Labile Organic Matter—
Chemical Compositions, Function, and Significance in Soil and the Environment

Labile Organic Matter—
Chemical Compositions, Function, and Significance in Soil and the Environment

Zhongqi He and Fengchang Wu, Editors

SSSA Special Publication 62

Soil Science Society of America, Inc.
5585 Guilford Road, Madison, WI 53711-5801 USA
soils.org

dl.sciencesocieties.org
societystore.org

ISBN: 978-0-89118-962-6 (print)
ISBN: 978-0-89118-963-3 (digital)
doi:10.2136/sssaspecpub62
Library of Congress Control Number: 2015949373

SSSA Special Publications
ISSN: 2165-9826 (online)
ISSN: 0081-1904 (print)

Cover design: Patricia Scullion
Cover image: Composite image adapted from Fig. 1 of "Composition of Whole and Water-Extractable Organic Matter of Cattle Manure Affected by Management Practices," by Zhongqi He and Heidi M. Waldrip. Photographs by Heidi M. Waldrip, USDA-ARS Conservation and Production Research Laboratory, Bushland, TX.

Printed in the United States of America.

Contents

Foreword

In recent years, our understanding of the natural organic matter in soils and dissolved organic matter in water has benefitted greatly from technological advances and novel and more sophisticated methods to examine elemental composition, structural and functional groups, and the ability to better compare the mobility of humic fractions in soils and the environment. In this compendium, renowned authors describe cutting-edge research in the field of organic matter and its significance to the ecosystem. The collection synthesizes recent advances in labile soil organic matter research for technical experts. Its utility extends to those who would simply like to learn more about labile organic matter.

This publication is timely given the need to better understand the whole of carbon stocks and what labile organic matter means in relation to environmental dynamics and the socioeconomic issues surrounding the present day emphasis on soil carbon sequestration.

The contents consist of 18 chapters arranged in three sections: characterization of labile organic matter, functions of labile organic matter, and ecosystem variability of labile organic matter. This book will serve as a needed resource for the latest in research on labile soil organic matter for years to come.

Carolyn Olson, 2015 SSSA President

Preface

Natural organic matter is ubiquitous in both terrestrial and aquatic environments. Although not strictly defined, natural organic matter can be differentiated as labile organic matter with a half life ranging from days to years, stable organic matter with a half-life between years and decades, and inert organic matter with a half-life between decades and centuries. Labile organic matter includes, but is not limited to, light carbon (C) fractions, easily extractable C pools, microbial biomass C, and labile humic materials. Because of their dynamic nature, labile organic matter fractions can respond rapidly to changes in C supply, thus releasing part of the nutrients for plant usage and providing energy and nutrients for microorganisms in soil and aquatic environments. Labile organic matter also plays roles in environmental dynamics and cycling of nutrients, as well as toxicity and fate of pharmaceutical, chemical (e.g., residues from herbicides and insecticides), metal, and organic pollutants in soil, sediments, water, and air. Therefore, labile organic matter attracts research attentions from a broad range of researchers in chemistry, biogeochemistry, agriculture, soil science, limnology, etc. In recent years, we have witnessed active research on chemical compositions, function, and significance of labile organic matter in soil and the environment. Thus, we organized this book project to highlight recent advances in labile organic matter studies. Subjects of interest include, but are not limited to, labile organic matter from agricultural soils, crop residues, animal manure, and sediments in freshwater, marshland, and coastal waters. The goals of this book are to (i) provide a timely synthesis of recent research and an appreciation for the remarkable range of wet chemistry and advanced instrumental techniques available for labile organic matter research and (ii) give a realistic evaluation of the current and potential impacts of labile organic matter on soil and the environment.

This book contains a total of 18 chapters divided into three parts. Part I is focused on characterization of labile organic matter. Five chapters in this part review and present case studies on the chemical composition and characteristics of dissolved organic matter in soils, whole and dissolved organic matter in livestock manure, and soil humic and fulvic acids, as revealed by wet chemistry and various advanced instrumental technologies. Part II is focused on the function of labile organic matter. Seven chapters in Part II cover issues related to effects of plant-derived organic matter on phosphorus sorption behaviors, structures, and functions of soil organic matter in diverse cropping systems, and nonideal interactions between organic contaminants and dissolved organic matter. Part III compiles and discusses variations of labile organic matter across different ecosystems. The first chapter in Part III discusses the influence of organic matter leached from leaf litter on streamwater composition. The other five chapters in Part III focus on composition and function of sediment organic matter from various freshwater and marine environments.

Chapter contribution for this publication was by invitation only and is a result of the combined efforts of a distinguished group of international contributors. We thank all authors for their timely and informative contributions. We profusely thank the unsung heroes (i.e., ad-hoc reviewers) for their time and effort in providing many valuable comments, which certainly improved the quality of this work. Special thanks go to Nicole Sandler and Lisa Al-Amoodi of the SSSA Headquarters staff for their excellent editorial assistance during the completion of this project.

Zhongqi He and Fengchang Wu, Editors

Contributors

Buyer, Jeffrey S.	USDA-ARS, Sustainable Agricultural Systems Lab., Beltsville Agricultural Research Center, 10300 Baltimore Ave., Bldg. 001, Rm. 118, Beltsville, MD 20705 (Jeffrey.buyer@ars.usda.gov)
Cavigelli, Michel A.	USDA-ARS, Sustainable Agricultural Systems Lab., Beltsville Agricultural Research Center, 10300 Baltimore Ave., Bldg. 001, Rm. 118, Beltsville, MD 20705 (michel.cavigelli@ars.usda.gov)
Cuss, C.W.	Environmental and Life Science Graduate Program, Trent Univ., 1600 West Bank Drive, Peterborough, Ontario, Canada, K9J 7B8 (chadcuss@trentu.ca)
DeLaune, Ronald D.	Dep. of Oceanography and Coastal Sciences, Louisiana State Univ., Baton Rouge, LA 70803 (rdelaun@lsu.edu).
Dodla, Syam K.	School of Plant, Environmental and Soil Sciences, 104 Sturgis Hall, Louisiana State Univ. Agricultural Center, Baton Rouge, LA 70803 (sdodla@agcenter.lsu.edu)
Duan, Dandan	State Key Lab. of Organic Geochemistry, Guangzhou Inst. of Geochemistry, Chinese Academy of Sciences, Yong No. 511 Kehuajie Tianhe, Guangzhou, Guangdong 510640, China; also Univ. of Chinese Academy of Sciences, Beijing 100049, China (dan113133@126.com)
Gerzabek, Martin H.	Inst. of Soil Research, Dep. of Forest and Soil Sciences, Univ. of Natural Resources and Life Sciences (BOKU), Peter Jordan Straße 82, 1190 Vienna, Austria
Guéguen, C.	Chemistry Dep., Trent Univ., 1600 West Bank Drive, Peterborough, Ontario, Canada, K9J 7B8 (celinegueguen@trentu.ca)
He, Zhongqi	USDA-ARS Southern Regional Research Center, 1100 Robert E. Lee Blvd., New Orleans, LA 70124 (zhongqi.he@ars.usda.gov)
Hunt, James F.	Dep. of Plant, Soil, and Environmental Sciences, Univ. of Maine, New England PSW Lab, Orono, ME 04469 (James_Hunt@umit.maine.edu)
Klepsch, Sabine	Inst. of Soil Research, Dep. of Forest and Soil Sciences, Univ. of Natural Resources and Life Sciences (BOKU), Peter Jordan Straße 82, 1190 Vienna, Austria, also Austrian Inst. of Technology GmbH (AIT), Konrad-Lorenz-Straße 24, 3430 Tulln, Austria
Martin-Neto, Ladislau	Empresa Brasileira de Pesquisa Agropecuária (EMBRAPA), Parque Estação Biológica – PqEB s/n, Brasília, DF, Brazil 70770-901 (ladislau.martin@embrapa.br)
Martins, Bruno Henrique	Instituto Agronômico do Paraná (IAPAR) – Área de Solos, Rodovia Celso Garcia Cid, km 375, Londrina, Paraná, Brazil 86047-902 (brn0321@gmail.com)
Maul, Jude E.	USDA-ARS, Sustainable Agricultural Systems Lab., Beltsville Agricultural Research Center, 10300 Baltimore Ave., Bldg. 001, Rm. 118, Beltsville, MD 20705 (Jude.maul@ars.usda.gov)
Monreal, Carlos M.	Eastern Cereal and Oilseed Research Center, Agriculture and Agri-Food Canada, Room 4046 K.W. Neatby Building, Central Experimental Farm, 960 Carling Ave., Ottawa, Canada, K1A 0C6 (carlos.monreal@agr.gc.ca)
Pan, Bo	Faculty of Environmental Science & Engineering, Kunming Univ. of Science & Technology, 727 S Jingming Rd., Kunming, Yunnan, China 650500 (panbocai@gmail.com)
Ran, Yong	State Key Lab. of Organic Geochemistry, Guangzhou Inst. of Geochemistry, Chinese Academy of Sciences, 511 Kehuajie Tianhe, Guangzhou 510640, China (yran@gig.ac.cn)
Reeves, James B., III	retired, USDA-ARS, Cropping Systems and Global Change Lab., Beltsville Agricultural Research Center, Beltsville, MD 20705
Reichenauer, Thomas	Austrian Inst. of Technology GmbH (AIT), Konrad-Lorenz-Straße 24, 3430 Tulln, Austria

Schnitzer, Morris I.	Eastern Cereal and Oilseed Research Center, Agriculture and Agri-Food Canada, Room 4046 K.W. Neatby Building, Central Experimental Farm, 960 Carling Ave., Ottawa, Canada, K1A 0C6
Senwo, Zachary N.	Dep. of Biological and Environmental Sciences, College of Agricultural, Life, and Natural Sciences, Alabama A&M Univ., 4900 Meridian Street, Normal, AL 35762 (zachary.senwo@aamu.edu)
Soja, Gerhard	Austrian Inst. of Technology GmbH (AIT), Konrad-Lorenz-Straße 24, 3430 Tulln, Austria
Spiegel, Heide	Austrian Agency for Health and Food Safety (AGES), Inst. for Sustainable Plant Production, Dep. for Soil Health and Plant Nutrition, Spargelfeldstraße 191, 1220 Vienna, Austria
Tatzber, Michael	Inst. of Soil Research, Dep. of Forest and Soil Sciences, Univ. of Natural Resources and Life Sciences (BOKU), Peter Jordan Straße 82, 1190 Vienna, Austria, also Austrian Inst. of Technology GmbH (AIT), Konrad-Lorenz-Straße 24, 3430 Tulln, Austria, also Federal Research and Training Centre for Forests, Natural Hazards and Landscape (BFW), Seckendorff-Gudent Weg 8, A-1131 Vienna, Austria and Austrian Agency for Health and Food Safety (AGES), Div. of Radiation Protection, Spargelfeldstraße 191, 1220 Vienna, Austria (michael.tatzber@boku.ac.at)
Tazisong, Irenus A.	Dep. of Biological and Environmental Sciences, College of Agricultural, Life, and Natural Sciences, Alabama A&M Univ., 4900 Meridian Street, Normal, AL 35762 (irenus.tazisong@aamu.edu)
Übner, Monika	Pärnu College, Univ. of Tartu, Pärnu 80010, Estonia (monika.ubner@ut.ee)
Waldrip, Heidi M.	USDA-ARS, Conservation and Production Research Lab., P.O. Drawer 10, 2300 Experiment Station Road, Bushland, TX 79012 (heidi.waldrip@ars.usda.gov)
Wang, Jim J.	School of Plant, Environmental, and Soil Sciences, 104 Sturgis Hall, Louisiana State Univ. Agricultural Center, Baton Rouge, LA 70803 (jjwang@agcenter.lsu.edu)
Wu, Fengchang	State Key Lab. of Environment Criteria and Risk Assessment, Chinese Research Academy of Environmental Sciences, 8 Dayangfang BeiYuan Rd., Chaoyang District, Beijing 100012, China (wufengchang@vip.skleg.cn)
Zhang, Di	Faculty of Environmental Science & Engineering, Kunming Univ. of Science & Technology, 727 S Jingming Rd., Kunming, Yunnan, China 650500 (zhangdi2002113@sina.com)
Zhang, Mingchu	Dep. of Agriculture and Horticulture, School of Natural Resources and Extension, Univ. of Alaska, 905 N. Koyukuk Drive, 303 O'Neill Bldg., Fairbanks, AK 99775 (mzhang3@alaska.edu)
Zhang, Yulong	State Key Lab. of Organic Geochemistry, Guangzhou Inst. of Geochemistry, Chinese Academy of Sciences, 511 Kehuajie Tianhe, Guangzhou 510640, China (gammalong@163.com)
Zhao, Aiqin	Editorial Dep. of the Transactions of the Chinese Society of Agricultural Engineering (TCSAE), No. 41 Maizidian Street, Chaoyang District, Beijing, China 100125 (zhaoaiqin2002@126.com)
Zhu, Yuanrong	State Key Lab. of Environment Criteria and Risk Assessment, Chinese Research Academy of Environmental Sciences, 8 Dayangfang BeiYuan Rd., Chaoyang District, Beijing 100012, China (zhuyuanrong07@mails.ucas.ac.cn)

Characteristics of Dissolved Organic Carbon Revealed by Ultraviolet–Visible Absorbance and Fluorescence Spectroscopy: The Current Status and Future Exploration

Mingchu Zhang* and Zhongqi He

Abstract

Dissolved organic carbon (DOC) is an important research subject for various disciplines. The objectives of this chapter are to review and summarize recent advancement in ultraviolet–visible (UV-Vis) and fluorescence spectroscopies and to identify the information gaps for future research. The search for causes of rising DOC in rivers and lakes in Europe and North America has led to no exclusive conclusion, especially from DOC exported from arable land. More studies are still needed in this area. Most of UV-Vis and fluorescence methods developed in the past are still widely used in current studies in the literature, such as specific ultraviolet absorbance (SUVA), spectral slope for UV-Vis, and excitation and emission matrix (EEM), fluorescence indexes (i.e., FI, F_{eff}) for fluorescence spectroscopy. New methods have been developed and tested, such as ratio of tryptophan-like to fulvic–humic-like, self-organizing map and classification/regression tree for EEM data analysis, and UV spectral deconvolution. Nevertheless, these new methods are tested in limited sites and studies. Critics on current fluorescence methods can also be found in the literature, such as quenching agents in DOC solutions and interaction of DOC fluorophores. Future research should include standardizing test conditions (e.g., pH, ionic strength, ion composition), as such parameters would affect fluorescence yield of DOC. In addition, new data analysis methods should be developed and tested.

Abbreviations: CRP, conservation reserve program; DOC, dissolved organic carbon; DON, dissolved organic nitrogen; EEM, excitation and emission matrix; FI, fluorescence index; SUVA, specific ultraviolet absorbance; UV-Vis, ultraviolet–visible.

Mingchu Zhang, Dep. of Agriculture and Horticulture, School of Natural Resources and Extension, Univ. of Alaska, 905 N. Koyukuk Drive, 303 O'Neill Bldg., Fairbanks, AK 99775. *Corresponding author (mzhang3@alaska.edu). Zhongqi He, USDA-ARS Southern Regional Research Center, 1100 Robert E. Lee Blvd., New Orleans, LA 70124 (zhongqi.he@ars.usda.gov).
doi:10.2136/sssaspecpub62.2014.0032

Labile Organic Matter—Chemical Compositions, Function, and Significance in Soil and the Environment. SSSA Special Publication 62. Zhongqi He and Fengchang Wu, editors.

Dissolved organic carbon has been an important research subject for soil and surface water for quite some time. Research efforts have been focused on at least four disciplinary areas for DOC: (i) impact of DOC on surface water (fresh and sea water) ecosystems, (ii) treatment efficiency of sewage water treatment facilities or processes, (iii) detecting trace DOC in drinking water, and (iv) DOC and dissolved organic nitrogen (DON) in soil system as affected by management and land use changes. Even though the common denominator for these four areas is DOC and much of the research methods are similar, such as UV-Vis and fluorescence spectroscopies, the connection of these four areas is limited, especially the link between surface water DOC and terrestrial DOC under certain management practices (Stanley et al., 2012; Diebel et al., 2008). The research on DOC on surface water emphasizes its impact on light transmission, carbon and nutrient cycling, and characterization of DOC and its impact on water quality. In comparison, for terrestrial DOC, many studies have reported on DOC export from forest land (Neff and Asner, 2001; Inamdar et al., 2012) and on quality and quantity of DOC under different land uses and management in arable land (Mattsson and Raike, 2005; Zhao et al., 2012; Sun et al., 2013). Most recently, more studies have been published on DOC runoff to surface water from manured land, which partially bridges the information gap (Singh and Inamdar, 2014; Old et al., 2012; Dutta et al., 2012a, 2012b). Nevertheless, these publications have not fully provided a comprehensive understanding of the relationship between terrestrial land management practice and DOC in surface water. Indeed, this information gap has made the causes of rising DOC in US and EU countries' fresh water systems an unanswered question (Sucker and Krause, 2010; Stanley et al., 2012). Further research in this area will likely continue until a full understanding of agriculture soil management practice and its impact on surface water is established. Even though UV-Vis and fluorescence spectroscopies are useful tools for assessing DOC properties (such as degree humification, component of DOC, etc.) in solution, they cannot provide information on the structure of DOC. Such limitations can be overcome by the use of other instruments such as FTIR (Fourier transform infrared spectroscopy), NRM (nuclear magnetic resonance), NIR (near infrared spectroscopy), MS (mass spectroscopy), and FTIR-MS. Research already demonstrates the use of such equipment to reveal the components and structures of DOC (Ohno et al., 2010; He et al., 2011, 2014). Pyrolysis-based methods such as Curie point pyrolysis-gas chromotograph/mass spectrometry (Cp-Py = GC/MS), pyrolysis-field ionization mass spectrometry (Py-FIMS), and pyrolysis-field-desorption mass spectrometry (Py-FDMS) are also widely used to determine the functional groups in DOC (Monreal and Schnitzer, 2012, 2015). The objectives of this chapter are to provide a review on the research methods of UV/Vis absorbance and fluorescence spectroscopy for soil and water DOC and identify the information gaps that require more study in the future. Selected case studies of DOC with these two methods can be found in the relevant chapters of this volume (Hunt and He, 2015; Zhang et al., 2015; Cuss and Guéguen, 2015).

Significance of Dissolved Organic Carbon Research in Soil and Surface Water Science

Dissolved organic C in the surface water primarily comes from two sources: carbon generated from the biological processes in the aquatic ecosystem system, and C imported from the terrestrial ecosystem. In the past, forest land carbon export was regarded as a primary source of terrestrial DOC in streams and lakes, especially in the spring flush (Neff and Asner, 2001; Strohmeier et al., 2013). It is only in recent years that terrestrial DOC contribution from other land use types has been studied. Wilson and Xenopoulos (2009) reported the contribution of agricultural land on DOC in streams in Ontario, Canada. But contrasting results have also been reported (Warrner et al., 2009). Indeed, the observations for rise of DOC in surface water have been reported by Freeman et al. (2001) based on a duration of 12-yr data (1988–2000). Since then, more evidence has been reported for rising DOC in surface water in different continents (Evans et al., 2005; Findlay, 2005; Monteith et al., 2007; Tian et al., 2013). With such observations, effort has been made in searching for the causes of such rises. However, no conclusive causes have been found (Sucker and Krause 2010). Nevertheless, evidence of DOC exported from agricultural fields to surface water are continuously reported in the literature (Naden et al., 2010; Old et al., 2012; Dutta et al., 2012a, 2012b).

For DOC exported from arable land, Naden et al. (2010) showed that liquid dairy manure applied at 46 m^3/ha on the soil surface in an intensively farmed grassland in England is linearly related to export of DOC sampled at each storm event to headwaters in a small catchment in the UK. The DOC from liquid dairy manure application can still be detected in storm water 5 wk after the application in the form of highly humified dissolved organic material and nitrate attributed to the decomposition of applied dairy slurry. In a similar climatic condition under beef, dairy, and sheep operation, Old et al. (2012) compared the key fluorescence properties of the drainage water from a livestock operation area and the ones without such an operation and concluded that livestock operation is contributing DOC to the headwaters in the catchment. Research results from the Delaware area of the United States show that poultry litter application on the land can lead to an increase in stream estrogen concentration (Dutta et al., 2012a). Moreover, the estrogen in the stream from the poultry application is related to the tillage practice. More runoff is found from the tilled land as compared with no-tillage land (Dutta et al., 2012a). This is, perhaps, the first report that links the contaminants with farm management practices. More research is needed in this direction to provide guidance for land management to minimize runoff contaminants from agricultural soils. Other research has also provided direct or indirect evidence that agricultural land use contributes to DOC in the surface water (Hernes et al., 2008; Williams et al., 2010; Graeber et al., 2012; Stanley et al., 2012; Singh and Inamdar, 2014). However, a contrasting report is also found in literature. In a study of stream water quality in an agricultural land use watershed, Warrner et al. (2009) found no link between the increase of DOC in stream water and drainage water DOC from agricultural land. Vidon et al. (2008) also reported no direct link of agricultural land use on stream DOC increase. The contrasting reports in DOC contribution from agricultural land uses to surface water has led to the need for more research in the area. This needed research at least should be in two areas: mechanisms of dissolved organic matter (DOM), DOC and/or DON generated from agricultural land and the pathways of such DOC transport.

Currently, there are limited publications in DOC and DON research in agricultural field. A recent publication by Zhang et al. (2011a) shows that DOC increased in manured land where tillage practices (tillage vs. no-tillage) were disregarded . Zhao et al. (2013) reported that whether the DOC is hot water extracted (80°C) or cold water extracted (23°C) all are biologically active (Zhao et al., 2012, 2013). The bioavailability of terrestrial DOC to microorganisms is also reported for nonagricultural land (Fellman et al., 2009; Roehm et al., 2009). The vertical movement of soil DOC from soil surface to subsurface is more likely for small size and more condensed DOC, but this condensed DOC may not correspond with a high-fluorescence humification index (HIX) (Zhang et al., 2011a). Hassouna et al. (2010) also reported that subsoil DOC is more humified, hydrophobic, but aliphatic. Land use change affects the soil DOC (Chantigny, 2003). Often, such comparison is made among different land uses on a local scale. A recent study (Sun et al., 2013) was conducted on the effect of land use change on soil DOC among two different countries on two separate continents (i.e., northern China and Canada) with similar climatic conditions; the results show no difference for quality and quantity of soil DOC among forest, grassland, and arable land . Sun et al. (2013) used air-dried soil samples, but air-dried soil samples could eliminate the difference among different land uses in DOC quality and quantity (Zhao et al., 2012). Among different land uses, the smaller size of DOC 1 kDa is dominated in arable land as compared with forest and land under conservation from agriculture (Zhao et al., 2012). There is limited research on the pathways of soil DOC to surface water. Understanding these processes will allow one to better manage DOC in runoff water.

Ultraviolet–Visible Spectroscopy in Dissolved Organic Carbon Research

Characterization of DOC in water includes two components: data acquisition and data analysis. For data acquisition, nearly all research on DOC either from soil or from surface water employ UV-Vis spectroscopy to characterize DOC properties. Figure 1 shows the UV-Vis spectra of water extracts of Texas Blackland Vertisol under different pasture managements. These spectra all look like monotonically decreasing curves with increasing wavelength. However, minor spectral differences, especially at the UV range, can be seen between these samples. These spectral features are typical for soil DOC samples (He et al., 2009; Hassouna et al., 2012). Thus, more specific parameters have developed to analyze the UV-Vis spectrum for DOC (Jaffrain et al., 2007, Zhang et al., 2011b). These parameters include specific ultraviolet absorbance (SUVA), absorbance ratio, and spectral slope of single exponential model. These methods are summarized in Table 1.

As illustrated in Table 1, these parameters are used in two areas: determination of properties of the DOC and identification of DOC sources. Specific UV absorbance at 254 and 280 nm ($SUVA_{254}$, $SUVA_{280}$), absorbance ratios of E2/E3 and E4/E6 are used for assessing the properties of DOC. The UV-Vis spectral slope is used to determine the sources of DOC. Table 2 lists some of these parameters in a pasture soil affected by a history of poultry litter application (Zhang et al., 2011b).

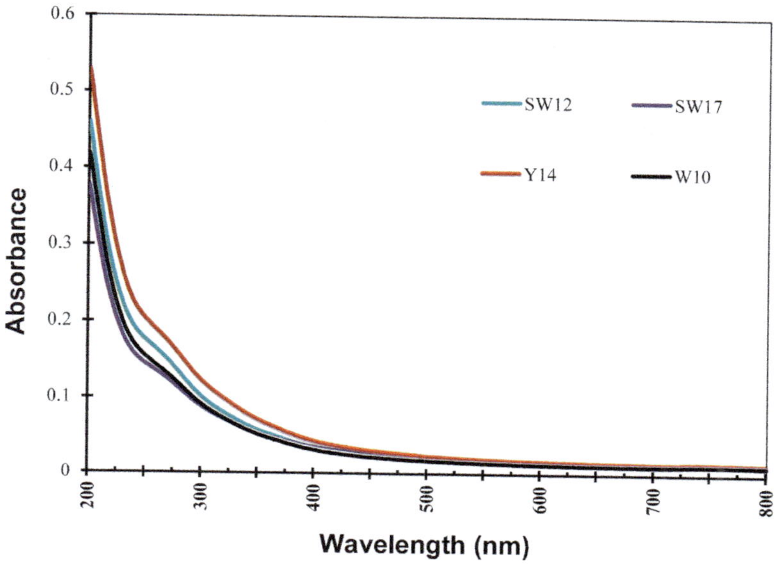

Fig. 1. Ultraviolet–visible (UV-Vis) spectra of the water extracts (1 g soil with 10 mL of water) of soil samples collected from Texas Blackland Vertisol with native ungrazed pasture (SW12), grazed pasture (SW17), hayed pasture (Y14), and grazed hay field (W10). The water-soil mixture was shaken for 30 min, followed by centrifuging for 30 min at 900 × *g*, and then filtrated with 0.45-μm filter paper. Unpublished data (Zhang, 2015).

These soil samples were collected from a long-term field experiment of land application of poultry litter to soils for pasture growth for nearly two decades in the Sand Mountain region of north Alabama (He et al., 2008a). The water extractable organic matter in these soil samples tends to be different in their spectral properties compared with the litter application years. Although the differences are not statistically significant ($P > 0.05$), the trend of changes is clearly demonstrated in Table 2. For all poultry litter–amended soils, the values of $SUVA_{254}$ and $SUVA_{280}$ increase compared with the soils with no poultry litter application, indicating increase in aromaticity in these litter-amended soils. Over the years of application, the spectral ratio of E2/E3 was not changed as obvious as the E4/E6 ratio. The E4/E6 ratio increases as the application of poultry litter increases over time. With the fact of the decrease in $SUVA_{254}$ and $SUVA_{280}$ the increase in the E4/E6 ratio seems to largely come from the aliphatic long chain compounds originated from poultry litter (Zhang et al., 2011b). During their evaluation of the concentration and composition of DOC in runoff from cropland (corn) receiving poultry manure, Singh and Inamdar (2014) reported that runoff DOC immediately following poultry manure application has a low $SUVA_{254}$ thus low aromaticity. Similarly, Li et al. (2014) evaluated multiple UV-Vis parameters of DOC during municipal solid waste composting. Among all parameters, they found that the values of $Area_{200-226}$, $Area_{226-400}$, and $Slope_{275-295}$ reflect the compost maturity best. Erlandsson et al. (2012) presented an analysis of UV-Vis absorbance spectra from 983 lakes throughout Sweden. They found that, relative to $SUVA_{254}$ the values of

Table 1. Selected ultraviolet–visible (UV-Vis) parameters used for dissolved organic carbon (DOC) studies.

Parameter	Description	Calculation	Application
SUVA254	UV specific absorbance at 254 nm normalized by DOC concentration	Absorbance at 254 nm/DOC concentration.	Linearly related to the size of DOC (Chin et al., 1994).
SUVA280	UV specific absorbance at 280 nm normalized by DOC concentration.	Absorbance at 280 nm/DOC concentration.	Apparent size of DOM, hydrophobic C content (He et al., 2006).
E2/E3	Ratio of absorbance at 250 and 365 nm or at 254 and 365 nm.	Absorbance at 250 nm/absorbance 365 nm, or absorbance at 254 nm/absorbance at 365 nm.	An indication of size of DOC. A high absorbance at the higher wavelength means the smaller the ratio, and existence large size of DOC (Yang and Xing, 2009).
E4/E6	Ratio of absorbance at 400 and 600 nm or at 465 and 665 nm.	Absorbance at 400 nm/absorbance at 600 nm, or absorbance at 465 nm/absorbance at 665 nm.	An indication for aromaticity and size of DOC. It reflects a well decomposed DOC when the ratio is low (Kang and Xing, 2008).
Slope (S)	Slope at single exponential model: $\alpha_\lambda = \alpha_{\lambda ref}\, e^{-S(\lambda-\lambda ref)}$, where S = slope; α_λ = wavelength (m^{-1}), $\alpha_{\lambda ref}$ = reference wavelength.	Use of linear transformation of the exponential function followed by linear regression from selective wavelength (i.e., 300–650 nm, by Stedmon and Markager, 2001); or use of nonlinear regression to directly obtain the S.	To find the sources of DOC in ocean (Stedmon and Markager, 2001). To differentiate DOC from different land uses (Zhang et al., 2006).

Table 2. The UV-Vis spectral properties of water extracts of poultry litter-amended soils. (Reprinted from Zhang et al. [2011b])

Litter application history	$SUVA_{254}$†	$SUVA_{280}$	E2/E3	E4/E6	Spectral slope (300–375 nm)
	m mg C^{-1} L^{-1}	m mg C^{-1} L^{-1}			nm^{-1}
0 yr	2.61 a‡	2.18 a	3.42 a	4.13 a	0.0109 a
5 yr	3.72 a	3.07 a	3.33 a	4.48 a	0.0109 a
10 yr	4.27 a	3.49 a	3.28 a	5.28 a	0.0112 a
15 yr	3.52 a	2.84 a	3.96 a	6.32 a	0.0131 a
20 yr	3.48 a	2.78 a	4.00 a	6.02 a	0.0132 a
Probability (*F* test)	0.170	0.218	0.134	0.062	0.093

† SUVA, specific ultraviolet absorbance.
‡ Values with different letters were significant with LSD at $P \leq 0.05$.

SUVA at longer wavelengths and metrics of spectral shape were more sensitive to acidity and less sensitive to the latitude of boreal lake waters. This observation confirms that the most informative UV-Vis range is between approximately 250 and 360 nm, where the between-lake variability is largest and absorbance can be precisely measured.

The $SUVA_{254}$ is widely used as a reliable indicator of a sample's average aromaticity with values <2.0 L mg C^{-1} m^{-1}, indicating low aromaticity (Weishaar et al., 2003; Hassouna et al., 2012). De Troyer et al. (2011) reported that $SUVA_{254}$ of the DOC in the soil samples amended with maize residues is significantly lower (around 6.6 cm g^{-1} L^{-1}) at the beginning and then increased to a similar value (19–29 cm g^{-1} L^{-1}) of the control samples after 2 wk of incubation, suggesting that most of the DOC released by maize residues mainly consists of hydrophilic (i.e., nonaromatic) compounds (98%) and is easily degradable. Dalzell et al. (2011) investigated the influence of subsurface drainage on the quantity and quality of DOC export from agricultural landscapes and found that subsurface drains with downstream ditch and stream site DOC decomposition at all sites from subsurface drains to downstream ditch and stream result in an increase in $SUVA_{254}$ values, showing that microbial decomposition preferentially removes nonaromatic compounds from the bulk DOC pool. Warrner et al. (2009) investigated the dynamics of DOC for 14 months in six headwater streams in a corn–soybean cropland with tile-drained landscape in the US Midwest. They found no relationship between DOC concentration and $SUVA_{254}$ in stream water, but an inverse relationship between DOC and $SUVA_{254}$ in water from tile drains. They attribute the observation to the fact that as tile flow increases, DOC low in aromaticity is likely mobilized and transported to the streams. On the other hand, the DOC pool in the streams is probably a mixture of DOC from a variety of precursor materials, which result in the lack of relationship between DOC concentration and $SUVA_{254}$ in the streams during nonstorm flows.

Fluorescence Spectroscopy in Dissolved Organic Carbon Research

Even with a shorter history of development than UV-Vis spectroscopy, fluorescence spectroscopy is widely used as a tool for DOC characterization (Ohno and He, 2011; Fellman et al., 2010). The most commonly used fluorescence methods include: EEM followed by parallel factor (PARAFAC) analysis, the humification index (HIX), fluorescence index (FI), and the ratio of β/α (Parlanti et al., 2000). The PARAFAC analysis after EEM is of particularly importance, for such analysis allows for the decomposition of EEM into the relevant components and calculation of the component distribution. The description of these parameters is summarized in Table 3.

Excitation and emission matrix spectra of DOC from soil and water are frequently decomposed into humic-like, fulvic-like, tryptophan-like, and tyrosine-like components by PARAFAC analysis (e.g., Hassouna et al., 2012; Ohno et al., 2009; He et al., 2010). Figure 2 shows a representative EEM fluorescence spectrum of water DOC and the four PARAFAC components (He et al., 2008b). The visual "peak-picking" indicates that the DOC probably contained three fluorophores: the strongest and broad peak had an excitation maxima of <240 nm and an emission maxima at ~440 nm (Ex < 240 nm/Em ~400 nm), a moderate peak at Ex ~315 nm/Em ~440

Table 3. Selected fluorescence spectroscopic methods for DOC characterization.

Parameter	Description	Calculation	Application
Humification index (HIX)	Ratio of emission intensity summation from 435 to 480 nm over intensity summation of 300 to 345 nm with an excitation wavelength at 254 nm.	$HIX = \Sigma I_{435\ to\ 480}/\Sigma I_{300\ to\ 345}$	Used as indicator content of humic materials in DOC, or indicator of degree of decomposition, i.e., more humified DOC (Zsolnay et al., 1999).
β: α ratio for freshness of DOC	First developed by Coble (1996) using C and M instead of α and β by Parlanti et al. (2000). α is the excitation maximum from 330 to 350 nm over the emission maximum from 420 to 480 nm. β is excitation maximum from 310 to 320 nm over the emission maximum from 380 to 420 nm.	β/α	Freshness of the DOC. The high value of the ratio indicates oldness of the DOC. α represents more decomposed (humic-like), while β represents more recent derived (marine humic-like) (Parlanti et al., 2000).
Fluorescence index (FI)	The ratio of fluorescence intensity at emission wavelength of 450 nm to that at 500 nm from the excitation wavelength of 370 nm.	Fluorescence intensity at 450 nm/fluorescence intensity at 500 nm.	The FI is used for differentiating sources of DOC; the high FI ~1.8 indicates DOC derived from microbes, and the low FI ~1.2 indicates DOC from terrestrial origin (McKnight et al., 2001).
Tryptophan index (TI)	Average fluorescence intensity in the region from excitation wavelength 275 to 285 and emission wavelength of 346 to 354 nm.	Average fluorescence intensity in the region.	The TI is used for tracking sources of DOC from various land uses (Old et al., 2012).

Parameter	Description	Calculation	Application
Fulvic/humic index	Average fluorescence intensity in the region from excitation wavelength 325 to 375 and emission wavelength of 450 to 500 nm.	Average fluorescence intensity in the region.	The index is used for tracking sources of DOC from various land uses (Old et al., 2012).
Excitation–emission matrix (EEM)	The method was first proposed by Christian et al. (1981). Its first biochemical application was done by Leiner and Wolfbeis (1988). First systematic research for DOC was by Coble (1996). Later, the parallel factor (PARAFAC) analysis was introduced by Bro (1997). The wavelength for EEM is excitation from 240 to 400 nm and emission from 300 to 500 nm.	Three dimensional contour curve is made with emission wavelength as *x* axis, excitation wavelength as *y* axis, and fluorescence intensity as *z* axis. The peaks in the contour curve can be identified by PARAFAC analysis.	The EEM method followed by PARAFAC analysis can give proximity components of DOC. These components include humic-like, tyrosine-like, and tryptophan-like DOC, etc. Detailed review for these components can be found from Leenheer and Croue (2003) and Fellman et al. (2010).

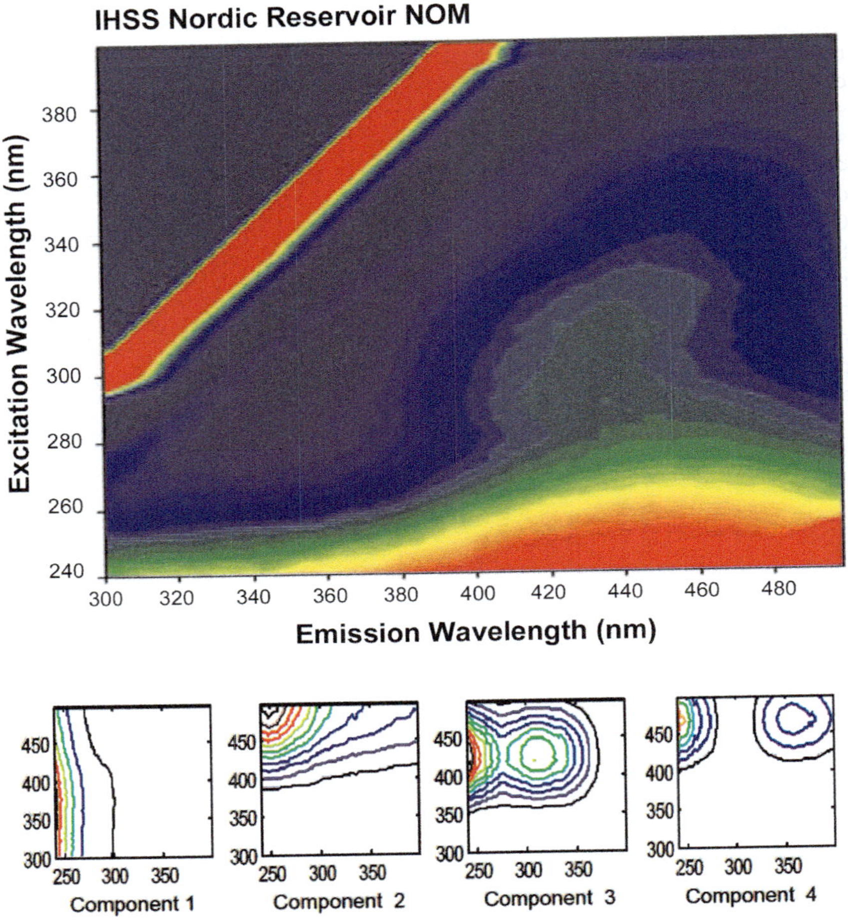

Fig. 2. Excitation–emission matrix spectrum of Nordic Reservoir natural organic matter and the spectral characteristics of the four parallel factor (PARAFAC) components. Reprinted from He et al. (2008b).

nm, and a weak peak at Ex ~305 nm/Em ~350 nm. Apparently, PARAFAC analysis revealed more information on the sample's fluorophoric nature with four fluorescent components. Component 1 is a strong UV humic-like peak. The classification of Component 2 is unknown with regard to the fluorophore types. Component 3 is assumed to be a weak peak from "humic-like" materials of terrestrial sources. Component 4's identity is ambiguous (He et al., 2008b). Jamieson et al. (2014) identified three humic-like components and one protein-like component biochar-derived DOC. They observed significant differences in the proportional distribution of PARAFAC components in different samples, and attributed the observation to differences in feedstocks and pyrolytic conditions.

Zhao et al. (2013) determined the bioavailability and spectral properties of soil DOC of three land uses (forest, agriculture, and land under a conservation reserve program [CRP]) in Alaska. They reported that DOC from all samples contains three PARAFAC components (Fig. 3); namely, Component 1: humic-like substances (Ex250 and 350 nm/Em 470 nm); Component 2: fulvic acid-like substances (Ex310 nm/Em 400 nm); and Component 3: tyrosine-like substances (Ex < 250 nm/Em < 300 nm). Zhang et al. (2011a) reported three similar components of soil DOC responding to poultry manure addition and tillage management of Georgia's Cecil sandy loam soil. In their study, the relative distribution of these three components shows that the humic-like fluorophore (Component 1) tends to increase over the years of poultry application. In contrast, the tyrosine-like fluorophore (Component 3) decreases with time. The fulvic-like fluorophore remains nearly constant over the years of the poultry application. Similarly, three fluorophoric components are also found in water-extracted DOC from field-moist and air-dried soil samples from subarctic Alaska under three land use managements (i.e., forest, agriculture, and grassland converted from agricultural use and under a CRP; Zhao et al., 2012). In contrast to the study of soil DOC, more components are decomposed from EEM from studying waste water and surface water (Chen et al., 2003; Stedmon and Markager, 2005). Chen et al. (2003) divided EEM into five regions to qualify the fluorophores in waste water and the data from literature. These five regions are aromatic protein (Region I, Ex 200–250, Em 280–330), aromatic protein II (Region II, Ex 200–250, Em 330–380), fulvic acid-like (Region III, Ex 200–250, Em 380–550), soluble microbial (Region IV, Ex 250–340, Em 280–380), and humic acid-like (Region V, EX 250–400, Em 380–550). However, Stedmon and Markager (2005) identified eight fluorescence components in studying DOC from a temperate estuary. There are four humic fluorophore groups (Ex < 250, Em 448; Ex < 250 or 305, Em 412; Ex 325, Em 428; Ex < 250 or 320, Em 400), two fulvic acid fluorophore groups (Ex < 250 or 385, Em 504; Ex < 250 or 360, Em 440), one tryptophan-like group (Ex 280, Em 344), and one tyrosine-like group (Ex 275, Em 304). Yamashita et al. (2011) also identified five fluorophore components in surface water DOC.

Whereas the EEM-PARAFAC method is used to identify the components of fluorophores for a semiqualification, FI is used to determine the sources of the DOC, and HIX is frequently used for assessing the properties of DOC. The fluorescence index was proposed by McKnight et al. (2001) (Table 3), and since then it has been widely used to differentiate the source of DOC. Typically, the FI is in the range of 1.3 to 1.8 for fresh surface water samples. The lower number within this range is regarded as terrestrial-derived DOC, and a high FI is indicative of microorganism-derived DOC (McKnight et al., 2001). Recently, the FI index has been used to differentiate the source of DOC in permafrost soils and lakes (Roehm et al., 2009), to characterize DOC from cave and spring waters (Birdwell and Engel, 2010) and the hydrological flow path of forested wetland (Johnson et al., 2011), and to identify the source of DOC in headwaters in central Europe (Graeber et al., 2012). The humification index (HIX) derived from fluorescence intensity has been another commonly used index for determination of DOC properties since it was proposed by Zsolnay et al. (1999). It is calculated as a ratio of summation of the emission intensity from 435 to 480 nm to the intensity summation of 300 to 345 nm at the excitation wavelength of 254 nm. The index was recently used

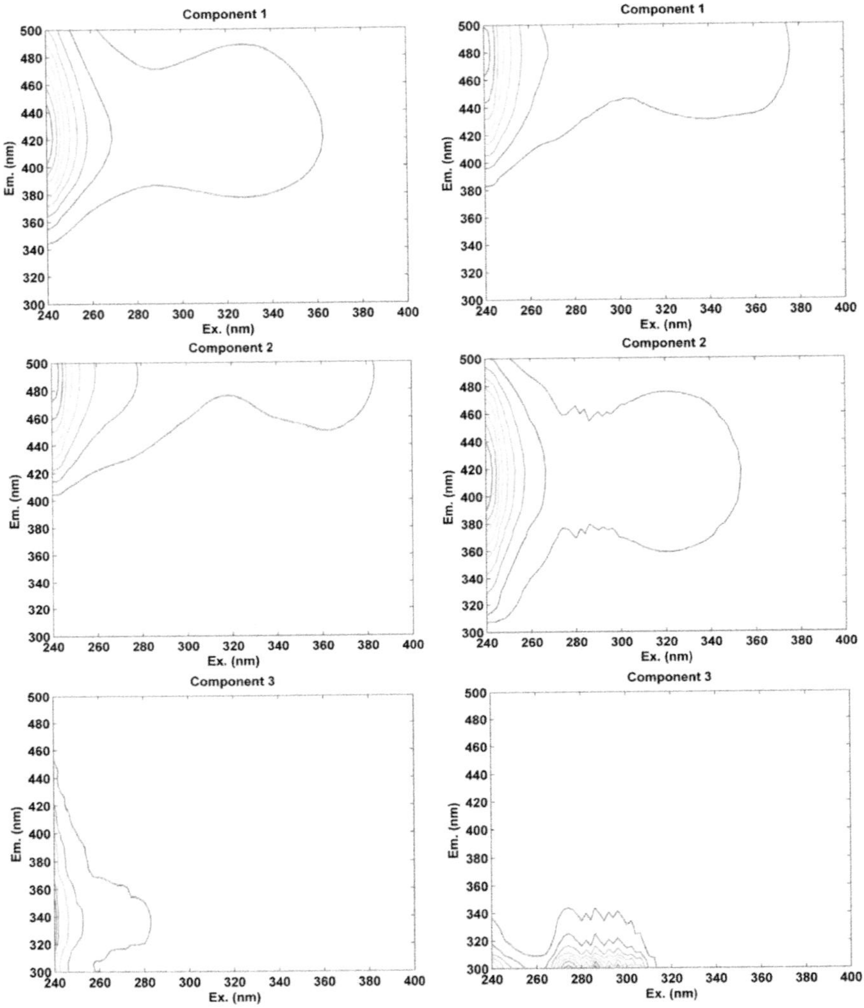

Fig. 3. The three PARAFAC fluorophore components for cold (22°C) (left) and hot (80°C) (right) water extractable organic matter from subarctic Alaska soils. For left column, component 1, humic-like compounds; component 2, fulvic-like compounds; component 3, tryptophan-like compounds. For right column, component 1, fulvic-like compounds; component 2, humic-like compounds; component 3, tyrosine-like compounds. Reprinted from Zhao et al. (2013).

for characterization of various sources of DOC (Roehm et al., 2009; Johnson et al., 2011; Graeber et al., 2012).

Other fluorescence parameters have also been developed, such as the ratio of tryptophan-like to fulvic/humic-like fluorescence (TI/FI) (Baker, 2002; Naden et al., 2010; Granger et al., 2010; Old et al., 2012; Graeber et al., 2012). Baker (2002) compared the ratio of river water samples with various samples from manure waste and found that the river water samples have the TI/FI ratio of less than 1, whereas those of the water samples from animal manure waste are from 1.8 to 25.1. In addition, the ratio decreases over time. Similarly, Naden et al. (2010) compared differences in DOC composition between manure and nonmanure agricultural sources for DOC fluorescence, and found that the ratio of TI/FI for animal manure is distinct from nonmanure sources. Whereas the TI/FI ratios of concentrated animal waste (direct extracts or samples collected close to the source) are fairly high (2–5 or higher), these ratios are relatively low (around 1 or <1) for diluted manure runoff or runoff collected at outlets of watersheds impacted by manure application (Naden et al., 2010; Old et al., 2012). Old et al. (2012) reported their values for TI/FI varied between 0.3 (premanure application) and 1.0 (immediately postmanure application). These values are lower than those reported by Baker (2002). They speculated that low TI/FI values are caused by the loss of the fluorescence signal of DOC due to the DOC sorption on soil particles and microbial degradation.

In addition, attempts have been made to predict DOC concentration by measuring the fluorescence or UV-Vis parameters. Dalzell et al. (2011) reported a negative linear relationship between FI and DOC concentration. In contrast, Petrone et al. (2011) reported an exponential decay function between FI and DOC concentration. Old et al. (2012) found a strong linear relationship between the tryptophan component and DOC concentration. A similar finding is also reported by Naden et al. (2010). Zhao et al. (2013) calculated five fluorescence parameters and two SUVA parameters of water-extracted DOC in 14 Alaskan soils (Table 4). For all soils, the hot water–extracted DOC had significantly ($P < 0.01$) lower HIX, $SUVA_{254}$, and $SUVA_{280}$ but significantly higher ($P < 0.01$) LQ (low quartile, summation of fluorescence emission intensity of 300 to 450 nm at excitation at 254 nm) and F_{eff} (ratio fluorescence maximum at 254 nm over the UV absorbance of 254 nm) than cold water–extracted DOC, indicating that hot water DOC consisted of smaller, less-humified molecules. Cold and hot DOCs show no significant ($P > 0.05$) differences in FI and high quartile (HQ). In addition, there were no significant differences in $SUVA_{254}$, F_{eff}, LQ, HQ, and HIX of cold and hot DOC among forest, CRP, and agricultural land.

Future Research Needs

Even though there is extensive use of UV-Vis and fluorescence spectroscopies in characterizing DOC, there are some fundamental issues with these methods worthy of more in-depth study. Some of these concerns are addressed in the recent review by Sharpless and Blough (2014). The unstable chromophores-fluorophores caused by interactions of various components of DOC can lead to the misassignment of a structural compound. For example, the underlying principle for PARAFAC analysis of EEM data is based on the assumption of no interference among fluorophores, that is, these fluorophores are independent from each other. However, no experiment ever proves such independence existed. In addition to the internal interaction of different molecules of DOC, the environmental variables also affect the absorbance and quantum yields. Metal concentrations in water, pH, and organic contaminants can affect the quantum yield of DOC (Yamashita and Jaffe, 2008; Aiken et al., 2011; Wang and Zhang, 2014). Ionic copper in water, for example, can decrease quantum yield of DOC as its concentration

Table 4. Effects of land use and extraction temperature on chemical characteristics of cold and hot water extractable organic matter of Alaskan soils. (Reprinted from Zhao et al. [2013].)

Soils	LQ†		HQ‡		HIX§		F_{eff}¶		FIX#		$SUVA_{254}$††		$SUVA_{280}$‡‡	
	Cold	Hot	Cold	Hot	Cold	Hot	Cold	Hot	Cold	Hot	Cold	Hot	Cold	Hot
1	0.06	0.23	1.19	1.14	19.61	5.58	0.08	0.48	1.07	1.07	1.90	0.31	0.76	0.38
2	0.16	0.32	1.63	1.48	12.90	4.77	0.08	0.50	1.05	1.05	1.65	0.65	1.50	0.36
3	0.22	0.27	1.42	1.46	9.81	5.41	0.07	0.93	1.11	1.09	1.57	0.29	1.80	0.30
Mean for CRP§§	0.15b¶¶	0.27a	1.41	1.36	14.10a	5.25b	0.08b	0.64a	1.08	1.07	1.71a	0.42b	1.35a	0.34b
4	0.08	0.26	1.43	1.26	18.66	4.92	0.04	0.62	1.13	1.10	3.51	0.41	2.51	0.15
5	0.20	0.21	1.67	1.38	9.78	6.60	0.08	0.33	1.06	1.08	1.77	0.70	1.03	0.19
6	0.10	0.32	1.44	1.79	15.04	5.23	0.07	0.85	1.12	1.08	2.01	0.38	1.58	1.06
Mean for forest	0.12b	0.26a	1.52	1.48	14.50a	5.58b	0.07b	0.60a	1.10	1.09	2.43a	0.50b	1.71a	0.47b
7	0.12	0.21	1.57	1.17	15.06	5.51	0.08	0.45	1.09	1.05	1.59	0.47	1.44	0.17
8	0.12	0.23	1.78	1.51	15.40	6.16	0.06	0.91	1.09	1.07	2.44	0.26	2.17	0.57
9	0.11	0.25	1.58	1.46	14.30	5.83	0.09	0.59	1.13	1.10	2.38	0.42	1.69	0.39
10	0.19	0.37	1.17	1.08	6.07	2.94	0.09	0.33	1.14	1.12	1.35	1.10	1.00	0.12
11	0.05	0.30	1.53	2.02	36.04	6.97	0.10	0.12	1.12	1.13	1.80	2.42	1.28	0.30
12	0.05	0.37	1.32	1.65	28.05	4.92	0.11	0.31	1.10	1.12	1.46	1.20	1.06	0.39
13	0.06	0.28	1.66	1.69	30.78	6.01	0.08	0.20	1.11	1.12	2.16	1.42	1.51	0.50
14	0.10	0.32	1.43	1.77	14.23	5.61	0.07	0.94	1.13	1.10	2.08	0.34	1.46	0.64
Mean for agriculture	0.07b	0.32a	1.47	1.70	24.35a	5.51b	0.09b	0.48a	1.11	1.11	1.90a	0.99b	1.34a	0.51b
Sig for land uses	NS##	NS	NS	NS	NS	NS	NS	NS	NS	NS	NS	NS	NS	NS

† Low quartile (LQ), sum of fluorescence intensity across 300–345 nm.
‡ High quartile (HQ), sum of fluorescence intensity across 435–480 nm.
§ Humification index (HIX).
¶ Fluorescence efficiency (Feff).
# Fluorescence index (FIX).
†† UV absorptivity at 254 nm (SUVA254).
‡‡ UV absorptivity at 280 nm (SUVA 280).
§§ CRP, conservation reserve program.
¶¶ Different letters in the column indicate significance at $P < 0.05$ between cold (22°C) and hot (80°C) water extractable organic matter for chemical characteristics.
No significant difference ($P > 0.05$).

increases (Yamashita and Jaffe, 2008). Research shows that divalent metals vary in their affinity to different sizes of DOC (Zhang et al., 2006), with most of Cu^{2+} associated with DOC particles of 0.45 um to 1 kDa disregarded with land uses, whereas Zn^{2+} is likely associated with DOC particles less than 1 kDa in undisturbed land uses. In addition, the frequently used wavelength for EEM excitation from 240 to 400 nm and emission from 300 to 500 nm (Table 3) as signal noises can be a significant problem at <230 nm on some instruments (Ohno and He, 2011). However, the excitation peak at a lower wavelength of 226 nm or nearby can provide an alternative or complementary characterization of the tryptophan fluorescence features (He et al., 2014). Using sequentially extracted water- and alkali-soluble cottonseed protein fractions (He et al., 2013), He et al. (2015) demonstrated that the fluorescence intensity of the peak at Ex 230 nm/Em 335 to 350 nm and the one at Ex 277 nm/Em 335 to 350 nm responded differently to the addition of denaturing reagents, as the former decreased continuously with increasing denaturant concentrations but the latter did not. In investigation of protein-like fluorescence intensity of river water of the River Tyne in northeast England, Baker and Inverarity (2004) demonstrated that tryptophan-like fluorescence intensity at Ex 280 nm/Em 350 nm correlates with both phosphate and nitrate, whereas tryptophan-like fluorescence intensity at Ex 220 nm/Em 350 nm correlates with biochemical oxygen demand, ammonia, and dissolved oxygen. Thus, if doable with the instrumental capability, extension of the excitation wavelength to <220 nm may provide more insight on DOC fluorescent features and its components and environmental influences.

In addition to the above mentioned fluorescence methods, Yu et al. (2014) reported a coupled method of EEM and self-organizing map for classification and regression tree analysis of waste water for monitoring DOC content changes. The self-organizing map is an unsupervised neural network that allows a low-dimensional view of high-dimensional data such as EEM. As such, the relationship between the organic fraction in DOC and the fluorescence group can be revealed. The application of such an algorithm in EEM is most often used in monitoring drinking water quality and waste water treatment efficiency (Bieroza et al., 2009; 2011). Since the method is used for quality and quantity monitoring, a use of such a method may also be helpful for tracing DOC movement in a watershed scale. The other advancement in fluorescence research is to compare the PARAFAC results from site-specific samples to large-scale samples (or prevalidated samples) (Fellman et al., 2009; Singh et al., 2013).

Ultraviolet spectral deconvolution (UVSD) is a multiwavelength deterministic or semideterministic procedure with an UV spectrum (Albrecht et al., 2011). Exploitation of this procedure is based on the fact that the UV spectrum of a sample is a linear combination of preselected representative absorption spectra, called reference spectra, and that the combination of the reference spectra predefined allows the restitution of the shape of the UV spectrum of any unknown sample. Ultraviolet spectral deconvolution takes into consideration the totality of the information present in a spectrum, making it more comprehensive and sensitive to global changes in a sample (Thomas et a., 1996). Ultraviolet spectral deconvolution has been used to quantify DOC of natural water or wastewater (Domeizel et al., 2004; Coulomb et al., 2006). Per Hassouna et al. (2007), UVSD could represent an appropriate alternative method for DOC quantification and fractionation, especially in on site measurements.

With the use of UV-Vis and fluorescence spectroscopies, knowledge of DOC both in surface water and from soil has been advanced. The connection, however, between these two ecosystems appears to be lacking. From a management perspective, farmers need to know which management practices can help to reduce DOC runoff from their field, when the vulnerable time for such DOC runoff from the field is, and if it is in the storm event, lasting small rains, or spring runoff. Most research attempting to connect those two systems is focused on storm events. For the storm event, the contact time between storm water and soil surface is short; therefore, the amount of DOC runoff is less. Dissolution of organic matter in soil requires time just like dissolution of most other chemicals. Future research may focus on lasting small rains or on spring runoff water. The other connection between these two systems is the changes of the soil DOC during transport and in surface water. Dissolved organic C from soil once in runoff will have a different ionic strength and pH than the place of origin. Knowing how the DOC will be transformed under different pH and ionic strength will help us to understand the fate of soil DOC in surface water so that the information gap between sources of DOC and its destination can be filled. It is well known that both pH and ionic strength affect the DOC fluorescence quantum yield. Yet, studies on the impact of these parameters on DOC fluorescence intensity are not commonly found in the DOC research papers. Standardizing the testing conditions (for example, pHs and ionic strengths) will provide insight of DOC properties and also allow cross comparison of different research papers in the literature. On the other hand, Jollymore et al. (2012) reported a methodology case study looking at the deployment of a submersible UV-Vis spectrophotometer to determine stream organic carbon dynamics within a headwater catchment located near Campbell River (British Columbia, Canada). These field-based absorbance measurements of DOC were made before and after forest harvest, highlighting the advantages of high temporal resolution compared with traditional grab sampling and laboratory measurements. This work proves that remote deployment strategies are a very promising tool for determining DOC transformations within aquatic ecosystems, so that it provides a means of gaining insight into dynamics at a temporal scale that is impossible to emulate through traditional approaches.

Future research may also call for use of a combination of UV-Vis and fluorescence spectroscopies and other advanced instruments, so that DOC molecular structure and properties can be revealed. With the assistance of these advanced instruments, molecular structure and the components of DOC are revealed (Ohno et al., 2010; He et al., 2011; He et al., 2015). More research based on the use of these instruments is expected to occur in the future. Such research will help to enrich our understanding of DOC and its changes under different land use and management practices. Also, such research will allow the connection of UV-Vis and fluorescence spectroscopies with DOC molecular structures and components, which will help to advance the methodologies of UV-Vis and fluorescence spectroscopies.

References

Aiken, G.R., H. Hsu-Kim, and J.N. Ryan. 2011. Influence of dissolved organic matter on the environmental fate of metals, nanoparticles and colloids. Environ. Sci. Technol. 45:3196–3201. doi:10.1021/es103992s

Albrecht, R., J. Le Petit, G. Terrom, and C. Perissol. 2011. Comparison between UV spectroscopy and NIRS to assess humification process during sewage sludge and green wastes co-composting. Bioresour. Technol. 102:4495–4500. doi:10.1016/j.biortech.2010.12.053

Baker, A. 2002. Fluorescence properties of some farm wastes: Implications for water quality monitoring. Water Res. 36:189–195. doi:10.1016/S0043-1354(01)00210-X

Baker, A., and R. Inverarity. 2004. Protein-like fluorescence intensity as a possible tool for determining river water quality. Hydrol. Processes 18:2927–2945. doi:10.1002/hyp.5597

Bieroza, M., A. Baker, and J. Bridgeman. 2009. Exploratory analysis of excitation-emission matrix fluorescence spectra with self-organizing maps as a base for determination of organic matter efficiency at water treatment works. J. Geophys. Res. 114:1–8. doi:10.1029/2009JG000940

Bieroza, M., A. Baker, and J. Bridgeman. 2011. Classification and calibration of organic matter fluorescence data with multiway analysis methods and artificial neural networks: An operational tool for improved drinking water treatment. Environmetrics 22:256–270. doi:10.1002/env.1045

Birdwell, J.E., and A.S. Engel. 2010. Characterization of dissolved organic matter in cave and spring waters using UV-Vis absorbance and fluorescence spectroscopy. Org. Geochem. 41:270–280. doi:10.1016/j.orggeochem.2009.11.002

Bro, R. 1997. PARAFAC. Tutorial and applications. Chemom. Intell. Lab. Syst. 38:149–171. doi:10.1016/S0169-7439(97)00032-4

Chantigny, M.H. 2003. Dissolved and water-extractable organic matter in soils: A review on the influence of land use and management practices. Geoderma 113:357–380. doi:10.1016/S0016-7061(02)00370-1

Chen, W., P. Westerhoff, J.A. Leenheer, and K. Booksh. 2003. Fluorescence excitation-emission matrix regional integration to quantify spectra for dissolved organic matter. Environ. Sci. Technol. 37:5701–5710. doi:10.1021/es034354c

Chin, Y.-P., G. Alken, and E. O'Loughlin. 1994. Molecular weight, polydispersity, and spectroscopic properties of aquatic humic substances. Environ. Sci. Technol. 28:1853–1858. doi:10.1021/es00060a015

Christian, G.D., J.B. Callis, and E.R. Davidson. 1981. Array detectors and excitation-emission matrices in multicomponent analysis. In: E.L. Wehry, editor, Modern fluorescence spectroscopy, 4. Plenum, New York. p. 111–165.

Coble, P.G. 1996. Characterization of marine and terrestrial DOM in seawater using excitation-emission matrix spectroscopy. Mar. Chem. 51:325–346. doi:10.1016/0304-4203(95)00062-3

Coulomb, B., Y. Richardson, C. Brach-Papa, J.-L. Boudenne, and F. Theraulaz. 2006. Rapid estimation of TOC in a marine urban sewage area by UV spectral deconvolution. Int. J. Environ. Anal. Chem. 86:1079–1093. doi:10.1080/03067310600687518

Cuss, C.W., and C. Guéguen. 2015. Characterizing the labile fraction of dissolved organic matter in leaf leachates: Methods, indicators, structure, and complexity. In: Z. He and F. Wu, editors, Labile organic matter—Chemical compositions, function, and significance in soil and the environment. SSSA Spec. Publ. 62. SSSA, Madison, WI. doi:10.2136/sssaspecpub62.2014.0043.

Dalzell, B.J., J.Y. King, D.J. Mulla, J.C. Finlay, and G.R. Sands. 2011. Influence of subsurface drainage on quantity and quality of dissolved organic matter export from agricultural landscapes. J. Geophys. Res. 116. doi:10.1029/2010JG001540

De Troyer, I., F. Amery, C. Van Moorleghem, E. Smolders, and R. Merckx. 2011. Tracing the source and fate of dissolved organic matter in soil after incorporation of a ^{13}C labelled residue: A batch incubation study. Soil Biol. Biochem. 43:513–519. doi:10.1016/j.soilbio.2010.11.016

Diebel, M.W., J.T. Maxted, P.J. Nowak, and M.J. Vander Zanden. 2008. Landscape planning for agriculture nonpoint source pollution reduction I: A geographical allocation framework. Environ. Manage. 42:789–802. doi:10.1007/s00267-008-9186-3

Domeizel, M., A. Khalil, and P. Prudent. 2004. UV spectroscopy: A tool for monitoring humification and for proposing an index of the maturity of compost. Bioresour. Technol. 94:177–184. doi:10.1016/j.biortech.2003.11.026

Dutta, S., S. Inamdar, J. Tso, D.S. Aga, and J.T. Sims. 2012a. Dissolved organic carbon and estrogen transport in surface runoff from agricultural land receiving poultry litter. J. Am. Water Resour. Assoc. 2:1–12.

Dutta, S., S. Inamdar, J. Tso, and D.S. Aga. 2012b. Concentration of free and conjugated estrogens at different landscape positions in an agricultural watershed receiving poultry litter. Water Air Soil Pollut. 223:2821–2836. doi:10.1007/s11270-011-1069-1

Erlandsson, M., M.N. Futter, D.N. Kothawala, and S.J. Kohler. 2012. Variability in spectral absorbance metrics across boreal lake waters. J. Environ. Monit. 14:2643–2652. doi:10.1039/c2em30266g

Evans, C.D., D.T. Monteith, and D.M. Cooper. 2005. Long-term increases in surface water dissolved organic carbon: Observations, possible causes and environmental impacts. Environ. Pollut. 137:55–71. doi:10.1016/j.envpol.2004.12.031

Fellman, J.B., E. Hood, D.V. D'Amore, R.T. Edwards, and D. White. 2009. Seasonal change in the chemical quality and biodegradability of dissolved organic matter exported from soils to streams in coastal temperate rainforest watersheds. Biogeochemistry 95:277–293. doi:10.1007/s10533-009-9336-6

Fellman, J.B., E. Hood, and R.G.M. Spencer. 2010. Fluorescence spectroscopy opens new windows into dissolved organic matter dynamics in freshwater ecosystems: A review. Limnol. Oceanogr. 55:2452–2462. doi:10.4319/lo.2010.55.6.2452

Findlay, S.E.G. 2005. Increased carbon transport in the Hudson River: Unexpected consequence of nitrogen deposition? Front. Ecol. Environ 3:133–137. doi:10.1890/1540-9295(2005)003[0133:ICTITH]2.0.CO;2

Freeman, C., C.D. Evans, D.T. Monteith, B. Reynolds, and N. Fenner. 2001. Export of organic carbon from peat soils. Nature 412:785. doi:10.1038/35090628

Graeber, D., J. Gelbrecht, M.T. Pusch, C. Anlanger, and D. von Schiller. 2012. Agriculture has changed the amount and composition of dissolved organic matter in Centre European headwater streams. Sci. Total Environ. 438:435–446. doi:10.1016/j.scitotenv.2012.08.087

Granger, S.J., R. Bol, L. Dixon, P.S. Naden, G.H. Old, J.K. Marsh, G. Bilotta, R. Brazier, S.M. White, and P.M. Haygarth. 2010. Assessing multiple novel tracers to improve the understanding of the contribution of agricultural farm waste to diffuse water pollution. J. Environ. Monit. 12:1159–1169. doi:10.1039/b915929k

Hassouna, M., C. Massiani, Y. Dudal, N. Pech, and F. Theraulaz. 2010. Changes in water extractable organic matter (WEOM) in a calcareous soil under field conditions with time and soil depth. Geoderma 155:75–85. doi:10.1016/j.geoderma.2009.11.026

Hassouna, M., F. Theraulaz, and C. Massiani. 2007. Direct estimation of nitrate, total and fractionated water extractable organic carbon (WEOC) in an agricultural soil using direct UV absorbance deconvolution. Talanta 71:861–867. doi:10.1016/j.talanta.2006.05.067

Hassouna, M., F. Théraulaz, and C. Massiani. 2012. Production and elimination of water extractable organic matter in a calcareous soil as assessed by UV/Vis absorption and fluorescence spectroscopy of its fractions isolated on XAD-8/4 resins. Geoderma 189-190:404–414. doi:10.1016/j.geoderma.2012.06.003

He, Z., H. Cao, H.N. Cheng, H. Zou, and J.F. Hunt. 2013. Effects of vigorous blending on yield and quality of protein isolates extracted from cottonseed and soy flours. Modern Appl. Sci. 7(10):79–88. doi:10.5539/mas.v7n10p79

He, Z., C.W. Honeycutt, and H. Zhang. 2011. Elemental and Fourier transformed-infrared spectroscopic analysis of water- and pyrophosphate-extracted soil organic matter. Soil Sci. 176:183–189. doi:10.1097/SS.0b013e318212865c

He, Z., J. Mao, C.W. Honeycutt, T. Ohno, J.F. Hunt, and B.J. Cade-Menun. 2009. Characterization of plant-derived water extractable organic matter by multiple spectroscopic techniques. Biol. Fertil. Soils 45:609–616. doi:10.1007/s00374-009-0369-8

He, Z., T. Ohno, B.J. Cade-Menun, M.S. Erich, and C.W. Honeycutt. 2006. Spectral and chemical characterization of phosphates associated with humic substances. Soil Sci. Soc. Am. J. 70:1741–1751. doi:10.2136/sssaj2006.0030

He, Z., T. Ohno, D.C. Olk, and F. Wu. 2010. Capillary electrophoresis profiles and fluorophore components of humic acids in Nebraska corn and Philippine rice soils. Geoderma 156:143–151. doi:10.1016/j.geoderma.2010.02.011

He, Z., T. Ohno, F.C. Wu, D.C. Olk, C.W. Honeycutt, and M. Olanya. 2008b. Capillary electrophoresis and fluorescence excitation-emission matrix spectroscopy for characterization of humic substances. Soil Sci. Soc. Am. J. 72:1248–1255. doi:10.2136/sssaj2007.0305

He, Z., I.A. Tazisong, Z.N. Senwo, and D. Zhang. 2008a. Soil properties and macro cations status impacted by long-term applied poultry litter. Commun. Soil Sci. Plant Anal. 39:858–872. doi:10.1080/00103620701880925

He, Z., M. Uchimiya, and H. Cao. 2014. Intrinsic fluorescence excitation-emission matrix spectral features of cottonseed protein fractions and the effects of denaturants. J. Am. Oil Chem. Soc. 91:1489–1497. doi:10.1007/s11746-014-2495-1

He, Z., M. Zhang, X. Cao, Y. Li, J. Mao, and H.M. Waldrip. 2015. Potential traceable markers of organic matter in organic and conventional manure using ultraviolet–visible and solid-state ^{13}C nuclear magnetic resonance spectroscopy. Org. Agric. doi:10.1007/s13165-014-0092-0 (in press).

Hernes, P.J., R.G.M. Spencer, R.Y. Dyda, B.A. Pellerin, P.A.M. Barchand, and B.A. Bergamashi. 2008. The role of hydrologic regimes on dissolved organic carbon composition in an agricultural watershed. Geochim. Cosmochim. Acta 72:5266–5277. doi:10.1016/j.gca.2008.07.031

Hunt, J.F., and Z. He. 2015. Characteristics of plant-derived water-extractable organic matter and its effects on phosphorus sorption behavior. In: Z. He and F. Wu, editors, Labile organic matter—Chemical compositions, function, and significance in soil and the environment. SSSA Spec. Publ. 62. SSSA, Madison, WI. doi:10.2136/sssaspecpub62.2014.0037.

Inamdar, S., N. Finger, S. Singh, M. Mitchell, D. Levia, H. Bais, D. Scott, and P. McHale. 2012. Dissolved organic matter (DOM) concentration and quality in a forested mid-Atlantic watershed, USA. Biogeochemistry 108:55–76. doi:10.1007/s10533-011-9572-4

Jaffrain, J., F. Gerard, M. Meyer, and J. Ranger. 2007. Assessing the quality of dissolved organic matter in forest soils using ultraviolet absorption spectrophotometry. Soil Sci. Soc. Am. J. 71:1851–1858. doi:10.2136/sssaj2006.0202

Jamieson, T., E. Sager, and C. Gueguen. 2014. Characterization of biochar-derived dissolved organic matter using UV–visible absorption and excitation–emission fluorescence spectroscopies. Chemosphere 103:197–204. doi:10.1016/j.chemosphere.2013.11.066

Johnson, M.S., E.G. Couto, M. Abdo, and J. Lehmann. 2011. Fluorescence index as an indicator of dissolved organic carbon quality in hydrologic flowpaths of forested tropical watersheds. Biogeochemistry 105:149–157. doi:10.1007/s10533-011-9595-x

Jollymore, A., M.S. Johnson, and I. Hawthorne. 2012. Submersible UV-Vis spectroscopy for quantifying streamwater organic carbon dynamics: Implementation and challenges before and after forest harvest in a headwater stream. Sensors 12:3798–3813. doi:10.3390/s120403798

Kang, S., and B.S. Xing. 2008. Humic acid fractionation upon sequential adsorption onto goethite. Langmuir 24:2525–2531. doi:10.1021/la702914q

Leenheer, J.A., and J.P. Croue. 2003. Aquatic organic matter: Understanding the unknown structure is key to better treatment of drinking water. Environ. Sci. Technol. 37:18A–26A. doi:10.1021/es032333c

Leiner, M.J.P., and O.S. Wolfbeis. 1988. Biochemical applications of 3-dimensional fluorescence spectrometry. In: J.R. Lakowicz, editor, Time-resolved laser spectroscopy in biochemistry. SPIE Proc., 909:134–138.

Li, D., X.S. He, B.D. Xi, Z.M. Wei, H.W. Pan, and D.Y. Cui. 2014. Study on UV-visible spectra characteristic of dissolved organic matter during municipal solid waste composting. Adv. Mater. Res. 878:840–849. doi:10.4028/www.scientific.net/AMR.878.840

Mattsson, T.P.K., and A. Raike. 2005. Export of DOM from boreal catchments: Impacts of land use cover and climate. Biogeochemistry 76:373–394. doi:10.1007/s10533-005-6897-x

McKnight, D.M., E.W. Boyer, P.K. Westerhoff, P.T. Doran, T. Kulbe, and D.T. Andersen. 2001. Spectrofluorometric characterization of DOM for indication of precursor material and aromaticity. Limnol. Oceanogr. 46:38–48. doi:10.4319/lo.2001.46.1.0038

Monreal, C.M., and M.I. Schnitzer. 2012. Biooil production from poultry litter: Potential energy, environmental and chemical opportunities. In: Z. He, editor, Applied research of animal manure: Challenges and opportunities beyond the adverse environmental concerns. Nova Sci. Publ., New York. p. 83–138.

Monreal, C.M., and M.I. Schnitzer. 2015. Labile organic matter in soil solution: II. Separation and identification of metabolites from plant–microbial communication in soil solutions of wheat rhizospheres. In: Z. He and F. Wu, editors, Labile organic matter—Chemical compositions, function, and significance in soil and the environment. SSSA Spec. Publ. 62. SSSA, Madison, WI. doi:10.2136/sssaspecpub62.2014.0074.

Monteith, D.T., J.L. Stoddard, C.D. Evans, H.A. de Wit, M. Forsius, T. Høgåsen, A. Wilander, B.L. Skjelkvåle, D.S. Jeffries, J. Vuorenmaa, B. Keller, J. Kopácek, and J. Vesely. 2007. Dissolved organic carbon trends resulting from changes in atmospheric deposition chemistry. Nature 450:537–540. doi:10.1038/nature06316

Naden, P.S., G.H. Old, C. Eliot-Laize, J.M.B. Hawkins, R. Bol, and P. Haygarth. 2010. Assessment of natural fluorescence as a tracer of diffuse agricultural pollutions from slurry spreading on intensely-farmed grasslands. Water Res. 44:1701–1712. doi:10.1016/j.watres.2009.11.038

Neff, J.C., and G.P. Asner. 2001. Dissolved organic carbon in terrestrial ecosystems: Synthesis and a model. Ecosystems 4:29–48. doi:10.1007/s100210000058

Ohno, T., and Z. He. 2011. Fluorescence spectroscopic analysis of organic matter fractions: The current status and a tutorial case study. In: Z. He, editor, Environmental chemistry of animal manure. Nova Sci. Publishers, New York. p. 83–103.

Ohno, T., Z. He, R.L. Sleighter, C.W. Honeycutt, and P.G. Hatcher. 2010. Ultrahigh resolution mass spectrometry and indicator species analysis to identify marker component of soil- and plant biomass-derived organic matter fractions. Environ. Sci. Technol. 44:8594–8600. doi:10.1021/es101089t

Ohno, T., Z. He, I.A. Tazisong, and Z.N. Senwo. 2009. Influence of tillage, cropping, and nitrogen source on the chemical characteristics of humic acid, fulvic acid, and water-soluble soil organic matter fractions of a long-term cropping system study. Soil Sci. 174:652–660. doi:10.1097/SS.0b013e3181c30808

Old, G.H., P.S. Naden, S.J. Granger, G.S. Bilotta, R.E. Bazier, C.J.A. Macleod, T. Krueger, R. Bol, J.M.B. Hawkins, P. Haygarth, and I. Freer. 2012. A novel application of natural fluorescence to understand the sources and transport pathways of pollutants from livestock farming in small headwater catchments. Sci. Total Environ. 417-418:169–182. doi:10.1016/j.scitotenv.2011.12.013

Parlanti, E., K. Worz, L. Geoffroy, and M. Lamotte. 2000. Dissolved organic matter fluorescence spectroscopy as a tool to estimate biological activity in a coastal zone submitted to anthropogenic inputs. Org. Geochem. 31:1765–1781. doi:10.1016/S0146-6380(00)00124-8

Petrone, K.C., J.B. Fellman, E. Hood, M.J. Donn, and P.F. Grierson. 2011. The origin and function of dissolved organic matter in agro-urban coastal streams. J. Geophys. Res. 116:1–13. doi:10.1029/2010JG001537

Roehm, C.L., R. Giesler, and J. Karlsson. 2009. Bioavailability of terrestrial organic carbon to lake bacteria: The case of a degrading subarctic permafrost mire complex. J. Geophys. Res. 114:G03006. doi:10.1029/2008JG000863

Sharpless, C.M., and N.V. Blough. 2014. The importance of charge-transfer interactions in determining chromophoric dissolved organic matter (CDOM) optical and photochemical properties. Environmental Sci. Environ. Sci. Processes Impacts 16:654–671. doi:10.1039/c3em00573a

Singh, S.S.D., and S. Inamdar. 2014. Land application of poultry manure and its influence on spectrofluoreometric characteristics of dissolved organic matter. Agric. Ecosyst. Environ. 193:25–36. doi:10.1016/j.agee.2014.04.019

Singh, S., S. Inamdar, and D. Scott. 2013. Comparison of two PARAFAC models of dissolved organic matter fluorescence for a mid-Atlantic forested watershed in the USA. J. Ecosyst. 2013:632424.

Stanley, E.H., S.M. Powers, N.R. Lottig, I.B. Buffam, and J.T. Crawford. 2012. Contemporary changes in dissolved organic carbon (DOC) in human-dominated rivers: Is there a role for DOC management? Freshwater Biol. 57:26–42. doi:10.1111/j.1365-2427.2011.02613.x

Stedmon, C.A., and S. Markager. 2001. The optic of chromophoric dissolved organic matter (CDOM) in the Greenland Sea: An algorithm for differentiation between marine and terrestrially derived organic matter. Limnol. Oceanogr. 46:2087–2093. doi:10.4319/lo.2001.46.8.2087

Stedmon, C.A., and S. Markager. 2005. Resolving the variability in dissolved organic matter fluorescence in a temperate estuary and its catchment using PARAFAC analysis. Limnol. Oceanogr. 50:686–697. doi:10.4319/lo.2005.50.2.0686

Strohmeier, S., K.H. Knorr, M. Reicher, S. Frei, J.H. Fleckenstein, S. Peitter, and E. Matzner. 2013. Concentration and flux of dissolved organic carbon in runoff from a forested catchment: Insights from light frequent measurements. Biogeosciences 10:905–916. doi:10.5194/bg-10-905-2013

Sucker, C. and K. Krause. 2010. Increasing dissolved organic carbon concentration in freshwaters: What is the actual driver? iForest, 3:106–108.

Sun, S., J. Liu, Y. Li, P. Jiang, and S. Chang. 2013. Similar quality and quantity of dissolved organic carbon under different land use systems in two Canadian and Chinese soils. J. Soils Sediments 13:34–42. doi:10.1007/s11368-012-0604-z

Thomas, O., F. Theraulaz, C. Agnel, and S. Suryani. 1996. Advanced UV examination of wastewater. Environ. Technol. 17:251–261. doi:10.1080/09593331708616383

Tian, Y.Q., Q. Yu, A.D. Feig, C. Ye, and A. Blunden. 2013. Effects of climate and land-surface processes on terrestrial dissolved organic carbon export to major U.S. coastal rivers. Ecol. Eng. 54:192–201. doi:10.1016/j.ecoleng.2013.01.028

Vidon, P., L.E. Wagner, and E. Soyeux. 2008. Changes in the character of DOC in streams during storms in two Midwestern watersheds with contrasting land uses. Biogeochemistry 88:257–270. doi:10.1007/s10533-008-9207-6

Wang, H.B., and Y.J. Zhang. 2014. Mechanisms of interaction between polycyclic aromatic hydrocarbons and dissolved organic matters. J. Environ. Sci. and Health 49:78–84.

Warrner, T.J., T.V. Royer, J.L. Tank, N.A. Griffiths, E.J. Rosi-Marshall, and M.R. Whiles. 2009. Dissolved organic carbon in streams from artificially drained and intensively farmed watershed in Indiana, USA. Biogeochemistry 95:295–307. doi:10.1007/s10533-009-9337-5

Weishaar, J., G. Aiken, B. Bergamaschi, M. Fram, R. Fujii, and K. Mopper. 2003. Evaluation of specific ultraviolet absorbance as an indicator of the chemical composition and reactivity of dissolved organic carbon. Environ. Sci. Technol. 37:4702–4708. doi:10.1021/es030360x

Williams, C.J., Y. Yamashita, H.F. Wilson, R. Jaffe, and M.A. Xenopoulos. 2010. Unraveling the role of land use and microbial activity in shaping dissolved organic matter characteristics in stream ecosystems. Limnol. Oceanogr. 55:1159–1171. doi:10.4319/lo.2010.55.3.1159

Wilson, H.F., and M.A. Xenopoulos. 2009. Effects of agricultural land use on the composition of fluvial dissolved organic carbon. Nat. Geosci. 2:37–41. doi:10.1038/ngeo391

Yamashita, Y., and R. Jaffe. 2008. Charactering the interactions between trace metals and dissolved organic matter using excitation-emission matrix and parallel factor analysis. Environ. Sci. Technol. 42:7374–7379. doi:10.1021/es801357h

Yamashita, Y., B.D. Kloeppel, J. Knoepp, G.L. Zausen, and R. Jaffe. 2011. Effects of Watershed history on dissolved organic matter characteristics in headwater streams. Ecosystems 14:1110–1122. doi:10.1007/s10021-011-9469-z

Yang, K., and B.S. Xing. 2009. Adsorption of fulvic acid by carbon nanotubes from water. Environ. Pollut. 157:1095–1100. doi:10.1016/j.envpol.2008.11.007

Yu, H., Y. Song, R. Liu, H. Pan, L. Xiang, and F. Qian. 2014. Identifying changes in dissolved organic matter content and characteristics by fluorescence spectroscopy coupled with self-organizing map and classification and regression tree analysis during wastewater treatment. Chemosphere 113:79–86. doi:10.1016/j.chemosphere.2014.04.020

Zhao, A., M. Zhang, and Z. He. 2012. Characteristics of soil water-soluble organic C and N under different land uses in Alaska. Soil Sci. 177:683–694. doi:10.1097/SS.0b013e31827e1fa8

Zhao, A., M. Zhang, and Z. He. 2013. Spectroscopic characteristics and biodegradability of cold and hot water extractable soil organic matter under different land uses in subarctic Alaska. Commun. Soil Sci. Plant Anal. 44:3030–3048. doi:10.1080/00103624.2013.829086

Zhang, M., Z. He, A. Zhao, H. Zhang, D. Endale, and H.H. Schomberg. 2011a. Water-extractable soil organic carbon and nitrogen affected by tillage and manure application. Soil Sci. 176:307–312. doi:10.1097/SS.0b013e31821d6d63

Zhang, M., Z. He, and A. Zhao. 2011b. Ultraviolet-visible absorptive features of water extractable and humic fractions of animal manure and relevant compost. In: Z. He, editor, Environmental chemistry of animal manure. Nova Sci. Publ., New York. p. 61–81.

Zhang, M., Z. He, and A. Zhao. 2015. Water extractable organic carbon and nitrogen affected by crop rotation and fertilizer management. In: Z. He and F. Wu, editors, Labile organic matter—Chemical compositions, function, and significance in soil and the environment. SSSA Spec. Publ. 62. SSSA, Madison, WI. doi:10.2136/sssaspecpub62.2014.0038

Zhang, M., S.D. Sparrow, and S. Seefeldt. 2006. Spectral characteristics of water extractable organic matter from soils of different land uses in a subarctic Alaska environment. Soil Science Society of American Annual Conference, Indianapolis, IN. 12–16 Nov. 2006. https://scisoc.confex.com/crops/responses/2006am/1253.pdf (accessed 12 Sept. 2014).

Zsolnay, A., E. Baigar, M. Jimenez, B. Steinweg, and F. Saccoumandi. 1999. Differentiating with fluorescence spectroscopy the sources of dissolved organic matter in soil subjected in drying. Chemosphere 38:45–50. doi:10.1016/S0045-6535(98)00166-0

Determination of Soil Organic Matter Features of Extractable Fractions Using Capillary Electrophoresis: An Organic Matter Stabilization Study in a Carbon-14-Labeled Long-Term Field Experiment

Michael Tatzber,* Sabine Klepsch, Gerhard Soja, Thomas Reichenauer, Heide Spiegel, and Martin H. Gerzabek

Abstract

Insights into the pathways of soil organic matter (SOM) turnover and its stabilization are important in soil science because soil fertility and quality are connected with the amount and quality of the formed SOM pool. Capillary electrophoresis (CE) is a valuable tool to determine the degree of SOM stabilization because it is based on a separation principle (separation of charged molecules in a very strong electric field), which is complementary to most other applied techniques in SOM analyses (e.g., pyrolysis gas chromatography–mass spectrometry [GC-MS], spectroscopic methods, and elemental analyses). The first part of this chapter reviews the current status on application of CE in SOM studies. The second part is a case study in which the potential of CE is evaluated as a method to detect SOM stabilization in a well-investigated ^{14}C-labeled long-term field experiment established in 1967. Referring to a time sequence in the investigated permanent bare fallow (BF) plot, in the case study, we observed four electrophoretic signals that significantly increased over time. Furthermore, extracted SOM from silt- and clay-sized particles was determined. Many signals in the electropherograms (two humps and six peaks) significantly reflected differences in SOM, which were attributed to various cropping systems or different particle size fractions. These results suggest that the signal intensities of hump 2 and the peaks 2, 4, and 6 represent subfractions of SOM that are clearly related to increasing SOM stabilization; hump 1 apparently included a considerable labile proportion of SOM. Based on the very consistent data for hump 2 and the peaks 2, 4, and 6, we conclude that CE is a highly applicable tool to investigate the stabilization of SOM in soil systems.

Abbreviations: BF, permanent bare fallow; CE, capillary electrophoresis; CIEF, capillary isoelectric focusing; C_{org}, organic C; CR, crop rotation; CSP, clay-sized particles; CZE, capillary-zone electrophoresis; FA, fulvic acid; GC-MS, gas chromatography–mass spectrometry; HA, humic acid; HPLC, high performance liquid chromatography; HS, humic substances; MC, monoculture; NMR, nuclear magnetic resonance; NOM, natural organic matter; PAGE, polyacrylamide gel electrophoresis; SDS, sodium dodecyl sulfate; SEC, size-exclusion chromatography; SOM, soil organic matter; SSP, silt-sized particles; UV-Vis, ultraviolet–visible.

Michael Tatzber, Sabine Klepsch, and Martin H. Gerzabek, Institute of Soil Research, Department of Forest and Soil Sciences, University of Natural Resources and Life Sciences (BOKU), Peter Jordan Straße 82, 1190 Vienna, Austria. Michael Tatzber, Sabine Klepsch, Gerhard Soja, and Thomas Reichenauer, Austrian Institute of Technology GmbH (AIT), Konrad-Lorenz-Straße 24, 3430 Tulln, Austria. Heide Spiegel, Austrian Agency for Health and Food Safety (AGES), Institute for Sustainable Plant Production, Department for Soil Health and Plant Nutrition, Spargelfeldstraße 191, 1220 Vienna, Austria. Michael Tatzber, also Federal Research and Training Centre for Forests, Natural Hazards and Landscape (BFW), Seckendorff-Gudent Weg 8, A-1131 Vienna, Austria and Austrian Agency for Health and Food Safety (AGES), Division of Radiation Protection, Spargelfeldstraße 191, 1220 Vienna, Austria. *Corresponding author (michael.tatzber@boku.ac.at).

doi:10.2136/sssaspecpub62.2014.0033

Labile Organic Matter—Chemical Compositions, Function, and Significance in Soil and the Environment. SSSA Special Publication 62. Zhongqi He and Fengchang Wu, editors.

Analyses of SOM, combined with the determination of its responses to different agricultural treatments, are essential when C storage in such management systems must be predicted. In such cases, special attention is paid to characterizing and understanding the labile and stable proportions of SOM (Fang et al., 2005; Knorr et al., 2005; von Lützow et al., 2007). The fractions that represent the active SOM pool (with turnover rates of less than 10 yr [von Lützow et al., 2007]) well are the SOM fractions of microbial biomass and the light fraction obtained by density fractionation. These have turnover rates of <10 yr (von Lützow et al., 2007). Extractable SOM fractions are well analyzable, and some of them also correspond to labile and active SOM pools. Prominent examples are water-extractable organic C (C_{org}) (Kalbitz et al., 2000, 2003; Chantigny, 2003) and SOM fractions extracted with alkaline media (Cao et al., 2011; Abelenda et al., 2011). Soil organic matter extracted by 1 M sodium hydroxide (NaOH) yields the extracted fraction and ^{14}C labels (Tatzber et al., 2009b); moreover, subsequent determinations with spectroscopic methods such as mid-infrared, fluorescence, and ^{13}C nuclear magnetic resonance (NMR) spectroscopy significantly reflect management influences and site characteristics in SOM (e.g., Tatzber et al., 2007, 2008, 2009a, 2012). This chapter provides a general review of progress made in characterization of SOM with CE, and then a case study is presented where CE was used to evaluate extracted SOM from a ^{14}C-labeled long-term field trial.

Previous Applications of Capillary Electrophoresis to Characterize Soil Organic Matter

Electrophoresis has been shown to be an effective method of separating SOM into different fractions based on molecular charge. The basis of this method is the separation of charged molecules in an alkaline solution by application of a strong electrical field. This gives it considerable potential for valuable additional information about influences on SOM (e.g., under different soil management systems) in a soil system when used with frequently applied methods in SOM analyses (solid-state ^{13}C NMR, mid-infrared, fluorescence, or ultraviolet–visible (UV-Vis) spectroscopies; pyrolysis GC-MS; etc.). Numerous studies have used CE as a separation method, providing good evidence of its utility in providing information on the forms and quantity of several SOM fractions. Since the mid-1990s, electrophoresis has been used for SOM investigations (Kutsch and Schumacher, 1994; Garrison et al., 1995; Pompe et al., 1996; Trubetskoj et al., 1997), and numerous further studies provided information about usable and useful variations of this separation technique (Dunkelog et al., 1997; reviews of Schmitt-Kopplin and Junkers, 2003; Široká et al., 2011).

Early works on CE analyses of extracted SOM focused on several aspects. These included the obtained fractions and fingerprint information from electrophoretic separation of humic substances (HS) (Kutsch and Schumacher, 1994; Pompe et al., 1996), the use of different capillary electrophoretic techniques (e.g., free solution CE, capillary gel electrophoresis, isoelectric focusing in ultrathin layers, micellar electrokinetic chromatography, and sodium dodecyl sulfate (SDS) gradient gel electrophoresis; Dunkelog et al., 1997) and different buffers (Garrison et al., 1995; Dunkelog et al., 1997; Schmitt-Kopplin et al. 1998b,c). Another focus was on comparing different electroosmotic flow-suppressing additives in a hydrodynamically closed system (Kaniansky et al., 1997) and on studying the critical concentration of aggregation of HS with capillary-zone electrophoresis (CZE) (Fetsch et al., 1998). In another study, CE was used to develop a method separating acidic monosaccharides and their corresponding 1,4-lactones in a single run without derivatization (Schmitt-Kopplin et al., 1998a). The fractions separated by electrophoresis were, in most cases, detected by UV-Vis (Garrison et al., 1995; Pompe et al., 1996; Dunkelog et al., 1997; Schmitt-Kopplin et al., 1998b; Schmitt-Kopplin and Junkers, 2003), but mentioned are also detectors based on conductivity (Kaniansky et al., 1997; Široká et al., 2011), laser-induced fluorescence, and ^{11}B NMR (Schmitt-Kopplin et al., 1998c).

Janoš (2003) reviews different separation methods for HS and mentions various techniques based on a size-exclusion effect, namely GC-MS for identification of pyrolysis or thermochemolysis products of HS. In this review, electrophoretic methods were described as providing detailed characterization of HS by their fingerprints; however, these authors pointed out the difficulty of determining specific subfractions, structure, or other properties of HS with these methods (Janoš, 2003). The review also mentioned a limitation by the great diversity of procedures employed and by the low degree of harmonization because no data on reproducibility and between-laboratory comparability were available. Nonetheless, numerous further publications demonstrated the applicability of CE to analyses of humic acids (HAs) and also increased our knowledge about the methodology of electrophoresis to characterize SOM and evaluate the obtained electropherograms.

Polyacrylamide Gel Electrophoresis

Numerous publications employed size-exclusion chromatography (SEC) and polyacrylamide gel electrophoresis (PAGE) for separation of SOM components based on molecular size and electrophoretic mobilities. In Cavani et al. (2003), CZE was used to characterize HA fractions previously separated by SEC and PAGE (SEC-PAGE). Capillary zone electrophoresis was conducted using diverse buffers; the best resolution of electropherograms was obtained with a Tris-borate-EDTA (ethylenediaminetetraacetic acid)-buffer with urea and SDS (Cavani et al., 2003). Trubetskoj et al. (1997), in contrast, compared different SEC and ultrafiltration techniques and assayed them by PAGE. Their results indicated that SEC gave better separation of HA on fractions differing in electrophoretic mobility and molecular size than ultrafiltration for soil HA fractionation. This methodology was then used in further works; Trubetskoj et al. (1999) used SEC-PAGE to fractionate HA from four different soil types (podzuvisol, phasozem, chernozem, and ferralsol). They reported that the mass of the separated high-molecular size fraction was highest in podzuvisol and ferralsol and smallest in phasozem and chernozem, but the latter both contained considerably more of the lower molecular size fraction than HA from the former.

Trubetskaya et al. (2002) investigated HA extracted from three soil types (chernozem, ferralsol, ranker) with steady-state fluorescence spectroscopy in the emission mode after separating three different fractions by tandem SEC and PAGE (leading to fractions A, B, and C+D). Independent of the HA source, the main fluorophores were located in the polar and low-molecular size fractions C+D (Trubetskaya et al., 2002). A pH effect was observed and led to the assumption that aromatic structures bearing carboxylate and hydroxyl substituents are involved in the longer wavelength emissions (Trubetskaya et al., 2002). An andisol HA was investigated by Richard et al. (2007) by separation into three different fractions applying a SEC-PAGE approach and subsequent UV-Vis, mid-infrared, and 3D-fluorescence spectroscopies. Mid-infrared measurements of acidified samples revealed that aliphatic moieties such as polysaccharides and peptides were mainly localized in the high-molecular size fractions, while carboxylic groups were present in the low-molecular size fractions (Richard et al., 2007).

Trubetskaya et al. (2008) compared HS extracted from detritus and soil by using SEC-PAGE. Their results experimentally confirmed the hypothesis that the transformation of plant detritus into HS proceeds in the direction from the high-molecular compounds to low-molecular compounds. Another outcome was that the HS differed significantly in the color intensity of the electrophoretic fractions, molecular weight distribution and protein content (as evidenced by color of the electrophoretic fractions when treated with a protein-specific dye) (Trubetskaya et al., 2008). Francioso et al. (2009) separated Chernozem soil HA by SEC-PAGE into different fractions and investigated them with mid-infrared spectroscopy and thermal and isotopic analyses (isotopic signature $\delta^{13}C$). One result was that the lowest molecular size fraction contained the highest proportion of aromatic compounds and the lowest proportion of aliphatic, carboxylic, carbohydrate, and amino acid groups compared with the other fractions (Francioso et al., 2009). Gieguzynska et al. (2009) extracted HA by applying three different extraction procedures and analyzed the isolated HS for elemental composition, functional groups content, absorption and emission properties, electrophoretic characteristics, and the ability to produce singlet oxygen using furfuryl alcohol as a scavenger. Electrophoresis was conducted as PAGE without SEC because the experience gained in previous studies (Trubetskaya et al., 2002; Richard et al., 2007) could be used to assign the obtained fractions to molecular sizes and functional group characteristics (Gieguzynska et al., 2009). Comparing all the mentioned methods, photochemical and electrophoretic determinations turned out to be sensitive indicators to evaluate differences among extraction procedures of HA (Gieguzynska et al., 2009). Trubetskoj et al. (2010) used ^{13}C, ^{1}H, and dimethyl sulfoxide-*d6* NMR for HA characterization after separation by the SEC-PAGE approach. A ratio of aryl C to alkyl C was calculated and increased by a factor of five from the highest to the lowest molecular size fraction. This suggests significant heterogeneity in terms of molecular structure and functionality among the various fractions, differing in both electrophoretic mobility and molecular size (Trubetskoj et al., 2010).

Capillary Zone Electrophoresis

Krivácsy et al. (2000) combined SEC with CZE to investigate humic-like substances from fog and interstitial aerosol between the fog droplets. They then compared the results to those of reference HS. Capillary electrophoresis was applied to determine the protonation constants' range (Krivácsy et al., 2000). The scavenging ratio for the humic-like substances was calculated and found to be the same as those of ammonium and sulfate, implying that such compounds play a role in cloud condensation (Krivácsy et al., 2000). A methodology using capillary isoelectric focusing (CIEF) was published by Schmitt et al. (1997), who investigated HS from different natural environments. In their work, 50 mg of a soil fulvic acid (FA) were fractionated by CIEF and the obtained fractions were characterized with UV-Vis spectroscopy, gel permeation chromatography, and CZE. Capillary isoelectric focusing provided good information on the electrofocusing patterns of HS using ampholytes in the pH range 3 to 10; this allowed the focusing profile of various HS to be compared (Schmitt et al., 1997). The CZE approach has also often been employed for HS characterization. Schmitt-Kopplin et al. (1999) analyzed HS for their relative charge densities (described as an important parameter when studying their reactivity toward organic and metallic species) and obtained characteristic fingerprint patterns in the electropherograms. In their study, the measurements were performed in aqueous solutions when both pH and ionic strength could be controlled, and this allowed a description of the behavior of HS mixtures in aqueous conditions close to environment and biological media.

Capillary zone electrophoresis was successfully used in many other studies with various research foci. Parlanti et al. (2002) used three-dimensional fluorimetry, high performance liquid chromatography (HPLC), and CZE to differentiate between marine and freshwater samples. Their analyses of HPLC-fractionated dissolved organic matter classes using both three-dimensional fluorimetry and CZE with UV-Vis and laser-induced fluorescence detection showed that the isolated fractions were specific to water origin. Furthermore, the authors showed that CZE had considerable potential for fingerprinting and for better characterizing dissolved organic matter from different origins. Peuravuori et al. (2004) performed a comparative study separating aquatic substances by CZE with uncoated, polymer-coated, and gel-filled capillaries and employed different background electrolytes and organic modifiers. The gel-filled capillary including linear noncross-linked PAGE with 10% gel concentration and 5M urea–water as the sample modifier using 40 mM phosphate buffer at pH 6.8 produced the most reliable results. This reflects the adaptation of physical gels, especially if the interactions of humic solutes with the gel matrix are not critical (Peuravuori et al., 2004). The addition of SDS in the linear PAGE gel increased the interaction of humic solutes with the gel matrix but also improved the separative power and strengthened the chaotropic effect of the urea modifier (Peuravuori et al., 2004). He et al. (2008), in turn, used CZE and fluorimetry to investigate five FA, six HA, and two unprocessed natural organic matter (NOM) samples. The electropherograms of all samples revealed three peak features with sharp first and third peaks and a hump-shaped second peak (He et al., 2008). The differences among the FA, HA, and unprocessed NOM samples were in the pattern and shapes of these peaks, and the information on fractions obtained by electrophoresis along with fluorescence spectroscopy was useful for investigating NOM decomposition dynamics (He et al., 2008).

Work by de Moraes and Rezende (2008) evaluated the possibility of eliminating the purification steps involved in characterizing HA by CZE (voltage, 20kV; T, 30°C; detection, UV-Vis at 214 nm). Their data suggested that sample purification was unnecessary to characterize HA with this method and the authors concluded that CZE without purification is a promising tool to characterize HA from different origins. Vanifatova et al. (2008) fractionated and characterized HA from soddy-podzolic and chernozem soils and obtained at least three HA subfractions with different electrophoretic mobilities under appropriate fractionation conditions (i.e., pH 7). The potential of CZE in estimating the isoelectric point was demonstrated. Moreover, the changes in the humate charge resulting from the varying ionization degree of its functional groups as a function of pH could be estimated based on electrophoretic mobility values (Vanifatova et al., 2008). Takahashi et al. (2009) separated water-soluble HS by CE using a polyacrylamide-coated capillary and a phosphate buffer solution at pH 7 containing hydroxyethyl cellulose. They used this approach to study complexation of HS with metal ions. Adding ferric ions or EDTA to the sample solution of FA yielded a distinct change in the electropherogram pattern based on the conformational change of FA (Takahashi et al., 2009). Chen et al. (2010) used CZE with UV absorption detection for analyzing lignin-derived phenolic distribution in the degradation residues after alkaline Cu-oxide oxidation of samples with HS. For well waters and fertilizers, the content of phenolic fragments was in agreement with the findings obtained by the elemental and spectroscopic measurements (Chen et al., 2010). He et al. (2010) applied a combination of CZE and fluorescence excitation–emission matrix to investigate HA samples from soils taken from corn (*Zea mays* L.) and rice (*Oryza sativa* L.) fields in Nebraska and the Philippines, respectively. The plots from which the Nebraska soil was taken had been under differing fertilization regimes (inorganic fertilizer, beef cattle feedlot manure, and unfertilized control). The Philippine rice soil had been under different irrigation regimes (flood-irrigated and rain-fed) and fertilization (inorganic fertilizer and unfertilized control) before sampling. Greater differences were observed in the CZE profiles between soils than between management practices on the same soil, reflecting the distinct fingerprint CE features of these soil humic fractions (He et al., 2010). Furthermore, modeling analysis of excitation–emission data indicated that four fluorophore components were present in the sequentially extracted mobile and recalcitrant humic fractions; three of these were already known from the literature and the unknown one was proposed as being relevant to agricultural humic fractions (He et al., 2010).

Übner et al. (2010) used HS as a background electrolyte to characterize HS–reagent interactions. In their work, the addition of lead nitrate [$Pb(NO_3)_2$] to HS enabled the separation of specific, characteristic regions in electropherograms correlating with the salt concentration. Further, they obtained two complexes (Pb–HS fractions I and II) with different mobility and each of the HS fractions contained those complexes in a different ratio. In a previous study (Tatzber et al., 2011), 113 HA originating from different layers of diverse forest sites were investigated with UV-Vis, mid-infrared, and fluorescence spectroscopies, and elemental analyses; the obtained data were compared with the signals from CZE (two humps and three peaks). Many significant correlations (Spearman, $p < 0.01$) between the electrophoretic signals (two humps and three peaks) and the signal intensities of other spectroscopic methods (UV-Vis, mid-infrared, and fluorescence spectroscopies) as well as C and N contents were obtained, and assignments for the electrophoretic signals could be attributed based on these results (Tatzber et al., 2011). In summary, the authors proposed that the first hump in the electropherogram is representative for preferably labile SOM and that all other signals in the electropherogram are connected with more stabilized SOM compounds.

Work by d'Orlyé and Reiller (2012) probed the physicochemical properties of three different HS using CZE in alkaline carbonate buffers at pH 10 and drew special attention to the impact of electrolyte ionic strength and counter-ion nature on HS electrophoretic mobility. The extended Onsager model reasonably described the measured electrophoretic mobilities in the ionic strength range 1 to 50 mM, thus allowing estimation of limiting mobilities and ionic charge numbers for the different HS samples (d'Orlyé and Reiller, 2012). An unexpected HS electrophoretic mobility increase was observed in the order $Li^+ < Na^+ < K^+ < Cs^+$ (d'Orlyé and Reiller, 2012). He et al. (2012) reviewed several case studies in which CE was used to investigate components of cotton (*Gossypium hirsutum* L.) and HS. These researchers monitored changes in homogalacturonans and activities of pectin-degrading enzymes during cotton cotyledon expansion, examined peptide fragments from structural components of cell wall grown in suspension culture, quantified gossypol in cottonseed, and characterized the fingerprint of several HS. These studies demonstrated that CE could improve our understanding of agricultural and soil chemistry factors that affect crop growth, plant-derived products, and the soil nutrient status (He et al., 2012).

The quoted literature clearly shows that CE has a good potential for delivering information about SOM characteristics, including more- and less-humified SOM fractions. Such analyses represent a useful approach because of CE's unique separation principle based on the charge of molecules in a strong electric field. This method is therefore a valuable complementary technique in combination with the well-established spectroscopic methods for SOM analyses (^{13}C NMR, mid-infrared, fluorimetry, and UV-Vis spectroscopy, etc.) or methods such as pyrolysis GC-MS. Tatzber et al. (2011) built on the results obtained for the highly diverse soil samples investigated in previous works. This allowed them to use the very well-defined soil sample set of the ^{14}C-labeled long-term field experiment in Fuchsenbigl (Austria) to clarify a key question: is the separation by CZE sensitive enough to make detectable those changes that occur in similar soils experiencing different soil management effects?

The ^{14}C-Labeled Long-Term Field Experiment in Fuchsenbigl, Austria: Basis for the Case Study

The Background Information on the Field Experiment

In 1967, a long-term field experiment with ^{14}C-labeled wheat straw and farmyard manure was established with three different cropping systems, that is, crop rotation (CR; sugarbeet [*Beta vulgaris* L. subsp. *vulgaris*], winter wheat [*Triticum aestivum* L.], and spring barley [*Hordeum vulgare* L. subsp. *vulgare*]), monoculture (MC; spring wheat [*T. aestivum*]), and permanent bare fallow (BF). A scheme of the long-term field experiment is depicted in Fig. 1. It is located in Fuchsenbigl, Lower Austria, approximately 45 km east of Vienna, in a wheat-producing region. A detailed description of the experiment is given in Zeller et al. (1968) and Oberländer and Roth (1974), including the generation of farmyard manure and the design of the long-term field experiment. Results about the depletion of the different labeled organic manures were reported by Oberländer and Roth (1980). The production of the ^{14}C-labeled wheat straw was described by Oberländer and Roth (1968). Compiled depletion curves of the ^{14}C-labeled straw and farmyard manures plus ^{14}C contents in the SOM pools extracted with 1M NaOH can be found in Tatzber et al. (2009b). Results of spectroscopic investigations of the extractable SOM are presented in Tatzber et al. (2009a).

Before the experiment started, the field site was used as cropland. The average annual temperature at the field site is 9.4°C, and the mean annual precipitation is 529 mm. The soil is described as Haplic Chernozem according to WRB (2006)

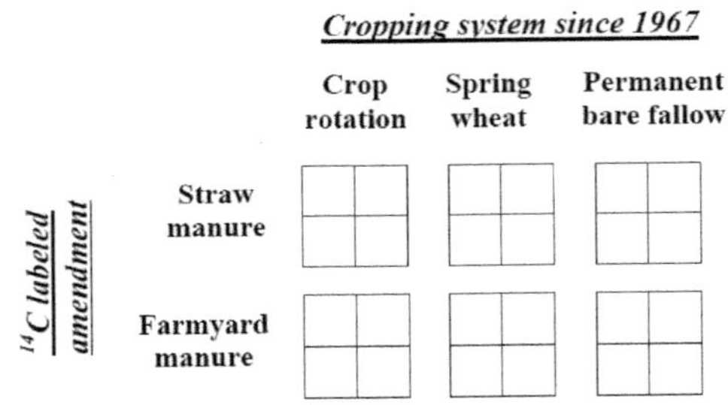

Fig. 1. Scheme of the long-term field experiment established in 1967 in Fuchsenbigl, Austria. The ^{14}C-labeled straw and farmyard manures were amended in 1967. The two exceptions are the straw manure on the plot experiencing crop rotation, where the amendment was carried out twice, and the straw manure on the plot with spring wheat monoculture, where the straw manure was added seven times.

(clay, 15%; silt, 41%; sand, 54%; pH ($CaCl_2$), 7.40; C_{org}, 1.53%; inorganic C, 1.61%; total N, 0.16%; values were measured before the beginning of the experiment; soil depth was 0–20 cm). The experimental area consisted of six plots (5.76 m^2 each), each divided into four subplots (1.44 m^2 each), which were separately tilled, cropped, harvested, and analyzed, leading to four replicates for each treatment. The ^{14}C-labeled material was applied as straw or farmyard manure to three plots each, treated with three different cropping systems each: CR, MC, and BF. The farmyard manure was produced by feeding a cow with ^{14}C-labeled plant material under veterinary supervision. The applied labeled straw (equivalent to 5 t dry matter ha^{-1}) was supplemented with 35 kg N ha^{-1} in all three cropping systems in August 1967. Labeled straw (plus mineral N) was amended once more for CR and six more times for MC. Labeled farmyard manure was applied once in November 1967 (equivalent to 5 t dry matter ha^{-1}) to all cropping systems. The radioactive straw was worked in by rotary hoeing and farmyard manure by hand spading followed by rotary hoeing. Immediately after the rotary hoeing, the first samples were taken and all plots were subsequently tilled to a depth of 20 cm. Organic residues were removed after harvest. Therefore, the annual plant input through plant biomass in CR and MC resulted from remaining roots and microorganisms.

All plots received 100 kg P_2O_5 ha^{-1} and 100 kg K_2O ha^{-1} in autumn. Cereals received additionally 40 kg N ha^{-1} in spring and 20 kg N ha^{-1} in the stage of ear emergence; sugar beet was fertilized with 160 kg N ha^{-1} plus additional boron. In BF, additional PK fertilization was stopped in 1977 and additional N fertilization in 1979. Fertilizer application rates were representative for on-farm practices, and plots were managed at high-yield level. The listed information plus more details can be accessed in Zeller et al. (1968), Oberländer and Roth (1968, 1974, 1980), and Tatzber et al. (2009b).

Previously published results about the ^{14}C-labeled experiment in Fuchsenbigl include soluble and nonsoluble SOM fractions (Zeller et al., 1968; Oberländer and Roth, 1974; Tatzber et al., 2009a,b, 2012), the depletion of the ^{14}C-labeled straw and farmyard manures (Oberländer and Roth, 1971, 1974, 1980; Oberländer, 1992; Stemmer et al., 2000; Tatzber et al., 2009a,b, 2012), microbial activities and rates of C mineralization (Stemmer et al., 2000) and yields, elemental contents (C and N), and the spectroscopic behavior of extractable SOM (Tatzber et al., 2009a,b, 2012). In two cases, the investigations included different particle size fractions, which were the silt (2–63 µm) and the clay (<2 µm) fraction (Stemmer et al., 2000; Tatzber et al., 2012).

This long-term experiment is unique due to its special layout. In particular, because of its well-defined and well-investigated plots, it provides an ideal basis for methodological developments in soil science. This was one of the main reasons why its samples were chosen for a second electrophoretic measurement series after successfully applying CE on the quite diverse set (Tatzber et al., 2011). This second series was designed to determine and illustrate how precisely the method developed in Tatzber et al. (2011) can detect changes in soil management systems.

Capillary Electrophoresis Features of Extracted Soil Organic Matter of the Fuchsenbigl Experiment

The purpose of this measurement series was to determine whether the signal assignments for the diverse sample set in Tatzber et al. (2011) could be verified for the data of the ^{14}C-labeled long-term field experiment in Fuchsenbigl. Basic data (C_{org}, total N, and basic data of extracted HAs) are provided in Table 1, and the data for SOM in the separated silt- and clay-size particles are listed in Table 2. An interesting fact is the highly significant behavior of the extraction yields of extracted HAs from the silt fraction (Table 2). The applied methodology for recording the electropherograms was identical to that in Tatzber et al. (2011). Briefly, the injection volume was 50 µL, the measurements were made with a stop time of 30 min, a voltage of 30 kV, and a performance of 0.6 W. The separated fractions were detected with a photometer at 254 nm. The electropherogram selection shown in Fig. 2 illustrates six evaluated signals of the HA. The biggest differences in the shape and size of peaks and humps are visible between the samples taken from the BF plot at the experiment's start in 1967 and in 2004 and between the HA sample from silt-sized particles and that from clay-sized particles. Figure 3 provides signal characteristics of selected electrophoretically separated HA from the BF plot until the year 2004. Exponential fitting equations describe the signal behavior over time. Finally, analyses of variance (ANOVAs) with the Scheffe post-hoc test were calculated; if the measured data were significantly different, it was indicated by respective letters in Fig. 3. Hump 2, peak 2, peak 4, and peak 6 showed significant signal increases with time, suggesting that they represent SOM fractions that are enriched during progressive humification of the bulk SOM (Fig. 3). For peak 3 and peak 5, such increases were indicated.

Table 1. Ranges of organic carbon (C_{org}) and total nitrogen (N_t) contents of the different bulk soil and humic acid (HA) samples together with the ranges of obtained extraction yields and ash contents for different farmyard manured cropping systems from 1967 to 2004. The values are given in arithmetic means (*n* = 3). (Compiled from Tatzber et al., 2009a,b).†

Cropping system‡	C_{org}	N_t	HA-C	HA-N	Yield of HA-C	Ash content
	g kg^{-1}		%		mg g^{-1}	%
CR	14.3–19.5	1.20–1.77	43.0–51.4	4.10–4.44	0.74–0.98	1.8–9.4
MC	14.9–20.6	1.23–1.65	44.8–50.5	3.99–4.36	0.69–0.99	1.7–3.8
BF	12.6–18.0	0.98–1.76	38.4–50.6	3.31–4.32	0.33–1.07	1.6–17.9

† See also http://www.publish.csiro.au/nid/84/paper/SR08231.htm.

‡ BF, bare fallow; CR, crop rotation; MC, monoculture.

Table 2. Selected properties of silt- and clay-size particles and extracted soil organic matter (from Tatzber et al., 2012).

Plot	Soil fraction in percentage of bulk soil	C_{org}†	N_t‡	Yield of HA-C§ in soil fraction	C_{org} in HA	N_t in HA	Ash content of HA
		g kg^{-1}			%		
Silt							
CR 2004	34.8 ± 0.9a¶	12.4 ± 1.3a	1.2 ± 0.1a	1.70 ± 0.02a	50.2 ± 0.1a	4.70 ± 0.24a	2.1 ± 0.6
MC 2004	34.1 ± 0.6a	11.6 ± 4.3a	1.4 ± 0.3a	1.55 ± 0.06b	50.7 ± 1.8a	4.37 ± 0.03ab	2.2 ± 0.7
BF 2004	34.8 ± 0.6a	9.2 ± 0.8a	0.9 ± 0.1a	0.75 ± 0.05c	48.4 ± 0.3a	4.20 ± 0.09b	3.8 ± 0.7
Clay							
CR 2004	9.6 ± 0.7a	45.5 ± 1.1a	4.6 ± 0.1a	0.74 ± 0.06a	44.3 ± 1.8a	4.70 ± 0.36a	2.8 ± 1.2
MC 2004	8.9 ± 0.6a	41.6 ± 0.3b	4.4 ± 0.1a	0.63 ± 0.04a	46.0 ± 0.4a	4.88 ± 0.07a	2.1 ± 1.6
BF 2004	9.2 ± 0.5a	35.9 ± 1.2c	3.4 ± 0.1b	0.49 ± 0.05b	40.0 ± 1.4b	4.01 ± 0.13b	12.5 ± 8.7

† C_{org}, total organic C.

‡ N_t, total N content.

§ Yield of humic acids (HAs) × C_{org} of HA.

¶ Different letters label significantly different groups ($P < 0.05$); a, b, and c identify groups in the cropping systems, x and y differences between particle sizes.

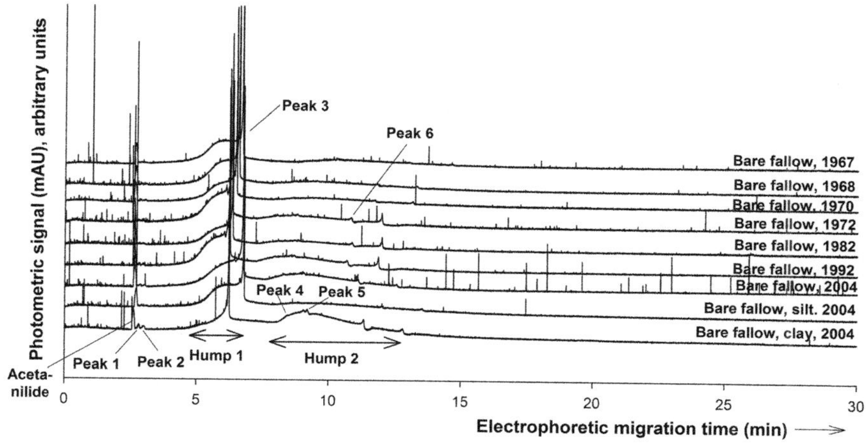

Fig. 2. Representative electropherograms of humic acids (HAs) of the time sequence extending from 1967 to 2004 and of HAs extracted from separated silt- and clay-sized fractions of samples from 2004.

Hump 1 (data not shown) decreased from 1967 until 1982 and then increased after that time point again; peak 1 also showed no constant increase or decrease. The signal deriving from acetanilide (data not shown), which is an internal standard and labels the beginning of the electropherogram, also showed no significant differences. This enabled excluding artifacts in the scope of the measurement procedure. Nonetheless, the behavior of the signal intensity of hump 1 was quite unexpected; based on the results of previous work (Tatzber et al., 2011), a continuous depletion of signal intensity with time was expected. Our explanation was that the SOM fraction reflected by hump 1 consists of more subfractions than solely labile ones and that, in this specific soil system, fragments of other depleted SOM pools entered this fraction over time; they might have compensated the initial loss of the labile proportions. The increase of signal intensities of hump 2 as well as of the peaks was in line with the observation of the aforementioned study (Tatzber et al., 2011). The suggested stabilized nature for most of these electrophoretic signals was confirmed by the data of the ^{14}C-labeled long-term field experiment.

The impression that hump 1 consists of several subfractions is supported by the data shown in Fig. 4, where HAs were extracted from silt- and clay-sized particles of samples taken in 2004. Accordingly, the signal intensities of hump 1 were significantly higher for SOM extracted from silt- than from clay-sized particles. The data presented in Figures 4 and 5 show that all other signals except hump 1 and peak 3 had the significantly higher intensities in the clay-sized fraction. As presented in Tatzber et al. (2012) the HA from silt-sized particles were less humified than HA from clay-sized particles. These results were based on many data (remaining farmyard manure ^{14}C, extraction yields, C_{org}, total N, and spectroscopic data from mid-infrared, fluorescence, and UV-Vis spectroscopy). Except for peak 3, the opposite behavior of hump 1 to the other signals of the electropherogram (Fig. 4, 5) was observed between the different particle fractions.

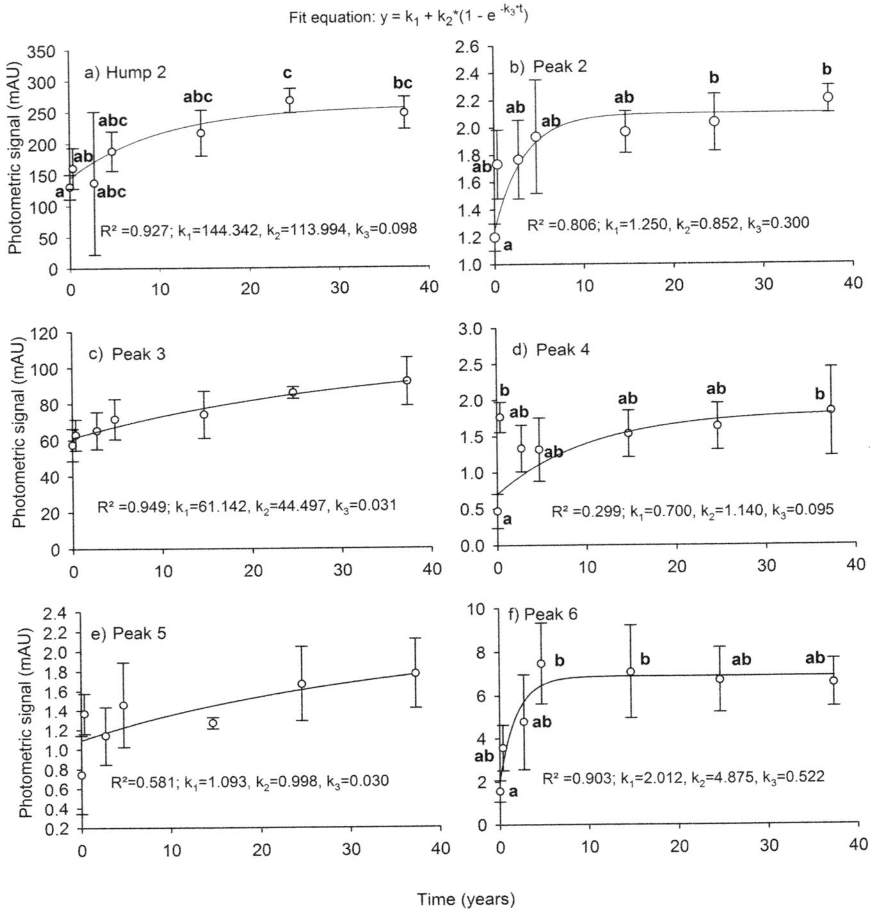

Fig. 3. Selection of integrated photometric signal areas of electrophoretically separated humic acids (HAs) plus calculated equations behind the shown graphs. Regression equations were calculated with Matlab (v. 7.0.1) applying the module fminsearchbnd. The location of the signals in the electropherogram is shown in Fig. 2. Different letters indicate significantly different groups (*n* = 3, ANOVA calculated with SPSS, the PostHoc test was the Scheffe's procedure).

Peak 3 already showed a quite different behavior compared with the other signals in Tatzber et al. (2011). This was explained as an artifact because of complex formation between borate ions and 1,2-diols, 1,3-diols, or both (Schmitt-Kopplin et al., 1998b; Schmitt-Kopplin and Junkers, 2003). The results depicted in Figures 3 through 5 confirm most of the conclusions drawn in Tatzber et al. (2011); with respect to determination of the labile character of SOM, all signals except hump 1 and peak 3 are important, that is, the lower their signal areas, the higher the labile character of the SOM sample. Based on the data of the ^{14}C-labeled long-term field experiment, the conclusions concerning hump 1 must be modified; this fraction was assigned to rather labile SOM (Tatzber et al., 2011), but the data from the time

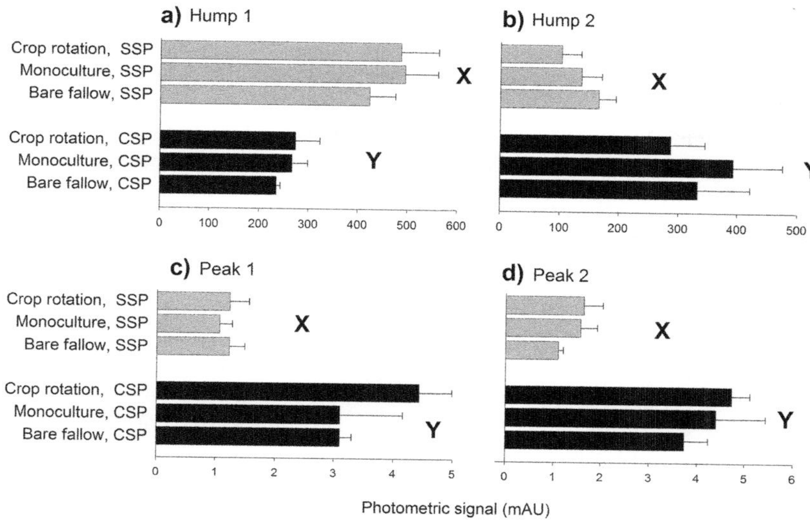

Fig. 4. Signal areas of the two humps and the first two peaks of the electropherograms for humic acids (HAs) from soil samples taken in 2004 from plots of the three different cropping systems (crop rotation, monoculture, and permanent bare fallow) and separated into silt-sized particles (SSP) and clay-sized particles (CSP). Different letters indicate significantly different groups ($n = 3$, ANOVA calculated with SPSS, the PostHoc test was the Scheffe's procedure).

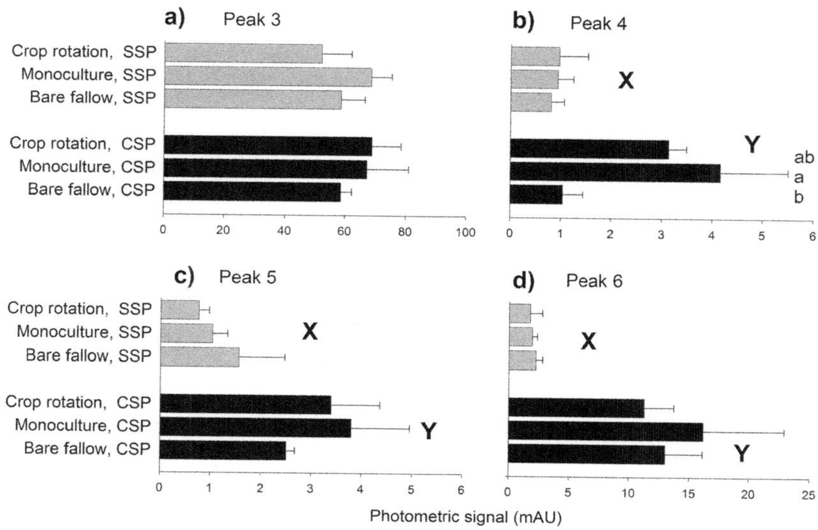

Fig. 5. Signal areas of the four peaks occurring at later migration times in electropherograms of humic acids (HAs) from soil samples taken in 2004 from plots of the three different cropping systems (crop rotation, monoculture, and permanent bare fallow) and separated into silt-sized particles (SSP) and clay-sized particles (CSP). Different letters indicate significantly different groups ($n = 3$, ANOVA calculated with SPSS, the PostHoc test was the Scheffe's procedure).

sequence of the BF plot of the ^{14}C experiment (Fig. 3) suggest that this fraction consists of several subfractions. These might not all be labile or could be depletion products of other fractions entering this fraction (and maybe also eventually leaving it). Nonetheless, the significantly enhanced signal intensity of hump 1 in the HA extracted from silt-sized particles (vs. clay-sized particles; see Fig. 4) still show that a considerable part of this fraction is recruited from rather labile SOM pools. Further investigations would be needed to validate the assignment of the SOM fraction represented by hump 1.

Curves for different signals have also been calculated based on mid-infrared spectroscopy on the extracted SOM (Tatzber et al., 2009a). These curves reflected mid-infrared signal areas from functional groups such as aromatics, carbonyl groups, and different amide bands and had different shapes. This was not surprising because the electrophoretically separated fractions contained different functional groups and hence, different SOM pools. In the ^{14}C-labeled long-term field trial, this priming effect was visible in several electrophoretic signals in the time sequence of the BF plot receiving ^{14}C-labeled farmyard manure in 1967 (see especially peaks 2, 4, and 5 in Fig. 3). Interestingly, most of the functional groups detected by infrared spectroscopy also reflected the priming effect that occurred in 1968; this was true for aromatics, carbonyl groups, and perhaps the amide III band (Tatzber et al., 2009a). Priming effects are well known and occur usually shortly after additions of organic matter to field experiments. Jenkinson (1966) describes four potential pathways leading to priming effects. From our point of view, the most probable causes for this effect in the ^{14}C-labeled long-term field trial are differences between the microbial environments in soils with and without organic amendments; these are associated with different mechanisms by which organic additions can directly increase the decomposition rate of the native organic matter (Jenkinson, 1966). The concordance between mid-infrared spectroscopy and UV-visible detection of electrophoretically separated SOM fractions regarding this priming effect allows the final conclusion that it occurred in the long-term field experiment and that both methodologies are sensitive enough to detect the influences of priming effects on SOM. As the aromatic and carbonyl groups showed a clear increase (Tatzber et al., 2009a) and the amide and sulfone bands decreased, we suggest that hump 2, peak 2, peak 4, and peak 6 are connected (among others) with aromatic and carbonyl groups.

Conclusions

Capillary electrophoresis has been used in characterization of soluble fractions of SOM samples and their changes due to various environmental and management impacts. The brief overview of the current status of CE determination of SOM features provides an introduction to apply various CE techniques in SOM studies. In the case study, significantly different electrophoretic signal intensities of extracted SOM were obtained for both a time sequence of the BF plot and particle size fractions of the ^{14}C labeled long-term field experiment. Hump 2, peak 2, peak 4, and peak 6 were significantly different in both the time-sequence and between particle size fractions. These signals are suggested to represent stabilized SOM fractions due to their increasing signal intensity over time in a BF plot since 1967. Hump 1 was significantly different between the separated fractions of silt and clay. Whether it really represents labile fractions must be clarified by further studies. The following electrophoretic signals detected priming effects: hump 2, peak 4, peak 5, and peak 6; for peak 4 the signal assignable to the priming effect (in 1968) was even significantly different compared with the initial signal in 1967. As a consequence, CE is very well applicable for detecting SOM stabilization. For future SOM studies using CE, focus on alternative detection principles than photometry (especially fluorimetry and mass spectroscopy) appears promising. Another important field for future research is the molecular build up of the obtained electrophoretic fractions; also investigations of them using specific detection systems (e.g., coupling with mass spectroscopy or application of pyrolysis GC-MS) would deliver valuable information. Further, there can be no doubt that CE on HA is a suitable tool for all kinds of scientific investigations that focus on SOM stabilization.

Acknowledgments

Parts of this work were financed by the project Modellierung des Stofftransportes in der ungesättigten Zone—Verbesserung bestehender Modelle, which was funded by the Austrian Institute of Technology and by the Austrian Science Fund (FWF), project number P16667-B06. We thank Gabriele Pusch for helpful information. The authors thank Leonhard Jaitz and Maria Fürhacker for helpful suggestions and Walter Frank for valuable technical support. We are much obliged to Ruth Pöll, Elisabeth Kopecky, and Ewald Brauner for performing the elemental analyses. Special thanks to Michael Stachowitsch for his detailed and conscientious linguistic input.

References

Abelenda, M.S., P. Buurman, M. Camps Arbestain, J. Kaal, A. Martinez-Cortizas, N. Gartzia-Bengoetxea, and F. Macías. 2011. Comparing NaOH-extractable organic matter of acid forest soils that differ in their pedogenic trends: A pyrolysis-GC/MS study. Eur. J. Soil Sci. 62:834–848. doi:10.1111/j.1365-2389.2011.01404.x

Cao, X., D.C. Olk, M. Chappell, C.A. Cambardella, L.F. Miller, and J. Mao. 2011. Solid-state NMR analysis of soil organic matter fractions from integrated physical-chemical extraction. Soil Sci. Soc. Am. J. 75:1374–1384. doi:10.2136/sssaj2010.0382

Chantigny, M.H. 2003. Dissolved and water-extractable organic matter in soils: A review on the influence of land use and management practices. Geoderma 113:357–380. doi:10.1016/S0016-7061(02)00370-1

Cavani, L., C. Ciavatta, O.E. Trubetskaya, O.I. Reznikova, G.V. Afanas'eva, and O.A. Trubetskoj. 2003. Capillary zone electrophoresis of soil humic acid fractions obtained by coupling size-exclusion chromatography and polyacrylamide gel electrophoresis. J. Chromatogr. A 983:263–270. doi:10.1016/S0021-9673(02)01737-5

Chen, M.Y., Y.Z. Chang, F.J. Lu, and J.L. Chen. 2010. Capillary electrophoretic determination of selected phenolic compounds in humic substances of well waters and fertilizers. Anal. Sci. 26:561–567. doi:10.2116/analsci.26.561

de Moraes, S.L., and M.O.O. Rezende. 2008. Capillary electrophoresis (CE): A powerful tool to characterize humic acid (HA). J. Braz. Chem. Soc. 19:24–28. doi:10.1590/S0103-50532008000100005

d'Orlyé, F., and P.E. Reiller. 2012. Contribution of capillary electrophoresis to an integrated vision of humic substances size and charge characterizations. J. Colloid Interface Sci. 368:231–240. doi:10.1016/j.jcis.2011.11.046

Dunkelog, R., H.H. Rüttinger, and K. Peisker. 1997. Comparative study for the separation of aquatic humic substances by electrophoresis. J. Chromatogr. A 777:355–362. doi:10.1016/S0021-9673(97)00330-0

Fang, C., P. Smith, J.B. Moncrieff, and J.U. Smith. 2005. Similar response of labile and resistant soil organic matter pools to changes in temperature. Nature 433:57–59. doi:10.1038/nature03138

Fetsch, D., M. Hradilová, E.M. Peña Méndez, and J. Havel. 1998. Capillary zone electrophoresis study of aggregation of humic substances. J. Chromatogr. A 817:313–323. doi:10.1016/S0021-9673(98)00369-0

Francioso, O., D. Montecchio, P. Gioacchini, L. Cavani, C. Ciavatta, O. Trubetskoj, and O. Trubetskaya. 2009. Structural differences of Chernozem soil humic acids SEC-PAGE fractions revealed by thermal (TG-DTA) and spectroscopic (DRIFT) analyses. Geoderma 152:264–268. doi:10.1016/j.geoderma.2009.06.011

Garrison, A.W., P. Schmitt, and A. Kettrup. 1995. Capillary electrophoresis for the characterization of humic substances. Water Res. 29:2149–2159. doi:10.1016/0043-1354(95)00031-F

Gieguzynska, E., A. Amine-Khodja, O.A. Trubetskoj, O. Trubetskaya, G. Guyot, A. ter Halle, D. Golebiowska, and C. Richard. 2009. Compositional differences between soil humic acids extracted by various methods as evidenced by photosensitizing and electrophoretic properties. Chemosphere 75:1082–1088. doi:10.1016/j.chemosphere.2009.01.047

He, Z., T. Ohno, F. Wu, D.C. Olk, C.W. Honeycutt, and M. Olanya. 2008. Capillary electrophoresis and fluorescence excitation-emission matrix spectroscopy for characterization of humic substances. Soil Sci. Soc. Am. J. 72:1248–1255. doi:10.2136/sssaj2007.0305

He, Z., T. Ohno, D.C. Olk, and F. Wu. 2010. Capillary electrophoresis profiles and fluorophore components of humic acids in Nebraska corn and Philippine rice soils. Geoderma 156:143–151. doi:10.1016/j.geoderma.2010.02.011

He, Z., H.M. Waldrip, and Y. Wang. 2012. Applications of capillary electrophoresis in agricultural and soil chemistry research. In: Z. He, editor, Capillary electrophoresis. Nova Science Publishers, Inc., Hauppauge, NY. p. 131–151.

Janoš, P. 2003. Separation methods in the chemistry of humic substances. J. Chromatogr. A 983:1–18. doi:10.1016/S0021-9673(02)01687-4

Jenkinson, D.S. 1966. The priming action. In: The use of isotopes in soil organic matter studies. Report of the FAO/IAEA Technical Meeting, Symposium Publications division, Pergamon Press, Oxford. p. 199–208.

Kalbitz, K., J. Schmerwitz, D. Schwesig, and E. Matzner. 2003. Biodegradation of soil-derived dissolved organic matter as related to its properties. Geoderma 113:273–291. doi:10.1016/S0016-7061(02)00365-8

Kalbitz, K., S. Solinger, J.H. Park, B. Michalzik, and E. Matzner. 2000. Controls on the dynamics dissolved organic matter in soils: A review. Soil Sci. 165:277–304. doi:10.1097/00010694-200004000-00001

Kaniansky, D., M. Masár, and J. Bielčíková. 1997. Electroosmotic flow suppressing additives for capillary zone electrophoresis in a hydrodynamically closed separation system. J. Chromatogr. A 792:483–494. doi:10.1016/S0021-9673(97)00743-7

Knorr, W., I.C. Prentice, J.I. House, and E.A. Holland. 2005. Long-term sensitivity of soil carbon turnover to warming. Nature 433:298–301. doi:10.1038/nature03226

Krivácsy, Z., G. Kiss, B. Varga, I. Galambos, Zs. Sárváribb, A. Gelencséra, Á. Molnára, S. Fuzzic, M.C. Facchinic, S. Zappolid, A. Andracchiod, T. Alsberge, H.C. Hansson, and L. Persson. 2000. Study of humic-like substances in fog and interstitial aerosol by size-exclusion chromatography and capillary electrophoresis. Atmos. Environ. 34:4273–4281. doi:10.1016/S1352-2310(00)00211-9

Kutsch, H., and B. Schumacher. 1994. Isoelectric focusing of humic substances on ultrathin polyacrylamide gels: Evidence of fingerprint performance. Biol. Fertil. Soils 18:163–167. doi:10.1007/BF00336464

Oberländer, H.E., and K. Roth. 1968. The use of a growth chamber for labelling plants with ^{14}C. Atompraxis 14:1–6.

Oberländer, H.E., and K. Roth. 1971. Der Umsatz von ^{14}C-markiertem Stroh und Stallmist im Boden. In: Vortragstagung Chemie und landwirtschaftliche Produktion: 100 Jahre Landwirtschaftlich-Chemische Bundesversuchsanstalt, 1870–1970. Proceedings, Landwirtsch.-Chem. Bundesversuchsanst., Vienna. p. 143–150.

Oberländer, H.E., and K. Roth. 1974. A small scale field trial on the decomposition and humification of ^{14}C labelled straw and farm-yard manure. J. Landwirtsch. Forsch. 25:111–129.

Oberländer, H.E., and K. Roth. 1980. Turnover of ^{14}C labelled organic manures in soil. Landwirtsch. Forsch. 33:179–188.

Oberländer, H.E. 1992. Humusbildung und Humuserhaltung in einer nachhaltig betriebenen Landwirtschaft. In: Riesenschilfgras, Proc. Worksh. 11–12 Mar. 1992. Ministry of Sci. and Res., Vienna. p. 47-70.

Parlanti, E., B. Morin, and L. Vacher. 2002. Combined 3D-spectrofluorometry, high performance liquid chromatography and capillary electrophoresis for the characterization of dissolved organic matter in natural waters. Org. Geochem. 33:221–236. doi:10.1016/S0146-6380(01)00154-1

Peuravuori, J., V. Lepane, T. Lehtonen, and K. Pihlaja. 2004. Comparative study for separation of aquatic humic substances by capillary zone electrophoresis using uncoated, polymer coated and gel-filled capillaries. J. Chromatogr. A 1023:129–142. doi:10.1016/j.chroma.2003.09.054

Pompe, S., K.-H. Heise, and H. Nitsche. 1996. Capillary electrophoresis for a "fingerprint" characterization of fulvic and humic acids. J. Chromatogr. A 723:215–218. doi:10.1016/0021-9673(95)00815-2

Richard, C., G. Guyot, A. Rivaton, O. Trubetskoj, L. Cavani, and C. Claudio Ciavatta. 2007. Spectroscopic approach for elucidation of structural peculiarities of Andisol soil humic acid fractionated by SEC-PAGE setup. Geoderma 142:210–216. doi:10.1016/j.geoderma.2007.08.019

Schmitt, Ph., A.W. Garrison, D. Freitag, and A. Kettrup. 1997. Capillary isoelectric focusing (CIEF) for the characterization of humic substances. Water Res. 31:2037–2049. doi:10.1016/S0043-1354(97)00021-3

Schmitt-Kopplin, Ph., K. Fischer, D. Freitag, and A. Kettrup. 1998a. Capillary electrophoresis for the simultaneous separation of selected carboxylated carbohydrates and their related 1,4-lactones. J. Chromatogr. A 807:89–100. doi:10.1016/S0021-9673(98)00056-9

Schmitt-Kopplin, Ph., D. Freitag, A. Kettrup, N. Hertkorn, U. Schoen, R. Klöcking, B. Helbig, F. Andreux, and A.W. Garrison. 1999. Analysis of synthetic humic substances for medical and environmental applications by capillary zone electrophoresis. Analusis 27:390–395. doi:10.1051/analusis:1999270390

Schmitt-Kopplin, Ph., A.W. Garrison, E.M. Perdue, D. Freitag, and A. Kettrup. 1998b. Capillary electrophoresis in the analysis of humic substances facts and artifacts. J. Chromatogr. A 807:101–109. doi:10.1016/S0021-9673(98)00077-6

Schmitt-Kopplin, Ph., N. Hertkorn, A.W. Garrison, D. Freitag, and A. Kettrup. 1998c. Influence of borate buffers on the electrophoretic behavior of humic substances in capillary zone electrophoresis. Anal. Chem. 70:3798–3808. doi:10.1021/ac971223j

Schmitt-Kopplin, P., and J. Junkers. 2003. Capillary zone electrophoresis of natural organic matter. J. Chromatogr. A 998:1–20. doi:10.1016/S0021-9673(03)00636-8

Široká, J., P. Jáč, and M. Polášek. 2011. Use of inorganic, complex-forming ions for selectivity enhancement in capillary electrophoretic separation of organic compounds. Trends Analyt. Chem. 30:142–152. doi:10.1016/j.trac.2010.08.006

Stemmer, M., K. Roth, and E. Kandeler. 2000. Carbon mineralization and microbial activity in a field site trial used for ^{14}C turnover experiments over a period of 30 years. Biol. Fertil. Soils 31:294–302. doi:10.1007/s003740050659

Takahashi, T., J. Kawana, and H. Hoshino. 2009. Capillary electrophoretic separation of humic substances using hydroxyethyl cellulose as a buffer additive and its application to characterization of humic substances in a river water sample. Anal. Bioanal. Chem. 393:761–766. doi:10.1007/s00216-008-2476-z

Tatzber, M., F. Mutsch, A. Mentler, E. Leitgeb, M. Englisch, and M.H. Gerzabek. 2011. Capillary electrophoresis characterisation of humic acids: Application to diverse forest soil samples. Environ. Chem. 8:589–601. doi:10.1071/EN11054

Tatzber, M., M. Stemmer, H. Spiegel, C. Katzlberger, G. Haberhauer, and M.H. Gerzabek. 2008. Impact of different tillage practices on molecular characteristics of humic acids in a long-term field experiment– An application of three different spectroscopic methods. Sci. Total Environ. 406:256–268. doi:10.1016/j.scitotenv.2008.07.048

Tatzber, M., M. Stemmer, H. Spiegel, C. Katzlberger, G. Haberhauer, A. Mentler, and M.H. Gerzabek. 2007. Development and application of advanced FTIR characterization methods for extracted and residual humic substances. J. Plant Nutr. Soil Sci. 170:522–529. doi:10.1002/jpln.200622082

Tatzber, M., M. Stemmer, H. Spiegel, C. Katzlberger, C. Landstetter, G. Haberhauer, and M.H. Gerzabek. 2012. ^{14}C-labeled organic amendments: Characterization in different particle size fractions and humic acids in a long-term field experiment. Geoderma 177-178:39–48. doi:10.1016/j.geoderma.2012.01.028

Tatzber, M., M. Stemmer, H. Spiegel, C. Katzlberger, F. Zehetner, G. Haberhauer, E. Garcia-Garcia, and M.H. Gerzabek. 2009a. Spectroscopic behaviour of ^{14}C-labeled humic acids in a long-term field experiment with three cropping systems. Aust. J. Soil Res. 47:459-469. doi:10.1071/SR08231

Tatzber, M., M. Stemmer, H. Spiegel, C. Katzlberger, F. Zehetner, G. Haberhauer, K. Roth, E. Garcia-Garcia, and M.H. Gerzabek. 2009b. Decomposition of ^{14}C-labeled organic amendments and humic acids in three cropping systems of a long-term field experiment. Soil Sci. Soc. Am. J. 73:744–750. doi:10.2136/sssaj2008.0235

Trubetskaya, O.E., O.A. Trubetskoi, B.A. Borisov, and N.F. Ganzhara. 2008. Electrophoresis and size-exclusion chromatography of humic substances extracted from detritus and soils of different geneses. Eurasian Soil Sci. 41:171–175. doi:10.1134/S1064229308020087

Trubetskaya, O., O. Trubetskoj, G. Guyot, F. Andreux, and C. Richard. 2002. Fluorescence of soil humic acids and their fractions obtained by tandem size exclusion chromatography-polyacrylamide gel electrophoresis. Org. Geochem. 33:213–220. doi:10.1016/S0146-6380(01)00153-X

Trubetskoj, O.A., P.G. Hatcher, and O.E. Trubetskaya. 2010. ^{1}H-NMR and ^{13}C-NMR spectroscopy of chernozem soil humic acid fractionated by combined size-exclusion chromatography and electrophoresis. Chem. Ecol. 26:315–325. doi:10.1080/02757541003785825

Trubetskoj, O.A., O.E. Trubetskaya, G.V. Afanas'eva, O.I. Reznikova, and C. Saiz-Jimenez. 1997. Polyacrylamide gel electrophoresis of soil humic acid fractionated by size-exclusion chromatography and ultrafiltration. J. Chromatogr. A 767:285–292. doi:10.1016/S0021-9673(97)00019-8

Trubetskoj, O., O. Trubetskaya, O. Reznikova, and G. Afanas'eva. 1999. Weight and optical differences between soil humic acids fractions obtained by coupling SEC-PAGE. Geoderma 93:277–287. doi:10.1016/S0016-7061(99)00070-1

Übner, M., M. Kaljurand, and M. Lopp. 2010. Humic substances as a background electrolyte in capillary electrophoresis. Chem. Ecol. 26:157–165. doi:10.1080/02757540.2010.494157

Vanifatova, N.G., A.G. Zavarzina, and B.Ya. Spivakov. 2008. Potential of capillary zone electrophoresis for estimation of humate acid-base properties. J. Chromatogr. A 1183:186–191. doi:10.1016/j.chroma.2008.01.004

von Lützow, M., I. Kögel-Knabner, K. Ekschmitt, H. Flessa, G. Guggenberger, E. Matzner, and B. Marschner. 2007. SOM fractionation methods: Relevance to functional pools and to stabilization mechanisms. Soil Biol. Biochem. 39:2183–2207. doi:10.1016/j.soilbio.2007.03.007

WRB. 2006. World reference base for soil resources. A framework for international classification, correlation and communication. World soil resources report No. 103, Food and Agriculture Organisation of the UN, Rome.

Zeller, A., H.E. Oberländer, and K. Roth. 1968. A field experiment on the influence of cultivation practices on the transformation of ^{14}C-labelled farmyard manure and ^{14}C-labelled straw into humic substances In: Isotopes and Radiation in soil organic-matter studies. Proc. Ser. Int. Atomic Energy Agency, Vienna. p. 265–274.

Composition of Whole and Water-Extractable Organic Matter of Cattle Manure Affected by Management Practices

Zhongqi He and Heidi M. Waldrip*

Abstract

Organic matter (OM) is a major component of animal manure. In this chapter, we present two case studies on the multiple spectral features of whole and water-extractable organic matter (WEOM) of cattle manure affected by differing management practices. Using wet chemistry and Fourier-transform infrared (FT-IR) and ultraviolet–visible (UV-Vis) spectroscopic techniques, Case Study I demonstrated that the composition and contents of OM and nutrients in beef cattle manure differ depending on the location on a feedyard (i.e., unconsolidated surface manure and the underlying manure pack in feedyard pens, settling basin, and retention pond sediments), with higher concentrations of available nutrients in recently excreted materials and increased molecular weight and humification as manure aged during its on-feedyard life cycle. Using multiple spectroscopic techniques, Case Study II compared structural and functional groups in dairy manure between dairies under conventional and organic management. Primary findings were that cutin- and cutan-related ^{13}C nuclear magnetic resonance (NMR) peaks were useful for tracing dietary management to distinguish grass-fed from grain-fed animals. Furthermore, pharmaceutical-related pyrolysis gas chromatograph–mass spectroscopy (GC-MS) peaks can monitor use of antibiotics and other pharmaceuticals in cattle. Thus, spectral features are traceable markers that may be useful to certify that dairy and beef cattle are managed according to standards for grass-fed or organic products. Confirmation of the relationships between NMR and pyrolysis GC-MS features with forage ingredients and pharmaceutical additives using manure from controlled studies (i.e., specific diet and management information are known) will provide scientific basis for authentication of organic and alternative cattle management practices.

Abbreviations: CD, conventional dairy; CP/TOSS, cross-polarization/total sideband suppression; DM, dry matter; DN, dissolved N; DOC, dissolved organic carbon; EEM, fluorescence excitation–emission matrix; FT-IR, Fourier-transform infrared; GC-MS, gas chromatograph–mass spectroscopy; NMR, nuclear magnetic resonance; OD, organic dairy; OM, organic matter; UV-Vis, ultraviolet–visible; WEOM, water-extractable organic matter; WW, wet weight.

Zhongqi He, USDA-ARS, Southern Regional Research Center, 1100 Robert E. Lee Blvd., New Orleans, LA 70124, USA (zhongqi.he@ars.usda.gov). Heidi M. Waldrip, USDA-ARS, Conservation and Production Research Laboratory, P.O. Drawer 10, 2300 Experiment Station Road, Bushland, TX 79012. *Corresponding author (heidi.waldrip@ars.usda.gov).

doi:10.2136/sssaspecpub62.2014.0034

Labile Organic Matter—Chemical Compositions, Function, and Significance in Soil and the Environment. SSSA Special Publication 62. Zhongqi He and Fengchang Wu, editors.

Approximately 1.1 billion Mg (wet weight, WW) manure are produced annually in the United States by domestic livestock, with cattle manure (beef and dairy) accounting for >80% of total manure production (~0.92 billion Mg yr^{-1} [WW]) (Zhang and Schroder, 2014). Animal manure is a mixture of urine, feces, bedding (e.g., straw, wood shaving), scurf, feed residue, and water and contains high concentrations of OM (Powell, 2012). Manure has been traditionally used as crop fertilizer but also has applications as a soil amendment, bioenergy feedstock, and as a means to remediate contaminated soils (He, 2012; Waldrip et al., 2011, 2012a; Jensen and Sommer, 2013). In *Environmental Chemistry of Animal Manure* (He, 2011), five chapters were dedicated to review and discussion of the chemical composition and structural environments of OM in animal manure and relevant compost. Techniques employed included pyrolysis mass spectrometry (Wang et al., 2011), infrared spectroscopy (He et al., 2011a), solid-state ^{13}C NMR spectroscopy (He and Mao, 2011), UV-Vis spectroscopy (Zhang et al., 2011), and fluorescence spectroscopy (Ohno and He, 2011). The first three spectroscopic techniques (pyrolysis mass spectrometry, infrared, and NMR spectroscopies) can be used for characterization of both whole OM and WEOM. The latter two techniques (UV-Vis and fluorescence spectroscopies) are used only for analysis of WEOM. Numerous reports have been published on the composition and functional group contents of OM of different manure types and other organic compounds as well as OM stability, decomposition characteristics, and fate in soil and the environment (e.g., He et al., 2011a; Harwood, 2014; Luo et al., 2013; Singh et al., 2014). However, studies on the influence of management practices, including animal diet and manure handling, on properties of manure OM are very limited. In this chapter, we present two case studies on the multiple spectral features of whole OM and WEOM of cattle (beef and dairy) manure affected by management practices. Information on the variance of manure OM components derived from these spectroscopic investigations could be used to assist beef cattle and dairy producers in making management decisions that influence manure fertilizer value and environmental impact in the event of runoff or leaching from animal housing or manure storages. In addition, this type of information could be used by regulatory agencies to certify that organic production regulations are met and improve recommendations for manure management to producers by extension and conservation agency personnel.

Case Study I: Comparison of Manure Organic Matter Collected from Different Areas of Open-Lot Beef Cattle Feedyards

Beef Cattle Feedyard Background Information

This case study is part of the work reported in Waldrip et al. (2014). In the southern High Plains (Texas Panhandle), open-lot feedyards are typical for beef cattle production. The commercial feedyard used for this study contained approximately 20,000 cattle (one-time capacity) in pens at a stocking density of 15 m^2 $head^{-1}$. The cattle were fed high-concentrate finishing diets that were based on steam-flaked corn (*Zea mays* L.) and contained 7 to 10% alfalfa (*Medicago sativa* L.). Dietary crude protein concentration was generally 13.0 to 14.5% of dry matter;

however, this varied depending on available feedstuffs. The feedyard is sloped to the west, and most of the manure solids in runoff are captured in three settling basins (185-m long by 12-m wide) (Fig. 1a,b). Approximately two to three times per year, when finished cattle are removed for slaughter, pens at this feedyard are scraped almost to bare soil and manure harvested. A thin layer of manure is usually left in the pens to avoid disrupting the soil–manure interface layer that prevents leachate losses of nitrate (NO_3^-). The liquid fraction of the runoff is diverted from settling basins into a 1.33-km circumference earthen retention pond lined with a layer of bentonite clay (Fig. 1a,b). Four types of manure samples were collected from this feedyard: (i) surface manure collected from five representative pens of finishing cattle by gently scraping the loose, unconsolidated manure from pen surfaces with a garden trowel; (ii) manure pack, large chunks (~12-cm diam.) of dry–wet pack were collected from stockpiled manure harvested from three recently scraped pens (Fig. 1c); (iii) settling basin sediment (comprised largely of solids from pen runoff) collected from three locations with a garden trowel to a depth of 25 cm (Fig. 1d); and (iv) retention pond sediment (comprised of material originating from overflow or diversion from settling basins), three samples taken with a garden trowel to a depth of 12 cm from three locations during a period of low precipitation when the pond was dry (Fig. 1e). Each sample was air-dried, ground to pass 2 mm, and stored at –20°C until use.

Chemical Composition of Manure Samples from Different Feedyard Sources

The chemical compositions of feedyard manures are presented in Table 1. There were no significant differences in total Kjehldahl nitrogen (N) and organic N concentrations between samples taken from surface manure, manure pack, or retention pond sediment. However, changes were observed in total ammoniacal N (NH_3 + NH_4^+; NH_x–N), with significantly ($P < 0.05$) higher NH_x–N concentrations in surface manure than manure pack and samples taken from other locations at the feedyard. Specifically, loose surface manure contained 91% more NH_x–N than stockpiled pack manure. These data indicate a high N loss due to ammonia (NH_3) volatilization during the on-farm manure life cycle. The highest concentration of total phosphorus (P) was in the settling basin and retention pond sediments, and P concentrations decreased in the order of settling basin ≥ retention pond > manure pack > surface manure. In soils, P mobility is usually limited by its propensity to interact with mineral and organic components; however, the high pH of feedyard manure (~7.5–8.0 [Waldrip et al., 2012b]) could lead to greater net negative charge of OM and thus a decrease in sorption of phosphate anions (Bolan et al., 1988; Fernandes and Coutinho, 1999) and higher concentrations of soluble P in runoff. In contrast, concentrations of total potassium (K) were highest in the manure pack and decreased in the order manure pack ≥ surface manure ≥ retention pond > settling basin. The inconsistent orders of N, P, and K levels in these manures indicate that when used as fertilizer, settling basin sediment, in addition to posing a risk of NO_3^- leaching, could potentially lead to excessive soil P accumulation or conditions that are limiting in sufficient K for crop growth.

There was 35% less OM in the manure pack samples than surface manure. Further, even lower content of OM was found in the samples collected from the settling basin and retention pond than more freshly excreted pen manure.

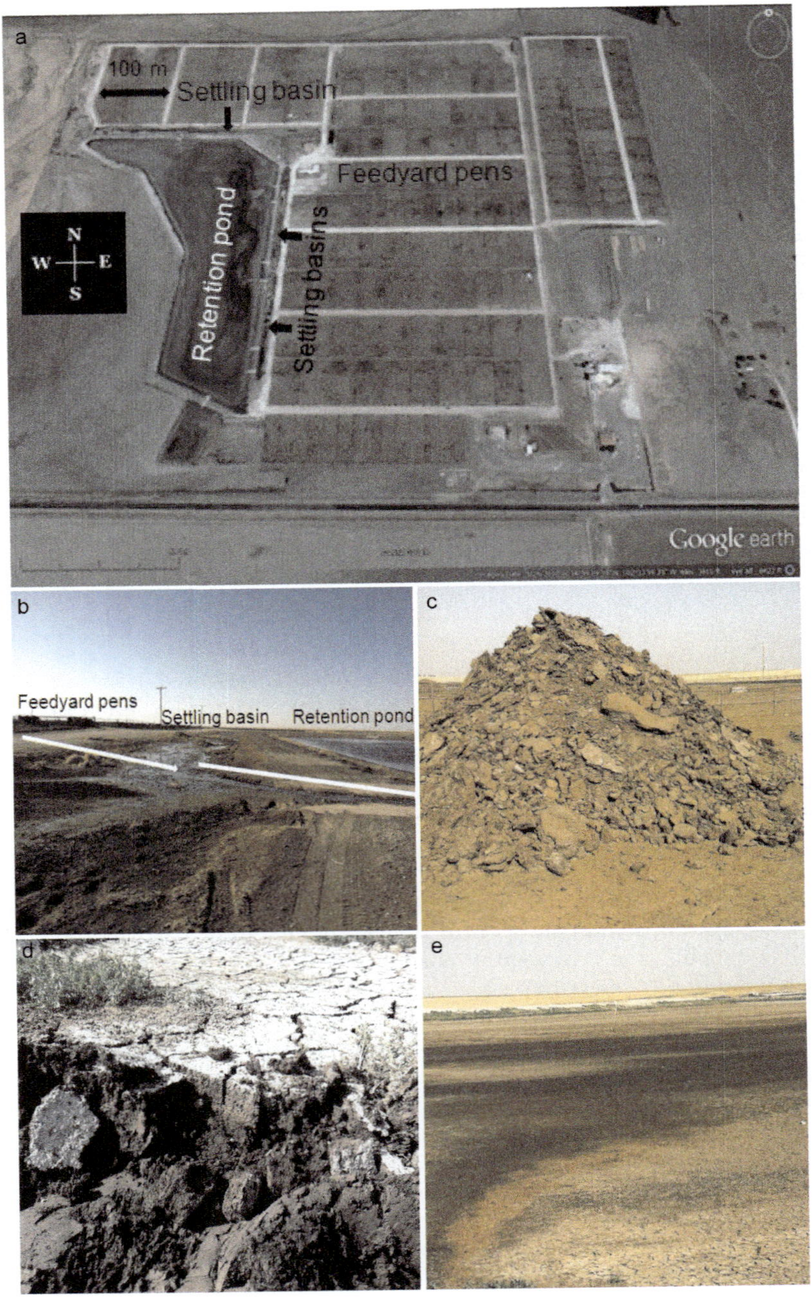

Fig. 1. (a) Aerial view of commercial feedyard showing key features, (b) flowpath of runoff from feedyard pens through settling basin to retention pond, (c) stockpiled pen manure after scraping, (d) accumulated sediment in settling basin, and (e) sediment accumulated in dry retention pond

Table 1. Selected properties of beef cattle manure collected from different feedyard sources. Data expressed on a dry matter (DM) basis as average ($n = 3$). Values within a column proceeded by different letters indicate significant difference ($P < 0.05$). Adapted from Waldrip et al. (2014).

Sample type	Moisture	TKN†	Organic N	NH_x–N‡	NO_3 + NO_2–N	Total P	Total K	OM§	DOC¶††	DN#††
	%	% DM							mg g^{-1} DM	
Surface manure	5.8a	2.6a	2.5a	0.259b	0.003a	0.88a	1.78b	79.9d	45.2d	9.73d
Manure pack	14.1b	2.6a	2.5a	0.022a	0.022a	1.41b	1.97b	45.5c	8.70c	2.12c
Settling basin	83.1d	3.6b	3.4b	0.003a	0.118b	1.82c	1.07a	27.5a	1.13a	0.72b
Retention pond	48.4c	2.7a	2.7a	0.040a	0.029a	1.56bc	1.72b	34.8b	2.11b	0.51a

† TKN, total Kjeldahl N.

‡ NH_x–N, total ammoniacal N.

§ OM, organic matter.

¶ DOC, dissolved organic carbon.

DN, dissolved nitrogen.

†† DOC and DN were extracted with 1:100 (g:mL) sample-to-water extraction ratio.

Similarly, surface manure had the highest dissolved organic carbon (DOC) and dissolved N (DN) concentrations. The underlying manure pack had a much lower DOC (81% less) and DN (78% less) contents than surface manure. Settling basin and retention pond sediments had very low DOC and DN contents, with 1.13 and 2.11 mg g^{-1} (DOC) and 0.72 and 0.51 mg g^{-1} (DN) dry matter (DM), respectively. Changes in whole and dissolved OM concentrations and chemical composition during the manure on-farm lifecycle demonstrate that the C-containing compounds in beef cattle manure are decomposing and NH_x–N is being lost, most likely as volatilized NH_3. Thus, manure from different feedyard sources are not necessarily equal in regard to potential for gaseous emissions of methane (CH_4), nitrous oxide (N_2O), and NH_3. In addition, when applied to cropland, these materials would have different fertilizer values and effects on soil microbial communities.

Fourier-Transform Infrared Characteristics of Beef Cattle Feedyard Manures

The FT-IR spectra of these beef cattle feedyard manures and their freeze-dried water extracts are shown in Fig. 2. These spectra are similar to those of animal manures and plant-derived OM (Calderón et al., 2006; He and Ohno, 2012; He et al., 2009; Soong et al., 2014), but less similar to those of soil OM and humic substances (Du et al., 2013; He et al., 2006, 2011b). The key spectral regions and their functional group assignments of these manure samples are also listed in Fig. 2. These assignments reveal a higher proportion of carboxyl groups, fats, and lipids in the surface manure than samples taken from the manure pack, settling basin, or retention pond, as evidenced by distinct peaks at 2917 and 2845 cm^{-1} and a shoulder at 1720 cm^{-1}. Furthermore, these samples exhibited more absorption within the 1640 to 1650 cm^{-1} region, which is evidence of the presence of amides (Calderón et al., 2006; Grube et al., 2006; He et al., 2013). Of the samples collected from the commercial feedyard, the intensity of the bands in these regions decreased in the order of surface manure > manure pack > settling basin > retention pond, indicating decomposition of fatty acids, polysaccharides, and proteins in manures from different sources.

The FT-IR spectra of WEOM from surface manure showed a shoulder and broad band from around 1740 to 1620 cm^{-1}; whereas, the spectra of the WEOM from the manure pack showed a wide strong band around 1700 to 1650 cm^{-1} with no visible shoulder at 1740 cm^{-1}. This band was even narrower in the spectra of water extracts from settling basin and retention pond sediments due to decrease in absorbance at 1650 and 1620 cm^{-1}. For the surface manure WEOM, the two aliphatic bands normally detected in the 2900 cm^{-1} region were obscured; however, in WEOM spectra of manure pack, settling basin, and retention pond sediments, there were decreases in absorbance around 3000 cm^{-1}, and the two bands were more prominent.

The specific FT-IR band intensity can be used as a semi-quantitative indicator of differences in OM quality based on proportional contributions of the various functional groups (Calderón et al., 2006; He et al., 2012; Soong et al., 2014). Specifically, intensity changes around 1080 (polysaccharides) and 2930 cm^{-1} (fats and lipids, polysaccharides, proteins, and humic substances) are indicative of the decomposition of organic components, and the relative intensi-

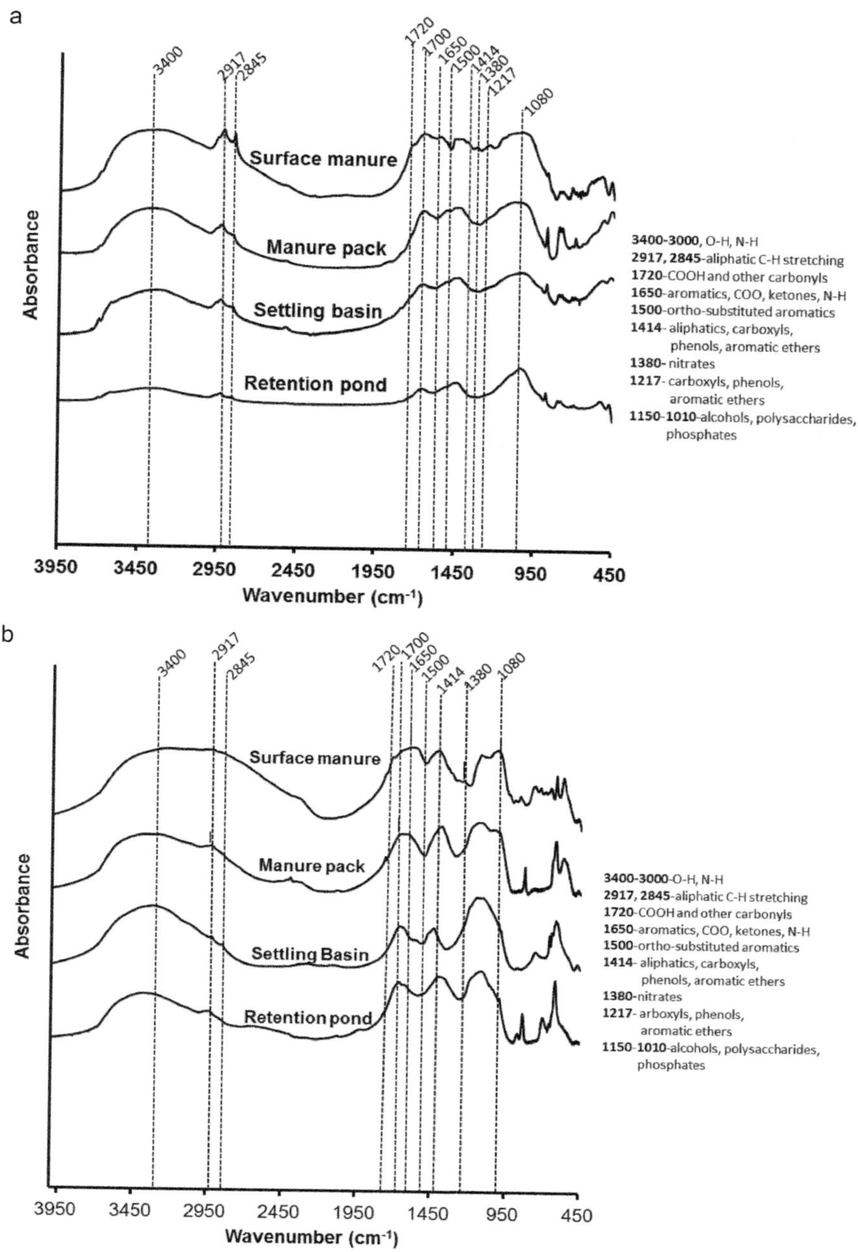

Fig. 2. Fourier-transform infrared (FT-IR) spectra of (a) whole feedyard manures and (b) their freeze-dried water-extractable organic matter (WEOM). Adapted from Waldrip et al. (2014).

ties of bands in these regions have been used to evaluate decomposition during composting (Grube et al., 2006; Smidt et al., 2002). Similarly, a decrease of the ratio of peak heights at 2925 and 1034 cm^{-1} has been reported to be indicative of decomposition of readily degradable components of OM, including aliphatics, polysaccharides, alcohols, and proteins (Grube et al., 2006; Hsu and Lo, 1999; Smidt et al., 2002).

Analysis of peak heights revealed considerable changes (relative to height at 1080 cm^{-1}) of samples taken from different locations on the commercial feedyard (Table 2). Specifically, 3400:1080 cm^{-1} ratios indicated decreases of as much as 41 and 36% in the proportion of amines and hydroxyl (OH) groups in whole manures and water extracts, respectively. Shown by the difference between absorbance due to aliphatic groups, polysaccharides, or proteins at 2930 and 1030 cm^{-1}, there was 30% less content in the manure pack than surface manure and up to 54% less in retention pond sediments. However, WEOM from retention pond sediment had 62% lower content in this region than surface manure. This finding suggests the preferential decomposition of fatty acids, polysaccharides, and proteins relative to cellulose and is supported by the work of Calderón et al. (2006), who also noted decreases in N-H and C-H bonding bands following a 10 wk incubation of dairy manure in soil. Relative intensity of bands in the 1650 cm^{-1} region show the decomposition of proteins and aromatic groups in both whole manure and water extracts of feedyard manures. The relative intensity of the band at 1380 cm^{-1}, which is usually attributed to NO_3^-, but could be due to carboxylate C-O symmetrical stretching (Ellerbrock et al., 1999), is interesting for the difference between whole samples and WEOM in retention pond and settling basin sediments. There was notably more absorbance in this region in whole samples of settling basin sediment than retention pond sediment; however, the band at 1380 cm^{-1} was not detected in WEOM from either settling basin or retention pond sediments. This indicates that much of the readily soluble NO_3^- and OM in runoff was lost by either leaching, utilized by microbial population, or transformed and

Table 2. The Fourier–infrared transform (FT-IR) band heights of organic matter in whole-beef cattle manures collected from different feedyard sources and their freeze-dried water-extractable organic matter (WEOM). Values within a column by different letters are significantly different ($P < 0.05$). Adapted from Waldrip et al. (2014).

Sample type	Approximate wavenumber (cm^{-1})					
	3400	2917	2845	1650	13800	1080
Whole-beef manures						
Surface manure	0.774c	0.178c	0.193c	0.699c	0.449c	0.668
Manure pack	0.612c	0.096b	0.072b	0.509bc	0.365c	0.631
Settling basin sediment	0.457b	0.089b	0.080b	0.334b	0.178b	0.381
Retention pond sediment	0.087a	0.082a	0.023a	0.099a	0.059a	0.283
WEOM						
Surface manure	0.474	0.019a	0.007	0.406a	0.197	0.339a
Manure pack	0.632	0.046b	0.041	0.710b	nd†	0.801b
Settling basin sediment	0.641	0.022a	0.027	0.344a	nd	1.043c
Retention pond sediment	0.530	0.038b	0.030	0.719b	nd	1.110c

† nd, not detected

volatilized as carbon dioxide (CO_2), CH_4, NH_3, N_2, or N_2O. However, this finding is consistent with the measured NO_3^- concentration in the whole samples (Table 1), in which NO_3^- was greatest in the settling basin sediment.

Ultraviolet–Visible Spectral Features of Feedyard Manures

The UV-Vis spectra of WEOM from the four sources of beef feedyard manure show monotonically decreasing curves with increasing wavelengths (data not shown). Consistent with swine and dairy manure samples (He et al., 2003, 2009), the UV-Vis spectra of WEOM from these feedyard manures are more complicated than of soil water extracts (He et al., 2003; Zhang et al., 2015, this volume), but simpler that plant WEOM (He et al., 2009). The spectra of WEOM from different organic materials reflects the abundance and complexity of OM (Zhang et al., 2011), which decreased in the order of plant material > feedyard manure > soil.

The quantitative values of E2/E3 (A_{254}/A_{365}), E2/E4 (A_{280}/A_{472}), E2/E6 (A_{280}/A_{664}), and E4/E6 (A_{472}/A_{664}) were calculated from the UV-Vis absorbances at specific wavelengths and are presented Table 3. The E2/E3 ratio of the surface manure sample was 7.24 and was significantly greater ($P < 0.05$) than the other three samples, which had values around 3.55. As a lower E2/E3 reflects higher average molecular weight (Peuravuori et al., 2005), a higher E2/E3 ratio for loose surface manure indicates recently excreted material with a lower molecular weight than older material in the manure pack, retention pond, and settling basin sediments. The ratio of E2/E4 has been used as an indicator of the relative abundance of lignin at the beginning of humification, while E2/E6 ratios reflect the relation between nonhumified and highly humified materials (Sellami et al., 2008). According to calculated E2/E4 values, there was significantly ($P < 0.05$) less lignin in the manure pack, settling basin, and retention pond sediments than in surface manure. The E2/E6 values point to a high proportion of humified OM in the settling basin sediment, where the proportion of humified organic matter was in the order of settling basin > retention pond ≥ manure pack > surface manure. The E4/E6 ratios of manure pack and retention pond sediment were >5.0, and that of surface manure and settling basin sediment was ~3.3. Typical E4/E6 ratios for humic acids extracted from soils are 3.3 to 5.0, whereas fulvic acids can range from 6.0 to 8.0 (Tan, 2003). Thus, it appears that humic substances in beef manures possess greater similarity to soil humic than fulvic acids.

Table 3. Ratios of ultraviolet–visible (UV-Vis) absorbance at specific wavelengths for water-extractable organic matter (WEOM) in beef cattle manure collected from different feedyard sources. Values within a column proceeded by different letters indicate significant difference (*P* < 0.05).

Sample type	E2/E3†	E2/E4	E2/E6	E4/E6
Surface manure	7.24b	30.5c	103c	3.34a
Manure pack	3.58a	14.6b	76.6b	5.26b
Settling basin sediment	3.55a	11.3a	37.1a	3.26a
Retention pond sediment	3.50a	13.7b	84.9b	6.18b

† Sample-to-water extraction ratio of 1:100 (g/mL). Samples were diluted so that absorbance at 254 nm was ~0.9.

Case Study II: Traceable Marker Components in Organic Dairy Manure

Background Information of Organic Dairy Farming Practices

Organic dairy (OD) production is drawing increasing attention because of public concerns about food safety, animal welfare, and the environmental impacts of conventional dairies (CDs) and other intensive livestock systems (Oudshoorn et al., 2011; Sato et al., 2005). In general, OD farming has significant differences from its conventional counterpart in respect to housing, grazing, animal nutrition, disease prevention, and veterinary treatment. Novak and Fiorelli (2010) listed requirements for ODs including, but not limited to, the following five management practices: (i) zero grazing is not permitted and the maximum density is two cows per hectare; (ii) at least half of the total housing floor is solid and not of slatted or grid construction; (iii) dry litter must be provided in the rest area; (iv) animals must be fed a diet comprised of only organically produced feedstuffs, with forage comprising at least 60% of the diet DM; and (v) the use of synthetic amino acids and growth promoters is forbidden, as well as the use of pharmaceuticals in the absence of illness. While the use of pharmaceuticals, including antibiotics, is restricted on OD farms, the USDA Organic Production and Handling Standards stipulate that appropriate medications and antimicrobial treatments be used temporarily to restore an ill animal to health (USDA, 2011). However, use of any treatment not within organic production standards will cause the animal to lose its organic status until treatment is complete (Sato et al., 2005). Compliance to these guidelines and maintenance of a USDA certification for OD farms is currently reliant on truthful reporting by management and annual inspections by third-party inspectors (Federal Register, 2012). Thus, there is often doubt in the authenticity of organic products and the adherence of producers to specific guidelines (Camin et al., 2007).

Identification of parameters that can differentiate organic and conventional agricultural products should help prevent misconduct and could provide a firm basis for comparative assessment of the two types of management and resultant products (Camin et al., 2007). Tres et al. (2012) conducted a feasibility study on authentication of organic feed for laying hens using near-infrared spectroscopy and chemometrics. Their study showed that the near-infrared spectroscopic approach provided a rapid and low-cost screening tool, while a more complicated fatty-acid fingerprinting model could be used for further confirmation of the organic identity of feed samples. These methods provided additional assurance to the administrative controls currently conducted in the organic feed sector. Capuano et al. (2014) evaluated the effect of feeding freshly harvested forage, pasture grazing, and organic or biodynamic farming on the milk triglyceride profile of dairy cows and discussed the implications for authentication of organic farms. Their work demonstrated that authentication of pasture grazing and of organic or biodynamic farming based on triacylglycerol profile is difficult during the grazing period. To address this inadequacy, we reviewed the relevant data (He and Ohno, 2012; He and Wang, 2012; He et al., 2015) and present a case study that analyzed and explored the difference between OD manure and CD manure. The goals of this work were to identify possible traceable marker components in dairy manure that could be used to verify that farms were managed under USDA OD guidelines.

General Characteristics of Dairy Manure Organic Matter

Both OD and CD manures were obtained from commercial dairy farms in Maine, USA (Table 4) (He et al., 2009; He and Ohno, 2012). The C and N contents ranged from 29.8 to 45.4% (C) and 0.88 to 2.79% (N) of manure DM for the OD manures, and from 12.4 to 41.2% (C), and 1.34 to 1.73% (N) of manure DM for the CD manures. The average N concentration for the 15 OD and seven CD manures were 1.70 ± 0.49% and 1.53 ± 0.13% of DM, respectively. These data indicated that C contents varied less in OD manure than manure from conventionally managed farms, which may reflect relative consistency of forage and bedding managements across OD farms. In contrast, the use of supplemental protein in the diets

Table 4. Selective management information and manure properties of 15 organic (OD) and seven conventional (CD) dairy manure samples. Adapted from He and Ohno (2012).

Sample	Forage†	Bedding	Storage	Type	Total C	Total N	DC‡	DN‡
					% dry matter		% total C or N	
OD1	1, 2	Hay sawdust	Stacking pad	Solid	38.0	2.79	6.6	12.3
OD2	1, 2, 3, 4	Sawdust sand	Pit	Solid	29.8	1.71	6.1	10.1
OD3	1, 3	Sawdust shaving	Pad	Solid	40.2	2.11	8.8	12.2
OD4	1, 3	NA	Pad	Solid	41.7	1.49	4.6	9.1
OD5	1, 2, 3	Green sawdust	Pit	Solid	41.5	1.46	3.0	7.3
OD6	1, 2	Sawdust	Roofed storage	Solid	34.9	1.82	7.1	9.6
OD7	1, 2, 3	Sawdust	Bedded pad	Solid	45.4	0.88	3.3	11.1
OD8	1, 3	Sawdust	Stacking pad	Solid	41.2	1.49	5.5	11.1
OD9	1, 2, 3	Sawdust	Stacking pad	Solid	41.5	1.41	1.6	3.2
OD10	1, 3	Sawdust shaving	Stack windrows	Solid	42.2	1.44	3.3	6.4
OD11	1, 3	Sawdust shaving	Stack windrows	Solid	43.2	1.28	2.8	6.9
OD12	1, 3	Sawdust	Pad	Solid	43.1	1.42	4.2	10.3
OD13	1, 3	Sawdust	Walled pad	Solid	38.7	2.16	11.3	18.6
OD14	NA§	NA	Pit	Semi-Solid	40.2	2.44	11.5	20.2
OD15	NA	NA	NA	Semi-Solid	42.1	1.57	4.5	10.5
CD1	NA	NA	NA	Solid	36.8	1.34	3.2	8.4
CD2	NA	NA	NA	Solid	41.2	1.53	3.8	7.8
CD3	NA	Sawdust	NA	Solid	40.2	1.56	4.8	12.0
CD4	NA	Sawdust	NA	Solid	40.1	1.45	4.3	7.8
CD5	NA	NA	NA	Solid	15.4	1.63	14.3	7.4
CD6	NA	NA	NA	Solid	22.1	1.73	17.7	34.4
CD7	NA	NA	Lagoon	Liquid	12.4	1.45	11.5	19.9

† 1 = hay; 2 = pasture; 3 = haylage; 4 = corn silage.

‡ NA, information is not available.

§ DC, dissolved C; DN, dissolved N. Measured in 1:100 (g/mL) extracts.

fed on CD farms was reflected by relatively small variations of manure total N as compared to total N content in OD manures (He and Ohno, 2012). About 1.6 to 11.5% of manure C in OD manure was water soluble, while dissolved C comprised 3.2 to 17.7% of the C in CD manure. Interestingly, there was a large range in dissolved N in the OD manures, which ranged from 3.2 to 20.2% of total manure N. In most samples, the extraction efficiency of N was approximately twice that of C. The average extraction efficiencies in these OD and CD samples were 5.6 and 8.5% of manure C, and 10.6 and 14.9% of manure N, respectively. A significant positive relationship was found between water-extractable C and N and total manure N content of OD manure; however, a similar relationship was not found between extractable nutrients and total manure C content. This could be explained by the fact that total manure C contents were not only affected by animal excreta but also by added bedding materials (e.g., sawdust), which contained less total N and soluble C than cattle excreta (He and Ohno, 2012).

Fourier-Transform Infrared, Fluorescence, and Ultraviolet–Visual Features of Dairy Manure Organic Matter

Both FT-IR and fluorescence excitation–emission matrix spectroscopies were used to characterize the C structural and bonding environments of whole and WEOM of these OD and CD manures (He and Ohno, 2012). The FT-IR features of manure OM varied mainly in the 1650 to 1550 cm^{-1} range, reflecting the presence of different N compounds in these manures. Fluorescence data revealed five fluorophore components present in the WEOM from the manures. Correlation analysis revealed a positive relationship between a tyrosine-like fluorophore component and the FT-IR band of amino-N compounds and a positive relationship between a humic-like fluorophore component and the FT-IR band attributed to polysaccharides. There were no clearly distinct differences observed between the two types of dairy manure with respect to C and N contents and FT-IR and fluorescence spectral features; however, based on the average values, whole manure from OD farms contained less soluble C and N compounds (DM basis) and more hydrophobic aliphatic groups than CD manure. Based on spectroscopic features, the authors (He and Ohno, 2012) reported that WEOM in OD manure contained more stable humic- and lignin-related components and less amino-protein-related components than WEOM from CD manure. They attributed these differences to the inclusion of more forage fed in the diets at farms under organic management and the inclusion of supplemental protein in CD diets.

In another study, He et al. (2015) were unable to discern specific border values in UV-Vis parameters that distinguished OM of OD manure from that of CD, which confirmed previous observations on manure FT-IR and fluorescence spectral characteristics (He and Ohno, 2012). However, the $SUVA_{254}$ and $SUVA_{280}$ (specific UV-Vis absorbances at 254 and 280 nm divided by the DOC concentration; used as an indicator of aromaticity), and average E4/E6 values of OD manure were significantly (P = 0.05) greater than those of CD manure. The higher average values of these three parameters of OD manure suggest the presence of more aromatic functional groups, and less aliphatic content, than OD manure (Heymann et al., 2005; Hunt et al., 2007; Shao et al., 2009).

Pyrolysis Gas Chromatography–Mass Spectrometric Analysis

The whole manure and WEOM of four CD manures (CD1–CD4) and five OD manures (OD1, OD2, OD3, OD7, and OD13) were characterized by pyrolysis gas chromatography–mass spectrometry (pyrolysis GC-MS) (He and Wang, 2012; Wang et al., 2011). Pyrolysis GC-MS analysis of whole manure samples showed the presence of 119 compounds, as indicated by peaks >0.1% of total ion intensity. These 119 compounds were segregated in eight general categories: aliphatics (20), benzenes (6), carboxylics (15), carbohydrates (11), lignin monomers (26), N-containing compounds (25), phenols (14), and sterols (2) (Wang et al., 2011). The relative abundance of these compounds was in the order of lignin monomers (~37%) > N-containing compounds (~18%) > aliphatics (~10%) > carbohydrates (~8%) > phenols (~6%) > carboxylics (~5%) > benzenes (~2%) > sterols (~0.4%). Generally, there were no remarkable differences in distribution of the eight groups of compounds between CD and OD manures.

Compared with that of whole dairy manures, the total ion chromatograms of pyrolysis GC-MS of WEOM were relatively simple (Fig. 3). Peaks related to benzenes (e.g., toluene [B1] and styrene [B4]), various phenols (P1, P2, P3, P5, P6, P8, etc.), and indole (N19) were more apparent in the chromatogram of WEOM than whole manure. This is due to the fact that lignin monomers, hydrophobic high

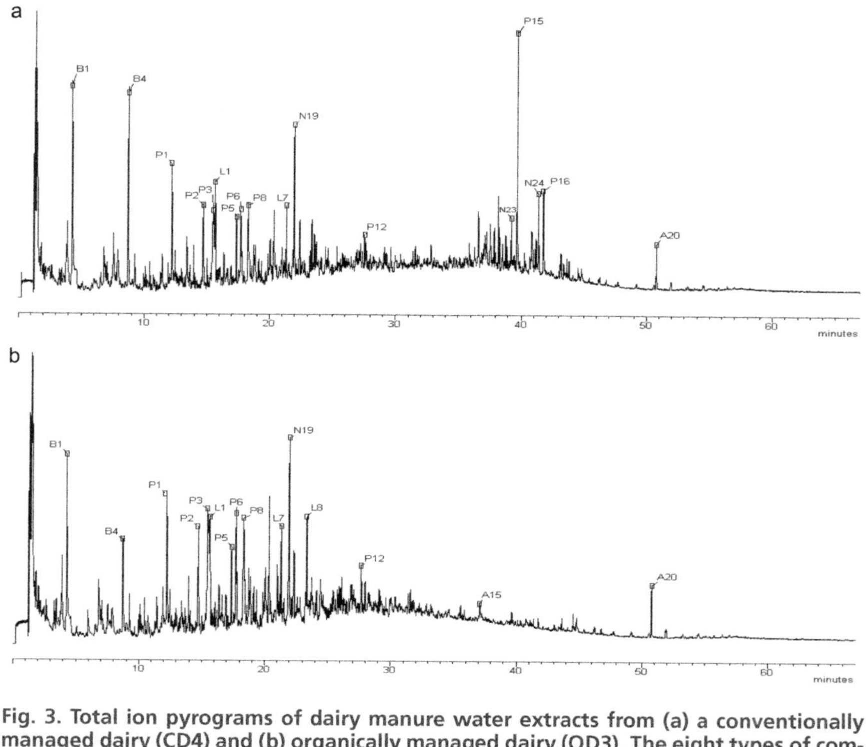

Fig. 3. Total ion pyrograms of dairy manure water extracts from (a) a conventionally managed dairy (CD4) and (b) organically managed dairy (OD3). The eight types of compounds identified are labeled as follows: A, aliphatics; B, benzenes; L, lignin monomers; N, N-containing compounds; P, phenols. Adapted from He and Wang (2012).

molecular mass compounds, and other insoluble forms of OM are not extractable with water; therefore, WEOM was enriched with soluble OM components (He and Wang, 2012).

Peaks eluted from the GC column at 35.5 to 42.5 min were found in WEOM of CD manure but not OD manure. Both aliphatic and carboxylic compounds were eluted at this retention time period. Two of the peaks were tentatively assigned to three-ring compounds: 1-methyl-7-(1-methyl)-1,2,3,4-tetrahydrophenanthrene (P15) and 2-isopropyl-10-methyl-phenanthrene (P16). He and Wang (2012) attributed these peaks to antibiotic-related metabolites or derivatives produced either by animal or ruminal microbial metabolism or during the pyrolysis process. Strong peaks of two amine-containing benzene ring structural compounds (N23 and N24) in CD manure could be from protein-based hormones (e.g., estradiol and androgenic agents, such as trembolone acetate). More than 400 active veterinary pharmaceutical ingredients have been used in animal production to control and prevent disease, increase growth rate, and improve feed/gain ratio (Song and Guo, 2014). The pyrolysis GC-MS peaks observed with the CD manure is consistent with the presence of residual pharmaceutical residues in dairy manure and wastewater, and has been suggested as a source for these compounds to be released into the environment (Jaffrezic et al., 2011; Song and Guo, 2014).

He and Wang (2012) noted that peaks attributed to pharmaceutical residues were absent in four of five OD manures tested but were present in all the four CD manures analyzed. The absence of these pharmaceutical-related peaks in OD manure chromatograms would be related to regulation against use of any chemical drugs (other than for treating illness) for dairies under certified USDA management (Sato et al., 2005). Although there was no clear rationale why the pyrolysis GC-MS chromatogram of one OD manure (OD7) contained peaks related to the presence of pharmaceutical residues, He and Wang (2012) proposed that the difference in the pyrolysis GC-MS chromatograms of WEOM of OD7 and four other OD samples could be due to (i) technical error, where a CD manure was mislabeled as OD; (ii) the manure was from an OD farm containing cows undergoing medical treatments; or (iii) the specific OD farm was not following the required organic farming practices. While the first cause (simple technical error) could be readily corrected, the latter two causes are related to OD management. Further confirmation of differences in pyrolysis GC-MS chromatograms between CD and OD manures is required to verify these findings; however, we predict that use of this technique could be used to prevent misconduct by management and provide a scientific basis for evaluation and authentication of pharmaceutical use in OD operations.

Solid-State ^{13}C Nuclear Magnetic Resonance Spectra of Whole Manure

The ^{13}C cross-polarization/total sideband suppression (CP/TOSS) NMR spectra of the four CD and four OD manure samples are shown in Fig. 4 (He et al., 2015). The relative and average abundances of C functional groups identified in these manures with ^{13}C CP/TOSS NMR are presented in Table 5. Overall, about 15% of the C functional groups in these manures were identified as aromatic, and 70% were aliphatic. The majority of the aliphatic C in these manures was ascribed to the O-alkyl group (60 to 94 ppm). There was not a statistically significant ($P \leq 0.05$)

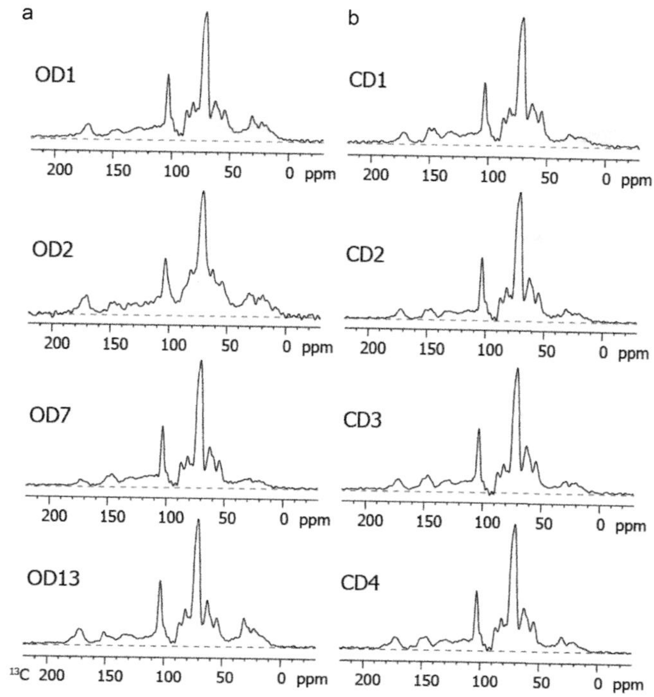

Fig. 4. Solid-state ^{13}C cross polarization–total sideband suppression nuclear magnetic resonance spectra of (a) organic (OD1, OD2, OD7, and OD13) and (b) conventional (CD1, CD2, CD3, and CD4) dairy manure samples. Adapted from He et al. (2015).

difference in the relative contents of these two major OM components (aliphatic and aromatic) between the OD and CD manures. In fact, the only statistically significant ($P = 0.003$) difference in the relative contents between the two types of manure were peaks found between 46 and 60 ppm, which were assigned to NCH & OCH_3 groups. The NCH group was attributed to amino acid and protein ingredients in manure (He and Mao, 2011); therefore, the greater peak content between 46 and 60 ppm in CD manure is consistent with the inclusion of more supplemental protein and amino acids in CD diets (He and Ohno, 2012).

Of the various peaks, the triple peaks or shoulders around 30 ppm showed a greater visual difference among the manure samples than any other peaks (Fig. 4). The triplet peak could be assigned to long chain $(CH_2)_n$ aliphatic biopolymers, such as cutin and cutan (Deshmukh et al., 2005; Shechter et al., 2010). Cutin is a high-molecular-weight, insoluble, polyester-like biopolymer, which is most often associated with waxes and cuticular polysaccharides. In some plant species, cutan (a base and acid hydrolysis-resistant polymethylene-like biopolymer) is associated with cutin (Shechter et al., 2010). It is evident that this plant cuticular matter, which is contained in high concentrations in forages, is less digestible than the readily degraded polysaccharides and proteins that are prevalent in grains and other starch-based feedstuffs commonly used in CD diets. As a higher amount of forage is typically required in OD diets, the relevant ^{13}C NMR peak

Table 5. Relative and average abundance (%) of carbon-containing functional groups in organic (OD) and conventional (CD) dairy manure obtained by integrations of ^{13}C cross polarization–total sideband suppression nuclear magnetic resonance spectroscopy (^{13}C CP/TOSS NMR) peak areas. Adapted from He et al. (2015).

Chemical shift range (ppm)	0–46	46–60	60–94	94–108	108–140	140–164	164–184	184–220	0–108	108–164
Assignment	Nonpolar aliphatics	NCH & OCH_3	O-Alkyl C	O-C-O anomerics	Aromatic C-C & C-H	Aromatic C-O	COO & N–C = O	ketone, quinone, aldehyde	Total aliphatic	Total aromatic
OD1	14.1	8.7	46.5	11.2	10.2	3.9	4.4	1.0	69.3	14.1
OD2	15.2	9.2	45.3	10.9	9.6	4.4	5.0	0.3	69.7	14.0
OD7	9.1	9.2	51.7	11.2	11.0	5.3	2.2	0.4	70.0	16.3
OD13	16.4	8.7	45.7	11.3	8.3	4.3	5.0	0.4	70.8	12.6
Average	13.7	9.0	47.3	11.2	9.8	4.5	4.2	0.5	70.0	14.3
Average (–OD7)†	15.2	8.9	45.8	11.1	9.4	4.2	4.8	0.6	69.9	13.6
SD	3.2	0.3	3.0	0.2	1.1	0.6	1.3	0.3	0.6	1.5
SD (–OD7)†	1.2	0.3	0.6	0.2	1.0	0.3	0.3	0.4	0.8	0.8
CD1	9.2	9.7	49.3	11.4	10.6	6.0	3.3	0.5	68.2	16.6
CD2	10.1	9.8	51.5	11.7	9.0	4.3	3.2	0.4	71.4	13.3
CD3	10.0	10.3	47.1	10.3	10.6	6.4	4.1	1.2	67.4	17.0
CD4	10.6	9.8	49.0	10.5	9.7	5.4	4.4	0.7	69.4	15.1
Average	10.0	9.9	49.2	11.0	10.0	5.5	3.8	0.7	69.1	15.5
SD	0.6	0.3	1.8	0.7	0.8	0.9	0.6	0.4	1.7	1.7
P value ($P > F$, OD vs. CD)	0.060	0.003	0.311	0.636	0.781	0.101	0.602	0.492	0.394	0.313
P value (OD [–OD7] vs. CD)†	<0.001	0.005	0.028	0.718	0.842	0.063	0.042	0.652	0.481	0.131

† Calculated without OD7 data.

at 30 ppm is much stronger in OD manure than in CD manure. The cuticular residues in manure were presumably from forage feedstuffs; thus, the apparent triple peaks or shoulders at 26, 29, and 33 ppm could serve as evidence for appropriate forage feeding under OD management.

Whereas the cutin and cutan ingredients could be traceable marker components in dairy manure for forage feed, the peaks at 30 ppm of OD7 were more similar to those of the four CD samples than those of other three OD samples, consistent with the results of pyrolysis GC-MS analysis. This evidence strongly suggests that OD7 was a mislabeled manure from a farm under CD management, or sampled from a dairy not complying with OD guidelines. This assumption is further supported by comparison of CD and OD data when the data of OD7 are excluded. When OD7 is excluded from the analysis, the significance of the difference between OD and CD manures in the 0 to 46 ppm peak area is greatly improved, as the *P*-value was reduced from 0.060 to <0.001 (Table 5). In contrast, the significance levels of the differences in other peak areas between the manures from two dairy management types are nearly identical with or without the exclusion of OD7. Confirmation of the differences in pyrolysis GC-MS and solid-state ^{13}C NMR spectroscopy chromatograms between CD and OD manures differences could help to develop evaluation methods with a scientific basis that are useful for preventing misconduct in the OD industry and authenticate OD farming programs.

Conclusions

Organic matter is a major component of animal manure. As the majority of manure OM is labile, both animal feeding and manure management and storage can affect the composition and content of manure OM. These effects can be monitored and evaluated by various spectroscopic techniques as demonstrated by two case studies of cattle manure in this chapter. Using wet chemistry and FT-IR and UV-Vis spectroscopic techniques, Case study I examined the composition and structural properties of whole manure and WEOM of beef cattle manure taken from different sources on a commercial feedyard (surface manure, manure pack, settling basin sediment, and retention pond sediment) in the Texas Panhandle. Results from this study demonstrated that the composition and content of OM and nutrients in beef cattle manure differ depending on the collecting location, with higher concentrations of available nutrients present in more recently excreted materials and increased molecular weight and degree of humification as manure ages during its on-farm life cycle.

Using multiple spectroscopic techniques, Case Study II compared the structural and functional groups in dairy manure between CD and OD farms. Based on the average values, the general patterns on the impact of organic farming are observed and can be attributed to more forage feedstuffs in OD farming management and more protein and other additives in CD diets. In particular, the cutin- and cutan-related ^{13}C NMR peaks may be used for tracing the dietary management on dairy farms distinguishing grass-fed cows from grain-fed cows, and veterinary pharmaceutical-related pyrolysis GC-MS peaks may be used for monitoring the usage of antibiotics, hormones, and other synthetic treatments in cattle. These observations revealed the possibility of developing a technical procedure to analyze traceable marker components in dairy manure for OD farm certification and authentication. The manure samples used in the Case Studies presented were collected from real world farms where specific management practices were sometimes unknown. Further research using manure from controlled studies with known feeding is needed to establish undisputed feed–manure linkages for such purposes.

References

Bolan, N.S., J.K. Syers, and R.W. Tillman. 1988. Effect of pH on the adsorption of phosphorous and potassium in batch and in column experiments. Aust. J. Soil Res. 26:165–170. doi:10.1071/SR9880165

Calderón, F.J., G.W. McCarty, and J.B. Reeves, III. 2006. Pyrolysis-MS and FT-IR analysis of fresh and decomposed dairy manure. J. Anal. Appl. Pyrolysis 76:14–23. doi:10.1016/j.jaap.2005.06.009

Camin, F., A. Moschella, F. Miselli, B. Parisi, G. Versini, P. Ranalli, and P. Bagnaresi. 2007. Evaluation of markers for the traceability of potato tubers grown in an organic versus conventional regime. J. Sci. Food Agric. 87:1330–1336. doi:10.1002/jsfa.2853

Capuano, E., R. Boerrigter-Eenling, A. Elgersma, and S. van Ruth. 2014. Effect of fresh grass feeding, pasture grazing and organic/biodynamic farming on bovine milk triglyceride profile and implications for authentication. Eur. Food Res. Technol. 238:573–580.

Deshmukh, A.P., A.J. Simpson, C.M. Hadad, and P.G. Hatcher. 2005. Insights into the structure of cutin and cutan from *Agave americana* leaf cuticle using HRMAS NMR spectroscopy. Org. Geochem. 36:1072–1085. doi:10.1016/j.orggeochem.2005.02.005

Du, C., Z. He, and J. Zhou. 2013. Characterization of soil humic substances using mid-infrared photoacoustic spectroscopy. In: J. Xu et al., eds. Functions of natural organic matter in changing environment. Springer. Amsterdam, The Netherlands, and Zhejiang Univ. Press. Hangzhou, China. p. 43–47.

Ellerbrock, R.H., A. Höhn, and J. Rogasik. 1999. Functional analysis of soil organic matter as affected by long-term manurial treatment. Eur. J. Soil Sci. 50:65–71. doi:10.1046/j.1365-2389.1999.00206.x

Federal Register. 2012. National organic program: Notice of draft guidance for accredited certifying agents, certified operations, and non-certified handlers of certified organic products. Vol. 77, no. 23. USDA Agricultural Marketing Service, Washington, DC. p. 5415–5416. http://www.gpo.gov/fdsys/pkg/FR-2012-02-03/html/2012-2377.htm (accessed 13 Mar. 2015)

Fernandes, M.L.V., and J.F. Coutinho. 1999. Effect of liming and phosphate application on sudangrass growth and phosphorus availability in two temperate acid soils. Commun. Soil Sci. Plant Anal. 30:855–871. doi:10.1080/00103629909370251

Grube, M., J.G. Lin, P.H. Lee, and S. Kokorevicha. 2006. Evaluation of sewage sludge-based compost by FT-IR spectroscopy. Geoderma 130:324–333. doi:10.1016/j.geoderma.2005.02.005

Harwood, J.J. 2014. Molecular markers for identifying municipal, domestic and agricultural sources of organic matter in natural waters. Chemosphere 95:3–8. doi:10.1016/j.chemosphere.2013.09.104

He, Z., editor. 2011. Environmental chemistry of animal manure. Nova Science Publishers, New York.

He, Z., editor. 2012. Applied research of animal manure: challenges and opportunities beyond the adverse environmental concerns. Nova Science Publishers, New York.

He, Z., H. Cao, H.N. Cheng, H. Zou, and J.F. Hunt. 2013. Effects of vigorous blending on yield and quality of protein isolates extracted from cottonseed and soy flours. Mod. Appl. Sci. 7:79–88. doi:10.5539/mas.v7n10p79

He, Z., C. Du, and J. Zhou. 2011a. Structural and bonding environments derived from infrared spectroscopic studies. In: Z. He, editor, Environmental chemistry of animal manure. Nova Science Publishers, New York. p. 23–42.

He, Z., C.W. Honeycutt, and T.S. Griffin. 2003. Comparative investigation of sequentially extracted P fractions in a sandy loam soil and a swine manure. Commun. Soil Sci. Plant Anal. 34:1729–1742. doi:10.1081/CSS-120021308

He, Z., C.W. Honeycutt, O.M. Olanya, R.P. Larkin, J.H. Halloran, and J.M. Frantz. 2012. Comparison of soil phosphorus status and organic matter composition in potato fields with different crop rotation systems. In: Z. He, et al., editors, Sustainable potato production: global case studies. Springer, Amsterdam, The Netherlands. p. 61–79.

He, Z., C.W. Honeycutt, and H. Zhang. 2011b. Elemental and Fourier-transform infrared spectroscopic analysis of water and pyrophosphate extracted soil organic matter. Soil Sci. 176:183–189. doi:10.1097/SS.0b013e318212865c

He, Z., and J. Mao. 2011. Functional groups identified by solid state ^{13}C-NMR spectroscopy. In: Z. He, editor, Environmental chemistry of animal manure. Nova Science Publishers, New York. p. 41–59.

He, Z., J. Mao, C.W. Honeycutt, T. Ohno, J.F. Hunt, and B.J. Cade-Menun. 2009. Characterization of plant-derived water extractable organic matter by multiple spectroscopic techniques. Biol. Fertil. Soils 45:609–616. doi:10.1007/s00374-009-0369-8

He, Z., and T. Ohno. 2012. Fourier transform infrared and fluorescence spectral features of organic matter in conventional and organic dairy manure. J. Environ. Qual. 41:911–919. doi:10.2134/jeq2011.0226

He, Z., T. Ohno, B.J. Cade-Menun, M.S. Erich, and C.W. Honeycutt. 2006. Spectral and chemical characterization of phosphates associated with humic substances. Soil Sci. Soc. Am. J. 70:1741–1751. doi:10.2136/sssaj2006.0030

He, Z., and J.J. Wang. 2012. Characterization of plant nutrients and traceable marker components in dairy manure for organic dairy farming management evaluation. In: Z. He, editor, Applied research of animal manure: challenges and opportunities beyond the adverse environmental concerns. Nova Science Publishers, New York. p. 3–19.

He, Z., M. Zhang, X. Cao, Y. Li, J. Mao, and H.M. Waldrip. 2015. Potential traceable markers of organic matter in organic and conventional dairy manure using ultraviolet-visible and solid-state ^{13}C nuclear magnetic resonance spectroscopy. Org. Agric. 5(2):113–122. doi:10.1007/s13165-014-0092-0

Heymann, K., H. Mashayekhi, and B.S. Xing. 2005. Spectroscopic analysis of sequentially extracted humic acid from compost. Spectrosc. Lett. 38:293–302. doi:10.1081/SL-200058702

Hsu, J.H., and S.L. Lo. 1999. Chemical and spectroscopic analysis of organic matter transformations during composting of pig manure. Environ. Pollut. 104:189–196. doi:10.1016/S0269-7491(98)00193-6

Hunt, J.F., T. Ohno, Z. He, C.W. Honeycutt, and D.B. Dail. 2007. Influence of decomposition on chemical properties of plant- and manure-derived dissolved organic matter and sorption to goethite. J. Environ. Qual. 36:135–143. doi:10.2134/jeq2006.0133

Jaffrezic, A., E. Jardde, A.M. Pourcher, M. Gourmelon, M.P. Caprais, D. Heddadj, P. Cottinet, M. Bilal, M. Derrien, and R. Marti. 2011. Microbial and chemical markers: Runoff transfer in animal manure-amended soils. J. Environ. Qual. 40:959–968. doi:10.2134/jeq2010.0355

Jensen, L.S., and S.G. Sommer. 2013. Manure organic matter-characteristics and microbial transformations. In: S.G. Sommer, et al., editors, Animal manure recycling: Treatment and management. John Wiley & Sons, Ltd, New York. p. 67–90.

Luo, Y., D. Zhang, G. Yu, and Q. Shen. 2013. Aromatic moieties from matured chicken manure and agriculture residues compost suppress growth of *lepidium sativum* L. and *trichoderma harzianum*. Pedosphere 23:826–834. doi:10.1016/S1002-0160(13)60074-0

Novak, S.M., and J.L. Fiorelli. 2010. Greenhouse gases and ammonia emissions from organic mixed crop-dairy systems: A critical review of mitigation options. Agron. Sustainable Dev. 30:215–236. doi:10.1051/agro/2009031

Ohno, T., and Z. He. 2011. Fluorescence spectroscopic analysis of organic matter fractions: The current status and a tutorial case study. In: Z. He, editor, Environmental chemistry of animal manure. Nova Science Publishers, New York. p. 83–103.

Oudshoorn, F.W., C.A.G. Sorensen, and I.J. de Boer. 2011. Economic and environmental evaluation of three goal-vision based scenarios for organic dairy farming in Denmark. Agric. Syst. 104:315–325. doi:10.1016/j.agsy.2010.12.003

Peuravuori, J., A. Monteiro, L. Eglite, and K. Pihlaja. 2005. Comparative study for separation of aquatic humic-type organic constituents by DAX-8, PVP and DEAE sorbing solids and tangential ultrafiltration: Elemental composition, size-exclusion chromatography, UV-vis and FT-IF. Talanta 65:408–422. doi:10.1016/j.talanta.2004.06.042

Powell, J.M. 2012. Alteration of dairy cattle diests for enhanced recycling of manure nutrients. In: Z. He, editor, Applied research of animal manure–challenges and opportunities beyond the adverse environmental concerns. Nova Science Publishers, New York. p. 21–42.

Sato, K., P.C. Bartlett, R.J. Erskine, and J.B. Kaneene. 2005. A comparison of production and management between Wisconsin organic and conventional dairy herds. Livest. Prod. Sci. 93:105–115. doi:10.1016/j.livprodsci.2004.09.007

Sellami, F., S. Hachicha, M. Chtourou, K. Medhioub, and E. Ammar. 2008. Maturity assessment of composted olive mill wastes using UV spectra and humification parameters. Bioresour. Technol. 99:6900–6907. doi:10.1016/j.biortech.2008.01.055

Shao, Z.H., P.J. He, D.Q. Zhang, and L.M. Shao. 2009. Characterization of water-extractable organic matter during the biostabilization of municipal solid waste. J. Hazard. Mater. 164:1191–1197. doi:10.1016/j.jhazmat.2008.09.035

Shechter, M., B. Xing, and B. Chefetz. 2010. Cutin and cutan biopolymers: Their role as natural sorbents. Soil Sci. Soc. Am. J. 74:1139–1146. doi:10.2136/sssaj2009.0313

Singh, S., S. Dutta, and S. Inamdar. 2014. Land application of poultry manure and its influence on spectrofluorometric characteristics of dissolved organic matter. Agric. Ecosyst. Environ. 193:25–36. doi:10.1016/j.agee.2014.04.019

Smidt, E., P. Lechner, M. Schwanninger, G. Haberhauer, and M.H. Gerzabek. 2002. Characterization of waste organic matter by FT-IR spectroscopy: Application in waste science. Appl. Spectrosc. 56:1170–1175. doi:10.1366/000370202760295412

Song, W., and M. Guo. 2014. Residual veterinary pharmaceuticals in animal manures and their environmental behaviors in soils. In: Z. He and H. Zhang, editors, Applied manure and nutrient chemistry for sustainable agriculture and environment. Springer, Amsterdam, the Netherlands. p. 23–52.

Soong, J.L., F.J. Calderón, J. Betzen, and M.F. Cotrufo. 2014. Quantification and FT-IR characterization of dissolved organic carbon and total dissolved nitrogen leached from litter: A comparison of methods across litter types. Plant Soil 385:125–137. doi:10.1007/s11104-014-2232-4

Tan, K.H. 2003. Humic matter in soil and the environment. Marcel Dekker, Inc., New York.

Tres, A., G. Van Der Veer, M. Perez-Marin, S. Van Ruth, and A. Garrido-Varo. 2012. Authentication of organic feed by near-infrared spectroscopy combined with chemometrics: A feasibility study. J. Agric. Food Chem. 60:8129–8133. doi:10.1021/jf302309t

USDA. 2011. Organic production and handling standards. USDA, Washington, DC. http://www.ams.usda.gov/AMSv1.0/getfile?dDocName=STELDEV3004445 (accessed 13 Mar. 2015)

Waldrip, H.M., Z. He, and M.S. Erich. 2011. Effects of poultry manure amendment on phosphorus uptake by ryegrass, soil phosphorus fractions and phosphatase activity. Biol. Fertil. Soil. 47:407-418.

Waldrip, H.M., Z. He, and T.S. Griffin. 2012a. Effects of organic dairy manure on soil phosphatase activity, available soil phosphorus, and growth of sorghum-sudangrass. Soil Sci. 177:629–637. doi:10.1097/SS.0b013e31827c4b78

Waldrip, H.M., Z. He, R.W. Todd, J.F. Hunt, M.B. Rhoades, and N.A. Cole. 2014. Characterization of organic matter in beef feedyard manure by UV-visible and Fourier-transform infrared spectroscopies. J. Environ. Qual. 43:690–700. doi:10.2134/jeq2013.09.0358

Waldrip, H.M., R.W. Todd, and N.A. Cole. 2012b. Characterization of ammonium sorption by beef cattle feedyard manure. Trans. ASABE 55:1609–1619. doi:10.13031/2013.42236

Wang, J.J., S.K. Dodla, and Z. He. 2011. Application of analytical pyrolysis-mass spectrometry in characterization of animal manure. In: Z. He, editor, Environmental chemistry of animal manure. Nova Science Publishers, New York. p. 3–24.

Zhang, H., and J. Schroder. 2014. Animal manure production and utilization in the U.S. In: Z. He and H. Zhang, editors, Applied manure and nutrient chemistry for sustainable agriculture and environment. Springer, Amsterdam, The Netherlands. p. 1–21.

Zhang, M., Z. He, and A. Zhao. 2011. Ultraviolet-visible absorptive features of water extractable and humic fractions of animal manure and relevant compost. In: Z. He, editor, Environmental chemistry of animal manure. Nova Science Publishers, New York. p. 61–81.

Zhang, M., Z. He, and A. Zhao. 2015. Water extractable organic carbon and nitrogen affected by crop rotation and fertilizer management. In: Z. He and F. Wu, editors, Labile organic matter—Chemical compositions, function, and significance in soil and the environment. SSSA Special Publ. 62. SSSA, Madison, WI. doi:10.2136/sssaspecpub62.2014.0038

Forms and Lability of Phosphorus in Humic and Fulvic Acids

Zhongqi He,* Irenus A. Tazisong, and Zachary N. Senwo

Abstract

Humic substances are involved in many biological and ecological processes in soils and natural waters. Characterization of P associated with humic substances may shed light on the function of humic substances in P cycling and nutrition. In this chapter, we review and discuss the forms and lability of humic-bound P based on our own and colleagues' research on various humic fractions obtained from the International Humic Substances Society and extracted from different agricultural soils. Solution ^{31}P nuclear magnetic resonance (NMR) spectroscopy reveals that the majority of humic-bound P are present in the forms of orthophosphate, monoesters, and diesters. Spiking with phytate confirms no phytate present in the examined humic fractions. The biotic and abiotic lability of humic-bound P was evaluated by enzymatic hydrolysis and ultraviolet irradiation. The wet chemistry analysis shows that 10 to 63% of humic-bound P is labile. The impact on lability of humic-bound P is in the order: sample origin > humic fraction type > management factors. Application of the knowledge gained from the comprehensive characterization by solution NMR spectroscopy and wet chemistry analysis in future research may lead to improved understanding of the dynamics and mechanisms of transformations of humic-bound P from unavailable to bioavailable forms in soil and the environment.

Abbreviations: AN, inorganic fertilizer; BF, control; CaHA, recalcitrant Ca humate; FA, fulvic acid; HA, humic acid; IHSS, International Humic Substances Society; MHA, mobile humic acid; NMR, nuclear magnetic resonance; UV, ultraviolet.

Zhongqi He, USDA-ARS Southern Regional Research Center, 1100 Robert E. Lee Blvd., New Orleans, LA 70124. Irenus A. Tazisong (Irenus.tazisong@aamu.edu) and Zachary N. Senwo (zachary.senwo@aamu.edu), Dep. of Biological and Environmental Sciences, College of Agricultural, Life and Natural Sciences, Alabama A&M Univ., Normal, AL 35762. *Corresponding author (zhongqi.he@ars.usda.gov).

doi:10.2136/sssaspecpub62.2014.0035

Labile Organic Matter—Chemical Compositions, Function, and Significance in Soil and the Environment. SSSA Special Publication 62. Zhongqi He and Fengchang Wu, editors.

Humic matter is a mixture of amorphous, polydespersed substances with yellow to brown-black color (Tan, 2003). They originate from the decomposed organic matter by "new formation" called humification and are usually obtained from soils and other environmental samples by extraction, fractionation, and isolation procedures with alkaline and acidic solutions. Humic acid (HA) and fulvic acid (FA) are two major fractions of humic matter. Humic acid is the humic fraction that is soluble in alkaline solution but precipitate on acidification of the alkaline extracts. Fulvic acid is soluble in alkaline solution and remains soluble on acidification of the alkaline extracts. Related to procedures of separation and fractionation, the HA fractions could be further grouped as gray HA (soluble in alkaline media and low ionic strength values) and brown HA (soluble in alkaline conditions independent of ionic strength values) (Baigorri et al., 2009), or standard HA (alkali-soluble, acid-insoluble) and nitro HA (extracted from HNO_3–nitrated residues after the standard HA extraction) (Nasir et al., 2011), mobile humic acid (MHA, alkali-soluble, acid-insoluble), and recalcitrant Ca humate (CaHA, extracted from HCl-washed residues after the MHA extraction) (He et al., 2013; Mao et al., 2008).

Phosphorus is an essential element for all forms of life. It is involved in the biosynthesis of diverse cellular components, including nucleic acids, proteins, lipids, and sugars. Phosphorus has long been known to be present in humic fractions collected from both terrestrial and aquatic samples (Nissenbaum, 1979; Swift and Posner, 1972). Thus, information on the forms and lability of P associated with humic fractions is required to enhance our understanding of the role of humic fractions in P cycling and nutrition. The forms of P in humic fractions are routinely characterized by ^{31}P NMR spectroscopy. There are at least 18 ^{31}P NMR studies of humic fractions extracted from peat (Bedrock et al., 1995; Francioso et al., 1998, 1996; He et al., 2006c; Makarov et al., 1997; Shand et al., 1999), forest floor humus (Ogner, 1983), mineral soils under forest and natural vegetation (Makarov, 1997; Guggenberger et al., 1996; Makarov et al., 1996, 1997; Shand et al., 1999; Tikhova et al., 2000, 2002; He et al., 2006c), and cultivated soils (Mahieu et al., 2000, 2002; Busato et al., 2005; Barančíková et al., 2007; He et al., 2011a).

There are only a few studies on the lability of humic-bound P, although labile P species in humic fractions is sometimes evaluated by the forms identified by ^{31}P NMR spectroscopy (Makarov et al., 1997). Evidence of lability of humic-bound P can be obtained by biotic and abiotic release of humic-bound P through phosphatase hydrolysis and ultraviolet (UV) irradiation, respectively (He et al., 2011a). The relevant data show that not all P forms associated with humic fractions were equally labile. On the other hand, fungal phytase can release humic-bound P to soluble inorganic P even though no phytate is found in humic samples (He et al., 2011a, 2009a). Therefore, more studies are needed to link the P forms detected by ^{31}P NMR to lability results from phosphatase hydrolysis and UV irradiation. More data are also needed to provide insights on how the lability of humic-bound P is affected by sample origins and environmental factors. Thus, based on our published and unpublished data (He et al., 2011a, 2009a, 2006c; Tazisong, 2007), this chapter presents and discusses P forms and lability in humic fractions from the International Humic Substances Society (IHSS) and agricultural soils with different cropping histories and management practices. Knowledge and insights derived from this chapter are helpful in promoting the research on the role of humic-bound P in P cycling and nutrition in agricultural ecosystems as well as in the environment.

Phosphorus and Selected Elemental Contents in Humic and Fulvic Acids

Table 1 lists P and four elements in HA and FA samples from IHSS and Alabama agricultural soils (Decatur silt loam). The IHSS Elliott soil is typical of the fertile prairie soils of the US states of Indiana, Illinois, and Iowa. The IHSS Waskish peat consists of deep very poorly drained organic soils that were formed mostly in slightly decomposed material from Sphagnum moss on raised bogs on glaciated terrain. The crop management practices at the Alabama soil site consisted of various combinations of tillage systems (conventional till, mulch till, and no-till), cropping systems, and management practices [cotton in summer and rye (*Secale cereale* L.) cover crop in winter], inorganic nitrogen fertilization [ammonium nitrate (0 and 100 kg N ha^{-1}), and poultry litter (0, 100, and 200 kg N ha^{-1})] (Ohno et al., 2009; Tazisong et al., 2013, 2014). The P contents of the four IHSS HA and FA samples are moderate with lower metal contents. On the other hand, the P

Table 1. Selected elemental contents of humic acid (HA) and fulvic acid (FA) from the International Humic Substances Society (IHSS) and Alabama agricultural soils (He et al., 2006c; Tazisong, 2007).

Sample ID and source	P		Ca		Fe		Al		Mn	
	HA	FA	HA	FA	HA	FA	HA	FA	HA	FA
	g kg^{-1} dry matter									
IHSS										
Elliott soil	2.4	1.2	<DL†	<DL	0.9	0.1	<DL	<DL	7×10^{-4}	<DL
Waskish peat	3.1	1.2	<DL	<DL	0.9	1.7	<DL	<DL	<DL	<DL
Alabama Soil‡										
BF	0.2	9.3	ND§	ND	1.8	0.7	8.7	21.9	<0.01	9.5
CT-C-AN	19.8	0.2	ND	ND	1.3	1.0	24.0	2.1	<0.01	0.3
CT-CR	14.4	7.7	ND	ND	10.7	3.7	5.5	8.4	<0.01	2.9
CT-CR-AN	12.9	0.2	ND	ND	0.02	0.8	0.4	4.7	<0.01	2.6
CT-CR-P	35.9	24.0	ND	ND	3.8	2.3	66.0	15.4	<0.01	10.1
MT-CR-AN	17.5	0.2	ND	ND	0.08	0.9	2.2	1.9	<0.01	0.4
MT-CR-P	24.8	20.8	ND	ND	0.5	1.4	36.6	12.4	<0.01	12.2
NT-C	13.9	16.4	ND	ND	0.1	1.2	6.4	4.2	<0.01	3.4
NT-C-AN	0.2	3.2	ND	ND	1.7	1.5	4.5	13.1	<0.01	0.6
NT-CR-AN	13.7	2.0	ND	ND	0.6	1.5	17.8	3.2	<0.01	0.4
NT-CR-P	17.8	13.7	<DL†	<DL	0.5	1.9	16.5	4.7	0.4	2.7
NT-CR-PP	13.2	8.9	<DL	<DL	0.06	4.8	0.4	4.4	<0.01	1.8

† Less than detection limit.

‡ BF, bare fallow; CT, conventional till; MT, mulch till; NT, no-till; C, cotton winter fallow; CR, cotton–rye cropping; AN, 100 kg N ha^{-1} from ammonium nitrate; P and PP, 100 and 200 kg N ha^{-1} from poultry litter, respectively.

§ Not determined.

content varies greatly among the 12 HA (0.2–35.9 g kg^{-1}) and 12 FA (0.2–24.0 g kg^{-1}) samples of Alabama soils. Previously, Makarov and Leoshkina (2009) reported the P content in a range from 1.5 to 15.4 g kg^{-1} in 19 soil HA samples and from 0.3 to 2.3 g kg^{-1} in 19 corresponding soil FA samples. They observed in their samples that the P content in FA isolated according to the IHSS method is much lower than that in HA. However, the content of P in only 9 of 12 FA samples is lower than that in their corresponding HA samples. The effects of tillage and cropping management on the P content are not obvious. However, the P content in five FA samples with the inorganic fertilizer (AN) is much lower than that of the control (BF). Whereas the P content in all but one of the HA samples increased compared with BF, which has very low (0.2 g P kg^{-1}) P content in its HA. The greatest increase of P content is HA (and also FA) with poultry litter due to the extra P input into soil from poultry litter application (He et al., 2012). In most HA and FA fractions, the contents of Al and Fe are relatively high with ≥1.9 g Al and 0.5 g Fe kg^{-1} dry matter; however, the content of Mn is higher in FA samples (0.3–10.1 g kg^{-1}), but lower in all but one HA samples (<0.01 g kg^{-1}).

Table 2 lists P and four elements in sequentially extracted labile MHA and recalcitrant CaHA samples from two agricultural soils (He et al., 2010; Olk et al., 2009). Detailed information on the agricultural importance and/or relevance of the two fractions can be found in this volume (He and Zhang, 2015). The MHA and CaHA samples reported here were obtained from rice soil fields that have been under continuous (>25 yr) cropping of irrigated and rainfed (dryland) lowland rice and were located at the International Rice Research Institute in Los Baños, Laguna, the Philippines. The second set of soil samples was collected from a corn field under continuous irrigation from a private farm in Hamilton County near Phillips, Nebraska, USA. The P content ranged from 1.7 to 3.7 g kg^{-1} for the Philippine MHA samples and from 1.5 to 2.3 g kg^{-1} for the CaHA samples. The P content ranged from 1.5 to 1.8 g kg^{-1} for the Nebraskan MHA samples and from 2.8 to 3.2 g kg^{-1} for the corresponding CaHA samples. The P content in these humic fractions is similar to each other and to the IHSS HA and FA samples in Table 1.

Table 2. Selected elemental contents of mobile humic acid (MHA) and calcium humate (CaHA) of Philippine rice soils and Nebraskan corn soils (He et al., 2011a; 2009a).

Sample ID and Source	P		Ca		Fe		Al		Mn	
	MHA	CaHA	MHA	CaHA	MHA	CaHA	MHA	CaHA	MHA	CaHA
	g kg^{-1} dry matter									
Philippine soils										
2-crop control	2.1	2.1	0.09	0.09	0.35	6.60	<0.10	0.35	<0.10	<0.10
2-crop-NPK†	2.1	2.3	0.07	0.12	0.43	6.36	0.16	0.28	<0.10	<0.10
3-crop-highN	1.7	1.5	<0.05	0.08	0.60	5.80	0.21	1.10	<0.10	<0.10
Dryland	3.7	1.8	0.26	0.40	0.54	1.07	<0.10	0.24	<0.10	<0.10
Nebraskan soil										
Control	1.5	2.8	0.26	0.17	0.36	0.85	<0.10	<0.10	<0.10	<0.10
N fertilizer	1.7	2.9	0.12	0.19	0.42	1.07	<0.10	<0.10	<0.10	<0.10
Beef manure	1.8	3.2	0.08	0.18	0.43	0.93	<0.10	<0.10	<0.10	<0.10

† NPK, inorganic fertilizer.

The contents of Al and Fe are higher in both the HA and FA fractions of Alabama soils, but the content of Al is lower than the detection limit in the IHSS HA and FA samples (Table 1). The content of Mn is greater in the FA fractions than the HA fractions of Alabamian soils. Whereas Ca and Al are detected in some samples, Fe is the major metal in MHA and CaHA fractions (Table 2). The content of Fe is especially greater in CaHA than its corresponding MHA in all samples. Olk et al. (1995) also report that the Fe content of the CaHA was eight-fold greater than that of the MHA fraction from a calcareous vermiculitic soil obtained from a cotton field in California. These data indicate that humic-Fe-phosphate could be major metal-humate complexes in CaHA, whereas humate-Ca-phosphate and humate-Al-phosphate are also present in these humic samples (Baigorri et al., 2013; He et al., 2011b; Urrutia et al., 2014).

Forms of Phosphorus in Humic and Fulvic Acids

The solution ^{31}P NMR spectra of the four IHSS HA and FA samples (Table 1) are relatively simple, indicating few P species in these samples (He et al., 2006c). Based on the ^{31}P NMR peak analysis and spectral integration, the two major P forms are orthophosphate and orthophosphate monoesters. Phosphate diesters and phosphonates may be present, but cannot be reliably distinguished from the background noise. There are 21% of the P as orthophosphate and 79% as orthophosphate monoesters in Elliott soil HA, and 40% of the P as orthophosphate and 60% as orthophosphate monoesters in Elliott soil FA. These results are similar to most studies of HA (Makarov, 1996; Guggenberger et al., 1996; Makarov et al., 1996, 1997; Tikhova et al., 2000, 2002) in that there is only one broad peak orthophosphate in the monoester NMR signal region. Only orthophosphate was detected in Waskish peat HA and FA spectra.

The solution ^{31}P NMR spectra of MHA and CaHA samples are more complicated and are more rigorously analyzed (Fig. 1) (He et al., 2011a). The P forms in both humic fractions were predominantly organic (87%), with no difference between MHA and CaHA (Table 3). These data are consistent with earlier observations that organic P forms comprised more than 80% of P in CaHA and MHA (Mahieu et al., 2000). Whereas there are no obvious trends in P forms for the different fertilizer treatments, some statistically significant differences can be seen between MHA and CaHA (Table 3). Of the seven orthophosphate monoester peaks or regions identified in these samples, significant differences are observed for α- and β-glycerophosphate, *scyllo*-inositol P, the monoester 2 region, and total orthophosphate monoesters. For all but *scyllo*-inositol P, the percentage of P is significantly higher in CaHA samples than in MHA. No previous studies have identified peaks in the orthophosphate monoester region in such detail. The widespread presence of *scyllo*-inositol P in HA samples is a very interesting observation. The origins of *scyllo*-inositol P in soil are unknown. Soil microbes could be responsible for its presence in soil because it was detected after a sand–clay mixture containing organic and inorganic nutrients was incubated with soil microbes (Caldwell and Black, 1958). The data in Table 3 shed no clarity on the source of the *scyllo*-inositol P in soil; however, the fact that it has been incorporated into humic materials supports the role of soil microbes in its formation.

Whereas phytate is a major organic P form in the environment (Giles and Cade-Menun, 2014; He et al., 2006b), none of the peaks for phytate appeared in the

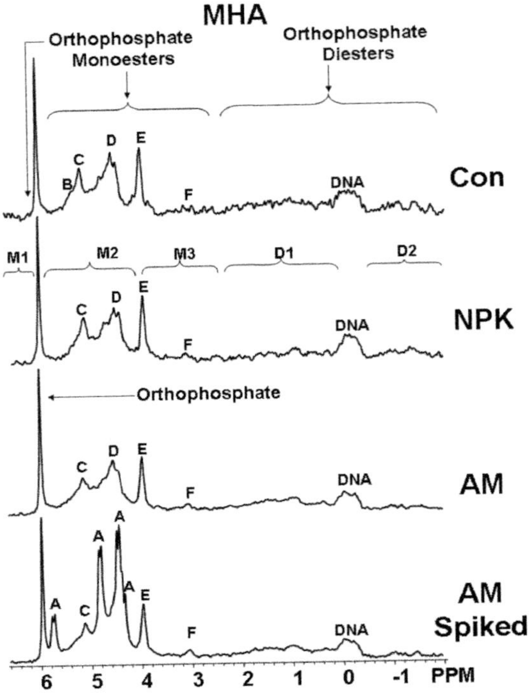

Fig. 1. Solution ^{31}P nuclear magnetic resonance spectra of mobile humic acid (MHA) extracted from the control (Con), inorganic fertilizer (NPK), and beef cattle manure (AM) soils. The AM-spiked spectrum was obtained after the addition of a phytate spike (spiked) into the AM sample. The main orthophosphate sharp peak is labeled in spectrum AM. Labeled peaks are for phytate (A), α-glycerophosphate (B), β-glycerophosphate (C), nucleotide P (D), scyllo-inositol phosphate (E), and an unidentified peak (F). M1, monoester 1; M2, monoester 2; M3, monoester 3; D1, diester 1; D2, diester 2. Please refer to the text for more details about these P species assignments (He et al., 2011a).

spectra of the humic fractions (MHA and CaHA), indicating that phytate was not part of humic-bound P in these fractions. This observation is further confirmed by the appearance of the typical four phytate P peaks in the spiked samples (Fig. 1). For earlier ^{31}P-NMR studies of humic fractions, the orthophosphate monoester region was either dominated by one broad peak so that it was impossible to determine the presence of phytate, or no peaks appeared in the expected NMR signal regions for phytate. Therefore, further studies that include spiking with phytate are required to confirm if the absence of phytate in humic fractions is a universal trait or not. Currently, the absence of phytate may indicate that it is not incorporated into soil humic fractions, or it may be a function of the procedure used to extract the HA fractions. However, the latter scenario is less possible as Mahieu et al. (2002), studying the MHA and CaHA fractions of tropical rice soils, significantly correlated ^{31}P NMR spectral proportions with other chemical properties of the fractions, including the H concentration, visible light absorption, and free radical concentration, which in turn are significantly correlated with the spectral proportions of ^{13}C and ^{15}N NMR. In other words, these correlations suggest

Table 3. Relative peak area (%) of ^{31}P nuclear magnetic resonance spectra of MHA and CaHA fractions extracted from the control (Con), inorganic fertilizer (NPK), and beef cattle manure (AM) soils.

	Orth†	Total phn	G6P	α-glc β-glc	Nucl	Scyllo	Mono 1	Mono 2	Mono 3	DNA	Dies 1	Dies 2	Total mono	Total dies	CTot mono	CTot dies	Total P_o
									MHA								
Con	12.1	2.7	7.6	13.6	7.6	10.6	1.5	16.7	3.8	4.4	15.9	3.5	61.4	23.8	52.6	32.6	87.9
NPK	10.9	4.7	8.2	16.4	8.2	9.1	0.9	12.8	4.6	11.2	9.3	3.7	60.2	24.2	57.0	27.4	89.1
AM	13.8	1.6	8.1	11.3	8.9	8.1	1.6	15.4	4.1	9.0	14.8	3.3	57.5	27.1	55.4	29.2	86.2
Mean‡	12.2	3.00	7.97	13.8*	8.23	9.27*	1.33	15.0*	4.17	8.20*	13.3*	3.50*	59.7*	25.0*	55.0*	29.7*	87.7
SD	1.46	1.57	0.32	2.55	0.65	1.26	0.38	1.99	0.40	3.47	3.54	0.20	2.00	1.80	2.23	2.64	1.46
									CaHA								
Con	12.5	1.7	10.2	18.1	14.7	5.7	1.1	26.1	5.7	1.7	2.5	0.0	81.6	4.2	48.8	37.0	87.5
NPK	11.1	1.6	8.3	17.8	14.2	5.9	1.2	28.4	3.6	2.4	4.0	0.8	79.4	7.2	47.5	39.1	88.2
AM	15.5	2.3	13.3	21.0	9.2	6.6	1.1	21.0	3.3	1.5	3.7	0.8	76.2	6.0	50.2	32.0	84.5
Mean	13.3	1.87	10.6	19.0*	12.9	6.07*	1.13	25.2*	4.20	1.87*	3.40*	0.53*	79.1*	5.80*	48.8*	36.0*	86.7
SD	1.97	0.38	2.52	1.77	2.64	0.47	0.06	3.79	1.31	0.47	0.79	0.46	2.72	1.51	1.35	3.65	1.97

* Statistically significant differences between MHA and CaHA at $P < 0.05$.

† Orth, orthophosphate; phn, phosphonates; G6P, glucose 6-phosphate; a-Glyc and b-Glyc, a and b glycerophosphate; nucl, mononucleotides; mono1, mono2, and mono3, orthophosphate monoesters, regions 1, 2, and 3; scyllo, scyllo-inositol P; Dies1 and Dies2, orthophosphate diesters regions 1 and 2; CTot mono and CTot dies, corrected totals; Tot Po, total of organic P forms from NMR (He et al., 2011a).

‡ Means calculated with single NMR samples, not field replicates; $n = 3$.

that any chemical alteration of P forms during humic fraction extraction does not overly distort the spectral proportions measured by ^{31}P NMR spectroscopy.

Enzymatic Hydrolysis of Phosphorus in Humic and Fulvic Acids

Enzymatic Hydrolysis of Phosphorus in HA and FA Fractions

Although solution ^{31}P NMR analysis is an important tool for characterizing P forms, it does not provide information on how labile humic-bound P is converted to free soluble bioavailable inorganic phosphate. On the other hand, enzymatic hydrolysis of organic P in a sample by external phosphatases can provide an estimate of hydrolyzable (labile) organic P fractions in the sample (He and Honeycutt, 2001, 2011; He et al., 2004b). When 3-phytase is included in the reaction mixture, 16 and 18% of total P are released from the HA and FA fractions of Elliott soil humic substances, respectively. In contrast, only 4% (HA) and 2% (FA) of total P associated with Waskish peat humic fractions are released by 3-phytase hydrolysis (He et al., 2006c). This portion of 3-phytase hydrolyzable P should be considered general labile monoester P. Furthermore, the inhibitory influence of Al and Fe additives on 3-phytase release of humic- or fulvic-associated P is negligible, suggesting that these two metals may have preferential interactions with functional groups other than phosphate esters in humic and fulvic acids, rather than forming enzyme-inaccessible metal-humic complexes or precipitates (He et al., 2006a; Tazisong et al., 2008).

To determine the bioavailability of P in HA and FA fractions of Alabama soils, these humic fractions are incubated with three acid phosphatases and phytase from wheat, respectively (Fig. 2). The P_i concentration in the control is around 12 μM for HA fractions and 2 μM for FA fractions. The presence of soluble P_i in these fractions is consistent with observations in He et al. (2006c). Whereas the procedure used to obtain humic substances would have removed soluble P_i from humic substances, this portion of soluble P_i in HA and FA fractions might be associated with the core structures of organic matter through weak van der Waals interactions or metalorganic matter complexes, which could be partially hydrolyzed under acid conditions (He et al., 2006c). It is clearly seen from Fig. 2 that P_i in all enzyme-humic reaction mixtures is greater than the control (no enzyme), suggesting hydrolysis of humic-bound compounds. The P_i released from these HA and FA fractions would reflect the possible types of P forms present in these humic fractions as these enzymes possess different substrate specificity (He et al., 2004a; Tazisong et al., 2013; Zhu et al., 2013). The hydrolysis by acid phosphatases is greater than by phytase as shown in Fig. 2, suggesting the hydrolysis of less complex phosphate monoesters. The low phytase activity may suggest that phytate which is the main substrate for this enzyme was low in concentration compared with the other phosphate compounds, or if present in appreciable quantities may have been inaccessible to phytase. The preferred hydrolysis of simple monoester phosphate and low phytate availability in these HA and FA fractions are further confirmed by enzymatic incubation of model P compounds in the presence of these HA and FA fractions. Approximately 70 to 90% of the simple monoester in the model P compound-HA reaction mixtures are hydrolyzed while less than 10% of phytate in the phytate-HA reaction mixtures are hydrolyzed (data not shown).

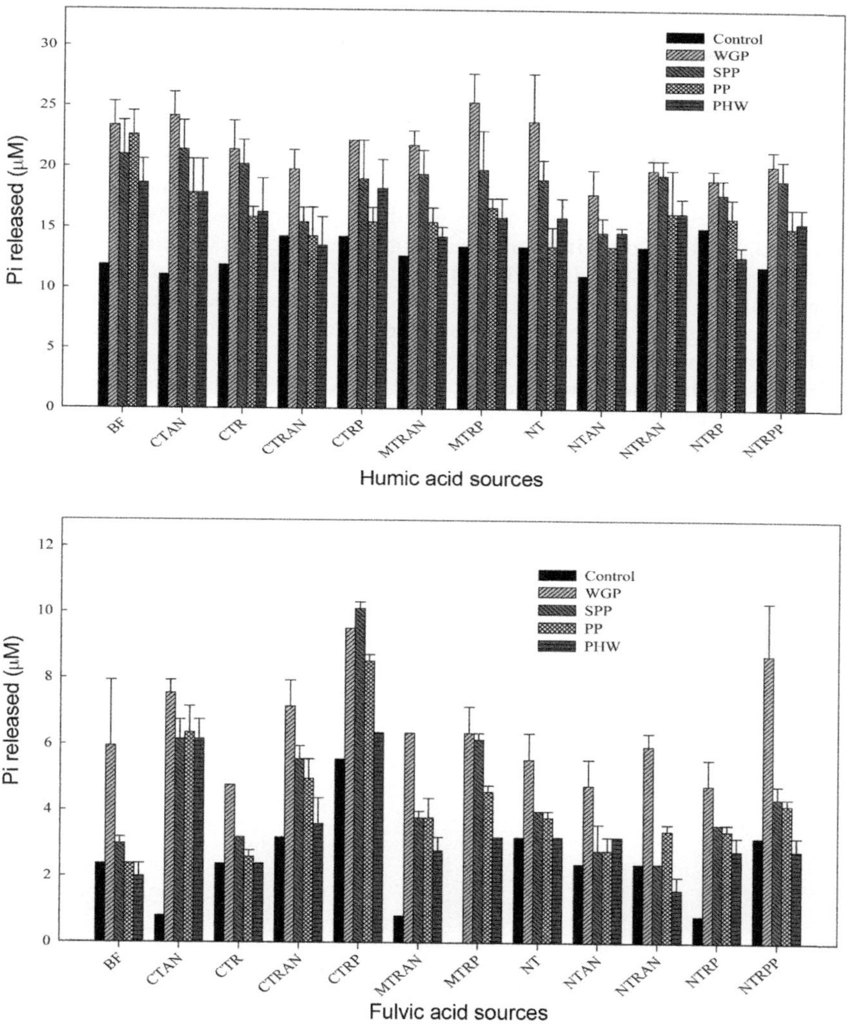

Fig. 2. Inorganic P (Pi) released from humic and fulvic acids obtained from soils with various treatments in the absence and presence of enzymes. WGP, wheat germ acid phosphatase; SPP, sweet potato acid phosphatase; PP, potato acid phosphatase; PHW, wheat phytase; BF, bare fallow; CT, conventional till; MT, mulch till; NT, no-till; C, cotton winter fallow; CR, cotton/rye cropping; AN, 100 kg N ha^{-1} from ammonium nitrate; P and PP, 100 and 200 kg N ha^{-1} from poultry litter, respectively. Data are the averages of laboratory tetraplicates with standard deviation (Tazisong, 2007).

A similar trend is observed for FA, except that greater than 50% of P in FA-phytate reaction mixture is hydrolyzed. The low P hydrolysis in phytate-HA reaction mixtures is noteworthy. The high aromacity and large molecular size of the HA may have created conditions for phytate to be tightly bound by metals present, or held covalently by nonmetalic moieties, especially the alkyl-C present in the surface and inner core of the HA structure, rendering it inaccessible to phytase.

Hydrolysis of Phosphorus in MHA and CaHA Fractions by Combination of Multiple Enzymes

The P in MHA and CaHA fractions has been investigated with a more comprehensive hydrolysis with multiple enzymes (He et al., 2011a, 2009a). The principle of this enzymatic hydrolysis is well described in literature (He and Honeycutt, 2001, 2005, 2011) and used in characterization of organic P in various environmental samples (Johnson and Hill, 2010, 2011; Pagliari, 2014; Pagliari and Laboski, 2012; Young et al., 2013; Zhu et al., 2013). In this procedure, the initial soluble P_i without enzymatic incubation is designated as labile P_i. Potato acid phosphatase releases most phosphomonoesters tested except for phytate. Fungal phytase catalyzes the hydrolysis of phosphomonoesters and phytate. Nuclease P1 endonucleolytically cleaves polynucleotide bonds in RNA and DNA. Then, the P-O bond in nuclease-cleaved mononucleotides is hydrolyzed by phosphomonoesterases to produce P_i. The typical combined enzymatic hydrolysis and results are shown in Fig. 3 and

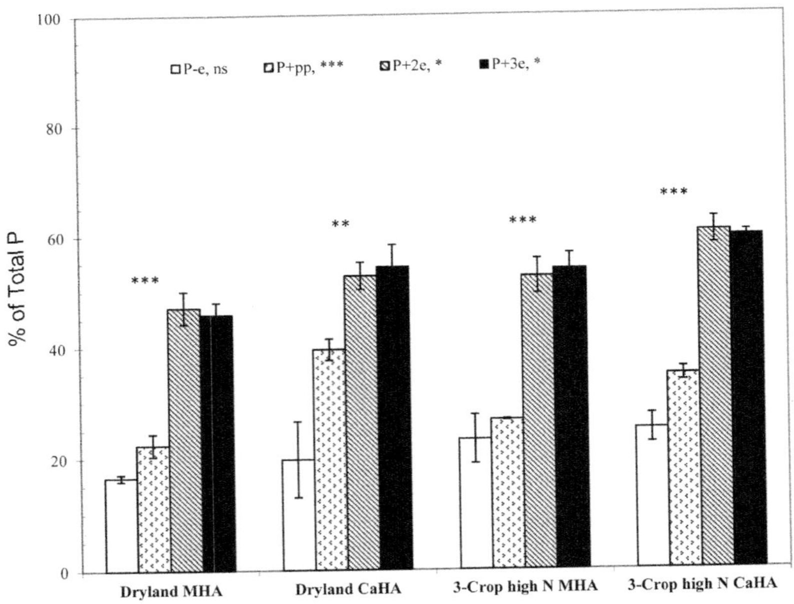

Fig. 3. Increase in soluble P_i for humic fractions from 3-crop and dryland soils after combined enzymatic hydrolysis. P_{-e}, control in the absence of enzyme; P_{+pp}, hydrolysis by potato phosphatase; P_{+2e}, hydrolysis by potato phosphatase + phytase; P_{+3e}, hydrolysis by potato phosphatase + phytase + nuclease P1. Data are the averages of laboratory duplicates with standard deviation. Symbol *, **, and *** represent significance of enzymatic treatments of soluble P_i measurement for a specific humic material, or significance of soluble P_i measurement among the four humic materials with the same enzymatic treatment at $P = 0.05$, 0.01, and 0.001, respectively. Symbol "ns" represents no significant difference at P = 0.05 (He et al., 2009a).

Table 4. Without any enzymatic hydrolysis, the content of soluble inorganic P (P_i) is 16.5 and 23.2% of total P in MHA fractions from dryland and 3-crop high N rice soils, and 19.7 and 25.1% of total P in the CaHA fractions, respectively. Incubation with potato phosphatase increased soluble P_i to 23.4 and 26.7% of total P in the MHA fractions for the dryland and 3-crop high N soils, respectively, and 39.5 and 34.8% of total P in the two CaHA fractions. Soluble P_i for the dryland and 3-crop high N humic fractions increased greatly after incubation with the combination of potato phosphatase and fungal phytase (P+2e); the percentage of total P measured as soluble P_i reached 47.1 and 52.4% for the MHA fractions and 54.2 and 60.5% for the CaHA fractions, respectively. Little difference in soluble P_i is observed between the two (P+2e)- and three (P+3e)-enzyme incubations, indicating only small amounts (0–4.2%) of polynucleotide-like P compounds in these humic fractions.

It is noteworthy that, unlike HA and FA fractions of the Alabamian soil, most of the enzymatically hydrolyzable P is phytate-like P in these MHA and CaHA fractions. The results are also in contrast to the fact that no phytate P was found in these fractions by solution ^{31}P NMR spectroscopy (Fig. 1). Therefore, the term "phytate-like P," rather than "phytate P" proposed by the original work in the enzymatic approach (He and Honeycutt, 2001) is in fact correct. Whereas the molecular mass of phytate is in the hundreds of Daltons, release of P from high molecular mass organic matter has been observed previously. For example, Pant et al. (1994) report that fungal phytase hydrolyzed about 80% of the organic P in the high molecular (>45 KD) fractions of root exudates of spring barley during a 24-h incubation. Working with solutions of sandy soils, He et al. (2006c) observe that incubation with fungal phytase hydrolyzed P in both low (<3 KD) and high (>3 KD) molecular mass humic and fulvic fractions from the Elliot soil in amounts that were 8 to 17% of total P. This observation suggested that the bonding environment in humic fractions is similar to the complicated intramolecular multiple-P structure of phytate, thus making humic P a substrate of phytase.

Table 4. Lability (%) of P in MHA and CaHA fractions extracted from the Con, NPK, and AM soils (He et al., 2011a).

P Form	Con		NPK		AM	
	MHA	CaHA	MHA	CaHA	MHA	CaHA
Labile P_i	10.9	8.5	9.2	8.7	9.0	9.0
Labile Po:	27.6:	18.1:	27.6:	17.8:	31.4:	19.5:
(Simple monoester)	(8.0)	(3.8)	(8.5)	(4.0)	(8.2)	(4.3)
(Phytate-like P)	(17.4)	(14.4)	(17.5)	(13.6)	(19.0)	(14.0)
(polynucleotide-like P)	(2.2)	(0.0)	(1.6)	(0.2)	(4.2)	(1.3)
UV labile P:	0.0:	7.8:	4.9:	10.6:	0.8:	9.2:
(P_i change)	(5.2)	(4.7)	(7.0)	(6.1)	(5.1)	(4.5)
(P_o change)	(−5.3)	(3.2)	(−2.1)	(4.5)	(−4.3)	(4.8)
Recalcitrant P	61.5	65.5	58.3	63.0	58.8	62.2

Ultraviolet Labile Phosphorus in Humic and Fulvic Acids

Ultraviolet degradation is an abiotic mechanism of releasing bioavailable P_i from complex P compounds in environmental samples (Francko and Heath, 1982; He et al., 2009b). Ultraviolet light can induce the release of P from Fe(III)-P compounds, suggesting that sunlight could play a role in maintaining labile P in soil (Pant and Warman, 2000). Ultraviolet light irradiation has been used to evaluate abiotic release of labile P in lake organic matter (Francko and Heath, 1982; Zhou, 1996; Zhou et al., 2004). Thus, He et al. (2006c, 2009b) propose that UV irradiation could be used to evaluate the lability of P associated with humic fractions, as UV irradiation may not only directly break the P-humic complexes to release soluble P_i, but also degrade other organic moieties of humic fractions, making the P_o groups less protected from phosphatase hydrolysis. This assumption is confirmed by the observation with the IHSS HA and FA fractions as more soluble P_i are observed in all four fractions after 4-h UV irradiation (Fig. 4). Ultraviolet irradiation releases 20 and 11% of P in Elliott soil HA and FA and 9 and 5% of P in Waskish peat HA and FA, respectively. Enzymatic hydrolysis of UV-irradiated Elliott soil HA and FA solutions releases 29 and 18% of P, higher than the P_i released by UV irradiation only. Enzymatic hydrolysis does not release more P_i from UV-irradiated

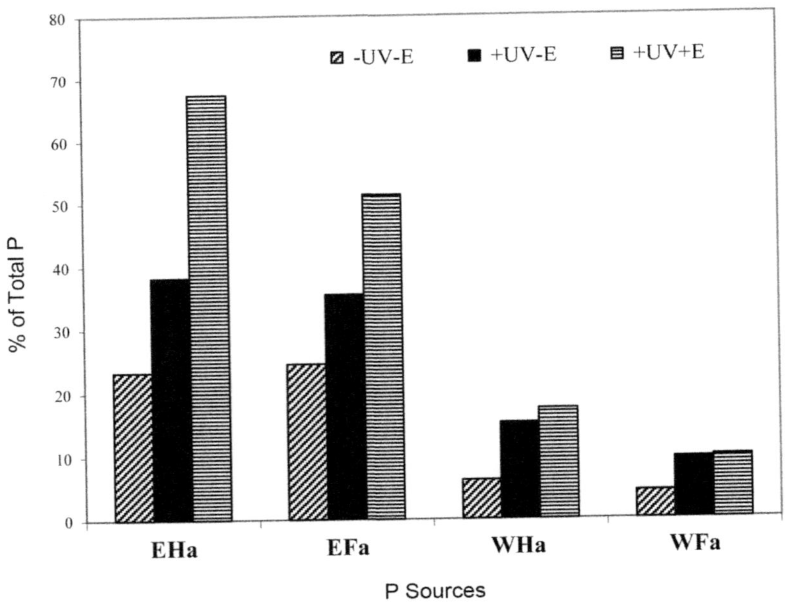

Fig. 4. Increase of soluble inorganic P (P_i) by ultraviolet (UV) irradiation. Reaction mixtures contain 0.322 mM P. The humic substance solutions were irradiated by UV light for 4 h (+UV) before incubation in the presence (+E), or absence (−E) of 3-phytase. The sample "−UV-E" is the control without irradiation. EHa, Elliott soil standard humic acid; EFa, Elliott soil standard fulvic acid; WHa, Waskish peat reference humic acid; WFa, Waskish peat reference fulvic acid. Adapted from He et al. (2006c).

Waskish peat HA and FA solutions, as ^{31}P NMR reveals little organic P in the two fractions (He et al., 2006c).

The similar effects of UV irradiation are also observed with MHA and CaHA fractions (He et al., 2011a, 2009a). Whereas not much difference in P lability is observed in the same type of humic fraction among the management treatments, there are some differences in the lability of humic-bound P between different types of humic fractions from the same source (Table 4). It is notable that the changes of UV labile P_o of the Nebraskan soil MHA fractions are negative, compared with those in unirradiated samples. The decreases in labile organic P with irradiation of the MHA fractions suggest that some hydrolyzable P_o associated with humic substances is directly broken down to soluble inorganic P, rather than to free organic P. In other words, part of UV labile P_o is also enzymatically hydrolyzable P_o. The portion of humic P (37–90%) is not characterized by either enzymatic hydrolysis or UV irradiation and is designated as recalcitrant P (Table 4). This portion of P could be either (i) tightly humic-bound inorganic P, (ii) organic P which was not a substrate of the enzymes used in this study, or (iii) protected from enzyme attacks.

To analyze the relative significance of the affecting parameters, the lability data of four sets of samples were further compared (Table 5). These data show remarkable differences between humic fractions from different origins. The labile P forms in either peat humic or fulvic acids are very low, with none of the three types of labile P more than 10% of total P, leaving 83 and 90% of total P in the two peat samples as recalcitrant P. The source of the IHSS soil humic and fulvic acids is a typical fertile soil found in the Midwest USA. The total labilities of P in the two IHSS soil humic and fulvic acids are more or less similar to the labilities of the rice soil humic fractions. However, their labile components are different. Rice soil humic fractions contained more labile P_o, whereas IHSS humic fractions contained more UV-sensitive labile P. The lability of humic-bound P in Nebraskan corn soils is moderate, as these humic fractions contained less labile P than the humic fractions from Philippine rice soils and the IHSS soil, but more than the peat humic fractions. Based on the data in Table 5, He et al. (2011a) concluded that the lability of humic-bound P is in the order: Sample origin > Humic type > Management factors.

Table 5. Comparison of lability (percent of total P in each humic fraction) for humic-bound P pools from different sources.

	Corn soil†		Rice soil‡		IHSS soil§		IHSS peat§	
	MHA	CaHA	MHA	CaHA	HA	FA	HA	FA
Labile inorganic P	9.7	8.7	20.0	21.6	23	25	6	4
Labile organic P	28.9	18.5	34.4	36.1	17	18	4	2
UV labile P	1.9	9.2	6.7	4.6	27	11	3	5
Total labile P	40.5	36.6	60.5	62.5	67	51	17	10

† Means of three MHA and three CaHA extracted from three fertilizer treatments in the Nebraska Hord soil planted to corn (He et al., 2011a).

‡ Means of four MHA and four CaHA from Philippine lowland soils under different rice cropping managements (He et al., 2009a).

§ Humic acid (HA) and fulvic acid (FA) fractions of the IHSS Elliott soil and Waskish peat (He et al., 2006c).

Conclusions

Solution ^{31}P NMR spectroscopy reveals various P forms present in humic fractions. Like soil and other environmental samples, monoester P is the major organic P in humic fractions. However, phytate, the typical organic P in many samples, is not present in the humic fractions. In contrast, *scyllo*-inositol P (a stereoisomer of phytate) is detected in the examined MHA and CaHA fractions. As there is only one such detailed NMR characterization of humic-bound P, it is not clear that the observations are due to the different extraction methods or in fact due to different soil type, ecological and environmental factors, and/or management practices. A systematic study involving spiking with phytate and other P forms is necessary to confirm whether the trends in phytate, *scyllo*-inositol P, and orthophosphate diesters observed in these MHA and CaHA are universal trends for humic fractions.

Enzymatic hydrolysis and UV irradiation can be used to evaluate the biotically and abiotically labile P in humic fractions. Such wet chemistry studies reveal that P in humic fractions is present in multiple forms with varying labilities. Data reviewed in this chapter demonstrate that neither assumption no P in humic fractions is labile or all P in humic fractions is labile is true, thus suggesting the need to experimentally evaluate the lability of humic-bound P. As the portion of labile P in humic fractions varies between 10 and 63% in the papers reviewed, the origin is the most important factor with respect to the forms and lability of humic-bound P. Cropping and fertilization management practices are least influential while the impact of humic type (i.e., HA, FA, MHA, CaHA) is modest.

References

Baigorri, R., M. Fuentes, G. Gonzalez-Gaitano, J.M. Garcia-Mina, G. Almendros, and F.J. Gonzalez-Vila. 2009. Complementary multianalytical approach to study the distinctive structural features of the main humic fractions in solution: Gray humic acid, brown humic acid, and fulvic acid. J. Agric. Food Chem. 57:3266–3272. doi:10.1021/jf8035353

Baigorri, R., O. Urrutia, J. Erro, M. Mandado, I. Perez-Juste, and J.M. Garcia-Mina. 2013. Structural characterization of anion-calcium-humate complexes in phosphate-based fertilizers. ChemSusChem 6:1245–1251. doi:10.1002/cssc.201300024

Barančíková , G., T. Liptaj, and N. Pronayova. 2007. Phosphorus fractions in arable and mountain soils and their humic acids. Soil Water Res. 2:141–148.

Bedrock, C.N., M.V. Cheshire, J.A. Chudek, A.R. Fraser, B.A. Goodman, and C.A. Shand. 1995. Effect of pH on precipitation of humic acid from peat and mineral soils on the distribution of phosphorus forms in humic and fulvic acid fractions. Commun. Soil Sci. Plant Anal. 26:1411–1425. doi:10.1080/00103629509369381

Busato, J.G., L.P. Canellas, V.M. Rumjanek, and A.C.X. Velloso. 2005. Phosphorus in an Inceptisol under long-term sugarcane: II. Humic acid analysis by NMR P-31. Rev. Bras. Cienc. Solo 29:945–953. doi:10.1590/S0100-06832005000600012

Caldwell, A.G., and C.A. Black. 1958. Inositol hexaphosphate: II. Synthesis by soil microorganisms. Soil Sci. Soc. Am. Proc. 22:293–296. doi:10.2136/sssaj1958.03615995002200040007x

Francioso, O., C. Ciavatta, V. Tuganoli, S. Sanchez-Cortes, and C. Gessa. 1998. Spectroscopic characterization of pyrophosphate incorporation during extraction of peat humic acids. Soil Sci. Soc. Am. J. 62:181–187. doi:10.2136/sssaj1998.03615995006200010024x

Francioso, O., S. Sanchez-Cortes, V. Tugnoli, C. Ciavatta, L. Sitti, and C. Gessa. 1996. Infrared, Raman, and nuclear magnetic resonance (^{1}H, ^{13}C, and ^{31}P) spectroscopy in the study of fractions of peat humic acids. Appl. Spectrosc. 50:1165–1174. doi:10.1366/0003702963905169

Francko, D.A., and R.T. Heath. 1982. UV-sensitive complex phosphorus: Association with dissolved humic material and iron in a bog lake. Limnol. Oceanogr. 27:564–569. doi:10.4319/lo.1982.27.3.0564

Giles, C.D., and B.J. Cade-Menun. 2014. Phytate in animal manure and soils: Abundance, cycling and bioavailability. In: Z. He and H. Zhang, editors, Applied manure and nutrient chemistry for sustainable agriculture and environment. Springer, Amsterdam, the Netherlands. p. 163–190.

Guggenberger, G., L. Haumaier, R.J. Thomas, and W. Zech. 1996. Assessing the organic phosphorus status of an Oxisol under tropical pastures following native savanna using ^{31}P NMR spectroscopy. Biol. Fertil. Soils 23:332–339. doi:10.1007/BF00335963

He, Z., X. Cao, J. Mao, T. Ohno, and H.M. Waldrip. 2013. Analysis of carbon functional groups in mobile humic acid and recalcitrant humate extracted from eight US soils. Pedosphere 23:705–716. doi:10.1016/S1002-0160(13)60063-6

He, Z., T.H. Dao, and C.W. Honeycutt. 2006a. Insoluble Fe-associated inorganic and organic phosphates in animal manure and soil. Soil Sci. 171:117–126. doi:10.1097/01.ss.0000187374.31026.ca

He, Z., and C.W. Honeycutt. 2001. Enzymatic characterization of organic phosphorus in animal manure. J. Environ. Qual. 30:1685–1692. doi:10.2134/jeq2001.3051685x

He, Z., and C.W. Honeycutt. 2005. A modified molybdate blue method for orthophosphate determination suitable for investigating enzymatic hydrolysis of organic phosphates. Commun. Soil Sci. Plant Anal. 36:1373–1383. doi:10.1081/CSS-200056954

He, Z., and C.W. Honeycutt. 2011. Enzymatic hydrolysis of organic phosphorus. In: Z. He, editor, Environmental chemistry of animal manure. Nova Science, New York. p. 253–274.

He, Z., C.W. Honeycutt, T. Ohno, F. Wu, and R. Zhang. 2009b. Characterization of phosphorus associated with natural organic matter. In: F. Wu and B. Xing, editors, Natural organic matter and its significance in the environment. Science Press, Beijing. p. 265–279.

He, Z., C.W. Honeycutt, and H. Zhang. 2011b. Elemental and Fourier transform infrared spectroscopic analysis of water and pyrophosphate extracted soil organic matter. Soil Sci. 176:183–189. doi:10.1097/SS.0b013e318212865c

He, Z., C.W. Honeycutt, T. Zhang, and P.M. Bertsch. 2006b. Preparation and FT-IR characterization of metal phytate compounds. J. Environ. Qual. 35:1319–1328. doi:10.2134/jeq2006.0008

He, Z., T.S. Griffin, and C.W. Honeycutt. 2004a. Enzymatic hydrolysis of organic phosphorus in swine manure and soil. J. Environ. Qual. 33:367–372. doi:10.2134/jeq2004.3670

He, Z., T.S. Griffin, and C.W. Honeycutt. 2004b. Phosphorus distribution in dairy manures. J. Environ. Qual. 33:1528–1534. doi:10.2134/jeq2004.1528

He, Z., T. Ohno, B.J. Cade-Menun, M.S. Erich, and C.W. Honeycutt. 2006c. Spectral and chemical characterization of phosphates associated with humic substances. Soil Sci. Soc. Am. J. 70:1741–1751. doi:10.2136/sssaj2006.0030

He, Z., T. Ohno, D.C. Olk, and F. Wu. 2010. Capillary electrophoresis profiles and fluorophore components of humic acids in Nebraska corn and Philippine rice soils. Geoderma 156:143–151. doi:10.1016/j.geoderma.2010.02.011

He, Z., D.C. Olk, and B.J. Cade-Menun. 2011a. Forms and lability of phosphorus in humic acid fractions of Hord silt loam soil. Soil Sci. Soc. Am. J. 75:1712–1722. doi:10.2136/sssaj2010.0355

He, Z., D.C. Olk, C.W. Honeycutt, and A. Fortuna. 2009a. Enzymatically and ultraviolet-labile phosphorus in humic acid fractions from rice soils. Soil Sci. 174:81–87. doi:10.1097/SS.0b013e3181981dc5

He, Z., Z.N. Senwo, and I.A. Tazisong. 2012. Long-term dynamics of labile and stable phosphorus following poultry litter application to pasture soils. Commun. Soil Sci. Plant Anal. 43:2835–2850. doi:10.1080/00103624.2012.728262

He, Z., and M. Zhang. 2015. Structural and functional comparison of mobile and recalcitrant humic fractions from agricultural soils. In: Z. He and F. Wu, editors, Labile organic matter: Chemical compositions, function, and significance in soil and the environment. SSSA Spec. Publ. 62. SSSA, Madison, WI. doi:10.2136/sssaspecpub62.2014.0036 (this volume).

Johnson, N.R., and J.E. Hill. 2010. Phosphorus species composition of poultry manure-amended soil using high-throughput enzymatic hydrolysis. Soil Sci. Soc. Am. J. 74:1786–1791. doi:10.2136/sssaj2009.0431

Johnson, N.R., and J.E. Hill. 2011. High-throughput measurement and classification of organic P in environmental samples. J. Vis. Exp. 52:1–3.

Mahieu, N., D.C. Olk, and E.W. Randall. 2000. Analysis of phosphorus in two humic acid fractions of intensively cropped lowland rice soils by 31P-NMR. Eur. J. Soil Sci. 51:391–402. doi:10.1046/j.1365-2389.2000.00329.x

Mahieu, N., D.C. Olk, and E.W. Randall. 2002. Multinuclear magnetic resonance analysis of two humic acid fractions from lowland rice soils. J. Environ. Qual. 31:421–430. doi:10.2134/jeq2002.0421

Makarov, M.I. 1996. Forms of P compounds in humic and fulvic acids in some types of soils. Moscow Univ. Soil Sci. Bull. 51:15–22.

Makarov, M.I. 1997. Phosphorus compounds of soil humic acids. Eurasian Soil Sci. 30:395–402.

Makarov, M.I., G. Guggenberger, W. Zech, and H.G. Alt. 1996. Organic phosphorus species in humic acids of mountain soils along a toposequence in the Northern Caucasus. Z. Pflanzenernaehr. Bodenkd. 159:467–470. doi:10.1002/jpln.1996.3581590509

Makarov, M.I., and N.A. Leoshkina. 2009. Phosphorus in the fulvate fraction of soil organic matter. Eurasian Soil Sci. 42:277–283. doi:10.1134/S1064229309030041

Makarov, M.I., T.I. Malysheva, L. Haumaier, H.G. Alt, and W. Zech. 1997. The forms of phosphorus in humic and fulvic acids of a toposequence of alpine soils in the northern Caucasus. Geoderma 80:61–73. doi:10.1016/S0016-7061(97)00049-9

Mao, J., D.C. Olk, X. Fang, Z. He, J. Bass, and K. Schmidt-Rohr. 2008. Influence of animal manure application on the chemical structures of soil organic matter as investigated by advanced solid-state NMR and FT-IR. Geoderma 146:353–362. doi:10.1016/j.geoderma.2008.06.003

Nasir, S., T.B. Sarfaraz, T.V. Verheyen, and A.L. Chaffee. 2011. Structural elucidation of humic acids extracted from Pakistani lignite using spectroscopic and thermal degradative techniques. Fuel Process. Technol. 92:983–991. doi:10.1016/j.fuproc.2010.12.020

Nissenbaum, A. 1979. Phosphorus in marine and non-marine humic substances. Geochim. Cosmochim. Acta 43:1973–1978. doi:10.1016/0016-7037(79)90009-7

Ogner, G. 1983. 31P-NMR spectra of humic acids: A comparison of four different raw types in Norway. Geoderma 29:215–219. doi:10.1016/0016-7061(83)90088-5

Ohno, T., Z. He, I.A. Tazisong, and Z.N. Senwo. 2009. Influence of tillage, cropping, and nitrogen source on the chemical characteristics of humic acid, fulvic acid, and water-soluble soil organic matter fractions of a long-term cropping system study. Soil Sci. 174:652–660. doi:10.1097/SS.0b013e3181c30808

Olk, D.C., M.M. Anders, T.R. Filley, and C. Isbell. 2009. Crop nitrogen uptake and soil phenols accumulation under continuous rice cropping in Arkansas. Soil Sci. Soc. Am. J. 73:952–960. doi:10.2136/sssaj2008.0069

Olk, D.C., K.G. Cassman, and T.W.-M. Fan. 1995. Characterization of two humic acid fractions from a calcareous vermiculitic soil: Implications for the humification process. Geoderma 65:195–208. doi:10.1016/0016-7061(95)94048-9

Pagliari, P.H. 2014. Variety and solubility of phosphorus forms in animal manure and their effects on soil test phosphorus. In: Z. He and H. Zhang, editors, Applied manure and nutrient chemistry for sustainable agriculture and environment. Springer, Amsterdam, the Netherlands. p. 141–161.

Pagliari, P.H., and C.A.M. Laboski. 2012. Investigation of the inorganic and organic phosphorus forms in animal manure. J. Environ. Qual. 41:901–910. doi:10.2134/jeq2011.0451

Pant, H.K., D. Vaughan, and A.C. Edwards. 1994. Molecular size distribution and enzymatic degradation of organic phosphorus in root exudates of spring barley. Biol. Fertil. Soils 18:285–290. doi:10.1007/BF00570630

Pant, H.K., and P.R. Warman. 2000. Phosphorus release from soils upon exposure to ultra-violet light. Commun. Soil Sci. Plant Anal. 31:321–329. doi:10.1080/00103620009370439

Shand, C.A., M.V. Cheshire, C.N. Bedrock, P.J. Chapman, A.R. Fraser, and J.A. Chudek. 1999. Solid-phase ^{31}P NMR spectra of peat and mineral soils, humic acids and soil solution components: Influence of iron and manganese. Plant Soil 214:153–163. doi:10.1023/A:1004720308820

Swift, R.S., and A.M. Posner. 1972. Nitrogen, phosphorus and sulfur contents of humic acids fractionated with respect to molecular weights. J. Soil Sci. 23:50–57. doi:10.1111/j.1365-2389.1972.tb01640.x

Tan, K.H. 2003. Humic matter in soil and the environment. Marcel Dekker, New York.

Tazisong, I.A. 2007. Enzymatic hydrolysis and P-31 nuclear magnetic resonance spectroscopy identification and quantification of organic phosphorus. Ph.D. diss., Alabama A&M Univ., Normal.

Tazisong, I.A., Z. He, and Z.N. Senwo. 2013. Inorganic and enzymatically hydrolyzable organic phosphorus of Alabama Decatur silt loam soils cropped with upland cotton. Soil Sci. 178:231–239. doi:10.1097/SS.0b013e31829cf464

Tazisong, I.A., Z.N. Senwo, B.J. Cade-Menun, and Z. He. 2014. Phosphorus forms and mineralization potentials of Alabama upland cotton production soils amended with poultry litter. In: Z. He and H. Zhang, editors, Applied manure and nutrient chemistry for sustainable agriculture and environment. Springer, Amsterdam, the Netherlands. p. 191–209.

Tazisong, I.A., Z.N. Senwo, R.W. Taylor, and Z. He. 2008. Hydrolysis of organic phosphates by commercially available phytases: Biocatalytic potentials and effects of ions on their enzymatic activities. J. Food Agric. Environ. 6(3&4):500–505.

Tikhova, V.D., M.M. Shakirov, V.P. Fadeeva, and M.I. Dergacheva. 2000. ^{31}P NMR study of soil humic acids. Russ. J. Appl. Chem. 73:1278–1281.

Tikhova, V.D., M.M. Shakirov, V.P. Fadeeva, M.I. Dergacheva, L.V. Samutenko, and L.V. Fedorova. 2002. NMR study of structural changes in humic acids of meadow soddy soils. Russ. J. Appl. Chem. 75:829–833. doi:10.1023/A:1020335301848

Urrutia, O., J. Erro, I. Guardado, S. San Francisco, M. Mandado, R. Baigorri, J. Claude Yvin, and J. Ma Garcia-Mina. 2014. Physico-chemical characterization of humic-metal-phosphate complexes and their potential application to the manufacture of new types of phosphate-based fertilizers. J. Plant Nutr. Soil Sci. 177:128–136. doi:10.1002/jpln.201200651

Young, E.O., D.S. Ross, B.J. Cade-Menun, and C.W. Liu. 2013. Phosphorus speciation in riparian soils: A phosphorus-31 nuclear magnetic resonance spectroscopy and enzyme hydrolysis study. Soil Sci. Soc. Am. J. 77:1636–1647. doi:10.2136/sssaj2012.0313

Zhou, Y. 1996. UV-sensitive P compounds: Release mechanism, seasonal fluctuation and inhibitory effects on alkaline phosphatase activity in a shallow Chinese freshwater lake (Donghu Lake). Hydrobiologia 335:55–62. doi:10.1007/BF00013683

Zhou, Y., J. Li, C. Song, and X. Cao. 2004. Variations and possible source of potentially available phosphorus in a Chinese shallow eutrophic lake. J. Freshwater Ecol. 19:87–96. doi:10.1080/02705060.2004.9664516

Zhu, Y., F. Wu, Z. He, J. Guo, X. Qu, F. Xie, J.P. Giesy, H. Liao, and F. Guo. 2013. Characterization of organic phosphorus in lake sediments by sequential fractionation and enzymatic hydrolysis. Environ. Sci. Technol. 47:7679–7687. doi:10.1021/es305277g

Structural and Functional Comparison of Mobile and Recalcitrant Humic Fractions from Agricultural Soils

Zhongqi He* and Mingchu Zhang

Abstract

Mobile humic acid (MHA) and Ca-bound humic acid (CaHA) are humic fractions sequentially extracted from soil samples. Mobile humic acid is extracted by dilute NaOH, and CaHA is subsequently extracted by dilute NaOH from the dilute HCl-washed soil residues of the first extraction. This chapter reviews the recent advance in characterization of the structural and functional groups of MHA and CaHA fractions of agricultural soils by sophisticated instrumental techniques. Capillary electrophoresis (CE) separates MHA and CaHA fractions into three or two major peaks, mainly superimposed on a characteristic humic "hump." The greater difference in the CE profiles between soils than between cropping management practices reflects the distinct "fingerprint" CE features of these soil humic fractions. Modeling analysis of fluorescence excitation-emission matrix spectral data reveals four general fluorophore components. Fourier transform infrared (FTIR) spectral data reveal a linear correlation between the aliphatic band (3020–2800 cm^{-1}) and the carboxyl band (1720–1600 cm^{-1}) in the same types (i.e., MHA or CaHA) of humic fractions from 10 agricultural soils, suggesting that these two functional groups may be associated with the same backbone or matrix in these humic fractions. Correlation analysis of ^{13}C nuclear magnetic resonance (NMR) peak area and intensity of fluorophore components of these humic samples suggests that the fluorophore Component 4 is enriched with nonpolar alkyl, NCH and OCH3 functional groups. Comparison of ^{13}C NMR and ultrahigh resolution mass spectral data reveals that proteinaceous material in these humic fractions is associated with the nonpolar alkyl groups, but competitively bound to the humic backbone structure against aromatic groups.

Abbreviations: CaHA, Ca-bound humic acid; CE, capillary electrophoresis; FTIR, Fourier transform infrared; MHA, mobile humic acid; NMR, nuclear magnetic resonance.

Zhongqi He, USDA-ARS Southern Regional Research Center, 1100 Robert E. Lee Blvd., New Orleans, LA 70124. Mingchu Zhang, Dep. of High Latitude Agriculture, School of Natural Resources and Agricultural Sciences, 303 O'Neill Building, Univ. of Alaska, Fairbanks, AL 99775 (mzhang3@alaska.edu). *Corresponding author (zhongqi.he@ars.usda.gov).

doi:10.2136/sssaspecpub62.2014.0036

Labile Organic Matter—Chemical Compositions, Function, and Significance in Soil and the Environment. SSSA Special Publication 62. Zhongqi He and Fengchang Wu, editors.

Natural organic matter refers to the nonliving organic material in the environment. Humification is the natural process that occurs in soil and other ecosystems by which organic material decomposes and breaks down to humic matter. As humic matter is a mixture of amorphous, polydispersed substances with yellow to brown-black color, it cannot be identified as belonging to an established group of organic compounds such as carbohydrates, peptides, and amino acids (Swift, 1996). Thus, humic matter is frequently fractionated by either physical or chemical means or both into different fractions for better characterization. Humic acid (HA) is a major fraction of humic matter that is soluble in alkaline solution but precipitates on acidification of the alkaline extracts.

An extraction method developed early in the 1960s further separates HA into MHA and recalcitrant CaHA fractions (Campbell et al., 1967; Mortensen, 1965). This fractionation procedure begins with a NaOH extraction without acid wash. This part of extracted HA is named MHA. Following the extraction of the MHA fraction, the soil residues are subject to dilute acid (HCl) wash to remove Ca^{2+} and perhaps other polyvalent cations, then a second NaOH extraction. The humic fraction recovered by the second NaOH extraction is named CaHA. Both HA fractions are then normally cleansed of soil and salt contaminants, often through HF washes and dialyzing, respectively, and finally dried, typically through freeze-drying (He et al., 2013; Olk, 2006). Mobile humic acid is considerably more labile than CaHA as it is richer in N and contains significantly greater quantities of hydrolzyable amino acids, whereas CaHA shows greater aromaticity and molecular weight compared to MHA (Olk and Cassman, 1995; Olk et al., 1995, 1998).

Whereas this fractionation approach has been applied mainly by Olk and his colleagues (Mahieu et al., 2002; Mao et al., 2008; Olk, 2006; Wingeyer et al., 2012), it has been adopted by several research groups (He et al., 2013; Kerek et al., 2003; Ohno et al., 2010; Slepetiene et al., 2010) in investigating soil humic fractions. Olk (2006) summarized and discussed the relevant studies that applied the MHA and CaHA fractionation procedure for exploring structure–function relations of soil organic matter impacted by cropping management practices. In particular, to demonstrate that characterization of the MHA and CaHA fractions could contribute toward a process-level understanding of nutrient cycling, Olk (2006) presented and analyzed three case studies: (i) a phenol accumulation in the MHA fraction was linked to an inhibition of N mineralization in tropical lowland soils under continuous rice (*Oryza sativa* L.) cropping; (ii) addition of the MHA to Californian cotton (*Gossypium hirstum* L.) soils in laboratory studies resulted in increased K availability and plant K uptake, reproducing the benefit of animal manure application in field conditions; and (iii) effects of straw management and winter flooding on N cycling in Californian rice soils were elucidated by studying a fraction comparable to the MHA fraction. Thus, this chapter focuses on the recent advances in applying multiple spectroscopic techniques in characterizing the structural and functional groups in MHA and CaHA fractions from agricultural soils. The forms and lability of phosphorus in MHA and CaHA fractions are reviewed and analyzed with other humic fractions in another chapter of this volume (He et al., 2015).

Capillary Electrophoresis Profiles of MHA and CaHA Fractions

Capillary electrophoresis (CE) is an attractive separation and analysis technique, providing automated, efficient high-resolution output coupled with short analysis times and minimal consumption of sample and reagents. In CE, the aqueous sample (1 to 50 nl) is first injected into a narrow-bore, fused-silica capillary, typically with a 25- to 75-μm inner diameter. Next, high voltage (10 to 30 kV) is applied to the capillary, and cationic and anionic compounds in the sample migrate to the cathode and anode, respectively, by electrostatic force. Electrophoretic mobility of the analyte is mainly based on the ratio of charge-to-size of the solvated ion; thus, CE separates analytes whose ratios of charge-to-size are different. Capillary electrophoresis has demonstrated potential and versatility for soil and environmental studies (He et al., 2012a; Wang et al., 2012). In particular, CE has been used in "fingerprint" characterization of humic acids and other natural organic matter (Campitelli et al., 2012; Cottrell et al., 2013; He et al., 2008; Tatzber et al., 2011). Another contribution to this volume provides more information on CE principles and applications (Tatzber et al., 2015).

The CE profiles of MHA and CaHA from three agricultural soils are presented in Fig. 1 and 2 (He et al., 2010). Because of the relative changes in the electroosmotic flow, the migration time of the peaks gradually shifted to longer times with repeated runs so that the peak pattern and shape rather than the precise migration time of each peak are critical when comparing these CE profiles. The CE profiles of the MHA fractions derived from the Nebraskan corn soils contain two major peaks (5.2 and 5.6 min) and a sharp buffer (borate) complexation peak at 7.7 min (Fig. 1a). The two major peaks are superimposed on a much broader "hump" from 3 to 7 min which is characteristic for humic fractions (Garrison et al., 1995; He et al., 2008). The CE profiles of the two MHA fractions of Philippine soils supporting two annual paddy rice crops show the similar three-peak pattern (Fig. 1b). The CE profile of the MHA of Philippine rice soil supporting three annual rice crops with high N fertilizer input (Fig. 1c) possessed the basic CE features of an MHA sample, namely the same three peaks described above. However, a strong and sharp peak appeared at the second half of the humic "hump." In the profile of the Philippine dryland rice soil MHA, the CE profile looked smoother and simpler, with a sharp peak superimposed on the broad humic hump and a relatively small peak at 8.7 min.

The three-peak pattern is also observed in the CE profiles of the CaHA fractions of corn soils with NPK and manure fertilization with a flatter overall look, as the humic hump is also flatter and the two peaks superimposed on the humic hump are weaker (Fig. 2a). The CE profile of the CaHA fraction from the control soil with two annual paddy rice crops is similar to that of the corresponding MHA fraction, but the two subpeaks of the sharp peak at 5.6 min in the CaHA fraction become more pronounced (Fig. 2b). The CE profile of the CaHA fraction of the NPK-amended rice soil is characterized by a sharp peak on a flatter hump and a smaller buffer-humic complex peak at 9.4 min (Fig. 2b). The CE profiles of the CaHA fractions of the other two rice soils (Fig. 2c) are smoother and simpler than their MHA counterparts, whereas the MHA of the rice soil supporting three annual rice crops with high N fertilizer input looks more different from the other three.

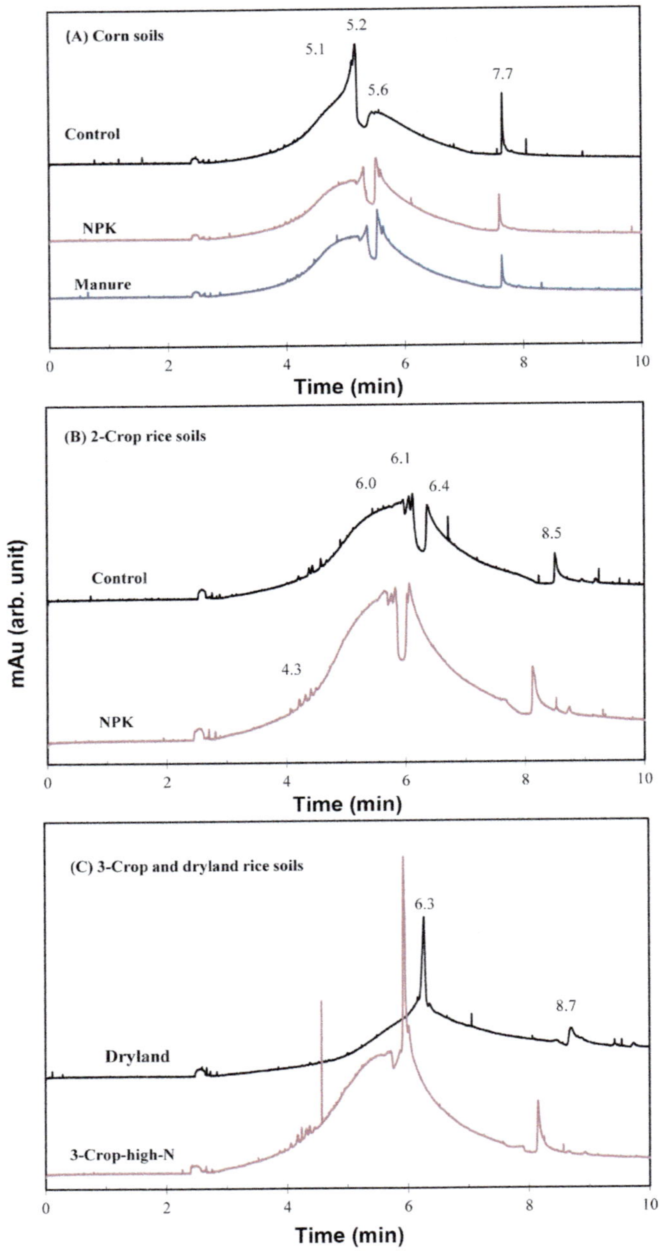

Fig. 1. The capillary electrophoresis (CE) profiles of the mobile humic acid (MHA) fractions. (a) Nebraskan (USA) corn soils with application of beef cattle feedlot manure (Manure), inorganic fertilizer (NPK), and an unfertilized control (Control). (b) Philippine double-cropped rice soil with NPK and an unfertilized control . (c) Philippine triple-cropped rice soil with inorganic fertilizer (3-crop-high-N) and Philippine rainfed rice soil (Dryland) (He et al., 2010).

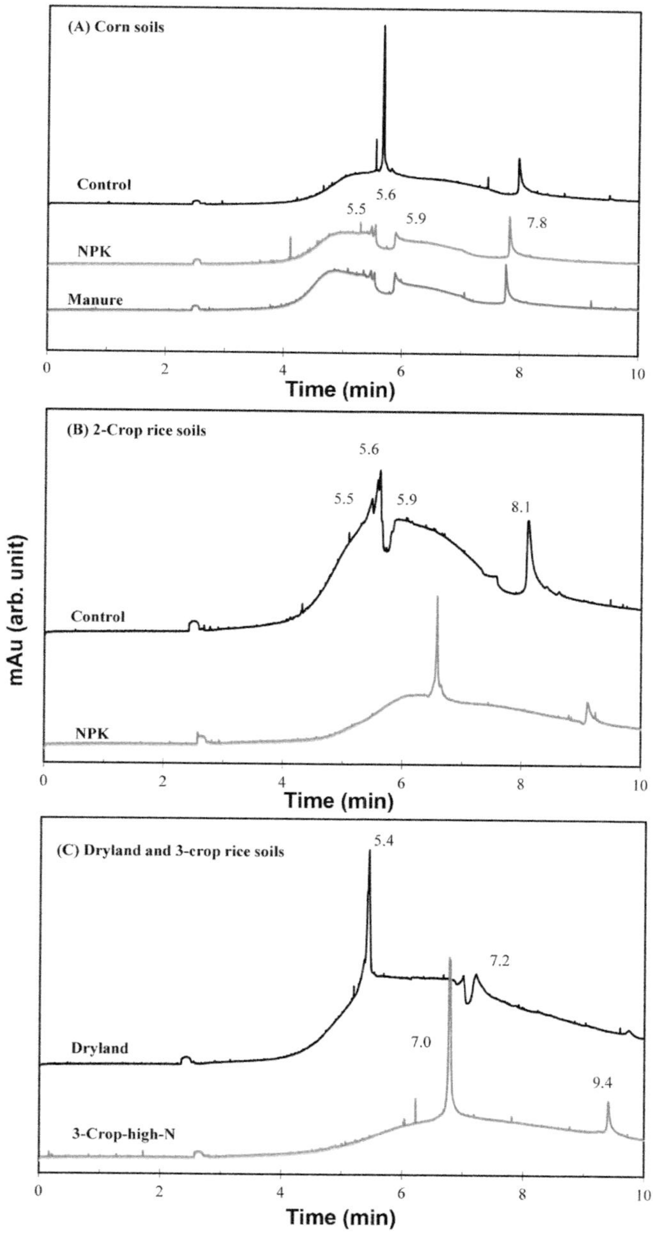

Fig. 2. The CE profiles of the Ca-bound humic acid (CaHA) fractions. (a) Nebraskan (USA) corn soils with application of beef cattle feedlot manure (Manure), inorganic fertilizer (NPK), and an unfertilized control (Control). (b) Philippine double-cropped rice soil with NPK and an unfertilized control. (c) Philippine triple-cropped rice soil with inorganic fertilizer (3-crop-high-N) and Philippine rainfed rice soil (Dryland) (He et al., 2010).

Of the CE profiles shown here, He et al. (2010) found only the MHA fraction of dryland soil resembled any of the profiles presented by Pompe et al. (1996) or He et al. (2008). The other CE profiles shown here were unique and were not similar to any other profiles in the literature, possibly reflecting the uniqueness of flooded rice soils.. Furthermore, there are more differences in the CE profiles between soils than between management practices on the same soil, confirming a fingerprint feature of these CE profiles.

Fluorophore Profiles of MHA and CaHA Fractions

Fluorescence excitation-emission matrix (EEM) spectroscopy has been one widely used tool in characterizing fluorophore components in agricultural products, byproducts, and plant nutrients (He et al., 2014; Ohno and He, 2011; Uchimiya et al., 2013; Zhao et al., 2013). Coupled with a multiple data analysis method, parallel factor analysis (PARAFAC), different fluorophore components (true or "quasiparticles") in humic fractions are derived from fluorescence EEM spectra (He et al., 2006, 2008; Orsetti et al., 2013; Santin et al., 2009). The characteristics of the four fluorophore components of MHA and CaHA fractions from two sets of soils are presented in Fig. 3. Components 1 and 2 represent amino acid–like pools and oxidized quinines in the humic fractions, respectively. They are also found in other humic and fulvic materials (He et al., 2006, 2008; Ohno and Bro, 2006). Components 3 and 4 are similar but not exactly the same in their respective counterparts in the two sets of samples. The differences are mainly on the contour shapes while the contour peak positions are basically the same. Thus, we believe that Components 3 and 4 in the two sets of soil samples represent the same fluorophores, but the fluorescence intensity is different. Componet 3 is likely a hydrophobic acid-related (Chen et al., 2003) or a lignin species (Djikanovic et al., 2007), and

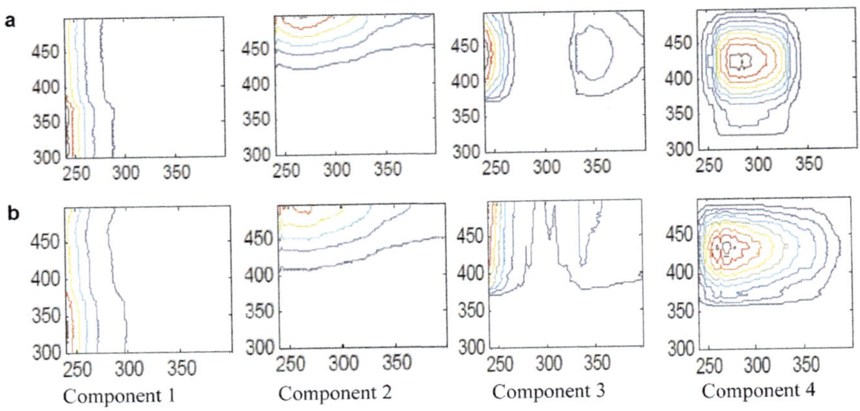

Fig. 3. Spectral characteristics of the four parallel factor (PARAFAC) fluorophore components of MHA and CaHA fractions. (a) MHA and CaHA fractions from Philippine rice and Nebraskan (USA) corn soils (He et al., 2010). (b) MHA and CaHA fractions from eight US agricultural soils; they are unpublished contour spectra from the study of Ohno et al. (2010).

Component 4 is likely a water-soluble fraction of soil organic matter (He et al., 2010). Indeed, both Components 3 and 4 are likely the major fractions of plant-derived water extractable organic matter (WEOM), accounting for 50 and 40% of total fluorescence on average, respectively (Ohno et al., 2010).

He et al. (2010) compared the relative abundance of the four fluorophore components with several previously measured properties of Philippine samples. The comparison reveals that Component 2 becomes more prominent with increasing degree of humification. It is more pronounced in CaHA (i.e., the more humified fraction) than in the MHA. Component 2 is also more pronounced in the dryland rice fractions than in all the fractions from double-cropped and triple-cropped rice fields that are less humified than the dryland humic samples (Olk et al., 2002). By contrast, Components 3 and 4 generally appeared inversely related to degree of humification, which is consistent with the observation that they are likely soluble soil organic matter of plant origin (Ohno et al., 2010).

In earlier work, He et al. (2008) collected CE fractions per three CE peak regions and analyzed the correlation between the CE fractions and fluorophore components by using 13 International Humic Substances Society (IHSS) humic, fulvic, and natural organic samples. They found that the same four fluorophore components in the 13 IHSS samples are present in all three CE fractions. However, the distributions of the four components differed among the three fractions. That is, CE Fraction 1 is rich in Fluorophore Components 1 and 4, sparse in Component 2, but less changed for Component 3 compared with the original samples. Fraction 2 is rich in Components 1 and 4, but sparse in Components 2 and 3. The distribution of the four components in Fraction 3 is similar to that of the original samples. Similarly, Cottrell et al. (2013) collected eight CE fractions of IHSS Suwannee River natural organic matter and analyzed their UV and 3D EEM fluorescence characteristics. Their results demonstrate that the CE fraction distributions have a unique fluorescent signature. For example, humic material-related fluorescence peak constitutes approximately 50% of the total fluorescence in all fractions except for Fraction 4 where it is approximately 35% of total fluorescence. On the other hand, Fraction 4 has the highest fluorescence intensity of protein-like material (>20% of the total fluorescence). Because only very small amounts (nL) of samples are applied to the CE column, the CE fractions must be repeatedly collected for 10 to 24 runs with extending loading times to get sufficient quantities for fluorescence spectroscopic analysis (Cottrell et al., 2013; He et al., 2008). Even so, such correlation of the CE peaks with changes in the four components of MHA and CaHA fractions would certainly provide insight into the chemical structures of the two types of humic acid fractions.

Fourier Transform Infrared Spectra of MHA and CaHA Fractions

Aliphatic and carboxyl groups are two important components in soil humic fractions. The absorbance bands in the ranges of 3020 to 2800 cm^{-1} and 1720 to 1600 cm^{-1} in Fourier transform infrared (FTIR) spectra have been attributed to the two groups, respectively (He et al., 2011a). Regarding the structural difference between MHA and CaHA of Nebraskan corn soil samples, Mao et al. (2008) reported that (i) the FTIR bands that represent aliphatic groups are more distinct in the MHA spectra than in the CaHA spectra, and (ii) these bands shift toward lower wave numbers in the CaHA spectra. These two observations suggest that the MHA fractions are more aliphatic and have a greater diversity of aliphatic fractions than do the CaHA fractions.

For quantitative analysis, three FTIR band regions are measured by band heights or areas (Capriel et al., 1995; He et al., 2011b, 2012b; Kaiser et al., 2007). Two bands in the range of 3020 to 2800 cm^{-1} are grouped and designated as Band A (aliphatic). Two bands in the range of 1740 to 1600 cm^{-1} are grouped and designated as Band B (carboxyl). A band at 1100 to 1030 cm^{-1} is named Band C and used as a reference for comparing the relative intensities of Bands A and B. The baselines were a tangent line drawn from near 2700 to 3020 cm^{-1} for band A, and a straight line from two low points near 900 and 1900 cm^{-1} for Bands A and B (Fig. 4). Bands A and B are also denoted as hydrophobic and hydrophilic, respectively, and the A/B ratio may be used as a measure for the potential wetability of a sample (Kaiser et al., 2007). Table 1 lists the relative band heights of both MHA and CaHA fractions of nine US agricultural soils (Fortuna et al., 2012) and one IHSS reference soil (Elliott) (International Humic Substances Society, 2011). Whereas in-depth analysis of these data is still under way, it is clear that the band heights vary among the 10 MHA and 10 CaHA samples, indicating the effects of the diversity of the soil sources of these humic samples. In general, Band B is stronger than Band A as the intrinsic nature of these samples extracted as a kind of carboxylic acid. In most samples, the ratio of Band A to Band B of an MHA fraction is greater than or about equal to that of the corresponding CaHA from the same soil. The exception is Newport soil samples with the ratio of 0.21 vs. 0.25. Correlation analysis shows that Band A is well correlated with Band B of a single type of humic acid. But no good correlations are observed between the data of MHA and CaHA samples. The general trend of wetability (i.e., ratio of Band A/Band B) is consistent with the qualitative FTIR observation of Nebraskan corn soil samples (Mao et al., 2008). We also measured the intensity of the three bands by band areas (data not

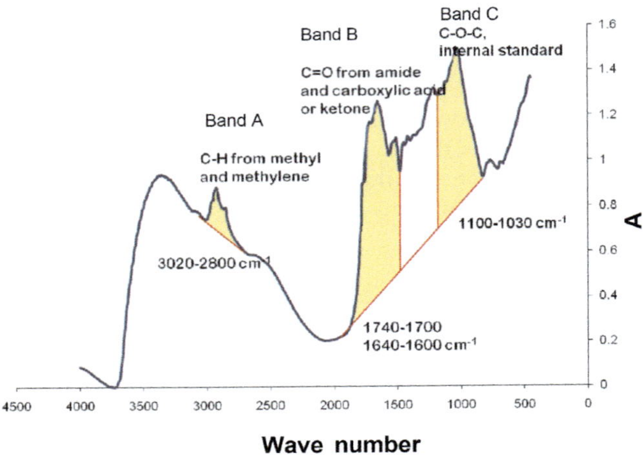

Fig. 4. Three band regions used for semiquantitative Fourier transform infrared (FTIR) analysis of MHA and CaHA fractions.

Table 1. Ratios of peak heights for FTIR Bands A, B and C of the MHA and CaHA fractions of 10 soils (He et al., unpublished data, 2015).†

Soil	Bands A/B and C		Bands B/B and C		Bands A/B and B	
	MHA	CaHA	MHA	CaHA	MHA	CaHA
Adkins	1.16	0.11	6.40	0.86	0.18	0.13
Brooksfield	0.39	0.35	3.54	4.41	0.11	0.08
Caribou	0.50	0.57	3.34	3.66	0.15	0.16
Catlin	0.27	0.08	3.61	0.91	0.07	0.08
Loyal	0.37	0.20	2.78	3.13	0.13	0.06
Newport	0.73	1.02	3.47	4.05	0.21	0.25
Rosholdt	0.31	0.33	3.57	3.79	0.09	0.09
Sharpsburg	0.28	0.29	2.37	3.86	0.12	0.08
Valentine	0.20	0.20	1.30	1.29	0.15	0.16
Elliott	0.24	0.14	3.71	3.26	0.06	0.04
Average	0.44	0.33	3.41	2.92	0.13	0.11
SD	0.29	0.28	1.29	1.37	0.05	0.06

† FTIR, Fourier transform infrared; MHA, mobile humic acid; CaHA, Ca-bound humic acid.

shown). The observations derived from band area measurements are similar to those derived from band height measurements as the coefficient values between the two sets of data are all high with significance levels at $P \leq 0.05$. This similarity can be shown by the correlation coefficients between Bands A, B, and C in the measurements of band height and area (Table 2). Both Tables 2 and 5 show that the best correlation is between Bands A and B in MHA samples. Bands A and B in CaHA samples are also somehow correlated although the coefficients are lower than those of the MHA samples. The close linear relationship between Bands A and B within the same type of humic fractions may imply that the methyl and ethylene groups represented by Band A and the carboxyl groups represented by Band B may come from the same components in these humic acid samples. Band C is not related to either Bands A or B of MHA samples or CaHA samples. No meaningful relationships are observed within the same band of different types of humic fractions.

In addition to the typical optical (transmittance or reflectance) detector, a photoacoustic device has been applied as the signal detector of FTIR (FTIR-PAS). During FTIR-PAS operation, a modulated infrared beam from the FTIR spectrometer strikes the sample. The light is absorbed and produces heat that alters the pressure of the surrounding gas. The changing pressure is detected as sound by a microphone and is then converted to an absorption spectrum, which identifies and quantifies functional groups in the sample. Recently, Du et al. (2013) compared the transmittance FTIR and FTIR-PAS spectra of HA, MHA, and CaHA of IHSS Leonardite. Their results demonstrated that the band heights of both aliphatic Band A and carboxyl Band B in the FTIR-PAS spectra were lower than those of transmittance spectra; however, the A/B ratios of these two bands were similar. Therefore, the FTIR-PAS technique can be used for investigating the functional groups in soil humic substances as the technique is relatively simpler and faster than optical FTIR. Based on their FTIR-PAS results, the ratios of Band A to Band B were 0.90, 0.67, and 0.85, respectively, for Leonardite HA, MHA, and

Table 2. Correlation coefficients between height and area for Bands A, B, and C of the MHA and CaHA fractions of 10 soil samples (He et al., unpublished data, 2015).

	MHA-A	CaHA-A	MHA-B	CaHA-B	MHA-C
Height					
CaA-A	0.511				
MHA-B	0.835**	0.209			
CaHA-B	0.018	0.594†	0.093		
MHA-C	0.505	0.194	0.570†	0.036	
CaHA-C	−0.268	−0.026	−0.343	−0.013	−0.137
Area	MHA-A	CaHA-A	MHA-B	CaHA-B	MHA-C
CaHA-A	0.579†				
MHA-B	0.784**	0.286			
CaHA-B	0.138	0.703*	0.144		
MHA-C	0.434	0.175	0.417	−0.012	
CaHA-C	−0.225	0.024	−0.399	0.072	0.400

* Significant at the 0.05 probability level.

** Significant at the 0.01 probability level.

† Significant at the 0.1 probability level.

CaHA fractions. Thus, among the three Leonardite samples, water repellence is in the order HA > CaHA > MHA. It is noted that the values of these three ratios are higher than those in Table 1, apparently due to the fact that Du et al. (2013) applied a different approach in measuring the band height than that used in the literature (Capriel et al., 1995; He et al., 2011b; Kaiser et al., 2007).

Solid-State ^{13}C NMR Spectra of MHA and CaHA Fractions

Similar to FTIR spectroscopy, NMR spectroscopy is a nondestructive technique that uses the magnetic resonance of nuclei to investigate surrounding chemical structural environments. Many organic matter samples are analyzed with solid-state ^{13}CNMR spectroscopy (Conte et al., 2004; He and Mao, 2011). By using this technique, Olk and his colleagues (Cao et al., 2011; Mahieu et al., 2002; Mao et al., 2008; Schmidt-Rohr et al., 2004) compared the C functional groups in MHA and CaHA fractions from same soils and also under different crop management treatments. For example, ^{13}C cross-polarization NMR and ultraviolet light absorption reveal that the MHA fractions from the Philippine rice soils are highly enriched in phenolic lignin residues (Olk et al., 2006). Saturation-pulse-induced dipolar exchange with recoupling NMR spectroscopy was used to analyze the MHA fractions from the Philippine rice soils (Schmidt-Rohr et al., 2004). The spectra of these MHA fractions show a sufficiently prominent signal for aromatic N (i.e., covalent binding of N by phenolic lignin residues) to enable estimation of the quantity of anilide N separated from that of heterocyclic N. The MHA fraction from the triple-cropped rice soil has a substantially larger content of anilide (25% of all MHA-N) than does the MHA extracted from the nearby dryland rice soil (11% of all MHA-N). These researchers believe that the magnitude of the difference between the two soils (converted to 55 kg N ha^{-1}) is of sufficient magnitude to potentially impact crop performance.

Whereas Olk and his collaborators (Cao et al., 2011; Olk et al., 2006) found that the MHA fraction is consistently composed of younger, more labile materials than the CaHA fraction from the same soil source, He et al. (2013) compared the structural differences between the two fractions extracted from eight US agricultural soils. The cross-polarization–total sideband suppression (CP-TOSS) ^{13}C NMR and corresponding CP-TOSS with dipolar dephasing spectra of MHA and CaHA fractions of Adkins soil are shown in Fig. 5. The spectra clearly show the different features between MHA and CaHA fractions as observed previously on the two fractions from a same soil (Cao et al., 2011; Mao et al., 2008). The distribution of the nine ^{13}C NMR-identified functional groups is listed in Table 3. Comparison of these data reveal that the SD values of five C functional groups (i.e., ketone, quinine, or aldehyde C; aromatic C-C; aromatic C-H; O-alkyl C; and nonpolar alkyl) in MHA fractions exceed 10% of the average values, implying these five C functional groups are more variable than the other four groups in the MHA fractions. In contrast, those five C functional groups displayed yet greater relative variability among the CaHA fractions than among the MHA fractions, and the SD values of all nine C functional groups in CaHA fractions are greater than 10% of their corresponding average values.

He et al. (2013) further analyzed the correlations between soil properties and abundance of C functional groups, and found that the abundance of COO and N-C = O in the MHA fractions was negatively correlated with the sand content, but positively correlated with the content of silt, total N, and soil organic C ($P \leq 0.05$). In contrast, the abundance of the functional group of COO and N-C = O in the CaHA fractions was positively correlated only with the content of clay ($P \leq$

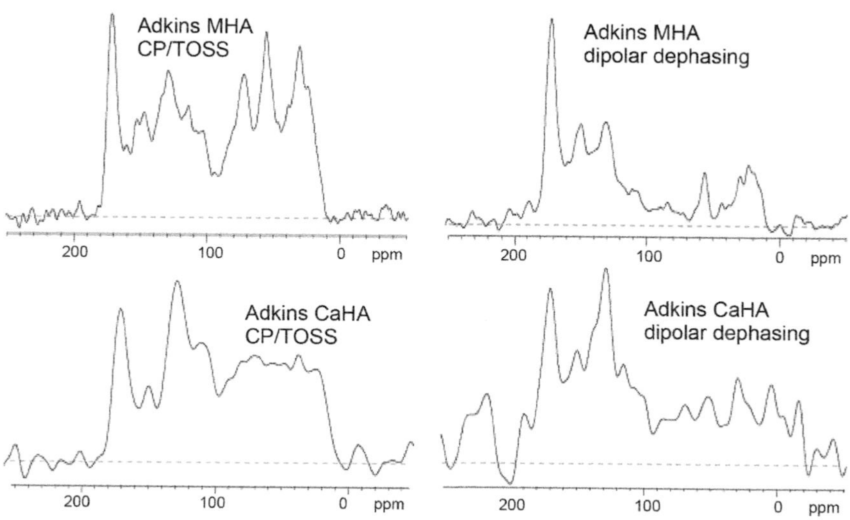

Fig. 5. Carbon-13 cross-polarization–total sideband suppression (CP-TOSS) spectra of MHA and CaHA fractions of Akins soil. Adapted from He et al. (2013).

Table 3. Relative and average abundance of C functional groups in MHA and recalcitrant CaHA fractions obtained by integrations of CP/TOSS ^{13}C nuclear magnetic resonance (NMR) peak areas (He et al., 2013).

Assignment	Chemical shift (ppm)									
	190–220	162–190	145–162	120–145	108–120	96–108	60–96	50–60	0–50	
	Ketone, quinone, or aldehyde C	COO and N-C = O	Aromatic C-O	Aromatic C-C	Aromatic C-H	O-C-O anomerics	O-alkyl C	NCH and OCH_3	Nonpolar alkyl	Aromaticity†
MHA										
Adkins	0.3 (−67)‡	12.0 (−14)	8.6 (2)	16.5 (−1)	7.0 (4)	5.0 (−6)	18.6 (−1)	9.0 (8)	23.1 (10)	42.1 (−4)
Brooksville	2.2 (144)	13.7 (−1)	9.7 (15)	13.9 (−17)	7.5 (12)	5.3 (0)	18.8 (1)	9.4 (13)	19.4 (−8)	43.4 (−1)
Caribou	1.0 (11)	13.5 (−3)	7.7 (−8)	13.6 (−19)	5.4 (−19)	4.9 (−8)	19.9 (6)	6.8 (−18)	27.2 (30)	36.9 (−16)
Catlin	0.1 (−89)	15.8 (14)	9.2 (10)	22.6 (35)	7.7 (15)	5.8 (9)	12.5 (−32)	8.1 (−2)	18.1 (−13)	53.9 (23)
Loyal	1.5 (67)	14.4 (4)	8.7 (4)	16.0 (−4)	6.6 (−1)	5.6 (6)	20.0 (7)	8.0 (−4)	19.1 (−9)	43.9 (0)
Rosholt	1.0 (11)	15.0 (8)	8.5 (1)	19.3 (16)	6.6 (−1)	5.1 (−4)	18.5 (−1)	7.7 (−7)	18.4 (−12)	46.9 (7)
Sharpsburg	1.3 (44)	15.0 (8)	7.2 (−14)	15.9 (−5)	5.9 (−11)	4.9 (−8)	20.1 (7)	9.1 (10)	20.6 (−2)	40.5 (−7)
Valentine	0.3 (−67)	12.1 (−13)	7.7 (−8)	16.2 (−3)	6.7 (0)	6.2 (17)	21.0 (12)	8.1 (−2)	21.8 (4)	41.9 (−4)
Average	0.9	13.9	8.4	16.7	6.7	5.3	18.7	8.3	21.0	43.7
SD	<u>0.7</u>§	1.4	0.8	<u>2.9</u>	<u>0.8</u>	0.5	<u>2.6</u>	0.8	<u>3.1</u>	5.0
CaHA										
Adkins	0.0 (−100)	12.3 (−21)	6.3 (−32)	19.5 (−9)	8.9 (16)	6.3 (26)	20.9 (22)	6.2 (3)	19.7 (18)	46.7 (−11)
Brooksville	2.5 (108)	19.0 (23)	10.7 (16)	16.9 (−21)	6.5 (−16)	5.1 (2)	15.5 (−10)	6.0 (0)	17.9 (7)	49.9 (−4)
Catlin	1.9 (58)	16.8 (8)	10.1 (10)	30.7 (43)	7.7 (0)	3.7 (−26)	12.5 (−27)	4.1 (−32)	12.5 (−25)	64.2 (23)
Loyal	0.0 (−100)	14.9 (−4)	10.2 (11)	19.6 (−8)	8.2 (6)	5.2 (4)	20.3 (18)	6.5 (8)	15.2 (−9)	50.8 (−3)
Rosholt	1.8 (50)	14.5 (−6)	8.9 (−3)	20.3 (−5)	7.4 (−4)	4.9 (−2)	16.9 (−2)	7.3 (22)	18.2 (9)	49.4 (−5)
Average	1.2	15.5	9.2	21.4	7.7	5.0	17.2	6.0	16.7	52.2
SD	<u>1.2</u>	<u>2.5</u>	<u>1.8</u>	<u>5.3</u>	<u>0.9</u>	<u>0.9</u>	<u>3.4</u>	<u>1.2</u>	<u>2.8</u>	6.9

† Aromaticty = peak area (108–162 ppm)/peak area (0–162 ppm) × 100.

‡ The values in parentheses are the change (%) related to the average abundance of the same functional groups.

§ The underlined standard deviation (SD) is greater than 10% of the average value.

Table 4. Correlation coefficients between ^{13}C NMR peak area and intensity of fluorophore components of 13 MHA and CaHA fractions of eight US soils (He et al. unpublished data, 2015).

Fluorophore component	Solid-state ^{13}C NMR peak (ppm)							
	190–220	162–190	145–162	108–145	96–108	60–96	50–60	0–50
Component 1	−0.16	−0.13	0.09	−0.13	−0.10	0.25	0.07	0.07
Component 2	0.31	0.43	0.14	0.28	−0.06	−0.43	−0.29	−0.30
Component 3	−0.35	−0.43	−0.14	−0.21	0.26	0.50	0.06	0.22
Component 4	−0.27	−0.65*	−0.57*	−0.35	0.26	0.28	0.67*	0.56*

* Significant at the 0.05 probability level.

0.01). Thus, the functional groups of COO and N-C = O of MHA fractions mainly come from the soil silt fraction, and the same functional groups in CaHA originate from the soil clay fraction. This observation is quite similar to that of total amounts of extractable amino N of these soils by Olk et al. (2008), who reported that total amounts of extractable amino N are most positively correlated with total soil N, silt content, and total soil C, and most negatively correlated with sand content. The correlations between the COO and N-C = O groups in the MHA fractions and the sand, total N, and soil organic carbon seemed simply due to the three soil parameters that were also closely tied with the silt contents in these soils (Fortuna et al., 2012).

Even though both EEM and ^{13}C NMR spectroscopic techniques are simultaneously used in the characterization of the same HA fractions in a work by Li et al. (2011), to our knowledge no report has been published on how the components identified by the two methods are related to each other. Thus, we analyzed the correlation coefficients of the four fluorophore EEM components and the nine ^{13}C NMR functional groups (Table 4). Fluorophore Component 4 is positively correlated with NMR peaks at 50 to 60 and 0 to 50 ppm, and negatively correlated with NMR peaks at 162 to 190 and 145 to 162 ppm. There are not statistically significant ($P \leq 0.05$) coefficients with other components. Therefore, Fluorophore Component 4, a water-soluble fraction of soil organic matter, is enriched with nonpolar alkyl, NCH, and OCH3 functional groups, but depleted in aromatic C-O, COO, and N-C = O groups.

Functional Components of MHA and CaHA Fractions Identified by Ultrahigh Resolution Mass Spectrometry

Ultrahigh resolution electrospray ionization Fourier transform ion cyclotron resonance mass spectrometry (ESI-FT-ICR-MS) is a powerful state-of-the-art method for molecular characterization of natural organic matter (Sleighter et al., 2012). With this technique, Ohno et al. (2010) determined the molecular composition of WEOM, MHA, and CaHA of the eight soils discussed above, two other soils (Maine Newport and Oregon Walla Walla series), and relevant WEOM from plant biomass. The negative-ion ESI-FT-ICR mass spectra for the samples within an OM type are visually similar across all samples, with thousands of peaks being detected across the m/z range of 200 to 800 and clusters of 10 to 20 peaks being detected at each nominal mass. To simplify the data analysis, these m/z peaks are grouped into six functional components: (i) lipids (H/C = 1.5–2.0, O/C = 0–0.3); (ii) proteins (H/C = 1.5–2.2, O/C = 0.3–0.67); (iii) lignins (H/C = 0.7–1.5, O/C = 0.1–0.67); (iv) carbohydrates (H/C = 1.5–2.4, O/C = 0.67–1.2); (v) unsaturated hydrocarbons (H/C = 0.7–1.5, O/C = 0–0.1); and (vi) condensed aromatic structures (H/C = 0.2–0.7, O/C = 0–0.67). The distribution of the six functional components in the alfalfa and Adkins soil derived OM fractions is shown in Fig. 6. The aqueous alfalfa extract components show a diverse distribution pattern, with the majority of the components in the lipid, protein, and lignin classifications. The soil WEOM, which represents the most labile form of SOM, has lignin components as the dominant type and decreased occurrence of lipid and protein molecules as compared with the alfalfa extract. The less refractory MHA fraction has two clusters of points: one in the lipid space and the other spanning the border of the lignin and condensed aromatic space. The more refractory CaHA fraction shows two clusters: one for lipid components and the other mostly in the condensed aromatic space.

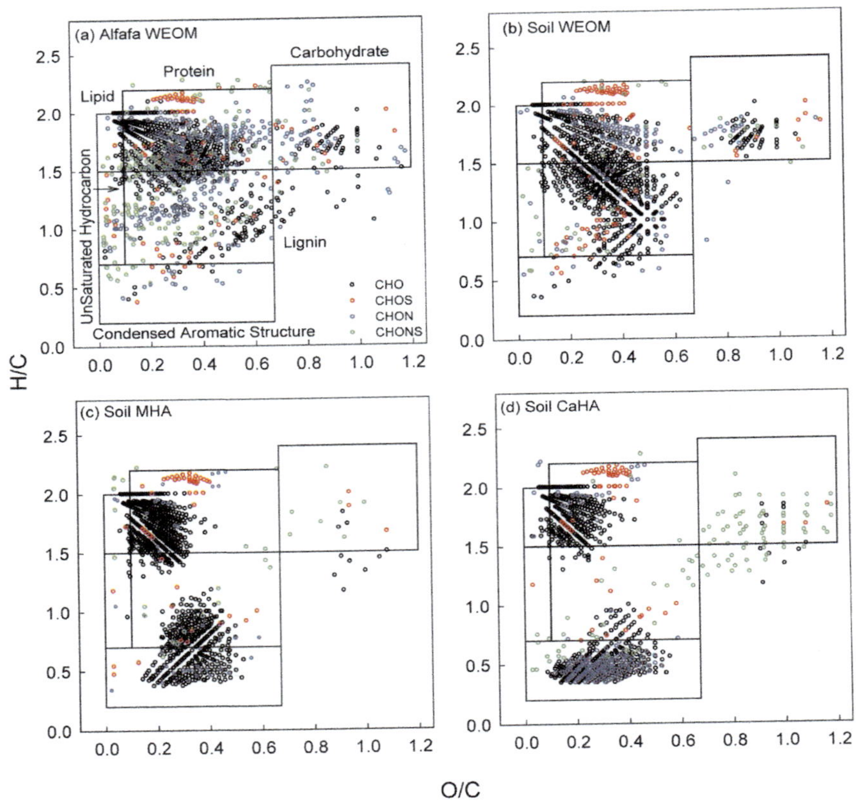

Fig. 6. van Krevelen diagrams from electrospray ionization Fourier transform ion cyclotron resonance mass spectra (ESI-FTICR-MS) of (a) water extractable organic matter (WEOM) from alfalfa biomass, (b) WEOM from Caribou soil, (c) MHA from Caribou soil, and (d) CaHA from Caribou soil (reprinted from Ohno et al., 2010).

The CaHA fraction also shows a minor cluster of elements CHONS components in the carbohydrate region (Fig. 6d).

The high data density of ultrahigh resolution mass spectrometry provides considerable chemical information, but the analysis and synthesis of that information is challenging with traditional approaches. Ohno et al. (2010) applied a multivariate indicator species analysis to identify marker components of the four organic matter types investigated. The analysis reveals that there are 772 marker components for plant WEOM, 237 for soil WEOM, 92 for MHA, and 418 for CaHA. The marker components for the plant WEOM are found in the spaces for lipids, proteins, carbohydrates, lignin, and unsaturated hydrocarbons, reflecting the diverse set of water-soluble biomolecules. The soil WEOM marker species are predominately concentrated in the higher H/C areas of the lignin space and in the carbohydrate space. The MHA indicator components are clustered in the lower H/C region of the lignin space, and the CaHA are clustered in the condensed aromatic area.

Whereas ESI-FT-ICR-MS has been utilized to characterize nearly all types of OM samples (i.e., aquatic, aerosol, soil, sediment, etc.), some studies have focused on correlating these data with ^{13}C NMR (Zhong et al., 2011) or EEM spectroscopies (Gonsior et al., 2011; Ohno et al., 2010). The information obtained from NMR and EEM may be qualitatively associated with specific types of peaks (i.e., molecular formulas) in FTICR-MS data, especially those that align in a certain region of the van Krevelen diagram (Sleighter et al., 2012). To better identify the links between NMR and FTICR-MS, He et al. (2013) quantitatively calculated the correlation coefficients between the C functional groups identified by the solid-state ^{13}C NMR spectroscopy and the C components identified by ESI-FT-ICR-MS in MHA, CaHA, and the combination of both fractions of the eight US agricultural soils (Table 5). For MHA fractions, the C components of lipid and carbohydrate showed no significant ($P > 0.05$) correlations with any of the nine functional groups classified by NMR spectroscopy. Lignin was positively correlated ($r = 0.74$) but unsaturated hydrocarbons were negatively correlated ($r = -0.77$) with the aromatic C-O functional group. Protein and condensed aromatic structures showed stronger correlations with the NMR-identified C functional groups. These statistically significant coefficients ($P \leq 0.05$) imply that protein is either associated with or bound to the nonpolar alkyl groups, and there was competition between proteinaceous and aromatic groups in the composition of MHA fractions. The aromatic components identified by the two methods are highly consistent, as four of the five coefficients between the condensed aromatic structures and the NMR-related aromatic parameters are statistically significant ($P \leq 0.05$). On the other hand, there is no statistically significant ($P > 0.05$) coefficient between the two sets of data for CaHA. He et al. (2013) are not sure if it is due to the lower sample number ($n = 5$) or it is a true indicator of the irrelevance between these functional groups in CaHA fractions. When the combination of both MHA and CaHA data was analyzed, the observations between the lipid and protein groups in MHA samples are enhanced. The observations on condensed aromatic structures derived from MHA data are generally applicable to the combination of MHA and CaHA data. The differences in the significance of coefficients of lignin, carbohydrate, and unsaturated hydrocarbons between MHA and the combination data imply that more data are needed to verify if any reliable correlations exist between these MS-assigned C components and NMR-assigned C functional

Table 5. Correlation coefficients between the nine C functional groups identified by ^{13}C NMR spectroscopy and the six C components identified by ESI-FT-ICR-MS (He et al., 2013).†

Functional groups determined by ^{13}C NMR spectroscopy	Functional groups determined by ESI-FT-ICR-MS					
	Lipid	Protein	Lignin	Carbohydrate	Unsaturated hydrocarbons	Condensed aromatic structures
	Mobile humic acid (MHA) (*n* = 8)					
Ketone, quinone, or aldehyde C	0.02	−0.02	0.31	−0.28	−0.21	−0.05
COO and N-C = O	−0.36	−0.47	−0.36	0.57	−0.32	0.61
Aromatic C-O	−0.66	−0.80*	0.74*	−0.53	−0.77*	0.72*
Aromatic C-C	−0.43	−0.68	−0.16	0.46	−0.28	0.74*
Aromatic C-H	−0.39	−0.81*	0.66	−0.23	−0.41	0.62
O-C-O anomerics	0.17	−0.30	0.11	0.17	0.24	0.13
O-alkyl C	0.61	0.64	−0.06	−0.07	0.52	−0.72*
NCH and OCH_3	0.11	−0.22	0.63	−0.11	0.05	−0.06
Nonpolar alkyl	0.27	0.86**	−0.26	−0.37	0.28	−0.71*
Total aromatic C	−0.50	−0.85**	0.18	0.20	−0.41	0.82*
Aromaticity	−0.53	−0.87**	0.14	0.25	−0.46	0.86*
	Recalcitrant Ca-bound humic acid (CaHA) (*n* = 5)					
Ketone, quinone, or aldehyde C	0.32	−0.27	0.76	−0.22	−0.85	−0.02
COO and N-C = O	0.28	−0.33	0.29	0.38	−0.38	0.13
Aromatic C-O	−0.20	−0.70	0.18	0.63	−0.07	0.59
Aromatic C-C	0.25	−0.01	0.14	0.06	−0.24	−0.09
Aromatic C-H	−0.08	0.45	−0.62	0.00	0.61	−0.22
O-C-O anomerics	−0.01	0.52	−0.49	−0.19	0.45	−0.32
O-alkyl C	−0.39	0.22	−0.59	0.03	0.72	0.05
NCH and OCH_3	−0.64	−0.29	0.17	−0.34	0.12	0.38
Nonpolar alkyl	0.05	0.40	0.06	−0.59	−0.09	−0.35
Total aromatic C	0.17	−0.08	0.01	0.25	−0.08	0.01
Aromaticity	0.26	−0.16	0.15	0.28	−0.25	0.02
	Combination of both MHA and CaHA (*n* = 13)					
Ketone, quinone, or aldehyde C	−0.03	−0.14	0.11	0.03	−0.22	0.09
COO and N-C = O	−0.34	−0.49	−0.39	0.54	−0.34	0.53
Aromatic C-O	−0.46	−0.61*	0.03	0.28	−0.49	0.54
Aromatic C-C	−0.42	−0.58*	−0.42	0.57*	−0.36	0.62*
Aromatic C-H	−0.48	−0.68**	−0.20	0.45	−0.45	0.65*
O-C-O anomerics	0.19	0.09	0.15	−0.20	0.24	−0.16
O-alkyl C	0.45	0.52	0.09	−0.24	0.45	−0.51
NCH and OCH_3	0.37	0.37	0.76**	−0.74**	0.35	−0.56*
Nonpolar alkyl	0.45	0.83***	0.34	−0.69**	0.42	−0.78**
Total aromatic C	−0.53	−0.75**	−0.37	0.60*	−0.47	0.75**
Aromaticity	−0.53	−0.76**	−0.39	0.64*	−0.49	0.77**

* Significant at the 0.05 probability level.

** Significant at the 0.01 probability level.

*** Significant at the 0.001 probability level.

† ESI-FT-ICR-MS, electrospray ionization Fourier transform ion cyclotron resonance mass spectrometry.

groups. Whereas reports of such correlation analysis are rare, Baigorri et al. (2009) calculated the correlation coefficients between pyrolysis GC–MS and ^{13}C NMR components of several gray HA, brown HA, and fulvic acids from different backgrounds. Their data show that the aromaticity measured by pyrolysis GC–MS was well correlated with aromatic C content ($r = 0.77$), but negatively correlated with the aliphaticity ($r = -0.81$) measured by ^{13}C NMR. There was also a high correlation between N compounds measured by pyrolysis GC–MS and O-alkyl-C and carbonyl-C by ^{13}C NMR ($r = 0.71$ and 0.82, respectively). Although two different mass spectroscopic methods were used by He et al. (2013) and Baigorri et al. (2009), the correlations revealed in the two studies are quite similar. Such correlation analysis may be used for better understating the functional groups in HA fractions and their changes subject to environmental and/or ecosystem impacts.

Conclusions

Capillary electrophoresis, EEM, and ESI-FT-ICR-MS have not been used extensively for these two types of humic fractions. On the other hand, earlier work with FTIR and ^{13}C NMR was more focused on the effects of cropping management practices on characteristics of MHA and CaHA fractions of the same soil. This chapter reviews and discusses the advance in characterization of soil MHA and CaHA fractions from multiple agricultural locations by using the five instrumental techniques. The multiple sets of data allow comparison and correlation analysis of the structural and functional groups of the two types of humic fractions identified by these complementary techniques. Data reviewed in this chapter indicate that MHA is mainly associated with the soil silt fraction, whereas CaHA could be associated with the clay fraction. Thus, soil type or location is more critical than cropping management practices to influence these spectral features of the two humic fractions. Whereas ESI-FT-ICR-MS provides marker components for the humification process from plant WEOM, soil WEOM, to MHA and CaHA, CE profiles may be used for "fingerprint" identification of these humic fractions from different sources.

References

Baigorri, R., M. Fuentes, G. Gonzalez-Gaitano, J.M. Garcia-Mina, G. Almendros, and F.J. Gonzalez-Vila. 2009. Complementary multianalytical approach to study the distinctive structural features of the main humic fractions in solution: Gray humic acid, brown humic acid, and fulvic acid. J. Agric. Food Chem. 57:3266–3272. doi:10.1021/jf8035353

Campbell, C.A., E.A. Paul, D.A. Rennie, and K.J. McCallum. 1967. Applicability of the carbon-dating method of analysis to soil humus studies. Soil Sci. 104:217–224. doi:10.1097/00010694-196709000-00010

Campitelli, P., M. Velasco, and S. Ceppi. 2012. Characterization of humic acids derived from rabbit manure treated by composting-vermicomposting process. J. Soil Sci. Plant Nutr. 12:875–891.

Cao, X., D.C. Olk, M. Chappell, C.A. Cambardella, L.F. Miller, and J. Mao. 2011. Solid-state NMR analysis of soil organic matter fractions from integrated physical-chemical extraction. Soil Sci. Soc. Am. J. 75:1374–1384. doi:10.2136/sssaj2010.0382

Capriel, P., T. Beck, H. Borchert, J. Gronholz, and G. Zachmann. 1995. Hydrophobicity of the organic matter in arable soils. Soil Biol. Biochem. 27:1453–1458. doi:10.1016/0038-0717(95)00068-P

Chen, W., P. Westerhoff, J.A. Leenheer, and K. Booksh. 2003. Fluorescence excitation-emission matrix regional integration to quantify spectra for dissolved organic matter. Environ. Sci. Technol. 37:5701–5710. doi:10.1021/es034354c

Conte, P., R. Spaccini, and A. Piccolo. 2004. State of the art of CPMAS ^{13}C-NMR spectroscopy applied to natural organic matter. Prog. Nucl. Magn. Reson. Spectrosc. 44:215–223. doi:10.1016/j.pnmrs.2004.02.002

Cottrell, B.A., W.R. Cheng, B. Lam, W.J. Cooper, and A.J. Simpson. 2013. An enhanced capillary electrophoresis method for characterizing natural organic matter. Analyst 138:1174–1179. doi:10.1039/c2an36144b

Djikanovic, D., A. Kalauzi, K. Radotic, C. Lapierre, and M. Jeremic. 2007. Deconvolution of lignin fluorescence spectra: A contribution to the comparative structural studies of lignins. Russ. J. Phys. Chem. A 81:1425–1428. doi:10.1134/S0036024407090142

Du, C., Z. He, and J. Zhou. 2013. Characterization of soil humic substances using mid-infrared photoacoustic spectroscopy. In: J. Xu et al., editors, Functions of natural organic matter in changing environment. Springer, Amsterdam, the Netherlands. 43–47.

Fortuna, A., C.W. Honeycutt, G. Vandemark, T.S. Griffin, R.P. Larkin, Z. He, B.J. Wienhold, K.R. Sistani, S.L. Albrecht, B.L. Woodbury, H.A. Torbert, J.M. Powell, R.K. Hubbard, R.A. Eigenberg, R.J. Wright, J.R. Alldredge, and J.M. Harsh. 2012. Links among nitrification, nitrifier communities and edaphic properties in contrasting soils receiving dairy slurry. J. Environ. Qual. 41:262–272. doi:10.2134/jeq2011.0202

Garrison, A.W., P. Schmitt, and A. Kettrup. 1995. Capillary electrophoresis for the characterization of humic substances. Water Res. 29:2149–2159. doi:10.1016/0043-1354(95)00031-F

Gonsior, M., B.M. Peake, W.T. Cooper, D.C. Podgorski, J. D'Andrilli, T. Dittmar, and W.J. Cooper. 2011. Characterization of dissolved organic matter across the Subtropical Convergence off the South Island, New Zealand. Mar. Chem. 123:99–110. doi:10.1016/j.marchem.2010.10.004

He, Z., X. Cao, J. Mao, T. Ohno, and H.M. Waldrip. 2013. Analysis of carbon functional groups in mobile humic acid and recalcitrant calcium humate extracted from eight US soils. Pedosphere 23:705–716. doi:10.1016/S1002-0160(13)60063-6

He, Z., C. Du, and J. Zhou. 2011a. Structural and bonding environments derived from infrared spectroscopic studies. In: Z. He, editor, Environmental chemistry of animal manure. Nova Science, New York. p. 23–42.

He, Z., C.W. Honeycutt, O.M. Olanya, R.P. Larkin, J.H. Halloran, and J.M. Frantz. 2012b. Comparison of soil phosphorus status and organic matter composition in potato fields with different crop rotation systems. In: Z. He et al., editors, Sustainable potato production: Global case studies. Springer, Amsterdam, the Netherlands. p. 61–79.

He, Z., C.W. Honeycutt, and H. Zhang. 2011b. Elemental and Fourier transform infrared spectroscopic analysis of water and pyrophosphate extracted soil organic matter. Soil Sci. 176:183–189. doi:10.1097/SS.0b013e318212865c

He, Z., T. Ohno, B.J. Cade-Menun, M.S. Erich, and C.W. Honeycutt. 2006. Spectral and chemical characterization of phosphates associated with humic substances. Soil Sci. Soc. Am. J. 70:1741–1751. doi:10.2136/sssaj2006.0030

He, Z., T. Ohno, D.C. Olk, and F. Wu. 2010. Capillary electrophoresis profiles and fluorophore components of humic acids in Nebraska corn and Philippine rice soils. Geoderma 156:143–151. doi:10.1016/j.geoderma.2010.02.011

He, Z., T. Ohno, F.C. Wu, D.C. Olk, C.W. Honeycutt, and M. Olanya. 2008. Capillary electrophoresis and fluorescence excitation-emission matrix spectroscopy for characterization of humic substances. Soil Sci. Soc. Am. J. 72:1248–1255. doi:10.2136/sssaj2007.0305

He, Z., and J. Mao. 2011. Functional groups identified by solid state ^{13}C NMR spectroscopy. In: Z. He, editor, Environmental chemistry of animal manure. Nova Science, New York. p. 41–59.

He, Z., I.A. Tazisong, and Z.N. Senwo. 2015. Forms and lability of phosphorus in humic and fulvic acids. In: Z. He and F. Wu, editors, Labile organic matter—Chemical compositions, function, and significance in soil and the environment. SSSA Spec. Publ. 62. SSSA, Madison, WI. doi:10.2136/sssaspecpub62.2014.0035 (this volume).

He, Z., M. Uchimiya, and H. Cao. 2014. Intrinsic fluorescence excitation-emission matrix spectral features of cottonseed protein fractions and the effects of denaturants. J. Am. Oil Chem. Soc. 91:1489–1497. doi:10.1007/s11746-014-2495-1

He, Z., H.M. Waldrip, and Y. Wang. 2012a. Application of capillary electrophoresis in agricultural and soil chemistry research. In: Z. He, editor, Capillary electrophoresis: Fundamentals, techniques and applications. Nova Science, New York. p. 131–151.

International Humic Substances Society. 2011. Source materials for IHSS samples. http://www.humicsubstances.org (accessed 18 Aug. 2011).

Kaiser, M., R.H. Ellerbrock, and H.H. Gerke. 2007. Long-term effects of crop rotation and fertilization on soil organic matter composition. Eur. J. Soil Sci. 58:1460–1470. doi:10.1111/j.1365-2389.2007.00950.x

Kerek, M., R.A. Drijber, and R.E. Gaussoin. 2003. Labile soil organic matter as a potential nitrogen source in golf greens. Soil Biol. Biochem. 35:1643–1649. doi:10.1016/j.soilbio.2003.08.011

Li, A., X. Zhao, H. Liu, and J. Qu. 2011. Characteristic transformation of humic acid during photoelectrocatalysis process and its subsequent disinfection byproduct formation potential. Water Res. 45:6131–6140. doi:10.1016/j.watres.2011.09.012

Mahieu, N., D.C. Olk, and E.W. Randall. 2002. Multinuclear magnetic resonance analysis of two humic acid fractions from lowland rice soils. J. Environ. Qual. 31:421–430. doi:10.2134/jeq2002.0421

Mao, J., D.C. Olk, X. Fang, Z. He, J. Bass, and K. Schmidt-Rohr. 2008. Influence of animal manure application on the chemical structures of soil organic matter as investigated by advanced solid-state NMR and FT-IR. Geoderma 146:353–362. doi:10.1016/j.geoderma.2008.06.003

Mortensen, J.L. 1965. Partial extraction of organic matter. In: C.A. Black, editor, Methods of soil analysis. Part 2: Chemical and microbiological properties. Agron. Monogr. 9. ASA, Madison, WI. p. 1401–1408.

Ohno, T., and R. Bro. 2006. Dissolved organic matter characterization using multiway spectral decomposition of fluorescence landscapes. Soil Sci. Soc. Am. J. 70:2028–2037. doi:10.2136/sssaj2006.0005

Ohno, T., and Z. He. 2011. Fluorescence spectroscopic analysis of organic matter fractions: The current status and a tutorial case study. In: Z. He, editor, Environmental chemistry of animal manure. Nova Science, New York. p. 83–103.

Ohno, T., Z. He, R.L. Sleighter, C.W. Honeycutt, and P.G. Hatcher. 2010. Ultrahigh resolution mass spectrometry and indicator species analysis to identify marker components of soil- and plant biomass-derived organic matter fractions. Environ. Sci. Technol. 44:8594–8600. doi:10.1021/es101089t

Olk, D.C. 2006. A chemical fractionation for structure-function relations of soil organic matter in nutrient cycling. Soil Sci. Soc. Am. J. 70:1013–1022. doi:10.2136/sssaj2005.0108

Olk, D.C., and K.G. Cassman. 1995. Reduction of potassium fixation by two humic acid fractions in vermiculitic soils. Soil Sci. Soc. Am. J. 59:1250–1258. doi:10.2136/sssaj1995.03615995005900050007x

Olk, D.C., K.G. Cassman, and T.W.-M. Fan. 1995. Characterization of two humic acid fractions from a calcareous vermiculitic soil: Implications for the humification process. Geoderma 65:195–208. doi:10.1016/0016-7061(95)94048-9

Olk, D.C., K.G. Cassman, N. Mahieu, and E.W. Randall. 1998. Conserved chemical properties of young humic acid fractions in tropical lowland soil under intensive irrigated rice cropping. Eur. J. Soil Sci. 49:337–349. doi:10.1046/j.1365-2389.1998.00163.x

Olk, D.C., K.G. Cassman, K. Schmidt-Rohr, M.M. Anders, J. Mao, and J.L. Deenik. 2006. Chemical stabilization of soil organic nitrogen by phenolic lignin residues in anaerobic agroecosystems. Soil Biol. Biochem. 38:3303–3312. doi:10.1016/j.soilbio.2006.04.009

Olk, D.C., M.C. Dancel, E. Moscoso, R.R. Jimenez, and F.M. Dayrit. 2002. Accumulation of lignin residues in organic matter fractions of lowland rice soils: A pyrolysis-GC-MS study. Soil Sci. 167:590–606. doi:10.1097/00010694-200209000-00004

Olk, D.C., A. Fortuna, and C.W. Honeycutt. 2008. Using anion chromatography-pulsed amperometry to measure amino compounds in dairy manure-amended soils. Soil Sci. Soc. Am. J. 72:1711–1720. doi:10.2136/sssaj2007.0420

Orsetti, S., J.L. Marco-Brown, E.M. Andrade, and F.V. Molina. 2013. Pb (II) binding to humic substances: An equilibrium and spectroscopic study. Environ. Sci. Technol. 47:8325–8333. doi:10.1021/es400999q

Pompe, S., K.-H. Heise, and H. Nitsche. 1996. Capillary electrophoresis for a "fingerprint" characterization of fulvic and humic acids. J. Chromatogr. A 723:215–218. doi:10.1016/0021-9673(95)00815-2

Santin, C., Y. Yamashita, X.L. Otero, M.A. Alvarez, and R. Jaffe. 2009. Characterizing humic substances from estuarine soils and sediments by excitation-emission matrix spectroscopy and parallel factor analysis. Biogeochemistry 96:131–147. doi:10.1007/s10533-009-9349-1

Schmidt-Rohr, K., J.-D. Mao, and D.C. Olk. 2004. Nitrogen-bonded aromatics in soil organic matter and their implications for a yield decline in intensive rice cropping. Proc. Natl. Acad. Sci. USA 101:6351–6353. doi:10.1073/pnas.0401349101

Sleighter, R.L., H. Chen, A.S. Wozniak, A.S. Willoughby, P. Caricasole, and P.G. Hatcher. 2012. Establishing a measure of reproducibility of ultrahigh-resolution mass spectra for complex mixtures of natural organic matter. Anal. Chem. 84:9184–9191.

Slepetiene, A., I. Liaudanskiene, J. Slepetys, and A. Velykis. 2010. The influence of reduced tillage, winter crops and ecologically managed long-term mono-and multi-component swards on soil humic substances. Chem. Ecol. 26:97–109. doi:10.1080/02757540.2010.501029

Swift, R.S. 1996. Organic matter characterization. In: D.L. Sparks, editor, Methods of soil analysis. Part 3. Chemical methods. Soil Sci. Soc. Am., Madison, WI. p. 1011–1084.

Tatzber, M., F. Mutsch, A. Mentler, E. Leitgeb, M. Englisch, and M.H. Gerzabek. 2011. Capillary electrophoresis characterization of humic acids: Application to diverse forest soil samples. Environ. Chem. 8:589–601. doi:10.1071/EN11054

Tatzber, M., S. Klepsch, G. Soja, T. Reichenauer, H. Spiegel, and M.H. Gerzabek. 2015. Determination of soil organic matter features of extractable fractions using capillary electrophoresis: An organic matter stabilization study on a ^{14}C-labeled long-term field experiment. In: Z. He and F. Wu, editors, Labile organic matter—Chemical compositions, function, and significance in soil and the environment. SSSA Spec. Publ. 62. SSSA, Madison, WI. doi:10.2136/sssaspecpub62.2014.0033 (this volume).

Uchimiya, M., T. Ohno, and Z. He. 2013. Pyrolysis temperature-dependent release of dissolved organic carbon from plant, manure, and biorefinery wastes. J. Anal. Appl. Pyrolysis 104:84–94. doi:10.1016/j.jaap.2013.09.003

Wang, Y., Z. He, and H.M. Waldrip. 2012. Capillary electrophoresis application in metal speciation and complexation characterization. In: Z. He, editor, Capillary electrophoresis: fundamentals, techniques and applications. Nova Science, New York. p. 117–130.

Wingeyer, A.B., D.T. Walters, R.A. Drijber, D.C. Olk, T.J. Arkebauer, S.B. Verma, D.A. Wedin, and C.A. Francis. 2012. Fall conservation deep tillage stabilizes maize residues into soil organic matter. Soil Sci. Soc. Am. J. 76:2154–2163. doi:10.2136/sssaj2012.0121

Zhao, A., M. Zhang, and Z. He. 2013. Spectroscopic characteristics and biodegradability of cold and hot water extractable soil organic matter under different land uses in subarctic Alaska. Commun. Soil Sci. Plant Anal. 44:3030–3048. doi:10.1080/00103624.2013.829086

Zhong, J., R.L. Sleighter, E. Salmon, G.A. McKee, and P.G. Hatcher. 2011. Combining advanced NMR techniques with ultrahigh resolution mass spectrometry: A new strategy for molecular scale characterization of macromolecular components of soil and sedimentary organic matter. Org. Geochem. 42:903–916. doi:10.1016/j.orggeochem.2011.04.007

Characteristics of Plant-Derived Water-Extractable Organic Matter and its Effects on Phosphorus Sorption Behavior

James F. Hunt* and Zhongqi He

Abstract

The incorporation of crop residues into soil is an important agricultural management technique for maintaining soil quality and nutrient availability through the replenishment of C to soil ecosystems. The water-extractable organic matter (WEOM) fraction is the most labile and mobile fraction of organic matter. In this chapter, we reviewed and discussed spectroscopic and wet-chemical work on the characterization, decomposition, and soil mineral P sorption effects of WEOM solutions derived from the roots and shoots of several agriculturally relevant crop plants. In contrast to its relatively minor proportion in soil, the content of WEOM in plant biomass is high. These plant-derived WEOM samples contained, on average, 27.9% lipids, 21.8% protein, 36.7% lignin, 7.1% carbohydrates, 3.1% unsaturated hydrocarbons, and 3.3% condensed aromatic compounds. In general, WEOM from decomposed materials inhibited P sorption to a greater extent than did WEOM derived from fresh materials. This increased inhibition highlights the contribution of WEOM decomposition (as well as relevant microbial processes) to the release of soluble soil P. Information in this chapter is helpful in providing a better understanding of the relationship between plant-derived WEOM composition and its behavior in soil environments and can contribute to the overall goal of better management of plant residues for agricultural sustainability.

Abbreviations: C_{TS}, total soluble organic matter; DBE, double-bond equivalents; DOC, dissolved organic carbon; EEM, excitation–emission matrix fluorescent spectroscopy; ESI-FT-ICR-MS, electrospray-ionization Fourier-transform ion-cyclotron resonance mass spectrometry; HIX, humification index; HPB, hydrophobic; HPL, hydrophilic; HS, humic substances; MW, molecular weight; MW_{AP}, apparent molecular weight; NMR, nuclear magnetic resonance; PARAFAC, parallel factor; UV-Vis, ultraviolet–visible; WEOM, water-extractable organic matter.

James F. Hunt, Dep. of Plant, Soil, and Environmental Sciences, University of Maine, New England PSW Lab, Orono, ME 04469. Zhongqi He, USDA-ARS, Southern Regional Research Center, 1100 Robert E Lee Blvd., New Orleans, LA 70124 (zhongqi.he@ars.usda.gov). *Corresponding author (James_Hunt@umit.maine.edu).

doi:10.2136/sssaspecpub62.2014.0037

Labile Organic Matter—Chemical Compositions, Function, and Significance in Soil and the Environment. SSSA Special Publication 62. Zhongqi He and Fengchang Wu, editors.

Organic matter is an important soil component. As a heterogeneous mixture of geomolecules and biomolecules, organic matter can be fractionated into distinct pools with different solubility and lability. Plant biomass is a major source of soil organic matter (Cotrufo et al., 2013; Gocke et al., 2013; Kaiser and Kalbitz, 2012). The WEOM fraction is the most labile and mobile fraction of organic matter in the soil (He et al., 2011; Ohno et al., 2009). Although WEOM accounts for a relatively small organic matter fraction, its high activity and fast turnover rates make it play a significant role in the regulation of nutrients, metals, and microbial activity in soil (Zhang et al., 2011; Zhao et al., 2012; 2013).

One of the extensively investigated fields of organic matter functions is its effect on P adsorption onto soil mineral particles (Ohno and Crannell, 1996; Ohno et al., 2007, Fu et al., 2013; Malik et al., 2013), as the majority of P in soils is poorly available because it is often present as insoluble precipitated, adsorbed or occluded forms (Sposito, 1989; Shang et al., 2013; Wang et al., 2013). As such, there is often little correlation between the total P present in a given soil and the available quantity (Havlin et al., 1991) with total P in surface soils ranging from 0.005 to 0.15% of soil (Gorgin et al., 2011). The influence of organic matter from plant residues and animal manures on the amount of plant-available P in soils has long held the interest of agronomists and soil scientists. As early as a century ago, an agricultural extension service paper tested the efficacy of cow manure for increasing the solubility of phosphate minerals in soils (Hartwell and Pember, 1912). Studies on organic matter involved in P sorption in soil include but are not limited to humic substances (HSs) (Sibanda and Young, 1986; Antelo et al., 2007; Hua et al., 2008; Du et al., 2013; Fu et al., 2013), WEOM derived from green manures (Ohno and Crannell, 1996; Erich et al., 2002: Schreeg et al., 2013), organic amendment (Ohno et al., 2005; Hosseinpur and Pashamokhtari, 2008), and organic acids (Bhatti et al., 1998; Strom et al., 2002; Schefe et al., 2011).

Whereas WEOM plays a critical role in soil fertility and quality, there are relatively few reports of the chemical nature of plant-derived WEOM or its role in P–mineral interaction. Our team research has focused on the chemical characteristics of WEOM fractions derived from the roots and shoots of agriculturally relevant crop plants. In this chapter, we have synthesized our work on the characteristics of plant-derived WEOM and its effects on P sorption onto soil minerals. We have also included some data on animal-manure-derived WEOM for the purpose of comparison. We hope this will help in providing a better understanding of the relationship between plant-derived WEOM composition and its behavior in soil environments, and eventually contribute to the overall goal of better management of plant residues for agricultural sustainability.

Characteristics of Plant-Derived Water-Extractable Organic Matter

Chemical Composition of Plant-Derived Water-Extractable Organic Matter

He et al. (2009) extracted WEOM from eight samples by 40:1 (v/w) water to sample ratio. These samples were plant shoots of field-grown alfalfa (*Medicago sativa* L.), corn (*Zea mays* L.), crimson clover (*Trifolium incarnatum* L.), hairy vetch (*Vicia villosa* Roth), lupin (*Lupinus albus* L.), soybean (*Glycine max* L. Merr.), and wheat (*Triticum aestivum* L.) air-dried, and ground to pass through a 1-mm sieve, as well as a dairy cattle (*Bos taurus*) manure sample (air-dried, and passed through a 2-mm sieve). These plant extracts contained total soluble organic matter (C_{TS}) levels ranging from 1100 to 3600 mg L^{-1} (Table 1). As all extractions were performed with the same water-to-sample ratio, the dissolved organic carbon (DOC) levels in the extracts reflected the abundance of soluble C in these plant materials. The lower C_{TS} (~450 mg L^{-1}) in dairy manure WEOM compared with the plant extracts suggested that the more-soluble C fraction of these materials had been preferentially removed through animal digestive processes. After decomposition incubation, the C_{TS} levels in all plant WEOM fractions decreased, but the level in manure WEOM fraction increased. This observation implied that the labile fresh plant organic matter was utilized by microorganisms for energy production and lost as CO_2 while recalcitrant manure organic matter was converted into more water soluble forms during the decomposition incubation (Hunt et al., 2007a).

Phosphorus was present in all WEOM solutions with the highest level in alfalfa and the lowest in wheat. Solution ^{31}P nuclear magnetic resonance (NMR) spectroscopic analysis was used to identify the P forms in the fresh WEOM fractions (He et al., 2009). The study revealed that five WEOM fractions (crimson clover, hairy vetch, lupin, wheat, and dairy manure) contained orthophosphate only. There were both orthophosphate (86%) and monoester P (14%) present in the alfalfa shoot WEOM. However, the distribution in soybean shoot WEOM was 68% orthophosphate and 32% monoester P. Within the monoester P of soybean shoot

Table 1. Total soluble organic C (C_{TS}) and selected elemental contents of water-extractable organic matter (WEOM) fraction derived from plant shoots (Hunt, unpublished data, 2014).

Sample†	C_{TS}	P	Ca	Mg	Al	Fe
	mg L^{-1} of extracts					
Alfalfa-F	2565 (2365)‡	62 (64)	37.7	40	0.2	0.4
Alfalfa-D	2049	39	56.2	46	0.8	6.1
Corn-F	1710 (1826)	56 (54)	39	45	0.1	0.2
Corn-D	1029	39	1	46	4.6	6.5
Crimson clover-F	3385 (3565)	39 (40)	162	63	0.7	0.3
Crimson-D	2325	18	109	56	1.5	0.3
Hairy vetch-F	2370 (2454)	45 (40)	150	41	1.6	0.4
Hairy vetch-D	1902	13	62	28	0.6	2.1
Lupin-F	2128 (2121)	43 (39)	58	38	0.3	1.6
Lupin-D	1148	17	42	36	1.4	0.2
Soybean-F	1158 (1153)	22 (22)	66	79	<0.05	0.1
Soybean-D	882	10	93	97	0.3	1.3
Wheat-F	1512 (1645)	4.6 (5.0)	25	19	<0.05	0.4
Wheat-D	1042	2.6	26	20	1.7	1.9
Dairy manure-F	424 (481)	39 (39)	11	9	<0.05	0.7
Dairy manure-D	726	43	19	14	0.1	2.1

† F, WEOM extracted from fresh samples; D, WEOM extracted from microbial decomposed samples.
‡ Data in the parenthesis are from He et al. (2009).

WEOM, 70% was phytate P while WEOM from corn shoot contained less monoester P (8% of total P). Ca and Mg were present in these WEOM samples. Two other measured elements (Al and Fe) were present in these WEOM in much lower contents. The difference in the content of the four metals in the plant WEOM fraction is consistent with the order of abundances of these metals in plants, where Ca and Mg are much higher in concentration than are Al and Fe (Demir et al., 2010; Tewolde et al., 2011; He et al., 2013b). It should be noted that the alfalfa, crimson clover, hairy vetch, and lupin samples were field grown and harvested at full flowering stage, while the corn, soybean, and wheat samples were crop residues left after harvest. It is not clear how the collecting conditions could have influenced the chemical compositions of these WEOM fractions. Noack et al. (2012) investigated P speciation in eight crop residues (wheat, barley [*Hordeum vulgare* L.], oat [*Avena sativa* L.], rye [*Secale cereal* L.], canola [*Brassica napus* L.], bean [*Phaseolus vulgaris* L.], lupin [*L. angustifolius* L.] and pea [*Pisum sativum* L.]). They found that the majority (65–95%) of P in stems was water-extractable, and most of this was detected as orthophosphate. Other P forms detected in stem and chaff were phospholipids (10–40%) and RNA (5–30%). Phytate was the dominant P species in seeds, and constituted up to 45% of total P in chaff but was only detected in minor amounts (<1%) in stem residues.

Ultrahigh resolution electrospray-ionization Fourier-transform ion-cyclotron resonance mass spectrometry (ESI-FT-ICR-MS) has remarkably advanced the definition of molecular composition in dissolved organic matter (Herzsprung et al., 2014). Using this technique, Ohno et al. (2010) determined the molecular composition of seven plant-derived WEOM fractions. Although no data was provided for the individual plant WEOM samples, the average and the range of the molecular mass was given, along with data from three additional types of organic matter fractions (WEOM, NaOH-extracted mobile humic acid [MHA], and calcium humate [CaHA], which was NaOH extracted after diluted HCl washing of the soil residues after MHA extraction) derived from 10 US soils with national and global significance (He et al., 2013a; Ohno et al., 2011). The ESI-FT-ICR-MS revealed that the average molecular weight (MW) of plant WEOM was 431 (i.e., the mass-to-charge ratio (m/z) value in Table 2). The value is at the same level of the apparent molecular weight (MW_{AP}) of these WEOM fractions as determined by high-performance size-exclusion chromatography. Except for soybean, the MW_{AP} of these plant WEOM fractions were reported to be mostly in a range from 200 to 600 (Hunt et al., 2007a).

On the other hand, the average highest MW peak was at 822 for plant WEOM, 683 for soil WEOM, 701 for MHA, and 665 for CaHA. This trend suggests that the higher MW biomolecules are transformed into lower MW components as a result of the humification process or through conversion into microbial biomass. The MWs when viewed as a bulk average of the components for the different organic matter types were similar to the MW range pattern, with the plant WEOM MW being significantly higher than the soil WEOM MW (Table 2). The average number of C atoms, which are the structural backbone of organic matter, was found in a narrow range between 20.0 and 23.6. The average O atom content, which is key to acidic functional groups of organic matter, was also found to be in a narrow range of 6.1 to 7.4 for the four organic matter types. The H content of plant and soil WEOM and the soil MHA fraction did not significantly differ from each other but was significantly different from the values found for the refractory

Table 2. Organic matter (OM) average formula atomic numbers, molecular weight, double-bond equivalents (DBE) for seven plant-derived water-extractable organic matter (WEOM), and three types of organic matter fractions (WEOM, mobile humic acid [MHA], and calcium humate [CaHA]) from 10 US soils. The ranges of the parameter values are shown in parentheses and the least significant difference values were calculated at the *p* = 0.05 level. Adapted from Ohno et al. (2010).

OM type	C	H	O	N	S	m/z†	DBE
Plant WEOM	22.7 (21.8–24.9)	33.3 (31.1–36.1)	7.43 (6.11–8.40)	0.79 (0.44–1.25)	0.22 (0.15–0.29)	431 (415–443)	7.38 (5.83–104)
Soil WEOM	20.0 (19.0–20.6)	29.5 (24.5–35.7)	7.36 (5.44–8.39)	0.24 (0.14–0.40)	0.16 (0.10–0.24)	395 (380–417)	6.35 (3.89–7.85)
Soil MHA	23.6 (20.8–23.9)	31.2 (19.6–37.1)	6.09 (4.72–7.01)	0.27 (0.14–0.57)	0.18 (0.10–0.35)	420 (381–429)	9.16 (5.42–14.5)
Soil CaHA	22.2 (19.9–24.2)	23.6 (16.6–33.3)	6.50 (5.78–7.62)	0.49 (0.19–1.11)	0.31 (0.14–0.62)	410 (389–448)	11.6 (6.71–16.2)
LSD	1.2	5.0	0.72	0.18	0.08	20	2.15

† m/z, mass-to-charge ratio.

CaHA fraction. The double-bond equivalents (DBE) for the two types of WEOM fractions were not significantly different, but increasing DBE values in the progression from soil WEOM to MHA to CaHA were significant, indicating greater aromaticity of the organic matter with humification. Interestingly, both N and S, which are macronutrients generally supplied to plants through mineralization of organic matter, increase in average content as the soil organic matter becomes more humified.

Carbon Functional Groups of Plant-Derived Water-Extractable Organic Matter

Ultraviolet–visible (UV-Vis), Fourier-transform infrared (FT-IR), solid-state ^{13}C NMR, and ESI-FT-ICR-MS spectroscopic techniques have been used to characterize the carbon functional groups in plant-derived WEOM samples. He et al. (2009) reported that the UV-Vis spectra of plant-derived WEOM is mainly composed of monotonically decreasing curves with increasing wavelength and an absorbance shoulder or peak between 260 and 280 nm. An apparent second absorbance shoulder or peak between 300 and 350 nm appeared in the spectra of the WEOM fractions of corn, hairy vetch, and alfalfa. The absorbance around 280 nm is mainly due to the –OH- and –COOH-substituted benzene rings of amino acids and proteins (He et al., 2003). The additional peak or shoulder between 300 and 350 nm could be due to some small molecules (e.g., NADH-nicotinamide adenine dinucleotide-reduced) and ring-fission products of phenolic carboxylic compounds present in some WEOM fractions (He and Spain, 1997, 2000). Recently, Khan et al. (2014) used UV absorbance to investigate the concentration of phenolic carboxylic compounds in WEOM derived from corn straw. The WEOM of corn straw was obtained using the method reported by He et al. (2009). Khan et al. (2014) demonstrated that absorption of the corn straw WEOM at 254 nm did not depend on pH in their tested normal working range, but was found to have a linear relationship with WEOM content ($R^2 = 0.998$). Thus, they concluded that the absorbance at 254 nm can be used as a surrogate for concentration of WEOM from corn straw.

Fourier transform infrared spectroscopy demonstrated that these plant- and manure-derived WEOM fractions contained more aliphatic and phenolic groups than other natural organic matter, but did not contain as many carboxylic groups as humic fractions (He et al., 2009). Some differences in FT-IR spectral features were also observed among the eight WEOM samples. Based on the strengths of the relevant FT-IR bands, these olefinic and N compounds decreased in the order of lupin > manure > crimson clover > hairy vetch > corn > wheat > alfalfa > soybean. The ^{13}C NMR spectra of plant-derived WEOM could be separated into three groups: (1) alfalfa, hairy vetch, and soybean; (2) crimson clover and lupin; and (3) corn and wheat. Group 1 contained very few aromatic functional groups, in contrast to terrestrial humic acids (He et al., 2013a; Mao et al., 2008). The ^{13}C NMR spectra of Group 3 were generally similar to those of Group 1, but with weaker amino acid or peptide-related peaks. This result was consistent with the fact that group 1 plants are leguminous. The ^{13}C NMR spectra of crimson clover and lupin (Group 2), with three sharp peaks at 36, 51, and 176 ppm for asparagine, appeared different from those of Group 1 and 3. However, if the three sharp peaks were removed, they would be quite similar to those of Group 1 since they are all leguminous plants. Therefore, WEOM fractions of Group 2 contained significant asparagine in addition to compounds found in Group 1.

The distribution of the six C functional components in the plant-derived WEOM fractions, identified by ESI-FT-ICR-MS, is listed in Table 3. Generally, the ESI-FT-ICR-MS data of the functional groups are similar to those of revealed by solid ^{13}C NMR spectroscopy. Unfortunately quantitative analysis is impossible due to the lack of detailed data from each sample. Using soil humic acid samples, He et al. (2013a) did find that the aromatic components identified by the two

Table 3. Average abundance (%) of six C functional components of seven plant-derived water-extractable organic matter (WEOM), and three types of organic matter fractions (WEOM, mobile humic acid [MHA], and calcium humate [CaHA]) from 10 US soils. The ranges of the parameter values are shown in parentheses and the least significant difference values were calculated at the $p = 0.05$ level. Adapted from the supplementary materials in Ohno et al. (2010)

Organic matter type	Lipid	Protein	Lignin	Carbohydrate	Unsaturated hydrocarbons	Condensed aromatic structures
Plant WEOM	27.9 (16.5–40.8)	21.8 (14.8 –29.3)	36.7 (28.6–46.4)	7.1 (5.1–8.4)	3.1 (1.6–5.8)	3.3 (1.1–6.6)
Soil WEOM	25.4 (14.3–56.1)	12.8 (4.5–24.4)	43.9 (13.5–59.6)	13.9 (3.5–28.1)	1.3 (0.8–2.4)	2.6 (0–7.4)
Soil MHA	49.0 (19.3–81.1)	1.9 (0.9–3.0)	19.0 (10.0–29.9)	2.5 (0.8–5.0)	2.1 (0.5–6.5)	25.5 (2.3–60.4)
Soil CaHA	33.1 (3.9–54.5)	2.2 (1.1–34.0)	9.3 (6.3–13.5)	5.4 (2.4–7.1)	0.6 (0.3–1.6)	49.5 (28.3–72.9)
LSD	14.0	3.5	8.5	3.3	1.2	12.0

methods are highly consistent. Comparison of the ESI-FT-ICR-MS data of the four types of organic matter, lignin, lipid, and protein were the major components. Comparison of the data in Table 3 reveals the changes in chemical composition of the organic matter as it undergoes humification from plant-released biomolecules to refractory HSs (Ohno et al., 2010). The initial phase of the humification process, as represented by the plant WEOM to soil WEOM transition, is marked by increases in the distribution of lignin and carbohydrate classes and decreases in the protein components. The formation of a moderately refractory MHA fraction is marked by a decrease in the lignin components with increases in the lipid and condensed aromatic components. The formation of the refractory CaHA fraction involves the continued decrease in the lignin and lipid components to increase the condensed aromatic structural compounds.

Fluorescence Components of Plant-Derived Water-Extractable Organic Matter

Excitation–emission matrix (EEM) fluorescent spectroscopy with three-way parallel factor (PARAFAC) modeling was used to characterize the WEOM from air-dried and ground samples of aboveground portions of oat, millet (*Panicum miliaceum* L.), soybean, corn, wheat, hairy vetch, crimson clover, alfalfa, canola, and lupin (Hunt and Ohno, 2007). For each fresh or decomposed WEOM sample, a full-scan fluorescence landscape was determined using an excitation range of 240 to 400 nm and an emission range of 300 to 500 nm. The representative spectra in the contour plots of fresh and decomposed corn WEOM fractions are shown in Fig. 1A. Visual examination suggests the presence of at least four fluorophores in the fresh sample at excitation/emission peaks of <240:440 nm, 250:300 nm, 270:355 nm, and 320:440 nm. The spectrum of decomposed corn WEOM differs as a new peak at 250:465 nm and substantial broadening of the peak at 320:440 nm.

Multiway PARAFAC modeling was used to analyze the resulting EEM spectra. In the absence of knowledge of specific quantum efficiencies these signal based concentration calculations could not be used to determine actual chemical concentrations (Ohno and He, 2011). Their usefulness, instead, was in assessing how decomposition of the plant residue affected the chemical nature of the resulting WEOM. A prior study used PARAFAC modeling on a similar fresh set of WEOM materials and identified five components (Ohno and Bro, 2006). When considered together, however, fresh and decomposed WEOM produced PARAFAC models with five to ten components. The most appropriate model as determined by its high (80.5) core consistency was found to be the seven-component model that explained >99% of the data variability in the samples. Microbial by-products of the decomposing amendments therefore added an additional two components to the model of the combined fresh and decomposed WEOM data set compared with the similar fresh-only WEOM set. The seven-component spectral loadings in the *x–y* plots for this WEOM set are shown in Fig. 1B.

Due to the ubiquitous presence of aromatic amino acids and humic-like substances in WEOM derived from other environments, previous studies have published qualitatively comparable fluorescence models (Cory and McKnight, 2005; Alberts and Takacs, 2004). In a study by Hunt and Ohno (2007), Components 2, 3, 4, and 6 had two excitation spectral peaks and a single emission peak

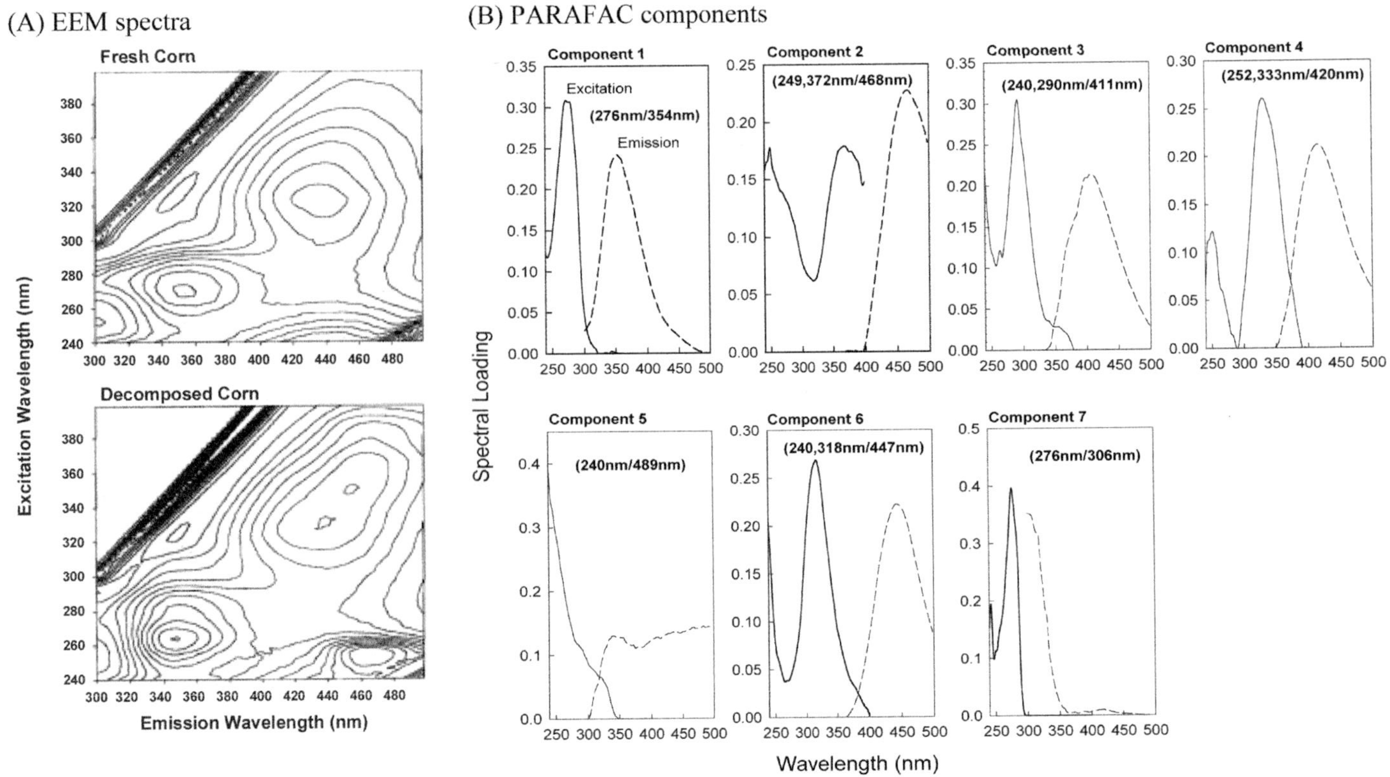

Fig. 1. (A) Fluorescence excitation–emission matrix (EEM) spectra of fresh and decomposed corn water-extractable organic matter (WEOM) and (B) the spectral characteristics of the seven three-way parallel factor (PARAFAC) components per the spectra of 20 plant and six manure WEOM samples. Adapted from Hunt and Ohno (2007).

in wavelength ranges that were characteristic of HSs (Sierra et al., 2005; Leenheer and Croue, 2003).

Components 1 and 7 are generally thought to be associated with proteins containing tryptophan and tyrosine (Yamashita and Tanoue, 2003), while Component 5 (240:489 nm) had a single excitation peak that was similar to Components 2 and 3, and emission characteristics spectra that that approximately corresponded to a sometimes observed, but poorly defined component, referred to by Stedmon and Markager (2005) as the A peak.

The effect of decomposition on changes in the concentration of components found in the plant-derived WEOM extracts normalized to 30 mg C_{TS} L^{-} was generally consistent. This allowed for a meaningful calculation of percentage changes in the WEOM fractions following decomposition. For all of the WEOM extracts, changes to the concentrations of Components 1, 3, and 5 due to decomposition was, on average, less than ~5%. Conversely, humic-like Components 2 and 4 increased, on average, by nearly 100 and 1000%, respectively, following the incubation at ambient temperature with the resident microbial population in the inoculating fresh field soil extracts. This indicated that the remaining WEOM extracts were more condensed (Tan, 2003), and that the simpler forms of organic compounds initially present were likely utilized as energy sources. The effect of decomposition on the animal-manure-derived WEOM components was not consistent among the sources, reflecting differences in both the digestive processes of the unrelated animals and the chemical properties of dissimilar materials used to feed them.

The PARAFAC decomposition results suggested either selective degradation of WEOM having lower MW and aromaticity or a pathway that produces molecules of greater MW and aromaticity compared with that extracted from fresh plant residue (Hunt and Ohno, 2007). The production of extracellular polymeric substances by microbial excretion following cell lysis might be responsible for the latter process, as these materials are known to contain large (MW_{AP} > 10,000) and highly aromatic substances (Sheng and Yu, 2006).

Hunt and Ohno (2007) further conducted Spearman rank–order correlations to relate selected chemical properties with the PARAFAC component scores (Table 4). Components 1 and 7, those related to tyrosine and tryptophan containing materials, had either negative or no correlation with UV–fluorescent indices such molar absorptivity (at 280 nm), humification index (HIX), or MW. These results were consistent with the presence of labile, low MW polypeptides and amino acids as revealed by the solid-state ^{13}C NMR and ESI-FT-ICR-MS data. Strong correlation between these components and the E2/E3 (absorbance at 250 nm to that at 365 nm) ratio index was observed, however, reinforcing the expectation of lower MW_{AP}.

Based on excitation and emission characteristics, Hunt and Ohno (2007) suggested that Component 4 corresponded closely with humic component "C" of Sheng and Yu (2006) and both components were correlated most strongly with MW_{AP}, molar absorptivity, and HIX, as would be expected of humic-like materials. The three humic-like Components 3, 4, and 5 displayed varying degrees of positive correlation with spectroscopic humification parameters. All of these components were negatively correlated with total soluble C content. On the other hand, Components 1 and 7 did not significantly correlate with this parameter, perhaps highlighting the importance of the role of decomposition in the formation of the humic-like fluorophores. Correlation analysis illustrated the

Table 4. Spearman rank correlation coefficients between selected water-extractable organic matter (WEOM) chemical characteristics and three-way parallel factor (PARAFAC) components (n = 52). Reprinted from Hunt and Ohno (2007).

PARAFAC component	Dissolved organic carbon concentration	E2/E3 ratio	Absorptivity at 280 nm	Apparent molecular weight	Humification index
Component 1	0.0316ns†	0.745***	−0.0512	−0.368**	−0.392**
Component 2	−0.663***	0.185ns	0.679***	0.580***	0.786***
Component 3	−0.294*	0.0686ns	0.377**	0.346**	0.493***
Component 4	−0.620***	0.311*	0.577***	0.443**	0.652***
Component 5	−0.858***	0.0867ns	0.284*	0.528***	0.399**
Component 6	−0.428**	−0.414**	0.372**	0.372**	0.288*
Component 7	0.238ns	0.382**	0.0503ns	0.157ns	−0.0294ns

* Significant at the 0.05 probability level.

** Significant at the 0.01 probability level.

*** Significant at the 0.001 probability level.

† ns, not significant.

relationship between the freshness of agricultural plant residues and the humic-like spectroscopic indices of the resulting WEOM. Other studies have found strong correlation between these spectroscopic parameters and sorption to common soil minerals (Ohno et al., 2007; Hunt et al., 2007a).

Stability of Plant-Derived Water-Extractable Organic Matter and its Effects

Sorption of Plant-Derived Water-Extractable Organic Matter on Soil Minerals

In the study of the effects of decomposition on the mineral sorption of WEOM (Hunt et al., 2007a), fresh and decomposed crop-residue-derived WEOM was first separated into hydrophilic (HPL) and hydrophobic (HPB) fractions using XAD-7 resins. Hydrophobicity was defined in that study as the percentage of WEOM C sorbed onto the XAD-7 resin relative to the total WEOM carbon. Whereas the content of C_{TS} in all decomposed WEOM samples decreased, their hydrophobicity (i.e., the percentage of the HPB fraction) increased, compared with the corresponding fresh samples. The overall mean percentage hydrophobicity for the fresh-plant-derived WEOM was found to be 24.9 ± 9.8% compared with 44.8 ± 14.6% for the WEOM derived from the decomposed residues of the same materials. The hydrophobicity for the fresh-crop-derived WEOM was close to the 30% found by Ohno and Cronan (1997) for XAD-8/XAD-4 fractionation and the ~20% found by Gigliotti et al. (2005) for XAD-8 fractionated leachate from early stage compost. The hydrophobicity increase of the WEOM fractions following decomposition of the residues was consistent with the high concentrations of readily metabolized amino acids, monosaccharides, and small chain fatty acids that constitute plant biomass extracts.

The mean amounts of fresh-extract-derived WEOM sorbed onto goethite using single-point initial concentrations of 5 mM C_{TS} for the HPL and HPB fractions were 28.0 (~47% C_{TS}) and 16.4 μg C m^{-2} (~27% of C_{TS}), respectively. Decomposition of the crop residues significantly increased the sorptivity of HPL fractions of the WEOM by 50 to 100%, and that of HPB fractions in lower rates (0–35%). Despite the overall increase in WEOM sorptivity, a decrease in overall C sorption was observed with WEOM derived from the shoots of corn, hairy vetch, and crimson clover following decomposition because of the significant loss of C concentration in these extracts.

Correlation analyses were done to relate sorption to the experimentally determined chemical characteristics such as MW_{AP}, carboxyl, and phenolic acidity and molar absorptivity. Sorption was strongly and significantly correlated with absorptivity, phenolic acidity, and MW_{AP} in all cases except in the fresh-extract-derived HPL fraction where MW_{AP} did not show significant correlation. Sorption of WEOM was better correlated to phenolic acidity than was carboxyl acidity content. In addition, phenolic acidity was more strongly correlated in the sorption of the HPL fraction than to that of the HPB WEOM fraction, suggesting ligand exchange alone might play a larger role in the sorption of the former than the latter. Decomposition increased the WEOM sorption correlation of UV molar absorptivity (280 nm), suggesting that hydrophobic effects complement the primary sorption mechanism of ligand exchange in the fresh HPB fractions as well as in both decomposed WEOM fractions. Significant sorptive fractionation was observed in the supernatant solutions following sorption, clearly illustrating how the heavier and more aromatic compounds preferentially sorbed to the goethite surface. For example, the MW_{AP} range for fresh and decomposed WEOM fractions before sorption ranged from 209 to 6740, while postsorption solutions ranged only from 138 to 315.

Hunt et al. (2007a) further looked at the sorptive characteristics of the WEOM fractions (solutions) on other common mineral surfaces such as kaolin and gibbsite. In this work, one-point sorption indices with 50 mg C L^{-1} solutions was used to determine the mean amount of C sorbed from WEOM solutions derived from corn, hairy vetch, and dairy manure and two model compounds (Aldrich Humic Acid and oxalate) to three mineral phases: (i) laboratory-prepared goethite, (ii) laboratory-prepared gibbsite, and (iii) commercial kaolin. The specific surface area of the synthetic goethite was 58.3 m^2 g^{-1}, which is comparable to the literature values in a range from 48.5 to 70.9 m^2 g^{-1} (Wang et al., 1997). The synthetic gibbsite, per Borggaard et al. (2005), was assumed a specific surface area of 264 m^2 g^{-1}. Commercial kaolin possesses a value of 17.7 m^2 g^{-1}. Corresponding to the order of the specific surface area, the two metal oxides had much greater affinity for C anions than did kaolin with average C sorptions of 14,200 μg C_{TS} g^{-1} for gibbsite, 5470 μg C_{TS} g^{-1} for goethite, and 322 μg C_{TS} g^{-1} for kaolin. Indeed, Saidy et al. (2013) applied goethite coating to increase the sorption capacity of kaolinitic clay with organic matter extracted from dried medic (*Medicago truncatula* Gaertn. 'Praggio') shoot residue. On average, decomposed WEOM solutions sorbed to a greater extent to goethite (7,892 μg C_{TS} g^{-1}) and gibbsite (15,569 μg C_{TS} g^{-1}) than did fresh WEOM solutions. It was observed that, following sorption to all of the minerals, significant fractionation of the WEOM solutions occurred with MW, absorptivity, and hydrophobicity of the WEOM solutions decreasing in value between 75 and 95%, indicating the preferential sorption of WEOM fractions with these characteristics (Hur and Schlautman, 2003; Oren and Chefetz, 2012).

Effects of Plant-Derived Water-Extracted Organic Matter on Phosphorus Sorption Behavior

When P was added (as KH_2PO_4) to the WEOM solutions immediately before exposure to the goethite, gibbsite, and kaolin slurries, C sorption to the metal oxides was reduced dramatically for all WEOMs (Hunt, unpublished data, 2014). For WEOM solutions, C sorption to goethite declined between 82 and 100%, while the decrease observed with gibbsite was between 56 and 78%. Thus with these mineral phases, P in the form of orthophosphate was clearly a much stronger competitor for sorption sites than was C from either the model or WEOM solutions. The affect of the presence of P on C sorption to kaolin was more confounded, a fact made less important by the very low overall sorption to that mineral. The reference organic acid solution of oxalate behaved much differently than the WEOM or Aldrich humic acid solutions, causing a net desorption of C from the goethite surface and better sorption to kaolin and gibbsite in the presence of P. This anomalous behavior was likely due to mineral dissolution in the former case and to organo-P bridging bonds in the latter.

The P sorption inhibition to the three minerals was examined by the same WEOM and reference solutions at low and high WEOM concentrations (i.e., 50 and 200 mg L^{-1} of C, respectively). Phosphorus sorption inhibition by the various WEOM solutions turned out to be highly mineral dependent. Fresh-plant-derived WEOMs showed little affect in inhibiting P sorption, which is consistent with their low aromatic contents (Table 3).

The P sorption results in Table 5 indicate no affect of C-containing solutions on P sorption to kaolin but significant ($p < 0.05$) P sorption inhibition to goethite was observed by Aldrich humic acid, oxalate, and decomposed clover and corn residue WEOM solutions at both concentrations of 50 and 200 mg C_{TS} L^{-1}. Antelo et al. (2007) found that competition between soil humic acid and phosphate for goethite sorption sites was highly pH dependent. In their study they observed phosphate adsorption decreases of up to 45% at solution pH values of 4.5, while solution pH values of 7.0 decreased humic acid adsorption between 20 and 35%. In our study, solution pH values ranged from approximately 6.0 before sorption to goethite to 6.4 after sorption. These relatively high pH values may therefore have played a role in the apparently superior affinity of phosphate for the mineral.

In the case of gibbsite, P sorption inhibition was significantly inhibited by 50 mg C_{TS} L^{-1} oxalate solution and WEOM derived from decomposed corn residue and fresh dairy manure residue (Table 5). Yet at the higher 200 mg C_{TS} L^{-1} concentration, significant P sorption inhibition to gibbsite was achieved by all WEOM solutions while in the case of kaolin, only oxalate solution inhibited P sorption. So even though the earlier WEOM sorption experiments had shown orthophosphate to be a much stronger competitor for metal-oxide binding sites, small but significant P sorption inhibition to both minerals still occurred with certain WEOM solutions.

The mechanisms involved in the WEOM-mediated P sorption inhibition were varied (Hunt et al., 2007b). Table 5 also illustrates the Fe and Al released into solution from the three minerals accompanying the phosphate-sorption inhibition by WEOM. These results indicate that WEOM may help release of Fe or Al into solution through organic acid dissolution of mineral surfaces. Thus, formation of soluble organo-Al or Fe-phosphate complexes may be responsible for both

Table 5. Percentage P sorption and associated metal release (in parenthesis) in the presence of water-extractable organic matter (WEOM). Low and high WEOM corresponding to 50 and 200 mg L^{-1} of C, respectively. Compiled from data in Hunt et al. (2007b) and Hunt (unpublished data, 2014).

WEOM sample	Goethite % P sorption (Fe released)		Gibbsite % P sorption (Al released)		Kaolin % P sorption (Al released)	
	Low WEOM	High WEOM	Low WEOM	High WEOM	Low WEOM	High WEOM
Control†	34.5 (1.02)	34.5 (1.02)	67.4 (<0.10)	67.4 (<0.10)	25.3 (1.97)	25.3 (1.97)
Crimson clover-F‡	33.8 (0.50)	33.8 (0.07)	66.3 (<0.10)	56.3* (0.54)	21.4 (0.13)	23.5 (2.05)
Crimson clover-D‡	29.8* (1.00)	29.2* (0.05)	65.1 (<0.10)	60.4* (0.37)	24.1 (1.93)	24.5 (2.28)
Corn-F	33.6 (1.00)	33.8 (0.41)	66.0 (<0.10)	52.4* (0.35)	25.8 (2.62)	24.6 (2.28)
Corn-D	28.4* (1.80)	28.6* (0.32)	58.9* (0.14)	56.6* (0.22)	22.9 (0.75)	24.7 (1.31)
Hairy vetch-F	34.3 (0.19)	32.5 (0.04)	68.9 (0.12)	61.8* (0.18)	22.4 (2.49)	22.7 (2.88)
Hairy vetch-D	33.7 (0.70)	35.3 (0.81)	66.1 |(0.11)	62.7* (0.40)	29.1 (1.56)	24.4 (3.14)
Dairy Manure-F	33.6 (3.03)	33.8 (1.73)	60.0* (0.08)	60.1* (0.30)	23.6 (0.97)	25.5 (2.71)
Dairy manure-D	35.7 (<0.15)	34.9 (0.59)	65.9 (0.12)	58.7* (0.33)	29.1 (1.20)	22.9 (1.88)
Oxalate	31.2* (0.73)	30.9* (0.70)	51.8* (1.31)	46.4* (3.81)	−2.0* (12.8)	−2.2* (13.9)
Aldrich humic acid	31.4* (5.91)	30.4* (4.95)	68.4 (0.02)	59.6* (0.32)	26.7 (1.15)	23.1 (3.15)

* Significant percentage P sorption inhibition at $P = 0.05$ using the student's t-test, compared with the control.

† In the absence of WEOM.

‡ F, WEOM extracted from fresh samples; D, WEOM extracted from microbial decomposed samples.

the observed mineral dissolution and the P sorption inhibition. In other words, WEOM may inhibit P sorption to mineral by dissolution of surface metal oxide sorption sites in addition to the competitive/negative surface charge effects.

Decomposed crimson-clover-derived WEOM appeared to inhibit P sorption to gibbsite through competitive effects alone (Hunt et al., 2007b). Yet in inhibiting P sorption to goethite by competitive effects, this particular WEOM seemed to also affect changes to that mineral's binding sites. Similarly, WEOM derived from decomposed corn residue inhibited P sorption to both goethite and gibbsite through a combination of competitive sorption and binding site alterations. In all cases, it is important to point out that even those WEOM solutions that significantly ($p < 0.05$) inhibited P sorption, they did not do so to a great extent. Sorption of P to the minerals was clearly a much stronger process than sorption of C from any of the residue-derived solutions. Based on these results it is doubtful that any of the WEOM solutions used here would have noticeable effects on P availability under realistic soil conditions. In fact, only pure oxalate solution, at concentrations no doubt thousands of times higher than would normally be found in soil solution, dramatically changed the sorption characteristics of P to any of the mineral surfaces. But again, it's important to not generalize results as soil systems are extremely diverse and complex. Perhaps it is best to say that for all of the positive reasons for adding green or animal manure residues to soils—such as improved soil structure, cation exchange capacity, nutrient content, water holding capacity—improved P availability may be yet another reason.

Modeling of Phosphorus Sorption Isotherms Affected by Plant-Derived Water-Extracted Organic Matter

For better understanding the inhibition mechanisms of plant-derived WEOM, P sorption isotherms were constructed with five P concentrations at 5, 10, 20, 50, and 50 mg P L^{-1} with WEOM at 50 mg C_{TS} L^{-1}. Tempkin (Hunt et al., 2007b) and Langmuir (Hunt, unpublished data, 2014) modeling analysis was used to interpret the resulting P isotherms. Both models have been used in P sorption studies (Lu et al., 2013). The Tempkin isotherm contains a factor that explicitly accounts for adsorbent–adsorbate interactions (Foo and Hameed, 2010), as the linear Tempkin form is P sorbed = a log[solution P] + b, where a represents the slope of the linear isotherm and b the y-intercept. With this model, increases or decreases in slope suggest fundamental changes in the affinity of binding sites to P, while changes in y-intercept are interpreted as competitive sorption effects (Table 6). The Langmuir adsorption model assumes monolayer adsorption (the adsorbed layer is one molecule in thickness) with adsorption only occurring at a finite (fixed) number of definite localized sites with no lateral interaction and steric hindrance between the adsorbed molecules (Foo and Hameed, 2010). The Langmuir isotherm can be expressed as P sorbed = bK[solution P]/(1 + K[solution P]) where b is Langmuir adsorption maximum and K is a constant related to binding energy (Shang et al., 2013). In this work, the values of Langmuir adsorption maximum and the correlation coefficients were calculated for comparison of the effects of WEOM on the P adsorption (Table 6). In the absence of WEOM, the b values of Langmuir maxima were 169, 667, and 9.63 μmol g^{-1} for goethite, gibbsite, and kaolin, respectively (Table 6), corresponding to the order of the specific surface area. The values and order differed from those in Shang et al. (2013) in which the specific surface area was 13.3 m^2 g^{-1} for goethite, 2.1 m^2 g^{-1} for gibbsite, and 23.6 m^2 g^{-1} for kaolin. Apparently, the different specific surface areas, along with other variables such as mineral particle size and adsorption conditions of pH and P concentration, contributed to the variations observed by different researchers.

The results from the two modeling isotherms are similar as far as the significance level ($p < 0.05$) observed in the data differences between treatments and controls. For example, with respect to P sorption onto goethite, significant difference in the b values of Langmuir adsorption maxima were found for six of the 10 WEOM solutions. Four of them were decreases in the b values (i. e. Aldrich humic acid, oxalate, and decomposed corn and clover). This observation is consistent with the P sorption inhibition to goethite observed for this C source using Tempkin plots. On the other hand, the two other significant changes were increases in the b values (i. e. decomposed dairy manure and hairy vetch). Tempkin plots showed a significant increase in the slope of the curve with decomposed dairy

Table 6. Adsorption modeling values for P adsorption by three minerals in the presence of 50 mg L^{-1} of C. Compiled from data in Hunt et al. (2007b) and Hunt (unpublished data, 2014.

	Goethite		Gibbsite		Kaolin	
Templin parameter	*y* intercept	Slope	*y* intercept	Slope	*y* intercept	Slope
Control†	1631	1288	7347	4870	−235	145
Crimson clover-F‡	1642	1169	7038	4913	−157	114
Crimson clover-D‡	1223*	1110**	7061**	4976	−159	118
Corn-F	1585	1234	7202	4870	−242	151
Corn-D	1517**	1025**	5896***	4527**	−201	135
Hairy vetch-F	1614	1295	7263	5184	−175	113
Hairy vetch-D	1597	1311	5010	5607**	−191	131
Dairy Manure-F	1576	1212	6753*	4872	−128	113
Dairy manure-D	1585**	13076**	6459	5385	−184	142
Oxalate	1490	1126**	4525***	3728***	−68	15*
Aldrich humic acid	1233	1313	7103	4954	−199	132
Langmuir parameter	Adsorption max., *b*	Correlation coefficient, r^2	Adsorption max., *b*	Correlation coefficient, r^2	Adsorption max., *b*	Correlation coefficient, r^2
Control†	169	0.998	667	0.997	9.63	0.923
Crimson clover-F	175	0.991	667	0.995	5.97	0.991
Crimson clover-D	160*	0.999	667	0.994	7.87	0.969
Corn-F	170	0.998	667	0.999	9.90	0.993
Corn-D	146*	0.998	607*	0.999	10.7	0.927
Hairy vetch-F	174	0.996	667	0.994	7.37	0.960
Hairy vetch-D	177*	0.994	667	0.996	8.44	0.974
Dairy manure-F	169	0.995	667	0.998	8.74	0.969
Dairy manure-D	187*	0.994	667	0.994	8.54	0.986
Oxalate	160*	0.998	541**	0.995	−0.49**	0.820
Aldrich humic acid	156*	0.995	667	0.993	8.86	0.975

*, **, *** Significant at the 0.05, 0.01, and 0.001 probability levels.

† In the absence of WEOM.

‡ F, WEOM extracted from fresh samples; D, WEOM extracted from microbial decomposed samples.

manure but not with decomposed hairy vetch. The WEOM inhibited P sorption to gibbsite in a manner similar to that observed with goethite. The WEOM fractions from decomposed clover and corn and fresh dairy manure significantly inhibited P sorption. Decomposed hairy vetch WEOM significantly increased the isotherm slope, indicating possible formation of new P sorption sites. Langmuir *b* data also revealed the same inhibition of P sorption onto gibbsite by oxalate and decomposed corn as did the Tempkin plots. Despite the mediocre fitting of the kaolin data to the Langmuir model, the results were the same as those of the Tempkin model, with oxalate also showing a strong inhibition effect on P sorption. Although there are no clear patterns of plant types on the parameters of the two P adsorption models, data in Table 6 demonstrate that the extent of inhibition by WEOM is closed related to the chemical structures of WEOM. This is evident by the differences in sorption observed between fresh and decomposed WEOM fractions and between mineral types.

Conclusions

In contrast to their relatively minor proportion in soil, the quantity of WEOM derived from plant biomass and animal manure is high. Studies on WEOM derived from seven ground and air-dried plant samples and dairy cattle manure found DOC contents that ranged between 1153 and 3566 mg L^{-1} for the plants and 481 mg L^{-1} for the manure. Five WEOM solutions (crimson clover, hairy vetch, lupin, wheat, and dairy manure) contained orthophosphate only, while alfalfa, corn, and soybean-shoot-derived WEOM also contained monoester P, primarily in the form of phytate. Calcium was not found in detectable levels in the manure WEOM but was present in all seven plant WEOM samples. The seven plant-derived WEOM samples possessed an average MW of 822 Da, and an average number of C and O atoms per molecule of 22.7 and 7.43, respectively. These plant-derived WEOM samples contained, on average, 27.9% lipids, 21.8% protein, 36.7% lignin, 7.1% carbohydrates, 3.1% unsaturated hydrocarbons, and 3.3% condensed aromatic compounds. Excitation–emission matrix fluorescent spectroscopy with three-way PARAFAC modeling revealed seven flurophoric components in these WEOM samples.

Decomposition of these plant-derived WEOM samples significantly increased their sorptivity to gibbsite and goethite minerals and correlation analyses showed that sorption was strongly and significantly correlated with absorptivity, phenolic acidity, and MW_{AP}. In the presence of equal concentrations of orthophosphate, all WEOM solutions showed greatly reduced C sorption to goethite and gibbsite, while P sorption inhibition by C was less dramatic. Certain decomposed WEOM solutions seem to inhibit P sorption to gibbsite through competitive effects alone, while others (i.e., decomposed corn) appeared to inhibit P sorption to both goethite and gibbsite through a combination of competitive sorption and binding-site alterations. However, more robust research is needed to confirm the different inhibition mechanisms involved in these WEOM fractions.

References

Alberts, J.J., and M. Takacs. 2004. Total luminescence spectra of IHSS standard and reference fulvic acids, humid acids and natural organic matter: Comparison of aquatic and terrestrial source terms. Org. Geochem. 35:243–256. doi:10.1016/j.orggeochem.2003.11.007

Antelo, J., F. Arce, M. Avena, S. Fiol, R. López, and F. Macías. 2007. Adsorption of a soil humic acid at the surface of goethite and its competitive interaction with phosphate. Geoderma 138:12–19. doi:10.1016/j.geoderma.2006.10.011

Bhatti, J.S., N.B. Comerford, and C.T. Johnson. 1998. Influence of oxalate and soil organic matter on sorption and desorption of phosphate onto a spodic horizon. Soil Sci. Soc. Am. J. 62:1089–1095. doi:10.2136/sssaj1998.03615995006200040033x

Borggaard, O.K., B. Raben-Lange, A.L. Gimsing, and B.W. Strobel. 2005. Influence of humic substances on phosphate adsorption by aluminum and iron oxides. Geoderma 127:270–279. doi:10.1016/j.geoderma.2004.12.011

Cory, R.M., and D.M. McKnight. 2005. Fluorescence spectroscopy reveals ubiquitous presence of oxidized and reduced quinones in dissolved organic matter. Environ. Sci. Technol. 39:8142–8149. doi:10.1021/es0506962

Cotrufo, M.F., M.D. Wallenstein, C.M. Boot, K. Denef, and E. Paul. 2013. The Microbial Efficiency-Matrix Stabilization (MEMS) framework integrates plant litter decomposition with soil organic matter stabilization: Do labile plant inputs form stable soil organic matter? Glob. Change Biol. 19:988–995. doi:10.1111/gcb.12113

Demir, K., O. Sahin, Y.K. Kadioglu, D.J. Pilbeam, and A. Gunes. 2010. Essential and non-essential element composition of tomato plants fertilized with poultry manure. Sci. Hortic. (Amsterdam) 127:16–22. doi:10.1016/j.scienta.2010.08.009

Du, Z.Y., Q.H. Wang, F.-C. Liu, H.L. Ma, B.Y. Ma, and S.S. Malhi. 2013. Movement of phosphorus in a calcareous soil as affected by humic acid. Pedosphere 23(2):229–235. doi:10.1016/S1002-0160(13)60011-9

Erich, M.S., C.B. Fitzgerald, and G.A. Porter. 2002. The effect of organic amendments on phosphorus chemistry in a potato cropping systems. Agric. Ecosyst. Environ. 88:79–88. doi:10.1016/S0167-8809(01)00147-5

Foo, K.Y., and B.H. Hameed. 2010. Insights into the modeling of adsorption isotherm systems. Chem. Eng. J. 156:2–10. doi:10.1016/j.cej.2009.09.013

Fu, Z., F. Wu, K. Song, Y. Lin, Y. Bai, Y. Zhu, and J.P. Giesy. 2013. Competitive interaction between soil-derived humic acid and phosphate on goethite. Appl. Geochem. 36:125–131. doi:10.1016/j.apgeochem.2013.05.015

Gigliotti, G., F. G. Erriquens, and D. Said-Pullicino. 2005. Changes in the chemical characteristics of dissolved organic matter during the composting process and their influence on compost stability and maturity. Geophys. Res. Abstr. 7:02416.

Gocke, M., Y. Kuzyakov, and G.L. Wiesenberg. 2013. Differentiation of plant derived organic matter in soil, loess and rhizoliths based on n-alkane molecular proxies. Biogeochem. 112:23–40.

Gorgin, N., M. Fekri, and L. Sadegh. 2011. Impact of Organic-matter application on phosphorus-desorption kinetics in two agricultural soils in southeastern Iran. Commun. Soil Sci. Plant Anal. 42:514–527. doi:10.1080/00103624.2011.546925

Hartwell, B.L., and F. R. Pember. 1912. Effect of cow dung on the availability of rock phosphate, R.I. Agr. Exp. Sta. Bul. 151. Agricultural Experiment Station of the Rhode Island State College, Kingston, RI.

Havlin, J.L., J.D. Beaton, S.L. Tisdale, and W.L. Nelson. 1991. Soil fertility and fertilizers: An introduction to nutrient management, 7th ed. Prentice Hall, Englewood Cliffs, N.J.

He, Z., X. Cao, J. Mao, T. Ohno, and H.M. Waldrip. 2013a. Analysis of carbon functional groups in mobile humic acid and recalcitrant humate extracted from eight US soils. Pedosphere 23:705–716. doi:10.1016/S1002-0160(13)60063-6

He, Z., C.W. Honeycutt, and T.S. Griffin. 2003. Comparative investigation of sequentially extracted P fractions in a sandy loam soil and a swine manure. Commun. Soil Sci. Plant Anal. 34:1729–1742. doi:10.1081/CSS-120021308

He, Z., C.W. Honeycutt, and H. Zhang. 2011. Elemental and Fourier transform infrared spectroscopic analysis of water and pyrophosphate extracted soil organic matter. Soil Sci. 176:183–189. doi:10.1097/SS.0b013e318212865c

He, Z., J. Mao, C.W. Honeycutt, T. Ohno, J.F. Hunt, and B.J. Cade-Menun. 2009. Characterization of plant-derived water extractable organic matter by multiple spectroscopic techniques. Biol. Fertil. Soils 45:609–616. doi:10.1007/s00374-009-0369-8

He, Z., M. Shankle, H. Zhang, T.R. Way, H. Tewolde, and M. Uchimiya. 2013b. Mineral composition of cottonseed is affected by fertilization management practices. Agron. J. 105:341–350. doi:10.2134/agronj2012.0351

He, Z., and J.C. Spain. 1997. Studies of the catabolic pathway of degradation of nitrobenzene by *Pseudomonas pseudoalcaligenes* JS45: Removal of the amino group from 2-aminomuconic semialdehyde. Appl. Environ. Microbiol. 63:4839–4843.

He, Z., and J.C. Spain. 2000. Reactions involved in the lower pathway for degradation of 4-nitrotoluene by *Mycobacterium* strain HL 4-NT-1. Appl. Environ. Microbiol. 66:3010–3015. doi:10.1128/AEM.66.7.3010-3015.2000

Herzsprung, P., N. Hertkorn, W. von Tumpling, M. Harir, K. Friese, and P. Schmitt-Kopplin. 2014. Understanding molecular formula assignment of Fourier transform ion cyclotron resonance mass spectrometry data of natural organic matter from a chemical point of view. Anal. Bioanal. Chem. 406:7977–7987. doi:10.1007/s00216-014-8249-y

Hosseinpur, A.R., and H. Pashamokhtari. 2008. Impact of treated sewage sludge application on phosphorus release kinetics in some calcareous soils. Environ. Geol. 55:1015–1021. doi:10.1007/s00254-007-1050-x

Hua, Q.X., J.Y. Li, J.M. Zhou, H.Y. Wang, C.W. Du, and X.Q. Chen. 2008. Enhancement of phosphorus solubility by humic substances in ferrosols. Pedosphere 18(4):533–538. doi:10.1016/S1002-0160(08)60044-2

Hunt, J.F., and T. Ohno. 2007. Characterization of fresh and decomposed dissolved organic matter using excitation–emission matrix fluorescence spectroscopy and multiway analysis. J. Agric. Food Chem. 55:2121–2128. doi:10.1021/jf063336m

Hunt, J.F., T. Ohno, Z. He, C.W. Honeycutt, and D.B. Dail. 2007a. Influence of decomposition on chemical properties of plant- and manure-derived dissolved organic matter and sorption to goethite. J. Environ. Qual. 36:135–143. doi:10.2134/jeq2006.0133

Hunt, J.F., T. Ohno, Z. He, C.W. Honeycutt, and D.B. Dail. 2007b. Inhibition of phosphorus sorption to goethite, gibbsite, and kaolin by fresh and decomposed organic matter. Biol. Fertil. Soils 44:277–288. doi:10.1007/s00374-007-0202-1

Hur, J., and M.A. Schlautman. 2003. Molecular weight fractionation of humic substances by adsorption onto minerals. J. Colloid Interface Sci. 264:313–321. doi:10.1016/S0021-9797(03)00444-2

Kaiser, K., and K. Kalbitz. 2012. Cycling downwards-dissolved organic matter in soils. Soil Biol. Biochem. 52:29–32. doi:10.1016/j.soilbio.2012.04.002

Khan, S., Y. Wu, X. Zhang, Y. Xu, J. Zhang, and S. Hu. 2014. Relationship for the concentration of dissolved organic matter from corn straw with absorbance by using UV–visible spectrophotometer. Intern. J. Environ. Pollut. Remediation 2:18–23.

Leenheer, J.A., and J.P. Croue. 2003. Characterizing dissolved aquatic organic matter. Environ. Sci. Technol. 37:2323–2331. doi:10.1021/es0264089

Lu, J., H. Liu, R. Liu, X. Zhao, L. Sun, and J. Qu. 2013. Adsorptive removal of phosphate by a nanostructured Fe-Al-Mn trimetal oxide adsorbent. Powder Technol. 233:146–154. doi:10.1016/j.powtec.2012.08.024

Malik, M.A., K.S. Khan, P. Marschner, and S. Ali. 2013. Organic amendments differ in their effect on microbial biomass and activity and on P pools in alkaline soils. Biol. Fertil. Soils 49:415–425. doi:10.1007/s00374-012-0738-6

Mao, J., D.C. Olk, X. Fang, Z. He, J. Bass, and K. Schmidt-Rohr. 2008. Influence of animal manure application on the chemical structures of soil organic matter as investigated by advanced solid-state NMR and FT-IR. Geoderma 146:353–362. doi:10.1016/j.geoderma.2008.06.003

Noack, S.R., M.J. McLaughlin, R.J. Smernik, T.M. McBeath, and R.D. Armstrong. 2012. Crop residue phosphorus: Speciation and potential bio-availability. Plant Soil 359:375–385. doi:10.1007/s11104-012-1216-5

Ohno, T., and R. Bro. 2006. Dissolved organic matter characterization using multiway spectral decomposition of fluorescence landscapes. Soil Sci. Soc. Am. J. 70:2028–2037. doi:10.2136/sssaj2006.0005

Ohno, T., and B.S. Crannell. 1996. Green and animal manure-derived dissolved organic matter effects on phosphorus sorption. J. Environ. Qual. 25:1137–1143. doi:10.2134/jeq1996.00472425002500050029x

Ohno, T., and C.S. Cronan. 1997. Comparative effects of ionic and nonionic-resin purification treatments on the chemistry of dissolved organic matter. Int. J. Environ. Anal. Chem. 66:119–136.

Ohno, T., T.S. Griffin, L. Liebman, and G.A. Porter. 2005. Chemical characterization of soil phosphorus and organic matter in different cropping systems in Maine, U.S.A. Agric. Ecosyst. Environ. 105:625–634. doi:10.1016/j.agee.2004.08.001

Ohno, T., and Z. He. 2011. Fluorescence spectroscopic analysis of organic matter fractions: The current status and a tutorial case study. In: Z. He, editor, Environmental chemistry of animal manure. Nova Science Publishers, NY. p. 83–103.

Ohno, T., Z. He, R.L. Sleighter, C.W. Honeycutt, and P.G. Hatcher. 2010. Ultrahigh resolution mass spectrometry and indicator species analysis to identify marker components of soil- and plant biomass-derived organic matter fractions. Environ. Sci. Technol. 44:8594–8600. doi:10.1021/es101089t

Ohno, T., Z. He, I.A. Tazisong, and Z.N. Senwo. 2009. Influence of tillage, cropping, and nitrogen source on the chemical characteristics of humic acid, fulvic acid, and water-soluble soil organic matter fractions of a long-term cropping system study. Soil Sci. 174:652–660. doi:10.1097/SS.0b013e3181c30808

Ohno, T., S. Hiradate, and Z. He. 2011. Phosphorus solubility of agricultural soils: A surface charge and phosphorus-31 NMR speciation study. Soil Sci. Soc. Am. J. 75:1704–1711. doi:10.2136/sssaj2010.0404

Ohno, T., B.R. Hoskins, and M.S. Erich. 2007. Soil organic matter effects on plant available and water soluble phosphorus. Biol. Fertil. Soils. 43:683–690.

Oren, A., and B. Chefetz. 2012. Sorptive and desorptive fractionation of dissolved organic matter by mineral soil matrices. J. Environ. Qual. 41:526–533. doi:10.2134/jeq2011.0362

Saidy, A.R., R.J. Smernik, J.A. Baldock, K. Kaiser, and J. Sanderman. 2013. The sorption of organic carbon onto differing clay minerals in the presence and absence of hydrous iron oxide. Geoderma 209–210:15–21. doi:10.1016/j.geoderma.2013.05.026

Schefe, C., P. Kappen, and P. Pigram. 2011. Carboxylic acids affect sorption and micro-scale distribution of phosphorus in an acidic soil. Soil Sci. Soc. Am. J. 75:35–44. doi:10.2136/sssaj2010.0068

Schreeg, L.A., M.C. Mack, and B.L. Turner. 2013. Leaf litter inputs decrease phosphate sorption in a strongly weathered tropical soil over two time scales. Biogeochemistry 113:507–524. doi:10.1007/s10533-012-9781-5

Shang, C., L.W. Zelazny, D.F. Berry, and R.O. Maguire. 2013. Orthophosphate and phytate extraction from soil components by common soil phosphorus tests. Geoderma 209–210:22–30. doi:10.1016/j.geoderma.2013.05.027

Sheng, G. P., and H. Q. Yu. 2006. Characterization of extracelluar polymeric substances of aerobic and anaerobic sludge using three-dimensional excitation and emission matrix fluorescence spectroscopy. Water Res. 40:1233–1239.

Sibanda, H.M., and S.D. Young. 1986. Competitive adsorption of humus acids and phosphate on goethite, gibbsite and two tropical soils. J. Soil Sci. 37:197–204. doi:10.1111/j.1365-2389.1986.tb00020.x

Sierra, M.M.D., M. Giovanela, E. Parlanti, and E.J. Soriano-Sierra. 2005. Fluorescence fingerprint of fulvic and humic acids from varied origins as viewed by single-scan and excitation/emission matrix techniques. Chemosphere 58:715–733. doi:10.1016/j.chemosphere.2004.09.038

Sposito, G. 1989. The chemistry of soils. Oxford Univ. Press, New York.

Stedmon, C.A., and S. Markager. 2005. Resolving the variability in dissolved organic matter fluorescence in a temperate estuary and its catchment using PARAFAC analysis. Limnol. Oceanogr. 50:686–697.

Strom, L., A.G. Owen, D.L. Godbold, and D.L. Jones. 2002. Organic acid mediated P mobilization in the rhizosphere and uptake by maize roots. Soil Biol. Biochem. 34:703–710. doi:10.1016/S0038-0717(01)00235-8

Tan, K.H. 2003. Humic matter in soil and the environment. Marcel Dekker, Inc., New York, N.Y.

Tewolde, H., A. Adeli, K.R. Sistani, and D.E. Rowe. 2011. Mineral nutrition of cotton fertilized with poultry litter or ammonium nitrate. Agron. J. 103:1704–1711. doi:10.2134/agronj2011.0174

Wang, L.L., Y. Chin, and S.J. Traina. 1997. Adsorption of (poly)-maleic acid and an aquatic fulvic acid by goethite. Geochim. Cosmochim. Acta 61:5313–5324. doi:10.1016/S0016-7037(97)00338-4

Wang, X., F. Liu, W. Tan, W. Li, X. Feng, and D.L. Sparks. 2013. Characteristics of phosphate adsorption-desorption onto ferrihydrite: Comparison with well-crystalline Fe (hydr) oxides. Soil Sci. 178:1–11. doi:10.1097/SS.0b013e31828683f8

Yamashita, Y., and E. Tanoue. 2003. Chemical characterization of protein-like fluorophores in DOM in relation to aromatic amino acids. Mar. Chem. 82:255–271. doi:10.1016/S0304-4203(03)00073-2

Zhang, M., Z. He, A. Zhao, H. Zhang, D.M. Endale, and H.H. Schomberg. 2011. Water-extractable soil organic carbon and nitrogen affected by tillage and manure application. Soil Sci. 176:307–312. doi:10.1097/SS.0b013e31821d6d63

Zhao, A., M. Zhang, and Z. He. 2012. Characteristics of soil water-soluble organic C and N under different land uses in Alaska. Soil Sci. 177:683–694. doi:10.1097/SS.0b013e31827e1fa8

Zhao, A., M. Zhang, and Z. He. 2013. Spectroscopic characteristics and biodegradability of cold and hot water extractable soil organic matter under different land uses in subarctic Alaska. Commun. Soil Sci. Plant Anal. 44:3030–3048. doi:10.1080/00103624.2013.829086

Water-Extractable Organic Carbon and Nitrogen Affected by Crop Rotation and Fertilizer Management

Mingchu Zhang,* Zhongqi He, and Aiqin Zhao

Abstract

The importance of soil water-extractable organic matter (WEOM) on soil and its environment has been recognized. In this chapter, the impact of crop rotation and fertilizer N management on soil water-extractable organic C (WEOC) and N (WEON) is demonstrated in two case studies. In studies on double (e.g., potato [*Solanum tuberosum* L.]–barley [*Hordeum vulgare* L.]) and triple (e.g., potato–timothy [*Phleum pratense* L.]–rape [*Brassica napus* L.]) crop rotations, both quantity and quality of WEOM were different among crop-rotation systems. The continuous-potato monocropping yielded a higher ($p < 0.05$) level of WEOC than most triple-crop rotations. This WEOC appeared to have a higher fluorescence humification index (HIX) but less double-bonded C as shown by the ultraviolet–visible (UV-Vis) spectra. The impact of a continuous-potato rotation on properties of WEOM diminished when other crops were added to the rotation, especially with the triple-crop rotations. In a study on N fertilizer management, some of the applied N in a pasture field remained in soil as WEON, which could be used by grasses in the subsequent year. Further studies on crop rotation are required to link WEOM with disease suppression mechanisms and with functional soil microbial communities. Furthermore, future studies on N management are needed to quantify fertilizer N retained as WEON and to include this fraction in N fertilizer recommendations and management.

Abbreviations: BP1, potato–barley; BP2, barley–potato; DOM, dissolved organic matter; DS, disease suppression; DS1, Sudan grass–potato–mustard; DS2, Potato–mustard–Sudan grass; DS3, mustard–Sudan grass–potato; EEM, excitation–emission matrix; HIX, humification index; POT, continuous potato; SC, soil conservation; SC1, timothy–barley–potato; SC2, potato–barley–timothy; SC3, barley–timothy–potato ; SI, soil improvement; SI1, timothy–potato–barley plus manure compost; SI2, potato–barley–timothy plus manure compost; SI3, potato plus compost; SUVA254, specific UV-Vis absorptivity at 254 nm; SUVA280, specific UV-Vis absorptivity at 280 nm; UV-Vis, ultraviolet–visible; WOEC, water-extractable organic C; WOEM, water-extractable organic matter; WOEN, water-extractable organic N.

Mingchu Zhang, Dep. of Agriculture and Horticulture, School of Natural Resources and Extension, University of Alaska Fairbanks, 905 N. Koyukuk Drive, Fairbanks, AK 99775. Zhongqi He, USDA-ARS Southern Regional Research Center, 1100 Robert E Lee Blvd., New Orleans, LA 70124 (zhongqi.he@ars.usda.gov). Aiqin Zhao, Editorial Dep. of the Transactions of the Chinese Society of Agricultural Engineering (TCSAE). No. 41 Maizidian Street, Chaoyang District, Beijing, China 100125 (zhaoaiqin2002@126.com). *Corresponding author (mzhang3@alaska.edu).

doi:10.2136/sssaspecpub62.2014.0038

Labile Organic Matter—Chemical Compositions, Function, and Significance in Soil and the Environment. SSSA Special Publication 62. Zhongqi He and Fengchang Wu, editors.

Dissolved organic matter (DOM) consists of water-soluble organic compounds found in the soil solution that are <0.45 μm in size by definition (Baldock and Nelson, 2000). This fraction consists of simple compounds including sugars, amino acids, and low-molecular-weight organic acids (e.g. citrate and malate) but may also include some larger molecules. Whereas indigenous soil solution (water) can be collected from fields for DOM (Singh et al., 2014), laboratory water extraction is frequently used to obtain the dissolved fraction of soil or other natural samples. The water extracts are termed either DOM (Hunt et al., 2007; Sun et al., 2013; Borisover et al., 2012) or, more frequently, WEOM (He et al., 2009; Ohno et al., 2009; Zhang et al., 2014). Subsequently, organic C and N in the WEOM fraction are referred to as WEOC and WEON, respectively.

Water-extractable organic matter is generally regarded as the most labile form of soil organic matter, which is susceptible to environmental change and management treatments (Kalbitz et al., 2000; Chantigny, 2003). In the comparison of the influence of tillage, cropping, and N source on the chemical characteristics of humic acid, fulvic acid, and WEOM fractions of a long-term cropping system study in an Alabama Decatur silt loam (fine, kaolinitic, thermic Rhodic Paleudults), Ohno et al. (2009) reported that the chemical composition of the WEOM fraction was significantly affected by the N source treatment. Zhang et al. (2011b) demonstrated that the quantity and spectral properties of WEOM responded to the types of nutrient application and tillage practices. Zhao et al. (2012, 2013) found that long-term agricultural land use in subarcitic Alaska lowered the WEOM levels in soils; however, the decreasing trend could be reversed by converting agricultural land to grassland under a conservation program. Similarly, Sun et al. (2013) studied the effects of various land-use systems (native forest, grassland, and arable land) in western Canada and northeast China on the quality and quantity of soil DOC. The Sun et al. (2013) study indicates that the main effect of land-use type on labile soil organic matter is on its concentration but not on labile C characteristics or organic matter functional group compositions.

Continuous potato can influence diseases and weed infestations. As such, double-crop rotation practices are often used to control soilborne plant diseases and weeds. Such double-crop systems as wheat (*Triticum aestivum* L.)–canola [*Brassica juncea* (L.) Czern. subsp. *Juncea*], barley–canola, and soybean [*Glycine max* (L.) Merr.]–wheat rotations have proven to suppress soilborne diseases and provide nutrients for subsequent crops (Piening and Orr, 1988; Haines et al., 1996; Cunfer and Rothrock, 1994; Rothrock and Cunfer, 1986). However, after a long period of time, such double-cropping systems can experience high frequency of soilborne plant diseases and increase in input cost and a flat crop yield (Cunfer and Phillips, 2006; Halloran et al., 2013). Introduction of a third crop into double-crop rotation has shown positive results (Cunfer and Phillips, 2006; Larkin et al., 2011). The triple-crop systems, especially with an introduction of canola crop in the rotation, substantially decrease plant disease incidences (Cunfer and Phillips, 2006; Larkin et al., 2011; Olanya et al., 2014). It has also been reported that with different cropping systems, the microbial populations and compositions vary (Larkin et al., 2011).

The fate of applied N has been studied for a long time; nevertheless, there exist plenty of information gaps that limit our full understanding of the outcomes of the applied N (Miller et al., 2012). We are still far from managing applied N to minimize production cost and the environmental impacts of N losses. Earlier studies show that the N remaining in soils from inorganic fertilizer applications is mainly in organic forms either as microbial biomass or as a fraction of soil organic matter (Harris et al., 1994; Reddy and Reddy, 1993). Microbial biomass N, in general, is regarded as part of the active soil organic N fraction (Doran, 1987). However, recent studies show that 12 to 15% of applied inorganic N can reside in soil organic matter for more than 25 yr (Sebilo et al., 2013). With studies involving inorganic N applied in pasture or turf fields, previous studies show that 36 to 47% of applied N resides on thatch and soils after application (Petrovic, 1990). Large quantities of applied N can be immobilized into the soil organic matter fractions (Power, 1972, Ledgard and Feyter, 1988). Even though past studies show that applied N is immobilized into soil organic matter, it is still not clear in what fraction of organic matter this N is immobilized and whether it can be mineralized in subsequent years.

While there have been several studies on WEOM, those linking soil soluble organic matter with soil microbial activities and with soil available N fraction seem limited. Also, little is known about the impact of double- and triple-crop rotations and of rotation entry point on WEOM. The objective of this chapter is to demonstrate how double- and triple-crop rotations affect WEOM based on the quality of WEOC and WEON. Also, this chapter attempts to link various experimental results to confirm the hypothesis that immobilized fertilizer N may remain in the water soluble fraction of soil organic matter.

Case Study I. Crop Rotation Impacts on Quantity and Quality of Soil Soluble Organic Carbon

Experiment and Sample Analysis

A long-term crop rotation experiment was established in May 2004 in a USDA–ARS field experimental site in Presque Isle, ME (He et al., 2010; 2011a). The experiment had five crop rotation treatments: continuous potato (POT) as a non-rotation control; 2-yr potato–barley rotation (BP), which represents status quo; 3-yr mustard [*Brassica nigra* (L.) W. D. J. Koch]–Sudan grass [*Sorghum bicolor* (L.) Moench nothosubsp. *drummondii* (Steud.) de Wet ex Davidse]–rye (*Secale cereal* L.)–potato for disease suppression (DS); 3-yr barley–timothy–timothy–potato for soil conservation (SC); and 3-yr barley–timothy–timothy–potato plus annual compost application for soil improvement (SI). Soil samples (0–20 cm) were taken in May 2009 after the completion of two 3-yr crop rotations (He et al., 2012). Depending on crop at the time of soil sampling, the crop rotation could be further divided into the following: barley (BP1) and potato (BP2); mustard (DS1), Sudan grass (DS2), and potato (DS3); barley–timothy (SC1), timothy (SC2), and potato (SC3); and barley–timothy plus compost (SI1), timothy plus compost (SI2), and potato plus compost (SI3) (He et al., 2011a). There were five replicates for each treatment; however, only four replicates of soil samples were used for this study. Soil sample SI3 was not included in this study due to limited sample quantity for a complete analysis. Each treatment was implemented in a 16-m by 4-m plot with four rows (0.95-m row spacing and 35 cm between potato plants within a row). The experiment was conducted on a Caribou sandy loam (fine-loamy, isotic, frigid Typic

Haplorthods). Primary tillage was performed using potato plots with a chisel plow in the fall (POT and BP) or spring (SC, SI, and DS), and secondary tillage with disks was also done before potato planting. Fertilizers were applied annually with 2240 kg ha^{-1} of commercial 10–10–10 (NPK) banded on both sides of the seed row and 5 cm below seeds. Seed potato cultivar Russet Burbank was hand planted, and straw mulch was used at a rate of 2 Mg ha^{-1} for SC and SI. Dairy manure compost was applied at 45 Mg ha^{-1} in the SI system. In the DS system, green manure (mustard and Sudan grass) was grown for 2 mo before mowing and incorporating into soil.

Soil samples were air-dried, and sieved (<2 mm). Moisture content in the air-dried soil samples was determined based on the oven-dried weight at 105°C to allow for calculation of mass-based concentration of WEOM analyzed in the laboratory. Soil water soluble organic C and N were extracted with deionized water at a soil/solution ratio of 1:10. The samples (soil + solution) were shaken for 30 min in a reciprocal shaker (Eberbach Corp.), centrifuged at 900 *g* for 30 min and filtered through a 0.45-μm filter (GN-6 Metrical Membrane Filter, Pall Life Science). The total soluble N and organic C in the extracts were determined using a Shimadzu TOC-V organic C and total N analyzer (Shimadzu Scientific Instrument, Inc.). The extracts were used for determining UV-Vis and fluorescence spectral properties. For UV-Vis, the absorbance of the solution from 200 to 700 nm was measured using a 10-mm (width) quartz cuvette and a Shimadzu UV-Vis near-infrared spectrometer. To avoid any inner filter effect, the extracts were diluted using deionized water in order for the absorbance to reach about 0.1 at wavelength 240 nm. Fluorescence intensity of the extracts was determined using a Fluomax-4 (Jobin Yvon Horiba) at excitation wavelength of 254 nm and emission wavelength of 300 to 480 nm. The wavelength increment was set at 3 nm for emission with the correction of instrument bias similar to that described by Zhang et al. (2011a,b). The fluorescence excitation–emission matrix (EEM) was also determined for the soluble organic matter with excitation wavelengths from 240 to 400 nm and emission wavelengths from 300 to 500 nm. The EEM were decomposed into different component by use of PARAFAC analysis in the N-way toolbox (Bro and Andersson, 2000). Detailed prescription of EEM acquisition and PARAFAC analysis can be found in Zhao et al. (2012).

Specific UV-Vis absorptivities at 254 and 280 nm (SUVA254 and SUVA280, respectively) were calculated by dividing absorbance at these wavelengths by the C concentration in the extracts. Absorption ratios E2/E3 and E4/E6 were calculated using absorbance at 250 to 365 nm, and 400 to 600 nm, respectively. The HIX was calculated as a ratio of integrated fluorescence emission intensity from 450 to 480 nm to that at 300 to 345 nm. Standard error was calculated for relative loading of EEM components, and analysis of variance (ANOVA) and least significant difference (LSD) at 5% were also used to compare the relative loadings among treatments. ANOVA was conducted using Statistix 10 (Analytical Software, 2013), while the LSD at 5% was used for mean comparisons.

Results and Discussion

The first experiment indicated that total WEOC with the dairy manure compost application (SI1 and SI2) was higher than other treatments (Table 1). However, compared among the various cropping systems without the compost application, POT had higher soluble organic C concentration than the triple-crop system, except SC2 (Table 1). This observation, which is similar to that of the same rainfed potato soils after the completion of the first 3-yr rotation (He et al., 2011b) indicates that POT led to accumulation of high WEOC in soils. However, for WEON, POT was greater than DS1 only (Table 1). The WEOC in soil could come from the potato root exudates, which might be recalcitrant. High WEOC in POT may be part of the reason that there are prevailing soilborne diseases infections and yield decreases as reported by Larkin et al. (2011).

The spectral properties of the WEOM among treatments were statistically different ($p < 0.05$). However, there was no discernable trend for either SUVA254 or SUVA280 (Table 1). High SUVA254 is an indicator for aromaticity of WEOM, whereas a high SUVA280 value indicates high molecular weight of the water soluble organic fraction. The rotation entry points in the barley–potato double-cropping system were statistically different ($p < 0.05$) from each other in SUVA254

Table 1. Impact of crop rotation on water-extractable organic C and N (WEOC and WEON, respectively).

Crop rotation	Symbol	Crop at soil sampling	WEOC	WEON	SUVA254†	SUVA280‡
					—— mg^{-1} m^{-1} L ——	
Continuous potato	POT	Potato	228.5bc§	23.2c	2.19cde	1.66bc
Potato–barley	BP1	Barley	180.9cd	24.6c	2.84ab	2.20a
Barley–potato	BP2	Potato	200.7cd	23.0cd	2.14de	1.62bc
Sudan grass–potato–mustard	DS1	Mustard	154.7d	17.1d	1.93de	1.45c
Potato–mustard–Sudan grass	DS2	Sudan grass	180.3d	20.8cd	1.92e	1.46c
Mustard–Sudan grass–potato	DS3	Potato	171.1d	20.4cd	2.57bc	1.92a
Timothy–potato–barley	SC1	Barley	175.5d	22.9cd	3.20a	2.45a
Potato–barley–timothy	SC2	Timothy	187.7cd	24.9c	2.33cd	1.76bc
Barley–timothy–potato	SC3	Potato	164.8d	21.4cd	2.82ab	2.12a
Timothy–potato–barley + manure compost application	SI1	Barley	250.3b	34.8b	3.19a	2.46a
Potato–barley–timothy + manure compost application	SI2	Timothy	302.2a	47.7a	2.33cd	1.80b
Probability (*F* test)			<0.0001	<0.0001	<0.0001	<0.0001

† SUVA254, specific UV-Vis absorptivity at 254 nm.

‡ SUVA280, specific UV-Vis absorptivity at 280 nm.

§ Different letter indicates least statistical difference (LSD) at 5% level.

and SUVA280, indicating that crop rotation entry point impacted WEOM. Among the triple-crop rotation systems, those with broad leaf mustard (DS1, DS2) were similar to POT for SUVA254, while for the rotations with both cereal crops (i.e., timothy and barley; SC1, SC2, SC3, SI1, and SI2), the SUVA254 tended to be higher than for POT. The trends were similar for SUVA280 and SUVA254. Thus, high aromaticity of WEOM corresponded with high molecular size of WEOM for potato–barley and triple-crop rotation with two-cereal crops (Table 1).

The UV-Vis spectral ratios are parameters that reflect molecular size of the dissolved organic matter. A high E2/E3 ratio represents the low molecular weight of dissolved organic matter since existence of large-molecular-weight soluble organic matter tends to shift the maximum absorbance toward the high-wavelength regions (Zhang et al., 2011b, Wang et al., 2009). A high E4/E6 indicates existence of smaller size molecules of WEOM, but since the wavelengths of E4 and E6 are at a much higher wavelength than for E2/E3, the low ratio of E4/E6 also serves as an indicator of high humic fraction in the solution. Numerically, the highest E2/E3 ratio was found with DS3, whereas the lowest was with BP1 (Table 2); however, the rest were rather similar (Table 2). In comparison, the highest E4/E6 ratio was found in the treatments with compost applications (SI1), while the lowest was in the triple-crop ration treatment (DS2). The HIX is another indicator suggesting the degree of humification of dissolved organic matter. Similar to UV-Vis, the shift of emission intensity to higher wavelengths indicates a high degree of humification and large molecule size (Zsolnay et al., 1999). The HIX clearly indicated that WEOM from POT had a higher degree of humification compared with any of the other crop rotation (Table 2). Although no trend was found for UV-Vis spectral properties, the soil WEOM had different HIX in POT than in all other crop rotations. As such, HIX from fluorescence appeared as a better method than that of UV-Vis measurements.

Table 2. Spectral ratio of water-extractable organic matter as affected by crop rotation.

Crop rotation	Symbol	Crop at soil sampling	E2/E3	E4/E6	Humification index
Continuous potato	POT	Potato	4.6ab†	7.0cdefg	13.0a
Potato–barley	BP1	Barley	4.1c	5.6fg	7.8bc
Barley–potato	BP2	Potato	4.9ab	6.0efg	7.5bc
Sudan grass–potato–mustard	DS1	Mustard	4.8ab	6.4defg	8.2bc
Potato–mustard–Sudan grass	DS2	Sudan grass	4.6bc	5.2g	7.7bc
Mustard–Sudan grass–potato	DS3	Potato	5.1a	9.0abc	9.0bc
Timothy–potato–barley	SC1	Barley	4.6b	7.6bcdef	7.7bc
Potato–barley–timothy	SC2	Timothy	4.9ab	7.9bcde	7.0c
Barley–timothy–potato	SC3	Potato	4.9ab	8.3abcd	10.0b
Timothy–potato–barley + manure compost application	SI1	Barley	4.7ab	10.0a	8.4bc
Potato–barley–timothy + manure compost application	SI2	Timothy	4.7ab	9.1ab	6.8c
Probability (*F* test)			0.02	0.0001	0.0048

† Different letter indicates least statistical difference (LSD) at 5% level.

The contour curve after decomposing the EEM using PARAFAC analysis showed three components of water soluble organic matter in soils for all the treatments (Fig. 1). These three components were (i) a terrestrial humic-like component (Ex: 240 and 348, Em: 480) (Fig. 1a) named as Component 1; (ii) a marine–freshwater humic-like component (Ex: 240 and 320, Em: 392–418) (Fig.

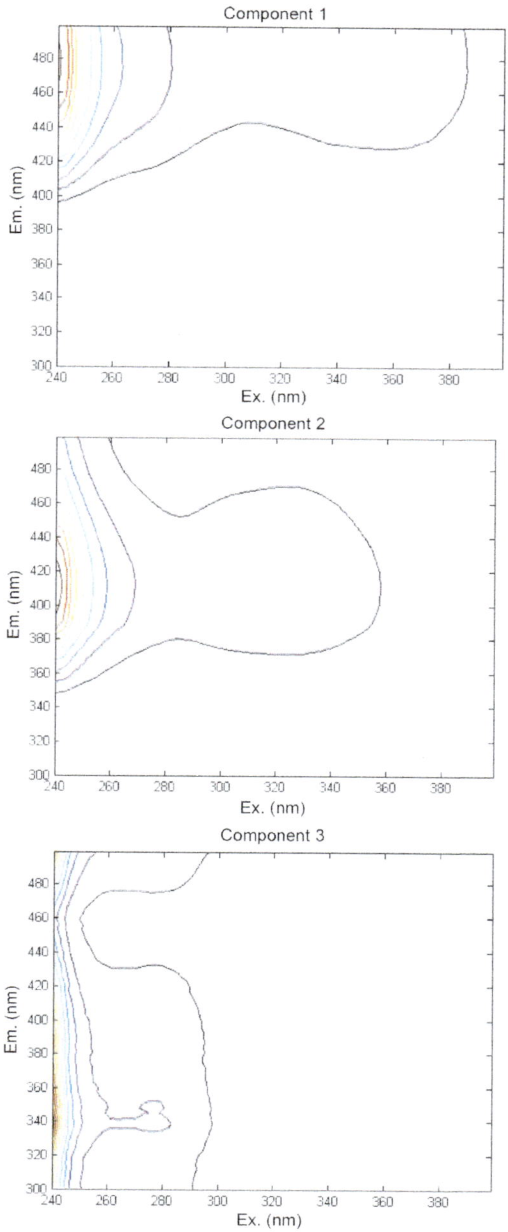

Fig. 1. Contour curves of fluorescence excitation–emission matrix (EEM) for three fluorescence components: Component 1 = a terrestrial humic-like fluorophore (Ex: 240 and 348, Em: 480); Component 2 = a marine–freshwater humic-like component (Ex: 240 and 320, Em: 392–418); and Component 3 = a tyrosine-like protein (Ex: 240, Em: 315–360). Ex, excitation; Em, emission.

1b) named as Component 2; and (iii) a tryptophan-like component (Ex: 240, Em: 315–360) (Fig. 1c) named as Component 3. The first two components have been reported in the literature for natural organic matter (Zhao et al., 2012; Ohno and He, 2011; Yamashita and Tanoue, 2004). Component 3, according to Chen et al. (2003), belongs to tryptophan-like protein. The existence of tryptophan-like protein indicates the presence of intact protein or less-degraded peptide materials in contrast to the more-degraded peptide materials such as tyrosine-like proteins. In studies of different land uses, no tryptophan-like protein is found in water soluble organic matter (Zhao et al., 2012). However, Fellman et al. (2008) reported tryptophan-like protein in forest and wet land soils. Because tryptophan-like component indicates less degraded peptide material (Yamashita and Tanoue, 2004), Fellman et al. (2008) concluded that WEON is of more recent origin. Thus, the existence of tryptophan-like protein in the soil in our potato study showed less decomposed protein. The relative loading of the three components indicated that more than 80% fluorophores were from the first two components, while the last component (i.e., the tryptophan-like protein of relatively recent origin) contributed less than 20% of the fluorophores, except BP1, which contributed 20% (Fig. 2). This showed that even though tryptophan-like protein occurred, it was a relatively small amount.

The relative loading of Component 1 was similar for POT as for SI2, SC3, DS3, and SC1, but lower ($p < 0.05$) than for SI1 and higher ($p < 0.05$) than for SC2, DS2, DS1, BP1, and BP2 (Fig. 2). The relative loading of Component 2 for POT was similar to most of the treatments, except it was higher ($p < 0.05$) than for BP1 and lower ($p < 0.05$) than for DS3 (Fig. 2). The relative loading of Component 3 of POT was similar to most treatments but lower ($p < 0.05$) than for BP1 and BP2 (the double-crop rotation treatments) (Fig. 2). There were large differences among treatments for Component 1 (i.e., the terrestrial humic-like fluorophores) than for Component 2, indicating that Component 2 may not be as sensitive as Component 1 to various crop rotations. For Component 3, the BP1 and BP2 of the double-crop rotation treatments had higher ($p < 0.05$) relative loading compared with the

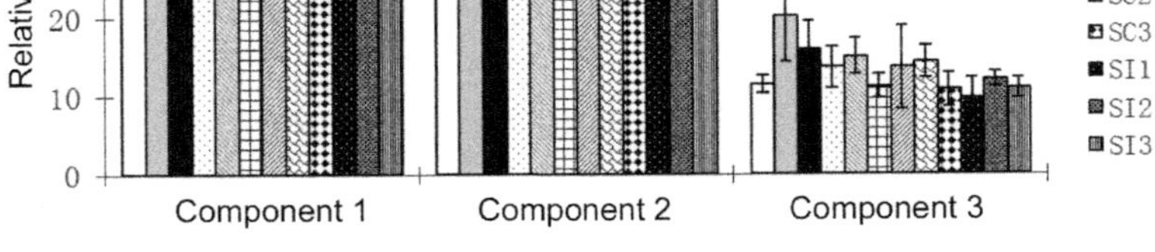

Fig. 2. Relative loading of the fluorescence components. Vertical bar indicates standard error (n = 5). POT, continuous potato; BP1, potato–barley; BP2, barley–potato; DS1, Sudan grass–potato–mustard; DS2, potato–mustard–Sudan grass; DS3, mustard–Sudan grass–potato; SC1, timothy–potato–barley; SC2, potato–barley–timothy; SC3, barley–timothy–potato; SI1, timothy–potato–barley + compost; SI2, potato–barley–timothy + compost.

other treatments, suggesting that the double-crop system resulted in more tryptophan-like protein in water-extractable organic fraction. This may suggest that Component 3 (tryptophan-like protein) might serve as an indicator for various crop rotations. However, the biological significance of tryptophan-like protein, if any, in relation to plant disease incidence is still unknown.

The full spectra from 200 to 700 nm for all the treatments are depicted in Fig. 3. For these, with no manure compost application, the rotational treatment tended to be different as compared with the POT (Fig. 3a–c). The difference between POT and SI1, SI2, and SI3 (Fig. 3d) obviously was due to manure compost applications in SI1, SI2, and SI3. For the double-crop rotation (i.e., potato–barley), the absorbance of POT from 200 to 230 nm overlapped with BP2 (Fig. 3a); however, from 230 to 300 nm, it overlapped with BP1. The only difference between BP1 and BP2 was the crop from which soil samples were taken in 2009, while with BP1 it was barley and with BP2 it was potato. Apparently, barley as a rotational crop affected the water soluble organic matter only in the short wavelength region. Absorbance in the short wavelength region is related to the C=C bonding (Shimazu Corporation, 2014). As the number of C=C bonding increased, the absorbance increases in the higher wavelength. The results suggested that WEOM from POT may contain two portions of soluble carbons, one aliphatic with single C=C (absorbance at short wavelengths) and the other multiple C=C, which might be aliphatic or aromatic (long wavelengths). When potato was the crop at the time of soil sampling (BP2), there were fewer C=C chains. In contrast, when barley was the crop at time of soil sampling (BP1), there were more C=C chains in soil WEOC; thus, introducing a rotational crop altered the properties of the WEOM.

For the potato–Sudan grass–mustard triple-crop rotation (i.e., DS), the UV-Vis spectral properties of the WEOM had lower absorbance at wavelengths from 228 to 290 nm than the POT (Fig. 3b). These differences were even more prominent for the wavelength ranging from 238 to 259 nm (Fig. 3b). Among the three treatments in DS, the difference in absorbance compared with POT depended on the period of time since potato was presented in the rotation. The largest difference compared with POT was with DS2, in which there had been 2 yr of no potato cultivation at the time of soil sampling. The smallest difference compared with POT was with DS1, in which the last crop present before soil sampling was potato. The results suggested that in the triple-crop rotation with the broad leaf, tap-rooted mustard as a rotational crop in combination with a grass species, the WEOM accumulated from growing potato lasts for 2 yr or less.

For the triple-crop rotation with only cereal crops in rotation with potato (i.e., SCs, barley–timothy–potato), the absorbance from 231 to 290 nm was very similar for the SC2, SC3, and POT, but SC1 was different from these (Fig. 3c). Differences were most pronounced between SC1 and SC2 or SC3, especially for wavelengths from 236 to 275 nm. The SC1 was in barley, SC2 was in timothy, and SC3 was in potato at time of soil sampling. With barley as the crop at the time of soil sampling, the WEOM seemed to contain more C=C structures. This echoed the results for barley–potato rotation illustrated in Fig. 3a.

It is unknown if there is a relationship between WEOM and diseases or phytotoxicity in continuous monocrop systems, especially with tube–root crops. Whether soil WEOM with more C=C structure was more or less phytotoxic or has no effect is not clear. Research on soilborne diseases shows that continuous cropping systems can increase disease inocula in soils and hence increase the

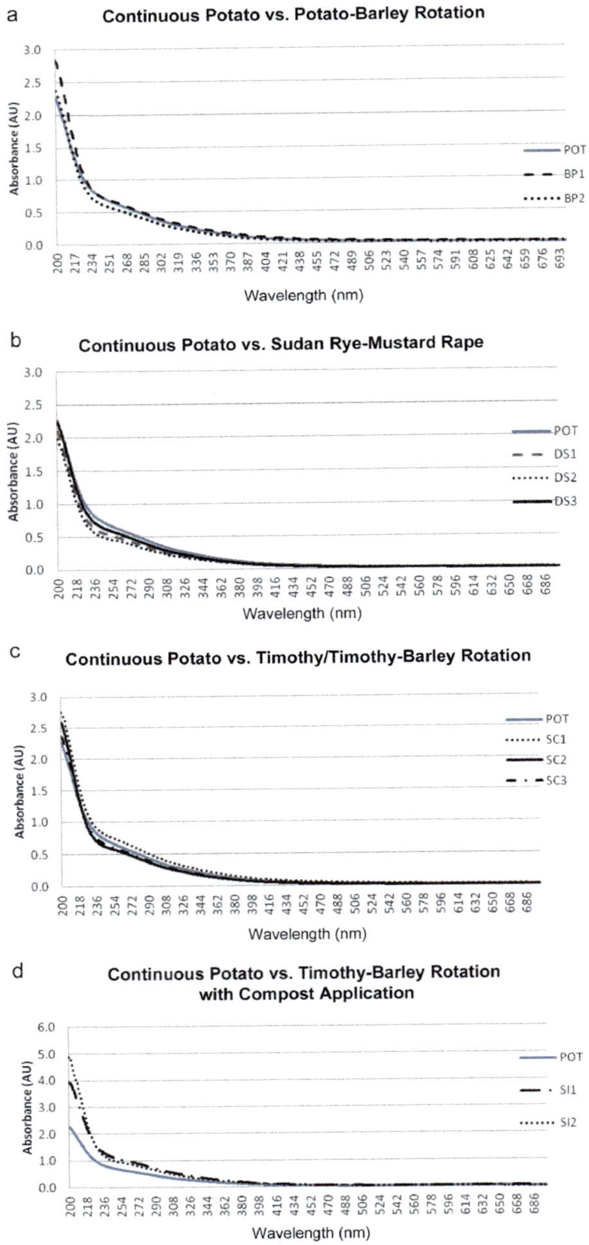

Fig. 3. Absorbance of water-extractable organic matter from the crop rotation study. (a) Continuous potato (POT) vs. potato–barley (BP1) and barley–potato (BP2); (b) POT vs. Sudan grass–potato–mustard (DS1), potato–mustard–Sudan grass (DS2), and mustard–Sudan grass–potato (DS3); (c) POT vs. timothy–potato–barley (SC1), potato–barley–timothy (SC2), and barley–timothy–potato (SC3) rotations; and (d) POT vs. timothy–potato–barley + compost (SI1), and potato–barley–timothy + compost (SI2).

chance of the crops being infected by soilborne disease organisms (Raaijmakers et al., 2009). Even with two crops in rotation, such as wheat–soybean or potato–barley, disease suppression from the rotational crop is not prominent compared with a triple-crop rotation that includes a rape crop (Cunfer and Phillips, 2006). There are at least two possible explanations for disease suppression when a third crop is introduced as a rotational crop. First, the accumulation of pathogens could decrease during a phase of the rotation where the crop is not a good host for the disease organism. This mechanism is most commonly cited. However, another possible mechanism of diseases suppression is modification of the soil microbial community (Larkin et al., 2012). From this perspective, continuous monocropping tends to support or favor a microbial community associated with disease infection. Our studies indicated that WEOM differed among different crop rotations; hence, we hypothesize that the soil water soluble organic matter resulting from rotational cropping might be able to suppress pathogen spore germination. The suppression could be achieved either by lack of chemical compounds that induce germination, such as cutin (biopolyester) (Roncero et al., 2003), possibly through chemical complexation or elicitor compound that can raise plant defense systems (Tamm et al., 2011). Either or both of these mechanisms may explain how WEOM could suppress plant diseases; however, more studies are needed to support such hypothesis.

Case Study II. Fate of Applied Inorganic Nitrogen in Soil

Mineral N is often the most limiting factor for crop production. Since the introduction of synthetic N fertilizer production, crop yield has significantly improved, supporting a substantially large world human population. Due to its importance, N in natural and farm ecosystem is one of the most intensively studied elements, especially with regards to agriculture. Nonetheless, despite tremendous effort and millions of dollars in investment, the N uptake by the major crops (maize [*Zea mays* L.], wheat, and rice [*Oryza sativa* L.]) is still below 50% on average (Cassman et al., 2002). Low fertilizer N use efficiency has led to N leaching to groundwater and running off into surface waters causing nitrate pollution in ground water and eutrophication in surface water systems. Most studies on improving N use efficiency or reducing N runoff or leaching have focused on mineral N (i.e., NH_4–N and NO_3–N), especially NO_3–N due to its mobility in soil. Little attention has been paid to WEON, particularly in agricultural research. We used two experiments to demonstrate the fate of applied fertilizer N. We hope more effort can be placed in this area so that we can better understand the fate of fertilizer N in agroecosystems and surrounding ecosystems.

Experiments

There were two experiments conducted at two different locations: one in Edmonton Alberta Canada (53.5° N, 113.5° W) and the other in Fairbanks, Alaska, USA (64.8° N, 147.7° W). Both areas are in the boreal forest ecosystem zone. The objectives of the experiment in Edmonton were determining the best time and rate of biosolids or municipal solid waste cocompost application in comparison with conventional N application rate in a turfgrass field for biomass production, soil nutrient status, and recovery of applied N. The treatments include (i) annual NPKS (100 kg N as urea + 20 kg P + 42 kg K + 20 kg S ha^{-1}) application for 3 yr (2002 to 2004); (ii) 50 t ha^{-1} cocompost annual application for 3 yr; (iii) 100 t ha^{-1} cocompost application in the first year plus 50 t ha^{-1} cocompost in the second year and no application in the third year; (iv) 150 t ha^{-1} cocompost application in the first year and no application in the second and third year; and (v) control with no fertilizer or cocompost application. On average, all compost treatments receive 50 t ha^{-1} annual application for 3 yr, which is equivalent of 297 kg total N ha^{-1} year^{-1}. Soil samples are taken at the beginning of each growing season. The experiment was conducted for 3 yr, and in the fourth year, neither N fertilizer nor compost was applied, but aboveground plant biomass and soil samples were taken. Thatch and soil samples were taken at thatch, 0 to 15 cm and 15- to 30-cm soil depth at two times (spring after snow thaw and fall before snow) in the fourth year. Mineral N in thatch and soil samples was determined using KCl extraction followed by colorimetric method determination. Aboveground plant samples were taken in June during the growing season and in September after the growing season. Biomass and plant tissue N concentration were determined, and N recovery was calculated (Zhang et al., 2010). The objectives of the experiment conducted in Fairbanks were to determine the time of N fertilizer application on a bromegrass (*Bromus inermis* Leyss.) pasture to improve N uptake by grasses and to illustrate relationship between WEON and N mineralization in soil. Fertilizer N application rates were 50 and 150 kg N ha^{-1} in 2004. Soil samples were taken at the end of the growing season and incubated in the laboratory for N release. Mineral N and WEON were also extracted before and after incubation. Regression analysis was conducted between water soluble organic N and mineral N released over the incubation time.

Results and Discussion

For the compost experiment, results showed that there were no differences ($p > 0.05$) among treatment in mineral N concentration in thatch, and at 0- to 15-cm and 15- to 30-cm soil depth at the beginning of the fourth year growing season (Table 3). However, higher ($p < 0.05$) leaf N concentration was found with the urea treatment than in the cocompost treatments in June but not in September ($p > 0.05$) (Table 3). Biomass was greater in both June and September ($p < 0.01$) for the urea fertilizer treatment compared with the other treatments (Table 3). Nitrogen uptake by the aboveground biomass was greater in the urea treatment ($p < 0.05$) than in the cocompost and control treatments (Table 3). Nitrogen recovered by plants in the urea treatment was 52 kg N ha^{-1} greater than the control. Since mineral N in soil was not different among treatments in May (before the growing season), the N recovered from the fertilizer treatment can be only attributed to soil N mineralization. The mineralizable soil organic N in the urea treatment but not in the control or cocompost treatment may indicate that part of the applied fertilizer remaining in soils is in active organic matter. It was not certain if the net 52 kg N ha^{-1} uptake in the fourth year in the urea treatment was a result from the previous 3-yr applied N or just from the N application in the third year. More research is needed in this area to quantify fertilizer N retention.

Curtin et al. (2006) reported that water soluble organic N is the fraction of soil N that can be mineralized during the growing season. The second experiment showed good correlation between water-extractable total organic N and

Table 3. Residual effect of urea fertilizers and cocompost applied at different times on soil mineral N, leaf N concentration, turfgrass biomass, and total N recovered measured in the fourth year (2005) after 3 yr (2002–2004) of application of urea in the spring of each year and differently timed cocompost application.

Treatment†	Mineral N in soil in May			Leaf N concentration		Biomass		Total N recovered
	Thatch	0–15 cm	15–30 cm	June 24	Sept. 29	June 24	Sept. 29	
	—— mg N kg^{-1} soil ——			—— % ——		——— kg ha^{-1} ———		
Control	15.0	7.8	8.3	12b‡	15	1206b	917b	29.7b
50 Mg cocompost annual	20.4	7.3	9.4	13b	14	1310b	989b	33.7b
100 Mg + 50 Mg split	28.2	7.3	11.7	14b	15	1459b	1000b	36.8b
150 Mg one time	24.9	14.1	10.7	13b	15	1629b	1089b	37.5b
NPKS annual	14.6	7.3	6.7	23a	15	2636a	1665a	81.9a
Probability (*F* test)	NS	NS	NS	<0.01	NS	<0.01	<0.05	<0.01

† Check, no cocompost nor chemical fertilizer application; 50 Mg cocompost annual, 50 Mg cocompost applied in the spring of each year from 2002 to 2004; 100 Mg + 50 Mg split, 100 Mg cocompost applied in the spring of the first year (2002), 50 Mg cocompost applied in the second year (2003), and no application in 2004 (the third year); 150 Mg one time, 150 Mg cocompost applied in the spring of the first year (2002) and no application in the second (2003) and third year (2004); NPKS annual, chemical fertilizer applied in the spring of each year from 2002 to 2004 at rate of (100 kg N as urea + 20 kg P + 42 kg K + 20 kg S) ha^{-1}.

‡ Different letter indicates least statistical difference (LSD) at 5% level.

mineral N released during a 2-wk soil incubation (Fig. 4a); the degree of correlation decreased as the incubation time increased to 4 wk (Fig. 4b). These results indicate that residual fertilizer N in soils was in a form that is quickly mineralized. These results also suggest that residual fertilizer N in the first experiment probably existed as WEON (Table 3). Thus, we hypothesize that fertilizer N retained in soil is in the active organic matter pool, which can be quickly mineralized. Similar research results have been summarized by Chantigny (2003) and Ros et al. (2009), although, different results have been reported (Ghani et al., 2003).

Efforts have been made to estimate soil mineralizable N to recommend a viable N fertilizer program. These efforts include acid hydrolyzable N (i.e., 1 N HCl), alkaline hydrolyzable N (i.e., 1 NaOH), and neutral solution (100°C KCl). Most recently, WEON was identified as an indicator for potential mineralizable N in soil (Curtin et al., 2006; Zhao et al., 2013). Soil WEON in combination with soil respiration test (i.e., Solvita test) is now used as a good indicator of estimating potential mineralizable C, N, and P in soil (Haney et al., 2008).

Conclusions and Future Research Directions

This chapter shows how crop rotation affects WEOC and WEON. Also both UV-Vis and fluorescence spectral properties of WEOM changed as a result of different crop rotations. Previous results from the same field experiment show that crop rotation affects soil microbial community (Larkin et al., 2011). Therefore, there may

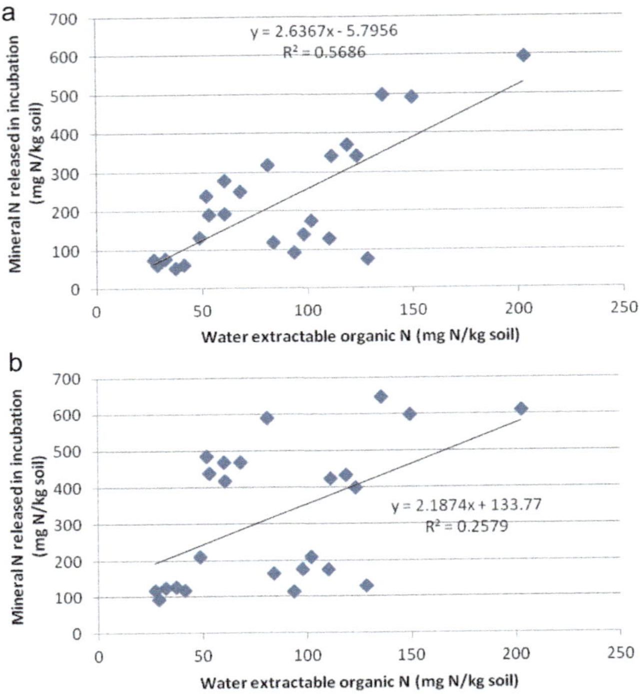

Fig. 4. Water-extractable total N before incubation and mineral N released in incubation in 20°C and 90% soil field capacity at (a) 2 wk and (b) 4 wk.

be a link between the microbial community and soil water soluble organic matter. Previous genomic method (ribosomal RNA) for microbial community analysis can only show the existence of species in soil, but it cannot indicate functional species in the community (Mettel et al., 2010). However, advance in extraction method shows the messenger RNA can be effectively extracted and enriched to reveal functional status of soil microbial communities (Mettel et al., 2010). A metatranscriptomic analysis of soil microbial community and functional group (Tveit et al., 2014) provides promises for linking active soil organisms and WEOM under different management practices. This information could be used to help to develop best management practices. Even though soil WEOM was different among the treatments, we still don't know if there is a link between WEOM and disease suppression in the multiple-crop rotation system. A multidisciplinary approach needs to be adopted to understand this completely.

In this chapter, the turnover rate of residual fertilizer N is not clear since the residual test was done after 3 yr of continuous application of N fertilizers. From the perspectives of improving N use efficiency and preventing N losses, quantifying annual fertilizer N carryover is meaningful. Solvita test (Haney et al., 2008) provides a practical way of soil respiration to estimate soil mineralizable N even though more research is still needed for the test. Since the soil water-extractable organic matter is the active fraction of soil organic matter, which is readily used by soil microorganisms, the soil respiration (i.e., Solvita test) should be related to the soil WEOM. In fact, some commercial laboratories have included soil water soluble organic N as part of their test for available N in soil. Despite numerous efforts, there are still significant amounts of N loss from the major agriculture area to the nation's surface water system. We suggest that including WEON in N management might provide solutions for N runoffs from agricultural fields.

References

Analytical Software. 2013. Statistix 10 analytical software for researchers. Analytical Software, Tallahassee, FL.

Baldock, J.A., and P.N. Nelson. 2000. Soil organic matter. In: M.E. Sumner, editor, Handbook of soil science. CRC Press, Boca Raton, FL. p. B25–B84.

Borisover, M., A. Lordian, and G.J. Levy. 2012. Water-extractable soil organic matter characterization by chromophoric indicators: Effects of soil type and irrigation water quality. Geoderma 179-180:28–37. doi:10.1016/j.geoderma.2012.02.019

Bro, R., and C.A. Andersson. 2000. The N-way toolbox for MATLAB, Chemometrics & Intelligent Laboratory Systems. 52:1–4. http://www.models.life.ku.dk/nwaytoolbox (accessed 19 Feb. 2015).

Cassman, K.G., A. Dobermann, and D.T. Walters. 2002. Agroecosystems, nitrogen-use efficiency, and nitrogen management. Ambio 31:132–140.

Chantigny, M.H. 2003. Dissolved and water-extractable organic matter in soil: A review on the influence of land use and management practices. Geoderma 113:357–380. doi:10.1016/S0016-7061(02)00370-1

Chen, J., E.J. LeBoeuf, S. Dai, and B. Gu. 2003. Fluorescence spectroscopic studies of natural organic matter fraction. Chemosphere 50:639–647. doi:10.1016/S0045-6535(02)00616-1

Cunfer, B.M., and D.V. Phillips. 2006. Effect of crop rotation on take-all of wheat in double-cropping systems. Plant Dis. 90:1161–1166. doi:10.1094/PD-90-1161

Cunfer, B.M., and C.S. Rothrock. 1994. The influence of conservation and double cropping practices on diseases of wheat in Georgia. Res. Bull. 418. Ga. Agric. Exp. Stn., Univ. GA.

Curtin, D., C.E. Wright, M.H. Beare, and F.M. McCallum. 2006. Hot water-extractable nitrogen as an indicator of soil nitrogen availability. Soil Sci. Soc. Am. J. 70:1512–1521. doi:10.2136/sssaj2005.0338

Doran, J.W. 1987. Microbial biomass and mineralizable nitrogen distribution in no-tillage and plowed soil. Biol. Fertil. Soils 5:68–75. doi:10.1007/BF00264349

Fellman, J.B., V. D'Amore, E. Hood, and R.D. Boone. 2008. Fluorescence characteristics and biodegradability of dissolved organic matter in forest and wetland soils from coastal temperate watershed in southeast Alaska. Biogeochemistry 88:169–184. doi:10.1007/s10533-008-9203-x

Ghani, A., M. Dexter, and K.W. Perrott. 2003. Hot-water extractable carbon in soils: A sensitive measurement for determining impacts of fertilisation, grazing and cultivation. Soil Biol. Biochem. 35:1231–1243. doi:10.1016/S0038-0717(03)00186-X

Haines, P.J., J.M. Curran, and D.R. Coventry. 1996. Rotation with canola and optimizing wheat yield in north-east Victoria Australia Society of Agronomy http://www.regional.org.au/au/asa/1996/contributed/281haines.htm (accessed 13 Mar. 2014).

Halloran, J.M., R.P. Larkin, S.L. DeFauw, O.M. Olanya, and Z. He. 2013. Economic potential of compost amendment as an alternative to irrigation in Maine potato production systems. Am. J. Plant Sci. 4:238–245. doi:10.4236/ajps.2013.42031

Haney, R.L., W.H. Brinton, and E. Evans. 2008. Estimating soil carbon, nitrogen and phosphorus mineralization from short-term carbon dioxide respiration. Comm. Soil Sci. and Plant Anal. 39:2706–2720. doi:10.1080/00103620802358862

Harris, G.H., O.B. Jesterman, E.A. Paul, S.E. Peters, and R.R. Janke. 1994. Fate of legume and fertilizer nitrogen-15 in a long-term cropping system experiment. Agron. J. 86:910–915. doi:10.2134/agronj1994.00021962008600050028x

He, Z., J. Mao, C.W. Honeycutt, T. Ohno, J.F. Hunt, and B.J. Cade-Menun. 2009. Characterization of plant-derived dissolved organic matter by multiple spectroscopic techniques. Biol. Fertil. Soils 45:609–616. doi:10.1007/s00374-009-0369-8

He, Z., C.W. Honeycutt, T.S. Griffin, R.P. Larkin, M. Olanya, and J.M. Halloran. 2010. Increases of soil phosphatase and urease activities in potato fields by cropping rotation practices. J. Food Agric. Environ. 8:1112–1117.

He, Z., C.W. Honeycutt, M. Olanya, R.P. Larkin, and J.M. Halloran. 2011a. Soil test phosphorus and microbial biomass phosphorus in potato fields. J. Food Agric. Environ. 9:540–545.

He, Z., C.W. Honeycutt, and H. Zhang. 2011b. Elemental and Fourier transform infrared spectroscopic analysis of water and pyrophosphate extracted soil organic matter. Soil Sci. 176:183–189. doi:10.1097/SS.0b013e318212865c

He, Z., H. Zhang, M.O. Olanya, J.M. Frantz, and R.P. Larkin. 2012. Differences in modified Morgan phosphorus levels determined by colorimetric and inductively coupled plasma methods. Open J. Soil Sci. 2:256–262. doi:10.4236/ojss.2012.23030

Hunt, J.F., T. Ohno, Z. He, C.W. Honeycutt, and D.B. Dail. 2007. Influence of decomposition on chemical properties of plant- and manure-derived dissolved organic matter and sorption to goethite. J. Environ. Qual. 36:135–143. doi:10.2134/jeq2006.0133

Kalbitz, K., S. Geyer, and W. Geyer. 2000. A comparative characterization of dissolved organic matter by means of original aqueous samples and isolated humic substances. Chemosphere 40:1305–1312. doi:10.1016/S0045-6535(99)00238-6

Larkin, R.P., C.W. Honeycutt, O.M. Olanya, J.M. Halloran, and Z. He. 2012. Impact of crop rotation and irrigation on soilborne diseases and soil microbial communities. In: Z. He, R. Larkin, and W. Honeycutt, editors, Sustainable potato production: Global case studies. Springer, New York. p. 23–41.

Larkin, R.P., C.W. Honeycutt, T.S. Griffin, O.M. Olanya, J.M. Halloran, and Z. He. 2011. Effects of different potato cropping system approaches and water management on soilborne diseases and soil microbial communities. Phytopathology 101:58–67. doi:10.1094/PHYTO-04-10-0100

Ledgard, S.F.K.W.S., and C. Feyter. 1988. Influence of time of application on the fate of 15N-labelled urea applied to dairy pasture. N.Z. J. Agric. Res. 31:87–91. doi:10.1080/00288233.1988.10421368

Mettel, C., Y. Kim, P.M. Shrestha, and W. Liesack. 2010. Extraction of mRNA from soil. Appl. Environ. Microbiol. 76:5995–6000. doi:10.1128/AEM.03047-09

Miller, J.J., B.W. Beasley, C.F. Drury, and B.J. Zebarth. 2012. Denitrification during the growing season as influenced by long-term application of composted versus fresh feedlot manure. Can. J. Soil Sci. 92:865–882. doi:10.4141/cjss2011-107

Ohno, T., Z. He, I.A. Tazisong, and Z.N. Senwo. 2009. Influence of tillage, cropping, and nitrogen source on the chemical characteristics of humic acid, fulvic acid, and water-soluble soil organic matter fractions of a long-term cropping system study. Soil Sci. 174:652–660. doi:10.1097/SS.0b013e3181c30808

Ohno, T., and Z. He. 2011. Fluorescence spectroscopic analysis of organic matter fraction: The current status and a tutorial case study. In: Z. He, editor, Environmental chemistry of animal manure. Nova Science Publishers, Hauppauge, NY. p. 83–103.

Olanya, O.M., R.P. Larkin, J.M. Halloran, and Z. He. 2014. Relationships of crop and soil management systems to meteorological variables and potato diseases on a Russet Burbank cultivar. J. Agric. Meteorol. (Tokyo) 70:91–104. doi:10.2480/agrmet.D-13-00022

Petrovic, A.M. 1990. The fate of nitrogenous fertilizer applied to turfgrass. J. Environ. Qual. 19:1–14. doi:10.2134/jeq1990.00472425001900010001x

Piening, L.J., and D. Orr. 1988. Effect of crop rotation on common root rot of barley. Can. J. Plant Pathol. 10:61–65. doi:10.1080/07060668809501766

Power, J.F. 1972. Fate of fertilizer nitrogen applied to a Northern Great Plains rangeland ecosystem. J. Range Manage. 25:367–371. doi:10.2307/3896548

Raaijmakers, J.M., T.C. Paulitz, C. Steinberg, C. Alabouvette, and Y. Moënne-Loccoz. 2009. The rhizosphere: A playground and battlefield for soilborne pathogens and beneficial microorganisms. Plant Soil 321:341–361. doi:10.1007/s11104-008-9568-6

Reddy, G.B., and K.R. Reddy. 1993. Fate of nitrogen-15 enriched ammonium nitrate applied to corn. Soil Sci. Soc. Am. J. 57:111–115. doi:10.2136/sssaj1993.03615995005700010021x

Roncero, M.I.G., C. Hera, M. Ruiz-Rubio, F.I.G. Maceira, M.P. Madrid, Z. Caracuel, F. Calero, J. Delgado-Jarana, R. Roldan-Rodriguez, A.L. Martinez-Rocha, C. Velasco, J. Rosa, M. Martin Urdiroz, D. Cordoba, A.D. Pietro. 2003. Fusarium as a model for studying virulence in soilborne plant pathogens. Physiol. Mol. Plant Pathol. 62:87–98. doi:10.1016/S0885-5765(03)00043-2

Ros, G.H., E. Hoffland, C. van Kessel, and E.J.M. Temminghoff. 2009. Extractable and dissolved soil organic nitrogen– A quantitative assessment. Soil Biol. Biochem. 41:1029–1039. doi:10.1016/j.soilbio.2009.01.011

Rothrock, C.S., and B.M. Cunfer. 1986. Absence of take-all decline in double-cropped fields. Soil Biol. Biochem. 18:113–114. doi:10.1016/0038-0717(86)90112-4

Sebilo, M., B. Mayer, B. Nicolardot, G. Pinay, and A. Mariotti. 2013. Long-term fate of nitrate fertilizer in agricultural soils. Proc. Natl. Acad. Sci. USA 110:18185–18189. doi:10.1073/pnas.1305372110

Shimazu Corporation. 2014. The relationship between UV-VIS absorption and structure of organic compounds. http://www.shimadzu.com/an/uv/support/uv/ap/apl.html (accessed 28 June 2014).

Singh, S., S. Dutta, and S. Inamdar. 2014. Land application of poultry manure and its influence on spectrofluorometric characteristics of dissolved organic matter. Agric. Ecosyst. Environ. 193:25–36. doi:10.1016/j.agee.2014.04.019

Sun, S., J. Liu, Y. Li, P. Jiang, and S. Chang. 2013. Similar quality and quantity of dissolved organic carbon under different land use systems in two Canadian and Chinese soils. J. Soils Sediments 13:34–42. doi:10.1007/s11368-012-0604-z

Tamm, L., B. Thurig, A. Fliessbach, A.E. Goltlieb, S. Karavani, and Y. Cohen. 2011. Elicitors and soil management to induce resistance against fungal plant diseases. NJAS-Wageningen Journal of Life Sciences 58:131–137.

Tveit, A., T. Urich, and M.M. Svenning. 2014. Metatranscriptomic analysis of Arctic peat soil microbiota. Appl. Environ. Microbiol. 80:5761–5772. doi:10.1128/AEM.01030-14

Wang, L., F. Wu, and R. Zhang. 2009. Characteristics of dissolved organic matter fractions from Lake Hongfen Southwestern China Plateau. J. Environ. Sci. (China) 21:581–588. doi:10.1016/S1001-0742(08)62311-6

Yamashita, Y., and E. Tanoue. 2004. Chemical characteristics of amino acid-containing dissolved organic matter in seawater. Org. Geochem. 35:679–692. doi:10.1016/j.orggeochem.2004.02.007

Zhang, M., Z. He, and A. Zhao. 2011a. Ultraviolet-visible absorptive features of water extractable and humic fraction of animal manure and relevant compost. In: Z. He, editor, Environmental chemistry of animal manure. NOVA Science Publishers, New York. p. 61–81.

Zhang, M., Z. He, A. Zhao, H. Zhang, D.M. Endale, and H.H. Schomberg. 2011b. Water extractable soil organic carbon and nitrogen as affected by tillage and manure application. Soil Sci. 176:307–312. doi:10.1097/SS.0b013e31821d6d63

Zhang, M., S.S. Malhi, R. Panasuik, and B. Henriquez. 2010. Response of turfgrass growth in a black chernozemic soil amended with municipal solid/biosolid waste compost. J. Plant Nutr. 34:183–205. doi:10.1080/01904167.2011.533321

Zhang, M., A. Zhao, and Z. He. 2014. Distribution and biodegradability of water soluble organic carbon and nitrogen in subarctic Alaskan soils under three different land uses. In: Z. He and H. Zhang, editors, Applied manure and nutrient chemistry for sustainable agriculture and environment. Springer, Amsterdam, the Netherlands. p. 313–332.

Zhao, A., M. Zhang, and Z. He. 2012. Characteristics of soil water-soluble organic C and N under different land uses in Alaska. Soil Sci. 177:683–694. doi:10.1097/SS.0b013e31827e1fa8

Zhao, A., M. Zhang, and Z. He. 2013. Spectroscopic characteristics and biodegradability of cold and hot water–extractable soil organic matter under different land uses in subarctic Alaska. Commun. Soil Sci. Plant Anal. 44:3030–3048. doi:10.1080/00103624.2013.829086

Zsolnay, W., E. Baigar, M. Jimenez, B. Steinweg, and F. Saccomandi. 1999. Differentiating with fluorescence spectroscopy the sources of dissolved organic matter in soils subjected in drying. Chemosphere 38:45–50. doi:10.1016/S0045-6535(98)00166-0

Elemental Composition and Functional Groups in Soil Labile Organic Matter Fractions

Irenus A. Tazisong,* Zachary N. Senwo, and Zhongqi He

Abstract

Labile organic matter fractions are major components involved in nutrient cycle in soil. In this chapter, we examine three labile organic matter fractions: light fraction (LF), humic acid (HA), and fulvic acid (FA) of Alabama cotton soils (Ultisol) amended with chemical fertilizer (NH_4NO_3) and poultry litter and managed as no-till, conventional-till, and mulch-till practices. Light fraction was significantly affected by management practices. On average, the LF in the no-till soils was slightly higher (42.02 ± 5.62 [SD, $n = 4$] g) than in the conventional-till (39.52 ± 1.58 g) and mulch-till (37.71 ± 0.61 g) per kilogram of dry soil. The C, N, and S contents of the LF from the amended and till soils were higher than the control (bare fallow [BF]) sample. Elemental analysis of HA and FA revealed that HA had greater aromatic functional groups and lower aliphatic functional groups than FA. Fourier transform infrared (FT-IR) and solid-state ^{13}C nuclear magnetic resonance (NMR) spectroscopy were applied to evaluate the effects of cropping management practices on the composition of soil humic substances (HS). These spectral data show that the impacts of management practices were mainly on the carbonyl group from amide and carboxylic acid or ketone compounds for HA and both carbonyl and aliphatic groups for FA. In HA, cropping management practices also increased ketone, quinone, or aldehyde, and decreased aromatic C-C and C-H groups in all treatments compared with those of the BF sample. The aromaticity of HA samples decreased with cropping managements and fertilizer applications, suggesting incorporation of aliphatic crop residues to the HAs.

Abbreviations: BF, bare fallow; DOM, dissolved organic matter; EC, electrical conductivity; FA, fulvic acid; FT-IR, Fourier transform infrared spectroscopy; HA, humic acid; HS, humic substance; LF, light fraction; SOM, soil organic matter.

Irenus A. Tazisong and Zachary N. Senwo (zachary.senwo@aamu.edu), Dep. of Biological and Environmental Sciences, College of Agricultural, Life and Natural Sciences, Alabama A&M University, 4900 Meridian Street, Normal, AL 35762. *Corresponding author (irenus.tazisong@aamu.edu). Zhongqi He, USDA-ARS Southern Regional Research Center, 1100 Robert E. Lee Blvd, New Orleans, LA 70124 (zhongqi.he@ars.usda.gov).

doi:10.2136/sssaspecpub62.2014.0039

Labile Organic Matter—Chemical Compositions, Function, and Significance in Soil and the Environment. SSSA Special Publication 62. Zhongqi He and Fengchang Wu, editors.

Since medieval times, it has been recognized that dark-colored soils with high organic matter content are more fertile than light-colored soils. The relationships between soil quality and sustainable productivity in agroecosystems are of great interest to scientists. To preserve and maintain soil quality, it is imperative to maintain the quantity and quality of soil organic matter (SOM) given its role in nutrient status and soil biological activity (Gigliotti et al., 2003). Soil organic matter is distributed into various C pools: the labile or active C, the slow or intermediate soil organic C, and the recalcitrant or inert soil organic C (Wander, 2004). According to Wander (2004), labile organic C can be characterized as (i) material with a half-life of days to a few years, (ii) material of recent origin or embodied living components, (iii) materials of high nutrient and energy value, or (iv) materials that lack physical protection, which makes them likely to participate in biological or chemically based reactions. Wander (2004) also classified the intermediate soil organic C as having a half-life of a few years to decades and by its physically protected status, which makes its decomposition moderately slow, whereas the recalcitrant SOM has the slowest decomposition rate with a half-life of decades to centuries. Any changes in these pools reflect the impact of the management practices on nutrient cycling.

This chapter examines labile C pools of an Alabama ultisol (highly weathered soil of warm and humic climate with a kandic or argillic horizon and a base saturation of less than 35%) because they serve as sources for readily metabolized C and because of their sensitivity to soil changes (Wander, 2004; Fynn et al., 2003). Labile C pools have been defined to include microbial biomass carbon; particulate organic C, which are organic matter suspended in water (Cambardella and Elliott, 1992); dissolved organic matter, defined as organic matter that can pass through 0.45-μm filter (Thurman, 1985); and LF C from partially decomposed plant tissue (Fynn, 2003). Studies (Gregorich and Ellert, 1993) have shown that the LF accounts for ~0.1 to 0.3% of total weight in cultivated soils, of which there is ~3 to 10% in grassland and forest soils, and are enriched in C and N, constituting approximately 30 to 40% of total soil organic C and N.

Light fraction of organic matter, HA, and FA can be considered as labile organic matter. Both HA and FA belong to HS that consist mainly of fully decomposed (humified) remains of plants and animal organic matter. About 60 to 80% of total C in terrestrial and aquatic media is made up of HS (Pena-Mendez et al., 2005). Humic substances have been shown to be comprised of extraordinary complex mixtures of organic compounds with no discrete identifiable molecular or structural formula, are yellow to black in color, and have high cation and anion exchange capacity (Aiken et al., 1985; Mayhew, 2004; Zhang and He, 2004).

Literature on the chemical composition of, preferential metal binding with, and the effect of tillage, cropping, and manure applications on labile organic matter in ultisol is scanty. Therefore, as a case study, this chapter presents and discusses unpublished data on LF, HA, and FA in Alabama cotton soils (ultisol) amended with chemical fertilizer (NH_4NO_3) and poultry litter and managed with no-till, conventional-till, and mulch-till practices.

Site and Treatment Description

The soil samples used were collected from sites that have been under experimental conditions since 1996 in an upland cotton production site at the Alabama Agricultural Experimental Station at Belle Mina, AL (34°41′ N, 86°52′ W). The soil is a Decatur silt loam (fine, kaolinitic, thermic Rhodic Paleudults). The site was continuously grown with cotton (*Gossypium hirsutum* L.) for 5 yr before the start of the experiment in 1996. Treatments (Table 1) consisted of various combinations of tillage systems (no-till, conventional-till, or mulch-till), cropping systems and management practices (cotton in summer, with or without rye [*Secale cereale* L.] cover crop in winter), and N fertilizer (ammonium nitrate, NH_4NO_3 [0 and 100 kg N ha^{-1}] or poultry litter [0, 100, and 200 kg N ha^{-1}]) (Tazisong et al., 2013; 2014).

Soils used in this study were collected in February 2006, with a hand probe (10 cm i.d.) from the 0- to 10-cm depth in four replicates in the central rows of the plots after removing the residue from the soil surface. The soils were air-dried, ground, sieved to pass a 2-mm sieve into plastic bags, and stored until used. Soil pH was measured in water at a soil/solution ratio of 1:2, and the filtered extract was used for electrical conductivity (EC) measurements, using an Orion

Table 1. Description of management practices, fertilizer source, and selected soil properties.

Management practice	Description of treatment	Cropping system	Fertilizer applied	pH (H_2O)	EC†	C	N	S	C/N
					μS cm^{-1}	—— g kg^{-1} ——			
BF	Bare fallow	No crop planted	None	5.21	38.85	9.97	1.1	0.03	9.3
CTAN	Conventional-till	Cotton, winter fallow	NH_4NO_3	5.59	55.05	13.5	1.8	0.16	7.7
CTR	Conventional-till	Cotton–rye cropping	None	5.74	43.08	12.2	1.5	0.14	8
CTRAN	Conventional-till	Cotton–rye cropping	NH_4NO_3	5.8	35.35	14	1.9	0.15	7.5
CTRP	Conventional-till	Cotton–rye cropping	Poultry litter	6.3	59.18	14.6	2	0.15	7.2
MTRAN	Mulch-till	Cotton–rye cropping	NH_4NO_3	5.81	55.23	14.8	2	0.16	7.5
MTRP	Mulch-till	Cotton–rye cropping	Poultry litter	5.96	53.25	13.5	1.4	0.15	9.6
NT	No-till	Cotton, winter fallow	None	5.97	30.9	11.3	1.4	0.12	8.3
NTAN	No-till	Cotton, winter fallow	NH_4NO_3	5.54	49.8	11.8	1.2	0.13	9.8
NTRAN	No-till	Cotton–rye cropping	NH_4NO_3	5.64	51.5	12.8	2.2	0.14	6.2
NTRP	No-till	Cotton–rye cropping	Poultry litter	6.17	62.15	16.4	2	0.18	8.2
NTRPP‡	No-till	Cotton–rye cropping	Poultry litter	6.22	81.05	15.1	2.1	0.18	7.1

† EC, electrical conductivity.

‡ Amount of poultry litter in this treatment was doubled.

conductivity meter (model 160, Thermo Orion). The pH and EC reported were temperature compensated at 25°C. Total soil C, N, and S were determined by dry combustion method using a Vario Max CNS analyzer (Elementar Germany).

Isolation and Composition of Light Fraction

The density gradient technique has been adopted by most investigators (Gregorich and Ellert, 1993; Janzen et al., 1992; Wander and Traina, 1996; Ding et al., 2002) to isolate organic matter LF. The LF was isolated using a modified method as described by Ding et al. (2002). Approximately 25 g of air-dried soil was weighed into 250-mL centrifuge bottles, to which was added 70 mL of NaI solution (1.7 g mL^{-1}). The mixture was shaken for 1 h and centrifuged at 120 *g* for 15 min. The suspended material was then filtered under suction through a 0.2-μm filter paper. The LF materials retained on the filter paper were rinsed with 100 mL of a 0.5 M $CaCl_2$ and 0.5 M $MgCl_2$ followed by a final rinse with 200 mL of deionized water. Rinsing of the LF with $CaCl_2$ and $MgCl_2$ was done to avoid any remnant biological toxicity because of Na saturation of the ion-exchange sites in the LF (Ding et al., 2002). The filter paper containing the LF was transferred into a drying pan and placed in an oven and dried at 70°C for 24 h. After 24 h of drying, the LF was weighed and then stored in a desiccator until analyzed. The LF was then analyzed for total C, N, and S using the Vario Max CNS analyzer. The total weight and elemental composition of the LF are shown in Table 2.

The studies indicated that the LF weight was significantly affected by management practices. On average, the LF in the no-till soils was slightly higher (42.02 ± 5.62 g) than in the conventional (39.52 ± 1.58 g) and mulch-till (37.71 ± 0.61 g) per kilogram of soil (Table 2). Similar findings were reported by Ding et al. (2002). Nascente et al. (2013) also reported that soils collected from no-till plots had more LF (average) than tilled plots irrespective of depth studied. Additionally, Assis et al. (2012) observed that LF increased in red Latosol under bermudagrass [*Cynodon dactylon* (L.) Pers. 'Tifton'] and was reported to be sensitive under soil cultivation. The LF weight in this study ranged from 37.78 ± 0.08 to 51.12 ± 0.23 g in the no-till soils, 37.32 ± 0.06 to 41.00 ± 1.41 g in the conventional-till soils, and 37.28 ± 0.06 to 38.14 ± 0.08 g in the mulch-till soils per kilogram of dry soil. In conventionally tilled soils, the LF elemental (C, N, and S) content from the CTRP (conventional till, cotton–rye cropping, poultry litter) treatment was greater than the other treatments (Table 2). Similarly, the LF elemental content from the NTRPP (no-till, cotton–rye cropping, double poultry litter) treatment was the highest among the no-tilled soils, whereas the MTRAN (mulch-till, cotton–rye cropping, NH_4NO_3) LF elemental content was higher than MTRP (mulch-till, cotton–rye cropping, poultry litter) except for the S content (Table 2). It has been reported that applying manure and chemical fertilizer to soils may increase SOM (Campbell et al., 1991; Biederbeck et al., 1994) due to increase in residue inputs into soils, which might have resulted from increase in shoots and roots. Manure (poultry litter) contains materials that are precursor to SOM. The positive effects of manure and inorganic fertilizer inputs on the LF were clearly observed in this study (Table 2). Based on our findings, the C, N, and S contents of the LF from the amended soils were significantly higher than the BF (control) sample. This finding was consistent to results reported by other investigators (Wu et al., 2005; Grandy et al., 2002). The LF content in

Table 2. Weight and elemental composition of light fraction (LF) organic matter. Values in parentheses are standard deviations.

Treatment†	Dry wt.	C	N	S	C/N
	g kg⁻¹ soil	mg kg⁻¹ LF			
BF	38.80 (1.13)	570 (0.01)	42.68 (0.01)	6.596 (0.001)	13.36
CTAN	37.32 (0.06)	900 (0.08)	62.50 (0.01)	6.138 (0.007)	14.4
CTR	40.20 (0.28)	1087 (0.20)	84.84 (0.00)	9.292 (0.014)	12.81
CTRAN	41.00 (1.41)	1294 (0.07)	95.58 (0.01)	10.30 (0.02)	13.53
CTRP	39.56 (0.00)	1525 (0.14)	116.42 (0.01)	13.266 (0.007)	13.1
MTRAN	37.28 (0.06)	1395 (0.03)	101.93 (0.01)	9.486 (0.007)	13.69
MTRP	38.14 (0.08)	1091 (0.07)	86.64 (0.01)	11.59 (0.007)	12.59
NT	51.12(0.23)	1108 (0.02)	92.04 (0.01)	10.92 (0.014)	12.03
NTAN	39.08 (0.06)	968 (0.01)	68.99 (0.01)	4.508 (0.007)	14.03
NTRAN	38.30 (0.14)	1083 (0.01)	85.25 (0.01)	9.792 (0.007)	12.7
NTRP	43.86 (0.20)	1214 (0.07)	104.28 (0.01)	9.24 (0.012)	11.65
NTRPP	37.78 (0.08)	1267 (0.0)	101.52 (0.01)	10.53 (0.02)	12.48

† Treatments described in Table 1.

poultry-litter-amended tilled soils was higher than in the mulch- or inorganic-fertilized tilled soils due to high organic matter content in the poultry litter and rapid LF mineralization in mulch and inorganic-fertilizated soils. The LF C contents ranged from 570 to 1525 mg kg^{-1} LF, the N value ranged from 42.68 to 116.42 mg kg^{-1} LF, and the S content ranged from 4.58 to 13.26 mg kg^{-1} LF in the various treatments (Table 2). There was tillage influence on the C, N, and S contents of the LF. The C, N, and S contents for the conventionally tilled soils were slightly higher than those of no-till soils, with most treatments almost equal to that of mulch-till soils. Similar trends were observed for the C/N ratio, though the difference was not significant. Nascente et al. (2013) observed higher C content in the no-till soil than conventionally tilled at the 0- to 0.05-m depth. Wander and Traina (1996) did not find significant impact of farming system on C and N content of the LF, but noticed a significant impact of the treatments on the LF C and N content ratios (Wander et al., 1994).

The latter report supports our results, which show effects of treatments on the LF C, N, and S contents. The difference in the LF contents among the treatments may be due to variations in such soil environmental factors, such as temperature and moisture, which strongly affect the decomposition rate (Janzen et al., 1992; Parton et al., 1987). Gosling et al. (2013) observed that climate, organic amendments, and inclusion of fallow periods were significant factors affecting LF organic matter. Among these factors, they reported that organic amendments and inclusion of fallow periods were the most important factors influencing LF, whereas climate had a larger influence on total SOM. Based on their findings, tillage practices and N fertilizer application were not significant variables. In the conventionally tilled soils, soil water content may be low, and the demand for water by crops maintains low moisture content, thus suppressing organic matter decomposition. In contrast, soil moisture is higher in no-till and mulch-till soils, thereby facilitating microbial decomposition. The LF has been suggested to be an important source of energy for soil microbial organisms due to strong correlations between LF and microbial respiration (Janzen et al., 1992). Significant correlations between the LF contents and microbial biomass N observed in this study (Table 3) support the fact that LF influences microbial activity. Negative and no correlations existed between LF content and microbial biomass C (Table 3). Similar results were observed by Janzen et al. (1992), who explained the lack of such relationships to confounding effects of increasing C substrates on the measurement of biomass C using fumigation–incubation techniques.

Extraction and Elemental Composition of Humic and Fulvic Acid

Humic and fulvic acids were extracted from air-dried soils ground to pass a 2-mm sieve based on the procedure recommended by the International Humic Substances Society as described by Swift (1996). Elemental analysis (C, N, and H) of HA and FA was determined with a PerkinElmer 2400 CHN combustion analyzer (PerkinsElmer, Inc.) of which the O content was obtained by subtraction. The E4/E6 ratio was determined by dissolving 1 mg of HA or FA in 100 mM sodium acetate pH 5.0 and measuring optical densities at 400 and 600 nm with a Genesys 10 UV spectrophotometer (Thermo Scientific). Selected metal (Fe, Zn, Cu, Pb, Ni, Cd, Cr, and Al) and P contents were determined after persulfate digestion and analyzed using an inductively coupled plasma–optical emission spectrometer model 210 DV (PerkinsElmer, Inc.).

On average, the C, H, and N contents were higher for HA than FA, whereas the FA had higher O content and C/N ratio (Table 4). The C concentration in the HA fraction ranged from 49.9 to 54.7% and was similar to ranges reported by other investigators (Francioso et al., 1998; Olk et al., 1999; Reintam et al., 2000). The C concentration in the FA fractions ranged from 15.43 to 45.24% and is similar to ranges reported by Reintam et al. (2000). The high C concentration in HA than FA may indicate that fixation of C or C sequestration takes place during humic matter synthesis and goes up slightly with an increase rate of humification (Tan, 2003). Generally, few differences exist between treatments and tillage practices, as was also noted by Wander and Traina (1996) and McDonnell et al. (2001) in other studies. The most notable differences observed was the low C contents of FA from MTRP (mulch-till, cotton–rye cropping, poultry litter) and NT (no-till, cotton [winter fallow], no fertilizer) treatments, which were two and three times lower than the C contents of the other treatments. The N content in HA ranged from 3.66 to 9.98%, and 0.42 to 2.69% in FA. The N content in HA was two to three times higher than in FA, which may be an indication that N immobilization increased with humification to HA (Tan, 2003). The high differences between N content of HA from CTR (conventional till, cotton–rye cropping, no fertilizer) and the other treatments could not be explained, but it has been suggested that high N content in HA may be due to the incorporation of protein-type compounds into the humic molecules (Gomez and Lejuene, 1987; Hernandez et al., 1993).

The elemental analysis data (Table 4) were used to calculate atomic mass ratio. The average H/C, O/C, O/H, and C/N ratio was higher for FA than HA. This result is in agreement with that reported by McDonnell et al. (2001), except that they had a higher average H/C ratio in HA than FA. In HA, the H/C values ranged from 0.80 to 0.93, whiles the O/C values ranged from 0.49 to 0.64. These ranges are

Table 3. Correlation matrix of the relationship between light fraction (LF) content and other indices of labile organic matter in soil.

	Total C	Total N	Total S	LF C	LF N	LF S	MBN†	MBC‡
Total N	0.785**							
Total S	0.869***	0.719**						
LF C	0.667*	0.625*	0.604*					
LF N	0.730**	0.666*	0.655*	0.973***				
LFS	0.519ns§	0.569ns	0.398ns	0.681*	0.737**			
MBN	0.648*	0.363ns	0.547ns	0.681*	0.715**	0.692*		
MBC	–0.645*	–0.270ns	–0.524ns	–0.518ns	–0.509ns	–0.363ns	–0.661*	
PCM¶	–0.806***	–0.604*	–0.648*	–0.695*	–0.755**	–0.686*	–0.938***	0.628*

* Significant at the 0.05 probability level.

** Significant at the 0.01 probability level.

*** Significant at the 0.001 probability level.

† MBN, microbial biomass N.

‡ MBC, microbial biomass C.

§ ns, not significant.

¶ PCM, potential C mineralized.

Table 4. Elemental composition of humic and fulvic acids extracted from soil.

	Humic acid						Fulvic acid				
Treatment†	C	H	N	O	E4/E6‡	C/N	C	H	N	O	C/N
	%						%				
BF	53.13	3.53	4.13	39.21	5.07	15.0	34.97	3.36	2.69	58.98	15.2
CTAN	49.91	3.53	3.66	42.90	5.01	15.9	39.27	3.08	1.40	56.26	32.7
CTR	52.51	3.49	9.98	34.02	4.97	6.1	42.98	3.70	1.75	51.57	28.7
CTRAN	43.78	3.13	4.51	48.59	5.02	11.3	41.02	3.59	1.91	53.48	25.1
CTRP	52.99	3.63	4.27	39.11	5.03	14.5	40.12	3.43	1.89	54.57	24.8
MTRAN	52.73	3.56	4.06	39.66	4.91	15.2	43.69	3.61	1.73	50.97	29.5
MTRP	51.91	3.66	4.12	40.32	4.96	14.7	29.53	2.58	1.39	66.50	24.8
NT	51.39	3.44	3.85	41.32	4.99	15.6	20.43	1.06	1.10	77.41	21.7
NTAN	51.56	3.59	3.91	40.94	5.17	15.4	41.34	3.56	2.04	53.06	23.6
NTRAN	54.57	3.66	4.12	37.65	5.14	15.5	45.24	3.82	1.86	49.08	28.4
NTRP	51.49	3.94	4.25	40.32	5.06	14.1	44.45	4.00	2.10	49.44	24.7
NTRPP	52.43	4.08	4.45	39.03	5.13	13.7	41.93	3.96	1.88	52.23	26.0

† Treatments described in Table 1.

‡ E4/E6, optical density or absorbance of dilute aqueous humic and fulvic acid solutions at 465 and 665 nm. The E4/E6 values for fulvic acid were not determined.

similar to those reported in the literature (Agnelli et al., 2000; Ayuso et al., 1997; Song et al., 2005; Dodla et al., 2012; Barros Soares et al., 2013). The H/C ratio has been used to indicate the degree of aromaticity or aliphacity of a substance, with highly aromatic macromolecules having H/C ratio <1, whereas larger values indicate aliphacity (Agnelli et al., 2000; Rice and MacCarthy, 1991; Garcia et al., 1992; Ussiri and Johnson, 2003). An H/C ratio above 1.3 may be indicative of nonhumic substances (Steelink, 1985) or contamination by biologically un-decomposed materials, which mimics HS when extracted in alkaline (Bracewell et al., 1988). Reported data (Dodla et al., 2012) showed that HA obtained from forest swamp soil is mostly aliphatic in composition, whereas freshwater marsh HA has the highest percentage of aromatics. The H/C ratios of HA were lower than those of FA, suggesting that the HA had greater aromaticity and lower aliphacity than FA (Table 4). It has been suggested that aromaticity increases with humification (Zech et al., 1992). The aromaticity in HA from NTRP (no-till, cotton–rye cropping, poultry litter) and NTRPP (NTRP with double poultry litter) treatments was lower than the rest of the treatments, and this may suggest that humification processes may have been slower in these two treatments than the other treatments. This observation is supported by ^{13}C NMR studies where most (average) of the aromatic functional groups (C-O, C-C, and C-H) were lower in NTRP and NTRPP treatments (Table 5). The O/C and O/H ratios are indexes for the abundance of O-containing functional groups and also are indicative of oxidation (Agnelli et al., 2000; McDonnell et al., 2001). This index was higher in FA than HA, suggesting higher oxidation level in the FA. The O/C ratio in HA ranged from 0.49 to 0.83, which is within the range for soil HA reported (Aiken et al., 1985; Stevenson, 1982; Song et al., 2005; McDonnell et al., 2001), while O/C ratio in FA ranged from 0.83 to 2.84. The C/N ratio for FA was higher than for HA, suggesting that a low degree of N was incorporated into the FA fraction as opposed to the HA fraction, which was associated with a high level of N and indicative of consistent biological activity. The color of HS has been used to characterize humic fractions in soils (Tan, 1998). The color ratio E4/E6 (optical density or absorbance of dilute aqueous HA and FA solutions at 465 and 665 nm) is used as an indication of the molecular weight of HS (Anderson and Schoenau, 1993). A high color ratio of 6:8 is observed for HS with low molecular weight (e.g. FA), while a color ratio of 4:5 corresponds to HS with high molecular weight (e.g. HA) (Tan, 2003, Zhang et al., 2011). The E4/E6 ratio in the HA fraction in this study ranged from 4.91 to 5.17 (Table 4) suggesting that the HA fraction from this soil is mainly composed of high molecular weight (>30,000 Da) compounds.

The metal content and P adsorbed in the HA and FA fractions are shown in Table 6 and 7. The amounts of Al, Fe, P, were higher in both the HA and FA fractions. The concentrations of Mn, Cr, and Zn were greater in the FA fraction than the HA fraction, which had a Mn, Cr, and Zn concentration of less than 12 mg kg^{-1}. The concentrations of Cd, Ni, and Pb were small in both fractions, except that Ni in FA had an appreciable quantity in some treatments. The high metal contents in these fractions are not surprising because HS are excellent chelators due to the presence of various functional groups, especially the O-containing functional groups carboxylic and phenolic OH groups. The low concentrations of Pb, Cd, and Ni seem to be due to the low concentrations of these elements in this soil (Senwo and Tazisong, 2004). The presence of significant quantity of P from these HA and FA fractions (Table 6 and 7) suggested the presence of P compounds

Table 5. Relative abundance of C functional groups in humic acids calculated from integrations of ^{13}C NMR spectral peaks. Chemical shifts of peaks (ppm) are given in parentheses.

Treatment†	Ketone, quinone, or aldehyde	COO and N-C-O	Aromatic C-O	Aromatic C-C and C-H	O-C-O anomerics	o-Alkyl Cc	NCH and OCH_3	Non-polar alkyl
	(190–220)	(162–190)	(145–162)	(108–145)	(96–108)	(60–96)	(50–60)	(0–50)
BF	1.4	14.22	8.26	32.71	5.06	16.06	5.71	16.59
CTAN	2.5	13.47	7.89	30.2	4.63	17.46	6.02	17.75
CTRAN	1.46	14.42	8.38	29.59	4.98	16.19	6.49	18.68
CTRP	2.26	14.86	9.11	31.03	5.03	15.48	6.3	15.94
CTR	2.09	15.39	8.8	30.5	4.94	16.69	5.37	16.22
MTRAN	1.83	14.41	8.34	29.39	6.07	16.97	6.11	16.88
NTAN	2.31	13.98	8.14	29.31	5.62	17.51	6.34	16.8
NTRAN	2.64	15.58	9.43	29.67	5.85	15.18	5.92	15.72
NTRP	2.35	15.45	9.96	30.1	5.53	15.79	5.85	14.98
NTRPP	2.16	13.42	8.56	28.02	5.00	18.16	6.9	17.78
NT	2.16	16.04	9.6	32.4	4.44	14.08	5.55	15.73

† Treatments described in Table 1.

Table 6. Selected total metal contents of humic acid.

Treatment†	Al	Cd	Cr	Cu	Fe	Mn	Ni	P	Zn	Pb
	mg kg^{-1}									
BF	8,679	<2	<2	80	1,844	<10	<2	200	36	<4
CTAN	24,034	<2	<2	230	1,300	<10	<2	19,778	66	<4
CTR	5,482	<2	<2	270	10,692	<10	28	14,382	126	<4
CTRAN	353	<2	<2	30	20	<10	<2	12,876	26	<4
CTRP	65,977	<2	<2	564	3,788	<10	<2	35,878	116	<4
MTRAN	2,224	<2	<2	68	80	<10	<2	17,502	20	<4
MTRP	36,604	<2	<2	40	498	<10	<2	24,757	44	<4
NT	643	<2	<2	64	100	<10	<2	13946	28	<4
NTAN	4454	<2	<2	98	1696	<10	<2	200	10	<4
NTRAN	17835	<2	<2	78	570	<10	<2	13725	42	<4
NTRP	16544	<2	<2	128	548	400	<2	17782	22	<4
NTRPP	371	<2	<2	38	60	<10	<2	13157	28	<4

† Treatments described in Table 1.

Table 7. Selected total metal contents of fulvic acid.

Treatment†	Al	Cd	Cr	Cu	Fe	Mn	Ni	P	Zn	Pb
	mg kg^{-1}									
BF	21,861	<8	9,394	140	654	9,523	50	9,291	380	12
CTAN	2,094	<8	200	222	958	323	36	200	174	<2
CTR	8,412	<2	386	1176	3710	2,866	40	7,731	548	20
CTRAN	4,676	<8	1,846	230	832	2,575	156	200	220	<4
CTRP	15,449	<6	9,725	548	2346	10,104	32	23,997	230	14
MTRAN	1,878	<8	200	256	866	400	16	200	128	<6
MTRP	15,404	<8	12,658	530	1414	12,211	114	20,756	254	<4
NT	4,246	<8	2,775	216	1160	3,364	<8	16,378	204	<6
NTAN	13,108	<4	200	132	1480	570	<6	3,228	310	14
NTRAN	3,169	<8	200	54	1506	354	<6	2,003	88	<10
NTRP	4,685	<8	1,638	250	1908	2,735	16	13,714	236	<2
NTRPP	4,410	<8	200	388	4802	1,755	82	8,903	786	28

† Treatments described in Table 1.

attached to or as part of the humic structure (refer to another contribution to this volume for P associated with humic fractions, He et al., 2015).

Fourier Transform Infrared Spectroscopy of Humic and Fulvic Acids

Fourier transform infrared spectroscopy has been used as one of the most versatile analytical techniques to study HA and FA compositions because of its enhanced frequency and accuracy, high signal-to-noise ratios, and high data acquisition speed (Pistorius, 1995). The FT-IR spectra of HA and FA were obtained using KBr discs (He et al., 2006). Figure 1 depicts FT-IR spectra of HA from no-till management soils. These spectra are typical for soil HA fraction (He et al., 2006; Mao et al., 2008) but are different from water-extractable organic matter (He et al., 2009; 2011b; Deng and Che, 2012; Bernier et al., 2013) probably signifying difference in the structural composition between HA in soil and water. The broad band at 3347 to 3291 cm^{-1} was assigned to O-H (stretching of phenolic OH) and N-H stretching. The band at ~2927 cm^{-1} corresponds to aliphatic C-H stretching of CH_2. A prominent peak appearing at 1720 cm^{-1} can be attributed to C=O stretching of COOH and other carbonyl groups. The shoulder peak at ~1624 cm^{-1} was attributed to aromatic C=C stretch or asymmetric COO^- stretch. A weak shoulder band at 1414 cm^{-1} may be attributed to C-H deformation of CH_3 and C-H bending of CH_2. The weak band at ~1376 cm^{-1} is due to symmetric COO^- vibrations or C-H bending of aliphatics. The broad peak around 1222 to 1233 cm^{-1} was attributed C-O stretch and OH deformation of COOH, phenols and aromatic esters. A weak band centered ~1070 to 1032 cm^{-1} is attributed to C-O stretching of polysaccharides (cellulose and residual hemicellulose), Si-O of silicate impurities, or O-Al-OH, O-Fe-OH vibrations. Due to the high concentration of P in our samples

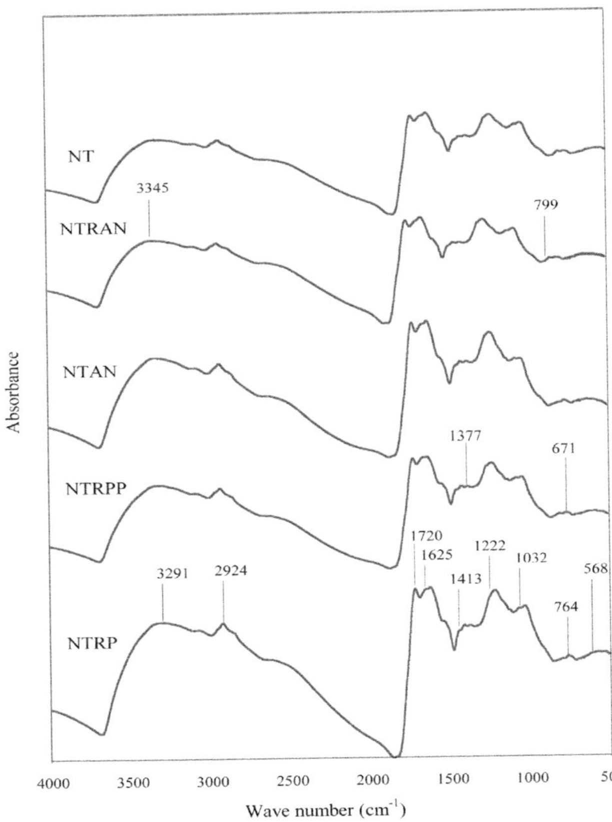

Fig. 1. Fourier transform infrared spectra of humic acids of no-tilled amended soils.

(Table 6 and 7), the bands around this region in this study can also be attributed to phosphate groups (He et al., 2006, 2011a).

To compare the impacts of different management practices, the FT-IR analysis is focused on two areas (He and Ohno, 2012; He et al., 2011a; Waldrip et al., 2014). The first area is ~3020 to 2800 cm^{-1}. The C-H vibration from methyl and methylene groups contributes absorbance around this area, that is, the aliphatic groups in SOM. The second area is ~1740 to 1700 and 1640 to 1600 cm^{-1}. These absorptions can be attributed to C=O from amide and carboxylic acid or ketone compounds. The absorbance at 1740 to 1700 cm^{-1} is due more to the COOH group, and the absorbance at 1640 to 1600 cm^{-1} is due more to the COO ion, amide, and aromatic groups. Based on the relative intensity, the infrared spectra of humic fractions can be classified into three types (He et al., 2011b). The characteristics of Type I spectra are equally strong bands at 1720 and 1600 cm^{-1} with no discernible absorption being evident at 1640 cm^{-1}. Type II spectra possess a very strong 1720 cm^{-1} band, a shoulder at 1650 cm^{-1}, and the absence of a 1600 cm^{-1} band. Type III spectra are similar to Type I with additional strong bands between 2900 and 2840 cm^{-1}. The aliphatic bands between 2900 and 2840 cm^{-1} are also observed, so that they are Type III spectra of HS. The 2900 and 2840 cm^{-1} features of these spectra look similar or identical. However, the impacts of the cropping management can be found in the relative intensity of the two bands at 1720 and 1624 cm^{-1} (Fig. 1; Table 8). Among the 12 HA samples, seven samples show equal FT-IR bands. Three samples show relatively smaller 1740-cm^{-1} bands and two possess a much greater 1640 cm^{-1} band. For example, the two band intensities are about the same for CTAN (conventional-till, cotton [winter fallow], NH_4NO_3) and CTRP (conventional-till, cotton–rye cropping, poultry litter) samples, but the 1720 cm^{-1} band is a little lower than that of 1624-cm^{-1} bands for CTR (conventional-till, cotton–rye cropping, no fertilizer) and BF (control) samples. The 1720-cm^{-1} bands are obviously lower for the CTRAN (conventional-till, cotton–rye cropping, NH_4NO_3) sample. Based on the nature of soil HA, a strong peak at 1624 cm^{-1} may represent more aromatic components in the humic samples.

Table 8. Effects of treatments on the Fourier transform infrared spectroscopy (FT-IR) features of humic acids.

Treatment†	FT-IR feature
BF	1720 >> 1624
CTAN	1720 >> 1624
CTR	1720 < 1624
CTRAN	1720 << 1624
CTRP	1720 >> 1624
MTRAN	1720 >> 1624
MTRP	1720 << 1624
NT	1720 < 1624
NTAN	1720 >> 1624
NTRAN	1720 < 1624
NTRP	1720 >> 1624
NTRPP	1720 >> 1624

† Treatments described in Table 1.

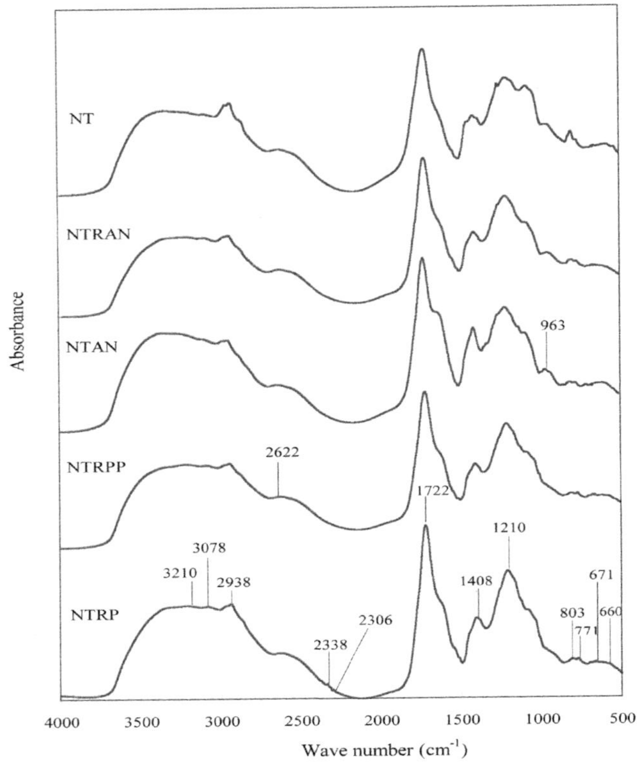

Fig. 2. Fourier transform infrared spectra of fulvic acids of no-tilled amended soils.

The FT-IR spectra of FA samples differ from the spectra of HA samples (Fig. 2). These spectra show a very strong 1720-cm^{-1} band and a weak shoulder at 1640 cm^{-1}, so that they are Type II spectra of HS. Whereas both bands at 1640 and 2900 cm^{-1} are weak in the BF (control) sample, cropping practices increase the intensities of both bands (Table 9). The impact of different management practices can be observed in the 1640-cm^{-1} shoulder and the 2900 cm^{-1} peaks as the CTRAN and CTRP samples show a strong 1640-cm^{-1} shoulder, and the CTAN and CTRAN show a strong 2900-cm^{-1} peak.

Solid-State Carbon-13 Nuclean Magnetic Resonance Spectra of Humic Acids

The ^{13}C NMR spectroscopy were performed on the HA samples only, as there were not enough amounts of the FA samples for the analysis. The spectra of three no-till samples are presented in Fig. 3. The main differences in spectral features between these samples are (i) ketone, quinone, or aldehyde groups, (ii) aromatic lignin group, (iii) NCH–OCH_3 groups, and (iv) nonpolar alkyl groups. Detail information on the ^{13}C NMR spectral peak intensities and their corresponding C

Table 9. Effects of treatments on the Fourier transform infrared spectroscopy (FT-IR) features of fulvic acids.

Treatment†	Band at 1640 cm^{-1}	Band at 2900 cm^{-1}
BF	Weak	Weak
CTAN	Moderate	Strong
CTR	Moderate	Weak
CTRAN	Strong	Strong
CTRP	Strong	Weak
MTRAN	Moderate	Strong
MTRP	Moderate	Strong
NT	Very weak	Very strong
NTAN	Very strong	Moderate
NTRAN	Weak	Moderate
NTRP	Moderate	Strong
NTRPP	Moderate	Moderate

† Treatments described in Table 1.

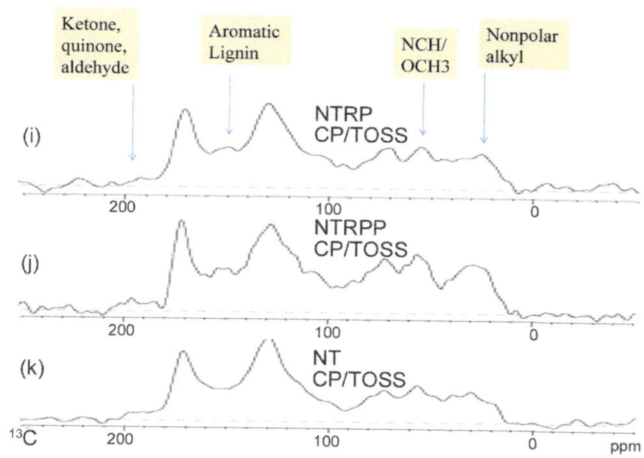

Fig. 3. Solid-state ^{13}C cross-polarization total sideband suppression (CP/TOSS) nuclear magnetic resonance spectra of humic acids of no-tilled soils with cotton and rye cropping +100 (NTRP, curve [i]) or +200 (NTRPP, curve [j]) kg N ha^{-1} from poultry litter or cotton, winter fallow (NT, curve [k]).

functional group assignments of these HA samples is listed in Table 5. Whereas the peak intensities vary among these HA samples, compared with the BF sample, the intensity of the peak in the 190 to 220 ppm of other cropping samples always increased, and the intensity of the peak in the 108 to 145 ppm region of cropping samples consistently lowered. In addition, we measured the relative area intensity in peaks from 100 to 162 ppm and that from 0 to 162 ppm as the ratio of these two intensities can be used as a measurement of aromaticity (He and Mao, 2011, He et al., 2013). Analysis of aromaticity of these HA samples reveals that cropping managements seem to increase aromaticity, compared with the BF sample. However, the higher the fertilizer applied, the lower the aromaticity. This observation may not be due to less aromatic components incorporated into the soil HA samples, but rather to more aliphatic components from crop residues incorporated into the SOM. Previously, based on their ^{13}C NMR observations, Barros Soares et al. (2013) reported that HA and FA from eucalypt plantation soils showed greater contributions of nonpolar alkyl groups and a smaller abundance of O-alkyl groups in comparison to the native vegetation soil, and sugarcane (*Saccharum officinarum* L.) cultivation increased HA aromatics in comparison to those from the native Atlantic Forest soil, but when sugarcane was substituted with eucalypt, the aromatics decreased and O-alkyl C recovered in HA and FA.

Conclusions

Organic matter LF represents a high proportion of total C and N in amended soils. Examination of the characteristics of the LF from Alabama cotton soils (ultisol) of a 9-yr agroecosystem study with different tillage, cropping system, and N source managements shows that tillage can significantly influence the weight and the C, N, and S content of the LF. On average, the C, N, and S content of the LF of the conventionally tilled soils were slightly higher than those of no-till soils in most treatments and almost equal to that of mulch-till soil. The LF has been used as a sensitive and reasonably precise measurement of organic matter changes, but it is just a transient property and only reflects short-term effects. Significant correlations between LF contents and microbial biomass N observed in this study support the fact that LF influences microbial activity.

The elemental composition of HS substances reported here is within the range of those reported in most literature. The H/C ratio of HA ranged from 0.80 to 0.93 while that of FA ranged from 0.82 to 1.15, suggesting that for each C atom, there is indeed one H atom. Differences of FT-IR band intensities were observed among the samples. Generally, the impacts of management practices reflected on the changes of band intensity in 3000 to 2900 cm^{-1}, (C-H of the aliphatic group), 1624 cm^{-1} (aromatic C=C, COO^-, H-bonded C=O of conjugated ketones), 1212 to 1233 cm^{-1} (C-O and O-H of carboxyls, phenols, and aromatic ethers), and 1074 to 1084 cm^{-1} (alcoholic and polysaccharide C-O, SiO_2, and P compounds). Cropping management practices increased the ^{13}C NMR peak at 190 to 220 ppm (ketone, quinone, or aldehyde) and decreased the peak at 108 to 145 ppm (aromatic C-C and C-H) compared with those of the BF (control) HA sample. The aromaticity of HA samples decreased with cropping system and fertilizer applications, suggesting incorporation of aliphatic crop residues to the HAs.

References

Agnelli, A., L. Celi, A. Degl'Innocenti, G. Corti, and F.C. Ugolini. 2000. Chemical and spectroscopic characterization of the humic substances from sandstone-derived rock fragments. Soil Sci. 165:314–327. doi:10.1097/00010694-200004000-00003

Aiken, G.R., D.M. McKnight, R.L. Wershaw, and P. MacCarthy. 1985. An introduction to humic substances in soil, sediment, and water. In: G.R. Aiken et al., editors, Humic substances in soil, sediment, and water: Geochemistry, isolation and characterization. Wiley Interscience, New York. p. 1–9.

Anderson, D.W., and J.J. Schoenau. 1993. Soil humus fractions. In: M.R. Carter, editor, Soil sampling and methods of analysis. Can. Soc. Soil Sci, CRC Press, Boca Raton, FL. p. 391–395.

Assis, C., I. Jucksch, E. Mendonca, J. Neves, L. Silva, and B. Wendling. 2012. Distribution and quality of the organic matter in light and heavy fractions of a Red Latosol under different uses and management practices. Commun. Soil Sci. Plant Anal. 43:835–846. doi:10.1080/00103624.2012.648469

Ayuso, M., J.L. Moreno, T. Hernandez, and C. Garcia. 1997. Characterization and evaluation of humic acids extracted from urban waste as liquid fertilisers. J. Sci. Food Agric. 75:481–488. doi:10.1002/(SICI)1097-0010(199712)75:4<481::AID-JSFA901>3.0.CO;2-K

Barros Soares, E.M., I.R. Silva, R.F. de Novais, Y.Y. Hu, and K. Schmidt-Rohr. 2013. Alterations in molecular composition of humic substances from eucalypt plantation soils assessed by ^{13}C-NMR spectroscopy. Soil Sci. Soc. Am. J. 77:293–306. doi:10.2136/sssaj2011.0070

Bernier, M.H., G.J. Levy, P. Fine, and M. Borisover. 2013. Organic matter composition in soils irrigated with treated wastewater: FT-IR spectroscopic analysis of bulk soil samples. Geoderma 209–210:233–240. doi:10.1016/j.geoderma.2013.06.017

Biederbeck, V.O., H.H. Janzen, C.A. Campbell, and Z.P. Zentner. 1994. Labile soil organic matter as influenced by cropping practices in an arid environment. Soil Biol. Biochem. 26:1647–1656. doi:10.1016/0038-0717(94)90317-4

Bracewell, J.M., G. Abbt-Braun, J.W. De Leeuw, M.H.M. Hayes, H.H. Nimz, E.M. Norwood, E.M. Perdue, M. Schnitzer, S.A. Visser, M.A. Wilson, and W. Zeichmann. 1988. The characterization and validity of structural hypotheses group report. In: F.H. Frimmel and R.F. Christman, editors, Humic substances and their role in the environment. Wiley Interscience, NEW YORK. p. 151–164.

Cambardella, C.A., and E.T. Elliott. 1992. Particulate soil organic matter changes across a grassland cultivation sequence. Soil Sci. Soc. Am. J. 56:777–783. doi:10.2136/sssaj1992.03615995005600030017x

Campbell, C.A., V.O. Biederbeck, R.P. Zentner, and G.P. LaFond. 1991. Effect of crop rotations and cultural practices on soil organic matter, microbial biomass and respiration in a thin Black Chernozem. Can. J. Soil Sci. 71:363–376. doi:10.4141/cjss91-035

Deng, L., and D. Che. 2012. Chemical, electrochemical and spectral characterization of water leachates from biomass. Ind. Eng. Chem. Res. 51:15710–15719. doi:10.1021/ie301468b

Ding, G., J.M. Novak, D. Amarasiriwardena, P.G. Hunt, and B. Xing. 2002. Soil organic matter characteristics as affected by tillage management. Soil Sci. Soc. Am. J. 66:421–429. doi:10.2136/sssaj2002.4210

Dodla, S.K., J.J. Wang, and R.L. Cook. 2012. Molecular composition of humic acids from coastal wetland soils along a salinity gradient. Soil Sci. Soc. Am. J. 76:1592–1605. doi:10.2136/sssaj2011.0346

Francioso, O., C. Ciavatta, V. Tugnoli, S. Sanchez-Cortes, and C. Gessa. 1998. Spectroscopic characterization of pyrophosphate incorporation during extraction of peat humic acids. Soil Sci. Soc. Am. J. 62:181–187. doi:10.2136/sssaj1998.03615995006200010024x

Fynn, R.W.S., R.J. Haynes, and T.G. O' Connor. 2003. Burning causes long-term changes in soil organic matter content of a South African grassland. Soil Biol. Biochem. 35:677–687. doi:10.1016/S0038-0717(03)00054-3

Garcia, C., T. Hernandez, F. Costa, B. Ceccanti, and A. Polo. 1992. A comparative chemical-structural study of fossil humic acids and extracted from carbon waste. Resour. Conserv. Recycl. 6:231–241. doi:10.1016/0921-3449(92)90033-X

Gigliotti, G., A. Macchioni, C. Zuccaccia, P.L. Giusquiana, and D. Businelli. 2003. A spectroscopic study of soil fulvic acid composition after six-year applications of urban waste compost. Agronomie 23:719–724. doi:10.1051/agro:2003048

Gomez, A., and C. Lejuene. 1987. Comparison of the physical and chemical properties of humic acids extracted from a podzolic soil and a mature city refuse compost. In: M. De Bertoldi et al., editors, Compost: Production, quality and use. Elsevier Appl. Sci. Publ., London, UK. p. 495–500.

Gosling, P., N. Parsons, and G.D. Bending. 2013. What are the primary factors controlling the light fraction and particulate soil organic matter content of agricultural soils? Biol. Fertil. Soils 49:1001–1014. doi:10.1007/s00374-013-0791-9

Grandy, A.S., G.A. Porter, and M.S. Erich. 2002. Organic amendment and rotation crop effects on the recovery of soil organic matter and aggregation in potato cropping systems. Soil Sci. Soc. Am. J. 66:1311–1319. doi:10.2136/sssaj2002.1311

Gregorich, E.G., and B.H. Ellert. 1993. Light fraction and macroorganic matter in mineral soils. In: M.R. Carter, editor, Soil sampling and methods of analysis. CRC Press, Boca Raton, FL. p. 397–407.

He, Z., X. Cao, J. Mao, T. Ohno, and H.M. Waldrip. 2013. Analysis of carbon functional groups in mobile humic acid and recalcitrant humate extracted from eight US soils. Pedosphere 23:705–716. doi:10.1016/S1002-0160(13)60063-6

He, Z., C. Du, and J. Zhou. 2011a. Structural and bonding environments derived from infrared spectroscopic studies. In: Z. He, editor, Environmental chemistry of animal manure. Nova Science Publishers, NEW YORK. p. 23–42.

He, Z., C.W. Honeycutt, and H. Zhang. 2011b. Elemental and Fourier transform infrared spectroscopic analysis of water and pyrophosphate extracted soil organic matter. Soil Sci. 176:183–189. doi:10.1097/SS.0b013e318212865c

He, Z., and J. Mao. 2011. Functional groups identified by solid state ^{13}C NMR spectroscopy. In: Z. He, editor, Environmental chemistry of animal manure. Nova Science Publishers, NEW YORK. p. 41–59.

He, Z., J. Mao, C.W. Honeycutt, T. Ohno, J.F. Hunt, and B.J. Cade-Menun. 2009. Characterization of plant-derived water extractable organic matter by multiple spectroscopic techniques. Biol. Fertil. Soils 45:609–616. doi:10.1007/s00374-009-0369-8

He, Z., and T. Ohno. 2012. Fourier transform infrared and fluorescence spectral features of organic matter in conventional and organic dairy manure. J. Environ. Qual. 41:911–919. doi:10.2134/jeq2011.0226

He, Z., T. Ohno, B.J. Cade-Menun, M.S. Erich, and C.W. Honeycutt. 2006. Spectral and chemical characterization of phosphate associated with humic substances. Soil Sci. Soc. Am. J. 70:1741–1751. doi:10.2136/sssaj2006.0030

He, Z., I.A. Tazisong, and Z.N. Senwo. 2015. Forms and lability of phosphorus in humic and fulvic acids. In: Z. He and F. Wu, editors, Labile organic matter—Chemical compositions, function, and significance in soil and the environment. SSSA Spec. Publ. 62. SSSSA, Madison, WI. doi:10.2136/sssaspecpub62.2014.0035

Hernandez, M.T., J.I. Moreno, and F. Costa. 1993. Infrared spectroscopic characterization of sewage sludge humic acids. Evidence of sludge organic matter–metal interactions. Agrochimica 37:12–17.

Janzen, H.H., C.A. Campbell, S.A. Brandt, G.P. Lafond, and L. Townley-Smith. 1992. Light-fraction organic matter in soils from long-term crop rotations. Soil Sci. Soc. Am. J. 56:1799–1806. doi:10.2136/sssaj1992.03615995005600060025x

Mao, J., D.C. Olk, X. Fang, Z. He, J. Bass, and K. Schmidt-Rohr. 2008. Influence of animal manure application on the chemical structures of soil organic matter as investigated by advanced solid-state NMR and FT-IR. Geoderma 146:353–362. doi:10.1016/j.geoderma.2008.06.003

Mayhew, L. 2004. Humic substances in biological agriculture. ACRES (USA), A voice for eco-agriculture. 34.

McDonnell, R., N.M. Holden, S.M. Ward, J.F. Collins, E.P. Farrell, and M.H.B. Hayes. 2001. Characteristics of humic substances in heathland and forested peat soils of the Wicklow Mountains. Biology and Environment: Proceedings of the Royal Irish Academy. 101B:187–197.

Nascente, A.S., Y.C. Li, and C.A.C. Crusciol. 2013. Cover crops and no-till effects on physical fractions of soil organic matter. Soil Tillage Res. 130:52–57. doi:10.1016/j.still.2013.02.008

Olk, D.C., G. Brunetti, and N. Senesi. 1999. Organic matter in double-cropped lowland rice soils: Chemical and spectroscopic properties. Soil Sci. 164:633–649. doi:10.1097/00010694-199909000-00003

Parton, W.J., D.S. Schimel, C.V. Cole, and D.S. Ojima. 1987. Analysis of factors controlling soil organic matter levels in Great Plains grasslands. Soil Sci. Soc. Am. J. 51:1173–1179. doi:10.2136/sssaj1987.03615995005100050015x

Pena-Mendez, E.M., J. Havel, and J. Patocka. 2005. Humic substances—compounds of still unknown structure: Application in agriculture, industry, environment, and biomedicine. J. Appl. Biomed. 3:13–24.

Pistorius, A.M.A. 1995. Biochemical applications of FT-IR spectroscopy. Spectrosc. Eur. 7:8–14.

Reintam, L., J. Kann, T. Kailas, and R. Kahrik. 2000. Elemental composition of humic and fulvic acids in the epipedon of some Estonian soils. Proc. Estonian Acad. Sci. Chem. 49:131–144.

Rice, J.A., and P. MacCarthy. 1991. Statistical evaluation of the elemental composition of humic substances. Org. Geochem. 17:635–648. doi:10.1016/0146-6380(91)90006-6

Senwo, Z.N., and I.A. Tazisong. 2004. Metal contents in soil of Alabama. Commun. Soil Sci. Plant Anal. 35:2837–2841. doi:10.1081/CSS-200036470

Song, J., P. Ping, and W. Huang. 2005. Characterization of humic acid-like material isolated from the humin fraction of a topsoil. Soil Sci. 170:599–611. doi:10.1097/01.ss0000180065.91004.19

Steelink, C. 1985. Implications of elemental characteristics of humic substances. In: G.R. Aiken et al., editors, Humic substances in soil, sediment, and water. Geochemistry, isolation, and characterization. Wiley Interscience, New York. p. 457–476.

Stevenson, F.J. 1982. Humus chemistry: Genesis, composition, reactions. Wiley-Interscience, New York.

Swift, R.S. 1996. Organic matter characterization. In: D.L. Sparks et al., editors, Method of soil analysis. Part 3. Chemical method. Soil Sci. Soc. Am. Book series. SSSA, Madison, WI. p. 1018–1020.

Tan, K.H. 1998. Principles of soil chemistry, 3rd ed. Marcel Dekker, Inc., New York.

Tan, K.H. 2003. Humic matter in soil and the environment. Marcel Dekker, Inc., New York.

Tazisong, I.A., Z. He, and Z.N. Senwo. 2013. Inorganic and enzymatically hydrolyzable organic phosphorus of Alabama Decatur silt loam soils cropped with upland cotton. Soil Sci. 178:231–239. doi:10.1097/SS.0b013e31829cf464

Tazisong, I.A., Z.N. Senwo, B.J. Cade-Menun, and Z. He. 2014. Phosphorus forms and mineralization potentials of Alabama upland cotton production soils amended with poultry litter. In: Z. He and H. Zhang, editors, Applied manure and nutrient chemistry for sustainable agriculture and environment. Springer, Amsterdam, the Netherlands. p. 191–209.

Thurman, E.M. 1985. Organic geochemistry of natural waters. Nijhoff–Junk Publishers, Dordrecht, The Netherlands.

Ussiri, D.A.N., and E. Johnson. 2003. Characterization of organic matter in a northern hardwood forest soil by ^{13}C NMR spectroscopy and chemical methods. Geoderma 111:123–149. doi:10.1016/S0016-7061(02)00257-4

Waldrip, H.M., Z. He, R.W. Todd, J.F. Hunt, M.B. Rhoades, and N.A. Cole. 2014. Characterization of organic matter in beef feedyard manure by UV–visible and Fourier transform infrared spectroscopies. J. Environ. Qual. 43:690–700. doi:10.2134/jeq2013.09.0358

Wander, M. 2004. Soil organic matter fraction and their relevance to soil function. In: F. Magdoff and R.R. Weil, editors, Soil organic matter in sustainable agriculture. CRC Press LLC, Boca Raton, FL. p. 67–102.

Wander, M.M., and S.J. Traina. 1996. Organic matter fractions from organically and conventionally managed soils: I. Carbon and nitrogen distribution. Soil Sci. Soc. Am. J. 60:1081–1087. doi:10.2136/sssaj1996.03615995006000040017x

Wander, M.M., S.J. Traina, B.R. Stinner, and S.E. Peters. 1994. The effects of organic and conventional management on biologically active soil organic matter pools. Soil Sci. Soc. Am. J. 58:1130–1139. doi:10.2136/sssaj1994.03615995005800040018x

Wu, T., J.J. Schoenau, F. Li, P. Qian, S.S. Malhi, and Y. Shi. 2005. Influence of fertilizer and organic amendments on organic-carbon fractions in Heilu soil on the loess plateau of China. J. Plant Nutr. Soil Sci. 168:100–107. doi:10.1002/jpln.200321295

Zech, W., F. Zeigler, I. Kogel-Knabner, and L. Haumaier. 1992. Humic substances distribution and transformation in forest soils. Sci. Total Environ. 117–118:155–174. doi:10.1016/0048-9697(92)90084-6

Zhang, M., and Z. He. 2004. Long-term changes in organic carbon and nutrients of an ultisol under rice cropping in southeast China. Geoderma 118:167–179. doi:10.1016/S0016-7061(03)00191-5

Zhang, M., Z. He, and A. Zhao. 2011. Ultraviolet–visible absorptive features of water extractable and humic fractions of animal manure and relevant compost. In: Z. He, editor, Environmental chemistry of animal manure. Nova Science Publishers, New York. p. 61–81.

Labile Organic Matter in Soil Solution: I. Metabolites of Chemical Signaling Pathways from Plant–Microbe Interactions

Carlos M. Monreal

Abstract

Labile organic matter in the soil solution of crop rhizospheres includes hundreds of small organic molecules, but little is known about their origins and ecological functions. This chapter presents a functional model for pools of labile soil organic matter on the basis of extracellular metabolites resulting from chemical signaling pathways of plant roots and soil microbe interactions in rhizospheres. The chemistry and dynamics of labile organic matter in soil solutions are closely associated with metabolic products of planned cellular responses during the main stages of chemical communication that include signal recognition, transduction, and transcription. The composition and dynamics of labile organic matter in soil solutions are important metabolites associated with the genome, transcriptome, proteome, and metabolome of living cells interacting continuously in the rhizospheres of soil–crop systems. Identifying the metabolites and signaling components resulting from root–microbe communication is essential for determining ecological function of compounds in labile soil organic matter pools in support of a sustainable agriculture.

Abbreviations: AHL, N-acyl homoserine lactone; AI-2, autoinducer-2; AIP, autoinducing peptide; AR, adhesion receptor; GPCR, G-protein coupled receptor; QS, quorum-sensing; ROS, reactive oxygen species.

Carlos M. Monreal, Eastern Cereal and Oilseed Research Center, Agriculture and Agri-Food Canada, 960 Carling Ave., Ottawa, Canada, K1A 0C6 (carlos.monreal@agr.gc.ca).

doi:10.2136/sssaspecpub62.2014.0073

Labile Organic Matter—Chemical Compositions, Function, and Significance in Soil and the Environment. SSSA Special Publication 62. Zhongqi He and Fengchang Wu, editors.

Temporal and spatial relationships are established between physical, chemical, and biological domains in soil–plant systems. In agroecosystems, interactions between crop roots, soil living organisms, and abiotic soil components are of increasing complexity, due in part to root–soil aggregate architecture; the chemical, physical, and biological heterogeneity of existing plant and soil components and their properties; and the influence of human interventions (Monreal and Schnitzer, 2013).

In rhizospheres, roots interact with other roots, soil solutions, soil atmosphere, soil fauna and microorganisms, humic substances, and inorganic colloids. Root interactions are dynamic and have neutral, antagonistic, beneficial, or competitor effects (Hirsch et al., 2003). Many of the organic compounds in soil solutions are secondary plant metabolites that represent key weapons in the continuous struggle of plant and soil microorganisms for ensuring vital space and access to nutrients and energy resources (Monreal and Schnitzer, 2013). The variability in response to interactions between plants and soil microorganisms can vary considerably depending on whether they are observed at the ecological community or whole organism levels or whether events are considered at the subcellular, biochemical, chemical, and DNA molecular levels (Korth and Thompson, 2006).

In vivo soils, the interactions of plants with living soil organisms, are influenced, to a large extent, by plant root exudates. Research advances made on roots has helped understand their development in relation to anthropogenic influences and some of their impact on soil environments. New molecular and genetic tools have also permitted researchers to study the expression of specific root genes or soil microbial genes (Xia et al., 2011). The reader is referred to extensive reviews published in the past decade on the various interactions of root exudates with soil organisms and their resulting effects by Bais et al. (2008) and Berg (2009).

The interactions between living cells and species use chemical and physical communication to accomplish many of their vital functions in the ecosystem. Advances in the understanding of chemical communications between plants and microorganisms and between species in terrestrial ecosystems has been limited due to inherent complexities of soil–plant systems and lack of adequate methods and analytical tools. In heterogeneous soil–plant systems, little is known about the ecological function of soluble organic components associated with chemical signaling between roots and soil microorganisms responding to both intraspecific and interspecific signals.

The main aim of this chapter is to give an overview of the influence that chemical signaling components of interacting plants and microbial cells have on the chemical composition and dynamics of labile organic compounds in soil solutions. This chapter provides an overview of chemical communication signaling pathways for the application of such knowledge in studies interpreting the significance of metabolites in the solution of terrestrial ecosystems. A model framework is presented for better understanding and interpreting ecological functions of metabolites in soil solutions.

Communication Signaling

Under the soil surface, plant roots and active microbial components need to obtain basic resources to survive, grow, reproduce, and evolve. The underground signals may be abiotic (e.g., pH, temperature, and osmolarity), biotic, or both. Biotic chemical communication is the earliest language that enabled life, biological reproduction, evolution, and diversity on earth. The communication between plants and other organisms has recently been accepted as a true ecological phenomenon (Adler, 2011).

During chemical communication, there is usually a movement and diffusion of small chemical molecules from senders to receivers. Single cells and multicellular organisms send, receive, and process chemical messages from other neighboring cells and organisms by integrating signals into a unified action plant that ultimately responds to changes in their surrounding environments (Nealson and Hastings, 1979). A signaling molecule is considered to act at low concentrations and is believed not to be involved in primary metabolism (Keller and Surette, 2006). In nature, communication exists among, within, and between animals, insects, plants, and microorganisms (Witzany, 2010). Secreted chemical signals, their diffusion, and signal detection components facilitate highly productive intra- and interspecies relationships (Bassler, 2002).

Communications are signaling mediated interactions and are not simply an exchange of information. Many chemical molecules can function as signals, messenger substances, information carriers, and memory medium in solid, liquid, or gaseous form (Witzany, 2006). In living cells, a group of individual signals complement and modify other signals to form signal sequences, much like words in sentences. The molecular syntax of molecular languages and codes determines the correct sequence and combination of signal molecules. Disrupting or deforming the syntactic rules can cause incomplete transmission of the message, triggering faulty interpretations and responses in the receiver (Witzany, 2006). In multicellular organisms, growth factors, hormones, neurotransmitters, and extracellular matrix components are some of the many types of chemical signals cells use. If necessary, chemical signals can travel short or long distances to reach their cell receptor targets.

Besides the communication between plants and soil microorganisms, further interactions between diverse populations of soil microorganisms and invertebrates increase the complexity of communication in ecosystems. Microorganisms such as soil-dwelling bacteria and fungi produce a multitude of low molecular weight molecules. Many of these compounds are produced as chemical signals or for defending the surrounding habitat, however, their true functions in their biological context are largely unknown (Yim et al., 2007). In nature, microorganisms can sense signal molecules that are produced by other organisms in their immediate environment. The identity of organic metabolites from chemical signaling pathways and cellular responses by plants and soil microorganisms to adapt to their changing environment remain largely underexplored. In the meantime, the repertoire of labile organic molecules in soils that may be associated with signaling systems continues to grow (Monreal and Schnitzer, 2013).

Communication in Eukaryotes—Plants

An intricate plant signaling network coordinates critical processes during their life cycle: photosynthesis, dark respiration and photorespiration, activities linked in terms of electron transfer, substrate, reducing power, and energy (Pfannschmidt et al., 2009). Plants are continuously challenged by biotic and abiotic stresses, such as heterotrophic organisms, which depend on the energy and carbon fixed by autotrophic plants. Plant communication occurs (i) between different plant species as well as between members of the same plant species (ii) between cells and in cells of the plant organism; and (iii) between plants and microorganisms, fungi, insects, and other animals (Witzany, 2006).

Plants use a number of strategies to communicate. Under biotic threat, plants recognize attackers with a high degree of sophistication in combination with intracellular signaling (Maffei et al., 2007). The latter signals are converted into appropriate biochemical, physiological, and cellular responses established locally on the site of infestation as well as systemically throughout the whole plant (Mithöfer and Boland, 2012). Most of the communication strategies are based on the chemical diversity of organic and inorganic compounds synthesized by plant cells or taken up from soils or the atmosphere. Plants have the ability to synthesize more than 200,000 compounds, or specialized metabolites, that have evolved in response to particular ecological challenges (Pichersky and Lewinsohn, 2011). These specialized metabolites include cyanogenic glycosides, glucosinolates, alkaloids, terpenoids, and many others, such as latex or proteinase inhibitors. Their modes of action include membrane disruption, inhibition of nutrient and ion transport, inhibition of signal transduction processes, inhibition of metabolism, or disruption of the hormonal control of physiological processes (Mithöfer and Boland, 2012).

An example of chemical signaling used by plants under stress include pathways that try to prevent increased production of reactive oxygen species (ROS) (Suzuki et al., 2012). In plants and other multicellular organisms, endocytosis and endosomes are important for auxin-mediated cell-to-cell communication as well as for gravitropic responses, stomatal movements, cytokinesis, and cell wall morphogenesis (Šamaj et al., 2005). Synaptic cell-to-cell communication is based on rapid endocytosis and vesicular recycling in plants (Baluska et al., 2005). Other examples of communication in plants involve organelles that synthesize and store chemical compounds in cells (i.e., plastids) and chemical signals that regulate nuclear gene expression to reorganize macromolecules in response to environmental influences (Strand, 2004).

The cellular organization of roots is determined during the plant's embryonic development and is controlled by intercellular communication via local signaling centers and the complex relationship between numerous different signaling pathways (Bonke et al., 2005). Communication in plants and roots requires and uses microscopic channels in the cell walls to communicate intracellularly (Tassetto et al., 2005).

Communication in Prokaryotes—Bacteria

Bacteria use quorum-sensing (QS) communication systems for cell-to-cell communication (Taga and Bassier, 2003). Bacteria use a wide variety of organic signaling molecules, signal-detection systems, and signal-transduction mechanisms to convert the information contained in the signal into changes in gene regulation at some threshold signal concentration (Lazdunski et al., 2004). Quorum sensing has been described in different bacterial species (Taga and Bassier, 2003). The QS system is a diffusion-based communication system that allows entire bacterial populations to collectively control gene expression and synchronize group behavior (Waters and Bassler, 2005). Bacterial interactions through QS chemicals involve cooperative signals and other interactions such as cues and chemical manipulations. Chemical signals can also be involved in conflicts within and between species (Keller and Surette, 2006). For further reading on QS systems, the reader is referred to an article by Kolenbrander et al. (2005).

Three well-defined classes of molecules are used for chemical cell-to-cell signaling in bacteria: oligopeptides, N-acyl homoserine lactone (AHLs), and the LuxS/autoinducer-2 (AI-2) class (Keller and Surette, 2006). The oligopeptide signaling systems are highly specific and used predominantly by Gram-positive bacteria (Ji et al., 1997). The active signaling peptide is exported from the cell (Lyon and Novick, 2004). The metabolic cost of the oligopeptide signal production is high (i.e., 184 ATP). The chemical structure of the signal is defined by its amino acid sequence, which can be modified to give more complex peptide structures, such as glycoproteins. The peptide signaling has a role in intraspecific competition, as well as in coordinated, cooperative behavior.

The AHL signaling system is produced in Gram-negative bacteria (Lazdunski et al., 2004). In these bacteria, R- and I-proteins are involved in the regulation of QS (Fuqua et al., 1996). The R-protein is a transcriptional regulator and the I-protein is the enzyme that synthesizes the signaling molecule (AHL). The production of the AHL signal system has in intermediate energy cost (i.e., 8 ATP) (Keller and Surette, 2006). The specificity of the AHL signaling system is moderate. In general, the AHL signal is detected by a member of the LuxR family of transcriptional regulators. In some cases, the LuxR protein will recognize a wide range of AHLs. An example involves *Rhizobium leguminosarum* biovar *viciae*, which has a QS system with five R-proteins, four I-proteins, and a cocktail of six different acyl-HSL molecules that are used for signaling (González and Marketon, 2003). The frequency distribution of AHL-based signaling within bacteria challenged the widespread view that QS is very common (Manefield and Turner, 2002).

The autoinducer LuxS/AI-2 pathway signaling system is found in many Gram-negative and Gram-positive bacteria (Vendeville et al., 2005). Autoinducers are molecules responsible for positive-feedback regulatory mechanisms and enabling intraspecies communication. For interspecies communication, the autoinducer AI-2 has been proposed as a universal signal (Xavier and Bassler, 2003). In this signaling system, the chemical signal appears to be 4,5-dihydroxy-2,3-petanedione, a molecule that is in chemical equilibrium with several furanones (Schauder et al., 2001). This pathway appears to lack specificity and so cannot convey precise information. This autoinducer pathway is associated with the lowest cost of ATP production (0–1 ATP) (Keller and Surette, 2006). The chemical signal is generated from the degradation of the key metabolic compound *S*-adenosylhomocysteine (also involved in the production of AHLs), which was considered a primary metabolite rather than a signal involved in cell-to-cell communication (Vendeville et al., 2005).

Besides the above three chemical signaling systems in bacteria, intimate physical bacterial–fungal interactions also contribute to the communication of microorganisms in nature. For example, intimate cocultivation of the fungus *Aspegillus nidulans* with dozens of soil-dwelling actinomycetes was necessary to trigger the activation of fungal genes controlling secondary metabolism, such as in the biosynthesis of secondary archetypal polyketides in *A. nidulans* (Schroeckh et al., 2009). The vast majority of publications have focused on the molecular aspects of cell-to-cell communication, with little attention paid to the ecological context of why bacteria and plants produce signaling molecules and respond to both intraspecific and interspecific signals.

Stages of Chemical Signaling

Chemical signaling between living cells involves the following steps: recognition, transduction, transcription, and response (Fig. 1). During recognition and after signal binding, most receptors transmit the signal through a sequence of molecular switches to internal signaling metabolic pathways that, in the end, result in a cell response to the environmental disturbance. For in-depth reviews of the four stages, the reader is directed to published books and selected research and review articles (Taiz and Zeiger, 2010).

Recognition of Chemical Signals

In any environment, chemical signals (or ligands) are recognized by protein receptors that bind to signaling molecules mostly on the surface of cell membranes and initiate a cell's physiological response (Pedersen and Ross, 1982). The detection of chemical signals by cell receptors produce electrical impulses to transmit the information through sensory systems within the living cells. Receptors undergoing conformational changes on signal binding induce specific biochemical intracellular signaling pathways (i.e., signal transduction) and amplify the message, producing multiple intracellular signals for every one receptor that is bound (Lefkowitz, 2007).

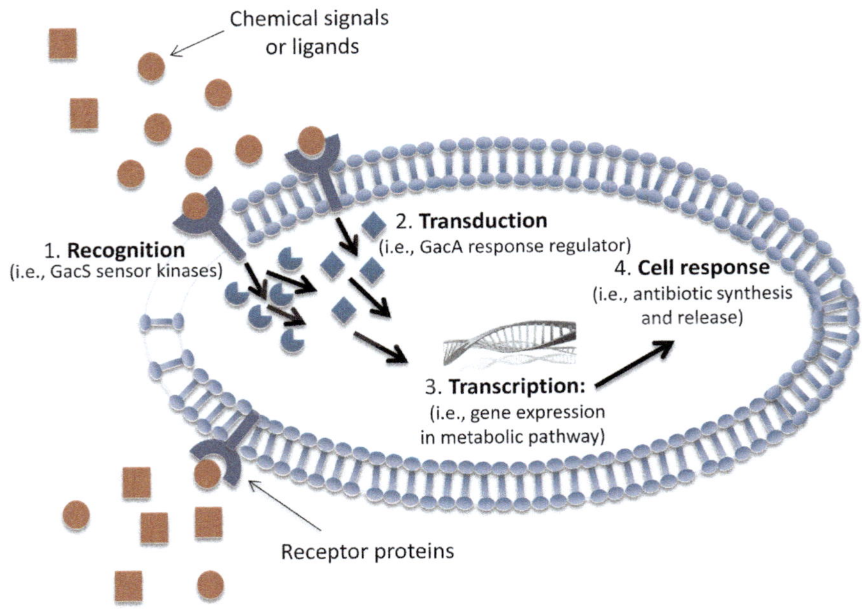

Fig. 1. Simplified diagram depicting the main components and steps of chemical signaling in a living cell.

The receptor proteins were first noted as receptive substances on cell tissue surfaces of eukaryotes (Dale, 1906). Work conducted in the 1970s had confirmed the existence of membrane receptors (Lefkowitz, 2007). Receptors can also be located inside the cell to bind molecules that pass through the membrane, such as nitrous oxide, ammonia, or hormones like estrogen. Decadal research efforts have helped to elucidate, in part, the molecular nature of chemical signal-receptor systems in eukaryotes and prokaryots. The 2012 Nobel Prize in Chemistry was awarded to B. Kobilka and R. Lefkowitz for having mapped how a family of G-protein coupled receptors (GPCRs) works in eukaryotes.

Each receptor has high affinity for a small set of similar agonists that bind to a specific site in its extracellular part. Agonists are ligands or signals that produce a biological response, and those that block the action are antagonists (Rosenbaum et al., 2011). In eukaryotes, binding of agonists favors the active receptor conformation and increases the affinity for the G-protein on the inside of the cell. Riboswitches constitute another family of receptors that are used to sense compounds fundamental to nearly all living organisms (Winkler et al., 2004). Riboswitches bind specifically to microbial metabolites and regulate cell gene expression without the need for protein factors.

Receptors in Plants and Bacteria

In plants and animals, signal receptors are called G-proteins due to their guanine nucleotide-binding proteins (Urano et al., 2013). In eukaryotes, the G-receptors encompass a large family of membrane proteins that are activated by a spectrum of structurally diverse ligands and have been shown to modulate the activity of different signaling pathways in a ligand specific manner. G-proteins have a high level of structural diversity and different chain lengths of amino acids in plants and animals (Trusov et al., 2012). In plants, G-proteins transmit signals by atypical mechanisms and effector proteins to control growth, cell proliferation, defense, stomate movements, channel regulation, sugar sensing, and some hormonal responses (Urano et al., 2013). The structure and activation of GPCRs have been reviewed in detail by Kobilka (2007).

Gaining knowledge on the mechanistic GPCR activity that detects the chemical stimuli in the environment and relays this information within the cell has been represented mathematically through a two-component signal system, where the receptor is in equilibrium between a single inactive state and a single active state (Galperin, 2005). The two-component signaling system is the most widely used signaling paradigm in living organisms (Wright et al., 2004). The two-component system, however, did not provide mechanistic details about events occurring at the level of the receptor; that is, the early two-component models are ambiguous as to what the agonist is doing to the receptor molecule to generate the cellular response (Park, 2012). Recent observations, however, indicate that the GPCRs can function via multiple active conformational substates rather than a single active state (Park 2012). For example, it has now been shown that G-proteins exhibit some promiscuity and a single GPCRs can couple to and signal via G-proteins from multiple classes to propagate the signals through multiple biochemical pathways to achieve different cellular responses instead of a single metabolic pathway and single cellular response as done by the two-model system (Park, 2012; Park and Ronald, 2012).

Bacterial receptors are transmembrane proteins that involve more than a dozen related families of periplasmic binding proteins. Despite the diversity of chemical signals or ligands, these proteins have virtually identical backbone structures surrounding the ligand-binding cleft, implying that residue side chains within the binding pocket are critical determinants of specificity (Johnson and Church, 2000). In Gram-positive bacteria, autoinducing peptides (AIPs) regulate virulence, the induction of competence for transformation, and the production of bacteriocins (Novick and Muir, 1999). In Gram-negative bacteria, AIPs are *N*-acyl homoserine lactones, which are freely diffusible and interact directly with intracellular regulatory proteins. In Gram-positive eubacteria, the known AIPs are peptides processed from larger propeptides. The receptors are ubiquitous in the bacterial cells, this is, they may be bound to the cytoplasmic membrane but concentrated at the poles of the cell instead of evenly distributed throughout the cytoplasmic membrane (Rayo et al., 2011).

Transduction of Chemical Signals

Signal transduction refers to how chemical signals are received and interpreted and the ways in which signals are used to respond to environmental cues (Alberts et al., 2002). Different pools of receptors act as integral parts for downstream signaling transduction cascades (von Zastrow and Sorkin, 2007). Much of the ongoing research on signal transduction focuses on elucidating the structures and mechanisms of action of functional chemical signal–receptor systems. For further reading on transduction processes in living cells, the reader is referred to an in-depth treatment of the theory of biocommunication among cells, tissues, and organisms in bacteria, fungi and plants in Witzany (2010).

Transduction in Plants and Bacteria

In plants, adhesion receptors (ARs) appear to generate and transduce chemical signals. The ARs contribute to cell proliferation, differentiation, and survival (Legate et al., 2009). Adhesion receptors, contribute to the onset of developmental genetic programs that occur synchronously (Pardi, 2010) by assembling large molecular networks at adhesion sites, behaving as scaffolds for many intermediates and effectors in major signaling pathways (Wang et al., 2009). The adhesion sites contain either integrins (in areas called focal adhesions) or cadherins proteins (at adherence junctions) (Maher et al., 1985). Integrins are protein receptors that pass information about the chemical composition and mechanical status of the extracellular matrix. Cadherins play important roles in cell adhesion forming adherence junctions to bind cells and are dependent on calcium to function.

In plants, a cascade reaction on the inside of the cell starts with nucleotide exchange and dissociation of the G-protein into three subunits (α, β, and γ) (Ross and Gilman, 1980). The α- subunit binds to and stimulates enzymes, such as adenylate cyclase, producing the cyclic nucleotide cyclic adenosine monophosphate, which diffuses easily and acts as a secondary messenger. Other proteins may interact with the β and γ subunits to further modulate the signal. The activated G-protein conformation lasts long enough to allow one bound agonist molecule to activate several G-proteins, which amplify the signal (Pedersen and Ross, 1982). The complex receptor–signal complex turns over very fast during cell growth, allowing G-proteins to reassociate after nucleotide hydrolysis and permit signal integration within minutes to produce the full cellular response (Plopper et al., 1995). Progress has been made in the identification and localization of immediate downstream signaling components of activated receptors during transduction in plants (Geldner and Robatzek, 2008), such as the role for the plant hormone abscisic acid in controlling the closure of stomatal pores during drought (Li et al., 2006).

In bacteria, the two-component signal system consists of a sensor kinase and a cognate response regulator. Gram-negative bacteria have the sensor kinase GacS and response regulator GacA. In plant-beneficial bacteria, the GacS–GacA system controls the expression of biocontrol factors that protect plants from pathogenic fungi (Heeb and Haas, 2001). The GacS–GacA system is also involved in the modulation of bacterial QS system and the expression of extracellular products (Kay et al., 2006). Signaling pathway cascades might sometimes have evolved for reasons other than cell-to-cell signaling or might have acquired new functions once they were established (Keller and Surette, 2006).

A census of signal-transduction proteins in bacterial and archaeal genomes showed uneven distribution of most signaling proteins among bacterial and archaeal phyla and that transduction proteins grew approximately as the square of genome size. While histidine kinases were found in representatives of all phyla and were distributed according to the power law, other signal transducers were abundant in certain phylogenetic groups but virtually absent in others (Galperin, 2005). The information on signaling systems is poorly represented in public databases. For example, while histidine kinases and response regulators are covered in the KEGG database (Kanehisa et al., 2004), other signaling systems are not (Galperin, 2005).

Transcription of Chemical Signals

Processes of transcription and translation are necessary for the expression of genes responsible for the synthesis of proteins and products that satisfy the physiological response planned by the living cells cell to organic compounds acting as chemical signals in soils. In transcription, a portion of the double-stranded DNA template gives rise to a single-stranded RNA molecule that can serve as a finished product to accomplish an important function within the cell. Often, the transcription of an RNA molecule is followed by a translation step, which ultimately results in the production of a protein molecule (Clancy, 2008). In translation, a protein is synthesized from the information contained in a molecule of messenger RNA. The transcription process has proofreading mechanisms that are less effective than the controls for copying DNA, resulting in a lower copying fidelity than DNA replication (Berg et al., 2006).

In transcription, a specific segment of the DNA (promoter region) is copied by the enzyme RNA polymerase into an RNA transcription unit that encodes for at least one or more genes. The RNA polymerase binds to the DNA promoter region in the presence of proteins (transcription factors) that signals which gene is to be transcribed, thus controlling the flow of genetic information from DNA to messenger RNA (Latchman, 1997). Eukaryotic cells have three different types of RNA polymerase, whereas bacteria have only one. In addition, bacterial promoters are usually composed of three sequence elements, whereas in eukaryotes, there are as many as seven elements (Clancy, 2008). During transcription, the rate of nucleotide addition ranges from about 15 to 20 amino acids per second in bacteria (Dennis and Bremer, 1974) and about five to eight amino acids per second in eukaryotes (Izban and Luse, 1992). In root radial and epidermal cell patterning, as well as in shoot organogenesis, the direct transmission of transcription factors in cell-to-cell communication plays a central role (Kurata et al., 2005).

The transcription initiation on protein-encoding genes represents a major control point for gene expression. Other proteins that participate in transcription that do not have DNA-binding domains include, among other, coactivators, acetylases, deacetylases, kinases, and methylases, which also play essential roles in gene regulation (Brivanlou and Darnell, 2002). For a detailed discussion of the role of coactivators in transcription and transcription step, the reader is referred to publications by Taiz and Zeiger (2010).

The termination of the transcription is caused by information stored in a sequence of DNA nucleotides that are found close to the ends of coding sequences. Bacteria have two types of the terminator sequences, and eukaryotes use different processes for the termination of transcription depending on the exact polymerase utilized (Clancy, 2008). On-going research in the area of prokaryotic and eukaryotic transcription has a large focus on unraveling the molecular details of this complex process (Nabavi and Nazar, 2008).

The cellular response by plants or bacteria will vary according to stressors or effectors exerted by abiotic and biotic factors. Therefore, many parallel signaling processes can take place at the same time within living cells of eukaryotes and prokaryotes. The latter results in concomitant responses that may see the action of many signaling components in cells and excreted metabolites that will affect the chemical composition of labile organic compounds found in soil solutions of crop rhizospheres.

Signaling in Soil–Plant Systems under Environmental Stress

To a large extent, the chemical composition of soil solutions reflects anthropogenic disturbances plus past and present interactions of plant and microorganisms associated with chemical signaling pathways. For example, researchers have uncovered a number of distinct interconnected signaling networks that are under both positive and negative control. These pathways operate, at least partly, through the action of small signaling molecules such as salicylate, jasmonate, and ethylene. The interplay of signals allows the plant to fine-tune defense responses in both local and systemic tissue (Feys and Parker, 2000). Plants and microorganisms also evoke local and systemic signaling systems through sensing of nutrient status or toxic levels of essential nutrients in the surrounding environment (Liu et al., 2014).

Roots are able to respond via changes in morphology and architecture and through molecular signaling components to local deficiencies or high levels of P (Niu et al., 2013) and N (Peng et al., 2008). Secondary messengers in many signal transduction pathways, such as ROS, play a role in plant P and N sensing and signaling (Nanjareddy et al., 2014). The ROS components have been shown to increase in N-, P-, and K-deprived roots (Shin et al., 2005). Under N deficiency, sorghum [*Sorghum bicolor* (L.) Moench] promotes the synthesis and exudation of deoxystrigol to facilitate the root infection by arbuscular mycorrhizal fungi in cell culture media (Yoneyama et al., 2007). For micronutrients, the regulation of plant and yeast Zn transporter gene expression was studied in relation to changes in Zn status. A GATA transcription factor was shown to play a role in Zn-independent regulation of Zn transporter gene expression in yeast and raises the possibility that GATA, as well as E2F, transcription factors may play roles in Zn homeostasis in plants (Milner et al., 2014).

It has been suggested that plant hormones such as auxin, cytokinins, abscisic acid, gibberellins, ethylene, brassinosteroids, and strigalactones are involved in reprogramming root development and architecture for increasing the efficiency of nutrient uptake, mobilization, and redistribution (Nagarajan and Smith, 2012). Systemic signaling becomes important and predominant when plants experience prolonged nutrient deficiency or toxicity over long periods (Kochian and Lucas, 2014).

Important products of the detection, transduction, and transcription processes of chemical signaling pathways responding to changes in the surrounding environments are the continuous synthesis and excretion of many different organic metabolites, many of them being secondary metabolites from plant and microbial cells that interact in soil systems.

Metabolomics of Labile Soil Organic Matter

There are key areas of science that will assist to better understand the underlying chemical, biochemical, and biological mechanisms and components in complex soil–plant systems. Many compounds of labile soil organic matter are metabolites from plant and microbial activities. Metabolomics is part of the *omics* cascade, which can be defined as the systematic, comprehensive, and quantitative analysis of metabolites in cellular systems and their surrounding growth media (i.e., soil solutions). For example, the knowledge and information obtained from the omics cascade that involve genomics (what can happen), transcriptomics (what appears to be happening), and proteomics (what makes it happen) can analytically integrate the response of a living organism to environmental disturbance (Dettmer et al., 2007). Nature uses the term omics to denote a broad discipline of science and engineering for analyzing the interactions of biological information objects in various domains (http://www.nature.com/omics/about/index.html). For further information on the omics cascade in soils or mass spectrometry-based metabolomics for potential application to labile soil organic matter in soil–plant systems, the reader is directed to publications by Nannipieri et al. (2014), and Dettmer et al. (2007), respectively.

Under the soil surface, soil solutions acts as reservoirs of inorganic and organic chemical compounds essential for growth and reproduction of all living cells. The model of Fig. 2 indicates that the soil solution is the liquid phase that receives metabolites synthesized and excreted by crop roots and soil microbial cells. The synthesis and excretion of metabolites by plant and soil microorganisms become inputs of labile organic matter in soils, thus continuously altering the chemical composition of soil solutions. As a consequence, continuous feedback loops are established between living plants and microbes defined by signal information, the surrounding growing environment, abiotic factors, and physiological responses of living cells. This model indicates that the integration of chemical data together with data obtained from the soil–plant omics cascade and the application of multivariate statistical analysis will help enhance our

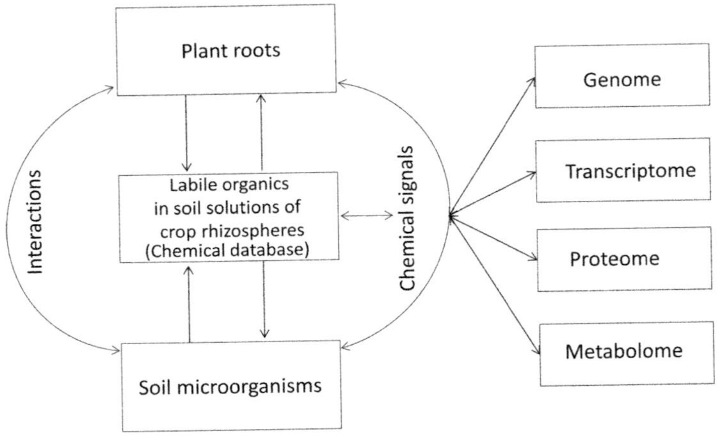

Fig. 2. A model for metabolites in labile soil organic matter of soil solutions associated with chemical signaling networks resulting from plant–microbe interactions in the rhizosphere.

knowledge on ecological functions of labile organic matter in soils. The model framework of Fig. 2 warrants testing with experimental data obtained from studies conducted in the rhizosphere of soil–plant systems.

The Future

In the future, defining ecological function for metabolites of labile soil organic matter pools in soil–plant systems will need multidisciplinary and integrated research studies on chemical signaling. Information obtained from genome, transcriptome, and proteome studies in rhizospheres will need to be integrated with robust chemical databases of metabolites in soil solution for rigorous interpretation and elucidation of the exact ecological function of an identified metabolite or group of metabolites. The multidisciplinary approach will help determine the influence of anthropogenic activities such as the addition of xenobiotics or the use of selected soil–crop and land-use-change practices on chemical signaling systems of living cells and their associated cellular response metabolic products. The characterization of signaling components in the soil–plant omics cascade will permit researchers to gain a deeper understanding of plant–microbe interactions, such as plant infection–defense systems and the supply and availability of soil nutrients controlled by crop demand within rhizospheres. This integrated, deeper knowledge of labile soil organic matter pools will also facilitate the development of novel soil reclamation practices and soil–crop management systems to enhance crop security, ecological function, and environmental health.

Conclusions

The chemical composition of labile soil organic matter in soil solutions reflects the multiple interactions of roots and soil microorganisms. The labile soil organic matter includes metabolites synthesized and excreted into soil solutions by living cells of plant and microorganisms. Many of these metabolites are end products of signaling pathways resulting from chemical communications between roots and microorganisms. Chemical signaling pathways involve the stages of recognition, transduction, transcription, and response. Each of these stages involves diverse cellular components that may carry out different and multiple functions in the cells. The identification of ecological function of labile soil organic matter metabolites requires parallel and complimentary studies on the genome, transcriptome, and proteome of crop rhizospheres, which will need to be integrated with robust chemical databases of metabolites in soil solution.

Acknowledgments

Support provided by Agriculture and Agri-Food Canada to the author for completing the writing of this chapter and the editorial changes indicated by reviewers are appreciated.

References

Adler, F.R. 2011. Plant signalling: The opportunities and dangers of chemical communication. Biol. Lett. 7:161–162. doi:10.1098/rsbl.2010.0790

Alberts, B., A. Johnson, J. Lewis, M. Raff, K. Roberts, and P. Walter. 2002. Molecular biology of the cell. 5th ed., Garland Science, Taylor and Francis Group, LLC., New York.

Bais, H.P., C.D. Broeckling, and J.M. Vivanco. 2008. Root exudates modulate plant-microbe interactions in the rhizosphere. In: P. Karlovsky, editor, Secondary metabolites in soil ecology. Soil Biology 14:241–252.

Baluska, F., D. Volkmann, and D. Menzel. 2005. Plant synapses: Actin based domains for cell-to-cell communication. Trends Plant Sci. 10:106–111. doi:10.1016/j.tplants.2005.01.002

Bassler, B. 2002. Small talk: Cell-to-cell communication in bacteria. Cell 109:421–424. doi:10.1016/S0092-8674(02)00749-3

Berg, G. 2009. Plant-microbe interactions promoting plant growth and health: Perspectives for controlled use of microorganisms in agriculture. Appl. Microbiol. Biotechnol. 84:11–18. doi:10.1007/s00253-009-2092-7

Berg, J., J.L. Tymoczko, and L. Stryer. 2006. Biochemistry. 6th ed. W.H. Freeman and Company, New York.

Bonke, M., S. Tähtiharju, and Y. Helariutta. 2005. Lessons from the root apex. In: A.J. Fleming, editor, Intercellular communication in plants. Ann. Plant Rev. 16:199–223.

Brivanlou, A.H., and J.E. Darnell. 2002. Signal transduction and the control of gene expression. Science 295:813–818. doi:10.1126/science.1066355

Clancy, S. 2008. RNA transcription by RNA polymerase: Pro-karyotes vs. eukaryotes. Nature Educ. 1:1.

Dale, H.H. 1906. On some physiological actions of ergot. J. Physiol. 34:163–206. doi:10.1113/jphysiol.1906.sp001148

Dennis, P.P., and H. Bremer. 1974. Macromolecular composition during steady-state growth of Escherichia coli B/r. J. Bacteriol. 119:270–281.

Dettmer, K., P.A. Aronov, and B.D. Hammock. 2007. Mass spectrometry-based metabolomics. Mass Spectrom. Rev. 26:51–78. doi:10.1002/mas.20108

Feys, B.J., and J.E. Parker. 2000. Interplay of signaling pathways in plant disease resistance. Trends Genet. 16:449–455. doi:10.1016/S0168-9525(00)02107-7

Fuqua, C., S.C. Winans, and E.P. Greenberg. 1996. Census and consensus in bacterial ecosystems: The LuxR-LuxI family of quorum-sensing transcriptional regulators. Annu. Rev. Microbiol. 50:727–751. doi:10.1146/annurev.micro.50.1.727

Galperin, M.Y. 2005. A census of membrane-bound and intracellular signal transduction proteins in bacteria: Bacterial IQ, extroverts and introverts. BMC Microbiol. 5:35–54. doi:10.1186/1471-2180-5-35

Geldner, N., and S. Robatzek. 2008. Plant receptors go endosomal: A moving view on signal transduction. Plant Physiol. 147:1565–1574. doi:10.1104/pp.108.120287

González, J.E., and M.M. Marketon. 2003. Quorum sensing in nitrogen-fixing rhizobia. Microbiol. Mol. Biol. Rev. 67:574–592. doi:10.1128/MMBR.67.4.574-592.2003

Heeb, S., and D. Haas. 2001. Regulatory roles of the GacS/GacA two-component system in plant-associated and other Gram-negative bacteria. Mol. Plant-Microbe Interact. 14:1351–1363. doi:10.1094/MPMI.2001.14.12.1351

Hirsch, A.M., W.D. Bauer, D.M. Bird, J. Cullimore, B. Tyler, and J.I. Yoder. 2003. Molecular signals and receptors: Controlling rhizosphere interactions between plants and other organisms. Ecology 84:858–868. doi:10.1890/0012-9658(2003)084[0858:MSARCR]2.0.CO;2

Izban, M.G., and D.S. Luse. 1992. Factor-stimulated RNA polymerase II transcribes at physiological elongation rates on naked DNA but very poorly on chromatin templates. J. Biol. Chem. 267:13647–13655.

Ji, G., R. Beavis, and R.P. Novick. 1997. Bacterial interference caused by autoinducing peptide variants. Science 276:2027–2030. doi:10.1126/science.276.5321.2027

Johnson, J.M., and G.M. Church. 2000. Predicting ligand-binding function in families of bacterial receptors. Proc. Natl. Acad. Sci. USA 97:3965–3970. doi:10.1073/pnas.050580897

Kanehisa, M., S. Goto, S. Kawashima, Y. Okuno, and M. Hattori. 2004. The KEGG resource for deciphering the genome. Nucleic Acids Res. 32:D277–D280. doi:10.1093/nar/gkh063

Kay, E., B. Humair, V. Dénervaud, K. Riedel, S. Spahr, L. Eberl, C. Valverde, and D. Haas. 2006. Two GacA-dependent small RNAs modulate the Quorum-sensing response in Pseudomonas aeruginosa. J. Bacteriol. 188:6026–6033. doi:10.1128/JB.00409-06

Keller, L., and M.G. Surette. 2006. Communication in bacteria: An ecological and evolutionary perspective. Nat. Rev. Microbiol. 4:249–258. doi:10.1038/nrmicro1383

Kobilka, B.K. 2007. G protein coupled receptor structure and activation. Biochim. Biophys. Acta, Biomembr. 1768:794–807. doi:10.1016/j.bbamem.2006.10.021

Kochian, L.V., and W.J. Lucas. 2014. Plant mineral nutrient sensing and signalling. J. Integr. Plant Biol. 56:190–191. doi:10.1111/jipb.12180

Kolenbrander, P.E., P.G. Egland, P.I. Diaz, and R.J. Jr. Palmer. 2005. Genome–genome interactions: Bacterial communities in initial dental plaque. Trends Microbiol. 13:11–15. doi:10.1016/j.tim.2004.11.005

Korth, K.L., and A.G. Thompson. 2006. Chemical signals in plants: Jasmonates and the role of insect-derived elicitors in responses to herbivores. In: S. Tuzun and E. Bent, editors, Multigenic and induced systematic resistance in plants. Springer Science+Business Media Inc., New York. p. 259–278.

Kurata, T., K. Okada, and T. Wada. 2005. Intercellular movement of transcription factors. Curr. Opin. Plant Biol. 8:600–605. doi:10.1016/j.pbi.2005.09.005

Latchman, D.S. 1997. Transcription factors. Int. J. Biochem. Cell Biol. 29:1305–1312. doi:10.1016/S1357-2725(97)00085-X

Lazdunski, A.M., I. Ventre, and J.N. Sturgis. 2004. Regulatory circuits and communication in Gram-negative bacteria. Nat. Rev. Microbiol. 2:581–592. doi:10.1038/nrmicro924

Lefkowitz, R.J. 2007. Seven transmembrane receptors: A brief personal retrospective. Biochim. Biophys. Acta 1768:748–755. doi:10.1016/j.bbamem.2006.11.001

Legate, K.R., S.A. Wickstrom, and R. Fassler. 2009. Genetic and cell biological analysis of integrin outside-in signaling. Genes Dev. 23:397–418. doi:10.1101/gad.1758709

Li, S., S.M. Assmann, and R. Albert. 2006. Predicting essential components of signal transduction networks: S dynamic model of guard cell abscisic acid signaling. PLoS Biol. 4:e312. doi:10.1371/journal.pbio.0040312

Liu, J., M.A. Piñeros, and L.V. Kochian. 2014. The role of aluminum sensing and signaling in plant aluminum resistance. J. Integr. Plant Biol. 56:221–230. doi:10.1111/jipb.12162

Lyon, G.J., and R.P. Novick. 2004. Peptide signaling in Staphylococcus aureus and other Gram-positive bacteria. Peptides 25:1389–1403. doi:10.1016/j.peptides.2003.11.026

Maffei, M.E., A. Mithöfer, and W. Boland. 2007. Before gene expression: Early events in plant-herbivore interactions. Trends Plant Sci. 12:310–316. doi:10.1016/j.tplants.2007.06.001

Maher, P.A., E.B. Pasquale, J.Y. Wang, and S.J. Singer. 1985. Phosphotyrosine-containing proteins are concentrated in focal adhesions and intercellular junctions in normal cells. Proc. Natl. Acad. Sci. USA 82:6576–6580. doi:10.1073/pnas.82.19.6576

Manefield, M., and S.L. Turner. 2002. Quorum sensing in context: Out of molecular biology and into microbial ecology. Microbiology 148:3762–3764.

Milner, M.J., N.S. Pence, J.L. Liu, and L.V. Kochian. 2014. Identification of a novel pathway involving a GATA transcription factor in yeast and possibly plant Zn uptake and homeostasis. J. Integr. Plant Biol. 56:271–280. doi:10.1111/jipb.12169

Mithöfer, A., and W. Boland. 2012. Plant defense against herbivores: Chemical aspects. Annu. Rev. Plant Biol. 63:431–450. doi:10.1146/annurev-arplant-042110-103854

Monreal, C.M., and M. Schnitzer. 2013. The chemistry and biochemistry of organic components in the soil solutions of wheat rhizosphere. Adv. Agron. 121:179–251. doi:10.1016/B978-0-12-407685-3.00004-9

Nabavi, S., and R.N. Nazar. 2008. Nonpolyadenylated RNA polymerase II termination is induced by transcript cleavage. J. Biol. Chem. 283:13601–13610. doi:10.1074/jbc.M710125200

Nagarajan, V.K., and A.P. Smith. 2012. Ethylene's role in phosphate starvation signaling: More than just a root growth regulator. Plant Cell Physiol. 53:277–286. doi:10.1093/pcp/pcr186

Nanjareddy, K., L. Blanco, M.K. Arthikala, X. Alvarado, F.S. Affantrange, and M. Lara. 2014. Nitrate regulates rhizobial and mycorrhizal symbiosis in common bean (Phaseolus vulgaris L.). J. Integr. Plant Biol. 56:281–298. doi:10.1111/jipb.12156

Nannipieri, P., G. Pietramellara, and G. Renella. 2014. Soil as a biological system. Chapter 1. Omics in soil science. Caister Academic Press. Poole, U.K.

Nealson, K.H., and J.W. Hastings. 1979. Bacterial bioluminescence: Its control and ecological significance. Microbiol. Rev. 43:496–518.

Niu, Y.F., R.S. Chai, G.L. Jin, H. Wang, C.X. Tang, and Y.S. Zhang. 2013. Responses of root architecture development to low phosphorus availability: A review. Ann. Bot. (Lond.) 112:391–408. doi:10.1093/aob/mcs285

Novick, R.P., and T.W. Muir. 1999. Virulence gene regulation by peptides in staphylococci and other Gram-positive bacteria. Curr. Opin. Microbiol. 2:40–45. doi:10.1016/S1369-5274(99)80007-1

Pardi, R. 2010. Signal transduction by adhesion receptors. Nature Educ. 3:38.

Park, C.J., and P.C. Ronald. 2012. Cleavage and nuclear localization of the rice XA21 immune receptor. Nat. Commun. 3:1–6.

Park, P.S.H. 2012. Ensemble of G protein-coupled receptor active states. Curr. Med. Chem. 19:1146–1156. doi:10.2174/092986712799320619

Pedersen, S.E., and E.M. Ross. 1982. Functional reconstitution of beta-adrenergic receptors and the stimulatory GTP-binding protein of adenylate cyclase. Proc. Natl. Acad. Sci. USA 79:7228–7232. doi:10.1073/pnas.79.23.7228

Peng, M., D. Hudson, A. Schofield, R. Tsao, R. Yang, H. Gu, Y.M. Bi, and S.J. Rothstein. 2008. Adaptation of Arabidopsis to nitrogen limitation involves induction of anthocyanin synthesis which is controlled by the NLA gene. J. Exp. Bot. 59:2933–2944. doi:10.1093/jxb/ern148

Pfannschmidt, T., K. Brautigam, R. Wagner, L. Dietzel, Y. Schroter, S. Steiner, and A. Nykytenko. 2009. Potential regulation of gene expression in photosynthetic cells by redox and energy state: Approaches towards better understanding. Ann. Bot. (Lond.) 103:599–607. doi:10.1093/aob/mcn081

Pichersky, E., and E. Lewinsohn. 2011. Convergent evolution in plant specialized metabolism. Annu. Rev. Plant Biol. 62:549–566. doi:10.1146/annurev-arplant-042110-103814

Plopper, G.E., H.P. McNamee, L.E. Dike, K. Bojanowski, and D.E. Ingber. 1995. Convergence of integrin and growth factor receptor signaling pathways within the focal adhesion complex. Mol. Biol. Cell 6:1349–1365. doi:10.1091/mbc.6.10.1349

Rayo, J., N. Amara, P. Krief, and M.M. Meijler. 2011. Live cell labeling of native intracellular bacterial receptors using aniline-catalyzed oxime ligation. J. Am. Chem. Soc. 133:7469–7475. doi:10.1021/ja200455d

Rosenbaum, D.M., C. Zhang, J.A. Lyons, R. Holl, D. Aragao, D.H. Arlow, S.G. Rasmussen, H.J. Choi, B.T. Devree, R.K. Sunahara, P.S. Chae, S.H. Gellman, R.O. Dror, D.E. Shaw, W.I. Weis, M. Caffrey, P. Gmeiner, and B.K. Kobilka. 2011. Structure and function of an irreversible agonist-b_2 adrenoceptor complex. Nature 469:236–240. doi:10.1038/nature09665

Ross, E.M., and A.G. Gilman. 1980. Biochemical properties of hormone-sensitive adenylate cyclase. Annu. Rev. Biochem. 49:533–564. doi:10.1146/annurev.bi.49.070180.002533

Šamaj, J., N.D. Read, D. Volkmann, D. Menzel, and F. Baluška. 2005. The endocytic network in plants. Trends Cell Biol. 15:425–433. doi:10.1016/j.tcb.2005.06.006

Schauder, S., K. Shokat, M.G. Surette, and B.L. Bassler. 2001. The LuxS family of bacterial autoinducers: Biosynthesis of a novel quorum-sensing signal molecule. Mol. Microbiol. 41:463–476. doi:10.1046/j.1365-2958.2001.02532.x

Schroeckh, V., K. Scherlach, H.-W. Nützmann, E. Shelest, W. Schmidt-Heck, J. Schuemann, K. Martin, C. Hertweck, and A.A. Brakhage. 2009. Intimate bacterial-fungal interactions triggers biosynthesis of archetypal polyketides in Aspergillus nidulans. Proc. Natl. Acad. Sci. USA 106:14558–14563. doi:10.1073/pnas.0901870106

Shin, R., R.H. Berg, and D.P. Schachtman. 2005. Reactive oxygen species and root hairs in Arabidopsis root response to nitrogen, phosphorus and potassium deficiency. Plant Cell Physiol. 46:1350–1357. doi:10.1093/pcp/pci145

Strand, A. 2004. Plastid-to-nucleus signalling. Curr. Opin. Plant Biol. 7:621–625. doi:10.1016/j.pbi.2004.09.004

Suzuki, T., M. Kusakabe, K. Nakayama, and E. Nishida. 2012. The protein kinase MLTK regulates chondrogenesis by inducing the transcription factor Sox6. Development 139:2988–2998. doi:10.1242/dev.078675

Taga, M.E., and B.L. Bassier. 2003. Chemical communication among bacteria. Proc. Natl. Acad. Sci. USA 100:14549–14554. doi:10.1073/pnas.1934514100

Taiz, L., and E. Zeiger. 2010. Plant physiology. 5th ed. Sinauer Associates Inc., Publishers, Sunderland, MA.

Tassetto, M., A. Maizel, J. Osorio, and A. Joliot. 2005. Plant and animal homeodomains use convergent mechanisms for intercellular transfer. EMBO Rep. 6:885–890. doi:10.1038/sj.embor.7400487

Trusov, Y., D. Chakravorty, and J.R. Botella. 2012. Diversity of heterotrimeric G-protein g subunits in plants. BMC Res. Notes 5:608. doi:10.1186/1756-0500-5-608

Urano, D., J.G. Chen, J.R. Botella and A.M. Jones. 2013. Heterotrimeric G protein signalling in the plant kingdom. Open Biol. 3:120186. doi:10.1098/rsob.120186

Vendeville, A., K. Winzer, K. Heurlier, C.M. Tang, and K.R. Hardie. 2005. Making 'sense' of metabolism: Autoinducer-2, LuxS and pathogenic bacteria. Nat. Rev. Microbiol. 3:383–396. doi:10.1038/nrmicro1146

von Zastrow, M., and A. Sorkin. 2007. Signaling on the endocytic pathway. Curr. Opin. Cell Biol. 19:436–445. doi:10.1016/j.ceb.2007.04.021

Wang, N., J.D. Tytell, and D.E. Ingber. 2009. Mechanotransduction at a distance: Mechanically coupling the extracellular matrix with the nucleus. Natl. Rev. 10:75–82. doi:10.1038/nrm2594

Waters, C.M., and B.L. Bassler. 2005. Quorum sensing: Cell-to-cell communication in bacteria. Annu. Rev. Cell Dev. Biol. 21:319–346. doi:10.1146/annurev.cellbio.21.012704.131001

Winkler, W.C., A. Nahvi, A. Roth, J.A. Collins, and R. Breaker. 2004. Control of gene expressions by natural metabolite-responsive ribozyme. Nature 428:281–286. doi:10.1038/nature02362

Witzany, G. 2006. Plant communication from biosemiotic perspective. Plant Signal. Behav. 1:169–178. doi:10.4161/psb.1.4.3163

Witzany, G. 2010. Biocomunication and natural genome editing. Springer, New York.

Wright, J.S., G.J. Lyon, E.A. George, T.W. Muir, and R.P. Novick. 2004. Hydrophobic interactions drive ligand–receptor recognition for activation and inhibition of staphylococcal quorum sensing. Proc. Natl. Acad. Sci. USA 101:16168–16173. doi:10.1073/pnas.0404039101

Xavier, K.B., and B.L. Bassler. 2003. LuxS quorum sensing: More than just a numbers game. Curr. Opin. Microbiol. 6:191–197. doi:10.1016/S1369-5274(03)00028-6

Xia, W., C. Zhang, X. Zeng, Y. Feng, J. Weng, X. Lin, J. Zhu, Z. Xiong, J. Xu, Z. Cai, and J. Zhongjun. 2011. Autotrophic growth of nitrifying community in an agricultural soil. ISME J. 5:1226–1236.

Yim, G., H.H. Wang, and J. Davies. 2007. Antibiotics as signalling molecules. Philos. Trans. R. Soc. Lond., B 362:1195–1200. doi:10.1098/rstb.2007.2044

Yoneyama, K., K. Yoneyama, Y. Takeuchi, and H. Sekimoto. 2007. Phosphorus deficiency in red clover promotes exudation of orobanchol, the signal for mycorrhizal symbionts and germination stimulant for root parasites. Planta 225:1031–1038. doi:10.1007/s00425-006-0410-1

Labile Organic Matter in Soil Solution: II. Separation and Identification of Metabolites from Plant–Microbial Communication in Soil Solutions of Wheat Rhizospheres

Carlos M. Monreal* and Morris I. Schnitzer

Abstract

Chemical data obtained by pyrolysis-field ionization mass spectrometry (Py-FIMS) analysis of soil solutions from wheat (*Triticum aestivum* L.) rhizospheres were organized in a database involving 3608 values of total ion intensities (TII). Models of multivariate analysis (i.e., principal component/factorial analysis) were used to understand the temporal variability, interpret ecological functions of discriminant metabolites, and test a model metabolomics framework for chemical signaling pathway components. The discriminant m/z signals involved four heterocyclic-N compounds, namely two alkaloids (m/z 383 and 347), a precursor of antibiotics and alkaloids (m/z 81), and an antibiotic (m/z 113); a polyphenol (m/z 473); an unsaturated fatty acid (m/z 304); a plant hormone or alkylaromatic fatty acid (m/z 346); and a fatty acid amide (m/z 345). Testing of the model framework indicated that the discriminant metabolites of labile soil organic matter reflect factors and signaling pathways involved in plant defense and pathogen root infection mechanisms. The chemical and statistical approach used herein, together with future complementary omics cascade studies, provide new opportunities to identify ecological functions for metabolites of labile soil organic matter and the health of soil–crop systems.

Abbreviations: AA, arachidonic acid; BP, benzyl palmitate; CA, chicoric acid; CM, classification matrix; DA, discriminant analysis; DAS, days after seeding; GA, gibberellic acid, GAs, ; PA, pyrrolizidine alkaloid; PCA-FA, principal component analysis–factor analysis ; Py-FIMS, pyrolysis-field ionization mass spectrometry; SOM, soil organic matter; TA, tropane alkaloid; TII, total ion intensities.

Carlos M. Monreal and Morris I. Schnitzer, Eastern Cereal and Oilseed Research Center, Agriculture and Agri-Food Canada, Room 4046 K.W. Neatby Building, Central Experimental Farm, 960 Carling Ave., Ottawa, Canada, K1A 0C6. *Corresponding author (carlos.monreal@agr.gc.ca).

doi:10.2136/sssaspecpub62.2014.0074

Labile Organic Matter—Chemical Compositions, Function, and Significance in Soil and the Environment. SSSA Special Publication 62. Zhongqi He and Fengchang Wu, editors.

Microbial metabolites and root exudates are important constituents of the labile soil organic matter (SOM) pool residing in soil solutions during the growth and senescence of plants (Waksman, 1961; Rovira and McDougall, 1967). Root exudates take part in signaling events that initiate the execution of complex biological interactions in soil–plant systems (Bais et al., 2006). Soil solutions in wheat rhizospheres are reservoirs containing hundreds of organic and inorganic compounds during the growing season, and 346 of those organic compounds were identified and assigned into 11 chemical classes (Monreal and Schnitzer, 2013). Up to 70% of the photosynthesized $^{13}CO_2$ is exuded by wheat roots at tilling, and the presence of wheat plants increased the content of carbon in soil solution three times relative to unplanted soil (Matus et al., 2014).

Plants and microorganisms in the environment and in vitro systems produce a bewildering array of low molecular weight organic molecules that have many biological activities (Goh et al., 2002). Most metabolites appear to modulate gene transcription at low concentrations, and the primary effect of small organic compounds appears to be the maintenance of microbial communities in the environment (Yim et al., 2007). Functional roles for some chemical components of labile SOM in soil solutions have been reported (Monreal and Schnitzer, 2013).

The exact functions of labile SOM associated with plant and microbial metabolites derived from their continuous communication are largely undefined, although many roles have been suggested (Wenke et al., 2010). Identifying and determining key ecological functions for metabolites in labile SOM residing in soil solution become essential knowledge and cornerstones for crop security, the sustainable production of food, and management of cultivated land. We used chemical data obtained by Py-FIMS together with multivariate statistical models to associate key metabolites of labile SOM with ecological functions in wheat rhizospheres. We also interpreted ecological functions within the model framework for metabolites and chemical signaling presented by Monreal (2015).

Metabolomics for Identifying Signaling Components in Soil Solutions

Knowledge and information obtained from the so-called "omics" cascade, which involve genomics (what can happen), transcriptomics (what appears to be happening), proteomics (what makes it happen), can analytically integrate the response of a living organism to environmental disturbance and help define ecological function (Dettmer et al., 2007). Metabolomics is part of the omics cascade that is defined as the systematic, comprehensive, and quantitative analysis of metabolites in cellular systems and their surrounding growth media. Metabolomics often use various complementary analytical techniques involving mass spectrometry in conjunction with liquid and gas chromatography and other spectroscopic techniques to identify compounds in heterogeneous sample matrices (Daviss, 2005; Bino et al., 2004). Thus, different analytical techniques can identify and quantify intra- and extracellular metabolites to compose databases that are then associated with specific metabolic pathways, and the genome and proteomes to understand their biological significance. Among the functional genomics technologies, metabolomics may better assist in understanding the physiology at the system level, because the metabolome is the ultimate outcome of biochemical networks or the omics cascade and close to phenotypes (Barupal et al., 2012). The temporal and spatial variability of metabolomes and biological diversity can complicate the analysis and interpretation of data in living systems (Vigneau-Callahan et al., 2001).

No comprehensive databases have been established or published yet to help explain the presence of certain metabolites in labile organic matter residing in soil solutions, or to understand physiological states or phenotypes of plants and/or soil microorganisms. In soil–plant systems, metabolic fingerprinting may be used as a diagnostic tool to shift in system or component health. The use of identified metabolites per se does not contribute alone to biochemical understanding of underlying mechanisms (Dettmer et al., 2007).

For further reading on the omics cascade in soils or mass spectrometry–based metabolomics for potential application in soil–plant systems, the reader is directed to publications by Nannipieri et al. (2014) and Dettmer et al. (2007), respectively.

Studying the Chemical Composition of Metabolites in Soil Solutions

Soil

A bulk soil sample (1000 kg) was taken from the 0- to 20-cm depth of a fallowed field (uncropped for the last 15 yr) located at the Central Experimental Farm of Agriculture and Agri-Food Canada in Ottawa, Ontario. The soil is sandy loam in texture and belongs to the Manotick series of the Haplorthods Great Group. The concentration of total carbon in the 0 to 20 cm of soil is 19.1 mg C g^{-1} soil, and that of total nitrogen is 1.8 mg g^{-1} soil. Details on the processing of the bulk soil sample for the preparation of unfertilized soil pots planted to wheat and grown to maturity in the greenhouse under controlled conditions of light, soil moisture, and temperature are reported by Monreal and Schnitzer (2013).

Collection of Soil Solutions

Planted soil pots were destructively sampled at four different wheat growth stages to obtain soil solution samples from the wheat rhizospheres. Roots and soil samples obtained from the wheat rhizosphere were handled and separated with gloves to prevent contact of soil with hands. Soil solutions were obtained after placing 50 g of moist rhizosphere soil in a Vivacell 70 (Sartorius Stedim Biotech, USA), and then centrifuged at 2000 *g* for 10 min at 4°C. Before its use, the Vivacell 70 was conditioned as indicated by the manufacturer. The collected solution was placed in a 10-mL glass vial and kept frozen at −18°C until chemical analysis by mass spectrometry. Soil solution samples were collected 25 d after seeding (DAS) (four leaves unfolded, one tiller), 53 DAS (flag leaf, two tillers), 72 DAS (grain-early milk), and 102 DAS (grain-dry seed).

Pyrolysis-Field Ionization Mass Spectrometry

Each soil solution sample collected from the planted soil was analyzed by Py-FIMS in triplicate. Before Py-FIMS analysis, the volumes of the soil solution samples were reduced by evaporation from 1 mL to about 100 μL.

Five-microliter aliquots of the samples were transferred into quartz crucibles, which were dried in a freeze dryer and stored until analyses in a desiccator. Thereafter the crucibles were successively placed in micro-oven and introduced into the fore-vacuum chamber of a double-focusing Finnigan MAT 900 mass spectrometer (Finnigan, MAT, Bremen, Germany). When the pressure had dropped below 1.5×10^{-2} Pa, the quartz micro-oven with sample was inserted into the ion source and heated under high vacuum (10^{-6} Pa) from 50 to 500°C in steps of 10°C per min. During about 16 min of total measurement time, 46 magnetic scans were recorded for the mass range 15 to 900 Dalton (single spectra). All Py-FIMS data were normalized with respect to a 1-μL sample. Analysis by Py-FIMS, which is a soft ionization technique, tells us what individual chemical compounds are volatilized under these experimental conditions, that is, it yields data and information at the molecular level. For detailed descriptions of the Py-FIMS methodology and of statistical evaluations of sample weight and residue, volatilized matter, and TII see Schulten (1987) and Sorge et al. (1993). Three sources of data and information were used to assign m/z signals to specific metabolic compounds: (i) the US National Institute of Standards and Technology (NIST) Chemistry Webbook (http://webbook.nist.gov) , (ii) work already published by Monreal and Schnitzer (2013), and (iii) other research articles as identified in the later section "Assignment of Significant m/z Signals to Individual compounds."

Experimental Design and Statistical Models

The greenhouse pot study was established as a randomized experiment with sampling time as the main factor with three replicated pots for each time of sampling. Multivariate statistical techniques in Systat (2007) involving discriminant analysis (DA) and principal component analysis–factor analysis (PCA-FA) were applied for evaluating the temporal variation in chemical composition of the soil solution in the rhizosphere during the growing season of an unfertilized wheat crop. Discriminant functions and classification matrices were obtained from the stepwise modes of linear DA (Systat, 2007). Discriminant functions were established after using the 40 most significant m/z signals according to the ranking established by Wilks's lambda test. Wilks's lambda test tests the null hypothesis that the means of all of the independent variables are equal across groups of the dependent variable. If the means of the independent variables are equal for all groups, the means will not be a useful basis for predicting the group to which a case belongs. We used DA and PCA-FA to help interpret the complex chemical data matrix associated with the soil solutions collected from the rhizosphere during the growing season of wheat. Results from DA help us to better understand the temporal changes in chemical composition of organic compounds in the soil solution. Principal component analysis–factorial analysis helps to compare the compositional pattern between the soil solution samples analyzed by Py-FIMS and the factors influencing each growth stage. In addition, PCA helps to summarize a multivariate data set as accurately as possible by using a few components. Factor analysis helps interpret each factor according to the meaning of the variables.

Testing the Model Framework for Metabolites in Labile Organic Matter

The Chemical Database for the Composition of Soil Solutions

A chemical molecular database involving 3608 values of TII for m/z values ranging from m/z 15 to m/z 900 was established. This data set is constructed based on information obtained from the analyses of soil solution samples characterized by Py-FIMS. Soil solution samples were collected at four different growth stages of wheat from seeding to grain harvest. The information associated with the chemical characterization of the same soil solution samples by Py-FIMS, their spectra, and thermograms were published earlier by Monreal and Schnitzer (2013). From the Py-FIMS analysis, the latter authors identified 346 individual organic compounds that were classified into 11 chemical classes.

Ranking the Importance of Organic Compounds in Soil Solutions

The data set consisting of 3608 TIIs values obtained from the Py-FIMS analysis were analyzed by the univariate *F*-distribution Wilks's lambda (λ) to find out the most significant m/z signals capable of discriminating among the four growth stages of wheat. The $F(3,8)$ values for three different probabilities show that, at $\alpha = 0.01$, only 25 m/z variables had F values >7.59, indicating significant differences between the four sampling dates. Further, Table 1 shows the ranking of the top 10 m/z signals out of a total of the 25 m/z signals. The ranking of the top 10 m/z signals according to Wilks's lambda (λ) test was used as a first approach to determine the most significant single variables (m/z signals) capable of discriminating among the four growth stages of wheat, namely four leaves unfolded and one tiller (25 DAS), flag leaf and two tillers (53 DAS), grain-early milk (72 DAS), and grain-dry seed (102 DAS). Such basic information was also useful in subsequent steps used to determine the chemical species identity and interpret specific ecological functions for the significant compounds in soil solution. According to Wilks's lambda (λ) test in Table 1, the top 10 significant m/z signals were m/z 383, 345, 473, 304, 81, 347, 39, 346, 23, and 113. Interestingly, the signals at m/z 383, 345,

Table 1. Significant m/z signals in soil solution discriminating among four stages of wheat growth according to Wilks's Lambda test and discriminant analysis.

Rank	1	2	3	4	5	6	7	8	9	10
m/z	383	345	473	304	81	347	39	346	23	113
λ	0.050	0.063	0.069	0.088	0.094	0.099	0.111	0.117	0.13	0.131
Discriminant analysis: classical										
F-to-remove	15.1	5.9	0.7	1.6	0.4	0.06	0.3	0.07	0	0
Tolerance	0.040	0.02	0.05	0.12	0.22	0.29	0.50	0.39	0	0
Discriminant analysis: backward stepwise mode										
F-to-remove	17.5	11.6	2.0	1.8	2.9	0.3	0	0	0	0
Tolerance	0.03	0.01	0.05	0.21	0.11	0.37	0	0	0	0

and 473 were the most useful in discriminating among the four growth stage groups, and m/z 23 and 113 the least useful variables.

Discriminant Analysis of Metabolites in Soil Solutions

Discriminant analysis was conducted by the classical, forward and backward stepwise mode by using the top 25 m/z signals identified by Wilks's lambda test. In DA, the *F*-to-remove values help to determine the relative importance of all variables, and a tolerance value of 0 (zero) indicates that variables are highly correlated with one or more of the others, showing that these m/z signals are redundant for explaining the variability in the data set. According to DA, data in Table 1 shows that only six m/z signals were capable of discriminating among the four growth stages of wheat, namely four leaves unfolded (25 DAS), flag leaf (53 DAS), grain-early milk (72 DAS), and grain-dry seed (102 DAS). These signals were m/z 383, 345, 473, 304, 81, and 347.

Discriminant functions and classification matrices (CMs) were obtained in the classical, backward and forward stepwise modes of DA (Systat, 2007). Table 2 presents the coefficients for the six most significant m/z signals determined by the classical and backward stepwise mode by using the top discriminant variables identified by Wilks's lambda test. The classical DA yielded 100% of the cases correctly in the CMs by using the six discriminant m/z signals. The more robust cross-validation classification matrix (i.e., the Jackknife CM) decreased to 42% the number of cases classified correctly. The backward stepwise mode DA also assigned 100% of the cases correctly in the CM by using six discriminant m/z signals. The Jackknifed CM, however, classified 50% of the cases correctly by the backward stepwise DA mode, with little difference in the match for each growth stage compared with the classical DA. In comparison, the DA forward stepwise mode could classify cases only into three groups but not into the four sampled growth stages. The DA forward stepwise mode identified m/z 384 and m/z 429 as two important discriminant variables. The latter two m/z signals, however, ranked 19th and 23rd by Wilks's lambda test. Based on the latter, and the consistent results obtained by Wilks's lambda test and the DA backward stepwise mode, the signals m/z 383, m/z 345, m/z 473, m/z 304, m/z 81, and m/z 347 were identified as the six most significant discriminating parameters to explain the temporal variability in chemical composition of organic

Table 2. Function coefficients obtained from discriminant analysis in a backward stepwise mode used to classify the six most significant m/z signals into four growth stages.

	Coefficients for cases grouped according to the number of days after seeding (DAS)			
	25 DAS	53 DAS	72 DAS	102 DAS
Constant	173.2	−3,372.3	−120.3	−289.5
m/z 383	200.4	5695.5	747.8	1402.0
m/z 345	78.2	4172.6	529.7	1035.2
m/z 473	318.5	−21,786.8	−2,506.2	−5,380.5
m/z 304	424.3	−537.6	156.9	89.5
m/z 81	453.5	3584.3	721.0	1203.8
m/z 347	−563.9	566.8	−256.9	−233.8

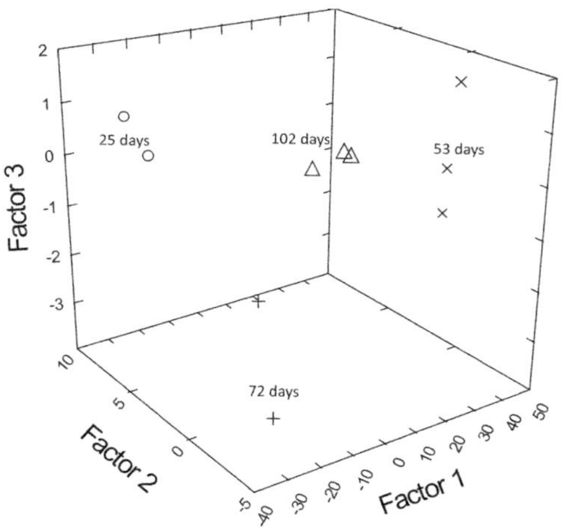

Fig. 1. Canonical 3-D scores plot of the discriminant analysis for the four growth wheat stages by using the 40 most significant m/z signals.

compounds in soil solutions during the four growth stages of wheat. Also, data from the DA backward stepwise mode was preferred because of the higher classification of cases obtained in the Jackknife CM relative to that obtained by the forward stepwise mode.

Figure 1 depicts the canonical 3-D scores plot from the DA. The axis represents the three canonical variables or factors, and the points represent the canonical variable scores. The DA shows that Factor 1 captured 98.6%, Factor 2 captured 1.2%, and Factor 3 captured 0.2% of the total data dispersion of the four growth stage groups as indicated by the very large first Eigen value (i.e., 1776) relative to the second (i.e., 22) and third values (i.e., 3). The Eigen values provide a measurement of the significance of the factor, with the largest value being the most significant. Eigen values $\geq$1.0 are considered significant (Kim and Mueller, 1987). The DA backward stepwise mode showed that the centroids (not shown in Fig. 1) for Day 72 and Day 102 are closest (3.4), and those for Day 25 and Day 53 are farthest apart (213.8). The distances between the centroids were significantly different at $P = 0.03$.

Most Significant Organic Compounds

Figure 2 shows a box and whisker plot representing the relative variability of the abundance (%TII) for the six most significant m/z signals during the four wheat growth stages. The largest variability is shown by signal m/z 345 and the smallest by m/z 473. In general, the relative distribution of TII abundances among the significant compounds was m/z 345 > m/z 383 > m/z 304 > m/z 81 > m/z 347 > m/z 473. Figure 3 shows that in general, the relative abundance (%TII) of most m/z signals increased from Day 25 (four leaves unfolded) to maximum relative abundances on Day 53 (flag leaf), steadily declining thereafter through the two grain stages (Days 72, early milk grain stage, and Day 102, grain-dry seed stage) until maturity.

The highest relative abundance of TIIs for m/z signals m/z 383, m/z 345, m/z 473, and m/z 347 was observed at the flag leaf stage or 53 DAS. In comparison, the relative abundance for m/z 304 in the soil solution was highest at Day 25, and it decreased steadily thereafter. Conversely, the relative abundance for m/z 81

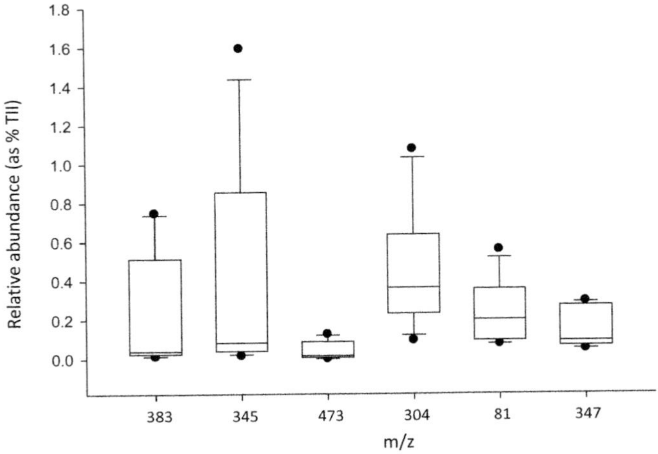

Fig. 2. Whisker and box plot for the temporal variability of six key m/z signals in the soil solution of wheat. Boundary of box closest to zero indicates the 25th percentile, line within the box marks the median, and the boundary of the box furthest from zero indicates the 75th percentile. The error bars (whiskers) above and below the box indicate the 90th percentiles. Solid circles represent outliers. TII, total ion intensities.

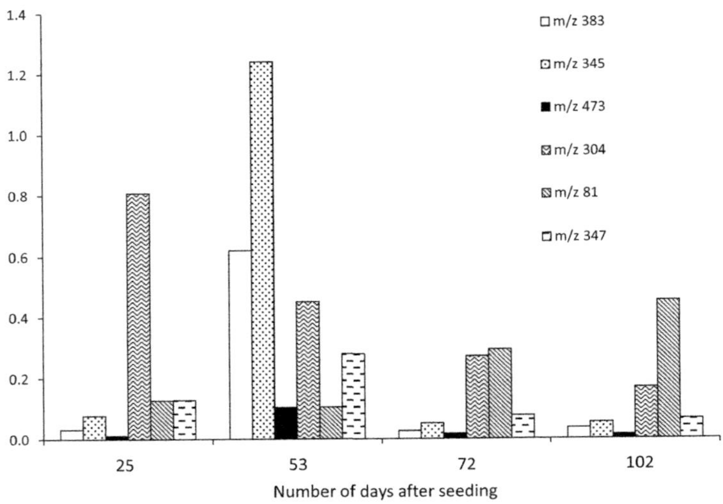

Fig. 3. The temporal variability of the six most significant m/z signals in soil solutions obtained at four growth stages of wheat.

was similar after 25 and 53 DAS, but it increased steadily from the four leaves unfolded to the milk and dry grain seed stages, 102 DAS. Monreal and Schnitzer (2013) discussed the temporal dynamics of more than 300 identified complementary organic compounds distributed in 11 chemical classes in the same soil solution samples of wheat rhizosphere.

Identification of Significant M/Z Signals in Soil Solutions

To establish the identity of the most significant discriminant m/z signals, we conducted an extensive search in the Py-FIMS soil database, the NIST webbook database, and also used information on metabolites from plant and microbial published studies. Table 3 shows the classes, subclasses, and the chemical identity for 10 of the most significant compounds in the soil solution samples of wheat rhizospheres. Table 3 includes information for six of the most significant m/z signals in soil solution samples as determined by DA, and four additional variables ranked by the Wilks lambda test. According to information retrieved from the mass spectrometry databases and other published works, we assigned the significant m/z signals to organic compounds involving four heterocyclic-N compounds (two alkaloids, m/z 383 and 347; an alkaloid precursor, m/z 81; and an antibiotic moiety, m/z 113), a polyphenol (m/z 473), an unsaturated fatty acid (m/z 304), a plant hormone (m/z 346), and a N-containing fatty acid (m/z 345). Two inorganic compounds identified as K (m/z 39) and Na (m/z 23) also contributed to discriminating the four growth stages of wheat, although to a lesser extent than most organic compounds.

Figure 4 shows the chemical structure for the nine significant organic compounds in soil solutions, with most of them having bioactive functions. Interestingly, natural bioactive products in the tissues of agronomic crops, root exudate, or soil solution studies receive little attention. In addition, crop breeding over centuries has resulted in decreased amounts of active compounds in their cells. As a result of the genetic changes during domestication, wheats are unable to survive wild in competition with better adapted species (Shewry, 2009).

Assignment of Significant m/z Signals to Individual Compounds

In the ensuing sections we provide supporting published evidence, rationale, and the function associated with the identification of 10 key plant and microbial metabolites in soil solution samples whose m/z signals were determined by Py-FIMS (Table 3).

Assignment of m/z 383 to Diacetyllycopsamine

The signal at m/z 383 was assigned to diacetyllycopsamine, a pyrrolizidine alkaloid (PA). The compound is produced by plants as a defense chemical product against insects, herbivores, bacteria, and fungi (Joosten and van Veen, 2011). More than 660 structures of PAs have been reported from >660 plant species in which the PAs also accumulate as N-oxides (Hartmann and Witte, 1995). Pyrrolizidine alkaloids are synthesized as N-oxides in the roots and are translocated to the aerial parts where they are converted into the species-specific alkaloids (Ober and Hartmann, 1999). Pyrrolizidine alkaloids are esters of hydroxylated methyl pyrrolizidines, consisting of a necine base and necic acid moiety, which are synthesized in plants during amino acid metabolism. The chemistry, occurrence, and significance of PAs have been presented and discussed by Colegate et al. (2012). The structure of 7-acetyllycopsamine and other PAs in *Amsinckia menziesii* were determined by nuclear magnetic resonance and mass spectrometry (Roitman, 1983).

Table 3. Classes, subclasses and assignment of 10 compounds in soil solutions identifying infection-defense factors and interactions of wheat roots with soil microorganisms.

m/z	Chemical			Metabolite		Reference
	Class	Subclass	Compound	Primary/Secondary	Function	
			Six most significant organic molecules in soil solutions			
383	Heterocyclic-N	Alkaloid	Diacetyllycopsa-mine	secondary	Plant defense†	Roitman (1983) and Mroczek and Glowniak (2002)
345	Alkylaromatic	N-benzylamide derivative	N-benzyl-palmitamide	secondary	Plant/microbial defense	Khalil (2006)
473	Phenol	Polyphenol	Chicoric acid	secondary	Plant defense‡	Chen et al. (2012) and Lee and Scagel (2009)
304	Fatty acid	Eicosanoid	Arachidonic acid	secondary	Fungal infection signal§	Savchenko et al. (2010)
81	Heterocyclic-N	Pyrrole derivative	Methylpyrrole	secondary	Microbial defense	Eich (2008) and Gross et al. (2003)
347	Heterocyclic-N	Alkaloid	Consabatine	secondary	Plant defense	Eich (2008)
			Additional four significant compounds as determined by Wilks's lambda test			
39	Potassium	Inorganic	K^+	primary	Crop nutrient	Wang and Wu (2013)
346	Alkylaromatic or hormone	Alkanoic acid or diterpene	Phenyl palmitate or Gibberellic acid	secondary secondary	Plant defense or growth/fungal infection	Mukku et al. (2002) and MacMillan (2002)
23	Sodium	Inorganic	Na^+	primary	Crop nutrient	Subbarao et al. (2003)
113	Heterocyclic-N	Antibiotic	Glutarimide	secondary	Microbial defense	Broadbent et al. (1971) and Kim et al. (1999)

† Plant defense mechanisms and synthesis of antimicrobial compounds.

‡ Plant signaling molecules central to various biological processes, especially during plant stress events caused by interactions with soil microorganisms and abiotic factors.

§ Infection signal of wheat roots by fungal pathogens.

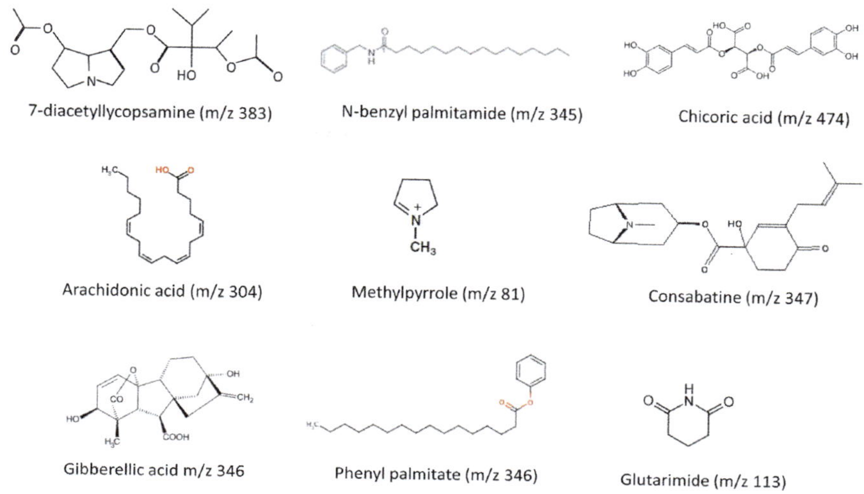

Fig. 4. Chemical structures for nine organic compounds in the soil solution that discriminate between the four growth stages of wheat.

Assignment of m/z 345 to Nitrogen-Benzyl Palmitamide

The signal at m/z 345 was assigned to N-benzyl palmitamide, an N-alkylamide (NA) derivative. Nitrogen-alkylamides are widely present in the whole biological kingdom and are a group of bioactive compounds with antimicrobial properties (Boonen et al., 2012). Nitrogen-benzyl palmitamide is responsible for the biochemical and toxicological effects on plant pathogens (Khalil, 2006). Most plant-derived alkylamides contain a polyunsaturated aliphatic fatty acid chain and a shorter substituent at the amine side. The fatty acid part is delivered by lauric (12:0), myristic (14:0), palmitic (16:0), or linoelaidic (18:2, *cis*,cis) acid. Hartmann et al. (1989) investigated the importance of the metabolic coupling of pyrrolizidine-alkaloid biosynthesis to polyamine metabolism.

Assignment of m/z 473 to Chicoric Acid

The signal at m/z 473 was assigned to chicoric acid (CA), a polyphenol derivative of caffeic and tartaric acids (Chen et al., 2012). Chicoric acid is involved in several plant responses to biotic and abiotic stress, and has antibacterial and antifungal properties (Fu et al., 2010). Chicoric acid is a hydroxycinnamic acid of the phenylpropanoid class and occurs in a variety of plant species (Vogt, 2010). This polyphenol derivative was identified in plant extracts employing both single quadrupole mass spectrometer with electrospray ionization negative mode and triple quadrupole tandem mass spectrometer with multiple-reaction monitoring mode (Chen et al., 2012; Lee and Scagel, 2009). Soil nitrogen and potassium, plant species, the developmental state of the plant, and the addition of phosphorus and root inoculation with arbuscular micorrhizal fungus enhanced the concentration and content of CA and caffeic acid derivatives (Scagel and Lee, 2012).

Assignment of m/z 304 to arachidonic acid

The signal at m/z 304 was assigned to arachidonic acid (AA, $C_{20:4(\omega-6)}$), a polyunsaturated omega-6 fatty acid, an important constituent of phospholipids that help maintain the fluidity of cell membranes. Fatty acids are also important as they act as signals between plants and associated microorganisms (Farmer, 1994). Species of soil *Phytophthora* and related oomycetes produce abundant amounts of AA which is released into plant tissue and the surrounding during the early stages of plant infection (Ricker and Bostock, 1994). Species of *Mortierella* (mostly saprophytic fungi) can also produce high amounts of AA (Higashiyama et al., 2002). An important family of regulatory molecules (i.e., jasmonic acid) in plant cells is derived from AA. Jasmonic acid and its derivatives can modulate a wide range of biological processes in plants, among them plant resistance to damage by insects and pathogens (Blée, 2002).

Assignment of m/z 81 to Methylpyrrole

The signal at m/z 81 was assigned to methylpyrrole (N-Methylpyrrole or N-Methyl-Δ^1-pyrrolinium cation), a derivative of pyrrole. Pyrroles are components of more complex macrocycles, including antibiotics, porphyrins, and chlorophyll, among others (Gross et al., 2003). Distamycin, a mixture of N-methylpyrrole containing antibiotic are synthesized by *Streptomyces* sp. and they exhibit predominantly antifungal activity (Arcamone et al., 1964). Methylpyrrole is also a central structural component of heteroaromatic polycycle antibiotics, such as distamycyn A, that preferentially binds to the nucleotide adenine-thymine–rich sites commonly found in bacterial promoters and replication origins (Gross et al., 2003). In addition, pyrrole tetra-amides are potent antibacterial against vancomycin resistant bacteria (Dyatkina et al., 2002).

Pyrroles can be produced via six different biosynthetic pathways from several precursors, such as, amino acids and organic acids (Siebenberg et al., 2011). Methylpyrrole is a specific precursor of ornithine-derived alkaloids and it abounds in soil solutions of wheat rhizospheres (Monreal and Schnitzer, 2013). The N-methyl-pyrrolinium cation (i.e., N-methylpyrrole) is also a precursor for the biosynthesis of simple plant alkaloids of the pyrrolidine, pyrrolidides, nicotinoids, tropanes, calystegines, and presumably indolizidines types (Eich, 2008).

Assignment of m/z 347 to Consabatine

The signal at m/z 347 was assigned to consabatine, a tropane alkaloid (TA), characterized by having an acyl residue provided by 2-deoxyconsiculic acid. Tropane alkaloids are produced and released by plant endophytes such as *Streptomyces* to defend against root pathogens. Methanol extracts of actinomycetes showed the presence of TAs, such as hyoscyamine and scopolamine, that were effective to inhibit soil bacteria and fungi (Christhudas et al., 2012). A detailed discussion on the chemistry and biosynthesis together with the economic and ecological significance of TAs has been published by Eich (2008).

Assignment of m/z 346

According to the NIST webbook and published references shown in Table 3, the signal at m/z 346 may be assigned either to benzyl palmitate (BP), a C_{23} phenyl terminated fatty acid, and/or gibberellic acid (GA). The potential presence of benzyl palmitate (plant roots or microbial) in soil solution warrants further research.

Alternatively, m/z 346 may also be assigned to GA ($C_{19}H_{22}O_6$), a natural fungal and plant hormone that was first found in rice plants infected with a fungus species complex of *Gibberella fujikuroi* (Sawada) (Brian et al., 1954). The chemical structure of GA was determined by Yabuta and Sumiki (1938) by liquid-chromatography-electrospray tandem mass spectrometry (Hou et al., 2008). There are currently more than 100 Gibberellins (GAs) identified from plants, fungi, and bacteria (MacMillan, 2001). In plants, GAs are a group of tetracyclic diterpenoids biosynthesized in three oxidation steps (Hayashi et al., 2010). Biosynthetic genes for GAs acquired by fungi might be evolutionally independent of those in plants (Kawaide, 2006; Yamaguchi, 2008). The *Gibberella fujikuroi* (Sawada) is a complex of *Fusarium* mating species that are responsible for important plant diseases (Nirenberg and O'Donnell, 1998). Species of these mating populations are commonly found in soils, asymptomatic and diseased wheat, maize, sorghum, and other plants (Nirenberg and O'Donnell, 1998).

The presence of m/z 346 strongly indicates the production and release of GA by a complex of *Fusarium* species during infection of wheat roots. The latter hypothesis is also consistent with the potential production and release of BP by the endophyte *Streptomyces* that is associated with wheat roots to protect against infection by species of the soil pathogen *Fusarium*.

Assignment of m/z 113 to Glutarimide

The signal m/z 113 was assigned to glutarimide, a structural moiety that is present in a large number of molecules with antimicrobial activity (Michalska et al., 1998). Several bioactive compounds of the glutarimide group of antibiotics have been isolated from various *Streptomyces* species (Betina, 1994). The biosynthesis of glutarimide antibiotics proceeds via polyketide pathway and these antibiotics have a strong antifungal activity by inhibiting protein synthesis. Examples of glutarimide antibiotics are cycloheximide and streptimidone (Kim et al., 1999). Coombs and Franco (2003) observed endophytic colonization of wheat from a very early stage of plant development with colonization of the embryo, endosperm, and emerging radicle. Conn and Franco (2004) reported that the highest diversity and level of endophytic colonization were found in the roots of wheat grown in a dark loam soil, and the lowest were found in water-repellent sandy soil from Australia. In their study, the dominant endophytic genera were *Mycobacterium*, *Streptomyces*, and *Kitasatospora*, and their relative abundance was dependent of soil type. The signal at m/z 113 (glutarimide) is a useful chemical fingerprint indicating the presence of *Streptomyces* in wheat planted soils.

Assignment of m/z 39 and 23

Potassium (K, m/z 39) and sodium (Na, m/z 23) are essential nutrients that affect biochemical and physiological processes responsible for plant growth and the content of bioactive compounds. The importance of K fertilizer as an essential macronutrient necessary for completing the life cycle of a crop and its quality is well known (Black, 1968). Sodium, on the other hand, is now considered a functional nutrient as it promotes maximal biomass yield and/or helps reduce the requirement of an essential element (Subbarao et al., 2003). An emerging role has been attributed to K in defending against a number of biotic and abiotic stresses, including diseases, pests, drought, salinity, cold and frost, and waterlogging (Wang et al., 2013). Potassium also influences the synthesis and content of alkaloid in plants (Gremigni et al., 2001). In greenhouse studies conducted in Australia, seed alkaloid concentrations increased between 200 and 400% under severe K deficiency in sweet varieties of *L. angustifolius.* In the field, application of K to deficient soils decreased seed alkaloid concentration. High concentrations of K decreased the internal competition of pathogens for nutrient resources, and its addition to crops significantly decreased the incidence of fungi, bacteria, insects and mites, viruses, and nematodes (Perrenoud, 1990). Also, K increased the yield of plants infested by the latter organisms.

Main Interactions Between Wheat and Soil Microorganisms

Statistical Evidence

We used PCA-FA of the Py-FIMS data set for (i) providing an ecological interpretation of the main identified factors, and (ii) examining the functional relationships between the most significant m/z signals. Figure 5 shows that the points in the factor loading plot are the m/z variables, and the coordinates are the rotated loadings (Systat, 2007).

Table 4 shows strong loading values for Factors 1 and 2. Strong factor loading have values >0.75, moderate factor loading values range from 0.5 to 0.75, and loading values from 0.3 to 0.5 are considered weak (Liu et al., 2003). The rotated loading matrix for PCA shows an Eigen value of 4.4, explaining 73% of the variability, with strong positive loadings on m/z 383 (0.95), m/z 345 (0.97), m/z 473 (0.98), and m/z 347 (0.89). The Eigen value for Factor 2 was 1.3 and it explained 22% of the total variability, with strong factor loadings on m/z 304 (0.99) and m/z 81 (−0.83). Both factors explained 95% of the total temporal variance.

Interpreting the Interaction Factors

The interpretation basis for the meaning of Factor 1 and Factor 2 was based on information obtained by Py-FIMS analysis and work published by workers shown in Table 3. The eight discriminant organic structures have bioactive properties and are as secondary metabolites of plants and microbial origins. Factor 1 on the

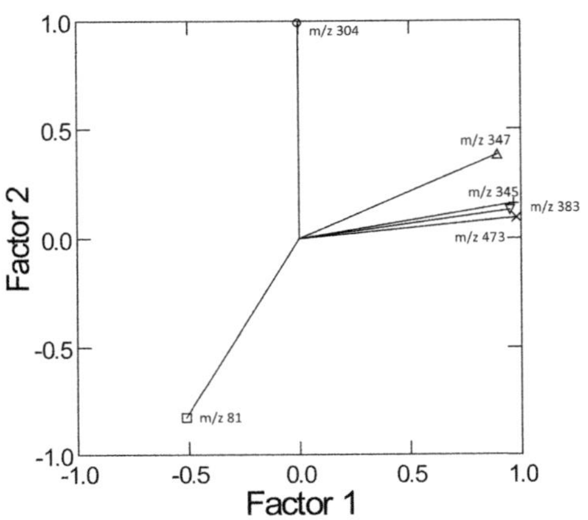

Fig. 5. Factor loading plot for the six most significant organic molecules in the soil solution of wheat rhizospheres.

Table 4. Organic compounds in soil solution with strong factor loadings explaining the total variance of the six most significant m/z signals as determined by discriminant analysis and identified by pyrolysis-field ionization mass spectrometry in the soil solution of wheat rhizosphere.

	Factor 1	Factor 2	Total variance
Variable (m/z)	383, 345, 473, 347	304, 81	95%
Factor loadings	0.95, 0.97, 0.98, 0.89	0.99, 0.83	
Eigen value	4.4	1.3	
Variance (%)	73	22	

x-axis of Fig. 5 represents a putative "plant defense" factor based on the identity of m/z 383 (diacetyllycopsamine), m/z 345 (N-benzylpalmitamide), m/z 473 (chicoric acid), and m/z 347 (consabatine), compounds derived preferentially through plant metabolism associated with plant defense pathways. That is, defense metabolites associated with Factor 1 involved two alkaloids, a polyphenol, and a fatty acid amine. The assignment of the *x*-axis to a defense factor was made also by considering the relevance of the metabolic intermediary methylpyrrole (m/z 81), the bioactive compound glutarimide (m/z 113), and the inorganic elements potassium (m/z 39) and sodium (m/z 23), as determined by Wilks's lambda test.

The positive vector of Factor 2 in Fig. 5 is interpreted as representing a putative "plant infection" factor as indicated by the presence of AA (m/z 304). This unsaturated fatty acid is a biochemical signal synthesized and released by soil-borne fungal pathogens of *Phytophtora* and *Fusarium* complexes during root infection of wheat. In potatoes, this infection signal elicits toxic antifungal terpenes (Bostock et al., 1981). The negative vector in Factor 2 is methylpyrrole, an intermediary metabolite that serves as a precursor for the synthesis of many plant alkaloids and of antibiotic moieties synthesized by Streptomyces, a wheat root endophyte (Eich, 2008).

Dynamics of Metabolites During Wheat Growth

Figure 6 shows a specific representation of the model framework to help understand the contextual function of the discriminant metabolites found in soil solution. Figure 6 depicts the main processes, metabolites, and their functions at the four growth stages of wheat. The model helps summarize and integrate information obtained from the chemical analysis by Py-FIMS, with that published in chemical databases and scientific articles, and data obtained from applying multivariate analyses models (i.e., Wilks's lambda test, DA, and PCA-FA).

The chemical information and statistical analyses of soil solution data point to components of a chemical signaling system resulting from the temporal interactions of wheat roots and associated endophytes with soil fungal pathogens. Our results point out that the presence of AA (m/z 304) indicates root infection by *Phytophtora* and/or *Fusarium* complexes likely occurring early in the growing season. Results show that the abundance of AA (m/z 304) in soil solution was highest at the four leaves and one tiller stage (25 DAS), and decreased continuously thereafter (Fig. 3). It is noteworthy that AA has been also identified as an evolutionary conserved signaling molecule that modulates plant stress signaling networks (Savchenko et al., 2010).

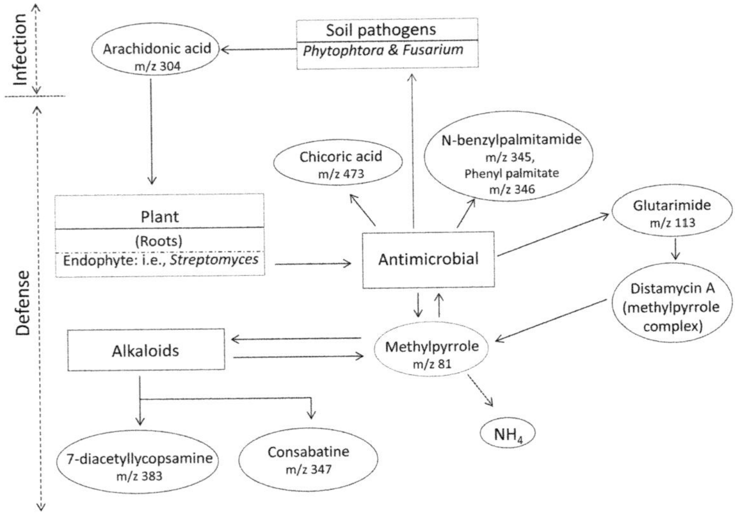

Fig. 6. A functional model for 10 metabolites of chemical signaling networks resulting from plant-soil microbial interactions. Metabolites from plant defense and soil microbial infection factors were identified in the soil solution of wheat rhizosphere.

According to Fig. 6, wheat continuously synthesized and released several defense alkaloids and other secondary metabolites (i.e., m/z signals 383, 345, 473, and 347) to control root infection by fungi. The dynamics of bioactive defense metabolites paralleled that of AA, the chemical fingerprint denoting the fungal infection AA metabolite. Sodium and potassium were also identified as key variables supporting the planned defense strategy by wheat against fungal root infection (Table 3). Plant PAs (m/z 383) also influence the diversity of fungal communities (Lotte, 2012). In our study, the predominance of plant defense chemical components during wheat growth may have been exacerbated by inadequate levels of soil nutrients for crop growth. Visual leaf color symptoms associated with N and P deficiencies were observed during the growing season of wheat.

Methylpyrrole (m/z 81) is the precursor of many antimicrobial compounds such as alkaloids and antibiotics (Eich, 2008). The abundance of methylpyrrole remained steady from Day 25 to 54 after seeding, but increased steadily thereafter. Considering the data of Fig. 3, we assumed that the steady abundance of methylpyrrole in soil solution during the first 54 d of wheat growth reflected a constant pool size. Figure 6 indicates that methylpyrrole may have served as a precursor to support the synthesis of alkaloids and antibiotics (i.e., m/z 383, 347, 113). Alternatively, methylpyrrole may be catabolized as C and N sources by soil microorganisms during soil N mineralization. It is noteworthy that antimicrobial N-containing compounds such as the alkaloids N-benzylpalmitade and glutarimide may also be used by a number of soil microbial species and strains, including those of the rhizospheres as a sole source of carbon and nitrogen (Aniszewski, 2007). The contribution of bioactive compounds containing N to soil N mineralization during periods of high N demand by crops has been largely ignored and warrants further research.

Our mass spectrometry data, plus published information, also indicate that *Streptomyces* and/or *Bacillus*, endophytes cohabiting roots beneath the epidermal cells of wheat, support the protection of wheat plants from invading *Phytophtora* or *Fusarium*. Endophytes are known to synthesize and release N-benzylpalmitamide and the antibiotics Distamycin A (methylpyrrole complex) and cycloheximide with glutarimide as the structural unit (Milewski et al., 2001). Endophytes are mutualistic to their host, receiving nutrition from the plant in exchange for producing substances such as secondary metabolites to protect the host from successful attack by fungi (Huang et al., 2007). Earlier studies have reported that isolates of *Bacillus* and one of *Streptomyces* effectively controlled damping-off caused by *Rhizoctonia solani* by reducing root infection (Broadbent et al., 1971).

The Future

Studies of biotic signaling pathways in terrestrial ecosystems need to consider the nature of the chemical and biochemical interactions when investigating ecological functions for labile SOM metabolites. New knowledge of signaling pathways components will have a significant impact on our ability to better understand plant defense mechanisms and components that fight off specific soil pathogen infection. On the other hand, knowledge of signaling components in soil–plant systems will also contribute, for example, to our understanding of the process of humification of SOM and the regulation of soil nutrient mineralization from SOM for plant growth.

In the future, studies on plant-microbial communication involving metabolomics of soil solutions will require parallel and complimentary studies of the omics cascade in crop rhizospheres. Information obtained from genome, transcriptome, and proteome studies in rhizospheres will need to be integrated with robust chemical databases of metabolites in soil solution for rigorous interpretation and elucidation of the exact ecological function of an identified metabolite or group of metabolites of the labile SOM pool.

Conclusions

Bioactive metabolites of labile SOM that discriminated between the four growth stages of wheat included the following compounds: diacetyllycopsamine, N-benzylpalmitamide, gibberellic acid, dicaffeoyl tartaric acid, arachidonic acid, methylpyrrole, consabatine, and glutarimide. Data obtained by Py-FIMS analyzed by multivariate statistical models (i.e., DA, PCA-FA) were used to test a model framework for chemical pathway components in labile SOM. Results indicated that the identified metabolites are consistent with factors and signaling pathways involved in plant defense and pathogen root infection mechanisms. Potassium and sodium in soil solutions were also involved in plant defense mechanisms.

The unambiguous identification of ecological functions of biotic metabolites from soil-plant signaling pathways require complimentary and parallel studies of genome, transcriptome, and proteome conducted in soil–plant systems. The latter knowledge is a prerequisite to make use of such information for human benefit.

Acknowledgments

We acknowledge financial support of Alberta Innovates Bio Solutions and Agriculture and Agri-Food Canada. We also thank Savka Orozovic, Terry Wu, and Bao-Luo Ma for their contribution in growing wheat in the greenhouse and for data collection. We are grateful to Dr. P. Leinweber of Rostock University in Germany for assistance with the pyrolysis-field ionization mass spectrometric analysis.

References

Aniszewski, T. 2007. Alkaloids—Secrets of life. Alkaloid chemistry, biological significance, applications and ecological role. Elsevier, Amsterdam, the Netherlands.

Arcamone, F., S. Penco, P. Orezzi, V. Nicolella, and A. Pirelli. 1964. Structure and synthesis of Distamycin *A*. Nature 203:1064–1065. doi:10.1038/2031064a0

Bais, H.P., T.L. Weir, L.G. Perry, S. Gilroy, and J.M. Vivanco. 2006. The role of root exudates in rhizosphere interactions with plants and other organisms. Annu. Rev. Plant Biol. 57:233–266. doi:10.1146/annurev.arplant.57.032905.105159

Barupal, D.K., K.H. Pradeep, G. Wohlgemuth, T. Kind, S.L. Kothari, K.E. Pinkerton, and O. Fiehn. 2012. MetaMapp: Mapping and visualizing metabolomics data by integrating information from biochemical pathways and chemical and mass spectral similarity. BMC Bioinformatics 13:99–114. doi:10.1186/1471-2105-13-99

Betina, V. 1994. Bioactive secondary metabolite of microorganisms. Progress in industrial microbiology. Elsevier, Amsterdam, the Netherlands.

Bino, R.J., R.D. Hall, O. Fien, J. Kopka, K. Saito, J. Draper, B.J. Nikolau, P. Mendes, U. Roessner-Tunali, M.H. Beale, R.N. Trethewey, B.M. Lange, E.S. Wurtele, and L.W. Sumner. 2004. Potential of metabolomics as a functional genomics tool. Trends Plant Sci. 9:418–425. doi:10.1016/j.tplants.2004.07.004

Black, C.A. 1968. Soil-plant relationships. 2nd ed. John Wiley & Sons, New York.

Blée, E. 2002. Impact of phyto-oxylipins in plant defense. Trends Plant Sci. 7:315–322. doi:10.1016/S1360-1385(02)02290-2

Boonen, J., A. Bronselaer, J. Nielandt, L. Veryser, G. DeTre, and B. De Spielgereer. 2012. Alkamid database: Chemistry, occurrence and functionality of plant N-alkylamides. J. Ethnopharmacol. 142:563–590. doi:10.1016/j.jep.2012.05.038

Bostock, R.M., J.A. Kuc, and R.A. Laine. 1981. Eicosapentaenoic and arachidonic acids from *Phytophthora infestans* elicit fungitoxic sesquiterpenes in the potato. Science 212:67–69. doi:10.1126/science.212.4490.67

Brian, P.W., G.W. Elson, H.G. Hemming, and M. Radley. 1954. The plant-growth-promoting properties of Gibberellic acid, a metabolic product of the fungus *Gibberella fujikuroi*. J. Sci. Food Agric. 5:602–612. doi:10.1002/jsfa.2740051210

Broadbent, P., K.F. Baker, and Y. Waterworth. 1971. Bacteria and actinomycetes antagonistic to fungal root pathogens in Australian soils. Aust. J. Biol. Sci. 24:925–944.

Chen, H.-J., B.S. Inbaraj, and B.-H. Chen. 2012. Determination of phenolic acids and flavonoids in *Taraxacum formosanum* Kitam by liquid chromatography-tandem mass spectrometry coupled with a post-column derivatization technique. Int. J. Mol. Sci. 13:260–285. doi:10.3390/ijms13010260

Christhudas, I.V.S.N., K.K. Kumar, and P. Agastian. 2012. Antimicrobial activity and HPLC analysis of tropane alkaloids in *Streptomyces* spp. Isolated from *Datura Stramonium* L. Asian J. Pharm. Clin. Res. 5(Suppl. 4):278–282.

Colegate, S.M., D.L. Gardner, R.J. Joy, J.M. Betz, and K.E. Panter. 2012. Dehydropyrrolizidine alkaloids, including monoesters with an unusual esterifying acid, from cultivated *Crotalaria juncea* (Sunn Hemp cv. 'Tropic Sun'). J. Agric. Food Chem. 60:3541–3550. doi:10.1021/jf205296s

Conn, V.M., and C.M.M. Franco. 2004. Effect of microbial inoculants on the indigenous actinobacterial endophyte population in the roots of wheat as determined by terminal restriction fragment length polymorphism. Appl. Environ. Microbiol. 70:6407–6413. doi:10.1128/AEM.70.11.6407-6413.2004

Coombs, J.T., and C.M.M. Franco. 2003. Visualization of an endophytic *Streptomyces* species in wheat seed. Appl. Environ. Microbiol. 69:4260–4262. doi:10.1128/AEM.69.7.4260-4262.2003

Daviss, B. 2005. Growing pains for metabolomics. Scientist 19:25–28.

Dettmer, K., P.A. Aronov, and B.D. Hammock. 2007. Mass spectrometry-based metabolomics. Mass Spectrom. Rev. 26:51–78. doi:10.1002/mas.20108

Dyatkina, N., C. Roberts, J. Keicher, Y. Dai, J. Nadherny, W. Zhang, U. Schmitz, A. Kongpachith, A. Fung, K. Novikov, L. Lou, M. Velligan, A. Khorlin, and M. Chen. 2002. Minor groove DNA binders as antimicrobial agents. 1. Pyrrole tetraamides are potent antibacterials against vancomycin resistant enterococci and methicillin-resistant *Staphylococcus aureus*. J. Med. Chem. 45:805–817. doi:10.1021/jm010375a

Eich, E. 2008. *Solanacea* and *convolvulaceae*: Secondary metabolites—Biosynthesis, chemotaxonomy, biological and economic significance. A Handbook. Springer, Berlin.

Farmer, E.E. 1994. Fatty acid signalling in plants and their associated microorganisms. Plant Mol. Biol. 26:1423–1437. doi:10.1007/BF00016483

Fu, J., K. Cheng, Z. Zhang, R. Fang, and H. Zhu. 2010. Synthesis, structure and structure–activity relationship analysis of caffeic acid amides as potential antimicrobials. Eur. J. Med. Chem. 45:2638–2643. doi:10.1016/j.ejmech.2010.01.066

Goh, E.B., G. Yim, W. Tsui, J. McClure, M.G. Surette, and J. Davies. 2002. Transcriptional modulation of bacterial gene expression by subinhibitory concentrations of antibiotics. Proc. Natl. Acad. Sci. USA 99:17025–17030. doi:10.1073/pnas.252607699

Gremigni, P., M.T.F. Wong, N.K. Edwards, D. Harris, and J. Hamblin. 2001. Potassium nutrition effects on seed alkaloid concentrations, yield and mineral content of lupins (*Lupinus angustifolius*). Plant Soil 234:131–142. doi:10.1023/A:1010576702139

Gross, M., R. Bürli, P. Jones, M. Garcia, B. Batiste, J. Kaizerman, H. Moser, V. Jiang, U. Hoch, J.-X. Duan, R. Tanaka, and K.W. Johnson. 2003. Pharmacology of novel heteroaromatic polycycle antibacterials. Antimicrob. Agents Chemother. 47:3448–3457. doi:10.1128/AAC.47.11.3448-3457.2003

Hartmann, T., A. Ehmke, U. Eilert, K. von Borstel, and C. Theuring. 1989. Sites of synthesis, translocation and accumulation of pyrrolizidine alkaloid N-oxides in *Senecio vulgaris* L. Planta 177:98–107.

Hartmann, T., and L. Witte. 1995. Chemistry, biology and chemoecology of the pyrrolizidine alkaloids. In: S.W. Pelletier, editor, Alkaloids: Chemical and biological perspectives. Vol. 9. Pergamon Press, Oxford, UK. p. 155–233.

Hayashi, K.-I., K. Horie, Y. Hiwatashi, H. Kawaide, S. Yamaguchi, A. Hanada, T. Nakashima, M. Nakajima, L.N. Mander, H. Yamane, M. Hasebe, and H. Nozaki. 2010. Endogenous diterpenes derived from ent-kaurene, a common gibberellin precursor, regulate protonema differentiation of the moss *Physcomitrella patens*. Plant Physiol. 153:1085–1097. doi:10.1104/pp.110.157909

Higashiyama, K., S. Fujikawa, E.Y. Park, and S. Shimizu. 2002. Production of arachidonic acid by *Mortierella* fungi. Biotechnol. Bioprocess Eng. 7:252–262. doi:10.1007/BF02932833

Hou, S., J. Zhu, M. Ding, and L. Guohua. 2008. Simultaneous determination of gibberellic acid, indole-3-acetic acid and abscisic acid in wheat extracts by solid-phase extraction and liquid chromatography-electrospray tandem mass spectrometry. Talanta 76:798–802. doi:10.1016/j.talanta.2008.04.041

Huang, W.Y., Y.Z. Cai, J. Xing, H. Corke, and M. Sun. 2007. Potential antioxidant resource: Endophytic fungi isolated from traditional Chinese medicinal plants. Econ. Bot. 61:14–30. doi:10.1663/0013-0001(2007)61[14:APAREF]2.0.CO;2

Joosten, L., and J.A. van Veen. 2011. Defensive properties of pyrrolizidine alkaloids against microorganisms. Phytochem. Rev. 10:127–136. doi:10.1007/s11101-010-9204-y

Kawaide, H. 2006. Biochemical and molecular analyses of Gibberellin biosynthesis in fungi. Biosci. Biotechnol. Biochem. 70:583–590. doi:10.1271/bbb.70.583

Khalil, A.T. 2006. Benzylamides from *Salvadora persica*. Arch. Pharm. Res. 29:952–956. doi:10.1007/BF02969277

Kim, B.S., S.S. Moon, and B.K. Hwang. 1999. Isolation, antifungal activity, and structure elucidation of the glutarimide antibiotic, Streptimidone, produced by *Micromonospora coerulea*. J. Agric. Food Chem. 47:3372–3380. doi:10.1021/jf981259s

Kim, J.-O., and C.W. Mueller. 1987. Introduction to factor analysis: What it is and how to do it. Quantitative Applications in the Social Sciences Series, No. 07-013. Sage Univ. Press, Newbury Park, CA.

Lee, J., and C.F. Scagel. 2009. Chicoric acid found in basil (*Ocimum basilicum* L.) leaves. Food Chem. 115:650–656. doi:10.1016/j.foodchem.2008.12.075

Liu, C.W., K.H. Lin, and Y.M. Kuo. 2003. Application of factor analysis in the assessment of groundwater quality in a Blackfoot disease area in Taiwan. Sci. Total Environ. 313:77–89. doi:10.1016/S0048-9697(02)00683-6

Lotte, J. 2012. Pyrrolizidine alkaloid composition of the plant and its interaction with the soil microbial community. Ph.D. thesis, Leiden University, Leiden, the Netherlands.

MacMillan, J. 2001. Occurrence of gibberellins in vascular plants, fungi and bacteria. J. Plant Growth Regul. 20:387–442. doi:10.1007/s003440010038

Matus, F., C. Monreal, M. Lefebvre, S.-S. Wu, R. Desjardins, and M. Derosa. 2014. Producing isotopically enriched plant, soil solution, and rhizosphere soil materials over a few hours. Commun. Soil Sci. Plant Anal. 45:865–880. doi:10.1080/00103624.2013.875190

Michalska, D., B. Morzyk, D.C. Bienko, and W. Wojciechowski. 1998. Glutarimide: A carrier transporting drug through cell membranes. Med. Hypotheses 64:472–474.

Milewski, S., F. Mignini, R. Prasad, and E. Borowski. 2001. Unusual susceptibility of a multidrug-resistant yeast strain to peptidic antifungals. Antimicrob. Agents Chemother. 45:223–228. doi:10.1128/AAC.45.1.223-228.2001

Monreal, C.M., and M. Schnitzer. 2013. The chemistry and biochemistry of organic components in the soil solutions of wheat rhizosphere. Adv. Agron. 121:179–251.

Monreal, C.M. 2015. The chemistry of labile organic matter in soil solution: I. A model for metabolites of chemical signaling pathways associated with plant–microbial interactions. In: Z. He and F. Wu, editors, Labile organic matter: Labile organic matter—Chemical compositions, function, and significance in soil and the environment. SSSA Spec. Publ. 62. SSSA, Madison, WI. doi:10.2136/sssaspecpub62.2014.0073

Mroczek, T., and K. Glowniak. 2002. Terrestrial natural products—Isolation, structure characterization, analytical methods. Chemodifferentiation. Pyrrolizidine alkaloids. Their chemistry, occurrence and significance as natural compounds. In: A.P. Rauter et al., editors, Natural products in the new millennium: Prospects and industrial applications. Kluwer Academic Publishers, Boston, MA. p. 1–46.

Mukku, V.J.R.V., R.P. Maskey, P. Monecke, I. Grün-Wollny, and H. Laatsch. 2002. 5-(2-methylphenyl)-4-pentenoic acid from a terrestrial Streptomycete. Z. Naturforsch. 57b:335–337.

Nannipieri, P., G. Pietramellara, and G. Renella. 2014. Soil as a biological system. Chapter 1. Omics in soil science. Caister Academic Press, Poole, UK.

Nirenberg, H.I., and K. O'Donnell. 1998. New *Fusarium* species and combinations within the *Gibberella fujikuroi* species complex. Mycologia 90:434–458. doi:10.2307/3761403

Ober, D., and T. Hartmann. 1999. Homospermidine synthase, the first pathway-specific enzyme of pyrrolizidine alkaloid biosynthesis, evolved from deoxyhypusine synthase. Proc. Natl. Acad. Sci. USA 96:14777–14782. doi:10.1073/pnas.96.26.14777

Perrenoud, S. 1990. Potassium and plant health. 2nd ed. Int. Potash Inst. Bern, Switzerland.

Ricker, K.E., and R.M. Bostock. 1994. Eicosanoids in the *Phytophthora infestans*-potato interaction: Lipoxygenase metabolism of arachidonic acid and biological activities of selected lipoxygenase products. Physiol. Mol. Plant Pathol. 44:65–80. doi:10.1016/S0885-5765(05)80095-5

Roitman, J.N. 1983. The pyrrolizidine alkaloids of *Amsinckia menziesii*. Aust. J. Chem. 36:769–778. doi:10.1071/CH9830769

Rovira, A.D., and M. McDougall. 1967. Microbiological and biochemical aspects of the rhizosphere. In: A.D. Mclaren and G.H. Peterson, editors, Soil biochemistry. Marcel Dekker, New York. p. 417–463.

Savchenko, T., J.W. Walley, E.W. Chehab, Y.X. Xiao, R. Kaspi, M.F. Pye, M.E. Mohamed, C. M. Lazarus, R.M. Bostock, and K. Dehesha, 2010. Arachidonic acid: An evolutionarily conserved signaling molecule modulates plant stress signaling networks. Plant Cell 22:3193–3205.

Scagel, C.F., and J. Lee. 2012. Phenolic composition of basil plants is differentially altered by plant nutrient status and inoculation with mycorrhizal fungi. HortScience 47:660–671.

Schulten, H.-R. 1987. Pyrolysis of soft ionization mass spectrometry of acquatic/terrestrial humic substances and soils. J. Anal. Appl. Pyrolysis 12:149–187. doi:10.1016/0165-2370(87)85063-5

Shewry, P.R. 2009. Wheat. J. Exp. Bot. 60:1537–1553. doi:10.1093/jxb/erp058

Siebenberg, S., N. Burkard, A. Knuplesch, B. Gust, S. Grond, and L. Heide. 2011. Two pathways for pyrrole formation in Coumermycin A1 biosynthesis: The central pyrrole moiety is formed from L-threonine. ChemBioChem 12:2677–2685. doi:10.1002/cbic.201100494

Sorge, C., R. Müller, P. Leinweber, and H.-R. Schulten. 1993. Pyrolysis-mass spectrometry of whole soils, soil particle size fractions, litter materials and humic substances: Statistical evaluation of sample weight, residue, volatilized matter and total ion intensity. Fresenius' J. Anal. Chem. 346:697–703. doi:10.1007/BF00321275

Subbarao, G.V., O. Ito, W.L. Berry, and R.M. Wheeler. 2003. Sodium—A functional plant nutrient. Crit. Rev. Plant Sci. 22:391–416.

Systat. 2007. Discriminant analysis. In: Chapter 11. SYSTAT 12 Statistics I. Systat Software, San Jose, CA.

Vigneau-Callahan, K.E., A.I. Shestopalov, P.E. Milbury, W.R. Matson, and B.S. Kristal. 2001. Characterization of diet-dependent metabolic serotypes: Analytical and biological variability issues in rats. J. Nutr. 131:924S–932S.

Vogt, T. 2010. Phenylpropanoid biosynthesis. Mol. Plant 3:2–20. doi:10.1093/mp/ssp106

Waksman, S.A. 1961. The role of antibiotics in nature. Perspect. Biol. Med. 4:271–286. doi:10.1353/pbm.1961.0001

Wang, Y., and W.-H. Wu. 2013. Potassium transport and signaling in higher plants. Ann. Rev. Plant Biol. 64:451–476.

Wang, M., Q.S. Zheng, and S. Guo. 2013. The critical role of potassium in plant stress response. Int. J. Mol. Sci. 14:7370–7390. doi:10.3390/ijms14047370

Wenke, K., M. Kai, and B. Piechulla. 2010. Belowground volatiles facilitate interactions between plant roots and soil organisms. Planta 231:499–506. doi:10.1007/s00425-009-1076-2

Yabuta,T., and Y. Sumiki. 1938. The crystallization of Gibberellins A and B. J. Agr. Chem. Soc. (Japan) 14:1526.

Yamaguchi, S. 2008. Gibberellin metabolism and its regulation. Annu. Rev. Plant Biol. 59:225–251. doi:10.1146/annurev.arplant.59.032607.092804

Yim, G., H.H. Wang, and J. Davies. 2007. Antibiotics as signalling molecules. Philos. Trans. R. Soc. Lond., B 362:1195–1200. doi:10.1098/rstb.2007.2044

Chemical Evaluation of Soil Organic Matter Structure in Diverse Cropping Systems

Bruno Henrique Martins,* Michel A. Cavigelli,* Jeffrey S. Buyer, Jude E. Maul, James B. Reeves III, and Ladislau Martin-Neto

Abstract

Soil organic matter (SOM) improves soil structure, nutrient and water retention, and biodiversity while reducing susceptibility to soil erosion. Soil organic matter also represents an important pool of C that can be increased to help mitigate global climate change. Our understanding of how agricultural management practices foster soil organic C (SOC) sequestration, however, is limited by gaps in our knowledge of the relationships among SOM structure, composition, and sequestration. Unraveling these connections is challenging because SOM is comprised of a complex mix of substances with various functional groups in various stages of decomposition, usually bonded with the soil mineral fraction. Recent application of infrared (IR) spectroscopy and pyrolysis–gas chromatography–mass spectrometry (PY-GC-MS) analysis to soil C fractions is helping uncover some of these relationships. In this chapter, we review these methods and our evolving understanding of SOM structure and composition. We use IR and PY-GC-MS data from a long-term agricultural research (LTAR) site in Maryland, USA, to explore management impacts on SOC structure and composition. The diverse cropping systems of the Farming Systems Project (FSP) LTAR differ in tillage, crop rotation, and source of fertility (mineral fertilizers vs. poultry litter [PL] and legume cover crops) as the factors that affect SOC structure and composition. Results from mid-IR analysis showed that differences in the C structure of occluded and free particulate organic matter (oPOM and fPOM, respectively) seem related to PL inputs in that degree of humification to the 20-cm depth was greater in systems receiving PL. At 20 to 30 cm, the degree of humification of oPOM and fPOM were greater in systems that were conventionally tilled than in a no-till system, suggesting that SOC sequestration at these depths may be facilitated by tilling organic matter into soils. The PY-GC-MS analysis suggests that fPOM composition is related to particular crops included in a crop rotation. The FSP cropping system that includes a perennial forage (alfalfa [*Medicago sativa* L.]) had the most unique set of pyrolysates; a system that included hairy vetch (*Vicia villosa* Roth) had the second-most unique set of pyrolysates. Combined use of IR spectroscopy and PY-GC-MS revealed different aspects of SOC structure and composition.

Abbreviations: CA, conservation agriculture; CI, condensation index; CT, conventional tillage; FA, fulvic acid; fPOM, free particulate organic matter; FSP, Farming Systems Project; HA, humic acid; HI, hydrophobicity index; HS, humic substances; HU, humin; IR, infrared; LTAR, long-term agricultural research; MIR, mid-infrared; NIR, near infrared; NT, no-tillage; OAS, organic agricultural system; OM, organic matter; oPOM, occluded particulate organic matter; Org2, organic agriculture system with a 2-yr rotation; Org3, organic agriculture system with a 3-yr rotation; Org6, organic agriculture system with a 6-yr rotation; PL, poultry litter; PY-GC-MS, pyrolysis–gas chromatography–mass spectrometry; SOC, soil organic C; SOM, soil organic matter.

Bruno Henrique Martins, Instituto Agronômico do Paraná (IAPAR) – Área de Solos, Rodovia Celso Garcia Cid, km 375, Londrina, Paraná, Brazil 86047-902. Michel A. Cavigelli, Jeffrey S. Buyer (Jeffrey.buyer@ars.usda.gov), and Jude E. Maul (Jude.maul@ars.usda.gov), USDA-ARS, Sustainable Agricultural Systems Lab., Beltsville Agricultural Research Center, 10300 Baltimore Ave., Bldg. 001, Rm. 118, Beltsville, MD 20705. James B. Reeves III, USDA-ARS, Cropping Systems and Global Change Laboratory, Beltsville Agricultural Research Center, Beltsville, MD 20705 (retired). Ladislau Martin-Neto, Empresa Brasileira de Pesquisa Agropecuária (EMBRAPA), Parque Estação Biológica – PqEB s/n, Brasília, DF, Brazil 70770-901 (ladislau.martin@embrapa.br). *Co-corresponding authors (brn0321@gmail.com, michel.cavigelli@ars.usda.gov).

doi:10.2136/sssaspecpub62.2014.0041

Labile Organic Matter—Chemical Compositions, Function, and Significance in Soil and the Environment. SSSA Special Publication 62. Zhongqi He and Fengchang Wu, editors.

Soils are complex, living systems where energy and matter are captured and transformed by plants, animals, and microbes. Plant and animal residues are transformed by soil organisms into SOM, which is then stabilized by physiochemical processes in complex ways, such that understanding SOM dynamics in soil is very challenging (Six et al., 2002b; Kleber and Johnson, 2010). Soil organic matter is composed of a complex mixture of substances with various functional groups in various stages of decomposition usually bonded with the soil mineral fraction. Soil organic matter improves soil quality by increasing soil structure, retaining soil water and plant nutrients, and increasing biodiversity while decreasing risks of soil erosion (Lal, 2008).

Changes in SOM content due to agricultural management practices—including tillage, crop rotation, and organic amendment application—affect soil bulk density, organic C storage and content, aggregate stability, porosity, water infiltration rate, microbial biomass and activity, enzyme activities, and nutrient cycling. Management also affects SOM composition, but these impacts are poorly understood. Developing agricultural management strategies that increase SOC stocks to mitigate climate change is a global priority (Beniston et al., 2014); however, we still have a limited understanding of the relationships among SOM quantity, composition, and SOC sequestration.

In this chapter, we briefly summarize the importance of SOM structure and composition and its association with soil management. We illustrate management impacts with experimental data from a LTAR site. First, we briefly review diverse soil management systems.

Soil Management Systems

Soil management affects both the quantity and composition of SOC. Intensive and continuous tillage practices have caused a worldwide decline in SOC with serious environmental and agricultural implications. Much of the losses in soil C can be attributed to reduced inputs of organic matter; increased decomposition of crop residues; tillage, which increases soil aeration and decreases the amount of physical and chemical protection of SOM (Yan et al., 2012); and artificial soil drainage (Baker et al., 2007).

Conventional tillage (CT) systems use tillage equipment such as moldboard and chisel plows to prepare a seedbed and kill weeds. Conventional tillage systems are common because they are highly effective in removing annual and perennial weeds and they provide a clean soil surface, facilitating precise seeding. On the other hand, plowing contributes to degradation of soil structure, subsoil compaction, soil surface seals, erosion, and a decrease in organic matter content. Plowing can also adversely affect biodiversity and productivity (Van Capelle et al., 2012).

Problems arising from CT in agriculture led to the development and promotion of conservation and no-tillage (NT) systems that minimize or eliminate tillage. These systems typically rely on increased use of herbicides and advances in planter technology to control weeds and to establish good crop stands, respectively (Triplett and Dick, 2008). Conservation tillage encompasses NT, where the soil is left undisturbed, and minimum or reduced till, where the number or intensity of tillage operations is reduced between harvest and planting to maintain residue cover (Page et al., 2013). The residues that remain at the soil surface have a lower decomposition rate and help reduce fluctuations in soil moisture and temperature (Mari and Changying, 2006).

One of the main goals of conservation tillage is to create biophysical conditions that slow the rate of organic matter decomposition and stabilize and protect soil (Six et al., 2002a; Glenn et al., 2011). When compared with CT practices, conservation tillage provides a series of benefits such as reversing SOM losses, improving soil porosity, ameliorating temperature fluxes, increasing water infiltration, reducing evaporation, lowering production costs, protecting soils from erosion, and reducing C emissions through savings in machinery and energy use (Van Capelle et al., 2012).

Brazil and the United States are pioneers in the development and use of soil conservation techniques, especially NT systems. In 2012, about half of Brazilian cropland was planted using NT, while 20% of US cropland was planted using NT in 2004 (Triplett and Dick, 2008; Martins et al., 2012). Since the early 1960s, conservation tillage systems have increasingly been adopted throughout the world due to their ability to help control erosion, lower fuel requirements, and help conserve soil chemical, biological, and mostly physical attributes (Verhulst et al., 2010). Nonetheless, concerns about using herbicides and simple crop rotations, and not having living roots in the soil between cash cropping periods in NT systems, have led to the development and promotion of conservation agriculture (CA).

Conservation Agriculture

Conservation agriculture involves three basic practices: (i) significant reduction of soil tillage and disturbance; (ii) permanent, or at least semi-permanent, soil protection by using crop residues or selected cover crops; and (iii) diversification of crop rotations and intercropping (FAO, 2002).

Each of these CA practices can affect SOC levels, distribution, and quality. Eliminating tillage results in (i) C inputs, such as plant residues remaining at the soil surface; (ii) reduced SOC decomposition rates; and (iii) changes in the quality of some SOC fractions (Kögel-Knabner, 2002; Zhang et al., 2011). The cover crops and crop residues in these systems serve to protect the soil surface against rain drop impacts that can break soil aggregates and increase water erosion, maintain soil moisture, suppress weed growth, and provide shelter and food for the soil biota (Blanchart et al., 2006). Crop rotations or intercropping are considered essential in CA, as they can help control weed and other pests and can improve soil fertility (Calegari, 2001). In addition, different crops, including their roots (White et al., 2011), have different chemical composition such that more diverse crop rotations are likely to provide more diverse C inputs into the soil; more diverse C inputs may also impact C sequestration rates (Drinkwater et al., 1998).

Organic Agriculture

Problems arising from conventional management in agriculture (including frequent pesticide applications, excess inorganic fertilizer use, declining soil organic matter content, soil erosion, and the presence of pesticide residues in food) led to the promotion of organic agricultural systems (OAS) that avoid the use of synthetic pesticides and conventional fertilizers (Kuepper and Gegner, 2004). In 1995, the USDA, through the National Organic Standards Board, defined organic farming as "a tillage and production system that promotes and enhances biodiversity, biologic cycles and soil biological activity based on activities that restore, maintain and increase the ecological equilibrium without using synthetic supplements."

Organic systems are intended to help farmers adapt to climate changes, establishing conditions that raise the soil ecosystem resistance to intrinsic alterations, largely by increasing SOC stocks (FAO, 2008). Most studies show greater SOC in organic than CT systems (Cavigelli et al., 2013). However, among the few studies comparing OAS with NT, results are mixed. In an OAS using animal manure, SOC level was greater than in a NT system that does not include animal manure (Cavigelli et al., 2013). When no animal manure was used in either system, however, SOC was greater in NT than OAS (Robertson et al., 2000). Cover crops in OAS have been shown to reduce soil erosion rates (Reganold et al., 1987; Green et al., 2005). Despite rapid growth and increased attention by political, media, social, and research institutions in the last 15 yr or so, OASs still represent less than 5% of agricultural areas (de Ponti et al., 2012).

Soil Organic Matter

Early definitions of SOM described a heterogeneous set of substances (humic and nonhumic) originating from the microbial decomposition of plant and animal residues (Jenkinson, 1988). During the decomposition process, these residues are broken into smaller particles. Subsequent decomposition of these particles by soil microorganisms generates a highly stable, dark colored compound known as humus.

The nonhumic SOM fraction was divided into three different groups: (i) an active fraction with a decomposition rate of years, (ii) an intermediate fraction with a decomposition rate of decades, and (iii) a passive fraction with a period of decomposition of centuries or millennia comprising organic matter stabilized by physical and chemical processes (Trumbore, 1997).

Given that SOM includes fractions with different mean residence times in soil, it is critical to understand the mechanisms involved in SOM stabilization and how different soil management systems impact these different fractions. Physical protection mechanisms refer to occlusion of SOM in soil aggregates, which form a physical barrier that limits access by decomposers and enzymes to organic substrates and limits oxygen diffusion (Pulleman and Marinissen, 2004). Chemical protection processes involve organic matter compounds bonding to mineral particles thereby decreasing the decomposition potential by enzymes and microorganisms (Six et al., 2002a).

More recently, our understanding of SOM structure has improved, accounting for the intrinsic complexity and the specific characteristics of SOM components. In a more detailed and recent definition, SOM is described as comprising a complex mixture of partially decomposed substances (organic molecules such as polysaccharides, lignin, aliphatic biopolymers, tannins, lipids, proteins, and amino sugars) originating from plant and animal residues as well as microbial biomass (Stockmann et al., 2013).

Soil Organic Matter Physical Fractionation

Soil organic matter can be physically and chemically fractionated. Physical fractionation can be based on differences in aggregate size or density. Aggregate size fractionation uses a set of sieves with different size openings to separate soil samples into different size fractions. Fractionation by aggregate size is thought to create soil fractions with more homogeneous physicochemical properties and turnover rates than total SOM (Six et al., 2002a).

Density fractionation uses a high-density solution, usually sodium polytungstate, to separate fractions according to differences in density: free or interaggregate particulate organic matter (fPOM; <1.6 g cm^{-3}), occluded or intra-aggregate particulate organic matter (oPOM; <2.0 g cm^{-3}), and heavy fraction (>2.0 g cm^{-3}) (Christensen, 2001; Marriott and Wander, 2006). Particulate organic matter is one of several measurable SOM pools used to model SOM dynamics in soil (Skjemstad et al., 2004)

Macroaggregates (250–2,500 μm) are formed around fresh residue, which is transformed into coarse, loosely associated fPOM. The fresh residue induces the formation of macroaggregates by supporting microbial activity since it is a source of C and macronutrients for microorganisms. These microorganisms then produce aggregated binding agents (Six et al., 1999; 2000). Macroaggregates are easily disrupted by disturbance and wetting and drying events (Tisdall and Oades, 1982). The processes of macroaggregate formation and disruption usually result in a loss of C and N from the fPOM pool as fPOM is transformed into oPOM. With time, the fresh material content of macroaggregates is decomposed, the more recalcitrant material remains, and these materials are combined with microbial metabolites and mineral particles, generating soil microaggregates within macroaggregates (Six et al., 2000).

The composition of fPOM is similar to that of the raw plant material (Cambardella and Elliot, 1992). The stability of fPOM is determined primarily by the recalcitrance of its chemical bonds (Oades, 1984). The stability of fPOM does not depend on physical protection processes. The oPOM fraction is composed of a more limited variety of organic compounds that have undergone a higher degree of decomposition than the fPOM fraction. The structural differences observed between fPOM and oPOM can be used as indicators of changes in SOM resulting from decomposition processes (and consequent aggregate formation), incorporation into the microbial biomass, and mineral adsorption (Golchin et al., 1994).

The heavy SOM fraction that is associated with the mineral fraction has undergone an even greater degree of decomposition than oPOM, has stronger physical and chemical adhesion to clays and other soil mineral components, and is, thus, more recalcitrant. Generally, this fraction is the most strongly associated with long-term C storage mechanisms in soil.

Soil Organic Matter Chemical Fractionation

Chemical fractionation of SOM separates the humic substances (HS), which comprise about 80% of SOM (Ikeya et al., 2011). The most widely used chemical fractionation method for SOM is the alkaline–acid extraction process proposed by the International Humic Substances Society. In this method, the HS are separated according to their solubility at various pH levels: (i) fulvic acid (FA), which is soluble in both basic and acidic solution; (ii) humic acid (HA), which is soluble in basic solutions but insoluble in acidic conditions; and (iii) humin (HU), which is insoluble in both acidic and basic solutions. Each fraction (FA, HA, and HU) is considered to be comprised of a set of molecules of different size, with few having the same structural configuration or reactive functional group arrangements (Helal et al., 2011).

The molecular structure and configuration of HS have been important areas of research on SOM. Various models have been proposed to explain the behavior of HS in the environment (Muscolo et al., 2013; Schnitzer and Monreal, 2011). The earliest model, based on the macromolecular theory, described HS as one single polymeric macromolecule with molecular weight varying between 20,000 and 50,000 Da. Structural stretching or shrinkage in solutions depended on the combination of pH values and ionic strength (Schulten and Schnitzer, 1993; Swift, 1999). However, after a series of spectroscopic and chromatographic studies, Piccolo (2001) stated that HS comprises a supramolecular set of relatively smaller molecules (molecular weight < 1000 Da) generated from degradation and decomposition of biological materials. In this supramolecular (or molecular aggregate) model, it is assumed that HS structures are maintained by weaker dispersive forces (hydrogen bridges and Van der Waals bonds) instead of covalent bonds. According to Wershaw (2004) and Nebbioso and Piccolo (2012), the supramolecular model is considered the best SOM HS representation because it provides detailed structural features, allowing more comprehensive analysis and characterization. In the context of this latest model, the HA fraction is formed from associations of predominantly hydrophobic compounds stabilized under neutral conditions by dispersive hydrophobic forces, while the FA fraction is formed from the association of smaller hydrophilic molecules having sufficient acidic functional groups to maintain scattered aggregates at any pH level (Piccolo, 2001). The HS structural configuration depends not only on the molecular structure itself, but also on environmental conditions such as ionic strength, pH, and occurrence of metallic ions such as Zn^{2+}, Cu^{2+}, and Fe^{2+} (Baalousha et al., 2006).

Analytical Methods Illustrated with a Case Study

Given the structural complexity of SOM and the inherent limitations to any method of characterizing it, multiple techniques are required to completely characterize SOM. In this section we illustrate the type of SOM structure information gleaned using IR absorption spectroscopy and pyrolysis coupled to mass spectrometry–gas chromatography (PY-GC-MS).

We illustrate the impact of soil management on SOM structure using soil samples collected from five different cropping systems at the FSP, an LTAR in Beltsville, MD, USA. The cropping systems are described in detail in Cavigelli et al. (2008, 2009), and their primary defining characteristics, each of which is likely to affect SOM composition and potentially SOC sequestration, are outlined in Table 1. Briefly, the FSP is comprised of two conventional and three organic cropping systems. A conventional NT system meets all three criteria of CA mentioned earlier, as it is never tilled and includes a rye (*Secale cereal* L.) cover crop following corn (*Zea mays* L.) within a 3-yr corn–soybean [*Glycine max* (L.) Merr.]–wheat (*Triticum aestivum* L.) double-cropped soybean rotation. The crop rotation, including a rye cover crop, is identical in a CT system that receives periodic tillage and fewer

Table 1. Characteristics of cropping systems likely to affect SOM properties at the Beltsville Farming Systems Project.

Crop rotation†	NT	CT	Org2	Org3	Org6
	C-r/S-W/S	C-r/S-W/S	v/C-r/S	v/C-r/S-W	C-r/S-W-A
Tillage‡	0	0.67	1	1	0.5
Cover crops§	0.67	0.67	1	1	1
Manure¶	0	0	~13	~18	~9
Perennial crops#	0	0	0	0	0.5

† A, alfalfa; C, corn (maize); S, soybean; W, wheat; r, rye cover crop; v, hairy vetch cover crop; crops separated by slashes (/) are harvested (A, C, S, W) or killed (r and v cover crops) in the same year; crops separated by dashes (-) are harvested or killed in different years.

‡ Proportion of years in crop rotation that include tillage.

§ Proportion of winters in crop rotation with living covers (rye, vetch, winter wheat, or alfalfa).

¶ Typical cumulative poultry litter application amount (Mg ha^{-1}) during 6 yr of crop rotation.

Proportion of years in crop rotation that include a perennial crop.

herbicides. Both CT and NT rely on synthetic N fertilizer for the bulk of crop N needs. An OAS with a 3-yr rotation (Org3) has a similar crop rotation as CT and NT, but N inputs are organic (i.e., PL) and the double-cropped soybean is replaced with a legume cover crop. The other two OASs differ in crop rotation length (2-yr [Org2] and 6-yr [Org6] rotations) and frequency of PL application. In addition, Org6 has lower tillage intensity than the two shorter rotations. While primary tillage in all OAS rotations involves a moldboard plow before planting corn and a chisel plow before planting soybean and wheat, Org6 receives no tillage during the 3 yr in which a perennial alfalfa crop is harvested for forage. While total C inputs have been essentially identical among the five systems (M. Cavigelli and C. Chapman, unpublished data, 2012), soil C levels to 30-cm depth are highest in Org3 and lowest in CT (Cavigelli, unpublished data, 2012).

Infrared Absorption Spectroscopy

Infrared absorption spectroscopy is a classical nondestructive analytical technique widely used for soil sample analysis. The IR spectral region is divided into three ranges: (i) near IR (NIR; 1000–2500 nm or 10,000–4000 cm^{-1}); (ii) mid-IR (MIR; 2500–25,000 nm or 4000–400 cm^{-1}) (Reeves III, 2010); and (iii) far IR (25,000–100,000 nm or 400–10 cm^{-1}).

To obtain IR spectra of soil samples, radiation containing all the relevant frequencies in a predetermined range is directed at the sample. The applied radiation causes vibrations (stretching or deformation) in particular molecular bonds, leading to light absorption with a specific energy quantum corresponding to the difference in levels after and before applying radiation. Since the energy quantum is proportional to the frequency (and inversely proportional to the wavelength), the resulting absorption spectrum produces a characteristic band, which can be used for analytical purposes (Miller, 2001). The wavelength at which the absorption process occurs (the energy magnitude) depends on the sample's chemical matrix and environmental factors such as temperature and adjacent functional groups. Thus, a wide variety of molecules that may contain the same type of bonding can be detected.

The IR spectroscopy techniques are highly sensitive to the organic and inorganic soil phases, leading to wide applicability in the analysis of agricultural and environmental samples (Viscarra Rossell et al., 2006). The more intense molecular vibrations found in soil samples occur at the MIR region. In IR spectroscopy the molecule of interest absorbs IR radiation resulting in a molecular vibration, which is a bending or stretching of a bond between atoms. The main difference between the NIR and MIR regions is that in the MIR region the absorption observed corresponds to fundamental vibrations in the molecule being studied, while in the NIR region the absorption corresponds to harmonics or overtones of the fundamental vibrations (Bellon-Maurel and McBratney, 2011).

The IR spectrum for soil samples—whether whole-soil or SOM fractions—often presents a typical profile with characteristic bands and peaks, enabling identification and characterization of SOM bonds. Differences regarding the origin of samples analyzed (such as depth or soil management) can be perceived in an infrared spectrum.

Mid-Infrared Spectroscopy of Soil Organic Matter from Diverse Cropping Systems

In 2011, we sampled soil from the soybean phase of the crop rotations in each of the five FSP cropping systems described in Table 1. We extracted fPOM and oPOM fractions (Marriott and Wander, 2006) from five depth increments (0–5 cm, 5–10 cm, 10–20 cm, 20–25 cm, and 25–30 cm). We used 5-cm increments near the surface and plow layers as SOC concentration gradients were expected to be steepest at these depths. Samples were taken from four field replicates, resulting in 100 total fPOM and oPOM samples.

The principal bands and peaks observed for the fPOM and oPOM samples are summarized in Table 2, and some of the IR spectra are illustrated in Fig. 1. We did not observe significant differences in spectra among fPOM and oPOM samples from organic and conventional systems at any of the five depth increments analyzed; however, in all soils, the height of peaks decreased with depth, reflecting decreases in SOC with depth (Fig. 1). We did observe higher absorbance intensity in the fPOM than the oPOM samples for bands at 3400 cm^{-1}, which Calderon et al. (2011) attributed to O-H groups originating from plant residues. Freixo et al. (2002), analyzing fPOM and oPOM fractions extracted with sodium iodide 1.8 g cm^{-3}, showed a broad band between 3700 and 3000 cm^{-1}, which they attributed to vibration stretching from aromatic O-H or even N-H groups. They also observed higher absorbance intensity in the fPOM than the oPOM samples.

Some aliphatic group (CH_2 and CH_3) peaks at the 2900 to 2850 cm^{-1} range can be noted in some spectra (Fig. 1), mostly for fPOM samples. These singular peaks are often attributed to methylene groups (Tandy et al., 2010). In certain spectral regions (around 1612 and 1315 cm^{-1}), overlapping bands, especially in the fPOM samples, indicate that the fPOM fraction may have a more heterogeneous character than the oPOM fraction. In the region between 1800 and 1500 cm^{-1}, stretching vibration of C-O groups is expected, although a more pronounced contribution of aromatic C=C is also expected. Vergnoux et al. (2011), analyzing soil HS, noted the presence of two distinct bands: the first one at 1700 cm^{-1} and the second one around 1620 to 1640 cm^{-1}. The first band can be attributed to carboxylic C=O stretching vibration. Nevertheless, the identification and attribution for the

Table 2. Main MIR peaks and bands observed and their respective attributions.

Wave number cm^{-1}	Attribution	Reference
3695/3622	C–H groups stretching (clay content)	Smidt and Parravacini (2009)
3400	O–H groups stretching (crop residues)	Calderon et al. (2011)
2930/2850	Methylene C–H groups stretching	Tandy et al. (2010)
1991/1869/1782	Possible silicates–sand interference	Calderon et al. (2011)
1733	Carboxylic acid C=O group stretching	Stenberg et al. (2010)
1612	Peaks overlapping range, with aromatic C=C predominance	Calderon et al. (2011)
1315	Amines C–N groups	Tandy et al. (2010)
1158	Polysaccharides C–O groups	Tandy et al. (2010)

second band is more difficult and complex since, in this region, there is the possibility of aromatic C=C bands and conjugated carbonyls (quinone), carboxylic salts (COO–), and type I amide C=O groups (Richard et al., 2007; Vergnoux et al., 2011; Calderon et al., 2011). Also, according to Vergnoux et al. (2011), bands observed in the range between 1200 and 1150 cm^{-1} can be attributed to phenol stretching, while peaks in the range 1130 and 1000 cm^{-1} could be attributed to C-O groups stretching in ethers and alcohols, as well as carbohydrates and silicates and mineral impurities including phosphate (He et al., 2006; 2011).

Several studies (Capriel et al., 1995; Capriel, 1997; Freixo et al., 2002) suggest calculating two different indices (condensation and hydrophobicity) to better understand the structural features of the analyzed samples. The condensation index (CI) expresses the ratio between conjugated and aliphatic groups. Higher CI values indicate more condensed and, thus, humified SOM. The ratio between the nonpolar and polar groups' absorbance intensities is called the hydrophobicity index (HI) and provides information regarding the resistance of compounds to microbial degradation. Higher HI values indicate SOM more resistant to microbial degradation. Both the CI and HI values for the analyzed samples are shown in Table 3.

The results show that CI values for oPOM samples were greater in the three OASs than the two conventional systems to a depth of 20 cm (Table 3). These results suggest a higher degree of humification, likely due to PL addition in the organic but not the conventional systems. However, from 20- to 30-cm depth, CI was also high in CT, suggesting that burial of organic matter (whether plant residue or PL) via tillage contributed to humification of oPOM at these depths (Table 3). These results suggest that these soils may provide a sink for C at 20- to 30-cm depth that is only realized by tilling organic materials to those depths, as suggested by Angers et al. (1997). The fPOM results show similar patterns but not as consistently. The CI index generally increased with soil depth in all systems, indicating that SOC humification increases with depth (Table 3).

The HI data for oPOM exhibited the opposite pattern in that HI was greater in NT than in most tilled systems at all depths, indicating that HI provides different information than CI (Table 3). The HI values generally decreased with soil

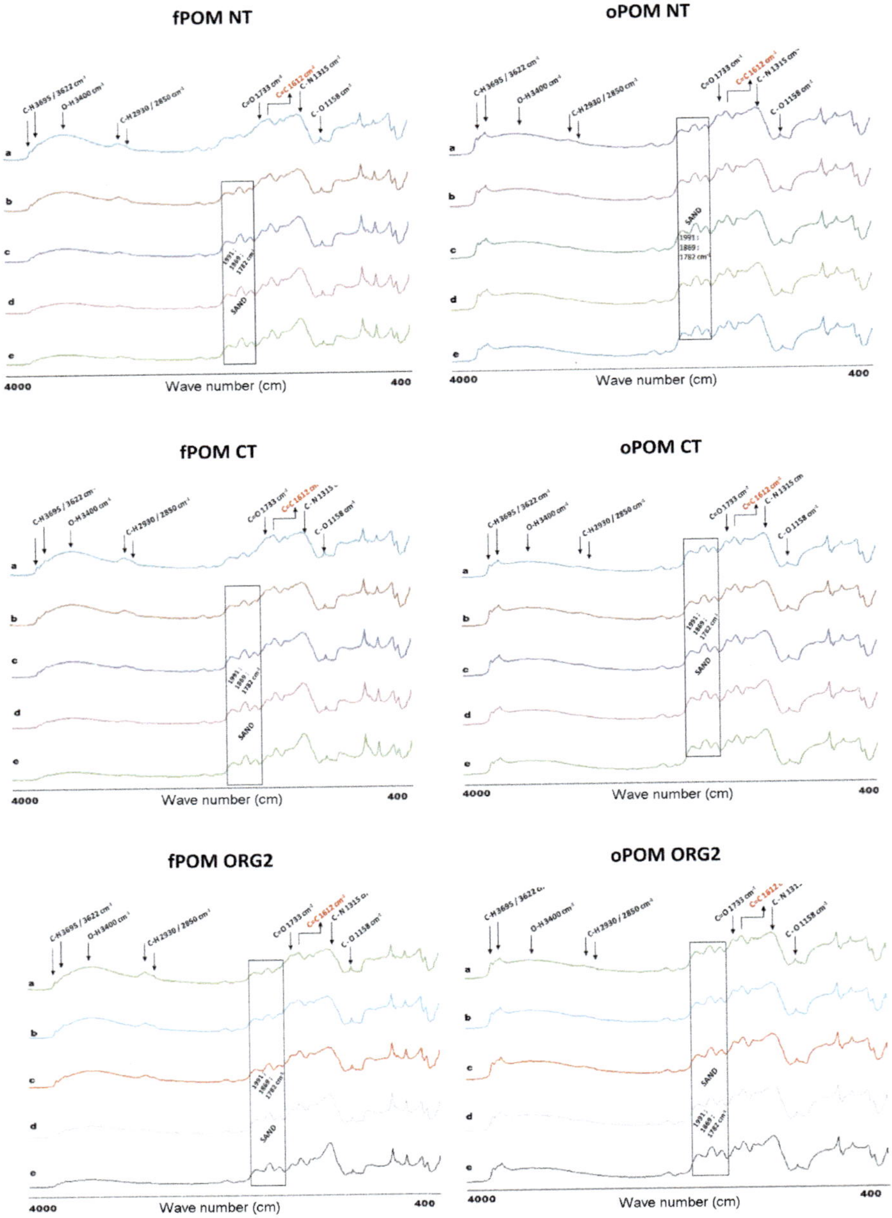

Fig. 1. Mid-infrared spectra for free particulate organic matter (fPOM) and occluded particulate organic matter (oPOM) samples for no-till (NT) and conventional till (CT). Graphs (a) to (e) represent soil depth increments 0 to 5, 5 to 10, 10 to 20, 20 to 25, and 25 to 30 cm, respectively. Org2, organic agricultural system with a 2-yr rotation.

Table 3. Condensation (CI) and hydrophobicity (HI) indexes calculated from infrared spectra for free particulate organic matter (fPOM) and occluded particulate organic matter (oPOM) samples in no-till (NT), conventional till (CT) and organic crop rotations at 2-, 3-, and 6-yr organic crop rotations (Org2, Org3, and Org6, respectively) at the Beltsville Farming Systems Project.

Depth	fPOM					oPOM				
	NT	CT	Org2	Org3	Org6	NT	CT	Org2	Org3	Org6
cm										
						CI†				
0–5	2.31	1.00	1.57	2.30	2.13	1.07	2.04	3.63	3.36	3.54
5–10	2.00	1.40	2.26	2.60	2.47	2.29	2.73	3.15	5.38	4.30
10–20	8.33	1.59	2.67	1.65	3.19	1.91	1.07	1.67	6.14	7.14
20–25	2.30	2.86	4.44	5.13	3.60	1.70	7.50	5.86	5.67	6.17
25–30	6.33	9.71	11.57	8.11	7.70	3.14	12.30	10.18	10.82	8.33
						HI‡				
0–5	1.85	2.16	2.30	2.22	2.30	1.64	1.09	1.60	1.57	1.22
5–10	2.00	1.88	1.90	2.00	2.11	2.12	1.38	1.60	1.14	1.11
10–20	1.00	1.61	1.50	1.24	1.78	1.14	2.14	1.03	1.00	1.00
20–25	0.81	1.08	0.82	0.80	0.91	1.43	0.67	1.00	0.86	0.75
25–30	0.60	0.33	0.40	0.47	0.50	0.86	0.53	0.55	0.58	0.67

† Ratio between absorbance intensity at 1612 and 2930 cm^{-1}.

‡ Ratio between absorbance intensity at 2930 and 1158 cm^{-1}.

depth for both fPOM and oPOM fractions in all systems, although this pattern was not statistically significant (ANOVA, $p > 0.05$). These results are unexpected, as greater resistance to microbial decomposition is expected with depth as shown with the CI data. The HI values were generally higher for the fPOM than the oPOM fraction to about 20 cm in tilled systems, while the opposite was true for NT except at the 0- to 5-cm depth (Table 3).

Given the structural complexity of SOM, more complete information regarding SOM structure can be gained by subjecting the samples to additional analytical techniques.

Pyrolysis Coupled to Gas Chromatography–Mass Spectrometry

Pyrolysis is a method that uses thermal breakdown to characterize SOM structure. In the pyrolytic process, macromolecules break into fragments and form more volatile products that undergo secondary and tertiary chemical reactions forming volatile products (pyrolysates), which, in turn, can be separated by means of gas chromatography and identified and quantified by mass spectrometry (Rahmonov et al., 2010). Although PY-GC-MS does not provide information regarding organic matter decomposition processes, the distribution of the different compounds identified following pyrolysis can be related to their origin (Muñoz et al., 2008).

The application of PY-GC-MS allows determination of the following groups of chemical compounds: normal-chain and branched-chain, saturated and unsaturated aliphatics; cycloaliphatics; aromatics; carbohydrates; N-containing compounds; sulfur containing compounds; and phenol, furan, and pyran derivatives. Relative contributions (in the peak area percentage) for the pyrolysis products can be assessed by calculating the area of the gas chromatographic peaks and assuming a linear relationship between the detector response and the quantity of a given analyte (Rahmonov et al., 2010). The wide use of gas chromatography to analyze pyrolysis fragments is due to the high incidence and the chemical complexity of small fragments generated by the pyrolysis process. The use of mass spectrometry provides the capability to identify fragments with high sensitivity—which is critical because sample size is in the milligram range—by their mass spectra.

According to Pognani et al. (2011), alkene compounds identified in soil samples analyzed with PY-GC-MS may represent peaks of lipids generated from bacterial activity, while the alkane compounds may be related to peaks of lipids contained in recent organic matter inputs. The furan peaks originate from polysaccharides, while the cyclicketones and pyrans—despite also being related to polysaccharides—are generated from decomposed and nondecomposed cellulose (Dignac et al., 2005). The aromatic molecules can be divided into two different categories according to their chemical properties (such as chain features, structural configuration, condensation degree) and origin: methoxy-phenols and methylketones (originating from lignin, and benzene and phenols (unknown origin) (González-Vila et al., 2009). According to de Assis et al. (2012), some phenols and alkylphenols can be attributed to lignin, tannins, proteins, and carbohydrates.

Cellulosic molecules undergoing pyrolysis generate carbonyls, acids, furans, pyranones, and phenols. Fatty acid molecules can be decarboxylated during the pyrolysis process (Kögel-Knabner, 2000), although the decarboxylation process does not occur readily.

In research conducted by Rumpel et al. (2009), it was verified that PY-GC-MS soil data from agricultural systems revealed peaks considered to be polysaccharide derivatives, lignin, and some unknown compounds. The main polysaccharide-derived compounds included structures similar to furans, cyclopentene, and cyclopropane, while the lignin-derivative compounds included higher incidence of methoxyphenol structures.

Pyrolysis–Gas Chromatography–Mass Spectrometry of Soil Organic Matter from Diverse Cropping Systems

The same fPOM and oPOM fractions described above were analyzed by PY-GC-MS. The analysis was performed in a 5200 Pyroprobe (CDS Analytical Inc.), at 600°C by 20 s, coupled to a PerkinElmer Clarus 500 (PerkinElmer Inc.) GC-MS, using He as mobile phase and a DB-35MS chromatographic column, with 20 m length, 0.180 mm internal diameter, and 35% phenol as stationary phase. The spectral scans were background subtracted, deconvoluted, and compounds identified using AMDIS software (National Institute of Standards and Technology [NIST]; http://chemdata.nist.gov/dokuwiki/doku.php?id=chemdata:amdis) and a customized database created by Chemical Data System Analytical (CDS) designed to characterize plant material derivatives. Compounds not identified with the CDS database were then identified using the NIST–EPA–NIH Mass Spectral Library (NIST). Pyrolysate compounds with the highest peak intensities in these samples, their chemical structures, and likely origins are listed in Table 4. The peaks were selected using the information-gain technique (Hall et al., 2009).

Table 4. Compounds with high intensity peaks for free particulate organic matter (fPOM) and occluded particulate organic matter (oPOM) fractions using pyrolysis gas chromatography–mass spectrometry.

Compound	oPOM	fPOM	Chemical structure	Classification; origin†
1,2,4-Trimethoxybenzene		x	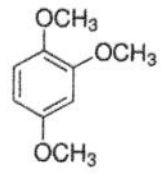	Aromatic ether; lignin derivative
1,3-benzenediol, 4-ethyl-		x	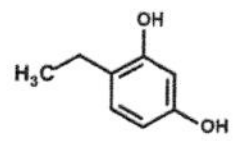	Aromatic; lignin, protein, tannin
1HIndene3-methyl-		x	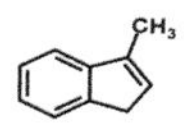	Aromatic hydrocarbon; lignin derivative
2(5H)Furanone		x	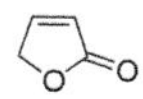	Furan derivative; polysaccharide
2′,4′-Dihydroxypropio-phenone		x	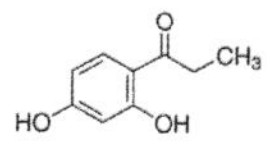	Phenolic ketone, soil lipid; lignin derivative
2-Cyclopenten1-one2-hydroxy-3-methyl-		x	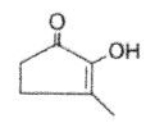	Methyl cyclo ketone; polysaccharide
2-Cyclopenten1-one, 2-methyl-		x	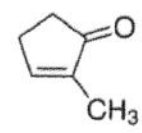	Methyl cyclo ketone; polysaccharide
2-Cyclopenten1-one, 3-methyl-		x	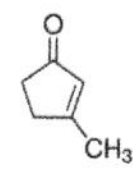	Methyl cyclo ketone; polysaccharide
2-Furancarboxaldehyde, 5-methyl-	x		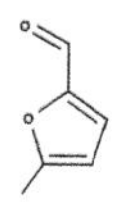	Furan derivative; polysaccharide
2-Methoxy-4-vinyl, Phenol	x		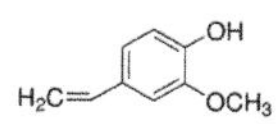	Phenolic ether; lignin derivative
2-Methoxy-5-methyl, Phenol	x	x	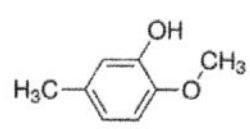	Phenolic ether; lignin derivative
2-Propanone, 1-(4-hydroxy-3-methoxyphenyl)-		x	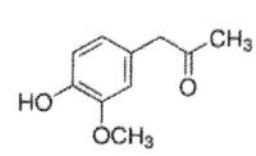	Ether ketone; lignin derivative
2-Propanone, 1-hydroxy-		x	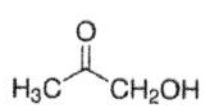	Ketone; polysaccharide
2-Propenoic acid, 3-(4-hydroxy-3-methoxyphenyl)-methylester		x	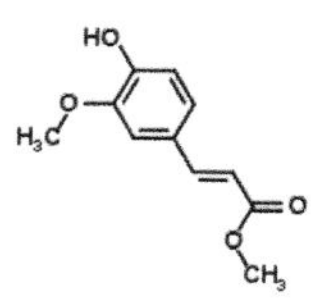	Aromatic ether acid; polysaccharide

Table continued.

Table 4. Continued.

Compound	oPOM	fPOM	Chemical structure	Classification; origin†
3′,5′-Dimethoxyaceto-phenone	x	x	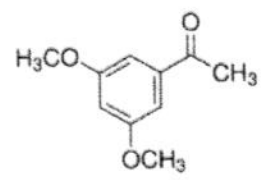	Aromatic ether-ketone; polysaccharide
3-Hydroxy-4-methoxymandelic acid	x		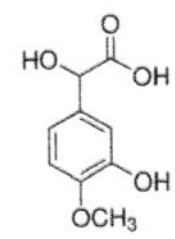	Aromatic ether-acid; polysaccharide
Acetophenone, 4′-hydroxy-		x	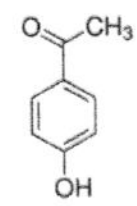	Phenolic ketone; polysaccharide, soil lipid
2-methyl-benzofuran		x	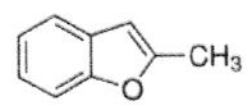	Furan derivative; Polysaccharide
1,3,5-Trimethylbenzene	x		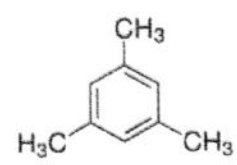	Aromatic; lignin
2,3-dihydro- benzofuran	x		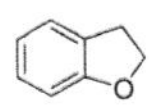	Furan derivative; polysaccharide
Ethanon1,-(3-hydroxy-4-methoxyphenyl)-	x		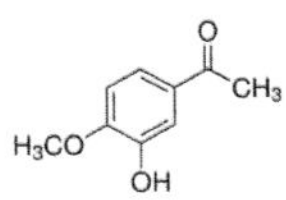	Ether ketone; lignin derivative
Furan		x		Cyclo hetero-hydrocarbon; polysaccharide
Furfural	x		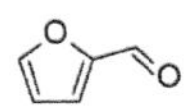	Furan derivative; polysaccharide
Maltol		x	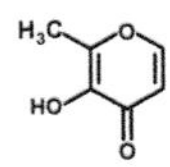	Pyran derivative; polysaccharide
Phenol	x		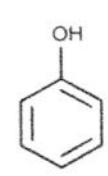	Phenol; lignin derivative
Phenol, 2,6-dimethoxy-	x	x	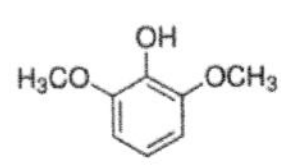	Phenolic ether; lignin derivative
Phenol, 2,6-dimethoxy-4-(2-propenyl)-		x	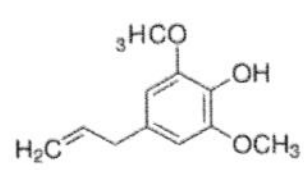	Phenolic ether; lignin derivative
Phenol, 2-ethyl-	x	x	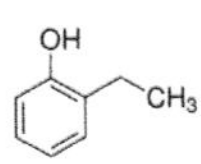	Phenolic hydrocarbon; lignin derivative

Table 4. Continued.

Compound	oPOM	fPOM	Chemical structure	Classification; origin†
Phenol, 2-ethyl-4-methyl-		x	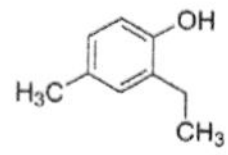	Phenolic hydrocarbon; lignin derivative
Phenol, 2-methoxy-	x	x	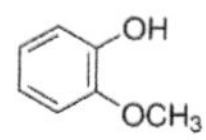	Phenolic ether; lignin derivative
Phenol, 4-ethyl-2-methoxy-		x	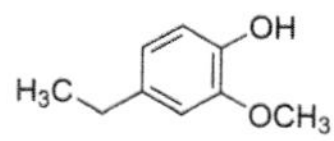	Phenolic ether; lignin derivative
Phenol, 4-methoxy-3-methyl-		x	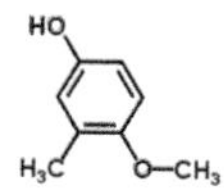	Phenolic ether; lignin derivative
Phenol, 4-methyl-	x	x	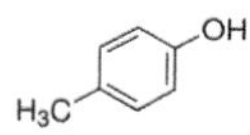	Phenol; lignin derivative
Styrene	x		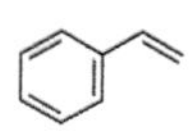	Aromatic; lignin, protein, tannin

† Classification according to Vancampenhout et al. (2009, 2012) and Pognani et al. (2011).

There were more peaks identified for the fPOM than oPOM fractions for all cropping systems. There were also more peaks related to furans and polysaccharides (such as methylfurans and furaldehyde), lignin aromatic derivatives (phenols and methoxyphenols) and some ketone-like soil lipids for the fPOM than the oPOM fraction. There was a higher incidence of structures with aromatic characteristics in the oPOM than the fPOM samples. In the fPOM samples, there were aromatic nuclei bonded to aliphatic chains. These two observations combined suggest that the oPOM fraction is more homogeneous, recalcitrant, and physically and chemically protected from microbial decomposition than the fPOM fraction. The lower number of peaks and fewer molecular structures identified for the oPOM than the fPOM samples may indicate a more biologically transformed material. The more heterogeneous character presented by the fPOM samples suggests a fraction more vulnerable to microbial attack but also reflects greater nutrient availability for crops and microbial biomass.

There were also more peaks identified for samples from organic than conventional systems for both fPOM and oPOM fractions. However, most compounds seemed to belong to similar chemical classes. According to Nierop et al. (2001), differences in SOM composition between organic and conventional agricultural systems are mostly due to the quantity and quality of organic material inputs. Since total C inputs were similar in all systems, differences seen here are likely due to differences in organic material composition consistent with management differences among systems listed in Table 1.

We used nonparametrical ANOVA to analyze the nonnormally distributed percentage data that are typical for PY-GC-MS analysis (Vancampenhout et al., 2012). Results for the oPOM samples showed no significant differences in peak intensities for compounds among the five cropping systems (data not shown). For the fPOM samples, there were statistically significant differences ($\alpha < 0.05$) among all systems for at least one compound for at least one depth, except between CT and Org2 (Table 5). For the Org2 and Org3 systems, we found significant differences for only two compounds (Table 5). These results indicate that CT, Org2, and Org3 had the most similar fPOM structure among the five cropping systems. Since C inputs in CT included only crop residues, while C inputs in Org2 and Org3 also included PL, these results suggest that PY-GC-MS is not able to distinguish impacts on SOC of PL additions. Also, there were few differences between CT and NT (only three of 48 significant pairwise comparisons), suggesting that tillage was not a significant factor influencing fPOM structure as defined with PY-GC-MS at this site.

By contrast, 30% of the 48 pairwise comparisons found to be significant were between Org6 and the other four cropping systems (Table 5). These results suggest that including a perennial crop in the rotation may have more influence on fPOM structure as defined by PY-GC-MS than tillage or PL inputs even 2 yr after the perennial crop was killed. These differences were especially striking at 0- to 5-cm and 20- to 25-cm depths where, respectively, 11 of 13 and nine of nine significant pairwise differences were between Org6 and the other four systems. The impact of plant inputs on PY-GC-MS results is consistent with other studies (Nierop et al., 2001; Dieckow et al., 2006).

At 10 to 20 cm, seven of 10 differences among systems were between Org3 and the other systems. Perhaps these differences were due to the inclusion of a hairy vetch cover crop, which would have been incorporated into the soil about 1 yr before soil samples were taken in this rotation. While Org2 would also normally have included a hairy vetch cover crop at this time, the vetch was not planted in fall 2009 in Org2 due to wet soil conditions. The previous vetch cover crop in these plots would have been killed 3 yr earlier. Our results are consistent with Marriott and Wander (2006), who found that fPOM quality reflected residue inputs, and with Drinkwater et al. (1998), who suggested that hairy vetch can have a strong impact on SOC pools due to the C quality of its tissues.

Conclusions

Long-term field experiments are crucial for understanding SOM dynamics. In this LTAR we observed distinct differences in SOM structure and composition between organic and conventional cropping systems. The use of spectroscopic techniques, specifically mid-IR and PY-GC-MS, to characterize POM fractions provided additional and complementary information to the bulk SOC data.

In general, the CI and HI (derived from mid-IR spectroscopy) revealed differences in the C structure of oPOM and fPOM that seem to be related to PL inputs and to tillage. These results suggest that the degree of humification (evaluated using CI) to 20 cm is greater in systems to which PL has been applied, possibly due to addition of more humified C sources in PL than in plant residues. At 20 to 30 cm, fPOM and oPOM showed greater humification in all tilled systems than NT, suggesting that soil C sequestration at these depths may be facilitated by tilling organic materials to these depths. On the other hand, difference in fPOM composition detected using PY-GC-MS seem to reflect the inclusion of a perennial

Table 5. Results of statistical analyses comparing the amounts of pyrolysate compounds for free particulate organic matter samples in no-till (NT), conventional till (CT) and organic crop rotations at 2-, 3-, and 6-yr organic crop rotations (Org2, Org3, and Org6, respectively) at the Beltsville Farming Systems Project.

Compound	Depth increment (cm)				
	0–5	5–10	10–20	20–25	25–30
2-Cyclopenten1-one2-hydroxy-3-methyl	NA†	NA	NA	Org3 < Org6‡	NT < Org3 Org3 < Org6
2-Cyclopenten1-one, 2-methyl	NA	NA	CT < Org3 Org2 < Org3	NA	NA
2-Methoxy-5-methyl, Phenol	NA	NA	NA	NA	CT < Org3 CT < Org6
2-Propanone, 1-(4-hydroxy-3-methoxyphenyl)-	NA	NA	NA	NA	NT < Org2 Org2 < Org6
2-Propanone, 1-hydroxy-	NA	NA	NA	NT < Org6 CT < Org6 Org2 < Org6	NA
1,3-benzenediol, 4-ethyl-	NT < Org6 Org2 < Org6 Org3 < Org6	NA	NA	NA	NA
2′,4′-Dihydroxypropiophenone	NA	NT < Org2 NT < Org6	NA	NA	NA
2(5H)Furanone	NA	NA	NT < Org3 NT < Org6 CT < Org3 CT < Org6	NT < Org6 CT < Org6 Org3 < Org6	NA
3′,5′-Dimethoxyacetophenone	NA	NA	NA	NA	NT < Org2 Org2 < Org6

Table continued.

Table 5. Continued.

Compound	Depth increment (cm)				
	0–5	5–10	10–20	20–25	25–30
1,2,4-Trimethoxybenzene	NA	NA	NA	NA	NT > CT NT < Org2 NT < Org3
Acetophenone, 4′-hydroxy-	NT > CT NT < Org6 CT < Org3 Org2 < Org6 Org3 < Org6	NA	NA	NA	NA
2-methyl-benzofuran	Org2 < Org6 Org3 < Org6	NA	NA	NA	NA
Furan	NA	NA	NT < Org3 CT < Org3 Org2 < Org3 Org3 < Org6	NT < Org6 CT < Org6	NA
Phenol, 2-ethyl-4-methyl-	NT < Org6 Org2 < Org6 Org3 < Org6	NA	NA	NA	NA
Phenol, 4-methoxy-3-methyl-	NA	NT > CT NT < Org2 NT < Org6	NA	NA	NA

† NA indicates there were no significant differences (a > 0.05) among systems at a given depth for a given compound.

‡ Only peaks with statistically significant differences ($\alpha = 0.05$) among systems are listed, that is, when the Kruskal–Wallis test was significant and pairwise differences were identified with the Mann–Whitney test for a given depth.

legume crop (alfalfa), and perhaps an annual legume cover crop (hairy vetch), in the rotations rather than either tillage or PL inputs.

A complete comprehension of SOM dynamics and structure is crucial to understanding SOM equilibrium in terrestrial ecosystems, especially when submitted to different soil management systems. Determining whether soils are sinks or sources of atmospheric C is critical to achieving environmental sustainability and climate change mitigation. Spectroscopic and PY-GC-MS techniques are important tools in characterizing and assessing SOM structural aspects and can lead to detailed understanding of how soil management practices can alter soil organic C sequestration mechanisms.

Acknowledgments

The authors would like to thank the Brazilian National Council of Research and Development (CNPq) for their financial support, Embrapa-Labex and BARC for the exchange opportunity, Kate White for the valorous help with the IR analysis, Ednaldo Ferreira for the support on the statistical analysis, and Chris Rasmann for coordinating and conducting soil core sampling and processing with the help of Gwen Bagley, Alice Chang, Kyle Cuilla, Grace Garst, Andrew Ghabranious, Sophie Jin, Victoria Lake, Clay Nadasen-Reynolds, Matt Stocker, Claire Teitelbaum, Taylor Throwe, and Alex Welch.

References

Angers, D.A., M.A. Bolinder, M.A. Carter, E.G. Gregorich, C.F. Drury, B.C. Liang, R.P. Voroney, R.R. Simard, G.G. Donald, R.P. Beyaert, and J. Martel. 1997. Impact of tillage practices on organic carbon and nitrogen storage in cool, humid soils of eastern Canada. Soil Tillage Res. 41:191–201. doi:10.1016/S0167-1987(96)01100-2

Baalousha, M., M. Motelica-Heino, and P. Le Coustumer. 2006. Conformation and size of humic substances: Effects on major cation concentration and type, pH, salinity, and residence time. Colloid Surf., A 272:48–55. doi:10.1016/j.colsurfa.2005.07.010

Baker, J.M., T.E. Ochsner, R.T. Venterea, and T.J. Griffis. 2007. Tillage and soil carbon sequestration: What do we really know? Agric. Ecosyst. Environ. 118:1–5. doi:10.1016/j.agee.2006.05.014

Bellon-Maurel, V., and A. McBratney. 2011. Near-infrared (NIR) and mid-infrared (MIR) spectroscopic techniques for assessing the amount of carbon stock in soils: Critical review and research perspectives. Soil Biol. Biochem. 43:1398–1410. doi:10.1016/j.soilbio.2011.02.019

Beniston, J.W., S.T. Dupont, J.D. Glover, R. Lal, and J.A.J. Dungait. 2014. Soil organic carbon dynamics 75 years after land-use change in perennial grassland and annual wheat agricultural systems. Biogeochemistry 120:37–49. doi:10.1007/s10533-014-9980-3

Blanchart, E., C. Villenave, A. Viallatoux, B. Barthès, C. Girardin, A. Azontonde, and C. Feller. 2006. Long-term effect of a legume cover crop (*Mucuna pruriens* var. *utilis*) on the communities of soil macrofauna and nematofauna, under maize cultivation, in southern Benin. Eur. J. Soil Biol. 42:S136–S144. doi:10.1016/j.ejsobi.2006.07.018

Calderon, F.J., J.B. Reeves, H.P. Collins, and E.A. Paul. 2011. Chemical differences in soil organic matter fractions determined by diffuse-reflectance mid-infrared spectroscopy. Soil Sci. Soc. Am. J. 75:568–579. doi:10.2136/sssaj2009.0375

Calegari, A. 2001. Cover crop management, p. 85–92. Madrid, Spain.

Cambardella, C.A., and E.T. Elliot. 1992. Particulate soil organic-matter changes across a grassland cultivation sequence. Soil Sci. Soc. Am. J. 56:777–783. doi:10.2136/sssaj1992.03615995005600030017x

Capriel, P. 1997. Hydrophobicity of the organic-matter in arable soils: Influence of management. Eur. J. Soil Sci. 48:457–462. doi:10.1046/j.1365-2389.1997.00098.x

Capriel, P., T. Beck, H. Borchert, J. Gronholz, and G. Zachmann. 1995. Hydrophobicity of the organic-matter in arable soils. Soil Biol. Biochem. 27:1453–1458. doi:10.1016/0038-0717(95)00068-P

Cavigelli, M.A., B.L. Hima, J.C. Hanson, J.R. Teasdale, A.E. Conklin, and Y.C. Lu. 2009. Long-term economic performance of organic and conventional field crops in the mid-Atlantic region. Renew. Agric. Food Syst. 24:102–119. doi:10.1017/S1742170509002555

Cavigelli, M.A., S.B. Mirsky, J.R. Teasdale, J.T. Spargo, and J. Doran. 2013. Organic management systems to enhance ecosystem services. Renew. Agric. Food Syst. 28:145–159. doi:10.1017/S1742170512000439

Cavigelli, M.A., J.R. Teasdale, and A.E. Conklin. 2008. Long-term agronomic performance of organic and conventional field crops in the mid-Atlantic region. Agron. J. 100:785–794. doi:10.2134/agronj2006.0373

Christensen, B.T. 2001. Physical fractionation of soil and structural and functional complexity in organic matter turnover. Eur. J. Soil Sci. 52:345–353. doi:10.1046/j.1365-2389.2001.00417.x

de Assis, C.P., J.A. González-Perez, J.M. Rosa, I. Jucksch, E.S. Mendonça, and F.J. González-Vila. 2012. Analytical pyrolysis of humic substances from a Latosol (Typic Hapludox) under different land uses in Minas Gerais, Brazil. J. Anal. Appl. Pyrolysis 93:120–128. doi:10.1016/j.jaap.2011.10.005

de Ponti, T., B. Rijk, and M.K. Van Ittersum. 2012. The crop yield gap between organic and conventional agriculture. Agric. Syst. 108:1–9. doi:10.1016/j.agsy.2011.12.004

Dieckow, J., J. Mielniczuk, F.J. Gonzales-Vila, H. Knicker, and C. Bayer. 2006. No-till cropping systems and N fertilisation influences on organic matter composition of physical fractions of a subtropical Acrisol as assessed by analytical pyrolysis (Py-GC/MS). Geoderma 135:260–268. doi:10.1016/j.geoderma.2006.01.002

Dignac, M.F., S. Houot, C. Francou, and S. Derenne. 2005. Pyrolitic study of compost and waste organic matter. Org. Geochem. 36:1054–1071. doi:10.1016/j.orggeochem.2005.02.007

Drinkwater, L.E., P. Wagoner, and M. Sarrantonio. 1998. Legume-based cropping systems have reduced carbon and nitrogen losses. Nature 396:262–265. doi:10.1038/24376

Food and Agricultural Organization of the United Nations (FAO). 2002. Conservation agriculture when agriculture is profitable and sustainable. Land and Water Development Division, FAO, Rome, Italy.

Food and Agricultural Organization of the United Nations (FAO). 2008. Conference on ecological agriculture: Mitigating climate change, providing food security and self-reliance for rural livelihoods in Africa. African Union Headquarters, Addis Ababa, Ethiopia, 26–28 Nov. 2008. ftp://ftp.fao.org/docrep/fao/011/aj993e/aj993e00.pdf (accessed 5 Mar. 2012).

Freixo, A.A., L.P. Canellas, and P.L.O.A. Machado. 2002. Propriedades espectrais da matéria organica leve-livre e leve intra-agregado de dois latossolos sob plantio direto e preparo convencional. (In Portuguese with English abstract.) Rev. Bras. Cienc. Solo 26:445–456 . doi:10.1590/S0100-06832002000200018

Glenn, A.J., B.D. Amiro, M. Tenuta, C. Wagner-Riddle, G. Drewitt, and J. Warland. 2011. Contribution of crop residue carbon to soil respiration at a northern Prairie site using stable isotope flux measurements. Agric. For. Meteorol. 151:1045–1054. doi:10.1016/j.agrformet.2011.03.008

Golchin, A., J.M. Oades, J.O. Skjemstad, and P. Clarke. 1994. Soil structure and carbon cycling. Aust. J. Soil Res. 32:1043–1065. doi:10.1071/SR9941043

González-Vila, F.J., J.A. González-Perez, K. Akdi, M.D. Gómis, F. Perez-Barrera, and T. Verdejo. 2009. Assessing the efficiency of urban waste biocomposting by analytical pyrolysis (Py-GC/MS). Bioresour. Technol. 100:1304–1309. doi:10.1016/j.biortech.2008.06.067

Green, V.S., M.A. Cavigelli, T.H. Dao, and D. Flanagan. 2005. Soil physical properties and aggregate-associated C, N, and P distributions in organic and conventional cropping systems. Soil Sci. 170:822–831. doi:10.1097/01.ss.0000190509.18428.fe

Hall, M., E. Frank, G. Holmes, B. Pfahringer, P. Reutemann, and I.H. Witten. 2009. The WEKA data mining software: An update. SIGKDD Explor. 11:10–18. doi:10.1145/1656274.1656278

Helal, A.A., G.A. Murad, and A.A. Helal. 2011. Characterization of different humic materials by various analytical techniques. Arabian J. Chem. 4:51–54. doi:10.1016/j.arabjc.2010.06.018

He, Z., C.W. Honeycutt, and H. Zhang. 2011. Elemental and Fourier transform-infrared spectroscopic analysis of water- and pyrophosphate-extracted soil organic matter. Soil Sci. 176:183–189. doi:10.1097/SS.0b013e318212865c

He, Z., T. Ohno, B.J. Cade-Menun, M.S. Erich, and C.W. Honeycutt. 2006. Spectral and chemical characterization of phosphates associated with humic substances. Soil Sci. Soc. Am. J. 70:1741–1751. doi:10.2136/sssaj2006.0030

Ikeya, K., T. Hikage, S. Arai, and A. Watanabe. 2011. Size distribution of condensed aromatic rings in various soil humic acids. Org. Geochem. 42:55–61. doi:10.1016/j.orggeochem.2010.10.006

Jenkinson, D.C. 1988. The turnover of organic matter in soil, Longman, Harlow, England. p. 564–607.

Kleber, M., and M.G. Johnson. 2010. Advances in understanding the molecular structure of soil organic matter: Implications for interactions in the environment. Academic Press, New York. p. 77–142.

Kögel-Knabner, I. 2000. Analytical approaches for characterizing soil organic matter. Org. Geochem. 31:609–625.

Kögel-Knabner, I. 2002. The macromolecular organic composition of plant and microbial residues and inputs to soil organic matter. Soil Biol. Biochem. 34:139–162. doi:10.1016/S0038-0717(01)00158-4

Kuepper, G., and I. Gegner. 2004. Organic crop production overview. ATTRA, Fayetteville, AR.

Lal, R. 2008. Carbon sequestration. Phylos. Trans., B 363:815–830. doi:10.1098/rstb.2007.2185

Mari, G.R., and J.I. Changying. 2006. Conservation tillage for the protection of soil quality and sustainability. Am.-Eurasian J. Sci. Res. 1:55–60.

Marriott, E.E., and M.M. Wander. 2006. Qualitative and quantitative differences in particulate organic matter fractions in organic and conventional farming systems. Soil Biol. Biochem. 38:1527–1536. doi:10.1016/j.soilbio.2005.11.009

Martins, M.R., D.A. Angers, and J.E. Corá. 2012. Co-accumulation of microbial residues and particulate organic matter in the surface layer of a no-till Oxisol under different crops. Soil Biol. Biochem. 50:208–213. doi:10.1016/j.soilbio.2012.03.024

Miller, C.E. 2001. Chemical principles of near-infrared spectroscopy. AACCI, Minnesota, MN. p. 19–37

Muñoz, C., C.M. Monreal, M. Schnitzer, and E. Zagal. 2008. Influence of *Acacia caven* (Mol) coverage on carbon distribution and its chemical composition in soil organic carbon fractions in a Mediterranean-type climate region. Geoderma 144:352–360. doi:10.1016/j.geoderma.2007.12.002

Muscolo, A., M. Sidari, and S. Nardi. 2013. Humic substance: Relationship between structure and activity. Deeper information suggests univocal findings. J. Geochem. Explor. 129:57–63. doi:10.1016/j.gexplo.2012.10.012

Nebbioso, A., and A. Piccolo. 2012. Molecular understanding of a humic acid by "humeomic" fractionation and benefits from preliminary HPSEC separation. IHSS, Beijing, China. p. 50–51.

Nierop, K.G.J., M.M. Pulleman, and J.C.Y. Marinissen. 2001. Management induced organic matter differentiation in grassland and arable soils: A study using pyrolysis techniques. Soil Biol. Biochem. 33:755–764. doi:10.1016/S0038-0717(00)00223-6

Oades, J.M. 1984. Soil organic matter and structural stability mechanisms and implications for management. Plant Soil 76:319–338. doi:10.1007/BF02205590

Page, K., Y. Dash, and R. Dalal. 2013. Impacts of conservation tillage on soil quality, including soil-borne crop diseases, with a focus on semi-arid grain cropping systems. Australas. Plant Pathol. 42:363–377. doi:10.1007/s13313-013-0198-y

Piccolo, A. 2001. The supramolecular structure of humic substances. Soil Sci. 166:810–832. doi:10.1097/00010694-200111000-00007

Pognani, M., R. Barrena, X. Font, F. Adani, B. Scaglia, and A. Sánchez. 2011. Evolution of organic matter in a full-scale composting plant for the treatment of sewage sludge and biowaste by respiration techniques and pyrolysis-GC/MS. Bioresour. Technol. 102:4536–4543. doi:10.1016/j.biortech.2010.12.108

Pulleman, M.M., and J.C.Y. Marinissen. 2004. Physical protection of mineralizable C in aggregates from long-term pasture and arable soil. Geoderma 120:273–282. doi:10.1016/j.geoderma.2003.09.009

Rahmonov, O., W.J. Kowalski, and R. Bednarek. 2010. Characterization of soil organic matter and plant tissues in an initial stage of plant succession and soil development by means of Curie-point pyrolysis coupled with GC-MS. Euras. Soil Sci. 43:1557–1568.

Reeves III, J.B. 2010. Near- versus mid-infrared diffuse reflectance spectroscopy for soil analysis emphasizing carbon and laboratory versus on-site analysis: Where are we and needs to be done? Geoderma 158:3–14. doi:10.1016/j.geoderma.2009.04.005

Reganold, J.P., L.F. Elliott, and Y.L. Unger. 1987. Long-term effects of organic and conventional farming on soil erosion. Nature 330:370–372. doi:10.1038/330370a0

Richard, C., G. Guyot, A. Rivaton, O. Trubetskaya, O. Trubetskoj, L. Cavani, and C. Ciavatta. 2007. Spectroscopic approach for elucidation of structural peculiarities of Andisol soil humic acid fractionated by SEC-PAGE set up. Geoderma 142:210–216. doi:10.1016/j.geoderma.2007.08.019

Robertson, G.P., E.A. Paul, and R.R. Harwood. 2000. Greenhouse gases in intensive agriculture: Contributions of individual gases to the radiative forcing of the atmosphere. Science 289:1922–1925. doi:10.1126/science.289.5486.1922

Rumpel, C., A. Chabbi, N. Nunan, and M. Dignac. 2009. Impact of landuse change on the molecular composition of soil organic matter. J. Anal. Appl. Pyrolysis 85:431–434. doi:10.1016/j.jaap.2008.10.011

Schnitzer, M., and C.M. Monreal. 2011. Quo vadis soil organic matter research? Academic Press, New York. p. 143–217.

Schulten, H.R., and M. Schnitzer. 1993. A state of art structural concept for humic substances. Naturwissenschaften 80:29–30. doi:10.1007/BF01139754

Six, J., T. Elliott, and K. Paustian. 1999. Aggregate and soil organic matter dynamics under conventional and no-tillage systems. Soil Sci. Soc. Am. J. 63:1350–1358. doi:10.2136/sssaj1999.6351350x

Six, J., T. Elliott, and K. Paustian. 2000. Soil macroaggregate turnover and microaggregate formation: A mechanism for C sequestration under no-tillage agriculture. Soil Biol. Biochem. 32:2099–2103. doi:10.1016/S0038-0717(00)00179-6

Six, J., C. Feller, K. Denef, S.M. Ogle, J.C.M. Sá, and A. Albretch. 2002a. Soil organic matter, biota and aggregation in temperate and tropical soils: Effect of no-tillage. Agronomie 22:755–775. doi:10.1051/agro:2002043

Six, J.C., E.A. Paul, and K. Paustian. 2002b. Stabilization mechanisms of soil organic matter: Implications for C-saturation of soils. Plant Soil 241:155–176. doi:10.1023/A:1016125726789

Skjemstad, J.O., L.R. Spouncer, B. Cowie, and R.S. Swift. 2004. Calibration of the Rothamsted organic carbon turnover model (RothC ver. 26.3) using measurable soil organic carbon pools. Aust. J. Soil Res. 42:79–88. doi:10.1071/SR03013

Smidt, E., and V. Parravacini. 2009. Effect of sewage sludge treatment and additional aerobic post-stabilization revealed by infrared spectroscopy and multivariate data analysis. Bioresour. Technol. 100:1775–1780. doi:10.1016/j.biortech.2008.10.003

Stenberg, B., R.A.V. Rossel, A.M. Mouazen, and J. Wetterlind. 2010. Visible and near infrared spectroscopy in soil science. Academic Press, Burlington, MA. p. 163–215

Stockmann, U., M.A. Adams, J.W. Crawford, D.J. Field, N. Henakaarchchi, M. Jenkins, B. Minasny, A.B. McBratney, V.R. Courcelles, K. Singh, I. Wheeler, L. Abbott, D.A. Angers, J. Baldock, M. Bird, P.C. Brookes, C. Chenu, J.D. Jastrow, R. Lal, J. Lehmann, A.G. O'Donell, W.J. Parton, D. Whitehead, and M. Zimmermann. 2013. The knowns, known unknowns and unknowns of sequestration of soil organic carbon. Agric. Ecosyst. Environ. 164:80–99. doi:10.1016/j.agee.2012.10.001

Swift, R.S. 1999. Macromolecular properties of soil humic substances: Fact, fiction, and opinion. Soil Sci. 164:790–802. doi:10.1097/00010694-199911000-00003

Tandy, S., J.R. Healey, N.A. Nason, J.C. Williamson, D.L. Jones, and S.C. Thain. 2010. FT-IR as an alternative method for measuring chemical properties during composting. Bioresour. Technol. 101:5431–5436. doi:10.1016/j.biortech.2010.02.033

Tisdall, J.M., and J.M. Oades. 1982. Organic matter and water-stable aggregates in soils. J. Soil Sci. 33:141–163. doi:10.1111/j.1365-2389.1982.tb01755.x

Triplett, G.B., Jr., and W.A. Dick. 2008. No-tillage crop production: A revolution in agriculture! Agron. J. 100:S153–S165. doi:10.2134/agronj2007.0005c

Trumbore, S.E. 1997. Potential responses of soil organic carbon to global environmental change. Proc. Natl. Acad. Sci. USA 94:8284–8291. doi:10.1073/pnas.94.16.8284

Vancampenhout, K., B. Vos, K. Wouters, R. Swennen, P. Buurman, and J. Deckers. 2012. Organic matter of subsoil horizons under broadleaved forest: Highly processed or labile and plant-derived? Soil Biol. Biochem. 50:40–46. doi:10.1016/j.soilbio.2012.03.005

Vancampenhout, K., K. Wouters, B. Vos, P. Buurman, R. Swennen, and J. Deckers. 2009. Differences in chemical composition of soil organic matter in natural ecosystems from different climate regions: A pyrolysis-GC/MS study. Soil Biol. Biochem. 41:568–579. doi:10.1016/j.soilbio.2008.12.023

Van Capelle, C., S. Schrader, and J. Brunotte. 2012. Tillage-induced changes in the functional diversity of soil biota: A review with a focus on German data. Eur. J. Soil Biol. 50:165–181. doi:10.1016/j.ejsobi.2012.02.005

Vergnoux, A., M. Giuliano, R. Rocco, M. Domeizel, F. Theraulaz, and P. Doumenq. 2011. Quantitative and mid-infrared changes of humic substances from burned soils. Environ. Res. 111:205–214. doi:10.1016/j.envres.2010.03.005

Verhulst, N., B. Govaerts, E. Verachtert, A. Catellanos-Navarrete, M. Mezzalama, P. Wall, A. Chocobar, J. Deckers, and K. Sayre. 2010 Conservation agriculture, improving soil quality for sustainable production systems. CRC Press, New York. p. 137–208.

Viscarra Rossell, R.A., D.J.J. Walvoort, A.B. McBratney, L.J. Janik, and J.O. Skjemstad. 2006. Visible, near infrared or combined diffuse reflectance spectroscopy for simultaneous assessment of various soil properties. Geoderma 131:59–75. doi:10.1016/j.geoderma.2005.03.007

Wershaw, R.L. 2004. Evaluation of conceptual models of natural organic matter (humus) from a consideration of the chemical and biochemical process of humification. USGS, Reston, VA.

White, K.E., J.B. Reeves, III, and F.J. Coale. 2011. Mid-infrared diffuse reflectance spectroscopy for the rapid analysis of plant root composition. Geoderma 167–168:197–203. doi:10.1016/j.geoderma.2011.08.009

Yan, Y., J. Tian, M. Fan, F. Zhang, X. Li, P. Christie, H. Chen, J. Lee, Y. Kuzyakov, and J. Six. 2012. Soil organic carbon and total nitrogen in intensively managed arable soils. Agric. Ecosyst. Environ. 150:102–110. doi:10.1016/j.agee.2012.01.024

Zhang, M., Z. He, A. Zhao, H. Zhang, D.M. Endale, and H.H. Schomberg. 2011. Water-extractable soil organic carbon and nitrogen affected by tillage and manure application. Soil Sci. 176:307–312. doi:10.1097/SS.0b013e31821d6d63

Nonideal Interactions Between Organic Contaminants and Dissolved Organic Matter

Di Zhang and Bo Pan*

Abstract

The nonideal interactions between organic contaminants and dissolved organic matter (DOM) are ubiquitous, which makes the environmental behavior and risk of organic contaminants complicated. We highlight the impacts of DOM heterogeneity on its nonideal interactions with organic contaminants as affected by environmental conditions, such as coexistent organic contaminants, pH, ionic strength, and temperature. The nonlinearity of binding isotherms of primary organic contaminants with DOM generally decreases due to other competitive organic contaminants binding with DOM. The decreased pH and increased ionic strength effectively neutralize negative charges, compress the electric double layer, and decrease the repulsion of DOM molecules, leading to aggregation and condensation of DOM. The nonlinear binding and desorption hysteresis result from the condensed domains, while the expanded domains contribute to the linear binding. The influences of pH, ionic strength, and temperature on nonideal interactions between organic contaminants and DOM may in large part depend on the change of DOM conformation. The complementary binding and temperature effects on phase transition of DOM should be examined in the future studies.

Abbreviations: AFM, atomic force microscopy; DOM, dissolved organic matter; HOC, hydrophobic organic chemical; PAH, polycyclic aromatic hydrocarbons; SOM, soil organic matter.

Di Zhang (zhangdi2002113@sina.com) and Bo Pan, Faculty of Environmental Science & Engineering, Kunming University of Science & Technology, 727 S Jingming Rd., Kunming, Yunnan, China 650500. *Corresponding author (panbocai@gmail.com).

doi:10.2136/sssaspecpub62.2014.0042

Labile Organic Matter—Chemical Compositions, Function, and Significance in Soil and the Environment. SSSA Special Publication 62. Zhongqi He and Fengchang Wu, editors.

Dissolved organic matter is ubiquitous in water, soil, and sediments and includes humic acid, fulvic acid, carbohydrate, and amino acid. Although the basic chemical nature, structure, and reactivity of DOM are still unclear due to chemical heterogeneity and geographical variability (Piccolo, 2002), DOM is recognized for its roles in controlling the fate of environmental pollutants and biogeochemistry of organic carbon in the global ecosystem (Mopper et al., 2007; Piccolo, 1996). For example, DOM can enhance the solubility of hydrophobic organic chemicals (HOCs); thus, the transport and environmental risk of HOCs may be enhanced (Huang and Lee, 2001). On the other hand, the binding of DOM and HOCs increases the molecule weight and hydrophilicity of HOCs, leading to their decreased bioavailability (McCarthy, 1989). Many models have been proposed to describe the interaction between organic contaminants and DOM. For example, the linear distribution model was first proposed because some researchers observed that the organic carbon normalized adsorption coefficient, K_{OC}, was independent of concentration of DOM and HOCs (Chiou et al., 1986). The interaction between HOCs and DOM was a linear distribution process, and K_{OC} was easily predicted based on HOC physicochemical properties, such as solubility and K_{OW} (Lee and Kuo, 1999). However, other studies found that it was difficult to establish a relationship between K_{OC} and a given DOM physicochemical property (Hur and Schlautman, 2003; Thomsen et al., 2002). Additionally, more and more nonideal interactions such as nonlinear binding (Borisover et al., 2006; de Perre et al., 2014; Hur et al., 2011; Laor and Rebhun, 2002; Pan et al., 2007, 2008a), desorption hysteresis (Akkanen et al., 2005; Pan et al., 2007, 2008a), and competitive binding (de Perre et al., 2014; Pan et al., 2007) of organic pollutants with DOM were generally observed. For solid organic matters, the two-stage sorption model (Loffredo and Senesi, 2006), three-site sorption model (Wehrhan et al., 2007), and dual mode sorption model (Pignatello and Xing, 1996) have been proposed to describe these nonideal interactions. These modeling concepts are applicable for DOM. According to the dual mode model, soil organic matter (SOM) was considered as a complex with condensed structure (glassy domain) and expanded structure (rubber domain). The nonlinear adsorption and desorption hysteresis resulted from the former, while the latter contributed to the linear adsorption. Hur et al. (2011) found that the aromatic condensed structures of DOM may play important roles in providing nonlinear strong binding sites for phenanthrene–DOM interactions. Pan et al. (2008a) also found that low pH and high ionic strength promoted the formation of larger aggregates and condensed structures of DOM, leading to enhanced nonlinear binding and desorption hysteresis of pyrene with DOM. These studies suggested that the dual mode sorption model could be adapted to DOM, although this concept was proposed based on solid SOM. Understanding these nonideal interactions between organic contaminants and DOM is critical for environmental fate modeling and risk assessment of organic contaminants.

This chapter summarizes current research progress on nonideal interactions between organic contaminants and DOM, with an emphasis on sorption-related topics. The environmental implications of interactions between organic contaminants and DOM are first introduced. The typical nonideal interaction mechanisms are the main focuses in the second section. In the third section, the influence factors for nonideal interaction are presented. Lastly, overall opinions and perspectives regarding nonideal interaction are summarized.

Environmental Implications of Dissolved Organic Matter

Soil structure, biological activity, and chemical bioavailability are three major soil properties affected directly by SOM (Piccolo, 2002). Once dissolved, SOM molecules became more flexible than solid phase. The inter- and intramolecular interactions of DOM molecules could result in various physical conformations (Pan et al., 2008b). In general, DOM refers to those dissolved SOM components that can pass through a 0.45-μm membrane. The chromophoric DOM is environmentally significant because of its influence on the optical properties of the water column, its role in photochemistry and photobiology, and its utility as a tracer of deep ocean biogeochemical processes and circulation (Nelson and Siegel, 2013). The heterogeneous DOM compositions could be operationally fractionated into hydrophobic acidic molecules, alkali molecules, neutral molecules, hydrophilic acidic molecules, alkali molecules, neutral molecules, and amphiphilic molecules (Leenheer and Croue, 2003). Because of the complicated components and structures, the interactions between DOM and organic pollutants are still unclear, although DOM is recognized for a role in affecting organic pollutants' speciation, solubility, mobility, and toxicity (Mopper et al., 2007; Piccolo, 1996). The effects of DOM on organic pollutants behavior and risks can be summarized according to three aspects of the influence of DOM: (i) direct interactions between DOM and organic pollutants, (ii) effects of DOM on uptake of organic pollutants by organisms, and (iii) effects of DOM coating on soil and other soil compositions (Fig. 1). For the first aspect, DOM can enhance the solubility of organic pollutants, thus enhancing the transport and environmental risk of organic pollutants (Chiou et al., 1987; Huang and Lee, 2001). On the other hand, the binding of DOM and organic pollutants increases the molecule weight and hydrophilicity of organic

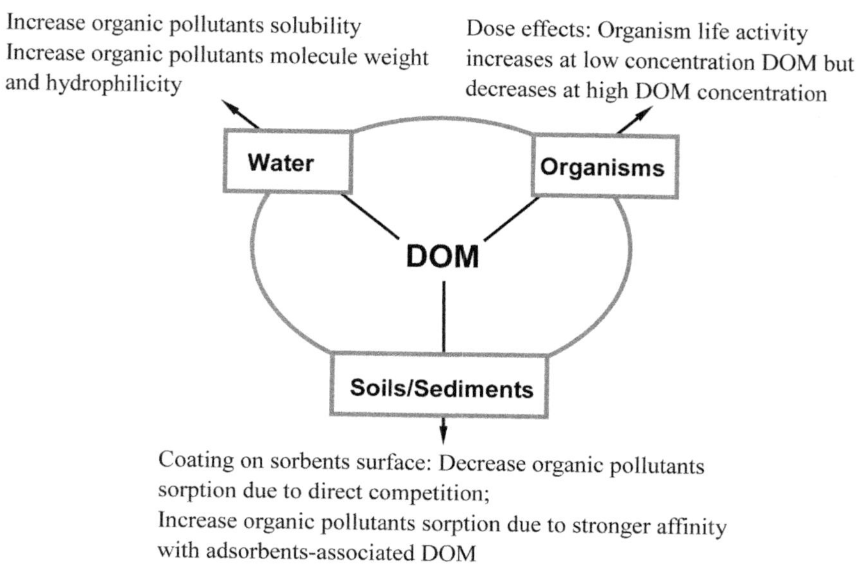

Fig. 1. The effects of dissolved organic matter (DOM) on organic pollutants behavior.

pollutants, leading to decreased bioavailability (McCarthy, 1989). For the second aspect, the response of organisms to DOM stress was significantly affected by dose effects. Low concentrations of DOM stimulated organism life activity, while the activity was inhibited at high DOM concentration (Steinberg et al., 2003). The DOM dose effects on organism life activity could indirectly affect organic pollutants uptake. For the third aspect, both enhanced and inhibited adsorption of organic pollutants have been observed as affected by DOM coating on soil or other soil portions in different studies. The enhanced adsorption of organic pollutants was attributed to the stronger affinity with adsorbents-associated DOM than adsorbents themselves (Ding et al., 2011; Navon et al., 2011). The direct competition and blockage of adsorption sites by DOM molecules led to the inhibited adsorption of organic pollutants (Haham et al., 2012). Even in a given system, the enhanced adsorption and inhibited adsorption of organic pollutants were simultaneously observed at different DOM concentrations. For example, Kulshrestha et al. (2004) found that low concentrations of DOM (1.0 mg L^{-1}) increased the adsorption of oxytetracycline on clay, while high concentrations of DOM (10 mg L^{-1}) resulted in an opposite effect. This result was attributed to the hydrophobic partitioning of oxytetracycline between clay-associated humic acid and "free" humic acids in the solution phase (Kulshrestha et al., 2004). Additionally, DOM could effectively disperse nanoparticles, leading to the stronger adsorption for organic contaminants. Thus the mobility and transport of dispersed nanoparticles and nanoparticles-adsorbed organic contaminants could subsequently be increased in natural aqueous environments, potentially enhancing the spread of various organic contaminants and their environmental risk (Pan et al., 2013). The general understanding of environmental functions of DOM, especially the effects of DOM on organic pollutants environmental behavior and risks, is still not sufficient. In the following section, the interactions between DOM and organic pollutants will be discussed.

Typical Nonideal Interaction between Organic Contaminants and Dissolved Organic Matter

The typical nonideal interactions between organic contaminants and DOM mainly include nonlinear binding, desorption hysteresis, and competitive binding. These nonideal interactions make the environmental behavior of organic contaminants more complicated, and thus the assessment for environmental risk of organic contaminants is difficult. In this section, the mechanisms of typical nonideal interactions will be demonstrated.

Nonlinear Binding

Nonlinear adsorption of organic contaminants in soil was first observed in 1983 (Mingelgrin and Gerstl, 1983). Similar phenomena were also observed for many other sorbents, including DOM (de Perre et al., 2014) and inorganic mineral and carbon-based particles (Zhang et al., 2010). The Freundlich model was widely used for describing the nonlinear binding isotherms because the factor of n could be easily obtained for the assessment of nonlinearity. Many studies indicated that in a small range of organic contaminants concentration, the binding basically was linear, while the binding became more and more nonlinear with increasing organic contaminants concentration, especially exceeding three orders of magnitude concentration range (Xing and Pignatello, 1997). For example, Borisover et al. (2006) found that the binding of pyrene on DOM was significantly concentration-dependent. The binding coefficient was up to sixfold at low pyrene concentration than at high pyrene concentration. Additionally, many studies suggested that the heterogeneous components of DOM played important roles in nonlinear binding. For example, the different components of DOM (weak binding components and strong binding components) resulted in the nonlinear binding of 3,4,3′,4′-tetrachlorobiphenyl and BaP (Akkanen et al., 2005). Pan et al. (2007) suggested that the nonlinear binding of HOCs with DOM resulted from the specific high energy binding sites (Pan et al., 2007). The aromatic condensed structures of DOM may play important roles in providing nonlinear strong binding sites for organic pollutants (Hur et al., 2011), as mentioned in the dual mode model, although it was proposed based on SOM. This concept has been applicable for DOM, because DOM was proved to contain micellelike regions due to the cross-linking of DOM molecules (Pan et al., 2007, 2008a). The dimension and compactness (condensed or expended) of the micellelike regions could be affected by DOM molecular structures, size, concentration, and solution chemistry (Pan et al., 2007, 2008a). Thus, the binding sites of DOM for organic contaminants are highly heterogeneous because of the complicated chemical components and alterable micellelike regions of DOM. Hernandez-Ruiz et al. (2012) also suggested that the binding of pharmaceuticals and personal care products with DOM was controlled by different components of DOM. For wastewater treatment plant DOM, the microbial byproduct-like and humic acid–like DOM components were main binding components with bisphenol A, whereas carbamazepine and ibuprofen mainly interacted with fulvic acid-like DOM components. For Suwannee River DOM, bisphenol A and carbamazepine mainly interacted with humic acid–like components, whereas fulvic acid–like components were main binding components with ibuprofen (Hernandez-Ruiz et al., 2012). Similarly, the aromatic and humic-like components would strongly bind propranolol, but high protein-like components interacted with propranolol more weakly (Peng et al., 2014). The highly heterogeneous binding sites and components result in the nonlinear binding of organic contaminants (Cheng et al., 2014; Ran et al., 2013; Sun et al., 2013).

Desorption Hysteresis

As a heterogeneous sorbent, another mostly significant characteristic of DOM is the desorption hysteresis of organic contaminants, namely, the binding curves do not overlap with the desorption curves, and even partly organic contaminants could not desorb from DOM (Pan et al., 2008a). It is easy to understand that the high binding but low desorption of organic contaminants will result in high retention in DOM and potential low environmental transport and risk and vice versa. Thus, understanding the mechanism of desorption hysteresis is vital for environmental fate modeling of organic contaminants and for biological remediation of polluted soil or water.

According to the literature, two important phenomenon are identified. One is that the stronger nonlinear binding results in more significant desorption hysteresis (Yuan and Xing, 2001). The other is that the strong desorption hysteresis is generally observed on condensed SOM or relatively compact DOM (Hur et al., 2011; Johnson et al., 2001; Pignatello and Xing, 1996; Ran et al., 2013;

Sun et al., 2013). For SOM, the penetration and pore filling were considered the major adsorption process for organic contaminants (Ran et al., 2013; Sun et al., 2013). Many research groups suggested that the inner pores of SOM might be irreversibly expanded, and some new pores might be formed due to the penetration of organic contaminants (Pignatello, 2012; Pignatello and Xing, 1996; Sun et al., 2013). These irreversibly expanded pores and new pores sequestrated more organic contaminants leading to desorption hysteresis of organic contaminants. The expansion of pores and formation of new pores were more effective for the condensed SOM; thus, the strong desorption hysteresis in condensed SOM was expected (Pignatello, 2012; Ran et al., 2013; Sun et al., 2013). For DOM, the possible conformation changes are shown in Fig. 2. The formation of micellelike structure was favorable for the nonlinear binding of HOCs (Pan et al., 2007). From Fig. 3, it is clear that more significant desorption hysteresis was observed on DOM micellelike structure (Pan et al., 2007). Ge et al. (2006) found that after the binding of naphthalene with DOM, the more condensed configuration of DOM was observed by atomic force microscopy (AFM) (Ge et al., 2006). The changed conformation of DOM from loosely coiled structures to condensed micellelike regions after binding with organic contaminants or under changed environmental conditions resulted in more sequestration of organic contaminants (Pan et al., 2007). As discussed above, the nonlinear binding of organic contaminants is

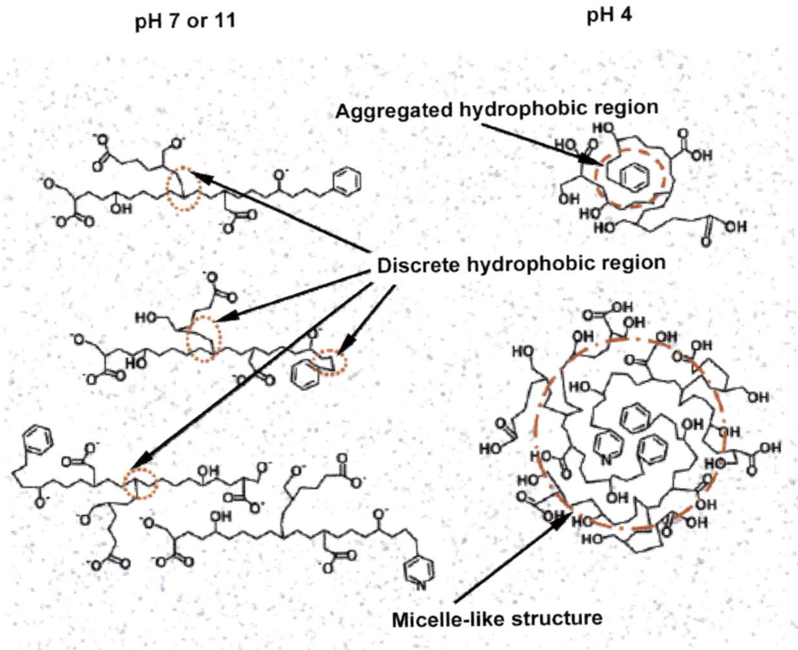

Fig. 2. Schematic diagram of possible conformation changes of dissolved organic matter (DOM). Discrete hydrophobic region (circle with dots), aggregated hydrophobic region (circle with dashes), and micellelike structure (circle with dots and dashes) are the possible binding region for HOCs. The area in color is the water molecules. From Pan et al. (2007), reprinted with permission from American Chemical Society.

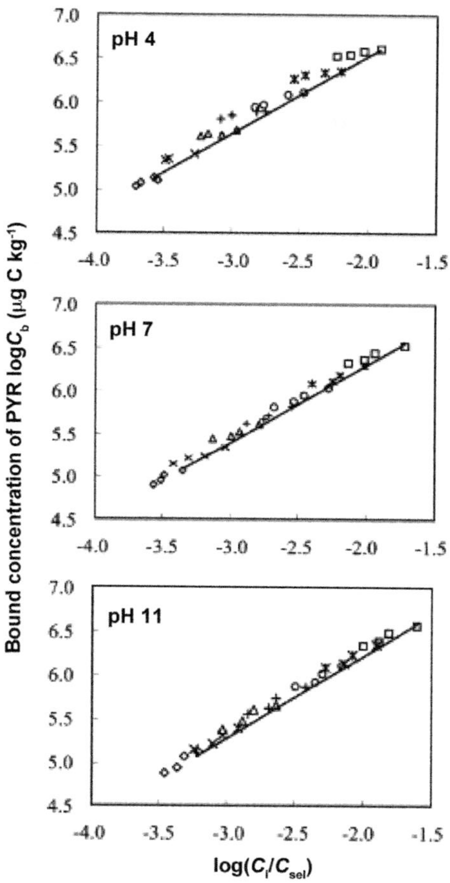

Fig. 3. Desorption isotherms of pyrene (PYR) from dissolved organic matter (DOM) at pH 4, 7, and 11 showing hysteresis. The solid lines are the original binding isotherms. The same symbol indicates the successive desorption from a single binding point. Considering Fig. 2, it was clear that more significant desorption hysteresis were observed on DOM micellelike structure. From Pan et al. (2007), reprinted with permission from American Chemical Society.

mainly attributed to the highly heterogeneous binding sites of DOM. The condensed conformation of DOM provided more nonlinear strong binding sites for organic pollutants (Hur et al., 2011), leading to more heterogeneous binding sites. Meanwhile, the condensed regions of DOM could effectively sequestrate organic contaminants (Pan et al., 2007). Thus, the more nonlinear binding was expected with the stronger desorption hysteresis. In addition, desorption hysteresis was also affected by environmental conditions, including pH, ionic strength, and temperature, which will be discussed later in the chapter.

Competitive Adsorption

The coexistence of various organic contaminants is very common in the real environment, and the interactions between these organic contaminants greatly alter their environmental behavior and health risks. Therefore, many efforts have been devoted to investigate the cosorption of two or three organic contaminants. The decreased adsorption of the primary sorbate due to the addition of competitor is general phenomenon (de Perre et al., 2014; Pan et al., 2007; Pelekani and Snoeyink,

1999; Yang et al., 2006, 2010). For example, benzene, toluene, and nitrobenzene could replace each other on charcoal, indicating the overlapping of adsorption sites of these compounds (Sander and Pignatello, 2005). Polycyclic aromatic hydrocarbons (PAH) could compete with each other for high energy adsorption sites via hydrophobic interactions and $\pi-\pi$ electron donor-acceptor interactions because of their similar aromatic structures (Yang et al., 2006). For a DOM–organic contaminants system, de Perre et al. (2014) found that the low molecular weight PAHs could effectively compete with high molecular weight PAHs because of its weak steric hindrance and strong interactions with DOM. Similarly, Pan et al. (2007) observed that phenanthrene showed significantly competitive binding for pyrene with DOM, especially at the low pyrene concentrations. In fact, the nonlinear binding and competitive binding essentially is consistent because the nonlinear binding of organic pollutants with DOM means that the binding sites of DOM should be heterogeneous, which would result in preferential binding on high-energy sites, and consequently causes competition between chemicals with similar properties.

As shown in Fig. 4, because of competitive adsorption or binding, the nonlinear degree of organic pollutants adsorption isotherms as identified by Freundlich nonlinear factor generally decreased no matter on DOM (de Perre et al., 2014; Pan et al., 2007) or other adsorbents (Shu et al., 2013; Xia and Ball, 2000). Researchers suggested that the addition of competitor inhibited or replaced the primary sorbate binding on low energy sites, leading to decreased energy distribution of binding sites and relatively homogeneous binding sites (Shu et al., 2013; Xia and Ball, 2000). The desorption and transport of a given organic contaminant may be enhanced by competitive adsorption of other organic contaminants, leading to increased environmental risk.

Influence Factors for Nonideal Interaction

In the real environment, the interactions between organic pollutants and DOM are more complicated because these interactions are generally affected by environmental conditions, such as pH, ionic strength, and temperature. Thus, investigating the effects of environmental factors on nonideal interaction between organic pollutants and DOM is vital for assessing the environmental risk of organic contaminants and environmental implication of DOM.

pH

pH is a widely discussed environmental factor affecting the nonideal interaction between organic contaminants and DOM. With increasing pH, both increased binding and decreased binding of organic contaminants with DOM have been observed in various studies. The authors suggested that at high pH, the increased binding was attributed to more exposed binding sites due to the stretch of DOM molecules (Jones and Tiller, 1999; Lassen and Carlsen, 1997), while the decreased binding resulted from increased polarity and decreased hydrophobicity of DOM due to dissociation of acidic functional groups (Marschner et al., 2005). Pan et al. (2008a) suggested that the binding of hydrophobic organic contaminants with DOM increased, nonlinearity increased, and desorption hysteresis was enhanced with decreasing pH. Many studies observed that the conformation of DOM could be changed with changing pH (Longstaffe et al., 2013; Pan et al., 2007, 2008a;

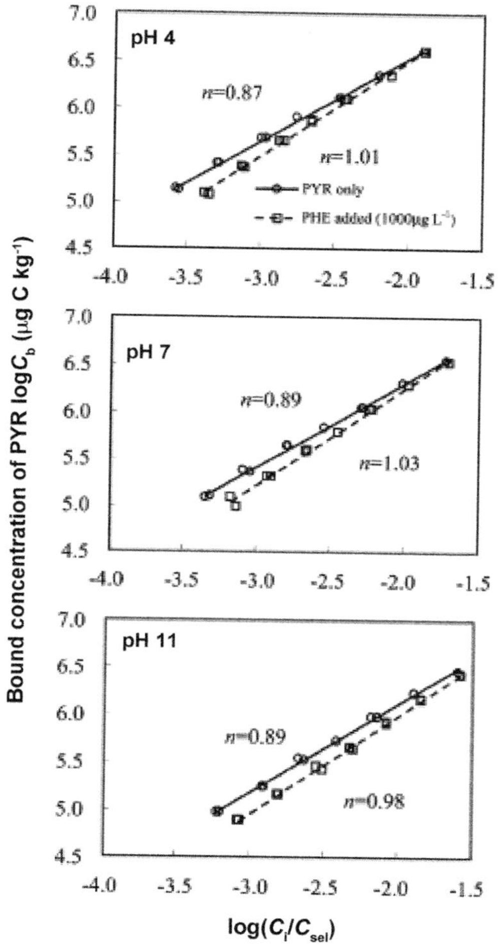

Fig. 4. Nonlinear isotherms of pyrene (PYR) on dissolved organic matter (DOM) and competitive effect after the addition of phenanthrene (PHE) (1000 μg L^{-1}) at pH 4, 7, and 11. The open circles are the measured sorption points for PYR only, and the open squares are those with PHE added as cosolute. The nonlinearity of PYR isotherms significantly decreased due to the competition binding of PHE with DOM. From Pan et al. (2007), reprinted with permission from American Chemical Society.

Piccolo, 2002). Using DOSY NMR, Longstaffe et al. (2013) found that the pH-dependence of the interactions between organofluorine and dissolved HA was correlated with changes in the conformation of the carbohydrate components of dissolved HA. At high pH, these structures opened up, making them more available for interactions with polar compounds. The direct evidence of DOM conformation changes could also be obtained by AFM (Fig. 5). Low pH was favorable for the formation of DOM aggregation. The possible process or mechanism of DOM conformation changes was shown in Fig. 6. At lower pHs, the negative charges on DOM would be neutralized, leading to the aggregation of DOM. The scattered binding sites, such as small hydrophobic regions aggregated together, leading to the formation of larger hydrophobic regions and relatively condensed structures (Pan et al., 2007, 2008a). The enhanced heterogeneous binding sites contributed to the nonlinear binding and desorption hysteresis (de Perre et al., 2014; Pignatello and Xing, 1996). On the contrary, at higher pHs, the DOM was negatively charged due to dissociation of acidic functional groups. The stretch and expanded conformation of DOM was expected due to electrostatic repulsion. The inner larger hydrophobic areas of DOM split into some smaller hydrophobic areas, leading to relatively uniform distribution of binding sites (Pan et al., 2007, 2008a). The enhanced homogeneous binding sites led to less nonlinear binding.

Ionic Strength

The presence of environmental ions makes the environmental behaviors of organic contaminants more complicated. The adsorption of organic contaminants as affected by ionic strength has been extensively studied, but the effects of ionic strength on the interaction between organic contaminants and DOM are still inconsistent in different studies. Both increased binding (Kang et al., 2010) and decreased binding of organic contaminants with DOM (Gondar et al., 2012) were observed with increasing ionic strength. Some studies also observed both increasing and decreasing effects in a single study, depending on ion types (Laegdsmand et al., 2004; Pan et al., 2012a). Summarizing the mechanisms of ionic strength effects, the increased binding was attributed to the decreased electrostatic repulsion due to binding of metal cations with DOM through cation-bridge. The decreased adsorption was attributed to the formation of a hydration shell of metal ions that decreased contact of organic contaminants with DOM. These studies mainly focused on the change of binding coefficient, so the nonideal interaction between organic contaminants and DOM as affected by ionic strength were not explored in detail. Pan et al. (2008a) observed that increased ionic strength could decrease the repulsion of DOM, leading to aggregation of DOM and formation of larger molecule cluster (Fig. 5 and 6). Taking Al as an example, the addition of Al could effectively neutralize negative charges, compress the electric double layer, and decrease the repulsion of DOM. Additionally, DOM could be flocculated by the Al cation-bridge effect (Kazpard et al., 2006). The same as for the low pH effect, the enhanced nonlinear binding and desorption hysteresis are expected at high ionic strength due to aggregated and condensed DOM structures (Pignatello and Xing, 1996).

Temperature

Temperature has been considered an important influencing factor for nonideal interactions between organic contaminants and SOM. Gunasekara and Xing (2003) observed that the HOCs' isotherms on soil humin with more condensed structures showed the highest adsorption nonlinearity and desorption hysteresis compared with soil humic acid and soil. The same nonlinear isotherm trend for naphthalene and 1, 2-dichlorobenzene adsorption on peat and its organic fractions was observed as: peat humic acid < peat < peat humin (Xing, 2001). This order of nonlinearity was consistent with the degree of condensed structures expected from organic matter extraction procedures (Cuypers et al., 2002). It is expected that the condensed structures of organic matters were contributed to nonideal interactions for organic contaminants (Pignatello and Xing, 1996). More importantly, the adsorption isotherms of organic contaminants became more linear at elevated temperatures. For example, the nonlinearity factor n of HOC adsorption on humin increased from 0.848 at 20°C to 0.955 at 85°C (Gunasekara and Xing, 2003). According to the dual mode model proposed by Pignatello

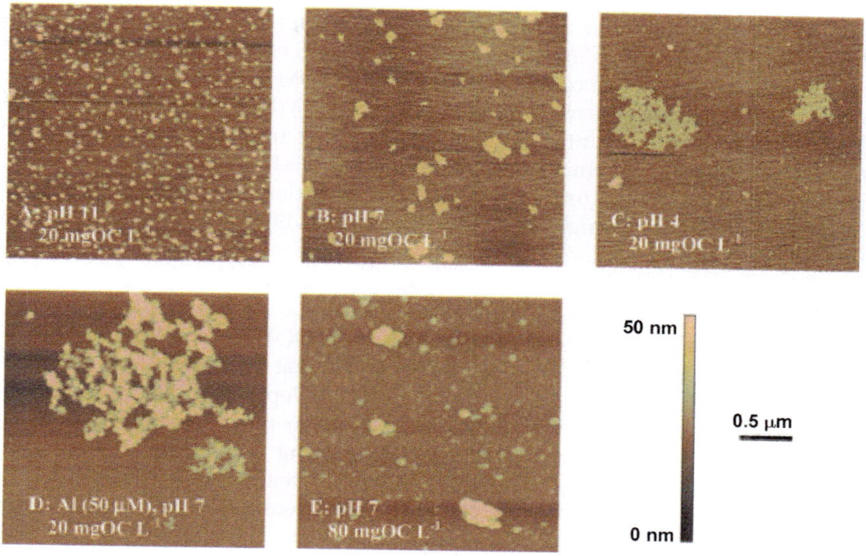

Fig. 5. Atomic force microscopy images showing different dissolved organic matter (DOM) conformations under various conditions. Aggregation increased significantly from (A) to (C) as pH decreased from 11 to 4. Sponge-like aggregates were observed for DOM at pH 4 (C) and with the presence of Al (D). Aggregates were also observed at high DOM concentration (E). From Pan et al. (2008a), reprinted with permission from American Chemical Society.

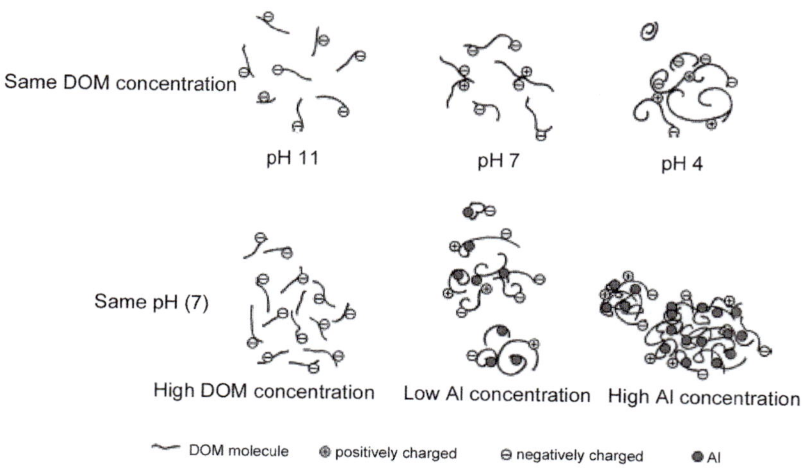

Fig. 6. Schematic diagram of dissolved organic matter (DOM) conformation under different pHs and ionic strength. At alkaline pH, the DOM molecules are dispersed and stretched due to high negative charges repulsion. As pH decreases, the negative charges are neutralized and electrostatic repulsive force decreases. The DOM molecules are aggregated and the condensed structures are formed. The addition of Al not only neutralizes negative charges, but also functions as bridging ions. Therefore, even larger aggregates and condensed structures are observed. From Pan et al. (2008a), reprinted with permission from American Chemical Society.

and Xing (1996), the SOM was considered as a complex with condensed structure (i.e., glassy domain) and expended structure (i.e., rubber domain). Many studies indicated that the condensed domains of SOM could become expanded domains above their glass transition temperatures (T_g) (LeBoeuf and Weber, 1997; Pignatello, 2012; Weber et al., 2001) (Fig. 7). The phase transition temperatures of different SOM are different. Generally, humic acid and fulvic acid have a lower phase transition temperature, while the phase transition temperatures of humin and lignin are relatively higher (LeBoeuf and Weber, 1997). The transition of condensed domain to expended domain of SOM with elevating temperature resulted in the decreased nonlinear adsorption of organic contaminants.

Dissolved organic matter is part of the dissolvable SOM, but so far, no study directly addressed the effects of temperature on DOM conformation. According to quenching kinetics, Pan et al. (2012b) found that the interactions between ofloxacin and DOM were significantly temperature dependent. The kinetics process could be identified as three quenching compartments, a very fast quenching compartment and two pseudo first-order quenching compartments. The fast quenching compartment had a positive relationship with temperature, while the two pseudo first-order quenching compartments were negatively related with temperature (Pan et al., 2012b). This study implied that the interactions between organic contaminants and DOM were highly heterogeneous. Dissolved organic matter conformation or properties might be changed with the changing temperature, and thus interactions were changed depending temperature. In this chapter, we propose that the conformation of DOM may also be changed with temperature alteration, consequently changing the nonideal interactions between DOM and organic contaminants. Are the phase transition temperatures for DOM similar to SOM? If yes, would DOM concentration be as high as micelle concentration?

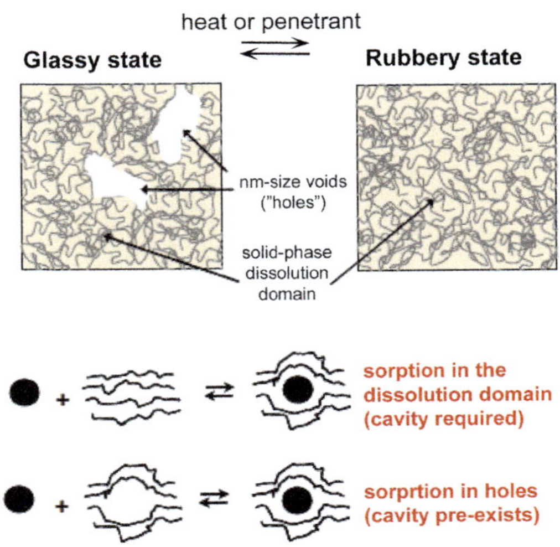

Fig. 7. Depiction of glassy and rubbery states showing the holes in the glassy phase and depiction of sorption in the dissolution domain, where a cavity has to open up, and sorption in the hole domain where the cavity preexists. Heating leads to the phase transition from condensed domains to expanded domains. From Pignatello (2012), reprinted with permission from Springer.

The answers to these questions are vital for understanding the nonideal interactions between DOM and organic contaminants and should be considered for new research areas and ideas in future studies.

According to the above discussions, we can point out that the nonideal interaction between organic contaminants and DOM may be controlled by DOM conformation, which can be changed by alterative pH, ionic strength, and temperature. Decreased pH, increased ionic strength, and reduced temperature are favorable to form more condensed structures, leading to the stronger nonideal interactions.

Conclusions and Perspectives

Conclusions

The nonideal interactions between organic contaminants and DOM are generally observed. Unlike the ideal interactions, such as linear partition adsorption, which are well understood, these nonideal interactions, such as nonlinear adsorption, desorption hysteresis, and competitive adsorption, make the environmental behavior of organic contaminants more complicated. Thus, the assessment for environmental risk of organic contaminants is more difficult. This chapter highlights the impacts of DOM heterogeneity on nonideal interactions of organic contaminants, along with environmental conditions, including coexistent organic contaminants, pH, ionic strength, and temperature.

The decreased pH and increased ionic strength effectively neutralize negative charges, compress electric double layer, and decrease the repulsion of DOM, leading to aggregation and condensation of DOM. The effects of temperature on SOM conformation should be extended to DOM. This point will be a new research idea for understanding DOM conformation and its environmental behavior. In summary, the nonideal interaction between organic contaminants and DOM may be controlled by DOM conformation, which can be changed by alterative pH, ionic strength, and temperature. However, the basic chemical nature, structure, and reactivity of DOM are still unclear due to chemical heterogeneity and geographical variability. Moreover, the different extraction procedures and different derivations result in the highly different properties of DOM, which requires that researchers pay special attention to different properties and components DOM in the future studies.

Perspectives

In the competitive adsorption studies of coexistent organic contaminants, most researchers have focused on the apparent adsorption behavior. The competitive adsorption was evaluated based on the adsorption isotherms or distribution coefficients of the primary sorbate in the absence or presence of competitor. This type of evaluation ignored that the competitor might adsorb on some adsorption sites that were not available for the primary sorbate adsorption, especially for those organic contaminants with different properties. Recently, a few studies observed that the complementary adsorption really existed for coexistent organic contaminants. An innovative method for data analysis was proposed, and the adsorption inhibition of primary sorbate and adsorption of competitor were calculated and compared. The results showed that the adsorption of competitors exceeded primary sorbate adsorption inhibition, indicating that these organic contaminants complementarily adsorbed on their preferred adsorption sites (Lerman et al., 2013; Wang et al., 2013; Zhang et al., 2012). These studies suggested that both competitive and complementary adsorption might simultaneously occur in a coexistent organic contaminants adsorption system. These results present new challenges for predicting the transport and fate of organic contaminants under the influence of cosolutes. On the one hand, the transport and risk of organic contaminants may be enhanced by competitive adsorption, which decreases the retention of organic contaminants. On the other hand, the availability of a given sorbent may be enhanced in binary adsorption system due to complementary adsorption. Thus, extend studies should be conducted to understand how to inhibit competitive adsorption and provoke complementary adsorption for decreasing the environmental risks of organic contaminants and increasing the availability of sorbents. These works will supply useful information for high efficient utilization of sorbent.

It is notable that these limited studies on complementary adsorption used carbon nanotubes as the probe due to its definite structures for simplifying the mechanism discussion. It is easy to expect that the complementary binding behavior should be ubiquitous for organic contaminants binding with DOM because DOM has more complicated structures and components and, consequently, more heterogeneous binding sites as preferred binding sites for different organic contaminants. However, the relevant information is still unavailable in literature. Thus, more investigations of complementary binding of different organic contaminants should be conducted and extended to DOM in the future.

Acknowledgments

Part of the research presented in this chapter is supported by the National Scientific Foundation of China (41303093, 41222025, 41273138), Scientific Foundation of Kunming University of Science and Technology (14118762), and Recruitment Program of Highly-Qualified Scholars in Yunnan (2010CI109).

References

Akkanen, J., A. Tuikka, and J.V.K. Kukkonen. 2005. Comparative sorption and desorption of benzo[α]pyrene and 3,4,3′,4′-tetrachlorobiphenyl in natural lake water containing dissolved organic matter. Environ. Sci. Technol. 39:7529–7534. doi:10.1021/es050835f

Borisover, M., Y. Laor, N. Bukhanovsky, and I. Saadi. 2006. Fluorescence-based evidence for adsorptive binding of pyrene to effluent dissolved organic matter. Chemosphere 65:1925–1934. doi:10.1016/j.chemosphere.2006.07.032

Cheng, C.-H., T.-P. Lin, J. Lehmann, L.-J. Fang, Y.-W. Yang, O.V. Menyailo, K.-H. Chang, and L. Lai. 2014. Sorption properties for black carbon (wood char) after long-term exposure in soils. Org. Geochem. 70:53–61. doi:10.1016/j.orggeochem.2014.02.013

Chiou, C.T., D.E. Kile, T.I. Brinton, R.L. Malcolm, J.A. Leenheer, and P. MacCarthy. 1987. A comparison of water solubility enhancements of organic solutes by aquatic humic materials and commercial humic acids. Environ. Sci. Technol. 21:1231–1234. doi:10.1021/es00165a012

Chiou, C.T., R.L. Malcolm, T.I. Brinton, and D.E. Kile. 1986. Water solubility enhancement of some organic pollutants and pesticides by dissolved humic and fulvic-acids. Environ. Sci. Technol. 20:502–508. doi:10.1021/es00147a010

Cuypers, C., T. Grotenhuis, K.G.J. Nierop, E.M. Franco, A. de Jager, and W. Rulkens. 2002. Amorphous and condensed organic matter domains: The effect of persulfate oxidation on the composition of soil/sediment organic matter. Chemosphere 48:919–931. doi:10.1016/S0045-6535(02)00123-6 de Perre, C., K. Le Menach, F. Ibalot, E. Parlanti, and H. Budzinski. 2014. Development of solid-phase microextraction to study dissolved organic matter-Polycyclic aromatic hydrocarbon interactions in aquatic environment. Anal. Chim. Acta 807:51–60. doi:10.1016/j.aca.2013.11.026

Ding, Q., H.L. Wu, Y. Xu, L.J. Guo, K. Liu, H.M. Gao, and H. Yang. 2011. Impact of low molecular weight organic acids and dissolved organic matter on sorption and mobility of isoproturon in two soils. J. Hazard. Mater. 190:823–832. doi:10.1016/j.jhazmat.2011.04.003

Ge, X.P., Y.M. Zhou, C.H. Lu, and H.X. Tang. 2006. AFM study on the adsorption and aggregation behavior of dissolved humic substances on mica. Sci. China Ser. B Chem. 49:256–266. doi:10.1007/s11426-006-0256-1

Gondar, D., R. López, J. Antelo, S. Fiol, and F. Arce. 2012. Adsorption of paraquat on soil organic matter: Effect of exchangeable cations and dissolved organic carbon. J. Hazard. Mater. 235–236:218–223. doi:10.1016/j.jhazmat.2012.07.044

Gunasekara, A.S., and B.S. Xing. 2003. Sorption and desorption of naphthalene by soil organic matter: Importance of aromatic and aliphatic components. J. Environ. Qual. 32:240–246. doi:10.2134/jeq2003.2400

Haham, H., A. Oren, and B. Chefetz. 2012. Insight into the role of dissolved organic matter in sorption of sulfapyridine by semiarid soils. Environ. Sci. Technol. 46:11870–11877. doi:10.1021/es303189f

Hernandez-Ruiz, S., L. Abrell, S. Wickramasekara, B. Chefetz, and J. Chorover. 2012. Quantifying PPCP interaction with dissolved organic matter in aqueous solution: Combined use of fluorescence quenching and tandem mass spectrometry. Water Res. 46:943–954. doi:10.1016/j.watres.2011.11.061

Huang, X.J., and L.S. Lee. 2001. Effects of dissolved organic matter from animal waste effluent on chlorpyrifos sorption by soils. J. Environ. Qual. 30:1258–1265. doi:10.2134/jeq2001.3041258x

Hur, J., B.M. Lee, and H.S. Shin. 2011. Microbial degradation of dissolved organic matter (DOM) and its influence on phenanthrene–DOM interactions. Chemosphere 85:1360–1367. doi:10.1016/j.chemosphere.2011.08.001

Hur, J., and M.A. Schlautman. 2003. Using selected operational descriptors to examine the heterogeneity within a bulk humic substance. Environ. Sci. Technol. 37:880–887. doi:10.1021/es0260824

Johnson, M.D., T.M. Keinath, and W.J. Weber. 2001. A distributed reactivity model for sorption by soils and sediments. 14. Characterization and modeling of phenanthrene desorption rates. Environ. Sci. Technol. 35:1688–1695. doi:10.1021/es001391k

Jones, K.D., and C.L. Tiller. 1999. Effect of solution chemistry on the extent of binding of phenanthrene by a soil humic acid: A comparison of dissolved and clay bound humic. Environ. Sci. Technol. 33:580–587. doi:10.1021/es9803207

Kang, J., H.J. Liu, Y.M. Zheng, J.H. Qu, and J.P. Chen. 2010. Systematic study of synergistic and antagonistic effects on adsorption of tetracycline and copper onto a chitosan. J. Colloid Interface Sci. 344:117–125. doi:10.1016/j.jcis.2009.11.049

Kazpard, V., B.S. Lartiges, C. Frochot, J.B.D. de la Caillerie, M.L. Viriot, J.M. Portal, T. Gorner, and J.L. Bersillon. 2006. Fate of coagulant species and conformational effects during the aggregation of a model of a humic substance with Al-13 polycations. Water Res. 40:1965–1974. doi:10.1016/j.watres.2006.03.014

Kulshrestha, P., R.F. Giese, and D.S. Aga. 2004. Investigating the molecular interactions of oxytetracycline in clay and organic matter: Insights on factors affecting its mobility in soil. Environ. Sci. Technol. 38:4097–4105. doi:10.1021/es034856q

Laegdsmand, M., L.W. de Jonge, P. Moldrup, and K. Keiding. 2004. Pyrene sorption to water-dispersible colloids: Effect of solution chemistry and organic matter. Vadose Zone J. 3:451–461.

Laor, Y., and M. Rebhun. 2002. Evidence for nonlinear binding of PAHs to dissolved humic acids. Environ. Sci. Technol. 36:955–961. doi:10.1021/es001996g

Lassen, P., and L. Carlsen. 1997. Solubilization of phenanthrene by humic acids. Chemosphere 34:817–825. doi:10.1016/S0045-6535(97)00010-6

LeBoeuf, E.J., and W.J. Weber. 1997. A distributed reactivity model for sorption by soils and sediments. 8. Sorbent organic domains: Discovery of a humic acid glass transition and an argument for a polymer-based model. Environ. Sci. Technol. 31:1697–1702. doi:10.1021/es960626i

Lee, C.L., and L.J. Kuo. 1999. Quantification of the dissolved organic matter effect on the sorption of hydrophobic organic pollutant: Application of an overall mechanistic sorption model. Chemosphere 38:807–821. doi:10.1016/S0045-6535(98)00228-8

Leenheer, J.A., and J.P. Croue. 2003. Characterizing aquatic dissolved organic matter. Environ. Sci. Technol. 37:19A–26A. doi:10.1021/es032333c

Lerman, I., Y. Chen, B. Xing, and B. Chefetz. 2013. Adsorption of carbamazepine by carbon nanotubes: Effects of DOM introduction and competition with phenanthrene and bisphenol A. Environ. Pollut. 182:169–176. doi:10.1016/j.envpol.2013.07.010

Loffredo, E., and N. Senesi. 2006. Fate of anthropogenic organic pollutants in soils with emphasis on adsorption/desorption processes of endocrine disruptor compounds. Pure Appl. Chem. 78:947–961. doi:10.1351/pac200678050947

Longstaffe, J.G., D. Courtier-Murias, and A.J. Simpson. 2013. The pH-dependence of organofluorine binding domain preference in dissolved humic acid. Chemosphere 90:270–275. doi:10.1016/j.chemosphere.2012.07.004

Marschner, B., R. Winkler, and D. Jödemann. 2005. Factors controlling the partitioning of pyrene to dissolved organic matter extracted from different soils. Eur. J. Soil Sci. 56:299–306. doi:10.1111/j.1365-2389.2004.00678.x

McCarthy, J.F. 1989. Bioavailability and toxicity of metals and hydrophobic organic contaminants. In: I.H. Suffet and P. MacCarthy, editors, Aquatic humic substances: Influence on fate and treatment of pollutants. Adv. in Chemistry 219. ACS Publications, Washington, DC. p. 263–277. doi:10.1021/ba-1988-0219.ch018

Mingelgrin, U., and Z. Gerstl. 1983. Reevaluation of partitioning as a mechanism of nonionic chemicals adsorption in soils. J. Environ. Qual. 12:1–11. doi:10.2134/jeq1983.00472425001200010001x

Mopper, K., A. Stubbins, J.D. Ritchie, H.M. Bialk, and P.G. Hatcher. 2007. Advanced instrumental approaches for characterization of marine dissolved organic matter: Extraction techniques, mass spectrometry, and nuclear magnetic resonance spectroscopy. Chem. Rev. 107:419–442. doi:10.1021/cr050359b

Navon, R., S. Hernandez-Ruiz, J. Chorover, and B. Chefetz. 2011. Interactions of carbamazepine in soil: Effects of dissolved organic matter. J. Environ. Qual. 40:942–948. doi:10.2134/jeq2010.0446

Nelson, N.B., and D.A. Siegel. 2013. The global distribution and dynamics of chromophoric dissolved organic matter. Annu. Rev. Mar. Sci. 5:447–476. doi:10.1146/annurev-marine-120710-100751

Pan, B., S. Ghosh, and B.S. Xing. 2007. Nonideal binding between dissolved humic acids and polyaromatic hydrocarbons. Environ. Sci. Technol. 41:6472–6478. doi:10.1021/es070790d

Pan, B., S. Ghosh, and B. Xing. 2008a. Dissolved organic matter conformation and its interaction with pyrene as affected by water chemistry and concentration. Environ. Sci. Technol. 42:1594–1599. doi:10.1021/es702431m

Pan, B., Y. Liu, D. Xiao, F. Wu, M. Wu, D. Zhang, and B. Xing. 2012b. Quantitative identification of dynamic and static quenching of ofloxacin by dissolved organic matter using temperature-dependent kinetic approach. Environ. Pollut. 161:192–198. doi:10.1016/j.envpol.2011.10.026

Pan, B., P. Ning, and B. Xing. 2008b. Part IV—Sorption of hydrophobic organic contaminants. Environ. Sci. Pollut. Res. Int. 15:554–564. doi:10.1007/s11356-008-0051-y

Pan, B., M.Y. Qiu, M. Wu, D. Zhang, H.B. Peng, D. Wu, and B.S. Xing. 2012a. The opposite impacts of Cu and Mg cations on dissolved organic matter-ofloxacin interaction. Environ. Pollut. 161:76–82. doi:10.1016/j.envpol.2011.09.040

Pan, B., D. Zhang, H. Li, M. Wu, Z. Wang, and B. Xing. 2013. Increased adsorption of sulfamethoxazole on suspended carbon nanotubes by dissolved humic acid. Environ. Sci. Technol. 47:7722–7728. doi:10.1021/es4008933

Pelekani, C., and V.L. Snoeyink. 1999. Competitive adsorption in natural water: Role of activated carbon pore size. Water Res. 33:1209–1219. doi:10.1016/S0043-1354(98)00329-7

Peng, N., K. Wang, G. Liu, F. Li, K. Yao, and W. Lv. 2014. Quantifying interactions between propranolol and dissolved organic matter (DOM) from different sources using fluorescence spectroscopy. Environ. Sci. Pollut. Res. Int. 21:5217–5226. doi:10.1007/s11356-013-2436-9

Piccolo, A. 1996. Humus and soil conservation. In: A. Piccolo, editor, Humic substances in terrestrial ecosystems. Elsevier, Amsterdam. p. 225–264. doi:10.1016/B978-044481516-3/50006-2

Piccolo, A. 2002. The supramolecular structure of humic substances: A novel understanding of humus chemistry and implications in soil science. Adv. Agron. 75:57–134. doi:10.1016/S0065-2113(02)75003-7

Pignatello, J.J. 2012. Dynamic interactions of natural organic matter and organic compounds. J. Soils Sediments 12:1241–1256. doi:10.1007/s11368-012-0490-4

Pignatello, J.J., and B.S. Xing. 1996. Mechanisms of slow sorption of organic chemicals to natural particles. Environ. Sci. Technol. 30:1–11. doi:10.1021/es940683g

Ran, Y., Y. Yang, B. Xing, J.J. Pignatello, S. Kwon, W. Su, and L. Zhou. 2013. Evidence of micropore filling for sorption of nonpolar organic contaminants by condensed organic matter. J. Environ. Qual. 42:806–814. doi:10.2134/jeq2012.0286

Sander, M., and J.J. Pignatello. 2005. Characterization of charcoal adsorption sites for aromatic compounds: Insights drawn from single-solute and Bi-solute competitive experiments. Environ. Sci. Technol. 39:1606–1615. doi:10.1021/es049135l

Shu, Y., P. Liu, Q. Zhang, and D. Wei. 2013. Competitive sorption between 1, 2, 4-trichlorobenzene/tetrachloroethene and 1, 2, 4, 5-tetrachlorobenzene by soils/sediments from South China. Sci. Total Environ. 463–464:258–263. doi:10.1016/j.scitotenv.2013.06.028

Steinberg, C.E.W., A. Paul, S. Pflugmacher, T. Meinelt, R. Klocking, and C. Wiegand. 2003. Pure humic substances have the potential to act as xenobiotic chemicals—A review. Fresenius Environ. Bull. 12:391–401.

Sun, K., Y. Ran, Y. Yang, B. Xing, and J. Mao. 2013. Interaction mechanism of benzene and phenanthrene in condensed organic matter: Importance of adsorption (nanopore-filling). Geoderma 204–205:68–74. doi:10.1016/j.geoderma.2013.04.008

Thomsen, M., S. Dobel, P. Lassen, L. Carlsen, B.B. Mogensen, and P.E. Hansen. 2002. Reverse quantitative structure-activity relationship for modelling the sorption of esfenvalerate to dissolved organic matter—A multivariate approach. Chemosphere 49:1317–1325. doi:10.1016/S0045-6535(02)00510-6

Wang, C., H. Li, S. Liao, H. Zheng, Z. Wang, B. Pan, and B. Xing. 2013. Coadsorption, desorption hysteresis and sorption thermodynamics of sulfamethoxazole and carbamazepine on graphene oxide and graphite. Carbon 65:243–251. doi:10.1016/j.carbon.2013.08.020

Weber, W.J., Jr., E.J. LeBoeuf, T.M. Young, and W. Huang. 2001. Contaminant interactions with geosorbent organic matter: Insights drawn from polymer sciences. Water Res. 35:853–868. doi:10.1016/S0043-1354(00)00339-0

Wehrhan, A., R. Kasteel, J. Simunek, J. Groeneweg, and H. Vereecken. 2007. Transport of sulfadiazine in soil columns- Experiments and modelling approaches. J. Contam. Hydrol. 89:107–135. doi:10.1016/j.jconhyd.2006.08.002

Xia, G., and W.P. Ball. 2000. Polanyi-based models for the competitive sorption of low-polarity organic contaminants on a natural sorbent. Environ. Sci. Technol. 34:1246–1253. doi:10.1021/es9812453

Xing, B., and J.J. Pignatello. 1997. Dual-mode sorption of low-polarity compounds in glassy poly (vinyl chloride) and soil organic matter. Environ. Sci. Technol. 31:792–799. doi:10.1021/es960481f

Xing, B.S. 2001. Sorption of anthropogenic organic compounds by soil organic matter: A mechanistic consideration. Can. J. Soil Sci. 81:317–323. doi:10.4141/S00-067

Yang, K., X.L. Wang, L.Z. Zhu, and B.S. Xing. 2006. Competitive sorption of pyrene, phenanthrene, and naphthalene on multiwalled carbon nanotubes. Environ. Sci. Technol. 40:5804–5810. doi:10.1021/es061081n

Yang, K., W.H. Wu, Q.F. Jing, W. Jiang, and B.S. Xing. 2010. Competitive adsorption of naphthalene with 2,4-dichlorophenol and 4-chloroaniline on multiwalled carbon nanotubes. Environ. Sci. Technol. 44:3021–3027. doi:10.1021/es100018a

Yuan, G., and B. Xing. 2001. Effects of metal cations on sorption and desorption of organic compounds in humic acids. Soil Sci. 166:107–115. doi:10.1097/00010694-200102000-00004

Zhang, D., B. Pan, M. Wu, H. Zhang, H. Peng, P. Ning, and B. Xing. 2012. Cosorption of organic chemicals with different properties: Their shared and different sorption sites. Environ. Pollut. 160:178–184. doi:10.1016/j.envpol.2011.08.053

Zhang, X., B. Pan, K. Yang, D. Zhang, and J.A. Hou. 2010. Adsorption of sulfamethoxazole on different types of carbon nanotubes in comparison to other natural adsorbents. J. Environ. Sci. Health Part A. Toxic/Hazardous Substances & Environmental Engineering 45:1625–1634.

Characterizing the Labile Fraction of Dissolved Organic Matter in Leaf Leachates: Methods, Indicators, Structure, and Complexity

C.W. Cuss and C. Guéguen*

Abstract

Leaves are an integral component of the worldwide C cycle. Leaching is the process through which soluble substances of varying character are removed from leaves in the early period following exposure to water. The dissolved organic matter leached from leaves, or leaf leachate dissolved organic matter (LLDOM), contributes substantial nutrients and energy in forested and aquatic ecosystems. The method used to produce leaf leachate can affect its properties and so should take account of the processes under study. The considerable proportion of LLDOM that is readily biotransformed by microorganisms is known as the labile fraction and has been defined in multiple ways. Many methods are used to characterize LLDOM and its lability; however, the accurate and absolute assessment of LLDOM lability is challenging due to its complex and dynamic nature and the consequentially high number of potentially influential factors. Nonetheless, the use of multiple indicators enables refined estimation of lability. Notwithstanding impressive progress, the extension of laboratory-based results to events in the natural world is still fraught with difficulty. Answering the call for further clarity, emerging analytical and investigative methods promise to shed new light on this important topic.

Abbreviations: AF4, asymmetrical flow field-flow fractionation; BDOC, biodegradable dissolved organic C; DOC, total dissolved organic C; DOM, dissolved organic matter; DON, dissolved organic N; DRP, dissolved reactive P; DUP, dissolved unreactive P; EEM, excitation–emission matrix; Ex/Em, excitation–emission; FI, fluorescence index; FlFFF, flow field-flow fractionation; FRP, filterable reactive P; HIX, humification index; HLF, humic-like fluorescence; LLDOM, leaf leachate dissolved organic matter; MRP, molybdate reactive P; MW, molecular weight; NMR, nuclear magnetic resonance; OP, orthophosphate; PARAFAC, parallel factor analysis; PCA, principal component analysis; PLF, protein-like fluorescence; SEC, size exclusion chromatography; SOM, self-organizing map; SRP, soluble reactive P; SUVA, specific ultraviolet absorbance; TDN, total dissolved N; TDP, total dissolved P; TFF, tangential flow filtration; TFP, total filterable P.

C.W. Cuss, Environmental and Life Science Graduate Program (chadcuss@trentu.ca), and C. Guéguen, Chemistry Dep., Trent University, 1600 West Bank Drive, Peterborough, Ontario, Canada, K9J 7B8.Trent University, 1600 West Bank Drive, Peterborough, Ontario, Canada, K9J 7B8. *Corresponding author (celinegueguen@trentu.ca).

doi:10.2136/sssaspecpub62.2014.0043

Labile Organic Matter—Chemical Compositions, Function, and Significance in Soil and the Environment. SSSA Special Publication 62. Zhongqi He and Fengchang Wu, editors.

Why Is Leaf Leachate Important?

Leaves contribute vital nutrients, energy, and organic matter to terrestrial and aquatic ecosystems worldwide. This contribution is so integral that the decomposition of leaf litter has been recommended as an indicator of ecosystem health in headwaters, wherein leaching is the first stage of litter decomposition (Gessner and Chauvet, 2002; Young et al., 2008). Leaching is the physical process through which soluble organic molecules such as carbohydrates, proteins, and tannins are extracted from the leaf into the surrounding liquid (Gessner, 1999; Bärlocher, 2005). Thus, leaching may occur wherever there is prolonged contact between water and senescing leaves, including during rainfall events as throughfall; after leaf drop, either through exposure to rainfall while on the ground or following inundation in streams; during freeze–thaw cycles in the autumn and spring; and while fully or partially immersed during spring freshet. Leaching may result in losses of up to 30% of leaf mass over the first 24 to 48 h of immersion in water, and water-soluble compounds can compose up to 50% of dried leaves by mass (Hättenschwiler et al., 2008). The amount and composition of leached material are both highly variable and depend on such factors as leaf age (pre- and post-abscission), dimensions, and integrity, and tree species (Gessner, 1999; Bärlocher, 2005).

The chemical composition of leaf leachate regulates the structure and function of microbial communities in lakes, headwaters, and streams of low order (McNamara and Leff, 2004; Docherty et al., 2006; Kritzberg et al., 2006; Hutalle-Schmelzer et al., 2010; Hall et al., 2012). It has been determined that the leaf leachate dissolved organic matter (LLDOM), which is readily degraded by microbes (i.e., labile), could support up to 54 and 51% of bacterial C demand and community respiration in a third-order stream, respectively, and may supply energy as far as 3.7 km downstream of its input (Wiegner et al., 2005). According to the river continuum concept (Vannote et al., 1980) and observed patterns in forested streams and navigable waters further downstream (Wipfli et al., 2007; Tank et al., 2010), communities of microbes and invertebrates as well as nutrient cycling are interrelated along the complete stream continuum. Thus, downstream ecosystems rely directly or indirectly on upstream inputs and processes so that the importance of LLDOM is palpable throughout freshwaters. In this chapter, we discuss the characterization of LLDOM, highlighting the wide range of analytical methods and approaches used to assess the labile fraction.

What Is Leaf Leachate Dissolved Organic Matter and How Is It Produced for Study?

Leaf leachate dissolved organic matter is the fraction of leaf leachate that passes through a 0.2-, 0.45-, 0.7-, or 1.0-μm filter. In practice, the most common method uses a sterilized, 0.45-μm glass-fiber filter. Leaching is defined as an abiotic process (Bärlocher, 2005); however, fungal colonization can be widespread in leaves and can occur even during early stages of growth (Osono, 2006). Decomposition by endophytic fungi may therefore contribute to leachate, but this contribution is not distinguished in leaching studies.

In studies of LLDOM lability, a wide variety of methods have been used for leaching (Table 1). However, the use of different methods can render results incomparable due to differences arising from the potential effects outlined below. Thus, in the absence of a standardized method, leaching procedures should be designed to reflect the natural processes under consideration. To extract leachate for laboratory-based experiments, leaves may be ground, cut into small pieces, or kept intact. Smaller pieces will release greater amounts of leachate, but mechanical cell lysis may release intracellular material in proportions that do not reflect natural processes. Exposure to water may be conducted via actual or simulated rainfall events (Hongve et al., 2000; Nishimura et al., 2012) or by immersion in water (Bärlocher, 2005; Cuss and Guéguen, 2012a, 2013). Leaching is considered complete within 24 to 48 h following immersion (Bärlocher, 2005); however, it has been found that the optical properties of LLDOM do not exhibit substantial change over the first 96 h (Cuss and Guéguen, 2013). Although warmer temperatures will release greater amounts of leachate, leaching temperature may also affect the size and optical properties of leached molecules (Shank et al., 2010).

How Is the Labile Fraction of Leaf Leachate Dissolved Organic Matter Defined?

Conventionally, the labile fraction of LLDOM is the fraction that undergoes biologically mediated transformation by microbes (biotransformation) relatively quickly. No single definition for lability has been specified along the temporal continuum (Table 1). LLDOM has been compartmentalized into labile (transformed in hours), semilabile (transformed over days), and recalcitrant fractions (undetermined transformation time) (Kaplan et al., 2008). Based on biodegradation experiments, the labile fraction alone has also been considered the fraction that is consumed over hours (McArthur and Richardson 2002; Castillo et al., 2003; Cuss and Guéguen, 2012a), days (Strauss and Lamberti, 2002; Maie et al., 2008; Hur et al., 2009), and months (Cleveland et al., 2004; Hur et al., 2011; Fellman et al., 2013). The relative labilities of different LLDOM compartments have also been estimated by extracting parameters from exponential decay models for DOC (Strauss and Lamberti, 2002; Scully et al., 2004; Vähätalo and Wetzel, 2004; Cuss and Guéguen, 2012a).

It is prudent to consider two types of experiments for assessing LLDOM biotransformation: a test of lability quantifies the amount of LLDOM that is biotransformed as a function of its initial properties, while a test of biotransformation assesses the changes that occur to LLDOM during biotransformation by microbes. There are relatively few published accounts of the former experiment type, which is understandable given the high number of influential variables and lack of a standard method. Since the microbe-mediated transformation of LLDOM tends to reduce its subsequent lability to microbes, it is tacitly assumed that LLDOM with properties similar to biotransformed LLDOM is relatively recalcitrant under similar operational conditions and vice-versa. Thus, the properties of LLDOM that tend to change during biotransformation are frequently considered indicators of lability. Accordingly, biotransformation experiments are often used to assess lability. For example, the amount of protein-like fluorescence (PLF) in LLDOM has been considered an indicator of lability because the concentration of proteins tends to decrease in concert with DOC during biotransformation (Hur et al., 2009; Cuss and Guéguen, 2012a).

Table 1. A summary of leaf leachate studies.

Study	Definition of lability	Leaf leachate	Inoculum origin	Inoculum pretreatment	Characterization of inoculum†
Attermeyer et al. (2013)	DOC turnover over 12 d	^{13}C-enriched beech leaf disks; leached at 4°C in dark in double-DI water; prefiltered (0.2 μm)	Lake	Filtered (0.8 μm); concentrated 25-fold by TFF,§ refrigerated and stored overnight	PLFA, DGGE
Beggs and Summers (2011)	DOC decrease over 61 d	Lodgepole pine needles; 3 types (undegraded, empty-bed degraded, established-bed degraded); litter dried then leached in lake water treated with activated C; leached in direct sunlight for 48 h at ambient temp. (August 2009, Boulder CO)	Lake	Filtered with granular activated carbon	None
Castillo et al. (2003)	Bacterial production over 36 h	Mixed leaves from forest floor; leached in distilled water for 3 d; prefiltered (0.2 μm)	Two rivers; multiple times of year	Storage in dark for 2–8 h during transport	None
Chen and Jaffé (2014)	DOC decrease over 28 d	Senescent plant material from multiple sources; leached at 21°C for 24 h in 0.22-μm filtered water from each site and filter sterilized (0.22 μm); diluted to 0.2–0.5 cm^{-1} absorbance at 280 nm with filtered site water	Surface water from respective sites where plant material collected	None; raw surface water	None
Cleveland et al. (2004)	DOC decrease over 100 d	Live foliage (i.e. green) and senesced litter from five temperate and tropical species of trees and herbaceous shrubs; 50 g cut into pieces <4 cm^2; leached in 1 L DI at 25°C; prefiltered (0.45 μm)	Soil from respective ecosystem	1 g soil + 9 mL DI water + vortex mixing; three sequential 1:10 dilutions in DI	None

Incubation conditions	DOC‡	Nutrients	Optical properties	Chemical Fraction	Polyphenol/ tannins	Bacterial biomass
Nutrients in excess (NO_3 and PO_4); inoculum added to leachate (7 mg C L^{-1}); different amounts of algae-derived DOC added incrementally in pulses; 24-h aeration, 18°C, 16:8 h light/dark at 72 μE; pH not reported	X		X			X
pH 7.2; incubation with constant agitation; nutrients, ambient levels after carbon filtration; leached in natural water containing ambient microbes (1.25 g L^{-1}; 10.06 ± 0.77 mg C L^{-1})	X		X			
Water chemistry was the same as ambient conditions in rivers (pH 4.1–5.4 and 5.4–6.6); incubation in the dark at 26–28°C; nutrient amendments varied (C, N, P, glucose) to reflect seasonal patterns of addition; leachates added to river water at six concentrations (0.9–7.6 mg L^{-1} above existing levels); algal leachates and nutrient amendments alone also added in separate experiments; different experiments used microbes collected at different times of the year	X	X				X
Inoculum added to leachate at volume ratio of 1:100; inoculum, leachate; nutrients, N and P added to 10 μM NH_4NO_3 and 1 μM KH_2PO_4; pH not reported; incubated in the dark	X		X			
10 mL of inoculum added to ~1.8 L of leachate; pH not reported; incubation in the dark (amber bottles) at 25°C; initial nutrient levels highly variable (80 < DOC < 1500 mg L^{-1}; 3 < TDN¶ < 45)	X	X		X		

Table 1. Continued.

Study	Definition of lability	Leaf leachate	Inoculum origin	Inoculum pretreatment	Characterization of inoculum†
Cuss and Guéguen (2012a)	DOC decrease over 7 d; modeled by two-compartment exponential decay to determine percentage labile	Senescent sugar maple leaves, frozen and thawed; whole leaves leached in Milli-Q water in the dark at 5°C for 5 h, with water decanted and refilled at 1 and 3 h; prefiltered (0.2 μm)	River	Filtered (5 μm); one inoculum stored overnight in the dark (5°C); one inoculum frozen and stored for 7 d in the dark (5°C)	None
Fellman et al. (2013)	DOC decrease over 28 d	Live foliage (i.e. green) from three different species; biodegradability tested for fresh litter and for litter aged in litter bags in full sunlight for 2, 4, and 6 mo; biodegradability of leachate DOC also tested following 48 h of irradiation (laboratory); 10 g cut into pieces <2 cm^2, leached in 500 mL milli-Q water for 2 h at room temp. (20–25°C) with stirring; prefiltered (0.7 μm)	Stream sediment	Leached with milli-Q for 24 h; Filtered (2.7 μm)	None
Hur et al. (2009)	Not specified (changes observed over 14 d)	Cut mixed fallen leaves into pieces ~1 cm^2 and shredded; leached in distilled DI water at solid/solution mass ratio 1:10 for 24 h with mixing; Leaf/water mixture then centrifuged at 10,000 rpm for 30 min and prefiltered (0.2 μm); humic substances also base-extracted using XAD-7 resin and mixed with various proportions of glucose (100%, 75:25, 50:50, 0)	River	Filtered (5 μm); concentrated 10-fold using 0.2-μm syringe filter	None
Hur (2011)	Not specified (changes observed over 28 d)	Three mixed samples (reeds, fallen leaves, and weeds collected in autumn); cut into pieces ~1 cm^2 and shredded; leached in distilled DI water at solid/solution mass ratio 1:20; leaching time not reported; leaf/water centrifuged at 10,000 rpm for 30 min. If necessary, and prefiltered (0.2 μm)	Lake and river (mixed)	See Hur et al. (2009)	None

Incubation conditions	DOC‡	Nutrients	Optical properties	Chemical Fraction	Polyphenol/ tannins	Bacterial biomass
10 mL of inoculum added to 200 mL of leachate; pH adjusted to 7; initial DOC conc. 26 and 65 mg L^{-1}; nutrients, ambient levels in leachate only	X		X			
Prefiltered (0.2 μm); 2 mL of inoculum added to 23 mL leachate; incubation in the dark at room temperature (20–25°C); initial nutrient levels highly variable (4 < DOC < 29 mg L^{-1}; 91 < DON < 290; 3 < SRP†† < 51); pH not reported	X	X	X			
Initial DOC conc. = 20–30 mg L^{-1}; 1 mL inoculum added to 60 mL leachate; nutrients NH_4NO_3 and K_2HPO_4 added to fix C/N/P = 30:10:3; ionic strength adjusted to 0.01 M; pH adjusted to 7; sealed flasks incubated in the dark at 25°C gently shaken daily	X	X	X	X		
Initial DOC conc. = 30 mg L^{-1}; 4 mL inoculum added to 250 mL leachate; nutrients NH_4NO_3 and K_2HPO_4 added to fix C/N/P = 30:10:3; ionic strength adjusted to 0.01 M; pH adjusted to 7; sealed flasks incubated in the dark at 20°C, gently shaken and opened daily	X		X	X		

Table 1. Continued.

Study	Definition of lability	Leaf leachate	Inoculum origin	Inoculum pretreatment	Characterization of inoculum†
Hur et al. (2012)	Not specified (changes observed over 28 d)	Mixed leaves from forest floor; shredded and leached in distilled DI water at solid/solution mass ratio 1:10 for 24 h; prefiltered (0.2 μm)	River	Concentrated 100-fold following Hur et al. (2009)	None
Kalbitz et al. (2003a)	CO_2 evolved over 90 d; fit with double exponential model to determine percentage labile	Litter layers of forest floor in spruce- and beech-dominated forests; cut into pieces ~1–2 cm^2, frozen, and thawed; leached in Ultrapure water at solid/solution mass ratio 1:10 for 24 h at 5°C, stirred three times; prefiltered (0.2 μm)	Mixed from soil sources beneath fallen litter	Preincubated for 2 wk at 20°C with moisture added; shaken for 30 min. with 4 nM $CaCl_2$; filtered (5 μm); enumerated and sources mixed with equal cell contributions	None
Kalbitz et al. (2003b)	As Kalbitz et al. (2003a)	CO_2; DOC; aromaticity; optical properties; humification; proportion of H associated with O-containing functional groups (carbohydrates); proportion of H associated with aromatics; polysaccharide/carboxylic ratio; ^{13}C isotope distributions			
Kaplan et al. (2008)	^{13}C-DOC uptake; fit with double exponential model to determine percentage labile	Ground leaves, stems, and roots from ^{13}C-enriched tulip poplar seedlings; 1–3 g of ground tissue leached in the dark at 4°C in 1 L of filter-sterilized DI water mixture then centrifuged, filtered, and heated to 70°C twice separated by 24 h at room temp. to ensure biological stability; stored in the dark at 4°C for ~1.5 mo; 120 L at 20 mg C L^{-1} used in stream addition; leachate added to a 1.27-km study reach over 2 h 3 wk before leaf fall (with Br^- tracer); samples collected from stream at eight stations over 8 h and filtered (0.7 μm); stored for 2 d	Stage 1: full, uncontrolled in-stream microbial community; Stage 2: river (same as Stage 1)	Stage 1: none (leachate added to stream); Stage 2: microbial community grown in the dark in a bioreactor; Stage 2: fed filtered (0.3 μm) stream water for 1 yr, colony regrown weekly	None

Incubation conditions	DOC‡	Nutrients	Optical properties	Chemical Fraction	Polyphenol/ tannins	Bacterial biomass
Initial DOC conc. = 548 mg L^{-1}; 2 mL inoculum added to 2 L leachate; nutrients NH_4NO_3 and K_2HPO_4 added to fix C/N/P = 30:10:3; pH not reported; covered with Parafilm, incubated in the dark at room temp. and gently shaken and opened daily	X		X	X		
Initial DOC conc. ≤ 20 mg L^{-1}; 7 mL inoculum added to 700 mL leachate; Nutrients at ambient levels in leachate only; pH not reported; five glass fiber filters added to facilitate biofilm formation; sealed and incubated in the dark at 20°C with daily shaking; opened four times to aerate, compressed air	X		X	X		
Stage 1: Leachate added to a 1.27-km study reach over 2 h 3 wk before leaf fall (with Br^- tracer); Stage 1: Samples collected from stream at eight stations over 8 h; Stage 2: in plug-flow bioreactor (chromatography column with glass beads) in dark (7 L river water, 8 m L leachate); Nutrients at ambient levels in leachate, stream, and bioreactor only; pH not reported	X					

Table 1. Continued.

Study	Definition of lability	Leaf leachate	Inoculum origin	Inoculum pretreatment	Characterization of inoculum†
Kiikkilä et al. (2011)	Mineralized DOC content over 110 (by CO_2 evolution) and 20 (by DOC) days; fit with double exponential model to determine percentage labile (110 days)	Humus layer under trees in 100% birch, 22–37% birch and spruce, and 1.6–7% birch and spruce stands from trees on plots: senescent birch leaves (green–yellow); needles from branches removed from spruce (mostly green); birch cut into 10 mm^2 pieces; needles removed and crushed, both stored at 4°C; humus samples adjusted to 100% ultrapure water-holding capacity, allowed to stand over night; 10 g leaves and needles leached in 100 mL ultrapure water overnight; water-saturated samples centrifuged at 5000 *g* for 20 min., filtered (0.45 μm), frozen, and thawed for 3–10 d (4°C); filtered (0.2 μm); 20 d frozen until further analysis	Humus layer under birch and spruce	10 g birch + spruce humus shaken (200 rpm, 1 h) in 100 mL water	Differentiated availability to bacteria and fungi
Maie et al. (2008)	DOC decrease over 7 and 28 d	Tannin extracts from near-senescent freeze-dried mangrove leaves	Freshwater marsh, mangrove swamp, estuary (7-d Exp.); Marine (7- and 28-d Exp.)	Filtered (0.7 μm) and stored at 4°C until use	None
McArthur and Richardson (2002)	Bacterial growth over 24 h	From senescent leaves and needles from five species of trees; 2 g of dried litter leached in 500 mL of sterilized DI water in the dark with mixing for 24 h and filtered (1 μm); leachate also separated into LMW (<10 kDa; two species only) and HMW fractions by ultrafiltration; stored for < wk at 4°C	River	Grown in-stream on glass beads; incubated in the dark on beads in stream water at 12°C for 24 h	None

Incubation conditions	DOC‡	Nutrients	Optical properties	Chemical Fraction	Polyphenol/ tannins	Bacterial biomass
For 110 d: 100 μL inoculate added to 30 mL leachate (25 ± 2.3 mg C L^{-1}); nutrient amendment (0.1 mM NH_4NO_3 + 0.1 mM K_2HPO_4) added to half the bottles; incubated in the dark at 20°C with continuous shaking (120 rpm) with semiregular aeration for 110 d; For 20 d: 0.65 mL inoculum added to 650 mL thawed leachate with glass fiber filter; incubated in the dark at 20°C with daily shaking by hand for 20 d; pH not reported	X		X			
1.2-mL aliquot of 2000 mg L^{-1} tannin solution (1060 mg C L^{-1}) mixed with 120 mL inoculum; covered with quartz plate (light conditions at 765 W m^{-2} solar simulator) or in HDPE amber (dark conditions); pH not reported; nutrients at ambient levels in tannin extract and inocula only	X		X		X	
Four beads added to 50 mg C L^{-1} low molecular weight and full leachates with micronutrient standard (NO_3, PO_4, Cu, Zn, Co, Mn, Fe); incubated in the dark at 12°C; pH not reported	X	X			X	X

Table 1. Continued.

Study	Definition of lability	Leaf leachate	Inoculum origin	Inoculum pretreatment	Characterization of inoculum†
Mineau et al. (2013)	DOC decrease over 30 d	Sugar maple leaves from N-enriched and control catchments (collected postabscission); leaves mechanically ground and leached in 1 L DI water at 2°C for 2 wk; leachates centrifuged, filtered (0.7 μm), and heated to 70°C twice separated by 24 h at room temp. to ensure biological stability; Stored at 2°C for <2 wk	1 from river in each catchment	100 mL filtered (0.7 μm) stream water + microbes scrubbed from rocks in stream	None
Pellerin et al. (2010)	DOC decrease over 21 d	Collected four crops and four aquatic macrophytes by cutting at soil–water surface (leaves and stems); rinsed with DI water and stored, then cut into <2.5 cm pieces and oven-dried (60°C); 15 g litter leached in 900 mL DI water in amber glass bottles for 4 h at 24°C with stirring, then filtered (0.3 μm) and diluted	River	None	None
Qualls (2005)	$^{14}CO_2$ evolved over 1 yr; fit with double exponential model to determine percentage labile	5 g of ^{14}C-labeled senescent poplar leaves; leaves decomposed on top of A horizon soil for 1 yr at 25°C, separated from soil by mesh; dried leaves ground to pass through 2-mm sieve and 0.8 g was leached in 30 mL DI water for 30 min at 3°C; leachate decanted, solids leached three more times at 25°C for 4 h; leachates filtered (0.45 μm), frozen, thawed, recombined; seven recombined fractions chemically fractionated into hydrophilic acid (HiA), neutral (HiN), base (HiB), fulvic acid (FA), humic acid (HA), weak hydrophobic acid (wHoA), and hydrophobic neutral (HoN); fractions used in inoculation (7): whole leachate, HA, FA, weak HiA, HiA, HiN, HiN + acids	Soil (7 fractions); sand (fulvic acid only)	Soil cores taken from upper 10 cm of A horizon (O horizon removed) in tubes; sterile, neutralized sand added to soil, adjusted to −20 kPa soil matric water; leachates and fractions added to soils; sand core for fulvic acid also mixed with micronutrients (C/N = 8) and inoculum prepared using soil	None

Incubation conditions	DOC‡	Nutrients	Optical properties	Chemical Fraction	Polyphenol/ tannins	Bacterial biomass
Leachate added to inoculum, separated in to four bottles; nutrients (N 350 μg L^{-1}, P 100 μg L^{-1} SRP, N+P, none) added to bottles (two inocula × two leachates × two amendments); DOC ± SE = 9.4 ± 0.38 mg L^{-1}; incubated in loosely capped bottles in the dark at 14°C; bottles inverted several times every 2 d; aerobic conditions ensured with probe; pH not reported	X	X	X			X
Raw inoculum added directly from river to leachate (3% v/v); DOC 40–50 mg C L^{-1}; incubated in the dark at 22°C, aerated continuously; pH not reported; nutrients at ambient levels in leachate	X	X			X	
Incubated in sealed mason jars with water added as necessary to maintain matric potential at 25°C; DOC not reported; pH not reported; Nutrients not reported for soils, C/N = 8 for sand	X			X		

Table 1. Continued.

Study	Definition of lability	Leaf leachate	Inoculum origin	Inoculum pretreatment	Characterization of inoculum†
Strauss and Lamberti (2002)	DOC decrease over 24 d; labile component determined from 6 d using single exponential fit; overall decomposition over 28 d using ln-transformed exponential model	Fallen leaves from 18 tree species ground into coarse fragments, leached for 24 h in RO-treated water at ~23°C; filtered (0.2 μm), stored at 4°C, and refiltered before use.	Mixed stream water and sediment	No water; sediment filtered (0.64 cm); stored on ice or the refrigerator until use (within 24 h)	None
Vähätalo and Wetzel (2008)	DOC decrease over 898 d	30 L of senescent and green leaves from common rush inundated with 38 L DI water for 19 wk filtered (0.7 μm) and lyophilized; 6.7 g dissolved in lake water and autoclaved, then diluted to 10 mg C L^{-1} with DI water	Creek	None	None
Wickland et al. (2007)	CO_2 evolved over 28- 31 d; fit with single and double exponential models to determine percentage labile	Leachates from senescent leaves and needles of two deciduous trees, one conifer (and twigs and bark) and one grass and two mosses; vegetation dried and leached (10–50 g) in 5–9 L of 0.001 N $NaHCO_3$ in clear foil-covered Pyrex for 7 d with aeration; filtered (0.45 μm)	Mixed soil from sites where vegetation collected	Soil mixed with DI, filtered (1.6 μm), and diluted serially to 10^{-3}	None
Wieder et al. (2008)	CO_2 evolved over 21 and 34 d	Recently senesced litter of six species collected from beneath at least four individuals, combined by species, and air dried; 25 g of each species leached in 500 mL DI water for 24 h at 25°C, filtered (1 μm), and diluted to 250 mg C L^{-1}	Soil cores from beneath eight trees of each species	Soils from under trees combined, homogenized, and stored until use (5°C); 1 g soil + 9 mL DI water + vortex mixing + three sequential 1:10 dilutions in DI (liquid Exp.)	None

Incubation conditions	DOC‡	Nutrients	Optical properties	Chemical Fraction	Polyphenol/ tannins	Bacterial biomass
Leachates added to unfiltered stream water + 1 mL sediment at 35 ± 4 mg C L^{-1} and total volume 200 mL; flasks loosely covered with foil, incubated in darkness at ~23°C; pH not reported; For nitrification Exp.: pH 7.1–7.8; Exp. 1, all leachates at 10 mg L^{-1}; Exp. 2, four leachates + glucose at 10, 30, and 50 mg C L^{-1}; incubated as other except 27°C for 72 h on orbital shaker (175 rpm); nutrients at ambient levels in stream water and leachates only	X	X	X			
5% v/v inoculum/leachate; DOC 10 mg L^{-1}; pH not reported; nutrients adjusted to 1.7 and 0.24 mg N and P L^{-1}; sealed in 20-mL glass ampoules with 5 mL of headspace and incubated in darkness at 20–25°C for 439 d	X		X			
1 mL inoculate added to 50 mL leachate diluted to ~10 mg C L^{-1} and incubated in sealed amber bottles at 22°C; 1/8 glass fiber filter added to inoculum, incubated in the dark with shaking; pH not reported; nutrients at ambient levels in leachates only	X		X	X		
For soil Exp.: 7 mL of 250 mg C L^{-1} leachate added to 25 g moist soil, incubated in the dark (23°C) for 34 d, routinely aerated; For liquid Exp.: 1 mL inoculate added to 70 mL of 20 mg C L^{-1} leachate, flasks sealed, incubated in darkness at 23°C with continuous mixing (orbital shaker) for 21 d; leachate pH differed by species (4.28–6.59); incubation pH not reported; For liquid Exp.: duplicated with nutrients added (NH_4NO_3 + K_2HPO_4 to C/N/P = 100:10:1)	X	X		X		

Table 1. Continued.

Study	Definition of lability	Leaf leachate	Inoculum origin	Inoculum pretreatment	Characterization of inoculum†
Wiegner et al. (2005)	^{13}C DOC uptake over 44 h; fit with double exponential model to determine percentage labile	Ground leaves, stems, and roots from ^{13}C-enriched tulip poplar seedlings; ~1 g of ground tissue leached in the dark at 4°C in 1 L of filter-sterilized DI water with stirring for 24 h; mixture then centrifuged, filtered, heated to 70°C twice separated by 24 h at room temp. to ensure biological stability; stored in the dark at 4°C for ~4 mo	Stream sediment	Incubated in an artificial stream fed with river water for 18 d; transferred to 15-L stream mesocosms fed with filtered stream water (0.3 μm); water recirculated in mesocosm at velocity about equal to river velocity, with level 0.04 m above sediments	Bacteria counts (PI epifluorescence)

† DGGE, denaturing gradient gel electrophoresis; PFLA, phospholipid fatty acid.

‡ DOC, dissolved organic carbon.

§ TFF, tangential flow filtration.

¶ TDN, total dissolved N.

†† SRP, soluble reactive.

What Methods Are Used to Characterize Leaf Leachate Dissolved Organic Matter and Its Lability?

The wide variety of direct measurements and proxies used to assess LLDOM and its lability may be broken up into three categories: quantitative measures, characterization of fractions, and spectroscopic indicators (Table 1; Fig. 1–3).

Quantitative Measures

The amount of nutrients and organic compounds in LLDOM are frequently quantified to assess its lability.

Measurements of nutrient content are primarily focused on C, N, and P. Nutrient levels have been favored in ecological studies of leaf leachate quality as determined by biodegradability because of their importance throughout global cycles (Hall et al., 2012; Fellman et al., 2013). Carbon is generally measured directly as total dissolved organic C (DOC). The labile fraction of LLDOM is often quantified by measuring the change in DOC concentration following inoculation with microbes, known as the biodegradable dissolved organic C content (BDOC; Cleveland et al., 2004; Fellman et al., 2013; Table 1). However, CO_2 production is also used to estimate biodegradation (Qualls, 2005; Wickland et al., 2007). The microbe-containing medium (e.g., concentrated pellets, cultured communities, or river water) may be added to the leachate or vice versa, and bioreactors may also be used, all with or without the inclusion of a microbial substrate (Table 1; Scully et al., 2004; Wickland et al., 2007; Maie et al., 2008; Hur et al., 2009; Cory and Kaplan, 2012; Cuss and Guéguen, 2012b). Since microbial communities are complex adaptive systems, it is not yet possible to control the microbial component of

Incubation conditions	DOC‡	Nutrients	Optical properties	Chemical Fraction	Polyphenol/ tannins	Bacterial biomass
For mesocosm: temperature-controlled (20.3 ± 2.1°C), light-modulated (14:10 h light/dark); 288 μmol photons m^{-2} s^{-1} PAR); 8 mL leachate (10 mmol C L^{-1}) injected into each of three identical mesocosms	X				X	X

biodegradation experiments. Consequently, the location from which inocula are collected and the conditions under which they are incubated should reflect the system, process, or mechanism under consideration. The N content of LLDOM may be measured as total dissolved nitrogen (TDN), nitrate (NO_3–N), and ammonia (NH_4–N), so that dissolved organic N (DON) is calculated as DON = TDN – (NO_3–N + NH_4–N) (Qualls et al., 1991; Yu et al., 2002; Castillo et al., 2003; Schreeg et al., 2013). Since the P content of DOM may be quantified in one of several ways, it is worthwhile clarifying the relationships between different measurements. Dissolved P may be measured as total dissolved or filterable (TDP or TFP), dissolved reactive or unreactive (DRP or DUP), molybdate reactive (MRP), or orthophosphate (OP). Phosphorous is considered reactive if it is detectable using colorimetric tests without having been hydrolyzed or subjected to oxidative digestion and consists of mainly OP (APHA, 1999). Dissolved reactive P is also referred to as soluble reactive (SRP), filterable reactive (FRP), or dissolved inorganic phosphorous, and MRP is equivalent to DRP in filtered leachates (Jarvie et al., 2002; Fig. 1A). Nutrient ratios (e.g., DOC/DON, DOC/DOP, DON/DOP) are also frequently used to characterize leachates (Qualls and Haines, 1991; Qualls and Richardson, 2003; Cleveland et al., 2004; Osborne et al., 2007; Wieder et al., 2008; Yoshimura et al., 2010; Fellman et al., 2013; Mineau et al., 2013; Schreeg et al., 2013).

Measurement of the wide range of organic molecules in leaf leachate tends to focus on major groups, including amino acids and proteins (Hernes et al., 2001; Yu et al., 2002; Maie et al., 2007), sugars and carbohydrates (Kaiser et al., 2001; Kalbitz et al., 2003a,b; Maie et al., 2006; Osborne et al., 2007; Hall et al., 2012), tannins (Hernes et al., 2001; Kraus et al., 2003; Maie et al., 2006; 2008), lignins (Ibrahima et al., 2008; Hernes et al., 2007; 2009; Spencer et al., 2008; Pellerin et al., 2010; Yoshimura et al., 2010; Hall et al., 2012), and phenolics (Maie et al., 2006; Osborne et al., 2007;

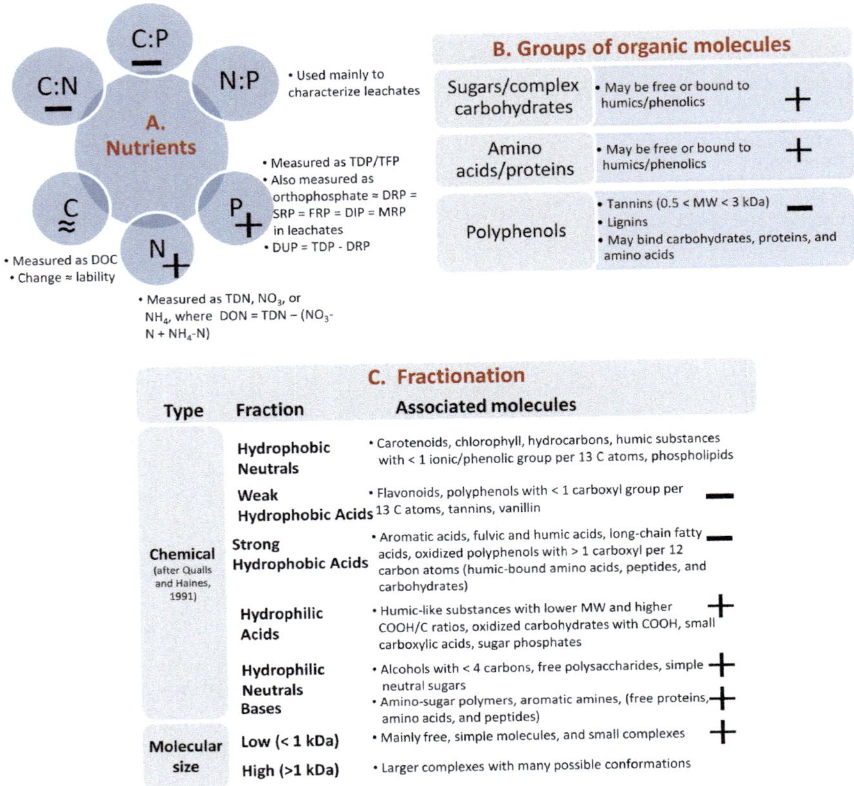

Fig. 1. Nonoptical methods for characterizing leaf leachate dissolved organic matter (LLDOM). Symbols (+ and –) have been added to indicators that tend to exhibit mainly positive and negative correlations with lability, respectively.

Kiikkilä et al., 2011; Nishimura et al., 2012) (Table 1; Fig. 1B). Phenolics are highly water-soluble aromatics with one or more hydroxyl substituents that vary widely in leaf leachates and include a broad spectrum of compounds (Hättenschwiler et al., 2008). Tannins are phenolics that precipitate proteins and have a molecular weight (MW) between 0.5 and 3 kDa and may be further classified as condensed or hydrolyzable (Hättenschwiler and Vitousek, 2000; Kraus et al., 2003). The latter distinction is important with regards to leaf leachates, as hydrolyzable tannins are limited to dicotelydonous plant species (Haslam, 1988). Lignins are operationally distinguished from tannins and other phenolics, and may be further separated into distinct species (Hernes et al., 2007); however, subclassifying phenolics may not produce mutually exclusive groups, results can be method-dependent, and misclassification is common (Hättenschwiler and Vitousek, 2000; Kraus et al., 2003; Khoddami et al., 2013). Notwithstanding these difficulties, the measurement of total phenolics, tannins, or lignin has been very useful for characterizing LLDOM and assessing lability (Hernes and Hedges, 2004; Hernes et al., 2007; Maie et al., 2008). Importantly, experiments that test the biodegradable DOC content and

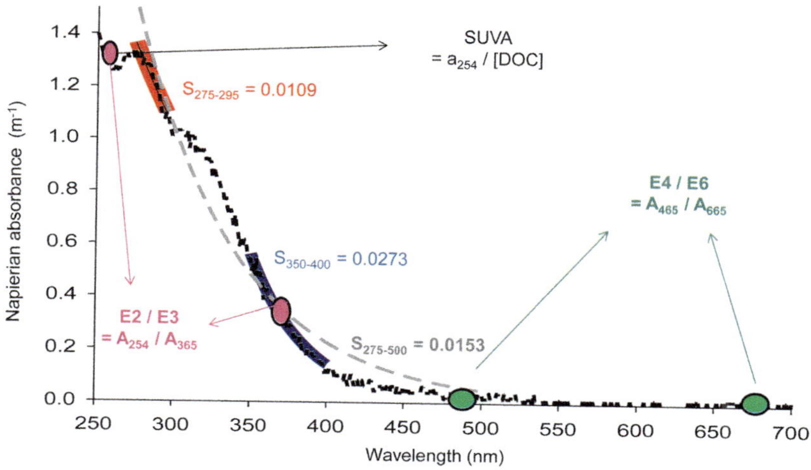

Fig. 2. Example of absorbance indicators for leaf leachate dissolved organic matter (DOM), using spectrum of *Salix bebbiana* Sarg. leachate (unpublished data, 2014).

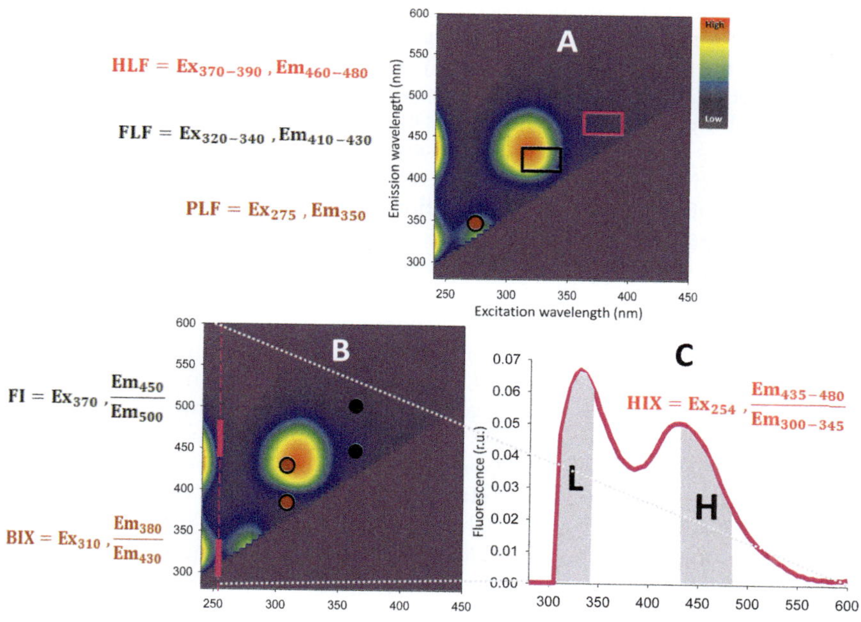

Fig. 3. (A, B) Fluorescence indicators for leaf leachate dissolved organic matter using excitation–emission matrices (EEMs) of *Acer saccharum* Marshall, (C) demonstrating the interference of protein-like fluorescence in the humification index (HIX) (unpublished, 2013).

lability of leaf leachates often do not test the concentrations of organic compounds directly, instead relying on proxies. For example, carbohydrate content is often attributed to the hydrophilic neutral fraction of leaf leachates, while protein content is frequently tested via fluorescence or considered to be reflected by DON. These surrogate analyses and their importance in the characterization of LLDOM and its lability are discussed below.

Characterization of Fractions

The lability of LLDOM is frequently tested by first fractionating it into chemical- and size-based fractions and testing the lability and associated properties thereof.

Most chemical fractionations of LLDOM rely on one of two chromatographic procedures developed for riverine DOM (Leenheer, 1981; Aiken et al., 1992; Aiken and Leenheer, 1993). These procedures use a series of resin-filled columns with acid–base treatments to separate DOM into hydrophobic and hydrohphilic acids, neutrals, and bases (Qualls and Haines, 1991; Kaiser et al., 2001; Qualls, 2005; Wickland et al., 2007; Wallace et al., 2008; Kiikkilä et al., 2011; Uselman et al., 2012). The resulting fractions and compounds typically attributed thereto are outlined in Fig. 1C. Similar chromatographic methods are used to extract humic and fulvic acids from LLDOM, which are both part of the hydrophobic acid fraction (Thurman and Malcom, 1981). Humic acids are operationally distinct from fulvic acids; they are insoluble at pH <2. However, total fulvic and humic acids compose only a fraction (~30–60%) of DOM in natural waters and a lower percentage of leaf leachates (~10–50%; Qualls, 2005; Wickland et al., 2007; Wallace et al., 2008). Chemical fractionation has proven useful for characterizing DOM according to the proportions of chemically reactive groups and for relating these to their functionalities; however, it has been shown that DOM can undergo large and irreversible changes at such extremes of pH (Chen and Kenny, 2007; Spencer et al., 2007; Cuss et al., 2014). Thus, it is likely that the properties and behavior of chemically fractionated LLDOM will not reflect natural conditions. Compounds typically attributed to each chemical fraction are far from a complete accounting and may vary between leachates.

Leaf leachate DOM has been characterized according to its MW using methods such as tangential flow filtration (TFF; Maie et al., 2006; Osborne et al., 2007; Kiikkilä et al., 2011), flow field-flow fractionation (FlFFF; Schimpf and Petteys, 1997; Cuss and Guéguen, 2012b; 2013), and size exclusion chromatography (SEC; Hongve et al., 2000; Scully et al., 2004; Hur et al., 2009, 2011; Hofmann et al., 2012) (Fig. 1C). Size exclusion chromatography is capable of assessing the entire continuum of LLDOM properties, while TFF produces discrete fractions, and small molecules may be lost through membrane pores in FlFFF. On the other hand, SEC uses carrier fluid with a high ionic strength and requires high concentrations of DOM (0.1 M NaCl; 30–50 mg C L^{-1}; Chin et al., 1994; Hur, 2011), which are known to alter its properties (Pan et al., 2008; Mei et al., 2009). Tangential flow filtration is conducted with the sample in its natural medium and frequently requires large volumes. On the other hand, FlFFF allows for pH and ionic strength to be adjusted to levels appropriate to the system of interest. Asymmetrical flow field-flow fractionation (AF4) has also been used with low volumes and environmentally relevant concentrations of DOM at high levels of precision to ensure that the measured properties reflect in situ conditions (<1 mL; 5–10 mg L^{-1}; ±100 Da; Guéguen and Cuss, 2011).

Spectroscopic Indicators

Owing to several attractive analysis properties (e.g., nondestructiveness, efficiency, cost-effectiveness, high information yield, and ease of insertion in coupled systems), LLDOM is increasingly characterized by its optical properties. Since the optical properties of DOM are indicative of its structure and composition, a wide variety of indicators have been developed both for direct estimation of lability via structural information and as proxies for organic molecules. The lability of DOM and LLDOM is of interest to terrestrial, fluvial, and marine scientists, giving rise to a wide variety of independently developed spectroscopic indicators. Thus, several indicators may be derived from each measurement of absorbance and fluorescence spectra.

The range of the absorbance spectrum from 250 to 700 nm is typically used for measuring LLDOM (Fig. 2). As for other DOM sources, absorbance at 254 nm (A_{254}) is commonly used as a proxy for DOM concentration, but this relationship differs with source. The specific ultraviolet absorbance (SUVA) is equal to the Napierian absorbance coefficient at 254 nm (A_{254}) or 280 nm measured in m^{-1}, divided by the concentration of DOC measured in mg L^{-1}. It has been shown that SUVA is highly correlated with aromaticity in the fulvic and hydrophobic acid extracts of DOM (Weishaar et al., 2003; Wickland et al., 2007; Spencer et al., 2008; Beggs and Summers, 2011). Absorbance properties are also used as proxies for MW including the E2/E3 and E4/E6 ratios (Chen et al., 1977; Chow et al., 2013; Fig. 2) and a suite of spectral slope parameters (e.g., $S_{275-295}$, $S_{350-400}$, S_R; Helms et al., 2008). The E4/E6 ratio has been shown to increase gradually in leachates with the increasing degradation of litter (Gressel et al., 1995) and to be correlated with MW (Osborne et al., 2007). However, E4/E6 has been unsuccessful at predicting other leachate characteristics (Osborne et al., 2007) and at characterizing LLDOM (Silveira et al., 2011). The S-parameter is calculated by fitting a 2- or 3-parameter exponential decay model to the absorbance curve of the Napierian absorbance coefficient according to the following relationship:

$$a_\lambda = a_{\lambda_r} e^{-S(\lambda-\lambda_r)} + K \qquad [1]$$

where S is the S-parameter, λ_r is a reference wavelength, and K is a background parameter that may be added to account for baseline shifts and background scatter (Markager and Vincent, 2000; Helms et al., 2008; Guéguen and Kowalczuk, 2013). Different ranges of the LLDOM absorbance spectrum have been fit to obtain the S-parameter, including 275 to 295 nm (Hur et al., 2009), 290 to 350 nm (Spencer et al., 2008), 350 to 400 nm (Helms et al., 2008), and 280/290 to 500 nm (Shank et al., 2010; Zhang et al., 2013). The ratio of $S_{275-295}/S_{350-400}$, or slope ratio (S_R), has also proven useful (Helms et al., 2008; Guéguen and Cuss, 2011; Inamdar et al., 2012; Chow et al., 2013; Zhang et al., 2013). The S-parameter has been highly correlated with MW for riverine DOM (Guéguen and Cuss, 2011); however, a peak near 275 nm and a shoulder near 310 nm are common in leaf leachates, which may interfere with $S_{275-295}$ and $S_{275-500}$ (Fig. 2; Zhang et al., 2013).

Several fluorescence-based indicators have been used to measure the quality and lability of leaf leachate DOM (Fig. 3). Single excitation–emission (Ex/Em) wavelength pairs and two-dimensional synchronous fluorescence scanning have both been used to distinguish leachates and assess lability (Hur et al., 2009; 2011). The fluorescence index (FI; McKnight et al., 2001) has been used to distinguish DOM age and end member (e.g., allochthonous vs. autochthonous; Wickland et al., 2007; Inamdar et al., 2012). Integrated areas of fluorescence scans have been efficaciously employed to distinguish leachates and measure changes in labile fractions (i.e., protein- or polyphenol-like fluorescence, PLF; fulvic-like fluorescence; and humic-like fluorescence, HLF; Hur et al., 2009; 2012; Hur, 2011; Fig. 3). Other measures of leaf leachate age and quality that only use portions of the fluorescence spectrum include the biological index (Huguet et al., 2009; Zhang et al., 2013), and the humification index (HIX; Zsolnay et al., 1999; Ohno, 2002). The HIX has been broadly applied to measure changes in the quality of leaf leachates during the leaching process and over biotransformation (Kalbitz et al., 2003b; Wickland et al., 2007; Hur et al., 2009; 2011; Inamdar et al., 2012). However, the application of structural indicators such as SUVA and HIX should be undertaken with care, as they have not been empirically tested for LLDOM. For example, the effectiveness of HIX was demonstrated for fumigated, shaken, centrifuged, and dried soil extracts analyzed at pH 2 (Zsolnay et al., 1999). These conditions are considerably different from those encountered in fresh, untreated leaf leachates at circumneutral or slightly acidic pH. Differences in structure and composition caused by altered physicochemical conditions can exert a significant impact on both lability and the properties reflected by indicators (Zsolnay, 2003). This effect is apparent in fresh LLDOM that contains a relatively large amount of proteins and >50% polyphenols (Uselman et al., 2012), in which the region of PLF (Ex/Em ≅ 250–300/300–350 nm) can be very large relative to the edge of the fulvic-like *A* peak (Ex/Em ≅ <240–260/375–500 nm; e.g., Cuss and Guéguen, 2012a, 2013). It has been demonstrated that several polyphenols also fluoresce in regions previously attributed to proteins (Aiken, 2014). Hence, large amounts of either protein- or polyphenol-like fluorescence may thereby interfere with the *L* region of HIX (Ex/Em = 254/300–445 nm), so that decreases in PLF alone will result in decreasing *L* with a corresponding increase in HIX, which may be mistakenly interpreted as increasing humification or aromaticity (Fig. 3C). High PLF in LLDOM may also interfere with indicators such as SUVA due to higher absorbance near 225 and 275 nm (Fig. 2).

Detailed fluorescence analysis of leaf leachate DOM can be achieved by scanning a range of emission wavelengths (~280–600 nm) corresponding to excitation along each of a series of wavelengths (~240–450 nm), which are aligned to form a three-dimensional excitation–emission matrix (EEM; Coble 1996; Fig. 3). The large volume of correlated data thereby yielded is a primary challenge in the study of DOM fluorescence, and the associated data analysis techniques are numerous. The interested reader is directed to the first chapter of this volume (Zhang and He, 2015) and to recent reviews in the literature (Fellman et al., 2010; Ishii and Boyer, 2012; Andrade-Eiroa et al., 2013a,b; Murphy et al., 2014b). In this chapter we will confine discussion to the most popular method for managing EEM data, parallel factor analysis (PARAFAC; Stedmon et al., 2003; Stedmon and Bro, 2008; Murphy et al., 2013; Fig. 4). However, two recent methods that are not included in the above reviews also deserve preliminary mention: self-organizing maps (SOMs; Bieroza et al., 2009, 2012; Carstea et al., 2010; Cuss et al., 2014; Cuss and Guéguen, 2015) and the finite mixture approach (Butturini and Ejarque, 2013). Self-organizing maps are unsupervised learning algorithms originally developed for pattern recognition in artificial intelligence research (Kohonen, 2001). As such, SOMs iteratively "learn" the primary patterns underlying collections of EEMs and project them on to a two- or three-dimensional map, whereby EEMs are self-organized into self-similar clusters. The finite mixture approach considers each fluorescence peak to be composed of a finite number of underlying peaks with a user-defined bivariate distribution. This method assumes noninteracting molecules, which theoretically produce distinct fluorescence peaks with Gaussian distributions; however, it is strongly suspected that the fluorescence of DOM results from interacting molecules (Del Vecchio and Blough, 2004), for which distributions are unknown and may differ both by peak and across samples. The PARAFAC algorithm uses alternating least-squares optimization to minimize the error in a trilinear fitting of correlated groups of Ex/Em wavelength pairs. However, PARAFAC components are not easily compared across studies (Murphy et al., 2014a). Parallel factor analysis also requires large numbers of relatively similar EEMs to cross validate a model capable of distinguishing samples. Despite these shortcomings, PARAFAC relies on relatively few assumptions and has been remarkably successful in detecting similar components across a wide swath of independent studies (Ishii and Boyer, 2012 and references therein; Cuss and Guéguen, 2013; Murphy et al., 2014b). Typical PARAFAC components and an example of sample differentiation are shown in Fig. 4.

Many studies have characterized LLDOM using more than one method or coupled instrumentation, leading to an improved understanding of properties and lability. Osborne et al. (2007) characterized LLDOM using several approaches: nutrient content, carbohydrates and phenolics, and E4/E6 and SUVA for four different size fractions separated using TFF. This multimethod approach revealed size-specific information about organic compound distributions. Principal component analysis (PCA) was used to distinguish leachates by species and expose relationships between MW and nutrient levels. Wickland et al. (2007) measured chemical fractions, SUVA, FI, HIX, and PARAFAC components for LLDOM and soil pore waters in Alaskan forests. Lability was strongly correlated with the proportion of hydrophilic DOC, chemical composition was distinct by species, and chemical fractions were also distinct in their PARAFAC component compositions, FI, and HIX. Finally, Cuss and Guéguen (2015) used on-line asymmetrical FlFFF coupled to ultraviolet–visible diode array and fluorescence detectors to measure the size, absorbance, and fluorescence properties of LLDOM from several tree species. Using PARAFAC and SOM, it was shown that the fluorescence composition of leachates differed significantly by species and size fraction, including the proportion of PLF, which was linked to lability in an earlier study (Cuss and Guéguen, 2012a; Fig. 4 and 5). Since all properties of the LLDOM samples were measured simultaneously, relationships between MW and fluorescence were revealed. Since lability is related to both MW and fluorescence properties (see below), measuring fluorescence at multiple points over the size continuum offers more information about structure and, hence, lability (Cuss and Guéguen, 2015; Romera-Castillo et al., 2014; Fig. 5). Thus, applying multiple measurements to assess LLDOM lability affords deeper insight into its structure, character, and hence functionality in the environment.

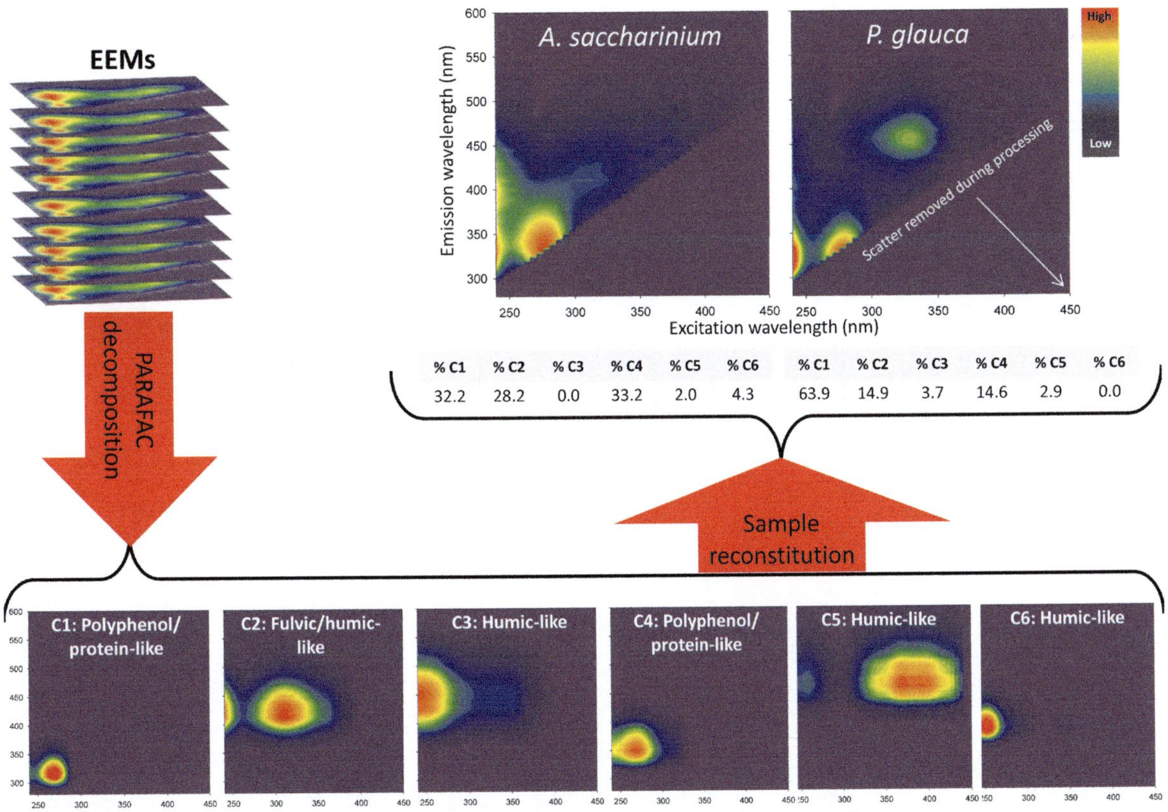

Fig. 4. Example of parallel factor analysis (PARAFAC) decomposition of leaf leachate excitation–emission matrices (EEMs) into typical components, followed by recomposition to achieve semiquantitative differentiation (unpublished data, 2012).

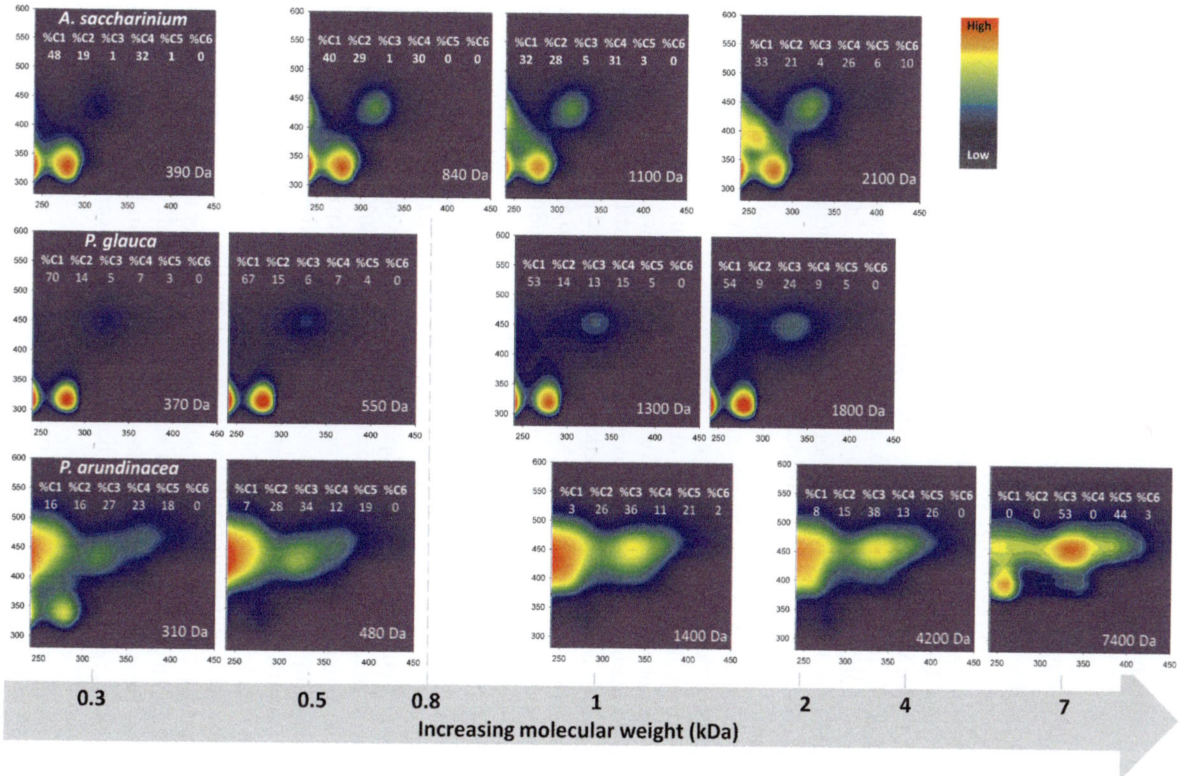

Fig. 5. Illustration of changes in optical properties with increasing molecular weight using excitation–emission matrices (EEMs) of different leaf leachates and reconstituted using parallel factor analysis (PARAFAC) components and loadings. Component identities follow Fig. 4 (unpublished data, 2014).

Factors Influencing the Measurement of Leaf Leachate Dissolved Organic Matter and Its Lability

The sheer breadth of experimental approaches and properties used to assess the lability of LLDOM is indicative of its complexity (Table 1). Here we discuss the major factors that have been found to affect the measurement of LLDOM and its lability.

Nutrient Content, Organic Compounds, and Indicators

Unsurprisingly, the lability of LLDOM is related to its value as a source of food and energy for microorganisms, which has been demonstrated in several studies relating lability and biotransformation to nutrient content and ratios (Strauss and Lamberti, 2002; Cleveland et al., 2004; Wieder et al., 2008; Fellman et al., 2013; Castillo et al., 2003; Mineau et al., 2013), carbohydrate content (Qualls and Haines, 1992; Kalbitz et al., 2003a; Qualls, 2005), and protein content or PLF (Scully et al., 2004; Maie et al., 2008; Hur et al., 2009; Cuss and Guéguen, 2012a). However, in some cases these indicators have not been good predictors of lability (Qualls and Haines, 1992; Cleveland et al., 2004; Mineau et al., 2013).

Importantly, in many cases, relationships between the organic molecules present in leaf leachate DOM and its lability are not tested directly, instead measuring (e.g., chemical fractions) (Cleveland et al., 2004; Qualls, 2005) or PLF (Wickland et al., 2007; Hur et al., 2009; Mineau et al., 2013). However, in LLDOM, such indirect proxies are not reliable indicators for labile organic molecules. For example, a recent study of freshly collected senescent leaves from five different tree species (Uselman et al., 2012) found that the fraction of bases contained substantial nonproteinaceous material, and protein-based N comprised only 6 to 65% of DON (Fig. 1). Thus, measurements of bases and DON in LLDOM may include more than 50% of nonproteinaceous material, making them indicators of primarily unknown nitrogenous material. It has also been demonstrated that polyphenols may fluoresce in the protein-like region, and this region has been shown to represent both labile and recalcitrant material of different MWs (Maie et al., 2007; 2008; Cuss and Guéguen, 2012a,b; Aiken, 2014). Further, the hydrophilic neutral fraction of forest throughfall was shown to be made up of only 50% carbohydrates, so that 50% was again not the assumed labile material (Fig. 1C; Qualls and Haines, 1991). Since indications may be so heavily biased by unknown materials or not composed of the assumed majority of labile compounds, and since the extent of this bias is quite variable, the relationships between quality indicators and lability are unreliable. In fact, the proportions of both hydrophobic and hydrophilic fractions, for both the acidic and neutral species as well as bases, have each been the strongest indicators of lability (Qualls and Haines, 1992; Qualls, 2005; Wickland et al., 2007; Wallace et al., 2008). These inconsistencies have been attributed to species-based differences in leaf-litter leachates, and in some cases, no chemical fraction has served as a significant lability indicator (Wallace et al., 2008).

Thus, indicators of DOM quality, organic compounds, and nutrient content can be used to estimate lability; however, they frequently do not reflect the majority of material in LLDOM, and hence may not reflect other influential factors. This is likely because the majority of tests and indicators have been transferred from studies of soil-derived, riverine, and marine DOM, which have considerably different compositions than LLDOM having been subjected to considerable microbial, photochemical, or adsorptive processing, or being borne of different end members. Hence, the proportions of key constituents may differ considerably with consequent differences in structure and intermolecular dynamics. These findings underscore the difficulty of accurately quantifying LLDOM lability and highlight the importance of employing multiple indicators while conducting experiments under in situ conditions, particularly since the structure of DOM depends on other physiochemical parameters such as concentration, pH, and ionic strength (Sutton and Sposito, 2005; Pan et al., 2008; Mei et al., 2009; Pace et al., 2012; Cuss et al., 2014).

Structural Considerations

The structure of LLDOM affects the accessibility of nutrients to microbes (i.e., bioavailability) (Sutton and Sposito, 2005; Moorhead and Sinsabaugh, 2006; Craine et al., 2007; Tomaszewski et al., 2011), so that quality alone is not a reliable indicator of lability. This hidden dimension explains the inconsistent behavior of quality indicators across studies. Thus, indicators of DOM structure have also proven useful for assessing lability in the complex mixture of fresh material characteristic of LLDOM.

Water-soluble, degradation-resistant, protein–polyphenol complexes are known to form during leaf senescence and may also form in the complex mixture of primarily polyphenols in leaf leachate following cell lysis (Hättenschwiler and Vitousek, 2000; Kraus et al., 2003; Maie et al., 2008). The existence of water-soluble, protein–polyphenol complexes created using natural polyphenols (Kusuda et al., 2006; Richard et al., 2006; Maie et al., 2008), coupled with evidence that polyphenols generate humic-like fluorescence following bio- and photo-degradation (Maie et al., 2008), supports prior evidence suggesting that the bioavailability of a sizeable portion of the proteins, amino acids, and carbohydrates in LLDOM may be inhibited through intimate association with more recalcitrant humic substances, hydrophobic acids, polyphenols, or hydrophilic acids (Wershaw, 2004 and references therein; Sutton and Sposito, 2005 and references therein; Chen and Kenny, 2007; Benner and Kaiser, 2011). Such associations have also been suggested by the appearance of two peaks in PLF in different size fractions of DOM separated by SEC (Romera-Castillo et al., 2014). This phenomenon was also detected in size-fractionated leaf leachates and aged DOM sources using AF4, where a single peak in PLF was found in the low- to medium-weight size fractions of fresh-leaf leachates and two peaks were detected in more aged sources (Cuss and Guéguen, 2015). The formation of phenolic glycosides in LLDOM through the complexation of carbohydrates with polyphenols or humic acids is also a strongly supported mechanism for limiting bioavailability (Kijima et al., 1997; Wershaw, 2004; Sutton and Sposito, 2005). Not surprisingly then, indicators of aromaticity and humification such as SUVA and HIX (Fig. 2 and 3) are often negatively correlated with lability in fresh and aged LLDOM as are the proportions of humic substances (via chemical fractionation) and HLF (Qualls and Haines, 1992; Kalbitz et al., 2003a; Wickland et al., 2007; Wieder et al., 2008; Hur, 2011; Hur et al., 2012). However, this negative correlation is not always present and requires deeper investigation (Strauss and Lamberti, 2002; Wickland et al., 2007; Pellerin et al., 2010). As an example of such investigative methods, Fellman et al. (2011) found a significant positive correlation between the amount of one humic-like PARAFAC component and both DOC and DON concentrations, while at the same time a significant negative correlation was found between both DOC and DON concentrations for a different humic-like component. Thus, different humic-like DOM constituents may be related to DON in different ways, potentially exerting different impacts on lability, which can be distinguished using PARAFAC. Humification index, SUVA, the fraction of humic-like substances, and HLF also often increase over the process of biotransformation (Kalbitz et al., 2003b; Hur et al., 2009; Beggs and Summers, 2011; Hur, 2011), indicating that LLDOM becomes more humic and aromatic as it is biotransformed. Thus, structural indicators such as aromaticity, humification, and fluorescence composition are useful indicators of both lability and biotransformation.

The MW of LLDOM is often monitored over the course of biotransformation to provide information about the preferential use of molecules of different sizes (Fig. 1C). Thus, it is also used to assess the impact of LLDOM structure on lability. In most instances, it has been found that microbes preferentially use smaller molecules such as free proteins and carbohydrates, thereby biotransforming LLDOM into larger, more aromatic and more humic DOM (Scully et al., 2004; Hur et al., 2009; Hur, 2011; Attermeyer et al., 2013). However, contrary findings exist, including the biotransformation of LLDOM has shifted MW distribution toward smaller sizes (Hur et al., 2012), the <10 kDa fraction has supported less microbial growth than whole leachates (McArthur and Richardson, 2002), and larger molecules have been preferentially used (Kritzberg et al., 2006). Hur et al. (2009) showed that the nominal MW of LLDOM increased following biotransformation but contrarily decreased in humic-substance extracts from the same leaf leachate, and no consistent trend in MW change was apparent when increasing proportions of glucose were added. Attermeyer et al. (2013) showed that LLDOM increased in nominal MW and aromaticity during biotransformation, while the opposite was true for algae-derived DOM that started with a higher MW, and mixtures again demonstrated intermediate results. Using three natural DOM samples with different MW (0.8, 1.1, and 1.9 kDa by SEC) and microbial colonies collected from each habitat, Young et al. (2005) found that the largest amount of DOM was degraded by two-thirds of the microbial inoculates when it came from their native habitat regardless of MW. Using ANOVA it was determined that microbial productivity was significantly affected by both DOM sample and microbial community ($p < 0.001$), but the 0.8 kDa DOM was only available to the microbial community from the same habitat while the 1.9 kDa DOM was available to all microbial communities. These results suggest that the microbial transformation of LLDOM is not only contingent on its structure, but also on the availability of alternate resources and microbial community composition with correspondingly variable relationships with MW. Thus, MW may also be useful for assessing lability when combined with other measurements.

The photodegradation of LLDOM has also provided insight into structural relationships between recalcitrant and labile portions. Traditionally, LLDOM lability has been tested in the absence of light (Strauss and Lamberti, 2002; Cleveland et al., 2004; Cory and Kaplan, 2012; Cuss and Guéguen, 2012a), as photodegradation and biodegradation can act synergistically, thereby increasing the labile proportion (Moran et al., 2000; Scully et al., 2004; Vähätalo and Wetzel, 2008; Benner and Kaiser, 2011; Fellman et al., 2013). Indeed, a recent study using biomass leachates showed that reductions in DOC and fluorescence caused by sequential photo- and biodegradation exceeded those caused by biodegradation alone, and PCA was effectively used to demonstrate that the fluorescence of sequentially degraded leachates was more similar to that of riverine DOM (Chen and Jaffé, 2014). In headwater streams where leaves are frequently the primary source of energy and nutrients (Abelho et al., 2001; Tank et al., 2010), photodegradation plays a minor role in the summer because of shade provided by forest canopy. However, photodegradation may exert substantial influence in shallow waters in late fall and spring freshet (Moran and Zepp, 1997; Vähätalo and Wetzel, 2004; Gonsior et al., 2013). Photoirradiation tends to increase the labile fraction of leaf leachate DOC and DON while decreasing the proportion of polyphenols, suggesting that photoirradiation may liberate labile constituents from polyphenolics (Qualls and Richardson, 2003; Scully et al., 2004; Maie et al., 2008; Fellman et al., 2013). Indicators of recalcitrance such as SUVA have also been shown to decrease following photoexposure (Hur et al., 2011; Fellman et al., 2013).

To date, molecular-level instrumental analysis has been sparingly applied to LLDOM degradation and lability, owing to high costs and scarcity; however, their application has provided insight into structural considerations for biodegradation pathways. For example, Benner et al. (1990) used ^{13}C nuclear magnetic resonance (NMR) and elemental analysis to compare the properties of senescent mangrove leaves on trees and those submerged in estuarine sediments and found that tannins remained in constant proportions during decomposition in the water corresponding with N immobilization. Kalbitz et al. (2003b) combined FTIR, ^{13}C NMR, and mass spectrometry to assess the behavior of various compounds before and after 90 d of LLDOM degradation (Table 1). The LLDOM was leached from beech (*Fagus sylvatica* L.) and spruce (*Picea* spp.) litter was collected from the Oi horizon. An increase in aromatic H was observed during incubation and was closely correlated with the proportion of mineralized DOC. The proportion of carbohydrate H also decreased substantially but was not correlated with any variables that described degradation. Interestingly, proportions of aromatic H and O-containing functional groups were closely but nonlinearly related to the percentage of total DOC mineralized. Such instrumental methods are clearly capable of providing detailed information about relationships between LLDOM composition and structure during biotransformation.

Microbial Considerations

In some experiments, a nutrient amendment that includes available N and P has been added to leachates to avoid the premature cessation of biological activity (e.g., Scully et al., 2004; Hur, 2011; Table 1). However, this is less common in recent studies (Wickland et al., 2007; Cuss and Guéguen, 2013; Fellman et al., 2013) and is unnecessary for stimulating bacterial production (Castillo et al., 2003). To assess differences in LLDOM lability arising from different sources, differences in structure and MW must be considered as they may affect bioavailability (Kraus et al., 2003). Differences in external nutrient availability may also lead to different degradation behavior, with correspondingly different impacts on structure and MW (e.g., N-mining; Craine et al., 2007). This phenomenon has been demonstrated by Wieder et al. (2008), in which SUVA was a highly significant ($p < 0.001$) predictor of total C mineralization from multiple leaf leachates (i.e., when more aromatic, recalcitrant DOM was present, less DOC was consumed) under conditions of excess nutrients (i.e., when more aromatic, recalcitrant DOM was present, less DOC was consumed); however, it was not a significant predictor in the absence of added nutrients. Thus, excess external nutrient levels may result in measurements of lability and biotransformation that do not represent natural consumption patterns, since the extraction of these nutrients from LLDOM is not required.

Tests of lability have generally failed to assess the potential effects of interactions between microbial community composition and different LLDOM sources, which may explain the differences in results observed across studies (Table 1). However, there is increasing evidence of intricate relationships between microbial community structure and DOM properties, including the following: community structure may differ quite significantly in freshwaters, the properties of DOM may govern the composition of associated microbial communities, and communities may evolve in response to changing DOM quality (Young et al., 2005; Docherty et al., 2006; Hahn, 2006; Kritzberg et al., 2006; Smart et al., 2008; McCarren et al., 2010; Portillo et al., 2012). It has also been found that microbial communities from different habitats may process DOM differently over the short term (Cottrell and Kirchman, 2000; Young et al., 2005). Hence, communities that are isolated for extended periods or subjected to harsh storage conditions before lability testing may not possess the genetic or species diversity required for normal behavior under ambient conditions.

Future Directions

In tests of lability, there are frequently too many unisolable governing variables and the effects of changing more than one variable are not simply additive, often leading to highly nonlinear, synergistic, and unexpected behavior, which would not be observable except with the highly sensitive measurement of multiple variables. Hence, the ability to detect significant differences currently exceeds the ability to isolate and identify the mechanisms through which these differences arise. Four broad approaches are therefore suggested for advancing understanding about the lability of leaf leachate DOM:

1. Continue assessments of lability using indigenous microbial inocula under controlled conditions that represent those found in situ, involving multiple measurements (e.g., absorbance, fluorescence, free and bound nutrients, and BDOC) and strategic control of key variables. The reduced costs of and easier access to previously rare instrumentation enables detailed molecular characterization while minimizing the disruption of DOM structure during sample introduction (e.g., electrospray ionization mass spectrometry and NMR spectroscopy). The reduction of these barriers allows for more information to be obtained about degradation pathways and may be especially useful when coupled to other instruments or used in combined analysis (e.g., Kalbitz et al., 2003b; Piccolo and Spiteller, 2003).
2. Establish a standard method for assessing lability with subsequent research into the effects of deviation from standard parameters that reflect conditions typically encountered in natural environments (e.g., pH, microbial community composition, temperature, ionic strength, degree of photoirradiation, abundance of external nutrients, etc.).
3. Advanced analytical methods that have not been extensively applied to LLDOM may provide novel information about the interactions between constituents, revealing unanticipated cause–effect mechanisms. For example, the characterization of microbial community structure has revealed novel information about the relationship between microbes and DOM degradation in soils and aquatic systems (Fernando and Li, 2012; Hall et

al., 2012; Attermeyer et al., 2013; Gabor et al., 2014) and may do the same if applied to LLDOM. The application of rapid biochemical characterization and gene-based fingerprinting methods such as fatty acid methyl ester profiling (Attermeyer et al., 2013), denaturing gradient gel electrophoresis (Docherty et al., 2006), terminal restriction fragment length polymorphism (Kritzberg et al., 2006; Hall et al., 2012), and pyrosequencing (McCarren et al., 2010; Eilers et al., 2012; Portillo et al., 2012; Gabor et al., 2014) facilitate the incorporation of microbial community characterization into studies of lability. Improved instrumentation for fluorescence measurement allows in situ monitoring of optical properties at high frequencies (e.g., Carstea et al., 2010; Cuss and Guéguen, 2015), allowing for the measurement of biotransformation at fine-grained time scales. Electrospray ionization Fourier transform ion cyclotron resonance mass spectrometry offers unprecedented chemical information about elemental ratios (H/C, O/C, and N/C) and their corresponding MWs and may be effectively combined with other methods to offer insight into the lability and biotransformation of LLDOM (e.g., Kim et al., 2006; Schmidt et al., 2011).

4. The interpretation and complete understanding of results obtained from data-rich, coupled instrumentation or experimental methods is challenging given the volume and dimensionality of information produced by measuring such a large number of correlated variables. Combined with the complexity of DOM cycling, these conceptual and data analysis challenges have led to remarks that the biogeochemical cycling and complexity of DOM are recalcitrant to available methods (Filella, 2009; Jiao et al., 2010). This complexity may be multiplied in studies of LLDOM lability due to the interaction of two complex systems (DOM and microbes) with a correspondingly enhanced number of relationships and governing variables. Combined with sophisticated analytical methods and experimental design, advanced statistical and mathematical methods for analyzing and interpreting high-dimensional and nonlinear phenomena may be useful for revealing information about interactions between LLDOM and microbes. Hence, the study of LLDOM lability may benefit from incorporating the methods and perspectives associated with the study of complex systems and systems science (Xiang et al., 2006; Newman, 2011; Fuchs, 2013).

References

Abelho, M. 2001. From litterfall to breakdown in streams: A review. Sci. World J. 1:656–680.

Aiken, G. 2014. Fluorescence and dissolved organic matter: A chemist's perspective. In: P. Coble, J. Lead, A. Baker, D. Reynolds, and R.G.M. Spencer, editors, Aquatic organic matter fluorescence. Cambridge Univ. Press, Cambridge. p. 35–74.

Aiken, G., and Leenheer, J., 1993. Isolation and chemical characterization of dissolved and colloidal organic matter. Chem. Ecology 8:135–151. doi:10.1080/02757549308035305

Aiken, G.R., D.M. McKnight, K.A. Thorn, and E.M. Thurman. 1992. Isolation of hydrophilic organic acids from water using nonionic macroporous resins. Org. Geochem. 18:567–573. doi:10.1016/0146-6380(92)90119-I

Andrade-Eiroa, Á., M. Canale, and V. Cerdá. 2013a. Environmental applications of excitation–emission spectrofluorimetry: An in-depth review I. Appl. Spectrosc. Rev. 48:1–49. doi:10.1080/05704928.2012.692104

Andrade-Eiroa, Á., M. Canale, and V. Cerdá. 2013b. Environmental applications of excitation–emission spectrofluorimetry: An in-depth review II. Appl. Spectrosc. Rev. 48:77–141. doi:10.1080/05704928.2012.692105

APHA. 1999. Method 4500-P phosphorous. Standard methods for the examination of water and wastewater, 20th ed. American Public Health Association, Washington, DC.

Attermeyer, K., T. Hornick, Z.E. Kayler, A. Bahr, E. Zwirnmann, H.-P. Grossart, and K. Premke. 2013. Increasing addition of autochthonous to allochthonous carbon in nutrient-rich aquatic systems stimulates carbon consumption but does not alter bacterial community composition. Biogeosciences Discuss. 10:14261–14300. doi:10.5194/bgd-10-14261-2013

Bärlocher, F. 2005. Leaching. In: M.A.S. Graça, F. Bärlocher, and M.O. Gessner, editors, Methods to study litter decomposition: A practical guide. Springer, Dordrecht, the Netherlands. p. 33–36.

Beggs, K.M.H., and R.S. Summers. 2011. Character and chlorine reactivity of dissolved organic matter from a mountain pine beetle impacted watershed. Environ. Sci. Technol. 45:5717–5724. doi:10.1021/es1042436

Benner, R., P.G. Hatcher, and J.I. Hedges. 1990. Early diagenesis of mangrove leaves in a tropical estuary: Bulk chemical characterization using solid-state ^{13}C NMR and elemental analysis. Geochim. Et Cosmochim. Acta 54:2003–2013.

Benner, R., and K. Kaiser. 2011. Biological and photochemical transformations of amino acids and lignin phenols in riverine dissolved organic matter. Biogeochemistry 102:209–222. doi:10.1007/s10533-010-9435-4

Bieroza, M., A. Baker, and J. Bridgeman. 2009. Exploratory analysis of excitation–emission wavelength matrix fluorescence spectra with self-organizing maps as a basis for determination of organic matter removal efficiency at water treatment works. J. Geophys. Res. 114. doi:10.1029/2009JG000940

Bieroza, M., A. Baker, and J. Bridgeman. 2012. Exploratory analysis of excitation–emission matrix fluorescence spectra with self-organizing maps: A tutorial. Educ. Chem. Engineers 7:e22–e31. doi:10.1016/j.ece.2011.10.002

Butturini, A., and E. Ejarque. 2013. Dissolved organic matter fluorescence: A finite mixture approach to deconvolve excitation–emission matrices. Biogeosciences Discuss. 10:4711–4732. doi:10.5194/bgd-10-4711-2013

Carstea, E.F., A. Baker, M. Bieroza, and D. Reynolds. 2010. Continuous fluorescence excitation–emission matrix monitoring of river organic matter. Water Res. 44:5356–5366. doi:10.1016/j.watres.2010.06.036

Castillo, M.M., G.W. Kling, and J.D. Allan. 2003. Bottom-up controls on bacterial production in tropical lowland rivers. Limnol. Oceanogr. 48:1466–1475. doi:10.4319/lo.2003.48.4.1466

Chen, M., and R. Jaffé. 2014. Photo- and bio-reactivity patterns of dissolved organic matter from biomass and soil leachates and surface waters in a subtropical wetland. Water Res. 61:181–190. doi:10.1016/j.watres.2014.03.075

Chen, Y., N. Senesi, and M. Schnitzer. 1977. Information provided on humic substances by E4:E6 ratios. Soil Sci. Soc. Am. J. 41:352–358. doi:10.2136/sssaj1977.03615995004100020037x

Chow, A.T., J. Dai, W.H. Conner, D.R. Hitchcock, and J.J. Wang. 2013. Dissolved organic matter and nutrient dynamics of a coastal freshwater forested wetland in Winyah Bay, South Carolina. Biogeochemistry 112:571–587. doi:10.1007/s10533-012-9750-z

Chen, H., and J.E. Kenny. 2007. A study of pH effects on humic substances using chemometric analysis of excitation–emission matrices. Ann. Envir. Sci. 1:1–9.

Cleveland, C.C., J.C. Neff, A.R. Townsend, and E. Hood. 2004. Composition, dynamics, and fate of leached dissolved organic matter in terrestrial ecosystems: Results from a decomposition experiment. Ecosystems 7:275–285. doi:10.1007/s10021-003-0236-7

Coble, P. 1996. Characterization of marine and terrestrial DOM in seawater using excitation–emission matrix spectroscopy. Mar. Chem. 51:325–346. doi:10.1016/0304-4203(95)00062-3

Cory, R.M., and A. Kaplan. 2012. Biological lability of streamwater fluorescent dissolved organic matter. Limnol. Oceanogr. 57:1347–1360. doi:10.4319/lo.2012.57.5.1347

Cottrell, M.T., and D.L. Kirchman. 2000. Natural assemblages of marine proteobacteria and members of the *cytophaga-flavobacter* cluster consuming low- and high-molecular-weight dissolved organic matter. Appl. Environ. Microbiol. 66:1692–1697. doi:10.1128/AEM.66.4.1692-1697.2000

Craine, J.M., C. Morrow, and N. Fierer. 2007. Microbial nitrogen limitation increases decomposition. Ecology 88:2105–2113. doi:10.1890/06-1847.1

Cuss, C.W., and C. Guéguen. 2012a. Impacts of microbial activity on the optical and copper-binding properties of leaf-litter leachate. Front. Microbiol. 3:166. doi:10.3389/fmicb.2012.00166

Cuss, C.W., and C. Guéguen. 2012b. Determination of relative molecular weights of fluorescent components in dissolved organic matter using asymmetrical flow field-flow fractionation and parallel factor analysis. Anal. Chim. Acta 733:98–102. doi:10.1016/j.aca.2012.05.003

Cuss, C.W., and C. Guéguen. 2013. Distinguishing dissolved organic matter at its origin: Size and optical properties of leaf-litter leachates. Chemosphere 92:1483–1489. doi:10.1016/j.chemosphere.2013.03.062

Cuss, C.W., and C. Guéguen. 2015. Relationships between molecular weight and fluorescence properties for size-fractionated dissolved organic matter from fresh and aged sources. Water Res. 68:487–497. doi:10.1016/j.watres.2014.10.013

Cuss, C.W., Y.X. Shi, S.M. McConnell, and C. Guéguen. 2014. Changes in the fluorescence composition of multiple DOM sources over pH gradients assessed by combining parallel factor analysis and self-organizing maps. J. Geophys. Res.: Biogeosciences 119:1850–1860 . doi:10.1002/2013JG002598

Del Vecchio, R., and N.V. Blough. 2004. On the origin of the optical properties of humic substances. Environ. Sci. Technol. 38:3885–3891. doi:10.1021/es049912h

Docherty, K.M., K.C. Young, P.A. Maurice, and S.D. Bridgham. 2006. Dissolved organic matter concentration and quality influences the structure and function of freshwater microbial communities. Microb. Ecol. 52:378–388. doi:10.1007/s00248-006-9089-x

Eilers, K.G., S. Debenport, S. Anderson, and N. Fierer. 2012. Digging deeper to find unique microbial communities: The strong effect of depth on the structure of bacterial and archaeal communities in soil. Soil Biol. Biochem. 50:58–65. doi:10.1016/j.soilbio.2012.03.011

Fellman, J.B., E. Hood, and R.G.M. Spencer. 2010. Fluorescence spectroscopy opens new windows into dissolved organic matter dynamics in freshwater systems: A review. Limnol. Oceanogr. 55:2452–2462. doi:10.4319/lo.2010.55.6.2452

Fellman, J.B., K.C. Petrone, and P.F. Grierson. 2011. Source, biogeochemical cycling, and fluorescence characteristics of dissolved organic matter in an agro-urban estuary. Limnol. Oceanogr. 56:243–256. doi:10.4319/lo.2011.56.1.0243

Fellman, J.B., K.C. Petrone, and P.F. Grierson. 2013. Leaf litter age, chemical quality, and photodegradation control the fate of leachate dissolved organic matter in a dryland river. J. Arid Environ. 89:30–37. doi:10.1016/j.jaridenv.2012.10.011

Fernando, W.G., and R. Li. 2012. Opening the black box: Understanding the influence of cropping systems and plant communities on bacterial and fungal population dynamics. Ceylon J. Sci. Bio. Sci. 41:89–110. doi:10.4038/cjsbs.v41i2.5380

Filella, M. 2009. Freshwaters: Which NOM matters? Environ. Chem. Lett. 7:21–35. doi:10.1007/s10311-008-0158-x

Fuchs, A. 2013. Nonlinear dynamics in complex systems: Theory and applications for the life-, neuro-, and natural sciences. Springer–Verlag, Berlin.

Gabor, R.S., K. Eilers, D.M. McKnight, N. Fierer, and S.P. Anderson. 2014. From the litter layer to the saprolite: Chemical changes in water-soluble soil organic matter and their correlation to microbial community composition. Soil Biol. Biochem. 68:166–176. doi:10.1016/j.soilbio.2013.09.029

Gessner, M.O. 1999. A perspective on leaf litter breakdown in streams. Oikos 85:377–384. doi:10.2307/3546505

Gessner, M.O., and E. Chauvet. 2002. A case for using litter breakdown to assess functional stream integrity. Ecol. Appl. 12:498–510. doi:10.1890/1051-0761(2002)012[0498:ACFULB]2.0.CO;2

Gonsior, M., P. Schmitt-Kopplin, and D. Bastviken. 2013. Depth-dependent molecular composition and photo-reactivity of dissolved organic matter in a boreal lake under winter and summer conditions. Biogeosciences 10:6945–6956. doi:10.5194/bg-10-6945-2013

Gressel, N., Y. Inbar, A. Singer, and Y. Chen. 1995. Chemical and spectroscopic properties of leaf litter and decomposed organic matter in the Carmel Range, Israel. Soil Biol. Biochem. 27:23–31. doi:10.1016/0038-0717(94)00139-R

Guéguen, C., and C.W. Cuss. 2011. Characterization of aquatic dissolved organic matter by asymmetrical flow field-flow fractionation coupled to UV-Visible diode array and excitation emission matrix fluorescence. J. Chromatog. A 1218:4188–4198. doi:10.1016/j.chroma.2010.12.038

Guéguen, C., and P. Kowalczuk. 2013. Colored dissolved organic matter in frontal zones. In: I.M. Belkin, editor, Chemical oceanography of frontal zones, the handbook of environmental chemistry. Springer–Verlag, Berlin. doi:10.1007/698_2013_244

Hahn, M.W. 2006. The microbial diversity of inland waters. Curr. Opin. Biotechnol. 17:256–261. doi:10.1016/j.copbio.2006.05.006

Hall, E.K., K. Besemer, L. Kohl, C. Preiler, K. Riedel, T. Schneider, W. Wanek, and T.J. Battin. 2012. Effects of resource chemistry on the composition and function of stream hyporheic biofilms. Front. Microbiol. 3:35. doi:10.3389/fmicb.2012.00035

Haslam, E. 1988. Plant polyphenols (syn. vegetable tannins) and chemical defense: A reappraisal. J. Chem. Ecol. 14:1789–1806. doi:10.1007/BF01013477

Hättenschwiler, S., B. Aeschlimann, M.M. Coûteaux, J. Roy, and D. Bonal. 2008. High variation in foliage and leaf litter chemistry among 45 tree species of a neotropical rainforest community. New Phytol. 179:165–175. doi:10.1111/j.1469-8137.2008.02438.x

Hättenschwiler, S., and P.M. Vitousek. 2000. The role of polyphenols in terrestrial ecosystem nutrient cycling. Trends Ecol. Evol. 15:238–243. doi:10.1016/S0169-5347(00)01861-9

Helms, J.R., A. Stubbins, J.D. Ritchie, E.C. Minor, D.J. Kieber, and K. Mopper. 2008. Absorption spectral slopes and slope ratios as indicators of molecular weight, source, and photobleaching of dissolved organic matter. Limnol. Oceanogr. 53:955–969. doi:10.4319/lo.2008.53.3.0955

Hernes, P.J., R. Benner, G.L. Cowie, M.A. Goñi, B.A. Bergamaschi, and J.I. Hedges. 2001. Tannin diagenesis in mangrove leaves from a tropical estuary: A novel molecular approach. Geochim. Cosmochim. Acta 65:3109–3122. doi:10.1016/S0016-7037(01)00641-X

Hernes, P.J., B.A. Bergamaschi, R.S. Eckard, and R.G.M. Spencer. 2009. Fluorescence-based proxies for lignin in freshwater dissolved organic matter. J. Geophys. Res. 114:3291. doi:10.1029/2009JG000938

Hernes, P.J., and J.I. Hedges. 2004. Tannin signatures of barks, needles, leaves, cones, and wood at the molecular level. Geochim. Cosmochim. Acta 68:1293–1307. doi:10.1016/j.gca.2003.09.015

Hernes P.J., A.C. Robinson, and A.K. Aufdenkampe. 2007. Fractionation of lignin during leaching and sorption and implications for organic matter "freshness". Geophys. Res. Lett. 34:L17401. doi:10.1029/2007GL031017

Hofmann, S., M.A. Timofeyev, A. Putschew, N. Saul, R. Menzel, and C.E.W. Steinberg. 2012. Leaf litter leachates have the potential to increase lifespan, body size, and offspring numbers in a clone of *Moina Macrocopa*. Chemosphere 86:883–890. doi:10.1016/j.chemosphere.2011.10.041

Hongve, D., P.A.W. Van Hees, and U.S. Lundström. 2000. Dissolved components in precipitation water percolated through forest litter. Eur. J. Soil Sci. 51:667–677. doi:10.1046/j.1365-2389.2000.00339.x

Huguet, A., L. Vacher, S. Relexans, S. Saubusse, J.M. Froidefond, and E. Parlanti. 2009. Properties of fluorescent dissolved organic matter in the Gironde Estuary. Org. Geochem. 40:706–719. doi:10.1016/j.orggeochem.2009.03.002

Hur, J. 2011. Microbial changes in selected operational descriptors of dissolved organic matter from various sources in a watershed. Water Air Soil Pollut. 215:465–476. doi:10.1007/s11270-010-0491-0

Hur, J., M.H. Lee, H. Song, and M.A. Schlautman. 2012. Microbial transformation of dissolved organic matter from different sources and its influence on disinfection byproduct formation potentials. Environ. Sci. Pollut. Res. 20:4176–4187. doi:10.1007/s11356-012-1384-0

Hur, J., M.H. Park, and M.A. Schlautman. 2009. Microbial transformation of dissolved leaf litter organic matter and its effects on selected organic matter operational descriptors. Environ. Sci. Technol. 43:2315–2321. doi:10.1021/es802773b

Hur, J., K.Y. Jung, and Y.M. Jung. 2011. Characterization of spectral responses of humic substances upon UV irradiation using two-dimensional correlation spectroscopy. Water Res. 45:2965–2974. doi:10.1016/j.watres.2011.03.013

Hutalle-Schmelzer, K.M.L., E. Zwirnmann, A. Krüger, and H.P. Grossart. 2010. Changes in pelagic bacteria communities due to leaf litter addition. Microb. Ecol. 60:462–475. doi:10.1007/s00248-010-9639-0

Ibrahima, A., P. Biyanzi, and M. Halima. (2008) Changes in organic compounds during leaf litter leaching: Laboratory experiment on eight plant species of the Sudano-guinea Savannas of Ngaoundere, Cameroon. iForest 1:27–33.

Inamdar, S., N. Finger, S. Singh, M. Mitchell, D. Levia, H. Bais, D. Scott, and P. McHale. 2012. Dissolved organic matter (DOM) concentration and quality in a forested mid-Atlantic watershed, USA. Biogeochemistry 108:55–76. doi:10.1007/s10533-011-9572-4

Ishii, S.K.L., and T.H. Boyer. 2012. Behavior of reoccurring PARAFAC components in fluorescent dissolved organic matter in natural and engineered systems. Crit. Rev. Environ. Sci. Technol. 46:2006–2017. doi:10.1021/es2043504

Jarvie, H.P., P.J.A. Withers, and C. Neal. 2002. Review of robust measurement of phosphorous in river water: Sampling, storage, fractionation, and sensitivity. Hydrol. Earth Syst. Sci. 6:113–132. doi:10.5194/hess-6-113-2002

Jiao, N., G.J. Herndl, D.A. Hansell, R. Benner, G. Kattner, S.W. Wilhelm, D.L. Kirchman, M.G. Weinbauer, T. Luo, F. Chen, and F. Azam. 2010. Microbial production of recalcitrant dissolved organic matter: Long-term carbon storage in the global ocean. Nat. Rev. Microbiol. 8:593–599. doi:10.1038/nrmicro2386

Kaiser, K., G. Guggenberger, L. Haumaier, and W. Zech. 2001. Seasonal variations in the chemical composition of dissolved organic matter in organic forest floor layer leachates of old-growth Scots pine (*Pinus sylvestris* L.) and European beech (*Fagus sylvatica* L.) stands in northeastern Bavaria, Germany. Biogeochemistry 55:103–143. doi:10.1023/A:1010694032121

Kalbitz, K., J. Schmerwitz, D. Schwesig, and E. Matzner. 2003a. Biodegradation of soil-derived dissolved organic matter as related to its properties. Geoderma 113:273–291. doi:10.1016/S0016-7061(02)00365-8

Kalbitz, K., D. Schwesig, J. Schmerwitz, K. Kaiser, L. Haumaier, B. Glaser, R. Ellerbrock, and P. Leinweber. 2003b. Changes in properties of soil-derived dissolved organic matter induced by biodegradation. Soil Biol. Biochem. 35:1129–1142. doi:10.1016/S0038-0717(03)00165-2

Kaplan, L.A., T.N. Wiegner, J.D. Newbold, P.H. Ostrom, and H. Gandhi. 2008. Untangling the complex issue of dissolved organic carbon uptake: A stable isotope approach. Freshwater Biol. 53:855–864. doi:10.1111/j.1365-2427.2007.01941.x

Khoddami, A., M.A. Wilkes, and T.H. Roberts. 2013. Techniques for analysis of plant phenolic compounds. Molecules 18:2328–2375. doi:10.3390/molecules18022328

Kiikkilä, O., V. Kitunen, and A. Smolander. 2011. Properties of dissolved organic matter derived from silver birch and Norway spruce stands: Degradability combined with chemical characteristics. Soil Biol. Biochem. 43:421–430. doi:10.1016/j.soilbio.2010.11.011

Kijima, H., T. Ide, I. Otsuka, C. Ogimi, E. Hirata, A. Takushi, and Y. Takeda. 1997. Water-soluble phenolic glycosides from leaves of *Alangium premnifolium*. Phytochemistry 44:1551–1557. doi:10.1016/S0031-9422(96)00760-1

Kim, S., L.A. Kaplan, and P.G. Hatcher. 2006. Biodegradable dissolved organic matter in a temperate and a tropical stream determined from ultra-high resolution mass spectrometry. Limnol. Oceanogr. 51:1054–1063. doi:10.4319/lo.2006.51.2.1054

Kohonen, T. 2001. Self-organizing maps. 3rd ed. Springer–Verlag, Berlin.

Kraus, T.E.C., R.A. Dahlgren, and R.J. Zasoski. 2003. Tannins in nutrient dynamics of forest ecosystems: A review. Plant Soil 256:41–66. doi:10.1023/A:1026206511084

Kritzberg, E.S., S. Langenheder, and E.S. Lindström. 2006. Influence of dissolved organic matter on lake bacterioplankton structure and function- implications for seasonal dynamics of community composition. FEMS Microbiol. Ecol. 56:406–417. doi:10.1111/j.1574-6941.2006.00084.x

Kusuda, M., T. Hatano, and T. Yoshida. 2006. Water-soluble complexes formed by natural polyphenols and bovine serum albumin: Evidence from gel electrophoresis. Biosci. Biotechnol. Biochem. 70:152–160. doi:10.1271/bbb.70.152

Leenheer, J. 1981. Comprehensive approach to preparative isolation and fractionation of dissolved organic matter from natural waters and wastewaters. Environ. Sci. Technol. 15:578–587. doi:10.1021/es00087a010

Maie, N., R. Jaffé, T. Miyoshi, and D.L. Childers. 2006. Quantitative and qualitative aspects of dissolved organic carbon leached from senescent plants in an oligotrophic wetland. Biogeochemistry 78:285–314. doi:10.1007/s10533-005-4329-6

Maie, N., O. Pisani, and R. Jaffé. 2008. Mangrove tannins in aquatic ecosystems: Their fate and possible influence on dissolved organic carbon and nitrogen cycling. Limnol. Oceanogr. 53:160–171. doi:10.4319/lo.2008.53.1.0160

Maie, N., N.M. Scully, O. Pisani, and R. Jaffé. 2007. Composition of a protein-like fluorophore of dissolved organic matter in coastal wetland and estuarine ecosystems. Water Res. 41:563–570. doi:10.1016/j.watres.2006.11.006

Markager, S., and W.F. Vincent. 2000. Spectral light attenuation and the absorption of UV and blue light in natural waters. Limnol. Oceanogr. 45:642–650. doi:10.4319/lo.2000.45.3.0642

McArthur, M.D., and J.S. Richardson. 2002. Microbial utilization of dissolved organic matter leached from riparian litterfall. Can. J. Fish. Aquat. Sci. 59:1668–1676. doi:10.1139/f02-135

McCarren, J., J.W. Becker, D.J. Repeta, Y. Shi, C.R. Young, R.R. Malmstrom, S.W. Chrisholm, and E.F. DeLong. 2010. Microbial community transcriptomes reveal microbes and metabolic pathways associated with dissolved organic matter turnover in the sea. Proc. Natl. Acad. Sci. USA 107:16420–16427. doi:10.1073/pnas.1010732107

McKnight, D.M., E.W. Boyer, P.K. Westerhoff, P.T. Doran, T. Kulbe, and D.T. Anderson. 2001. Spectrofluorometric characterization of dissolved organic matter for indication of precursor organic material and aromaticity. Limnol. Oceanogr. 46:38–48. doi:10.4319/lo.2001.46.1.0038

McNamara, C.J., and L.G. Leff. 2004. Response of biofilm bacteria to dissolved organic matter from decomposing maple leaves. Microb. Ecol. 48:324–330. doi:10.1007/s00248-003-1058-z

Mei, Y., L. Wang, and F. Wu. 2009. Effects of water chemistry and concentrations of dissolved organic matter on its fluorescence characteristics and molecular conformation. Chin. J. Geochem. 28:413–420. doi:10.1007/s11631-009-0413-2

Mineau, M.M., C.M. Rigsby, D.T. Ely, I.J. Fernandez, S.A. Norton, T. Ohono, H.M. Valett, and K.S. Simon. 2013. Chronic catchment nitrogen enrichment and stoichiometric constraints on the bioavailability of dissolved organic matter from leaf leachate. Freshwater Biol. 58:248–260. doi:10.1111/fwb.12054

Moorhead, D.L., and R.L. Sinsabaugh. 2006. A theoretical model of litter decay and microbial interaction. Ecol. Monogr. 76:151–174. doi:10.1890/0012-9615(2006)076[0151:ATMOLD]2.0.CO;2

Moran, M.A., W.M. Sheldon, Jr., and R.G. Zepp. 2000. Carbon loss and optical property changes during long-term photochemical and biological degradation of estuarine dissolved organic matter. Limnol. Oceanogr. 45:1254–1264. doi:10.4319/lo.2000.45.6.1254

Moran, M.A., and R.G. Zepp. 1997. Role of photoreactions in the formation of biologically labile compounds from dissolved organic matter. Limnol. Oceanogr. 42:1307–1316. doi:10.4319/lo.1997.42.6.1307

Murphy, K., R. Bro, and C. Stedmon. 2014b. Chemometric analysis of organic matter fluorescence. In: P. Coble, J. Lead, A. Baker, D. Reynolds, and R.G.M. Spencer, editors, Aquatic organic matter fluorescence. Cambridge Univ. Press, Cambridge. p. 339–376.

Murphy, K.R., C.A. Stedmon, D. Graeber, and R. Bro. 2013. Fluorescence spectroscopy and multi-way techniques. PARAFAC. Anal. Methods 5:6557–6566. doi:10.1039/c3ay41160e

Murphy, K.R., C.A. Stedmon, P. Wenig, and R. Bro. 2014a. OpenFluor: An online spectral library of auto-fluorescence by organic compounds in the environment. Anal. Methods 6:658–661. doi:10.1039/C3AY41935E

Newman, M.E.J. 2011. Complex systems: A survey. Am. J. Phys. 79:800–810. doi:10.1119/1.3590372

Nishimura, S., N. Maie, M. Baba, T. Sudo, T. Sugiura, and E. Shima. 2012. Changes in the quality of chromophoric dissolved organic matter leached from senescent leaf litter during the early decomposition. J. Environ. Qual. 41:823–833. doi:10.2134/jeq2011.0342

Ohno, T. 2002. Fluorescence inner filtering correction for determining the humification index of dissolved organic matter. Environ. Sci. Technol. 36:742–746. doi:10.1021/es0155276

Osborne, T.Z., P.W. Inglett, and K.R. Reddy. 2007. The use of senescent plant biomass to investigate relationships between potential particulate and dissolved organic matter in a wetland ecosystem. Aquat. Bot. 86:53–61. doi:10.1016/j.aquabot.2006.09.002

Osono, T. 2006. Role of phyllosphere fungi of forest trees in the development of decomposer fungal communities and decomposition processes of leaf litter. Can. J. Microbiol. 52:701–716. doi:10.1139/w06-023

Pace, M.L., I. Reche, J.J. Cole, A. Fernández-Barbero, I.P. Mazuecos, and Y.T. Prarie. 2012. pH change induces shift in the size and light absorption of dissolved organic matter. Biogeochemistry 108:109–118. doi:10.1007/s10533-011-9576-0

Pan, B., S. Ghosh, and B. Xing. 2008. Dissolved organic matter conformation and its interaction with pyrene as affected by water chemistry and concentration. Environ. Sci. Technol. 42:1594–1599. doi:10.1021/es702431m

Pellerin, B.A., P.J. Hernes, J.F. Saraceno, R.G.M. Spencer, and B.A. Bergamaschi. 2010. Microbial degradation of plant leachate alters lignin phenols and trihalomethane precursors. J. Environ. Qual. 39:946–954. doi:10.2134/jeq2009.0487

Piccolo, A., and M. Spiteller. 2003. Electrospray ionization mass spectrometry of terrestrial humic substances and their size fractions. Anal. Bioanal. Chem. 377:1047–1059. doi:10.1007/s00216-003-2186-5

Portillo, M.C., S.P. Anderson, and N. Fierer. 2012. Temporal variability in the diversity and composition of stream bacterioplankton communities. Environ. Microbiol. 14:2417–2428. doi:10.1111/j.1462-2920.2012.02785.x

Qualls, R.G. 2005. Biodegradability of fractions of dissolved organic carbon leached from decomposing leaf litter. Environ. Sci. Technol. 39:1616–1622. doi:10.1021/es049090o

Qualls, R.G., and B.L. Haines. 1991. Geochemistry of dissolved organic nutrients in water percolating through a forest ecosystem. Soil Sci. Soc. Am. J. 55:1112–1123. doi:10.2136/sssaj1991.03615995005500040036x

Qualls, R.G., and B.L. Haines. 1992. Biodegradability of dissolved organic matter in forest throughfall, soil solution, and stream water. Soil Sci. Soc. Am. J. 56:578–586. doi:10.2136/sssaj1992.03615995005600020038x

Qualls, R.G., B.L. Haines, and W.T. Swank. 1991. Fluxes of dissolved organic nutrients and humic substances in a deciduous forest. Ecology 72:254–266. doi:10.2307/1938919

Qualls, R.G., and C.J. Richardson. 2003. Factors controlling concentration, export, and decomposition of dissolved organic nutrients in the Everglades of Florida. Biogeochemistry 62:197–229. doi:10.1023/A:1021150503664

Richard, T., D. Lefeuvre, A. Descendit, S. Quideau, and J.P. Monti. 2006. Recognition characters in peptide-polyphenol complex formation. Biochim. Biophys. Acta 1760:951–958. doi:10.1016/j.bbagen.2006.01.005

Romera-Castillo, C., M. Chen, Y. Yamashita, and R. Jaffé. 2014. Fluorescence characteristics of size-fractionated dissolved organic matter: Implications for a molecular assembly based structure? Water Res. 55:40–51. doi:10.1016/j.watres.2014.02.017

Schimpf, M.E., and M.P. Petteys. 1997. Characterization of humic materials by flow field-flow fractionation. Colloids. Surf., A 120:87–100. doi:10.1016/S0927-7757(96)03679-5

Schmidt, F., B.P. Koch, M. Elvert, G. Schmidt, M. Witt, and K.W. Hinrichs. 2011. Diagenetic transformation of dissolved organic nitrogen compounds under contrasting sedimentary redox conditions in the Black Sea. Environ. Sci. Technol. 45:5223–5229. doi:10.1021/es2003414

Schreeg, L.A., M.C. Mack, and B.L. Turner. 2013. Nutrient-specific solubility patterns of leaf litter across 41 lowland tropical woody species. Ecology 94:94–105. doi:10.1890/11-1958.1

Scully, N.M., N. Maie, S.K. Dailey, J.N. Boyer, R.D. Jones, and R. Jaffé. 2004. Early diagenesis of plant-derived dissolved organic matter along a wetland, mangrove, estuary ecotone. Limnol. Oceanogr. 49:1667–1678. doi:10.4319/lo.2004.49.5.1667

Shank, G.C., R. Lee, A. Vähätalo, R.G. Zepp, and E. Bartels. 2010. Production of chromophoric dissolved organic matter from mangrove leaf litter and floating *Sargassum* colonies. Mar. Chem. 119:172–181. doi:10.1016/j.marchem.2010.02.002

Silveira, M.L., K.R. Reddy, and N.B. Comeford. 2011. Litter decomposition and soluble carbon, nitrogen, and phosphorous release in a forest ecosystem. Open J. Soil Sci. 1:86–96. doi:10.4236/ojss.2011.13012

Smart, K.A., H.L. Smart, and C.R. Jackson. 2008. The effects of fine scale environmental variation on microbial community structure and function in aquatic environments. In: G.V. Kurladze, editor, Environmental microbiology research trends. Nova Science Publishers, Inc., Hauppage, NY. p. 167–190.

Spencer, R.G.M., G.R. Aiken, K.P. Wickland, R.G. Striegl, and P.J. Hernes. 2008. Seasonal and spatial variability in dissolved organic matter quantity and composition from the Yukon River basin, Alaska. Global Biogeochem. Cycles 22. doi:10.1029/2008GB003231

Spencer, R.G.M., L. Bolton, and A. Baker. 2007. Freeze/thaw and pH effects on freshwater dissolved organic matter fluorescence and absorbance properties from a number of UK locations. Water Res. 41:2941–2950. doi:10.1016/j.watres.2007.04.012

Stedmon, C.A., and R. Bro. 2008. Characterizing dissolved organic matter fluorescence with parallel factor analysis: A tutorial. Limnol. Oceanogr. Methods 6:572–579. doi:10.4319/lom.2008.6.572

Stedmon, C.A., S. Markager, and R. Bro. 2003. Tracing dissolved organic matter in aquatic environments using a new approach to fluorescence spectroscopy. Mar. Chem. 82:239–254. doi:10.1016/S0304-4203(03)00072-0

Strauss, E.A., and G.A. Lamberti. 2002. Effect of dissolved organic carbon quality on microbial decomposition and nitrification rates in stream sediments. Freshwater Biol. 47:65–74. doi:10.1046/j.1365-2427.2002.00776.x

Sutton, R., and G. Sposito. 2005. Molecular structure in soil humic substances: The new view. Environ. Sci. Technol. 39:9009–9015. doi:10.1021/es050778q

Tank, J.L., E.J. Rosi-Marshall, N.A. Griffiths, S.A. Entrekin, and M.L. Stephen. 2010. A review of allochthonous organic matter dynamics and metabolism in streams. J. North Am. Benthol. Soc. 29:118–146. doi:10.1899/08-170.1

Thurman, E.M., and R.L. Malcom. 1981. Preparative isolation of humic substances. Environ. Sci. Technol. 15:463–466. doi:10.1021/es00086a012

Tomaszewski, J.E., R.P. Schwarzenbach, and M. Sander. 2011. Protein encapsulation in humic substances. Environ. Sci. Technol. 45:6003–6010. doi:10.1021/es200663h

Uselman, S.M., R.G. Qualls, and J. Lilienfein. 2012. Quality of soluble organic C, N, and P produced by different types and species of litter: Root litter versus leaf litter. Soil Biol. Biochem. 54:57–67. doi:10.1016/j.soilbio.2012.03.021

Vähätalo, A.V., and R.G. Wetzel. 2004. Photochemical and microbial decomposition of chromophoric dissolved organic matter during long (months-years) exposures. Mar. Chem. 89:313–326. doi:10.1016/j.marchem.2004.03.010

Vähätalo, A.V., and R.G. Wetzel. 2008. Long-term photochemical and microbial decomposition of wetland-derived dissolved organic matter with alteration of ^{13}C:^{12}C mass ratio. Limnol. Oceanogr. 53:1387–1392. doi:10.4319/lo.2008.53.4.1387

Vannote, R.L., G.W. Minshall, K.W. Cummins, J.R. Sedell, and C.E. Cushing. 1980. The river continuum concept. Can. J. Fish. Aquat. Sci. 37:130–137. doi:10.1139/f80-017

Wallace, T.A., G.G. Ganf, and J.D. Brookes. 2008. A comparison of phosphorous and DOC leachates from different types of leaf litter in an urban environment. Freshwater Biol. 53:1902–1913. doi:10.1111/j.1365-2427.2008.02006.x

Weishaar, J.L., G.R. Aiken, B.A. Bergamaschi, M.S. Fram, R. Fujii, and K. Mopper. 2003. Evaluation of specific ultraviolet absorbance as an indicator of the chemical composition and reactivity of dissolved organic matter. Environ. Sci. Technol. 37:4702–4708. doi:10.1021/es030360x

Wershaw, R.L. 2004. Evaluation of conceptual models of natural organic matter (humus) from a consideration of the chemical and biochemical process of humification. USGS Scientific Investigations Report 2004-5121. USGS, Reston, VA.

Wickland, K.P., J.C. Neff, and G.R. Aiken. 2007. Dissolved organic carbon in Alaskan boreal forest: Sources, chemical characteristics, and biodegradability. Ecosystems 10:1323–1340. doi:10.1007/s10021-007-9101-4

Wieder, W.R., C.C. Cleveland, and A.R. Townsend. 2008. Tropical tree species composition affects the oxidation of dissolved organic matter from litter. Biogeochemistry 88:127–138. doi:10.1007/s10533-008-9200-0

Wiegner, T.N., L.A. Kaplan, J.D. Newbold, and P.H. Ostrom. 2005. Contribution of dissolved organic C to stream metabolism: A mesocosm study using ^{13}C-enriched tree-tissue leachate. J. North Am. Benthol. Soc. 24:48–67. doi:10.1899/0887-3593(2005)024<0048:CODOCT>2.0.CO;2

Wipfli, M.S., J.S. Richardson, and R.J. Naiman. 2007. Ecological linkages between headwaters and downstream ecosystems: Transport of organic matter, invertebrates, and wood down headwater channels. J. Am. Water Resour. Assoc. 43:72–85.

Xiang, X., Y. Huang, G. Madey, S. Cabaniss, and L. Arthurs. 2006. Modeling the evolution of natural organic matter in the environment with an agent-based stochastic approach. Nat. Resour. Model. 19:67–90. doi:10.1111/j.1939-7445.2006.tb00176.x

Yoshimura, C., M. Fuji, T. Omura, and K. Tockner. 2010. Instream release of dissolved organic matter from coarse and fine particulate organic matter of different origins. Biogeochemistry 100:151–165. doi:10.1007/s10533-010-9412-y

Young, K.C., K.M. Docherty, P.A. Maurice, and S.D. Bridgham. 2005. Degradation of surface-water dissolved organic matter: Influences of DOM chemical characteristics and microbial populations. Hydrobiologia 539:1–11. doi:10.1007/s10750-004-3079-0

Young, R.C., C.D. Matthaei, and C.R. Townsend. 2008. Organic matter breakdown and ecosystem metabolism: Functional indicators for assessing river ecosystem health. J. North Am. Benthol. Soc. 27:605–625. doi:10.1899/07-121.1

Yu, Z., Q. Zhang, T.E.C. Kraus, R.A. Dahlgren, C. Anastasio, and R.J. Zasoski. 2002. Contribution of amino compounds to dissolved organic nitrogen in forest soils. Biogeochemistry 61:173–198. doi:10.1023/A:1020221528515

Zhang, M., and Z. He. 2015. Characteristics of dissolved organic carbon revealed by ultraviolet/visible absorbance and fluorescence spectroscopy: The current status and future exploration. In: Zhongqi He and Fengchang Wu, editors, Labile organic matter—Chemical compositions, function, and significance in soil and the environment. SSSA Spec. Publ. 62. SSSA, Madison, WI. doi:10.2136/sssaspecpub62.2014.0032

Zhang, Y., X. Liu, M. Wang, and B. Qin. 2013. Compositional differences of chromophoric dissolved organic matter derived from phytoplankton and macrophytes. Org. Geochem. 55:26–37. doi:10.1016/j.orggeochem.2012.11.007

Zsolnay, Á. 2003. Dissolved organic matter: Artefacts, definitions, and functions. Geoderma 113:187–209. doi:10.1016/S0016-7061(02)00361-0

Zsolnay, Á., E. Baigar, M. Jimenez, B. Steinweg, and F. Saccomandi. 1999. Differentiating with fluorescence spectroscopy the sources of dissolved organic matter in soils subjected to drying. Chemosphere 38:45–50. doi:10.1016/S0045-6535(98)00166-0

Bioavailability and Preservation of Organic Phosphorus in Freshwater Sediments and Its Role in Lake Eutrophication

Yuanrong Zhu, Fengchang Wu,* and Zhongqi He

Abstract

The dissolution of organic matter can result in the release of phosphorus from lake sediments, and organic phosphate (P_o) itself can also be an important component of lake organic matter. In this chapter, a case study with the sediments of a Chinese eutrophic lake is presented to demonstrate the bioavailability and preservation of P_o in the freshwater sediments and their role in lake eutrophication. Sequential extraction and phosphatase hydrolysis were used to identify the bioavailability of different P pools in these sediments. Those data show that the biogeochemical cycling of bioavailable P_o in sediments might be playing an important role in maintaining the eutrophic status of lakes after external input of P has been reduced. Sediment organic matter can affect the bioavailability and preservation of P_o in sediments by formation of P_o–metal-humic acid or P_o–metal complexes. Furthermore, the adsorption of P_o by minerals would be the other mechanism for some P_o preservation in sediments. The observations and insight derived from this chapter contribute to a better understanding of the biological cycles of sediments P_o in freshwater lakes and their contribution in lake eutrophication.

Abbreviations: BAP, bioavailable phosphorus; BAP_i, bioavailable inorganic phosphate; BAP_o, bioavailable organic phosphate; DOC, dissolved organic carbon; HA, humic acid; LOI, loss on ignition; NMR, nuclear magnetic resonance; P_i, inorganic phosphate; P_o, organic phosphate; P_t, total phosphate; TN, total nitrogen; TOC, total organic carbon; TP, total phosphorus.

Yuanrong Zhu (zhuyuanrong07@mails.ucas.ac.cn) and Fengchang Wu, State Key Laboratory of Environment Criteria and Risk Assessment, Chinese Research Academy of Environmental Sciences, 8 Dayangfang BeiYuan Rd., Chaoyang District, Beijing 100012, China. Zhongqi He, USDA-ARS Southern Regional Research Center, 1100 Robert E. Lee Blvd., New Orleans, LA 70124 (zhongqi.he@ars.usda.gov). *Corresponding author (wufengchang@vip.skleg.cn).

doi:10.2136/sssaspecpub62.2014.0044

Labile Organic Matter—Chemical Compositions, Function, and Significance in Soil and the Environment. SSSA Special Publication 62. Zhongqi He and Fengchang Wu, editors.

The dissolution of organic matter can result in release of P from lake sediments (Purushothaman and Chakrapani, 2014). As a matter of fact, P_o can be an important component of natural organic matter (Ged and Boyer, 2013; He et al., 2009b; Sannigrahi et al., 2006; Suzumura and Ingall, 2001). Knowledge of the distribution and forms of P_o in the environment has greatly increased since the beginning of this century, mainly because of the improved and combined wet chemistry and spectroscopic characterization. These characterization techniques include but are not limited to sequential fractionation (Pagliari and Laboski, 2012; Toth et al., 2011), enzymatic hydrolysis (He and Honeycutt, 2001, 2011; Johnson and Hill, 2010; 2011; Tazisong et al., 2014), solution ^{31}P nuclear magnetic resonance (NMR) spectroscopy (Ahlgren et al., 2006; Cade-Menun, 2011; Ozkundakci et al., 2014), solid state spectroscopy (Akinerimi et al., 2011; Brandes et al., 2007; He et al., 2009c; Manimel Wadu et al., 2013), and liquid chromatography (De Brabandere et al., 2008). In particular, the combination of ^{31}P NMR spectroscopy and enzymatic hydrolysis could provide complementary information on the forms and lability of P_o (Giles and Cade-Menun, 2014; He et al., 2011, 2008, 2007; Young et al., 2013). Although those techniques were developed for terrestrial samples, they have also been applied to water and lake sediment samples to shed light on the role of P_o in aquatic ecosystems and eutrophication (Ahlgren et al., 2007; Sannigrahi et al., 2006; Torres et al., 2014; Zhang et al., 2009; Zhu et al., 2013b). Based on previous studies (Bai et al., 2009; Ding et al., 2013; Zhu et al., 2013b), a large proportion of P_o is likely to remain intact in the sediments of lakes. Currently, little information has been reported regarding the mechanisms for P_o preservation in lake sediments even though understanding these mechanisms are important before one can start to understand the biogeochemical cycle of P_o in lake sediments.

Lake eutrophication in China is a serious environmental concern, especially in lakes from the middle and lower reaches of Yangtze River watershed and the Southwestern China Plateau (Le et al., 2010; Zhang et al., 2009). Most lakes in the two regions are under mesotrophic or eutrophic conditions because of rapidly expending township enterprises, large density of population, the overuse of agrochemical and chemical fertilizer, the discharge of the municipal sewage, and scale cultivation. Whereas exogenous nutrient input has been limited or reduced greatly in recent years, the eutrophic status of lakes has not improved accordingly (Zhu et al., 2013a, 2013b). This phenomenon could be attributed to the biological cycling of internal nutrients accumulated in sediments (Søndergaard et al., 2003). Thus, manipulation to decrease P recycling or promote the preservation of P within lakes would be one useful intervention for eutrophication reduction (Carpenter, 2005). For this purpose, more information on the chemical nature of sediment P is needed.

Thus, by using Lake Dianchi as a case study, this chapter first reviews the relationship of eutrophication and P levels in Lake Dianchi with historic data, and then presents and discusses our recent work on the species and bioavailability of P_o in the freshwater sediments and their roles in maintaining the eutrophic status of lakes after external input of P has been reduced. The information and insight derived from this chapter are helpful in better understanding the biological cycle of sediments P_o in freshwater lakes and their role in lake eutrophication.

Background Information and Eutrophication of Lake Dianchi

Description of Lake Dianchi

Lake Dianchi (24°40′–25°02′ N, 102°36′–102°47′ E), located in Kunming, Yunnan Province of China (Fig. 1), has an area of nearly 306 km^2 and an average depth of 5 m. The climate is subtropical plateau monsoon, with a long, warm, and humid summer and a short winter. The lake is divided into two sections: Caohai, the northern part, with a water area of 10.8 km^2, and Waihai, the southern part, with a water surface area of 298.2 km^2. Lake Dianchi is the sixth largest freshwater lake in China and is the largest lake in southwest China. The lake is of great importance for Kunming City because of its water supply, fisheries, water storage, aquaculture, flood regulation, recreation, climate, and biological diversity. However, since the 1970s, Lake Dianchi has received large loads of contaminants and nutrients from increasing human activities such as the rapid expansions of agriculture and industry and urbanization in the region; this has resulted in serious pollution issues with the fact that the lake is now seriously eutrophic (Gao et al., 2005; Wang et al., 2009). Furthermore, because there are many source rivers inflowed and only one outflowed, Lake Dianchi has a relatively long water residence period (3–8 yr) which exacerbates the retention and accumulation of

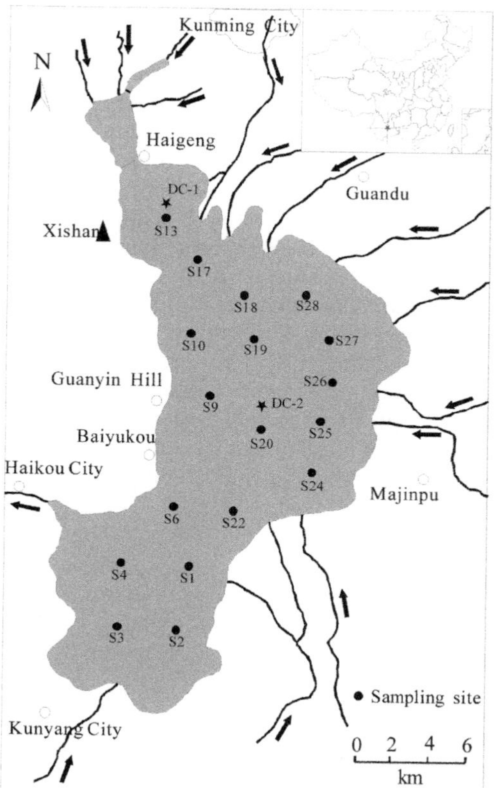

Fig. 1. Major inlets and the outlet of Lake Dianchi with sediment sampling sites.

nutriments in the lake (Fig. 1) (Xiong et al., 2010; Zhu et al., 2013b). In recent years, external contaminations and nutrient inputs have been reduced to a certain extent, but the water quality has not been improved markedly and algal blooms still outbreak seriously every year. The accumulation of nutrients in the sediments would be an important contributing factor to this phenomenon. Thus, this lake provides a good case study on the contribution of retaining nutrients (e.g., sediment P) to water quality (e.g., eutrophication) (Gao et al., 2005; Zhu et al., 2013b).

Changes of Nutrient Levels and Eutrophication of Lake Dianchi

The nutrients, such as concentrations of total phosphorus (TP) and total nitrogen (TN), in the overlying water of Lake Dianchi were increased gradually from the 1960s, which has resulted in serious eutrophication of Lake Dianchi since the 1970s (Gao et al., 2015; Lu et al., 2012). Historic data show that concentration of TP (average annual concentration) in the water column increased from 0.01 mg L^{-1} in 1960 to 0.14 mg L^{-1} in 1991, and further to a maximum of 0.33 mg L^{-1} in 1999. With the limited external input, TP concentrations decreased to an average value of 0.18 mg L^{-1} in 2012. Similarly, TN concentrations increased from 0.3 mg L^{-1} in 1960 to a maximum of 3.0 mg L^{-1} in 2007 and then decreased to 2.0 mg L^{-1} in 2012 (Fig. 2). The TN/TP mass ratio was less than 29, which possibly favors dominance by bloom-forming cyanobacteria in the early stage of eutrophication in lakes (Smith, 1983), such as Lake Dianchi. Since then, the TN/TP ratio decreased, indicating the blue-green algae blooming after the 1960s (Smith, 1983; Xie et al., 2003). From 1960 to 2000, the decreased values of the ratio also suggested that the exogenous input rate of TP increased faster relative to TN and release of internal P from sediments (Xie et al., 2003). Blue-green algae, including *Microcystis* sp. and *Aphanizomenon*, dominate the algal community of Lake Dianchi during the most eutrophic period (Dai et al., 2012). An enclosure experiment in a lake has shown that *Microcystis* blooms never occurred in the treatments with low P concentration in spite of the presence of sufficient N (Xie et al., 2003), which indicated that P is an important limiting nutrient for eutrophication in Lake Dianchi. Furthermore, *Aphanizomenon flos-aqua* isolated from Lake Dianchi, a main composition of cyanobacterial blooms, has a powerful nitrogen (N) fixation ability, which could cause N fixation even in a nitrogen-free medium (Wu et al., 2014). Phosphorus is much easier to control than N for eutrophication in Lake Dianchi. Therefore, the allochonous input of nutrients, especially P, should be reduced to control the eutrophication of Lake Dianchi. Unfortunately, even though the external input of P has been controlled to a large extent in recent years, serious algal blooms still occurs every year, which might be attributed to the internal recycling of P in sediments (Lu et al., 2012; Xie et al., 2003; Zhu et al., 2013a; 2013b).

Levels and Characteristics of Phosphorus in Sediments of Lake Dianchi

Because the importance of Lake Dianchi for southwestern China and the large sink of P in the sediments, numerous studies have been done to characterize P in the sediments (Table 1). For the top 5 cm of sediments from Lake Dianchi, concentrations of TP ranged from 4000 mg kg^{-1} to 6660 mg kg^{-1}, with an average value of 2283 mg kg^{-1}; for 5- to 10-cm sediments, concentrations of TP ranged from 250 mg kg^{-1} to 6150 mg kg^{-1}, with an average of 1958 mg kg^{-1}; for 10- to 20-cm sediments, concentrations of TP ranged from 440 mg kg^{-1} to 6410 mg kg^{-1}, with an average

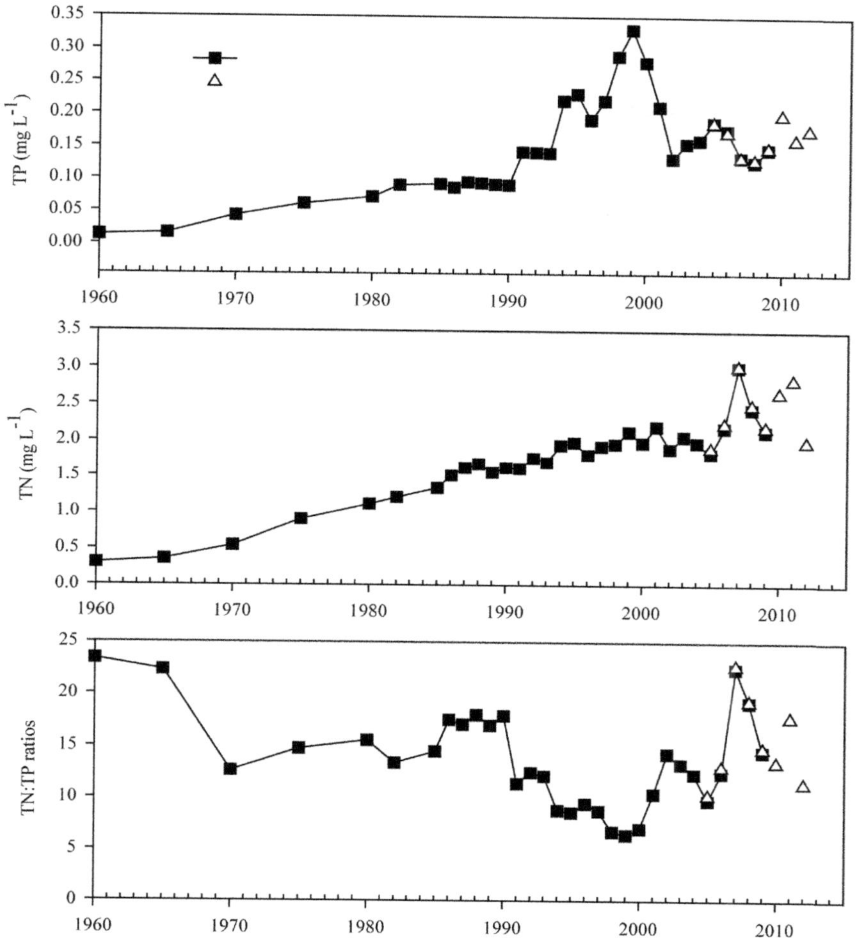

Fig. 2. Long-term changes in total phosphorus (TP), total nitrogen (TN), and TN/TP ratios of overlying water in Lake Dianchi. Compiled per data reported in Lu et al. (2012) (closed squares) and Zhang et al. (2014) (open triangles).

of 1823 mg kg^{-1} (Lu et al., 2005). The TP in the sediments was divided into loosely adsorbed P (NH_4Cl-P), redox-sensitive P (BD-P), metal oxides bound P (NaOH-P), calcium bound P (HCl-P), and residual P (Table 1). Based on the variation of P fractions in the sediments before cyanobacterial blooms and during cyanobacterial blooms, NH_4Cl-P, BD-P, and NaOH-P were labile and mobile (Hu et al., 2007). In particular, BD-P is intensively released from sediment to water and consequently promotes cyanobacterial blooms. Gao et al. (2005) indicate that NH_4Cl-P, BD-P, and NaOH-P potentially release from the sediment into the overlying water so that they could be the dominant factors in maintaining the trophic status of Lake Dianchi after the external P load is reduced.

Table 1. Selected studies on internal load of phosphorus in sediments from Lake Dianchi.

References	Sampling date	Samples	Analysis
Lu et al. (2005)	7 May to 30 July 2002	118 sediment samples from 0 to 5, 5 to 10, and 10 to 20 cm	Special variation of TP in sediments
Gao et al. (2005)	December 2002	6 sediment cores sliced into 0–2, 2–4, 4–6, 8–10, 10–15, 15–20, 20–25, 25–30, 30–35, and 35–40 cm	P fractions including NH_4Cl-P, BD-P, NaOH-P, HCl-P, and P_o in sediment profiles†
Hu et al. (2006)	March 2003	8 surface sediment samples (0–10 cm) were collected in the north and northeast part of Lake Dianchi	Potential bioavailable inorganic P forms including AAP, WSP, Olsen-P, LAP, and Inorganic P (P_i)‡
Hu et al. (2007)	April and August 2004	40 surface sediments (0–5 cm) were collected twice in April 2004 (before cyanobacterial bloom) and August 2004 (the climax of cyanobacterial bloom)	Variation of P fractions including NH_4Cl-P, BD-P, NaOH-P, HCl-P, and Residual P during cyanobacterial blooms†
Gao et al. (2010)	October 2007	33 surface sediments (0–1 cm) were collected in the whole lake	P_o fractions including labile P_o, moderately labile P_o, and nonlabile P_o were characterized in the sediments§
Ding et al. (2010)	September 2006	1 surface sediment sample (0–1 cm)	NaOH-EDTA extraction with ^{31}P-NMR analysis
Li et al. (2012)	February 2010	37 surface sediment samples (0–10 cm) were collected in the whole lake	P fractions including LAP, Al-P, Fe-P, Ca-P, occluded Al-P, occluded Fe-P, and residual P¶
Zhu et al. (2013b)	May 2010	18 surface sediments samples (0–10 cm) were collected in the lake	P_o and P_i fractions sequentially extracted by H_2O, $NaHCO_3$, NaOH, and HCl; Bioavailability of these P_o fractions was analyzed by enzymatic hydrolysis#
Xiong et al. (2014)	May 2013	4 sediment cores, selected from the sediment depths of 0–2, 2–5, 5–8, 8–12, 12–16, 16–20 cm	The same as previous study in Zhu et al. (2013b)

† NH_4Cl-P, BD-P, NaOH-P, and HCl-P were P fractions sequentially extracted by 1.0 M NH_4Cl-P, 0.11 M $NaHCO_3/Na_2S_2O_4$, 1.0 M NaOH, and 0.5M HCl; Residual P was calculated by total P (TP) minus total extracted reactive P (Hupfer et al., 1995).

‡ Algal available P (AAP), water soluble P (WSP), and Olsen-P were extracted by 0.1M NaOH, distilled water, and 0.5 M NaHCO3 (pH 8.5), described in Zhou et al. (2001); loosely adsorbed P (LAP) was extracted by 1M NH_4Cl (Hieltjes and Lijklema, 1980).

§ Procedure of organic phosphate (P_o) sequential extraction was based on the scheme of Ivanoff et al. (1998): Labile P_o including $NaHCO_3$–P_o extracted by 0.5 M NaHCO3 (pH 8.5); Moderately labile P_o including the organic P extracted by 1.0 mol/L HCl and fulvic acid P; Nonlabile P_o including humic acid P and residual organic P; Fulvic acid P and humic acid P was extracted by 0.5 mol/L NaOH, then the extractants was acidified to pH 0.2; Humic acid P was precipitated and fulvic acid P soluble in the acidified extractants.

¶ The scheme of inorganic P fractionation was based on Jin and Tu (1990): LAP, loosely adsorbed P extracted by 1M NH_4Cl; Al-P, aluminum bound P extracted by 0.5M NH_4F (pH 7.0); Fe-P, iron bound P extracted by 0.1M NaOH; Ca-P, calcium bound P extracted by 0.25 M H_2SO_4; occluded Al-P and Fe-P was extracted after removing oxidized iron film by 0.3 M sodium citrate, 1 M $NaHCO_3$ and $Na_2S_2O_4$; Residual P was the difference between TP and extracted inorganic P.

Bioavailable including labile monoester P (P_o hydrolysis by alkaline phosphatase), diester P (P_o hydrolysis by phosphodiesterase and alkaline phosphatase minus labile monoester P), phytate-like P (P_o hydrolysis by alkaline phosphatase, phosphodiesterase and phytase minus labile monoester P and diester P) (Zhu et al., 2013b).

Whereas most of these studies mainly focus on inorganic P (P_i) fractions and their bioavailability, few studies have focused on P_o in the sediments from Lake Dianchi. In fact, P_o was a major composition of TP in the sediments from the hypereutrophic Lake Dianchi, which can also be considered bioavailable for cyanobacterial blooms (Gao et al., 2010; Xiong et al., 2014; Zhu et al., 2013b). Concentrations of P_o ranged from 153 mg kg^{-1} to 1137 mg kg^{-1}, accounting for 10.5 to 52.2% of TP in the sediments (Gao et al., 2010). The P_o fractions include labile P_o ($NaHCO_3$–P_o), moderately labile P_o (HCl-P_o and fulvic acid P), and less and nonlabile P_o (humic acid P and residual P_o) in the sediments of Lake Dianchi (Table 1). Concentrations of labile P_o ranged from 18 mg kg^{-1} to 90 mg kg^{-1} and accounted for approximately 10% of P_o in the sediments. Concentrations of moderately labile and nonlabile P_o ranged from 56~567 mg kg^{-1} and 45~464 mg kg^{-1}, respectively (Gao et al., 2010). However, it should be pointed out that the bioavailability of P_o based on the conventional fractionation procedures should be interpreted with caution (Turner et al., 2005; Zhang et al., 2008). For better understanding of the bioavailability of P_o, enzymatic hydrolysis, as reported by He and Honeycutt (2011) and Bünemann (2008), should be applied to investigate the bioavailability of P_o in the sediments from Lake Dianchi.

Bioavailability and Preservation of Sediment Organic Phosphate in Lake Dianchi

Organic Phosphate Species and Bioavailability in the Sequential Fractions of Sediments

Zhu et al. (2013b) examined the bioavailability of P_o and the biogeochemical cycle of P_o in sediments per the sediment samples (~10 cm deep) collected from 18 locations in Lake Dianchi (Fig. 1). Sites identified as S1 to S24, S1, S2, S3, and S4 were located in the southwest part of the lake, near the Kunyang and Kunming Phosphate Fertilizer Factory. The site S13 was in the northern lake where, because of both nutrients and wind conditions, algal blooms aggregate. Sites S19 and S20 were in the deepest part of the lake, near the center where P has accumulated in recent years. The other sampling sites, including S6, S9, S10, S17, S18, S22, S24, S25, S26, S27, and S28 were near the lakeshore and are influenced by nonpoint agricultural inputs. Two sediment cores including DC-1 and DC-2 were also collected from the north and center of Lake Dianchi by use of a piston-percussion corer fitted with 5-cm internal Plexiglas cylindrical tubes (Fig. 1).

These sediment samples were subjected to sequential extraction with distilled H_2O, 0.5 mol L^{-1} $NaHCO_3$ (pH 8.5), 0.1 mol L^{-1} NaOH, and 1 mol L^{-1} HCl (He et al., 2004; Zhu et al., 2013b). Phosphorus in these fractions was separated into P_i and P_o portions. The concentrations of potentially bioavailable P_i (BAP_i), including P_i extracted by H_2O, $NaHCO_3$, and NaOH, varied from 283 to 850 mg kg^{-1} and accounted for 17.2 to 32.4% of TP in the sediments (Fig. 3d). Inorganic P extracted by HCl, primarily bound to calcium, was the main composition of P_i in the sediments, which would not be readily available to phytoplankton (Zhu et al., 2013a). Compared with BAP_i, P_o in the sediments was likely an important internal load of P for cyanobacterial blooms (Fig. 3d).

The bioavailability of P_o in the sediments and in their sequential extracts was then evaluated by enzymatic hydrolysis. Bioavailable P_o (BAP_o) is considered

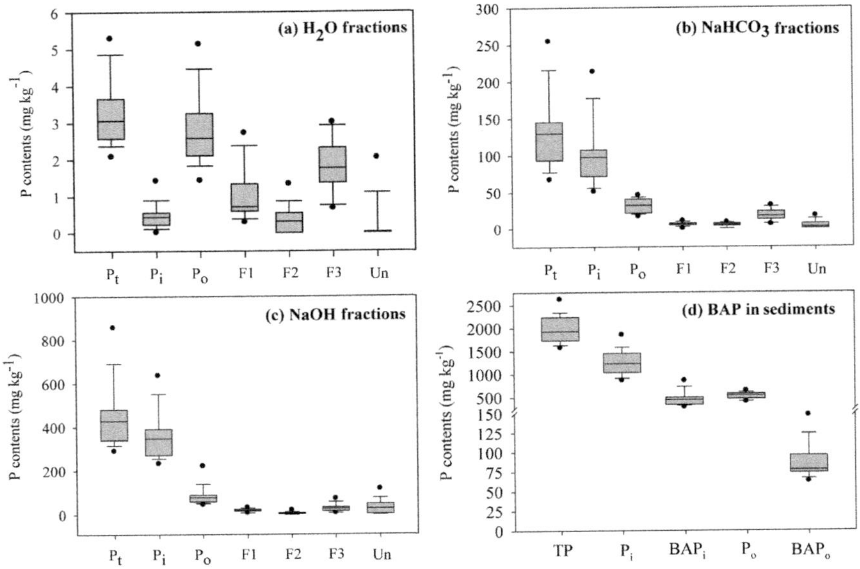

Fig. 3. Distribution of total phosphate (P_t), inorganic phosphate (P_i), and organic phosphate (P_o) species (mg kg^{-1}) in the (a) H_2O, (b) $NaHCO_3$, (c) and NaOH fractions of Lake Dianchi sediments characterized by phosphatase hydrolysis. F1, labile monoester P; F2, diester P; F3, phytate-like P; Un, unknown P_o. (d) Summarization of bioavailable P (BAP) including bioavailable P_i (BAP_i) and bioavailable P_o (BAP_o) in the sediments of Lake Dianchi. The lower boundary of each box indicates the 25th percentile, the full line within the box shows the median, the dotted line represents the mean, and upper boundary shows the 75th percentile. Error bars indicate the 95th and 5th percentiles (Zhu et al., 2013b).

as the sum of the three enzymatically hydrolysable P_o identified in the sequential extractions. There was little P_o in the HCl fraction of these samples (Zhu et al., 2013b) so that only the data in the first three (i.e., H_2O, $NaHCO_3$, and NaOH) fractions are discussed in this chapter. Based on specificity of substrates for phosphatase preparations and the sequential P_o fractions hydrolyzed by these enzymes, three types of labile P_o, monoester P, diester P, and phytate-like P were quantified in the H_2O, $NaHCO_3$, and NaOH fractions of sediments from Lake Dianchi (Fig. 3). For the H_2O-P_o, 36.7% (average) was labile monoester P, 14.8% was diester P, 69.9% was phytate-like P; in $NaHCO_3$–P_o, 19.9% was labile monoester P, 17.5% was diester P, 58.8% was phytate-like P; for NaOH-P_o, 25.6% was labile monoester P, 7.9% was diester P, and 35.9% was phytate-like P. Labile monoester P was active to support growth of algae to form blooms. Diester P mainly distributed in labile H_2O and $NaHCO_3$ fractions was readily available to cyanobacteria. Phytate-like P represents a major portion of the P_o in the NaOH fractions and also in the more labile H_2O and $NaHCO_3$ fractions. Based on the results of the sequential extraction of P_o and enzymatic hydrolysis, lability and bioavailability was in increasing order H_2O-P_o > $NaHCO_3$–P_o > NaOH-P_o, and bioavailable P_o accounted for only 12.1 to 27.2% of total P_o in sediments.

On the other hand, there was only 69 to 265 mg kg^{-1}, accounting for 15.1 to 48.8% of P_o extracted by H_2O, $NaHCO_3$, and NaOH in the sediments from Lake Dianchi. There were large proportions of P_o that could not be extracted by H_2O, $NaHCO_3$, and NaOH and further hydrolyzed by phosphatase (Fig. 3). The unknown P_o was ranked in order of highest concentration to lowest concentration as NaOH-P_o (31.5% average) > $NaHCO_3$–P_o (10.8% average) > H_2O-P_o (5.3% average). The analysis of bioavailability of P_o in the sediment profiles of Lake Dianchi also showed that a large proportion of P_o extracted by H_2O, $NaHCO_3$, and NaOH could not be hydrolyzed by phosphatase (Xiong et al., 2014). The unextracted P_o and unknown P_o was possibly nonbioavailable P_o and was in the sediments of Lake Dianchi. However, the mechanisms of P_o preservation in the sediments of lakes are still unknown, and clarification of these mechanisms is still need.

Relationship Between Hydrolyzable Organic Phosphate and Dissolved Organic Matter in Sequentially Extracted Fractions

Dissolved organic carbon (DOC) extracted by H_2O, $NaHCO_3$, and NaOH ranged from 5847 to 11026 mg kg^{-1}, which accounted for 10.7 to 18.8% of the total organic carbon (TOC) of sediments from Lake Dianchi. Distribution of DOC was similar to that of P_o and was ranked in order of highest concentration to lowest as NaOH (46.0–58.6% of the total extracted TOC) > $NaHCO_3$ (16.2–22.5%) > HCl (15.1–22.1%) > H_2O (8.7 –12.2%). Contents of hydrolyzable P_o in labile H_2O-P_o and $NaHCO_3$–P_o were significantly correlated with organic matter (Fig. 4). As a matter of fact, positive correlation of labile and moderately labile P_o in several freshwater lake sediments with organic matter has been reported previously (Ivanoff et al., 1998; Zhang et al., 2008). Furthermore, it is reported that phosphatase (e.g., APase) activity was proportional to organic matter and P_o in sediments (Su et al., 2005; Zhou et al., 2008). With greater phytoplankton biomass in eutrophic lakes, activities of APase would be expected to be greater in sediments (Zhou et al., 2002). Therefore, the bioavailable P_o would be released and continue to support eutrophication during periods when less orthophosphate is available. The biogeochemical cycle of bioavailable P_o might be an important process to self-regulate the nutrient status for eutrophic lake and maintain their eutrophic status longer, even after external sources of P have been controlled.

There were no statistically significant relationships ($P > 0.05$) between bioavailable P_o and organic matter in NaOH fractions. This could be due to the complexation of NaOH-P_o with polyvalent cations and/or organic matter (e.g., humic acid) that could render the NaOH-P_o, especially phytate-like P, less available for enzymatic hydrolysis. It is reported that phytate-like P in more complex forms, such as being associated with humic material or metals, is inaccessible to enzymatic hydrolysis (He et al., 2006a; Monbet et al., 2007; Reitzel et al., 2007). Therefore, the percentages of unknown P_o were the greatest in the NaOH fraction (0–60.2%, average 31.5%), leading to less significant correlation between hydrolyzable P_o and DOC in the NaOH fraction. Unlike the portion of hydrolyzable P_o, the level of total NaOH-P_o was still significantly correlated with organic matter ($P \leq 0.003$), suggesting that NaOH-P_o was associated with biogeochemical cycles of organic matter in sediments with a longer cycling period than H_2O- and $NaHCO_3$–P_o (Zhu et al., 2013b).

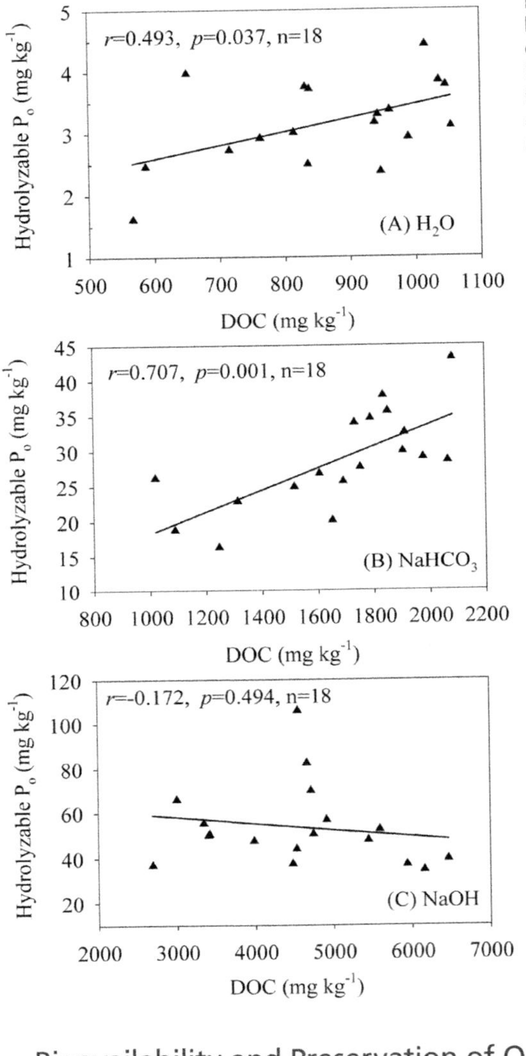

Fig. 4. Relationships between hydrolyzable P_o and dissolved organic matter (DOC, mg kg^{-1}) in the (a) H_2O, (b) $NaHCO_3$, and (c) NaOH fractions, respectively. Adapted from the supplementary material with Zhu et al. (2013b).

Bioavailability and Preservation of Organic Phosphate in the Sediment Profiles and Biogeochemical Cycling of Phosphorus

We further examined the relationship between P cycling and the trophic status of lakes before and after the reduction of the external P input, using the data of P_o from two sediment cores (DC-1 and DC-2) (unpublished data, 2015). Per the key parameters we measured [e.g., TOC, TN, loss on ignition (LOI), biogenic silica, the TOC/TN ratio, and P distribution], Lake Dianchi's evolution could be separated into three distinct phases: Section I (oligotrophic: 55–32 cm); Section II (mesotrophic: 32–12 cm); and Section III (eutrophic or hyper-eutrophic: 12–0 cm) (Fig. 5). The contents of BAP_o abruptly increased with TOC increasing in the sediment profiles when Lake Dianchi switched to eutrophic and hypereutrophic states. It was shown that Fe/Al-P, extracted by NaOH, is as an index of bioavailable P_i (Zhou

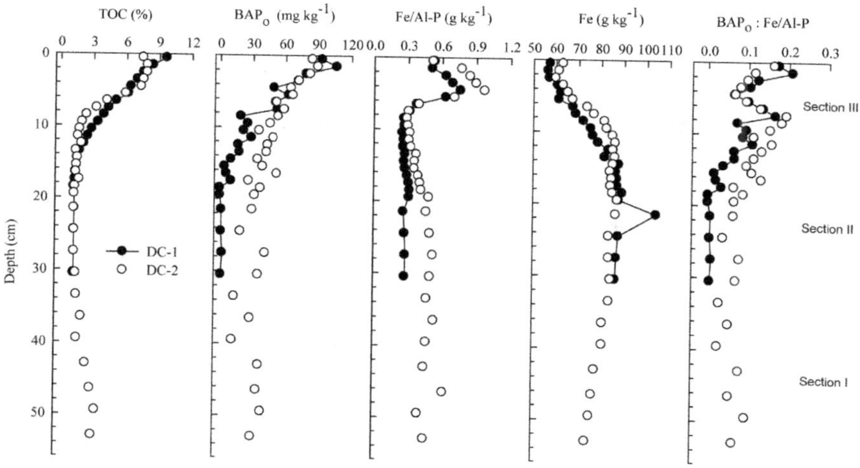

Fig. 5. The distribution of TOC (%), BAP$_o$, Fe/Al-P, Fe, and ratio (w/w) of BAP$_o$:Fe/Al-P in the sediment profiles. BAP$_o$ was the Po extracted by NaOH-EDTA and then hydrolyzed by alkaline phosphatase, phosphodiesterase, and phytase (unpublished data, 2015).

et al., 2001; Zhu et al., 2013a). Concentrations of Fe/Al-P decreased from sediment depth of 4 to 0 cm, which suggested that the release of Fe/Al-P from sediments to overlying water with frequently cyanobacterial blooms (Gao et al., 2005; Hu et al., 2007; Li et al., 2007). The distribution characteristics of Fe in the sediment profiles supported the release of Fe/Al-P in the process of eutrophication of Lake Dianchi. The reduction of Fe/Al-P observed in the sediments could also be related to the decrease of external input of P in recent years (Zhu et al., 2013a). The ratio (w/w) of BAP$_o$ and Fe/Al-P varied dramatically in the different evolutionary stages of Lake Dianchi (Fig. 5). The ratio of BAP$_o$ and Fe/Al-P increased in the eutrophic stage of Lake Dianchi (Section III), especially in the top 4 cm of sediment. The lack of bioavailable Fe/Al-P and increased accumulation of organic matter reinforces the important role of bioavailable P$_o$ cycling in maintaining the eutrophic status of Lake Dianchi, especially because the allochthonous input of P has been reduced to a large degree (as demonstrated by the top 4 cm of sediments).

The profiling study further demonstrated that the biogeochemical cycling of P was closely related to the ecological evolution of Lake Dianchi (Fig. 6). In the early stages, the biogeochemical cycling between P$_o$ and P$_i$ was in equilibrium (data not shown), during which superfluous P$_i$ was adsorbed by sediments, thereby maintaining the oligotrophic status of Lake Dianchi. In the middle stage, the biogeochemical cycle between bioavailable P$_i$ and P$_o$ became unstable. The external input of nutrients and release of bioavailable Fe/Al-P made Lake Dianchi's ecosystem unstable, resulting in transformations of Fe/Al-P to P$_o$ forms by phytoplankton, algae, and bacteria. During the eutrophic stage, external P input was beyond the sediments' capability to adsorb or control P cycling so that algal blooms occurred frequently. In recent years, now that the external input of P has been reduced and bioavailable sediment Fe/Al-P has been transformed to P$_o$, the biogeochemical cycle of P$_o$ would likely have played a more critical role in maintaining eutrophication in Lake Dianchi.

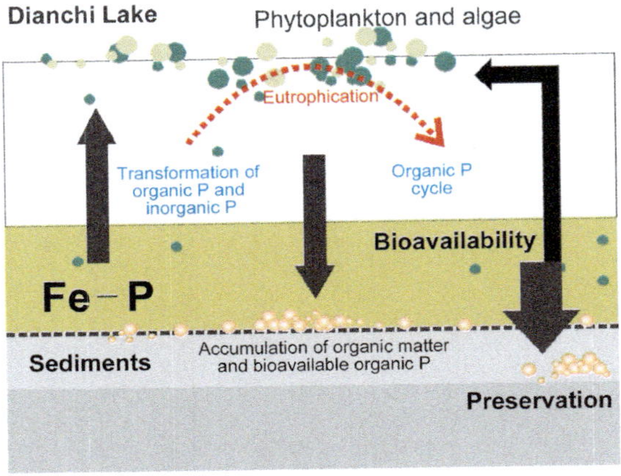

Fig. 6. Conceptual model of the biogeochemical cycle of phosphorus in the sediments from Lake Dianchi.

Additionally, part of the unextracted sediment P_o should be regarded as recalcitrant P_o, and even some labile P_o could be transformed to these recalcitrant P_o as the diagenetic processes progressed (Ding et al., 2013; Zhu et al., 2013a); it was 56 to 93% and 64 to 95% of P_o in the sediments from DC-1 and DC-2 cores, respectively, that could not be extracted by NaOH-EDTA solution (data not shown). Furthermore, 56 to 100% (DC-1) and 0 to 68% (DC-2) of P_o extracted by NaOH-EDTA could not be hydrolyzed by phosphatase in the sediments. This portion of P_o was thought to be nonbioavailable, thus possibly preserved in the sediments (Ding et al., 2013; Shinohara et al., 2012). The possible mechanisms for P_o resistant to enzymatic hydrolysis, thus the preservation in the sediments from lake, are further explored in the next section of this chapter.

The Mechanism for Preservation of Organic Phosphate of Sediments

It is reported that some P_o forms, (e.g., DNA), after association with humic acid (HA), could not be hydrolyzed by phosphatase (Crecchio and Stotzky, 1998; He et al., 2009a, 2011; Reitzel et al., 2007). Therefore, Zhu et al. (2015) studied the enzymatic hydrolysis of model P_o–HA, P_o–metal, and P_o–metal-HA complexes to shed light on the influence of natural organic matter on sediment P_o hydrolysis (Fig. 7) (Zhu et al., 2015). Compared with free P_o compounds, P_o associated with HA and metals (Al^{3+}, Ca^{2+}, Fe^{3+}, and Fe^{2+}) was protected against hydrolysis by enzymes, especially condensed P and phytate-like P. Enzymatic hydrolysis of PP was more affected by HA, Al^{3+}, Fe^{3+}, and Fe^{2+}, whereas Glu6P was less influenced by HA and metals. In a previous study, PP was found to be significantly correlated with the content of the LOI in surface sediments of Lake Tai ($R^2 = 0.301$, $P < 0.05$, $n = 18$) (Bai et al., 2009). These results indicate that the association with lake organic matter and metals is a potential mechanism for preservation of certain PP forms in the surface sediments of a lake. Glu6P, as a representative labile monoester P, formed the HA-M^{n+}-Glu6P model. HA-Ca^{2+}–Glu6P, HA-Fe^{2+}–Glu6P, and HA-Fe^{3+}–Glu6P

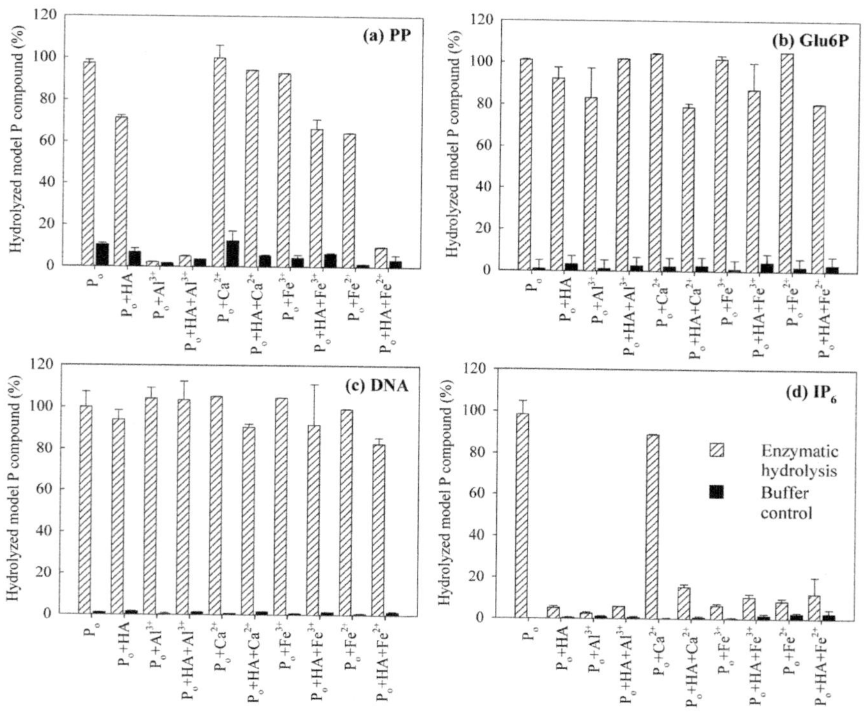

Fig. 7. Effect of humic acid (HA) and metals (M) on enzymatic hydrolysis of P_o compounds. Incubation solutions are (i) P_o only, (ii) mixture of P_o and HA (P_o + HA), and (iii) mixtures of P_o, HA, and M (P_o + HA + M). The solution was incubated in the presence (enzymatic hydrolysis) and absence (buffer control) of enzymes simultaneously. The incubation time was 1 h. PP, pyrophosphate (a); Glu6P, glucose-6-phosphate (b); DNA, deoxyribonucleic acid (c); and IP_6, inositol hexakisphosphate/phytate (d). Data are presented as the average value with standard deviation (*n* = 3). Reprinted from Zhu et al. (2015).

were hydrolyzed to a lesser extent by APase, compared with those of Glu6P only or corresponding Glu6P-M^{n+} (Fig. 7b). In other words, the complex formation could protect a small portion of labile monoester P from enzymatic hydrolysis even though Glu6P itself could hydrolyze spontaneously in the absence of enzymes. Only a small portion of DNA was likely to form complexes with HA or a combination of HA and metal ions that could resist enzymatic hydrolysis (Fig. 7c). In previous reports, maximal adsorption of DNA by HA was found to be ~0.23 μmol P/mg HA at pH 3.0 or 4.0 (Crecchio and Stotzky, 1998; Saeki et al., 2011). Hydrolysis data (Fig. 7c) indicated that metal ions tested in this case study cannot protect DNA hydrolyzed by enzymes, but it is possible to preserve DNA by the formation HA-M^{n+}-DNA complexes.

Enzymatic hydrolysis of IP_6, a representative phytate-like P, was obviously inhibited by the presence of HA and metal ions (Al^{3+}, Fe^{3+}, and Fe^{2+}), except for Ca^{2+} (Fig. 7d). It has been shown previously that concentrations of *myo*-IP_6 (characterized by ^{31}P-NMR) were significantly and positively correlated with total Al in the surface sediments from lakes (Jørgensen et al., 2011). Inositol phosphates are therefore thought to be stable and could be used as a paleo-indicator of P in sediments (Turner and Weckström, 2009). For the interaction of IP_6 with HA and Ca^{2+}, the HA-Ca^{2+}–IP_6 model were obviously more resistant to enzymatic hydrolysis than the Ca^{2+}–IP_6 model. Therefore, a large portion of phytate-like P would be resistant to enzymatic and chemical hydrolysis in the surface and profile sediments of lakes.

These data demonstrate that some P_o quantified by ^{31}P-NMR (including condensed P) cannot be hydrolyzed by phosphatase, and is not bioavailable when combined with HA with metal bridging or metal ions to form complex P_o compounds. This is probably a mechanism for the preservation of P_o, especially inositol phosphates, in sediments. Enzymatic hydrolysis and ultraviolet irradiation have shown that approximately 50% of P associated with soil humic and fulvic acids is labile; this percentage was lower (10–17%) for P after humification of peat (He et al., 2011, 2009a, 2006b). Besides natural organic matter, adsorption of P_o by minerals was also possibly an important mechanism for P_o resistant to enzymatic hydrolysis, and thus preservation in the sediments. In particular, the phytate-like P almost could not degrade by phytase after adsorption by goethite and montmorillonite (data not shown). Based on the results of the hydrolysis work, Zhu et al. (2015) proposed that free P_o is transformed to more recalcitrant P during the early stages of diagenesis, which can sequester some P_o in sediments of lakes for long periods (Ding et al., 2013).

Biogeochemical Cycling of Organic Phosphate and Its Role in Eutrophication

Unlike rivers, lakes always respond to changes in a nonlinear fashion (Fig. 8). Many lakes show a highly nonlinear response (hysteresis) to nutrient concentration (ILEC, 2005). In the case of P-limited lakes, external inputs of P result in the concentration of P in the overlying water increasing from A to B, the algal blooms may not be apparent until P concentrations are high and the lake abruptly switches its status. Take Dianchi, for example: it is likely that P contents in the water could be controlled by adsorption and desorption at the water-sediments interface, such as the formation of Ca-bonded P. With time process and P concentration increasing (from B to D), the plankton concentrations increased abruptly from B to D, which possibly, because of the superfluous P, could not be controlled in effect by water-sediment interface. The internal P accumulated in the sediments would be released, such as Fe-P in Lake Dianchi, with algal blooms and accumulation of organic matter (from C to D).

Even though the external inputs of P have reduced gradually or have been controlled, the lake cannot simply go from D back to B. There are likely to have been irreversible changes to the ecosystem, so recovery follows a path from D though F down to A. The accumulation of internal P from A to B to D would be an important mechanism for some irreversible changes. Based on our research in this chapter, bioavailability and preservation of P_o could be an important mechanism for the process from D though F to A. The semiclosed Lake Dianchi in this chapter is a good case in point, which possibly exhibits the process from C though D to E. For the process from D to F, the bioavailable P_o from frequent

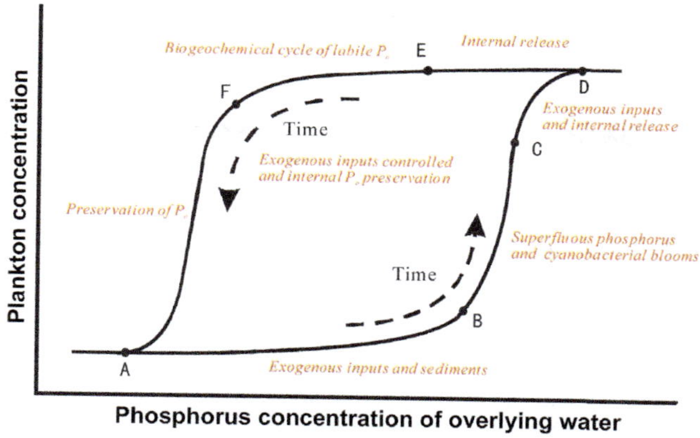

Fig. 8. Conceptual model of relationship between eutrophication and biogeochemical cycle of P_o in the P-limited lake. Per the model of ILEC (2005) with the observations in this study.

algal blooms accumulated in sediments would be released and continue to support eutrophication during periods when less phosphate is available (Fig. 6). With time, more and more P_o was deposited and preserved by combining naturally with organic matter or through adsorption by minerals, which would result in the decrease of the bioavailable internal P load. This would be an important mechanism for a eutrophic lake switching its status from F to A over a period of time. A further understanding of the bioavailability and preservation of P_o in the sediments of lakes is still needed, which would provide valuable information for the restoration of eutrophic lakes. For example, based on the information of the P_o biogeochemical cycle, it was possible to develop a scientific method to promote the preservation of P_o in the restoration process of a eutrophic lake.

References

Ahlgren, J., K. Reitzel, R. Danielsson, A. Gogoll, and E. Rydin. 2006. Biogenic phosphorus in oliotrophic lake sediments: Differences in composition measured with NMR spectroscopy. Water Res. 40:3705–3712. doi:10.1016/j.watres.2006.09.006

Ahlgren, J., H. DeBrabandere, K. Reitzel, E. Rydin, A. Gogoll, and M. Waldeback. 2007. Sediment phosphorus extractants for phosphorus-31 nuclear magnetic resonance analysis: A quantitative evaluation. J. Environ. Qual. 36:892–898. doi:10.2134/jeq2006.0235

Akinerimi, O.O., B. Ajiboye, and Z. He. 2011. Metal speciation of phosphorus derived from solid state spectroscopic analysis. In: Z. He, editor, Environmental chemistry of animal manure. Nova Science, New York. p. 301–324.

Bai, X., S. Ding, C. Fan, T. Liu, D. Shi, and L. Zhang. 2009. Organic phosphorus species in surface sediments of a large, shallow, eutrophic lake, Lake Taihu, China. Environ. Pollut. 157:2507–2513. doi:10.1016/j.envpol.2009.03.018

Brandes, J.A., E.D. Ingall, and D. Paterson. 2007. Characterization of minerals and organic phosphorus species in marine sediments using soft X-ray fluorescence spectromicroscopy. Mar. Chem. 103:250–265. doi:10.1016/j.marchem.2006.09.004

Bünemann, E.K. 2008. Enzyme additions as a tool to assess the potential bioavailability of organically bound nutrients. Soil Biol. Biochem. 40:2116–2129. doi:10.1016/j.soilbio.2008.03.001

Cade-Menun, B.J. 2011. Characterizing phosphorus in animal waste with solution ^{31}P NMR spectroscopy. In: Z. He, editor, Environmental chemistry of animal manure. Nova Science, New York. p. 275–299.

Carpenter, S.R. 2005. Eutrophication of aquatic ecosystems: Bistability and soil phosphorus. Proc. Natl. Acad. Sci. USA 102:10002–10005. doi:10.1073/pnas.0503959102

Crecchio, C., and G. Stotzky. 1998. Binding of DNA on humic acids: Effect on transformation of *bacillus subtilis* and resistance to DNase. Soil Biol. Biochem. 30:1061–1067. doi:10.1016/S0038-0717(97)00248-4

Dai, G., J. Li, L. Li, and L. Song. 2012. The spatio-temporal pattern of phytoplankton in the north basin of Lake Dianchi and related environmental factors. (In Chinese, with English abstract.) Acta Hydrobiol. Sinica 36:946–956. doi:10.3724/SP.J.1035.2012.00946

De Brabandere, H., N. Forsgard, L. Israelsson, J. Petterson, E. Rydin, M. Waldeback, and P.J. Sjoberg. 2008. Screening for organic phosphorus compounds in aquatic sediments by liquid chromatography coupled to ICP-AES and ESI-MS/MS. Anal. Chem. 80:6689–6697. doi:10.1021/ac8006335

Ding, S., D. Xu, X. Bai, S. Yao, C. Fan, and C.S. Zhang. 2013. Speciation of organic phosphorus in a sediment profile of Lake Taihu ii. Molecular species and their depth attenuation. J. Environ. Sci. (China) 25:925–932. doi:10.1016/S1001-0742(12)60137-5

Ding, S., D. Xu, B. Li, C. Fan, and C. Zhang. 2010. Improvement of ^{31}P NMR spectral resolution by 8-hydroxyquinoline precipitation of paramagnetic Fe and Mn in environmental samples. Environ. Sci. Technol. 44:2555–2561. doi:10.1021/es903558g

Gao, H., T. Liu, F. Zhu, and L. Deng. 2010. Characteristics of organic phosphorus in sediments of Lake Dianchi, China. Fresenius Environ. Bull. 19:3125–3130.

Gao, L., J.M. Zhou, H. Yang, and J. Chen. 2005. Phosphorus fractions in sediment profile and their potential contributions to eutrophication in Lake Dianchi. Environ. Geol. 48:835–844. doi:10.1007/s00254-005-0005-3

Gao, W., R.W. Howarth, D.P. Swaney, B. Hong, and H.C. Guo. 2015. Enhanced N input to Lake Dianchi Basin from 1980 to 2010: Drivers and consequences. Sci. Total Environ. 505:376–384. doi:10.1016/j.scitotenv.2014.10.016

Ged, E.C., and T.H. Boyer. 2013. Molecular weight distribution of phosphorus fraction of aquatic dissolved organic matter. Chemosphere 91:921–927. doi:10.1016/j.chemosphere.2013.01.113

Giles, C.D., and B.J. Cade-Menun. 2014. Phytate in animal manure and soils: Abundance, cycling and bioavailability. In: Z. He and H. Zhang, editors, Applied manure and nutrient chemistry for sustainable agriculture and environment. Springer, Amsterdam, the Netherlands. p. 163–190.

He, Z., B.J. Cade-Menun, G.S. Toor, A. Fortuna, C.W. Honeycutt, and J.T. Sims. 2007. Comparison of phosphorus forms in wet and dried animal manures by solution phosphorus-31 nuclear magnetic resonance spectroscopy and enzymatic hydrolysis. J. Environ. Qual. 36:1086–1095. doi:10.2134/jeq2006.0549

He, Z., T.H. Dao, and C.W. Honeycutt. 2006a. Insoluble Fe-associated inorganic and organic phosphates in animal manure and soil. Soil Sci. 171:117–126. doi:10.1097/01.ss.0000187374.31026.ca

He, Z., T.S. Griffin, and C.W. Honeycutt. 2004. Enzymatic hydrolysis of organic phosphorus in swine manure and soil. J. Environ. Qual. 33:367–372. doi:10.2134/jeq2004.3670

He, Z., and C.W. Honeycutt. 2001. Enzymatic characterization of organic phosphorus in animal manure. J. Environ. Qual. 30:1685–1692. doi:10.2134/jeq2001.3051685x

He, Z., and C.W. Honeycutt. 2011. Enzymatic hydrolysis of organic phosphorus. In: Z. He, editor, Environmental chemistry of animal manure. Nova Science, New York. p. 253–274.

He, Z., C.W. Honeycutt, B.J. Cade-Menun, Z.N. Senwo, and I.A. Tazisong. 2008. Phosphorus in poultry litter and soil: Enzymatic and nuclear magnetic resonance characterization. Soil Sci. Soc. Am. J. 72:1425–1433. doi:10.2136/sssaj2007.0407

He, Z., C.W. Honeycutt, T.S. Griffin, B.J. Cade-Menun, P.J. Pellechia, and Z. Dou. 2009c. Phosphorus forms in conventional and organic dairy manure identified by solution and solid state P-31 NMR spectroscopy. J. Environ. Qual. 38:1909–1918. doi:10.2134/jeq2008.0445

He, Z., C.W. Honeycutt, T. Ohno, F. Wu, and R. Zhang. 2009b. Characterization of phosphorus associated with natural organic matter. In: F. Wu and B. Xing, editors, Natural organic matter and its significance in the environment. Science Press, Beijing, China. p. 265–279.

He, Z., T. Ohno, B.J. Cade-Menun, M.S. Erich, and C.W. Honeycutt. 2006b. Spectral and chemical characterization of phosphates associated with humic substances. Soil Sci. Soc. Am. J. 70:1741–1751. doi:10.2136/sssaj2006.0030

He, Z., D.C. Olk, and B.J. Cade-Menun. 2011. Forms and lability of phosphorus in humic acid fractions of hord silt loam soil. Soil Sci. Soc. Am. J. 75:1712–1722. doi:10.2136/sssaj2010.0355

He, Z., D.C. Olk, C.W. Honeycutt, and A. Fortuna. 2009a. Enzymatically- and ultraviolet-labile phosphorus in humic acid fractions from rice soils. Soil Sci. 174:81–87. doi:10.1097/SS.0b013e3181981dc5

Hieltjes, A.H.M., and L. Lijklema. 1980. Fractionation of inorganic phosphates in calcareous sediments. J. Environ. Qual. 9:405–407. doi:10.2134/jeq1980.00472425000900030015x

Hu, J., Y. Liu, and J. Liu. 2006. The comparison of phosphorus pools from the sediment in two bays of lake Dianchi for cyanobacterial bloom assessment. Environ. Monit. Assess. 121:1–14. doi:10.1007/s10661-005-9036-1

Hu, J., Q. Shen, Y. Liu, and J. Liu. 2007. Mobility of different phosphorus pools in the sediment of Lake Dianchi during cyanobacterial blooms. Environ. Monit. Assess. 132:141–153. doi:10.1007/s10661-006-9509-x

Hupfer, M., R. Gächter, and R. Giovanoli. 1995. Transformation of phosphorus species in settling seston and during early sediment diagenesis. Aquat. Sci. 57:305–324. doi:10.1007/BF00878395

ILEC. 2005. Managing Lakes and their basins for sustainable use: A report for Lake basin managers and stakeholders. International Lake Environment Committee Foundation, Kusatsu, Japan.

Ivanoff, D.B., K.R. Reddy, and S. Robinson. 1998. Chemical fractionation of organic phosphorus in selected histosols. Soil Sci. 163:36–45. doi:10.1097/00010694-199801000-00006

Jin, X.C., and Q.Y. Tu, editors. 1990. The standard methods for observation and analysis in lake eutrophication. (In Chinese.) 2nd ed. Chin. Environ. Sci. Press, Beijing. p. 226–229.

Johnson, N.R., and J.E. Hill. 2010. Phosphorus species composition of poultry manure-amended soil using high-throughput enzymatic hydrolysis. Soil Sci. Soc. Am. J. 74:1786–1791. doi:10.2136/sssaj2009.0431

Johnson, N.R., and J.E. Hill. 2011. High-throughput measurement and classification of organic p in environmental samples. JoVE 52:1–3.

Jørgensen, C., H.S. Jensen, F.Ø. Andersen, S. Egemose, and K. Reitzel. 2011. Occurrence of orthophosphate monoesters in lake sediments: Significance of *myo*-and *scyllo*-inositol hexakisphosphate. J. Environ. Monit. 13:2328–2334. doi:10.1039/c1em10202h

Le, C., Y. Zha, Y. Li, D. Sun, H. Lu, and B. Yin. 2010. Eutrophication of lake waters in china: Cost, causes, and control. Environ. Manage. 45:662–668. doi:10.1007/s00267-010-9440-3

Li, H., Y. Wang, L.Q. Shi, J. Mi, D. Song, and X.J. Pan. 2012. Distribution and fractions of phosphorus and nitrogen in surface sediments from Dianchi Lake, China. Int. J. Environ. Res. 6:195–208.

Li, Q.-M., W. Zhang, X.-X. Wang, Y.-Y. Zhou, H. Yang, and G.-L. Ji. 2007. Phosphorus in interstitial water induced by redox potential in sediment of Dianchi Lake, China. Pedosphere 17:739–746. doi:10.1016/S1002-0160(07)60089-7

Lu, J., H. Wang, M. Pan, J. Xia, W. Xing, and G. Liu. 2012. Using sediment seed banks and historical vegetation change data to develop restoration criteria for an eutrophic lake in China. Ecol. Eng. 39:95–103. doi:10.1016/j.ecoleng.2011.11.006

Lu, J., H. Yang, L. Gao, and T. Yu. 2005. Special variation of P and N in water and sediments of Dianchi Lake, China. Pedosphere 15:78–83.

Manimel Wadu, M.C.W., V.K. Michaelis, S. Kroeker, and O.O. Akinremi. 2013. Exchangeable calcium/magnesium ratio affects phosphorus behavior in balcareous soils. Soil Sci. Soc. Am. J. 77:2004–2013. doi:10.2136/sssaj2012.0102

Monbet, P., I.D. McKelvie, A. Saefumillah, and P.J. Worsfold. 2007. A protocol to assess the enzymatic release of dissolved organic phosphorus species in waters under environmentally relevant conditions. Environ. Sci. Technol. 41:7479–7485. doi:10.1021/es070573c

Ozkundakci, D., D.P. Hamilton, R.W. McDowell, and S. Hill. 2014. Phosphorus dynamics in sediments of a eutrophic lake derived from P-31 nuclear magnetic resonance spectroscope. Mar. Freshwater Res. 65:70–80.

Pagliari, P.H., and C.A.M. Laboski. 2012. Investigation of the inorganic and organic phosphorus forms in animal manure. J. Environ. Qual. 41:901–910. doi:10.2134/jeq2011.0451

Purushothaman, P., and G. Chakrapani. 2014. Phosphorus biogeochemistry in sediments of high altitude lakes, Kumaun Himalayas, India. Arabian J. Geosci. 8:701–709. doi:10.1007/s12517-013-1234-5.

Reitzel, K., J. Ahlgren, H. DeBrabandere, M. Waldeback, A. Gogoll, L. Tranvik, and E. Rydin. 2007. Degradation rates of organic phosphorus in lake sediment. Biogeochemistry 82:15–28. doi:10.1007/s10533-006-9049-z

Saeki, K., Y. Ihyo, M. Sakai, and T. Kunito. 2011. Strong adsorption of DNA molecules on humic acids. Environ. Chem. Lett. 9:505–509. doi:10.1007/s10311-011-0310-x

Sannigrahi, P., E.D. Ingall, and R. Benner. 2006. Nature and dynamics of phosphorus-containing components of marine dissolved and particulate organic matter. Geochim. Cosmochim. Acta 70:5868–5882. doi:10.1016/j.gca.2006.08.037

Shinohara, R., A. Imai, N. Kawasaki, K. Komatsu, A. Kohzu, S. Miura, T. Sano, T. Satou, and N. Tomioka. 2012. Biogenic phosphorus compounds in sediment and suspended particles in a shallow eutrophic lake: A ^{31}P-nuclear magnetic resonance (^{31}P NMR) study. Environ. Sci. Technol. 46:10572–10578. doi:10.1021/es301887z

Smith, V.H. 1983. Low nitrogen to phosphorus ratios favor dominance by blue-green algae in lake phytoplankton. Science 221:669–671. doi:10.1126/science.221.4611.669

Søndergaard, M., J.P. Jensen, and E. Jeppesen. 2003. Role of sediment and internal loading of phosphorus in shallow lakes. Hydrobiologia 506–509:135–145. doi:10.1023/B:HYDR.0000008611.12704.dd

Su, Y., S. Ma, and S. Dong. 2005. Variation of alkaline phosphatase activity in sediments of shrimp culture ponds and its relationship with the contents of C, N and P. J. Ocean Univers. Chin. 4:75–79. doi:10.1007/s11802-005-0027-1

Suzumura, M., and E.D. Ingall. 2001. Concentrations of lipid phosphorus and its abundance in dissolved and particulate organic phosphorus in coastal seawater. Mar. Chem. 75:141–149. doi:10.1016/S0304-4203(01)00034-2

Tazisong, I.A., Z.N. Senwo, B.J. Cade-Menun, and Z. He. 2014. Phosphorus forms and mineralization potentials of Alabama upland cotton production soils amended with poultry litter. In: Z. He and H. Zhang, editors, Applied manure and nutrient chemistry for sustainable agriculture and environment. Springer, Amsterdam, the Netherlands. p. 191–209.

Torres, I.C., B.L. Turner, and K.R. Reddy. 2014. The chemical nature of phosphorus in subtropical lake sediments. Aquat. Geochem. 20:437–457. doi:10.1007/s10498-014-9228-9

Toth, J.D., Z. Dou, and Z. He. 2011. Solubility of manure phosphorus characterized by selective and sequential extractions. In: Z. He, editor, Environmental chemistry of animal manure. Nova Science, New York. p. 227–251.

Turner, B.L., B.J. Cade-Menun, L.M. Condron, and S. Newman. 2005. Extraction of soil organic phosphorus. Talanta 66:294–306. doi:10.1016/j.talanta.2004.11.012

Turner, B.L., and K. Weckström. 2009. Phytate as a novel phosphorus-specific paleo-indicator in aquatic sediments. J. Paleolimnol. 42:391–400. doi:10.1007/s10933-008-9283-6

Wang, F.S., C. Liu, M. Wu, Y. Yu, F. Wu, S. Lü, Z. Wei, and G. Xu. 2009. Stable isotopes in sedimentary organic matter from Lake Dianchi and their indication of eutrophication history. Water Air Soil Pollut. 199:159–170. doi:10.1007/s11270-008-9868-8

Wu, Y., L. Zheng, L. Li, D. Yin, G. Dai, and L. Song. 2014. Physiological response to low nitrogen level in Aphanizomenon flos-aquae from Lake Dianchi. (In Chinese, with English abstract.) J. Lake Sci. 26:131–136.

Xie, L., P. Xie, S. Li, H. Tang, and H. Liu. 2003. The low TN:TP ratio, a cause or a result of *Microcystis* blooms? Water Res. 37:2073–2080. doi:10.1016/S0043-1354(02)00532-8

Xiong, Q., L. Jiao, S. Wang, and X. Peng. 2014. Characteristics and bioavailability of organic phosphorus from different sources of sediments in Dianchi Lake. (In Chinese, with English abstract.) Environ. Sci. 35:4118–4126.

Xiong, Y., F. Wu, J. Fang, L. Wang, Y. Li, and H. Liao. 2010. Organic geochemical record of environmental changes in lake dianchi, china. J. Paleolimnol. 44:217–231. doi:10.1007/s10933-009-9398-4

Young, E.O., D.S. Ross, B.J. Cade-Menun, and C.W. Liu. 2013. Phosphorus speciation in riparian soils: A phosphorus-31 nuclear magnetic resonance spectroscopy and enzyme hydrolysis study. Soil Sci. Soc. Am. J. 77:1636–1647. doi:10.2136/sssaj2012.0313

Zhang, R., F. Wu, Z. He, J. Zheng, B. Song, and L. Jin. 2009. Phosphorus composition in sediments from seven different trophic lakes, China: A phosphorus-31 NMR study. J. Environ. Qual. 38:353–359. doi:10.2134/jeq2007.0616

Zhang, R., F. Wu, C. Liu, P. Fu, W. Li, L. Wang, H. Liao, and J. Guo. 2008. Characteristics of organic phosphorus fractions in different trophic sediments of lakes from the middle and lower reaches of Yangtze River region and Southwestern Plateau, China. Environ. Pollut. 152:366–372. doi:10.1016/j.envpol.2007.06.024

Zhang, T., W.H. Zeng, S.R. Wang, and Z.K. Ni. 2014. Temporal and spatial changes of water quality and management strategies of Dianchi Lake in southwest China. Hydrol. Earth Syst. Sci. 18:1493–1502. doi:10.5194/hess-18-1493-2014

Zhou, Q., C.E. Gibson, and Y. Zhu. 2001. Evaluation of phosphorus bioavailability in sediments of three contrasting lakes in China and the UK. Chemosphere 42:221–225. doi:10.1016/S0045-6535(00)00129-6

Zhou, Y., C. Song, X. Cao, J. Li, G. Chen, Z. Xia, and P. Jiang. 2008. Phosphorus fractions and alkaline phosphatase activity in sediments of a large eutrophic Chinese lake (Lake Taihu). Hydrobiologia 599:119–125. doi:10.1007/s10750-007-9185-z

Zhou, Y.Y., J.Q. Li, and M. Zhang. 2002. Temporal and spatial variations in kinetics of alkaline phosphatase in sediments of a shallow Chinese eutrophic lake (Lake Donghu). Water Res. 36:2084–2090. doi:10.1016/S0043-1354(01)00405-5

Zhu, Y., F. Wu, Z. He, J. Guo, X. Qu, F. Xie, J.P. Giesy, H. Liao, and F. Guo. 2013b. Characterization of organic phosphorus in lake sediments by sequential fractionation and enzymatic hydrolysis. Environ. Sci. Technol. 47:7679–7687. doi:10.1021/es305277g

Zhu, Y., F. Wu, Z. He, J.P. Giesy, W. Feng, Y. Mu, C. Feng, X. Zhao, H. Liao, and Z. Tang. 2015. Influence of natural organic matter on the bioavailability and preservation of organic phosphorus in lake sediments. Chem. Geol. 397: 51–60. doi:0.1016/j.chemgeo.2015.01.006"

Zhu, Y., R. Zhang, F. Wu, X. Qu, F. Xie, and Z. Fu. 2013a. Phosphorus fractions and bioavailability in relation to particle size characteristics in sediments from Lake Hongfeng, Southwest China. Environ. Earth Sci. 68:1041–1052. doi:10.1007/s12665-012-1806-9

Effect of Algal Organic Matter on Pollutant Accumulation in Reservoir Sediments of the Pearl River Delta, China

Dandan Duan and Yong Ran*

Abstract

Algae-derived organic matter plays an important role in biogeochemical cycling of pollutants in aquatic ecosystems, as algal organic matter can sequester hydrophobic organic pollutants and soluble heavy metals in a water column and bring these contaminants into sediments. An investigation of the increase of algal organic matter coupling with accumulation of pollutants may provide valuable insights on the "biological pump" effect of algal organic matter in aquatic ecosystems. In this chapter, we review and discuss the effects of algal organic matter on pollutants by using the sediment cores collected from three reservoirs in the Pearl River Delta. Rock-Eval pyrolysis and neutral sugar analysis were applied for characterization of algal organic matter. It was found that historical variations in the content of organic pollutants and heavy metals were strongly correlated with the concentration of algae-derived organic matter in the reservoirs (for most pollutants: $R^2 > 0.80$, $p < 0.01$). Increasing temperature in the subtropical region plays significant roles on both algal productivity and its co-settling of environmental pollutants. Detailed research in the future should take into account climate change associated with algal productivity and pollutants in fresh ecosystems.

Abbreviations: AOM, algae-derived organic matter; EOM, extracellular organic matter; HI, hydrogen index; HM, heavy metal; IOM, intracellular organic matter; LA, Lian'an; OCP, organochlorine pesticide; OM, organic matter; PAH, polycyclic aromatic hydrocarbon; PCB, polychlorinated biphenyl; PRD, Pearl River Delta; TOC, total organic carbon; XFJ, Xinfengjiang; ZT, Zengtang.

D. Duan and Y. Ran, State Key Laboratory of Organic Geochemistry, Guangzhou Institute of Geochemistry, Chinese Academy of Sciences, No. 511 Kehuajie Tianhe, Guangzhou, Guangdong 510640, China (dan113133@126.com); D. Duan also at University of Chinese Academy of Sciences, Beijing 100049, China. *Corresponding author (yran@gig.ac.cn).

doi:10.2136/sssaspecpub62.2014.0075

Labile Organic Matter—Chemical Compositions, Function, and Significance in Soil and the Environment. SSSA Special Publication 62. Zhongqi He and Fengchang Wu, editors.

Planktonic algae are ubiquitous in aquatic environments. They can produce and release large amounts of algae-derived organic matter (AOM) into water column during cell growth and lysis. Algae-derived organic matter fractions, including extracellular organic matter (EOM) and intracellular organic matter (IOM), are composed of a wide spectrum of chemical compounds such as oligosaccharides and polysaccharides, amino acids, peptides, proteins, organic acids (e.g., glycolic acid), lipids, and fatty acids (Henderson et al., 2008; Pivokonsky et al., 2006), all of which contain various functional groups (e.g., hydroxyl, carbonyl, and amino groups).

Planktonic algae and their AOM easily affect the settling of a number of toxic pollutants in aquatic ecosystems, such as polycyclic aromatic hydrocarbons (PAHs), organochlorine pesticides (OCPs), polychlorinated biphenyls (PCBs), and heavy metals (HMs) (Ko and Baker, 1995; Swackhamer and Skoglund, 1993; Taylor et al., 1991). Once the toxic pollutants are deposited in the water bodies, they could chemically and physically be absorbed and adsorbed by algae and their AOM, leading to a depletion of pollutants in the water phase (Richer and Peters, 1993; Skoglund et al., 1996). As the algal contents increase, the air–water exchange of the pollutants is not in equilibrium, reducing their atmospheric residence times and enhancing their diffusive flux in surface waters (Dachs et al., 2000, 1999). Eventually, AOM along with the pollutants are settled and preserved in sediments at the bottom of water bodies. This algal scavenging process of the pollutants, also called the "biological pump" effect (Fig. 1), has been extensively investigated and studied in recent years (Berrojalbiz et al., 2011; Galbán-Malagón et al., 2012; Outridge et al., 2007).

Employing sediment cores as historical archives for the long-term evaluation of the algal scavenging process can avoid the complicated calculations and multifactor analysis of environmental influences (e.g., diurnal variation and seasonal variation of algal abundance, meteorological conditions, and hydrological changes) that are frequently used in short-term investigations. One important step for long-term study is identification and characterization of AOM in sediment cores. To qualitatively and quantitatively analyze AOM

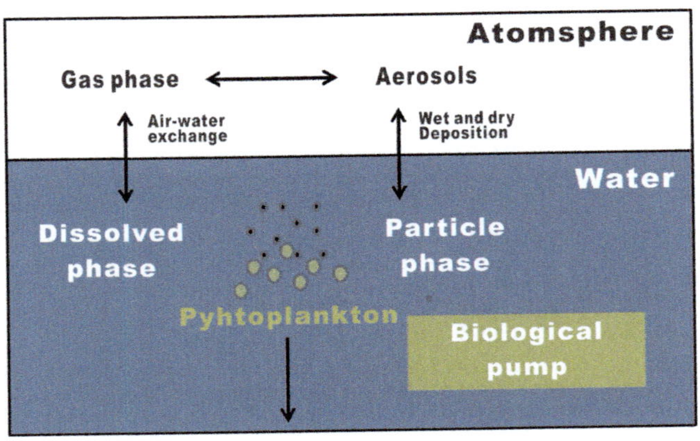

Fig. 1. Biological pump effects in natural waters.

and reflect its historical primary productivity in aquatic ecosystems, several analytical techniques have been proposed and developed, such as chlorophyll compound analysis, diatom count–biogenic silica measurement, and Rock-Eval pyrolysis (Keely and Brereton, 1986; Outridge et al., 2005; Sanei and Goodarzi, 2006). Along with AOM characterization and pollutant analysis of sediment cores, historical influences of algal productivity and AOM on the cycling of contaminants can be studied. Recent studies on the Arctic and sub-Arctic lakes indicated that significant portions of Hg in sediments could be partially explained by algal scavenging during the past decades and centuries (Carrie et al., 2010; Outridge et al., 2007). Similarly, significant associations between AOM and the distribution of ^{210}Pb, PAHs, and heavy metals were observed in the sediment cores of some Chinese lakes and reservoirs (Duan et al., 2014; Wu et al., 2012; Xu et al., 2011). However, this biological pump effect is disputed by Kirk et al. (2011) and Stern et al. (2009). Therefore, more in-depth studies with sediment cores are needed for the understanding of algal scavenging effects in the freshwater environment, especially in low-latitude areas.

In this chapter, three subtropical reservoirs in the Pearl River Delta (PRD) area with contrasting trophic levels were selected as a case study for showing the effects of AOM levels on pollutant accumulation in sediments. To do it, data from Rock-Eval pyrolysis combined with a molecular-level analysis of neutral sugars were presented for AOM characteristics. The historical variations in the concentrations of organic pollutants and heavy metals coupled with the AOM parameters and regional temperature trends were then analyzed and discussed. Information derived from this chapter would increase understanding on the internal mechanisms of interactions between algal productivity and pollutants in natural waters. These observations could also be helpful for shedding light on the importance of climate change on algal productivity in fresh ecosystems.

Study Areas

The Zengtang (ZT), Lian'an (LA), and Xinfengjiang (XFJ) reservoirs in PRD were chosen to collect the sediment cores (Fig. 2), and the water depths for the sampling sites at ZT, LA, and XFJ were 1, 17, and 36 m, respectively. Zengtang (23.22°N, 113.76°W) is a shallow, meso-eutrophic lake (surface area: 34 km^2; mean depth: 2.5 m; maximum depth: 5 m) located in the east of Zengcheng, where LA (23.4°N, 113.66°W) is the largest drinking water reservoir in Zengcheng (surface area: 43 km^2; maximum depth: 60 m). Xinfengjiang (23.77°N, 114.57°W), located near Heyuan, is the largest reservoir in Guangdong province (surface area: 5730 km^2; mean depth: 28.7 m; maximum depth: 93 m). They are mainly fed by precipitation and have no inflow. Zengtang reservoir was an oligo-mesotropic lake before 2003, then turned into the eutrophication status because of the fry misuse (Liu et al., 2011). By contrast, XFJ reservoir is an oligotrophic lake and has the lowest algal productivity. The trophic status of LA fell in between those of ZT and XFJ reservoirs. All reservoirs have been strictly protected from the direct input of nutrients (e.g., nitrogen and phosphorus) and hazardous compounds. No severe pollutions were detected since the dams were constructed (except in 2003 at ZT reservoir). Green algae, cyanophyta, and diatoms were the most abundant species in ZT, LA, and XFJ reservoirs. The sediment samples from three reservoirs were dated by ^{210}Pb radiometric dating and ^{137}Cs radiometric dating (Duan et al.,

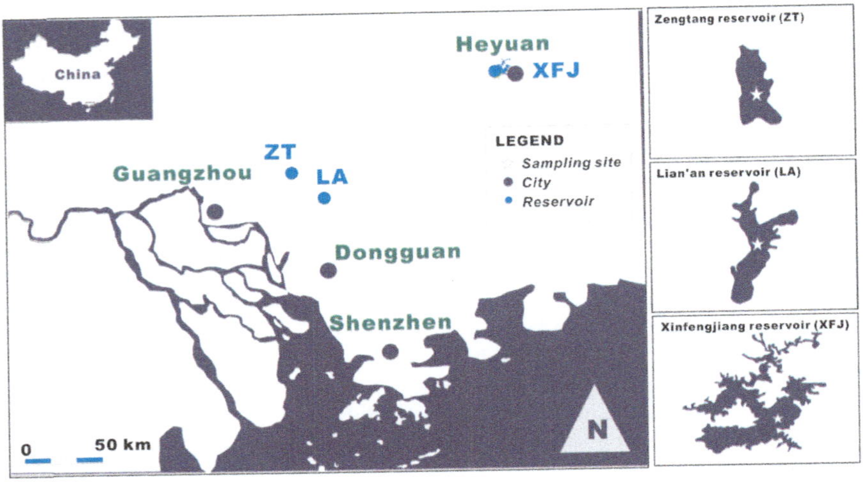

Fig. 2. Map of sampling sites.

2014). These dating data were used for the analysis of historical variation in pollutants and organic matter in three sediment core samples.

Rock-Eval Pyrolytic Analysis of Algae-Derived Organic Matter

Rock-Eval pyrolysis is a traditional analytical technique used in petroleum geochemistry that can determine the quantity and quality of the organic matter in the sample on the basis of pyrolytic degree and evolution of various organic compounds (Lafargue et al., 1998). Although this simplistic method for specific identification and assessment of particular organic matter (OM) sources (e.g., AOM) in recent sediments is controversial (Carrie et al., 2012; Wu et al., 2013), nevertheless, this rapid, efficient technique has been increasingly used for organic matter characterization during the last decades (Carrie et al., 2010; Duan et al., 2014; Outridge et al., 2007; Sanei and Goodarzi, 2006; Sanei et al., 2005; Stern et al., 2009).

The latest Rock-Eval product is the Rock-Eval 6, which was originally designed for oil and gas exploration and commercialized by Vinci Technologies. Bulk sediment was first heated in an inert, O_2–free oven (100–650°C), followed by combustion in an oxidation oven (400–850°C). The pyrolysis stage released two specific fractions of hydrocarbons, defined as S1 and S2, which were detected by a flame ionization detector. Simultaneously, CO and CO_2 detected by infrared (milligrams of CO and CO_2 per gram), representing mostly oxygen-bearing OM, was named as S3. The combustion stage burned the remaining OM, yielding the refractory carbon (weight percent) peak. Total organic carbon (TOC) (weight percent) was the sum of OM from the two steps (Fig. 3).

Signal S1 from the pyrolysis process represents labile organic matter (<500 Da), which is volatile and amorphous but may originate from part of the lower molecular weight of AOM. Signal S2 represents long-chain macromolecules,

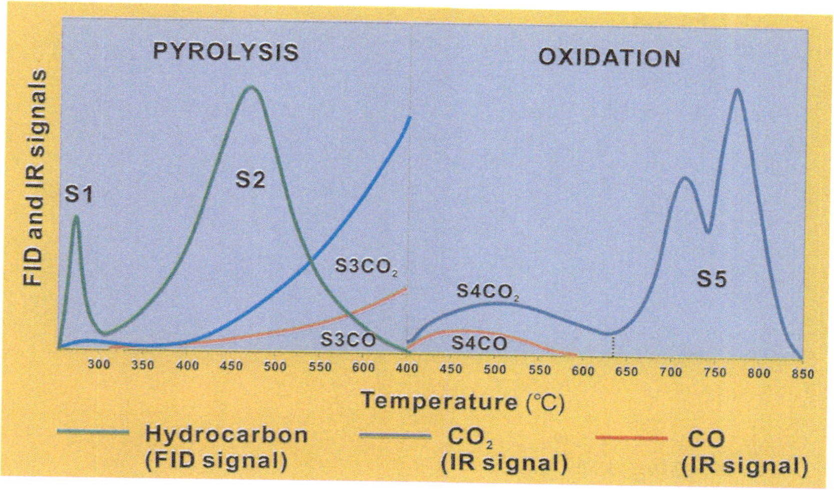

Fig. 3. Diagram of Rock-Eval pyrolysis with increasing temperature. FID, flame ionization detector; IR, infrared detector.

which could mainly contain hydrogen-rich aliphatic biomacromolecules derived from the cell walls of algal organic matter. To specifically indicate AOM, the fraction S2 was further separated into the S2a and S2b subtractions, which represent the integrations of S2 from 300 to 400°C and from 400 to 650°C, respectively. It was found that S2a is more reliable than S2 as a tracer for AOM in sediments of aquatic ecosystems (Carrie et al., 2012; Stern et al., 2009; Xu et al., 2011).

The TOC contents of sediments ranged from 0.81 to 3.38% at ZT, from 0.88 to 3.4% at LA, and from 0.64 to 1.3% at XFJ (Fig. 4). The highest TOC content occurs in the uppermost layers both in the ZT and LA reservoirs, whereas the XFJ reservoir displays a decreasing distribution, and the highest TOC level is at the deepest sections of this core. At ZT, the vertical profiles of S1 and S2 carbon increased by more than 8.7-fold and 7.41-fold in the sediment core, where the S1 and S2 profiles increased by more than 5.6-fold and 5.2-fold at LA, respectively (Fig. 4). The results indicate that ZT and LA may have similar historical primary productivity or undergo the same biogeochemical processes. In the core of XFJ, the S1 and S2 profiles decreased by 3.3-fold and 3.9-fold, respectively, from the top layer to the depth of 12 cm, (Fig. 4), which is consistent with the fact that XFJ is an oligotrophic reservoir.

Pearson correlation analysis demonstrated that all the organic matter fractions were significantly correlated with each other (Table 1) (Duan et al., 2014), indicating that the different organic matter fractions might have had the same source or have underwent similar geochemical changes. Considering the fact that different eutrophic conditions were present in the three reservoirs, it is reasonable to conclude that the organic matter composition in the three reservoirs had not been heavily influenced by biogeochemical processes. Furthermore, as planktonic algae was the primary producer of sediment organic matter in these reservoirs, as mentioned above, the algal production in the three reservoirs could be one of the most important factors contributing to the change of the organic matter compartment during the last six decades.

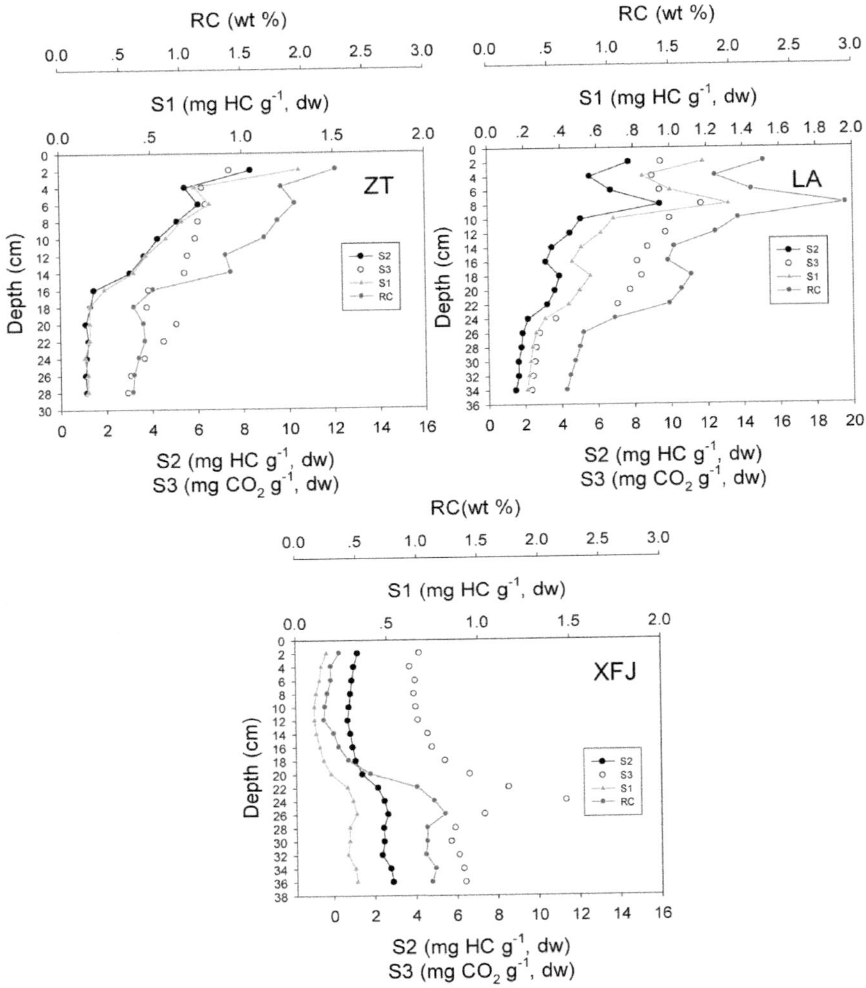

Fig. 4. Vertical distribution of Rock-Eval parameters of sediment cores of three reservoirs. RC, refractory carbon; ZT, Zengtang; LA, Lian'an; XFJ, Xinfengjiang.

The hydrogen index (HI) was calculated by normalizing the quantity of S2 to TOC and used as an indicator of primary productivity in aquatic environments (Bechtel and Schubert, 2009; Naeher et al., 2012). The HI values in the ZT and LA reservoirs were also similar (Fig. 5), suggesting their similar primary productivity, which was in accordance with their records of the S1 and S2 levels in the sediment profiles. For XFJ, the HI profile showed a different trend from the S1, S2, and TOC parameters, suggesting the relative increase of S2 (algal productivity) to the TOC in the sediment profile.

Table 1. Correlation between the Rock-Eval parameters in three reservoirs.

	S2a†	S1	S2	RC	TOC
			Zengtang reservoir		
S2a	1	0.995**	1.000**	0.974**	0.985**
S1		1	0.992**	0.954**	0.969**
S2			1	0.977**	0.988**
RC				1	0.998**
TOC					1
			Lian'an reservoir		
S2a	1	0.995**	0.999**	0.967**	0.975**
S1		1	0.997**	0.950**	0.961**
S2			1	0.963**	0.972**
RC				1	0.999**
TOC					1
			Xinfengjiang reservoir		
S2a	1	0.995**	0.987**	0.982**	0.986**
S1		1	0.995**	0.990**	0.988**
S2			1	0.987**	0.983**
RC				1	0.998**
TOC					1

* $P < 0.05$.

** $P < 0.01$.

† S1, the fraction of hydrocarbons released in an inert, O_2-free oven (100–650°C); S2, the fraction of hydrocarbons released in an oxidation oven (400–850°C); S3, CO and CO_2 released in pyrolysis stage; RC, refractory carbon; TOC, total organic carbon.

Relationship Between Neutral Sugar Concentrations and Algal Productivity

The vertical distribution of total concentrations of seven neutral monosaccharides (i.e., glucose, galactose, mannose, fucose, xylose, rhamnose, and arabinose) in the sediment cores of the three reservoirs is shown in Fig. 5. The concentrations at the ZT, LA, and XFJ reservoirs showed significant increases from the bottom to the top of the sediment cores ($P < 0.01$), ranging from 1.48–3.91, 0.31–4.41, and 0.56–1.62 mg/g, respectively (unpublished data, 2015) (Fig. 5). Total sugars were positively correlated with the content of S2a in the sediment cores at the ZT and LA reservoirs. The correlation coefficients for ZT and LA were 0.98 ($P < 0.01$) and 0.57 ($P < 0.01$), respectively, suggesting that the concentrations of total sugars were associated with the primary productivity in these reservoirs. The above result is consistent with that of principal component analysis (Table 2). For XFJ, the total sugar was poorly related to the content of S2a, but was well correlated with the HI index in the sediment core (Fig. 5). This observation provides further evidence that both the HI and total sugar could reflect the historical productivity in the oligotrophic reservoir although its TOC profile decreases from the bottom to the surface of the sediment core. The observed correlations suggest that total

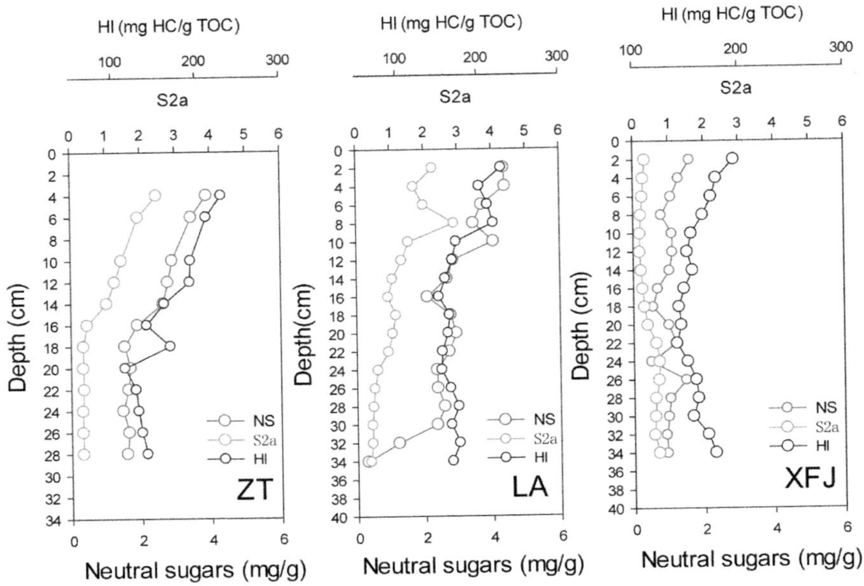

Fig. 5. Vertical profiles of neutral sugars (NS), S2a, and hydrogen index (HI) in sediment cores of three reservoirs.

Table 2. Loading of factors extracted by principal component analysis (PCA) in sediments from three reservoirs.

Factors	Zengtang	Lian'an	Xinfengjiang	
	PC1	PC1	PC1	PC2
Glucose	0.985	0.841	0.368	0.853
Galactose	0.988	0.964	0.950	0.041
Mannose	0.964	0.938	0.862	−0.398
Arabinose	0.984	0.984	0.975	0.056
Rahmnose	0.999	0.961	0.908	0.274
Fucose	0.953	0.974	0.883	−0.318
Xylose	0.989	0.751	0.013	0.788
S2a	0.995	0.809	−0.045	0.933

sedimentary carbohydrates originated from phytoplankton and are useful for indicating the historical primary productivity.

Molecular-level diagnostic parameters have often been applied to differentiate microbial, planktonic, and terrestrial sources (Cowie and Hedges, 1984; Guggenberger et al., 1994). For example, the ratios of mannose to xylose could indicate the OM sources derived from phytoplankton, bacteria, gymnosperm, and angiosperm tissues (Cowie and Hedges, 1984). Xylose% is a useful biomarker that is used to differentiate the inputs of aquatic and terrestrial inputs (Cowie and Hedges, 1984) and %(Ara + Gla)$_b$ ratios can indicate the presence of organic material derived from nonwoody angiosperm tissues, grasses, and phytoplankton (Cowie and Hedges, 1994).

As shown in Fig. 6, the values of mannose and xylose showed that most of the sediment core samples in ZT and LA were dominated with a phytoplankton source, suggesting the dominant input of algal organic matter in these two areas. However, the mannose/xylose ratio at XFJ was derived from gymnosperm tissues under the depth of 16 cm in the sediment core, which indicated the presence of terrestrial organic matter derived from angiosperm leaves and grasses. Low abundances of xyloses at ZT and LA also indicated the phytoplankton origin of neutral sugars. Further evidence in support of this conclusion can be obtained from the $\%(Ara + Gla)_b$. The ratios were around 40 in the sediment cores at ZT, LA, and XFJ (Fig. 6), indicating the important source of phytoplankton in these areas.

At the ZT and LA reservoirs, the main source for carbohydrates was phytoplankton, suggesting that the microbial degradation and decomposition of sedimentary sugars did not heavily contribute to OM during the deposition process, and some inert fractions were selectively preserved. In the oligotrophic reservoir, carbohydrate compounds might be transported and preserved without extensive degradation during the deposition process. The inputs of terrestrial plants and soil from the surrounding area were minimized if the sites were located in rural areas. In general, sedimentary carbohydrates were mainly derived from planktonic algae, and their concentration variations can reflect the historical primary productivity in the three investigated reservoirs.

Effects of Algae-Derived Organic Matter on Accumulation Of Heavy Metals in Reservoir Sediment Cores

Most of heavy metals were enriched in the upper layers of the sediment cores at the ZT and LA reservoirs (Table 3). Concentrations of Ni, Cu, Zn, Ag, Cd, and As at ZT significantly increased from the bottom to the top layers. For the XFJ reservoir, most metals varied slightly from the bottom to the top layers except As and Cd, and increased by 16.64-fold and 4.4-fold at the top layers compared with the bottom sections (Duan et al., 2014). Zengtang had a similar surface watershed area to LA. Since their location was close, they should have a similar atmospheric deposition of heavy metals. As ZT was shallower than LA, ZT was supposed to suffer from more serious heavy metal pollution. However, the levels of heavy metal contamination for the two reservoirs were similar. This may suggest that other internal factors in aquatic ecosystems were mainly responsible for the heavy metal pollution rather than anthropogenic activities.

To examine the internal factor (e.g., algal productivity) on the accumulation of heavy metals at the three reservoirs, Pearson correlation and principal component analysis were performed. Heavy metals were also normalized by Ti for removing the influences of geological and hydrodynamic processes. The increasing trends of S2a at ZT and LA were highly related to the variations of most metals, suggesting that AOM was likely to play an important role in accumulation of metals, especially in ZT reservoir (Duan et al., 2014).

As shown in Table 4, three principal components for ZT and LA, and four principal components for XFJ were extracted. At ZT, three principal components accounted for about 96% of the total variance. Based on the loading distribution of the variables, Ni, Cu, Zn, As, Ag, Cd, Mn, S1, S2, and TOC constituted a related

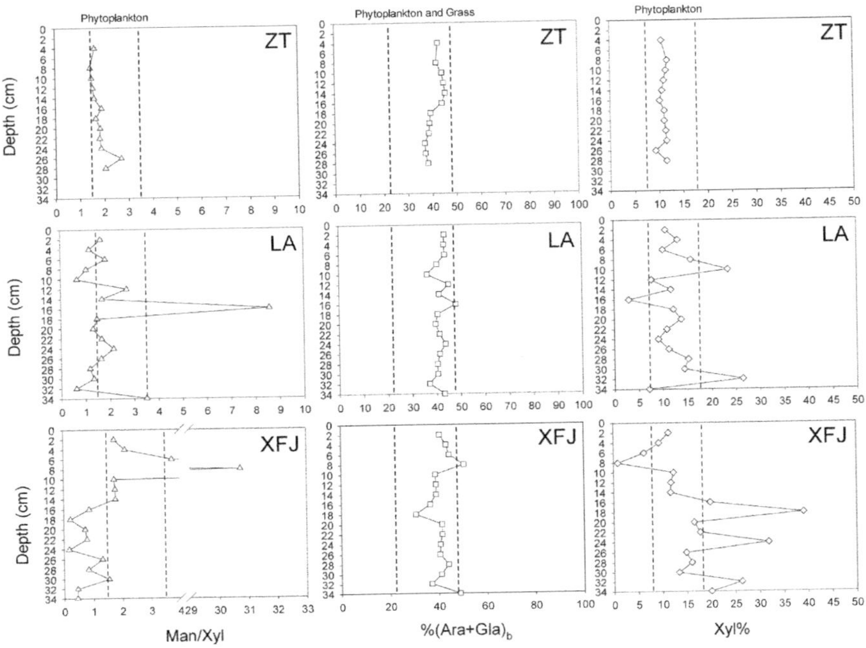

Fig. 6. Vertical profiles of mannose (Man) and xylose (Xyl), Xyl%, and %(Ara + Gla)$_b$ in sediment cores of three reservoirs.

group, while another group was composed of Cr, Co, and Ti. Moreover, after normalization with Ti, only two principal components were extracted in this reservoir. All of the trace elements (except for Pb) and S1, S2, and TOC were in the first principal component, which accounted for 72.41% of the total variance. The above results were consistent with the Pearson correlation analysis, indicating that heavy metals at ZT were accumulated by AOM during the process of settling.

The first principal component included Fe, Al, Cr, Co, Ni, Cu, Zn, As, Ag, Cd, Mn, S1, S2, and TOC at LA, accounting for 73.45% of the total variance and 72.41% of the variance after the Ti normalization, respectively (Duan et al., 2014). This observation showed that heavy metals may be strongly bound with algal organic matter and Fe, Al, and Mn oxides during the sedimentary process. Change in the oxidation and reduction of Fe, Al, and Mn oxides might also have an important effect on the bonding and releasing of heavy metals from sediments (Hübner et al., 2003; Lopez et al., 2006). For XFJ, the first principal component included Cr, Co, Ni, Cu, Zn, Cd, As, and Ag, accounting for 54.82% of the variance, whereas the second principal component included Mn, Fe, Al, Ag, S1, S2, and TOC, accounting for 18.85% of the variance. It confirmed the results from the Pearson correlation analysis, indicating that most heavy metal contents were not controlled by AOM.

Most of the heavy metals were strongly affected or controlled by AOM at the ZT and LA reservoirs in the last six decades. Concentrations of heavy metals were found to associate with the contents of AOM and Fe and Mn oxides at LA. On the contrary, most heavy metals at XFJ might mainly originate from the background input.

Table 3. Vertical distribution of heavy metals in sediment cores of reservoirs.

Depth	Cr	Co	Ni	Cu	Zn	As	Ag	Cd	Pb	Mn	Ti
cm	ng/g									μg/g	
Zengtang reservoir											
2	49.8	9.89	26.1	38.1	155	8.89	1.32	0.32	11.8	529	5.60
4	48.6	9.97	24.6	35.6	108	9.05	0.35	0.31	10.9	279	5.78
6	49.0	10.1	24.6	36.0	115	9.01	0.32	0.30	8.71	305	5.77
8	51.5	10.4	25.8	37.1	115	9.53	0.42	0.32	12.9	293	6.15
10	51.3	10.6	25.8	35.6	114	10.1	0.37	0.32	17.7	323	6.05
12	49.3	10.4	24.8	33.6	107	8.87	0.30	0.30	9.02	264	5.90
14	48.9	9.90	24.4	30.5	94.8	8.52	0.27	0.26	10.9	209	6.00
16	47.9	9.93	23.2	25.3	93.5	6.99	0.23	0.15	4.73	166	6.00
18	50.1	10.7	25.3	27.8	93.3	6.69	0.15	0.14	7.35	233	7.07
20	46.4	9.89	22.0	23.4	83.0	6.61	0.21	0.14	8.99	336	5.98
22	48.3	10.2	23.4	25.6	92.3	5.86	0.14	0.15	6.91	280	6.31
24	54.6	11.7	26.4	25.3	100	7.38	0.18	0.18	11.4	203	7.36
26	50.7	10.8	23.3	22.7	93.9	7.24	0.18	0.14	14.6	159	7.02
28	51.6	11.1	23.6	24.6	97.4	6.71	0.17	0.13	11.6	143	6.93
Lian'an reservoir											
2	81.9	18.3	45.4	49.8	149	45.4	0.22	0.46	12.3	336	4.81
4	70.5	18.5	40.4	43.6	143	45.8	0.19	0.45	16.4	328	4.18
6	73.0	15.1	38.2	41.9	140	46.6	0.22	0.50	14.2	262	4.70
8	65.4	17.0	35.4	44.3	139	59.5	0.25	0.50	11.5	444	3.89
10	74.5	17.3	38.8	43.0	124	51.9	0.22	0.28	9.98	362	4.80
12	77.6	17.0	40.5	44.7	121	46.4	0.22	0.26	8.19	276	4.78
14	73.6	15.3	38.1	39.9	119	39.5	0.25	0.22	10.0	272	4.72
16	76.2	14.5	39.0	40.0	112	34.3	0.22	0.19	7.82	269	4.87
18	77.5	16.4	39.8	42.0	118	36.9	0.25	0.22	6.11	271	5.08
20	73.1	17.0	40.0	40.6	122	31.3	0.24	0.23	8.13	324	4.75

Table continued.

Table 3. Continued.

Depth	Cr	Co	Ni	Cu	Zn	As	Ag	Cd	Pb	Mn	Ti
cm	ng/g									µg/g	
					Lian'an reservoir						
22	75.0	15.4	40.7	40.2	127	25.2	0.22	0.19	7.37	322	4.92
24	50.1	8.76	24.7	25.0	83.7	7.76	0.11	0.10	10.0	157	3.80
26	43.0	6.44	20.3	18.2	76.6	5.56	0.08	0.08	8.7	126	3.54
28	44.0	6.40	20.3	18.2	74.2	5.49	0.08	0.08	8.58	112	3.82
30	43.9	6.46	20.4	18.0	75.2	4.54	0.08	0.08	9.73	136	3.61
32	44.1	6.51	20.5	18.8	90.0	4.37	0.09	0.09	9.80	149	3.59
34	64.3	9.25	30.3	25.9	107	6.15	0.08	0.12	11.7	142	6.22
					Xinfengjiang reservoir						
2	31.6	18.7	20.3	33.9	130	69.3	0.11	0.44	19.0	549	10.9
4	28.8	17.7	18.1	28.7	117	45.9	0.09	0.41	18.9	380	11.4
6	36.4	15.2	22.5	20.5	91.9	33.7	0.08	0.28	8.93	317	10.6
8	35.0	14.8	20.0	22.4	97.9	32.6	0.10	0.26	13.1	270	10.2
10	35.1	14.9	20.1	22.5	98.1	32.7	0.10	0.26	13.1	270	10.3
12	37.1	15.4	19.5	19.6	88.1	25.0	0.09	0.18	15.7	188	9.06
14	38.6	13.1	19.1	20.0	97.3	25.7	0.09	0.25	14.9	209	5.81
16	35.8	13.4	18.2	18.0	89.8	28.2	0.09	0.16	11.3	276	6.50
18	33.5	12.6	17.7	17.6	83.0	25.3	0.08	0.12	12.3	299	6.15
20	33.8	14.0	18.7	19.0	88.0	23.7	0.09	0.13	18.8	384	5.63
22	38.3	15.8	21.1	21.8	95.4	21.1	0.12	0.16	20.6	637	5.22
24	35.3	14.9	20.1	20.4	92.3	20.1	0.12	0.16	17.6	1090	5.90
26	20.6	13.2	12.8	16.0	93.1	10.9	0.09	0.11	17.0	453	6.30
28	18.7	11.8	15.3	16.2	100	7.91	0.08	0.10	9.65	262	6.93
30	18.2	11.5	12.6	15.1	95.7	6.13	0.08	0.10	13.8	227	6.92
32	18.1	11.4	12.8	14.6	73.4	5.13	0.07	0.09	9.03	202	6.53
34	19.3	12.1	13.9	17.0	84.2	4.93	0.09	0.10	8.59	244	7.21

Table 4. Loading of factors extracted by principal component analysis in sediments from three reservoirs (reprinted from the supporting data of Duan et al. 2014).

	Zengtang reservoir					
Variable	PC1 (54.63%)	PC2 (24.33%)	PC3 (7.90%)	PC1^N (72.41%)	PC2^N (12.53%)	
Fe	0.34	−0.63	0.46	0.41	0.79	
Al	0.04	−0.74	0.37	0.14	0.88	
Cr	−0.07	0.95	0.15	0.97	0.11	
Co	−0.48	0.83	0.11	0.91	0.17	
Ni	0.56	0.66	0.17	0.98	0.09	
Cu	0.95	0.09	0.17	0.97	0.00	
Zn	0.89	0.28	−0.31	0.94	−0.20	
As	0.83	0.22	0.44	0.93	0.02	
Ag	0.82	0.06	−0.45	0.79	−0.26	
Cd	0.93	0.10	0.29	0.95	−0.04	
Pb	0.31	0.61	0.31	0.49	−0.38	
Mn	0.80	−0.16	−0.30	0.84	0.03	
Ti	−0.75	0.62	−0.05	−0.83	−0.42	
TOC	0.99	0.01	0.03	0.98	−0.14	
S2a	0.98	0.05	−0.10	0.96	−0.22	
	Lian'an reservoir					
Variable	PC1 (73.45%)	PC2 (13.88%)	PC3 (7.89%)	PC1^N (72.41%)	PC2^N (12.53%)	PC1 (54.82%)
Fe	0.91	−0.26	−0.23	0.89	−0.39	−0.05
Al	0.66	−0.57	−0.41	0.65	−0.61	−0.64
Cr	0.89	−0.29	0.33	0.96	−0.11	0.82
Co	0.98	−0.11	0.07	0.98	−0.06	0.90
Ni	0.93	−0.20	0.26	0.96	−0.08	0.87
Cu	0.98	−0.10	0.09	0.99	−0.04	0.87
Zn	0.96	0.14	0.19	0.90	0.37	0.69
As	0.97	0.05	−0.14	0.97	−0.03	0.92
Ag	0.92	−0.30	−0.11	0.92	−0.26	0.59
Cd	0.87	0.47	−0.02	0.90	0.37	0.85
Pb	0.28	0.87	0.24	0.20	0.94	0.56
Mn	0.94	−0.05	−0.21	0.93	0.03	0.26
Ti	0.41	−0.41	0.77	0.14	−0.63	−0.83
TOC	0.94	0.19	−0.22	0.97	0.02	−0.79
S2a	0.86	0.38	−0.19	0.92	0.21	−0.81
	Xinfengjiang reservoir					
Variable	PC2 (18.85%)	PC3 (10.59%)	PC4 (8.87%)	PC1^N (67.45%)	PC2^N (16.45%)	PC3^N (7.5%)
Fe	0.46	−0.42	0.74	0.02	0.67	0.68
Al	0.31	−0.17	0.59	−0.55	0.53	0.52
Cr	−0.08	−0.54	−0.12	0.91	0.11	−0.06
Co	0.28	0.21	0.00	0.98	0.15	−0.04
Ni	0.03	−0.39	0.05	0.95	0.15	0.02
Cu	0.24	0.37	0.13	0.95	0.10	0.04
Zn	0.31	0.52	0.22	0.96	0.13	0.05
As	0.05	0.21	0.25	0.89	−0.03	0.22
Ag	0.62	−0.16	−0.33	0.89	0.36	−0.21
Cd	0.03	0.40	0.27	0.84	−0.11	0.24
Pb	0.55	0.09	−0.40	0.81	0.34	−0.37
Mn	0.87	−0.28	−0.07	0.46	0.82	−0.21
Ti	0.02	0.49	0.06	−0.95	−0.12	0.05
TOC	0.59	0.02	−0.14	−0.78	0.57	−0.25
S2a	0.55	0.15	−0.07	−0.83	0.50	−0.17

Effects of Climate-Driven Algae-Derived Organic Matter on Accumulation of PAHs

Effects of Algae-Derived Organic Matter on the Cycling of Hydrophobic Pollutants in the Sediment Cores

The total PAH concentrations were estimated by the sum of 15 USEPA priority PAHs, except for naphthalene. Polycyclic aromatic hydrocarbons at ZT and LA were similarly enriched from the bottom to the top of the sediment cores (Fig. 7). However, PAHs at XFJ decreased since 1950, except for two intervals from 1969 to 1981 and from 1986 to 1995.

When PAHs enter aquatic ecosystems they are easily sorbed or absorbed by algal biomass because of their hydrophobicity. After algae die, their AOM together with PAHs eventually settle down into the sediments. The scavenging effect would be enhanced by the high algal biomass, and more hydrophobic compounds would be accumulated by AOM. To investigate the algal effect, PAHs normalized by Ti were compared with S2a in the three reservoirs. The vertical profiles of S2a and PAHs and Ti showed a significant correlation relationship in the sediment cores at both the ZT (R^2 = 0.832, P < 0.01) and LA reservoirs (R^2 = 0.792, P < 0.01) (Fig. 7) (unpublished data, 2015), suggesting that the inputs of PAHs are extensively affected by AOM. For XFJ, the vertical profiles of S2a and PAHs and Ti were also interrelated (R^2 = 0.60, P < 0.01), excluding the 15-cm sample. This abnormal data point at 15 cm was likely caused by redox change or terrestrial input change. Therefore, the above result indicated that the algal productivity was the major factor in controlling the deposition and preservation of PAHs in both eutrophic lakes and meso-trophic lakes

Effects of climate Change on Algae-Derived Organic Matter and Pollutants in Sediment Cores

Evidence recently obtained from some lakes in the Canadian High Arctic demonstrate that the dynamics and fate of Hg and persistent organic pollutants may have been influenced by climate-driven increases in algal productivity (Carrie et al., 2010; Outridge et al., 2007; Stern et al., 2009). However, these investigators also suggested that the climate effects would diminish at low-altitude areas because climate impacts on lake limnology would be more muted (Stern et al., 2009). But global warming could also take place on subtropical lakes, as observed for thermocline dynamics (Hambright et al., 1994). As few studies have been conducted on the climate warming effects in subtropical lakes and reservoirs, we applied the meteorological data of the China Meteorological Data Sharing Service System to estimate the climate change during the last six decades. When 5-yr moving temperature (T_5) and mean annual air temperature were estimated during this period, the results showed that the mean annual air temperature in the Guangzhou area had increased by about 1.5°C since 1970, but showed significant historical low values around 1970, 1986, and 1996 (Fig. 7). The mean annual air temperature in the Heyuan area had increased by about 1.52°C between 1957 and 2004 (unpublished data, 2015).

The changing trend of annual mean temperature and 5-yr moving average temperature over 60 yr showed significant correlations ($P \leq 0.05$) with the historical variations of PAHs and S2a in the three reservoirs (Table 5). At ZT and

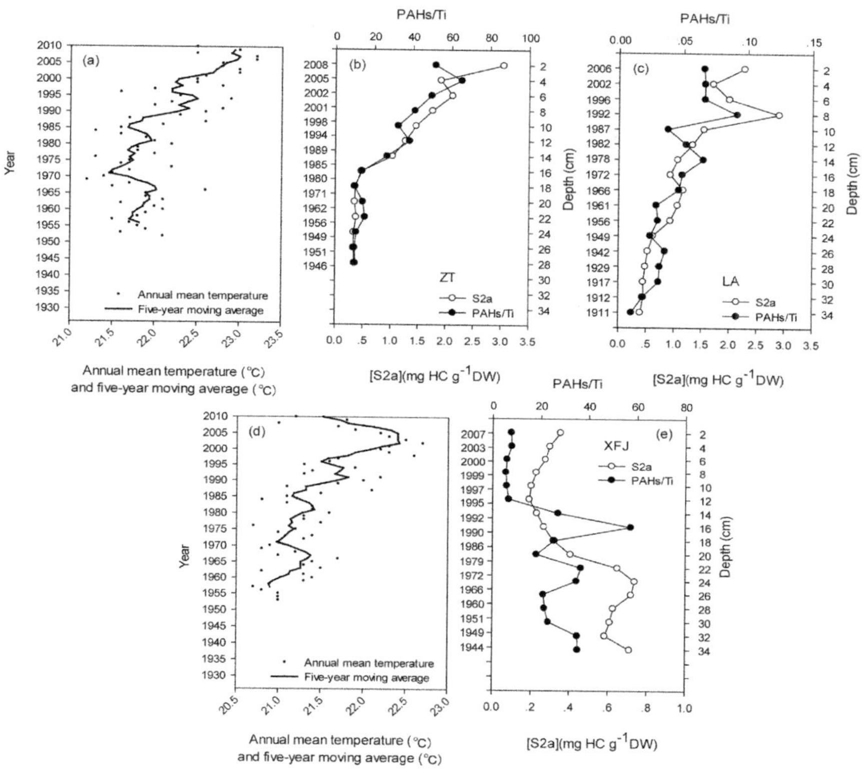

Fig. 7. (a) Annual mean temperature and 5-yr moving average (T_5) at Guangzhou area; (d) annual mean air temperature at Guangzhou area; (b, c, and e) Point-by-point comparison between polycyclic aromatic hydrocarbon [PAHs]/[Ti] and [S2a] carbon concentrations at ZT, LA, and XFJ, respectively. The normalized contents of PAHs by titanium are represented by [PAHs]/[Ti]. A significant increase in air temperature was observed in the period of 1971 to 2010 at Guangzhou area (y = 0.0378x – 52.827; R^2 = 0.89) and in the period of 1957 to 2004 at Heyuan area (y = 0.0249x + 20.817; R^2 = 0.72).

LA, PAHs and S2a were strongly associated with increasing T_5 during the past 60 yr, suggesting that climate warming also affected the deposition of hydrophobic contaminants by enhancing the scavenging of autochthonous algae in the reservoirs (Carrie et al., 2010); although, anthropogenic pollution undoubtedly contributed to the part of the increasing PAHs at ZT and LA. However, our findings suggested that the increasing PAHs were mainly attributed to the climate change in this area (unpublished data, 2015). As the XFJ reservoir was obviously affected by a limited phosphorus input, the primary productivity and algal biological pump effect on PAHs was very low. Hence, the profiles of PAHs and S2a showed a negative correlation with the increasing temperature during the last four decades. However, a strong relationship between T_5 and HI (S2/TOC × 100) at XFJ were observed, suggesting that S2a was still elevated relative to the decreasing TOC contents in the profile. Hence, the AOM fraction was significantly affected by the temperature change even though it was limited by nutrient condition. The HI index could be a more appropriate indicator for assessing the status of aquatic productivity in the reservoir regions. The above results suggest that climate warming plays a significant role in increasing primary productivity at the investigated reservoirs.

Conclusions

Sediment cores can provide valuable historical information for long-term research on the biological pump effect. Rock-Eval pyrolysis combined with the molecular-level analysis of biomarkers (neutral sugars) made the identification and characterization of AOM convincing. It was found that positive relationships between AOM and typical pollutants (PAHs and heavy metals) were highly significant in the investigated sediment cores in the three subtropical reservoirs, suggesting that historical algal productivity was primarily responsible for the deposition and fate of PAHs and heavy metals. The process of algal scavenging eventually led to the enhanced rates of pollutant deposition. Since the pollutants in sediment cores had been widely used as historical archives to estimate the impact of human activities, it was suggested the overestimation of anthropogenic deposition in the past investigations if this process of algal scavenging had not been taken into account.

Primary productivity in natural waters could be affected by many factors, including nutrient inputs, hydrological changes, temperature variations, etc. In this case study, it was demonstrated that the historical variations of primary productivity in subtropical reservoirs were significantly influenced by the increasing temperature during the last 60 yr. Simultaneously, the accumulation of AOM and pollutants (PAHs and heavy metals) was enhanced. The investigation in this chapter would help to understand the importance of the biological pump process compared with anthropogenic inputs in subtropical regions. Further research is needed on the effects of different algal populations–organic matter on pollutant accumulation by analysis of multiple biomarkers. More investigations are also needed for the demonstration of climate warming effects on primary productivity in aquatic environments at low latitude.

Acknowledgments

This study was supported by a key project of NNSFC-Guangdong (U1201235) and a "Team Project" of National Natural Science Foundation of China (41121063), and a Key Field Project, Chinese Academy of Sciences (Y234081A07). This is contribution No. IS-1700 from GIGCAS.

References

Bechtel, A., and C.J. Schubert. 2009. A biogeochemical study of sediments from the eutrophic Lake Lugano and the oligotrophic Lake Brienz, Switzerland. Org. Geochem. 40:1100–1114. doi:10.1016/j.orggeochem.2009.06.009

Berrojalbiz, N., J. Dachs, S. Del Vento, M.J. Ojeda, M.C. Valle, J. Castro-Jiménez, G. Mariani, J. Wollgast, and G. Hanke. 2011. Persistent organic pollutants in Mediterranean seawater and processes affecting their accumulation in plankton. Environ. Sci. Technol. 45:4315–4322. doi:10.1021/es103742w

Table 5. Correlation between the 5-yr average temperature and Rock-Eval parameters in three reservoirs.

	T_5†	HI	S2a	S2	PAHs	PAHs/Ti
			Zengtang reservoir			
T_5	1	0.950**	0.931**	0.937**	0.951**	0.948**
HI		1	0.947**	0.953**	0.878**	0.880**
S2a			1	0.999**	0.889**	0.898**
S2				1	0.895**	0.903**
PAHs					1	0.998**
PAHs/Ti						1
			Lian'an reservoir			
T_5	1	0.866**	0.702*	0.719*	0.503	0.513
HI		1	0.924**	0.937**	0.706*	0.733*
S2a			1	0.999**	0.717*	0.798**
S2				1	0.712*	0.785**
PAHs					1	0.967**
PAHs/Ti						1
			Xinfengjiang reservoir			
T_5	1	0.899**	−0.602*	−0.615*	−0.576*	−0.463
HI		1	−0.433	−0.435	−0.626*	−0.584*
S2a			1	0.996**	0.549	0.388
S2				1	0.541	0.367
PAHs					1	0.970**
PAHs/Ti						1

* $P < 0.05$

** $P < 0.01$.

† T_5, 5-yr moving temperature; HI, hydrogen index; PAH, polycyclic aromatic hydrocarbon.

Carrie, J., H. Sanei, and G. Stern. 2012. Standardisation of Rock-Eval pyrolysis for the analysis of recent sediments and soils. Org. Geochem. 46:38–53. doi:10.1016/j.orggeochem.2012.01.011

Carrie, J., F. Wang, H. Sanei, R.W. Macdonald, P.M. Outridge, and G.A. Stern. 2010. Increasing contaminant burdens in an Arctic fish, Burbot (*Lota lota*), in a warming climate. Environ. Sci. Technol. 44:316–322. doi:10.1021/es902582y

Cowie, G.L., and J.I. Hedges. 1984. Determination of neutral sugars in plankton, sediments, and wood by capillary gas chromatography of equilibrated isomeric mixtures. Anal. Chem. 56:497–504. doi:10.1021/ac00267a046

Cowie, G.L., and J.I. Hedges. 1994. Biochemical indicators of diagenetic alteration in natural organic matter mixtures. Nature 369:304–307.

Dachs, J., S.J. Eisenreich, J.E. Baker, F.C. Ko, and J.D. Jeremiason. 1999. Coupling of phytoplankton uptake and air-water exchange of persistent organic pollutants. Environ. Sci. Technol. 33:3653–3660. doi:10.1021/es990168o

Dachs, J., S.J. Eisenreich, and R.M. Hoff. 2000. Influence of eutrophication on air-water exchange, vertical fluxes, and phytoplankton concentrations of persistent organic pollutants. Environ. Sci. Technol. 34:1095–1102. doi:10.1021/es990759e

Duan, D., Y. Ran, H. Cheng, J.A. Chen, and G. Wan. 2014. Contamination trends of trace metals and coupling with algal productivity in sediment cores in Pearl River Delta, South China. Chemosphere 103:35–43. doi:10.1016/j.chemosphere.2013.11.011

Galbán-Malagón, C., N. Berrojalbiz, M.-J. Ojeda, and J. Dachs. 2012. The oceanic biological pump modulates the atmospheric transport of persistent organic pollutants to the Arctic. Nat. Commun. 3:862. doi:10.1038/ncomms1858

Guggenberger, G., B.T. Christensen, and W. Zech. 1994. Land-use effects on the composition of organic matter in particle-size separates of soil: I. Lignin and carbohydrate signature. Eur. J. Soil Sci. 45:449–458. doi:10.1111/j.1365-2389.1994.tb00530.x

Hambright, K.D., M. Gophen, and S. Serruya. 1994. Influence of long-term climatic changes on the stratification of a subtropical, warm monomictic lake. Limnol. Oceanogr. 39:1233–1242. doi:10.4319/lo.1994.39.5.1233

Henderson, R.K., A. Baker, S.A. Parsons, and B. Jefferson. 2008. Characterisation of algogenic organic matter extracted from cyanobacteria, green algae and diatoms. Water Res. 42:3435–3445. doi:10.1016/j.watres.2007.10.032

Hübner, A., G. De Lange, J. Dittmer, and P. Halbach. 2003. Geochemistry of an exotic sediment layer above sapropel S-1: Mud expulsion from the Urania Basin, eastern Mediterranean? Mar. Geol. 197:49–61. doi:10.1016/S0025-3227(03)00085-9

Keely, B.J., and R.G. Brereton. 1986. Early chlorin diagenesis in a recent aquatic sediment. Org. Geochem. 10:975–980. doi:10.1016/S0146-6380(86)80035-3

Kirk, J.L., D.C.M. Muir, D. Antoniades, M.S.V. Douglas, M.S. Evans, T.A. Jackson, H. Kling, S. Lamoureux, D.S.S. Lim, R. Pienitz, J.P. Smol, K. Stewart, X.W. Wang, and F. Yang. 2011. Climate change and mercury accumulation in Canadian high and subarctic lakes. Environ. Sci. Technol. 45:964–970. doi:10.1021/es102840u

Ko, F.-C., and J.E. Baker. 1995. Partitioning of hydrophobic organic contaminants to resuspended sediments and plankton in the mesohaline Chesapeake Bay. Mar. Chem. 49:171–188. doi:10.1016/0304-4203(94)00064-K

Lafargue, E., F. Marquis, and D. Pillot. 1998. Rock-Eval 6 applications in hydrocarbon exploration, production, and soil contamination studies. Rev. Inst. Fr. Pet. 53:421–437.

Liu, Y.F., J.Q. Qin, and Y.S. Liang. 2011. Analysis of aquatic ecosystem of Zeng Tang reservoir from virtuous cycle to eutrophication. Guangdong Chem. Indus. 7:53.

Lopez, P., E. Navarro, R. Marce, J. Ordoñez, L. Caputo, and J. Armengol. 2006. Elemental ratios in sediments as indicators of ecological processes in Spanish reservoirs. Limnetica 25:499–512.

Naeher, S., R.H. Smittenberg, A. Gilli, E.P. Kirilova, A.F. Lotter, and C.J. Schubert. 2012. Impact of recent lake eutrophication on microbial community changes as revealed by high resolution lipid biomarkers in Rotsee (Switzerland). Org. Geochem. 49:86–95. doi:10.1016/j.orggeochem.2012.05.014

Outridge, P., H. Sanei, G. Stern, P. Hamilton, and F. Goodarzi. 2007. Evidence for control of mercury accumulation rates in Canadian High Arctic lake sediments by variations of aquatic primary productivity. Environ. Sci. Technol. 41:5259–5265. doi:10.1021/es070408x

Outridge, P., G. Stern, P. Hamilton, J. Percival, R. McNeely, and W. Lockhart. 2005. Trace metal profiles in the varved sediment of an Arctic lake. Geochim. Cosmochim. Acta 69:4881–4894. doi:10.1016/j.gca.2005.06.009

Pivokonsky, M., O. Kloucek, and L. Pivokonska. 2006. Evaluation of the production, composition and aluminum and iron complexation of algogenic organic matter. Water Res. 40:3045–3052. doi:10.1016/j.watres.2006.06.028

Richer, G., and R. Peters. 1993. Determination of the short-term dynamics of PCB uptake by phytoplankton. Environ. Toxicol. Chem. 12:207–218. doi:10.1002/etc.5620120203

Sanei, H., and F. Goodarzi. 2006. Relationship between organic matter and mercury in recent lake sediment: The physical-geochemical aspects. Appl. Geochem. 21:1900–1912. doi:10.1016/j.apgeochem.2006.08.015

Sanei, H., L.D. Stasiuk, and F. Goodarzi. 2005. Petrological changes occurring in organic matter from Recent lacustrine sediments during thermal alteration by Rock-Eval pyrolysis. Org. Geochem. 36:1190–1203. doi:10.1016/j.orggeochem.2005.02.009

Skoglund, R.S., K. Stange, and D.L. Swackhamer. 1996. A kinetics model for predicting the accumulation of PCBs in phytoplankton. Environ. Sci. Technol. 30:2113–2120. doi:10.1021/es950206d

Stern, G.A., H. Sanei, P. Roach, J. Dalaronde, and P.M. Outridge. 2009. Historical interrelated variations of mercury and aquatic organic matter in lake sediment cores from a subarctic lake in Yukon, Canada: Further evidence toward the algal-mercury scavenging hypothesis. Environ. Sci. Technol. 43:7684–7690. doi:10.1021/es902186s

Swackhamer, D.L., and R. Skoglund. 1993. Bioaccumulation of PCBs by algae: Kinetics versus equilibrium. Environ. Toxicol. Chem. 12:831–838. doi:10.1002/etc.5620120506

Taylor, W., J. Carey, D. Lean, and D. McQueen. 1991. Organochlorine concentrations in the plankton of lakes in southern Ontario and their relationship to plankton biomass. Can. J. Fish. Aquat. Sci. 48:1960–1966. doi:10.1139/f91-233

Wu, F.C., L.B. Xu, H.Q. Liao, F. Guo, X.L. Zhao, and J.P. Giesy. 2013. Relationship between mercury and organic carbon in sediment cores from Lakes Qinghai and Chenghai, China. J. Soils Sediments 13:1084–1092. doi:10.1007/s11368-013-0694-2

Wu, F.C., L.B. Xu, Y.G. Sun, H.Q. Liao, X.L. Zhao, and J.Y. Guo. 2012. Exploring the relationship between polycyclic aromatic hydrocarbons and sedimentary organic carbon in three Chinese lakes. J. Soils Sediments 12:774–783. doi:10.1007/s11368-012-0498-9

Xu, L.B., F.C. Wu, G.J. Wan, H.Q. Liao, X.L. Zhao, and B.S. Xing. 2011. Relationship between $^{210}Pb_{ex}$ activity and sedimentary organic carbon in sediments of 3 Chinese lakes. Environ. Pollut. 159:3462–3467. doi:10.1016/j.envpol.2011.08.020

Characteristics and Functions of Labile Organic Carbon in Coastal Wetland Soils of the Mississippi River Deltaic Plain

Jim J. Wang,* Syam K. Dodla, and Ronald D. DeLaune

Abstract

The Mississippi River deltaic plain (MRDP) coastal wetlands are one of the largest in the United States. Adequate characterization of labile organic C (LOC) is essential for understanding the C transformation and overall biogeochemistry in these wetland soils. Among different MRDP wetlands, freshwater and brackish marsh soils contain higher LOC than saline marsh and forest swamp soils. Labile organic C estimated by aerobic incubation and salt and cold water extractions is generally less than 0.5% of soil organic C (SOC), whereas hot water extraction and especially acid hydrolysis yield 1.2 to 3.49% and 12.8 to 43% of SOC, respectively. Carboxyl C was the primary component of LOC by cold water and salt extractions, whereas both carboxyl C and O/N-alkyl C were released by aerobic mineralization. On the other hand, LOC by hot water extraction contains mainly carbonyl C, whereas acid hydrolysis includes the contribution of alkyl C besides O/N-alkyl C. In addition, higher LOC in freshwater marsh soils than in saline and forest swamp soils facilitates higher denitrification potential of the former, which is negatively correlated with the presence of phenolic aldehydes and ketones in these wetland soils. The addition of clay sediments reduce LOC degradation as is evident by lowered CO_2 emissions. Monovalent K^+ holds more DOC in forest swamp soils, whereas divalent Ca^{2+} retains more DOC in marsh soils. Selective degradation of syringyl lignin structure over guaiacyl occurs in both dissolved organic matter (DOM) and bulk soil organic matter (SOM) in the MRDP wetlands, which could enhance overall SOC lability and potentially contribute to the northern Gulf of Mexico hypoxia if exported offshore.

Abbreviations: AHC, acid hydrolyzable carbon; AMC, aerobically mineralizable carbon; BM, brackish marsh; CWEC, cold water extractable carbon; DOC, dissolved organic carbon; DOM, dissolved organic matter; FM, freshwater marsh; FS, forest swamp; HWEC, hot water extractable carbon; LOC, labile organic carbon; MRDP, Mississippi River deltaic plain; NOM, natural organic matter; PDR, potential denitrification rate; POM, particulate organic matter; SEC, salt extractable carbon; SM, saline marsh; SOC, soil organic carbon; SOM, soil organic matter.

Jim J. Wang and Syam K. Dodla, School of Plant, Environmental and Soil Sciences, 104 Sturgis Hall, Louisiana State Univ. Agricultural Center, Baton Rouge, LA 70803 (sdodla@agcenter.lsu.edu). Ronald D. DeLaune, Dep. of Oceanography and Coastal Sciences, Louisiana State Univ., Baton Rouge, LA 70803 (rdelaun@lsu.edu). *Corresponding author (jjwang@agcenter.lsu.edu).

doi:10.2136/sssaspecpub62.2014.0046

Labile Organic Matter—Chemical Compositions, Function, and Significance in Soil and the Environment. SSSA Special Publication 62. Zhongqi He and Fengchang Wu, editors.

Wetlands, despite comprising a relatively small portion of the total land area, consist of 20 to 25% of the terrestrial soil C and play a major role in the global C cycle (Amthor and Huston, 1998; Reddy and DeLaune, 2008). High rates of organic matter input coupled with slow mineralization rates have made wetland soils an important C sink for atmospheric CO_2. Low oxygen, acidic peat, and lack of nutrients are among the major factors contributing to the slow mineralization of SOM in wetlands (Bridgham and Richardson, 2003; Yavitt et al., 2004). On the other hand, the amount of C sequestered in these soils could be significantly impacted by relative distribution of SOC fractions with different lability for microbial mineralization (Rovira and Vallejo, 2007; Silveira et al., 2008; Wang et al., 2011). Therefore, adequate characterization of wetland soil LOC is essential to the understanding of the C transformation and overall biogeochemistry in these soils (Dodla et al., 2012b).

The MRDP coastal wetlands are one of the largest in the United States. These wetlands stretch over a land area of 11,000 km^2 from Vermillion Bay east to the Chandeleur Islands on the southern coast of Louisiana and account for 37% of the estuarine marsh in the conterminous United States (Fig. 1). The MRDP coastal wetlands represent one of most complex and anthropogenically altered coastal ecosystems in the world (Bianchi et al., 2011). While these wetlands were created over the past several thousand years through a series of delta lobes, recent history of both anthropogenic and natural effects has caused a significant loss of these wetlands at a rate of 26 to 30 km^2 yr^{-1} (Barras, 2006). The wetland loss is primarily a result of subsidence in the MRDP plus a similar rate of eustatic sea level rise. On the other hand, because of the construction of a levee along the Mississippi River for flood prevention, since the 1930s significant deterioration of these wetlands had begun because of the shortage of normal sediment inputs that would be otherwise historically associated with spring floods. Sediment inputs and organic

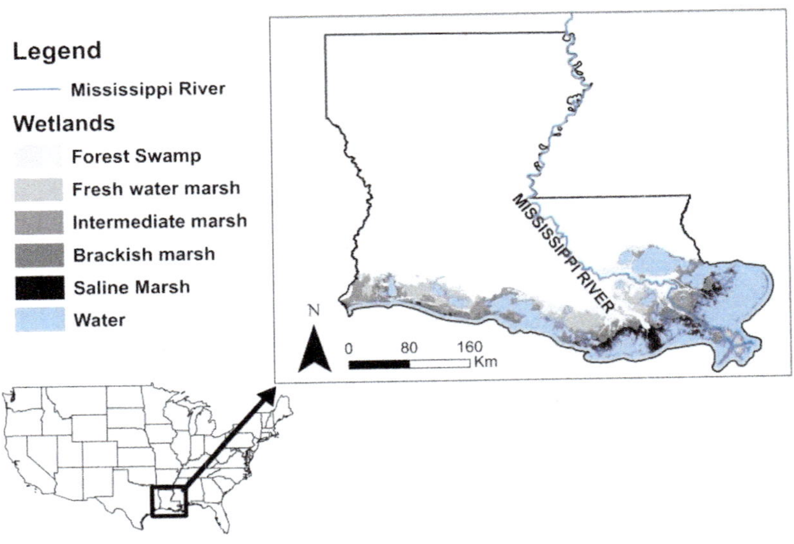

Fig. 1. The distribution of the Mississippi River deltaic plain (MRDP) wetlands.

matter accretion have been considered as the two most important factors for the stability of these wetlands (DeLaune and Pezeshki, 2003; Wang et al., 2011). For these reasons a series of diversion projects have been implemented to reintroduce freshwater and sediments from the Mississippi River into these targeted wetlands as part of a Louisiana coastal restoration process (Lane et al., 1999; Reddy and DeLaune , 2008). The MRDP wetlands are viewed as a case example of what may occur in other coastal areas in response to global warming and the accompanying increase in sea level (Reddy and DeLaune, 2008).

The MRDP coastal marshes are among the most productive ecosystems in the world. These coastal marshlands sequester large amounts of organic C through vertical accretion (Nyman et al., 1990). A significant portion of C previously sequestered in these soil profiles is, however, being lost through marsh deterioration (DeLaune and White, 2011). The DOM and sediment-associated particulate organic matter (POM) lost from marshes could be a significant source of organic C in fueling hypoxia in the northern Gulf of Mexico (Wilson and Allison, 2008; Bianchi et al., 2009, 2010). On the other hand, the majority of previous SOC characterizations, particularly LOC, were conducted on upland soils, and very little work has been done on wetland soils, especially at coastal zones, where soils are constantly influenced by different environmental factors such as redox fluctuations and salt water intrusions. Therefore, understanding these different organic C fractions is important for predicting and managing the contribution of these C sources in this highly dynamic region.

Soil Organic Carbon Pools and Lability

Extensive research has been performed characterizing the nature and transformation of SOC. Based on resistance to mineralization, SOM has been traditionally divided into labile, intermediate, and recalcitrant organic C pools (Torn et al., 2009). The soil LOC pool is the most active fraction of SOC, with rapid turnover rates from days to a few years, and it governs production and flux of CO_2 from soils to atmosphere (Silveira et al., 2008; Zou et al., 2005). While the lability of SOM has been found to be dependent on both molecular recalcitrance and physical protection from microbial degradation (McLauchlan and Hobbie, 2004; Rovira and Vallejo, 2007; Silveira et al., 2008), recent evidence has pointed primarily to the physicochemical stabilization processes through physical aggregation (Six et al., 2002) and chemical interaction with soil–sediment mineral surfaces of especially clay minerals and aluminum–iron oxides and with metal ions (Keil et al., 1994; von Lützow et al., 2007; Kleber 2010; Sanderman et al., 2014). Based on this concept shift, a labile molecule can become nonlabile because of physicochemical condition change regardless of its intrinsic nature of lability (Cuevas et al., 2011; Dodla et al., 2012a). Microbial activity is considered as the primary factor controlling the decomposition and stabilization of SOM, and all organic materials including those of chemically complex structures such as lignin can be decomposed if under "optimal" conditions (e.g., sufficient oxygen supply) (Chabbi and Rumpel, 2009; Schmidt et al., 2011; Stockmann et al., 2013). The perceived inherent resistance of some forms of organic matter is no longer considered as a material property but as a logistical problem constrained by microbial ecology, enzyme kinetics, environmental drivers, and matrix protection (Kleber, 2010). These conclusions appear to support findings of stable isotope research in that the chemical composition of the SOM has no correlation with the residence time of SOM (Amelung et al., 2008). In addition, molecular characterization of old and new SOM showed insignificant difference between their compositions (Paul et al., 2006). Nonetheless, it is well known that under oxic conditions, organic compounds with a simple structure such as polysaccharides followed by proteins and lipids are generally rapidly utilized by microbes (Harvey et al., 1995), whereas other compounds with high molecular weight, irregular structure, and aromatic structure, which were termed "recalcitrant," take longer times (Krull et al., 2003; Dodla et al., 2012a). These results suggest that the structural complexity of a given C compound may affect the kinetics of mineralization, but environmental conditions along with type and quantity of microbes present in an ecosystem determine the long-term persistence of SOC. Regardless of how it is operationally defined, the LOC fraction that has a turnover time of less than a few years, as compared with "recalcitrant" C fraction with a turnover time of several hundreds to thousands of years, can respond rapidly to changes in C supply, and, therefore, can be considered to be one of the most important indicators of soil quality (Parton et al., 1987; Haynes, 2005; Zhang et al., 2006).

Wetland Soil Organic Matter Characteristics

Research based on operationally defined selective extractions has shown that upland SOM, especially those extracted from cultivated soils, are generally more aromatic as compared with stream and marine natural organic matter (NOM) (Hatcher et al., 1980; Malcom, 1990; Mahieu et al., 1999). On the other hand, aquatic NOMs are richer in O and acid functional groups and lower in N content than terrestrial SOM (Senesi et al., 1989; González-Pérez et al., 2008). For wetland soils, different SOM molecular compositional characteristics have been reported (Mao et al., 2000; Gondar et al., 2005; Schellekens et al., 2011; Dodla et al., 2012a). While some noted that peat SOMs are high in aromatic carbon content (Mao et al., 2000), others observed a dominance of O-alkyl C content in SOM isolated from ombrotrophic peat bogs (Gondar et al., 2005). In addition, SOM formed under waterlogged conditions and vegetation colonization was likely to have different biochemical fractions of plants preserved because of the anoxia (Stevenson, 1994; Mendonca et al., 2004; Schellekens et al., 2011). Dodla et al. (2012a) reported that SOM from forest swamp of the MRDP region contained the higher percentage of aliphatics, whereas the freshwater marsh SOM of the same region had the highest percentage of aromatics. On the other hand, along the increasing salinity gradient toward the ocean, they found that there was generally an increase in the crystalline domain of alkyl C and more stable guaiacyl-type structural unit of lignin residue as well as an increased contribution of sulfur as a structural component in SOM, reflecting the effect of salinity on the compositional formation and degradation stage of SOM (Wang et al., 2011; Dodla et al., 2012a). The generally less degradation of guaiacyl units than syringyl units of lignin monomers were found in both DOM and humified SOM of these coastal wetland soils (Bianchi et al., 2009; Dodla et al., 2012a). Under high salinity conditions, a decreased mineralization of LOC was often observed because of the resistance of a more condensed form of DOM and the detrimental effect of salinity on type and quantity of microorganisms (Rietz and Haynes, 2003; Sparks, 2003). These results suggest that climate, environment, land use, and vegetation are important factors in controlling the degradation and composition of SOM.

While the lability of SOC is controlled primarily by physicochemical protection mechanisms, the hydrolysis–dissolution process has been shown to play an important role in the decomposition of SOM (Marschner and Kalbitz, 2003). In wetlands, the presence of appreciable amounts of water leads to the release of hydrophilic and polar compounds such as sugars, acids, and alcohols into the ecosystem and makes them more prone to microbial degradation, in other words more labile. On the other hand, because of water, wetlands are limited in available oxygen levels, which lead to the preservation of lignin-type compounds, as most of the lignin compounds are mineralized by aerobic microorganisms rather than by anaerobic ones (Van Soest, 1994).

A wide range of lability has been reported for DOC, depending on the degradation status of SOM that DOC is in equilibrium with (Kalbitz et al., 2003; Said-Pullicino et al., 2007; von Lützow et al., 2007). Analysis of DOC from a legume plant residue based on ^{13}C nuclear magnetic resonance spectroscopy showed the major presence of alkyl, O-alkyl and carboxyl C, indicating readily degradable polysaccharides, proteins, and organic acids in DOC (Saidy et al., 2012). For soils containing more of humified SOM, its DOC is more stable and less labile. For wetlands and surface waters, the concentration, composition, and chemistry of DOC are highly variable and depend on the sources of organic matter, ionic strength, major cation composition, pH, and temperature of the water and on the surface chemistry of sediment sorbents that act as solubility controls (Leenheer and Croue, 2003). While some showed that the majority of the water extractable C from marine sediments was labile (Keil et al., 1994; Bengtsson et al., 2012), others reported that on an average 14, 19, and 19% of the total DOC pools were found labile in the lakes, rivers, and seawater, respectively (Søndergaard and Middelboe, 1995). The dynamics of LOC are also influenced by the plant community and the individual plant species molecular composition (Breulmann et al., 2012). In addition, on a regional scale, the amount of LOC in wetland soils depends on the latitude, which affects temperature and type of in situ vegetation and ecosystem soil biogeochemical conditions. Wetlands in mid to high latitudes contain large soil C pools along with a large labile C fraction because of low climate temperature and poor nutrient status (Wang et al., 2012). Climate temperature affects both decomposition of labile SOM and the kinetic properties of different soil C fractions (Davidson and Janssens, 2006). Sunlight was also found to significantly affect the labile pool of DOC. An exposure of freshwater wetland organic matter to natural or artificial sunlight caused the bacteria to utilize 300% more organic matter as compared with unexposed organic matter (Bano et al., 1998). The exposure of DOC to sunlight led to the absorption of photo energy and the formation of new labile compounds (Bano et al., 1998).

Mississippi River Deltaic Plain Coastal Wetland Soil Labile Organic Carbon Characterization

The MRDP wetland soils contain variable amounts of organic C because of different net C inputs as affected by primary productivity, specific hydrological condition, and coastal influence. Freshwater marsh (FM) and brackish marsh (BM) soil profiles have higher organic C contents than saline marsh (SM) soils and forest swamp (FS) soils (Table 1). This trend is consistent with the generally high sediment C accumulation rate in these ecosystems (Reddy and DeLaune, 2008). Table 2 shows the LOC from different MRDP wetland soil profiles as estimated by several different methods. Aerobically mineralizable C (AMC), which measures the directly available LOC to microbes, is often considered as the basis for LOC characterization (Dodla et al., 2012b). Water extraction including both the cold and the hot along with salt extraction has also been commonly used as a quick method for estimating LOC (Gregorich et al., 2003; Fang et al., 2005; Silveira et al., 2008). As shown in Table 2, AMC as well as cold water extractable C (CWEC) and salt extractable C (SEC) were less than 2 g kg^{-1} in these soil profiles and were generally higher in the surface soil–sediment layers of these wetland sites than subsurface layers. The AMC, CWEC, and SEC all accounted for <0.50% of SOC, similar to those generally found for agricultural soils (Haynes, 2005). This result was surprising, considering the generally much higher SOC in wetlands. The SEC was 30 to 60% lower than CWEC, likely due to the coagulation effect of salt extraction on smaller labile organic compounds (Dou et al., 2008). On the other hand, hot water extractable C (HWEC) was found to be generally less than 10 g kg^{-1} and account for 1.2 to 3.49% of SOC and was 4 to 13 times greater than the CWEC in these wetlands. These results were generally similar to those reported for agricultural, pasture (Curtin et al., 2006), and forest soils (Wang and Wang, 2007), suggesting these simple characterization methodologies quantify similar fractions of LOC as they do for upland soils.

The three extraction-oriented measurements of LOC (CWEC, HWEC, and SEC) were strongly correlated among each other (R^2 = 0.79–0.84), whereas the incubation-oriented AMC was generally related, to a less extent (R^2 = 0.62- 0.69), with the three extraction-oriented measurements of LOC (Table 3). The later results likely suggest that the AMC measures slightly different fractions of LOC

Table 1. Physicochemical characteristics of the selected Mississippi River deltaic plain wetland soils.†

Site	Soil depth	pH	EC‡	BD§	C	N	C/N
	cm		dS m^{-1}	g cm^{-3}	—— g kg^{-1}	——	
Forest swamp	0–25	6.2	1.1	0.24	75.4	5.5	13.7
	25–50	6.1	1.3	0.44	56.3	3.7	15.2
	50–75	6.1	1.5	0.49	41.7	2.5	16.7
Freshwater marsh	0–50	6.1	2.7	0.12	332	17	16.2
	50–100	6.4	3.0	0.14	312	16	15.6
	100–150	6.5	3.3	0.11	333	17	14.9
Brackish marsh	0–75	6.8	2.6	0.08	386	23.0	16.8
	75–125	7.2	4.8	0.07	437	28.0	15.6
	125–150	7.4	6.1	0.08	387	24.0	16.1
	150–175	7.3	7.0	0.11	278	21.0	13.2
Saline marsh	0–50	7.4	23.6	0.21	151	8.4	18.1
	50–100	7.3	30.1	0.18	169	10.0	16.8
	100–150	7.7	27.7	0.63	42.2	3.3	13.8

† Data combined from Wang et al. (2011) and Dodla et al. (2012b).

‡ EC, electric conductivity.

§ BD, bulk density.

Table 2. Labile organic carbon estimated by different methods for selected sites of the Mississippi River deltaic plain wetland soils.†

Site	Soil depth	AMC‡	CWEC	HWEC	SEC	AHC
	cm	g kg^{-1} soil				
FS§	0–25	0.33 ± 0.01 (0.44)	0.29 ± 0.07 (0.38)	1.18 ± 0.03 (1.56)	0.13 ± 0.01 (0.17)	22.5 ± 1.8 (29.8)
	25–50	0.09 ± 0.01 (0.15)	0.18 ± 0.02 (0.31)	0.84 ± 0.01 (1.49)	0.09 ± 0.01 (0.15)	20.1 ± 1.1 (35.8)
	50–75	0.05 ± 0.01 (0.11)	0.13 ± 0.01 (0.32)	0.50 ± 0.08 (1.20)	0.09 ± 0.01 (0.22)	13.6 ± 0.4 (32.7)
FM	0–50	1.17 ± 0.05 (0.34)	1.11 ± 0.06 (0.36)	9.81 ± 0.41 (3.05)	0.62 ± 0.01 (0.19)	130.3 ± 6.0 (37.9)
	50–100	0.60 ± 0.11 (0.17)	1.06 ± 0.03 (0.32)	8.57 ± 0.48 (2.75)	0.57 ± 0.05 (0.17)	102.6 ± 1.7 (28.5)
	100–150	0.35 ± 0.01 (0.10)	0.82 ± 0.07 (0.25)	6.18 ± 0.04 (1.88)	0.49 ± 0.03 (0.15)	102.4 ± 3.1 (30.7)
BM	0–75	1.97 ± 0.11 (0.49)	1.39 ± 0.10 (0.34)	9.80 ± 0.01 (2.54)	0.59 ± 0.02 (0.15)	89.8 ± 1.9 (23.3)
	75–125	0.41 ± 0.04 (0.09)	1.59 ± 0.08 (0.35)	9.47 ± 0.64 (2.17)	0.64 ± 0.01 (0.14)	85.0 ± 4.7 (19.4)
	125–150	0.74 ± 0.11 (0.18)	1.28 ± 0.04 (0.32)	6.37 ± 0.19(1.65)	0.63 ± 0.07 (0.16)	61.4 ± 10.0 (15.9)
	150–175	0.55 ± 0.01 (0.14)	1.09 ± 0.10 (0.37)	9.71 ± 0.22 (3.49)	0.46 ± 0.02 (0.16)	48.5 ± 1.9 (17.4)
SM	0–50	0.30 ± 0.03 (0.20)	0.39 ± 0.01 (0.24)	4.76 ± 0.20 (3.12)	0.28 ± 0.02 (0.17)	36.8 ± 0.5 (12.8)
	50–100	0.34 ± 0.01 (0.16)	0.50 ± 0.03 (0.30)	2.83 ± 0.20 (1.78)	0.28 ± 0.01 (0.17)	46.9 ± 2.9 (29.0)
	100–150	0.06 ± 0.01 (0.30)	0.24 ± 0.03 (1.21)	0.77 ± 0.08 (2.88)	0.16 ± 0.02 (0.80)	16.6 ± 1.6 (43.3)

† Values in parentheses indicate labile organic C (LOC) as percentage of total soil organic C (SOC). Adapted from Dodla et al. (2012b).

‡ AMC, aerobically mineralizable C; CWEC, cold water extractable C; HWEC, hot water extractable C; SEC, salt extractable C; AHC, acid hydrolyzable C.

§ FS, forest swamp; FM, freshwater marsh; BM, brackish marsh; SM, saline marsh.

Table 3. Linear relationships between different labile organic carbon estimates for Mississippi River deltaic plain wetland soils.†

X\Y	CWEC‡	HWEC	SEC	AHC
AMC	$y = 1.54x + 0.123$ $R^2 = 0.62$***	$y = 12.36x + 0.41$ $R^2 = 0.69$***	$y = 0.72x + 0.101$ $R^2 = 0.67$***	$y = 107.5x + 20.1$ $R^2 = 0.47$***
CWEC		$y = 7.41x - 0.021$ $R^2 = 0.83$***	$y = 0.47x + 0.05$ $R^2 = 0.84$***	$y = 58.2x + 16.3$ $R^2 = 0.61$***
HWEC			$y = 14.1x - 0.063$ $R^2 = 0.79$***	$y = 7.80x + 19.08$ $R^2 = 0.70$***
SEC				$y = 120.3x + 12.4$ $R^2 = 0.71$***

*** Significant at $P < 0.001$.

† Adapted from Dodla et al. (2012b).

‡ AMC, aerobically mineralizable C; CWEC, cold water extractable C; HWEC, hot water extractable C; SEC, salt extractable C; AHC, acid hydrolyzable C.

in nature compared with CWEC, HWEC, and SEC, although all these methods yielded small quantities of LOC, especially CWEC, SEC, and AMC. On the other hand, the CWEC and SEC were highly correlated with each other ($R^2 = 0.84$), indicating that both SEC and CWEC approximated a similar type of wetland LOC. Also, the strong linear relationship between HWEC and CWEC ($R^2 = 0.83$) for these wetland soils of different profiles indicates a very close linkage between LOC values extracted by the two methods, even though the HWEC was much greater than the CWEC.

Besides simple water and salt extraction, 6 N HCl extraction has often been used to characterize acid hydrolyzable C (AHC). This fraction comprises both LOC and the slowly mineralizable C pool (Paul et al., 2006). The MRDP wetlands contain much higher AHC fractions than those of AMC, CWEC, SEC, and HWEC and generally accounted for 16 to 49% of SOC (Table 2). Previous studies have showed increase in AHC along depth in soil profiles of different land uses (Tan et al., 2004; Rovira and Vallejo 2007). However, as shown in Table 2, there was large variability in AHC in the majority of these wetland soil profiles, which was likely due to the disturbance caused by tidal movement in past history (Dodla et al., 2012b). In addition, Dodla et al. (2012b) observed a positive curve linear relationship between soil bulk density and AHC ($y = 10.7x^2 + 24.3x + 22.9$, $R^2 = 0.73$), which could indicate physicochemical protection of this fraction of acid hydrolyzable C compounds. They also found that following acid hydrolysis, C/N ratios of these wetland soils almost doubled from the initial range of 13.7 to 17.9 to after hydrolysis of 20.3 to 35.9. The increase in C/N ratio signifies that the majority of N-containing compounds has been hydrolyzed by the acid (McLauchlan and Hobbie, 2004). Despite the relatively large difference in quantity, AHC showed positive correlations ($R^2 = 0.61$–0.71) with the LOC estimates determined by the three extraction methods, as well as with the AMC ($R^2 = 0.47$) at a lesser degree (Table 3). In addition, all LOC estimates including AHC were correlated with the SOC of these wetlands with $R^2 \geq 0.72$ except for AMC, which had a slightly lower R^2 value of 0.56. These correlations support the important control of LOC by total SOC, a relation also commonly observed for upland soils (Dodla et al., 2012b).

Bulk Soil Organic Carbon Functionality Control of Labile Organic Carbon

Soil Organic Carbon Functionalities

Molecular moieties of SOM have significant influences on SOC transformation kinetics and effects on various soil process and functions (Kleber et al., 2007, Senesi et al., 2007; von Lützow et al., 2007). The distribution of C functional groups relative to total SOC in the MRDP wetlands showed that these soil profiles were dominated with O/N-alkyl C followed by approximately similar percentages of alkyl C and aromatic C followed by carboxyl C and carbonyl C (Wang et al., 2011; Dodla et al., 2008, 2012a, 2012b). This distribution of major C moieties in the order of polysaccharides, alkyls, total aromatics (aryl and phenolic), and carbonyls (carboxyl, and aldehyde and ketonic) was similar to that observed for the colder prairie wetland soils of Saskatchewan, Canada (Jokic et al., 2004), and for 300 various forest and cultivated agricultural soils (Mahieu et al., 1999). This broad similarity of general dominance of molecular C moieties in SOM among various soils suggests that overall SOM transformation in many ecosystems is likely controlled by similar processes or mechanisms (Wang et al., 2011; Dodla et al., 2008, 2012a, 2012b). Nonetheless, on average, there were no consistent differences observed in relative distribution of the C moieties among the soils along a salinity gradient toward ocean from forest swamp to freshwater marsh and saline marsh. In the soil profiles of all three wetland ecosystems, the surface soil horizons tended to have a higher percent of polysaccharides-C than the deeper horizons, whereas the opposite was true for aryl C (Dodla et al., 2008; Wang et al., 2011). The latter indicates a generally increased decomposition stage (or humification) of older plant materials with depth into the soil profile (Kögel-Knabner, 1997; Wang et al., 2011).

Relation between of Soil Organic Carbon Functionalities and Labile Organic Carbon

Soil organic matter molecular composition affects its turnover rates although the lability of SOM is primarily dependent of physicochemical protection mechanisms as discussed in previous sections. Dodla et al. (2012b) recently evaluated the relationships between LOC estimated by common methods and bulk SOM C moieties in the MRDP wetland soils and found that AMC, which represents the direct measurement of actual labile C pool available to microorganisms, was positively correlated with O/N-alkyl C and carboxyl C but a negative correlated with alkyl C (Table 4). This result indicates that polysaccharides and small molecular organic acids are the main constituents of LOC in these wetland soils, consistent with the general notion that these compounds are more labile than any other C moieties (Silveira et al., 2008). The hydrophilic nature of carboxylic functional group makes carboxyl compounds more susceptible to microbial degradation (von Lützow et al., 2007). The negative effect of alkyl C on AMC implies a more recalcitrant nature of aliphatic C in these wetland soils. Aliphatic C has also been shown to be less biodegradable and often concentrated as a result of extensive decomposition (or humification) of plant residues in upland soils (Baldock et al., 1997; Kögel-Knabner, 1997). On the other hand, both CWEC and SEC fractions only show the positive relations with carboxyl C, signifying strong contribution of soluble carboxyl C compounds such as organic acids to the cold water and salt

Table 4. Statistically significant relationships between labile organic carbon and carbon functionalities for Mississippi River deltaic plain wetland soils.†

Method	Relations	R^2	*P*
AMC	= 0.017 (O/N-alkyl C) + 0.148 (carboxyl C) – 0.094 (alkyl C) + 2.29	0.38	0.048
CWEC	= 0.105 (carboxyl C) – 0.046 (alkyl C) + 3.34	0.45	0.006
HWEC	= 0.165 (carbonyl C) + 1.46	0.28	0.017
SEC	= 0.056 (carboxyl C) – 0.029 (alkyl C) + 2.10	0.40	0.017
AHC	= 2.719 (alkyl C) + 1.287 (O/N-alkyl C) – 6.1 (aromatic C) + 342.8	0.65	0.001

† Adapted from Dodla et al. (2012b).

extractable C in these wetland soils. This difference may explain slightly low correlations between AMC and CWEC or SEC as compared with the high correlation between CWEC and SEC (Table 3). The lack of a significant effect of O/N-alkyl C fractions on CWEC and SEC indicates that the contribution from O/N-alkyl C to the cold water and salt extractable LOC was less important in the SOM of these wetland soil profiles even though they are generally hydrophilic and present as the dominant molecular C fraction in these wetlands (Wang et al., 2011; Dodla et al., 2012b). This observation was different from those often observed for upland soils (Fröberg et al., 2003; Haynes, 2005; von Lützow et al., 2007).

Labile organic C estimated by hot water extraction exhibited a completely different relationship with only positive contribution of carbonyl C, which was mainly ketonic and aldeyhyde compounds of SOM (Table 4). This was likely due to the ability of hot water to hydrolyze and cleave esters of various organic materials (Siskin and Katritzky, 1991; Stange et al., 2001). Research on upland soils has showed that the HWEC fraction consists of polysaccharides and N-containing compounds in soil solution (Ghani et al., 2003; Balaria et al., 2009), as well as an additional C fraction that was weakly adsorbed to mineral surfaces and humic acids (Ghani et al, 2003; von Lützow et al., 2007). Because of a generally much greater yield of LOC in the hot water extraction than the cold and salt extractions, the HWEC has been suggested as a valuable and rapid indicator of LOC (von Lützow et al., 2007). The LOC fraction in these wetland soils as extracted by hot water was different from those found in upland soils. As for slowly releasing AHC, the O/N-alkyl C and alkyl C were major positive contributing moieties, whereas the aromatic C groups were the negative control of this SOC fraction (Table 4). Besides O/N-alkyl C, the positive contribution of alkyl C to the AHC in these wetland soils indicates a different effect of HCl extraction from those reported for other soils. Comparing the residues of upland and wetland soils after 6 N HCl extractions, Silveira et al. (2008) noted that relative distribution of O/N-alkyl C was significantly reduced in both soils, whereas the aromatic C fraction was enriched in the wetland soil and the aliphatic C was enriched in the upland soil. Others also showed that the HCl extraction hydrolyzed nitrogenous compounds besides polysaccharides (Otto and Simpson, 2007) and nonhydrolyzable C fractions following HCl extraction were mainly organic materials consisting of aromatic and alkyl C groups such as lignin and wax (Ran et al., 2007; von Lützow et al., 2007). The results of Dodla et al. (2012b), however, indicated that alkyl C is likely extracted by HCl hydrolysis from these MRDP wetland soils.

The difference between TOC and AHC, or nonhydrolyzable C, has been suggested as a quick assessment of soil recalcitrant C because of the easy measurement, even though the nonhydrolyzable SOC could not separate functional SOM pools since HCl hydrolysis releases organic C stabilized by different mechanisms (Paul et al., 2006; von Lützow et al., 2007). The negative contribution of aromatic C to the AHC in these wetland soils appears to support the recalcitrant nature of nonhydrolyzable C. Dodla et al. (2012b) employed this assessment and estimated that the nonhydrolyzable C fraction ranged from 51 to 84% of total SOC in these MRDP wetland soil profiles, which was much greater than the 24 to 32% reported for other wetlands, such as Georgia hardwood forest wetland soils (Silveira et al., 2008). These results have important implications for the understanding of the liability and refractory character of SOC in these wetlands since such properties closely affect actual contribution of marsh SOC to the oxygen consumption variability as it relates to hypoxia in the Gulf of Mexico (Bianchi et al., 2010).

Influence of Labile Organic Carbon on Denitrification Potential

Denitrification as a tool has been used extensively to remove nitrate from nutrient rich surface waters through constructed and natural wetlands. While the Mississippi River has been diverted to targeted marshlands to provide sediments and nutrients to stabilize coastal marsh from further deterioration (Reddy and DeLaune, 2008), the process also has the function of removing nitrate from the river water before it reaches the ocean, one of the main causes for hypoxia in the Gulf of Mexico (Rabalais et al., 1996). The impact of elevated NO_3^- levels in the diverted Mississippi River water on the health of these coastal wetlands and their capacity to assimilate NO_3^- has become an increasingly important topic of study (DeLaune et al., 2005; Dodla et al., 2008, 2009). Understanding the relationship between the denitrification rate and the organic C quality could further elucidate the response of different wetlands to increased nitrogen loadings in these coastal wetlands.

As shown in Fig. 2, FM showed greater denitrification potential at all profile depths than for the SM and FS soils. The estimated potential denitrification rates (PDRs) were in the range of 3.77 to 18.46, 0.99 to 1.55, and 0.43 to 3.47 mg N kg^{-1} d^{-1} on weight basis and 0.72 to 1.28, 0.37 to 0.58, and 0.07 to 0.38 mg N m^{-3} d^{-1} on volume basis, for FM, FS, and SM soil profiles, respectively. The FM soil also exhibited faster kinetics of denitrification than the FS and SM soils. The lower PDR observed in SM soils among the three wetlands was attributable to the high salinity effect. On the other hand, the high PDR in FM was due to the high SOC in the profile. Overall, the PDR was positively correlated with TOC ($R^2 = 0.65$, $P < 0.01$) and particularly with the LOC fractions as estimated by AMC ($y = 256.7x + 1.46$, $R^2 = 0.94$, $P < 0.001$) (Dodla et al., 2008). Labile organic C has been considered as one of most important factors affecting denitrification besides the availability of nitrate (Burford and Bremner, 1975). These results suggest that the majority of LOC was evolved from the denitrification activity when nitrate was available. These results would be expected since polysaccharide-C was the dominant organic C within these soil profiles (Dodla et al., 2012b) and the major C moiety that contributes to AMC positively as discussed in the previous section.

Significant ($P \leq 0.04$) but negative relationships between PDR and percent phenolic C and between PDR and percent aldehyde/ketonic C were also observed (Fig. 3a and 3b). On the other hand, there was no statistically significant relation

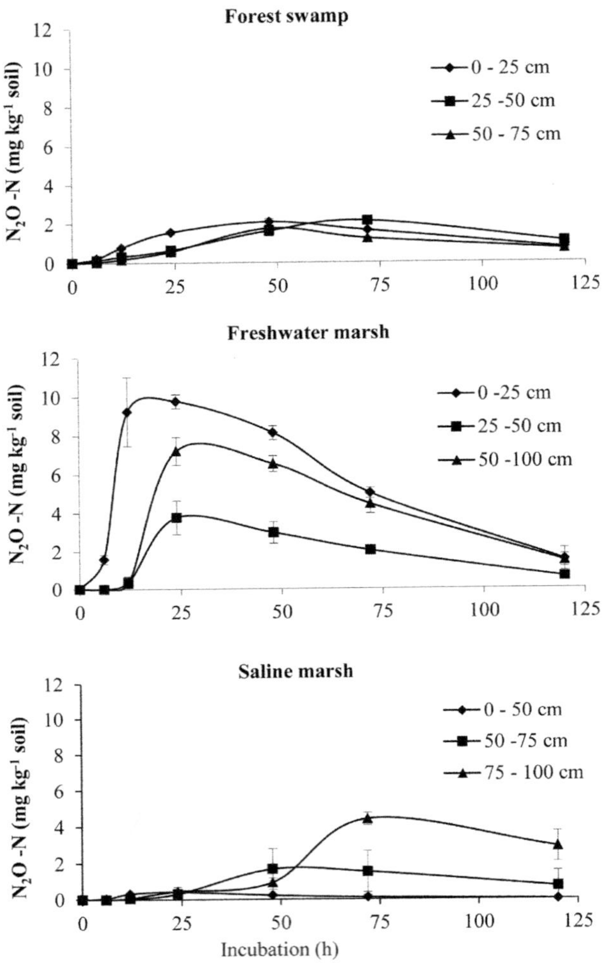

Fig. 2. Total N_2O–N released during the incubation of three MRDP wetland soils from different depths treated with 2 mg L^{-1} NO_3^-–N. Adapted from Dodla et al. (2008).

observed among PDR with percent alkyl C, aryl C, and carboxyl C (Dodla et al., 2008). Sirivedhin and Gray (2006) also showed that that addition of pore water containing a high level of phenolic C led to decreased denitrification rates as compared with the pore water containing a low level of phenolic C. Previously, certain strains of bacteria were found capable of denitrifying NO_3^- with aromatic C (including both phenolic and aryl moieties) under anaerobic condition (Nozawa and Maruyama, 1988). Also, addition of an alkyl C-rich soybean oil to wood chips significantly increased denitrification rates over woodchips alone (Greenan et al., 2006). The results of Dodla et al. (2008) revealed that phenolics not aryls, possibly of lignin, were likely associated with the inhibition of denitrification in these MRDP wetland soils. In addition, alkyl C likely was found to play a lesser role

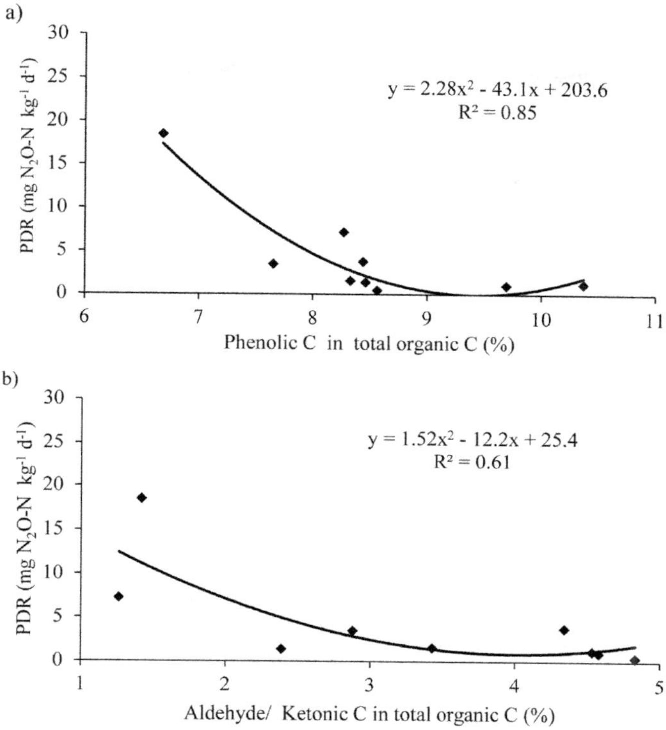

Fig. 3. The relationship between potential denitrification rate (PDR) and different C moieties for selected MRDP wetland soils: percent phenolic C (a) and aldehyde/ketonic C (b) under the treatment of 2 mg L^{-1} NO_3–N (b) . Modified from Dodla et al. (2012b).

in regulating denitrification in these wetland soils. Because of the fact that aldehyde and ketone C was significantly and positively correlated with phenolics (R^2 = 0.47, $P < 0.03$), but not related to any of other C moieties, Dodla et al. (2008) speculated that some of the aldehydes and ketones in these wetland systems were actually phenolic. As phenolic aldehydes are intermediates of lignin in the transformations of plant residues during organic matter decomposition, these results suggest that phenolic aldehydes and ketones could be the primary C source that inhibits denitrification in these wetland systems.

Role of Clay Mineral and Cation in Controlling Labile Organic Carbon Mineralization and Transformation

The unique situation of the Mississippi River diversion project, which carries rich sediments and nutrients, and constant seawater intrusion, which brings in large amounts of Na, K, Mg, and Ca ions, likely creates an integrated effect on LOC mineralization and transformation in the MRDP wetlands (Wang et al., 2011;

Dodla et al., 2012a). Mississippi River sediments contain smectite-rich clays from upper streams because of soil erosion. As shown in Fig. 4 in an incubation experiment, additions of clay and clay with K or Ca ions significantly decreased the mineralization of SOC in the MRDP freshwater marsh soil as evident by reduced CO_2–C emission, suggesting strong support of the river sediments to stabilize SOM in these wetlands. In addition, the addition of K and Ca ions to these wetlands significantly reduced DOC mobility as indicated by substantial reduction in the extractable amount of DOC by 10 mM K^+ and 5 mM Ca^{2+} solutions (Fig. 5). It was also interesting that Ca^{2+} was more effective in stabilizing the DOC in the FM soil than K^+, whereas it was opposite in the FS soil (Table 5). Solution analysis showed that a greater amount of added Ca^{2+} was sorbed in FM soil than in FS soil, whereas the opposite was true for added K^+ (Dodla, 2009). Since FS soil has a much higher content of clays than FM soil, these results could suggest a stronger K^+ effect on DOC stabilization in clayey wetlands, whereas Ca^{2+} has a stronger DOC stabilization effect in more organic wetlands.

The interaction of SOC with mineral surface and metals has been increasingly considered as an important mechanism of SOC stabilization in many soils (Saidy et al., 2012; Sanderman et al., 2014). Stability of clay absorbed organic C not only varies with types of mineralogy but is also correlated with the abundance of metals (Saidy et al., 2012; Sanderman et al., 2014). Based on laboratory incubation, Saidy et al. (2012) found that smectite, which has a high surface area, generally stabilized more LOC than kaolinite and illite. On the other hand, coating goethite to kaolinite significantly enhanced DOC sorption, whereas it did not affect DOC adsorption by illite and smectite (Saidy et al., 2012). They attributed this difference to the induction of surface charge change by coating hydrous Fe oxides to kaolinite surface but not to illite and smectite. Sanderman et al. (2014) also observed that allophone and metal oxides were more effective in retaining DOC than illite and smectite. In addition, they found that the DOC source, especially the proportions of aromatic, phenolic, and carboxylic C in the DOC, had a significant effect on the total amount of DOC to be retained by these minerals. Earlier studies also showed that clay minerals containing Ca such as smectites had more association with organic matter than the noncalcium-containing clays such as kaolinite (Furukawa, 2000). There has been tremendous evidence suggesting that divalent or trivalent cation bridging (e.g., Ca^{2+}, Al^{3+}, Fe^{3+}) between negatively charged clay

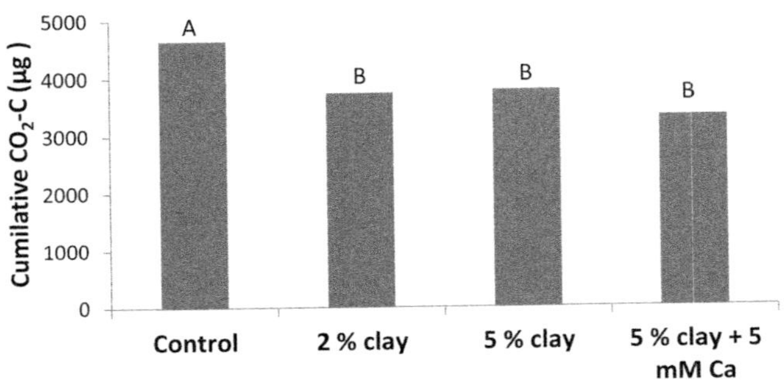

Fig. 4. Effect of smectite clay and cation addition on organic C mineralization from an MRDP fresh marsh soil.

mineral surfaces and dissociated carboxylic function groups of organic matter (Sparks, 2003). These divalent and trivalent cations can also precipitate or condense SOM, especially the aromatic and carboxylic C compounds (Scheel et al., 2008). While the literature has emphasized the role of multivalent cations in binding OM, the result shown in Fig. 5 clearly demonstrates that monovalent K^+ can also remove DOC from solution, especially in relatively high clayey wetland soils, which is likely associated with K^+ fixation and the closing up of interlayer spaces of smectite and vermiculite.

Fig. 5. Effect of Ca^{2+} and K^+ salt solution on dissolved organic C retention in an MRDP forest swamp soil (top) and a freshwater marsh soil (bottom). Control indicates deionized water. Ca^{2+} and K^+ solutions were added as chloride salts at 5 mM and 10 mM, respectively. Soil/water ratio was maintained at 1:1 on moist soil basis.

Table 5. Effect of K^+ and Ca^{2+} additions on dissolved organic carbon (DOC) and exchangeable cations from selected Mississippi River deltaic plain wetland soils.†

Soil	Treatment	DOC	K	Ca	Mg	Na	Al	Fe	Mn
		mg L^{-1}							
FS	Deionized H_2O	44.0	4.7	16.5	4.2	43.0	0.07	ND‡	ND
FS	10 mM K^+	15.0	78.9	69.6	24.3	60.5	0.04	0.62	2.1
FS	5 mM Ca^{2+}	18.3	9.5	67.5	21.1	66.4	ND	ND	1.6
FM	Deionized H_2O	118.3	3.3	5.3	2.2	41.7	0.29	ND	ND
FM	10 mM K^+	36.5	154.8	31.0	34.9	194.4	0.08	ND	ND
FM	5 mM Ca^{2+}	28.6	12.4	33.1	33.7	171.2	ND	ND	ND

† Results were based on 100 g of field moist soil samples extracted with 100 mL of treatment solution; K^+ and Ca^{2+} were added as chlorides salts.

‡ ND, not detectable.

Implication for Gulf Coast Hypoxia

The MRDP wetlands, especially the marshlands, provide the protection of Louisiana shorelines from storm surge; they also offer significant support for fishery species (Boesch and Turner, 1984). In addition, these wetlands sequester large amount of C because of their high primary productivity. On the other hand, while previously riverine C was considered as the main organic C source to fuel the hypoxia in the northern Gulf of Mexico, recent evidence has indicated that the DOM from estuarine wetlands also contributes significantly to this process (Bianchi et al., 2009, 2011). Analysis based on lignin biomarkers showed a source gradient of DOM from the upper marshes in the Terrebonne-Timbalier Bay estuary to the inner Louisiana shelf. The DOC concentrations in water of these wetlands were found to range from 500 to 700 μM, similar to those reported for other wetlands (Opsahl, 2005; Bianchi et al., 2009). On the other hand, there were significantly higher dissolved lignin concentrations in the water of marshlands than those in the inner shelf. In addition, the lignin monomer ratios of syringyl units to guaiacyl units (S/G) and cinnamyl to guaiacyl units (C/G) were low in the DOM of the upper marshes. According to Bianchi et al. (2009), these observations were likely a result of selective degradation of syringyl and cinnamyl over guaiacyl lignin phenols rather than the source C inputs from woody gymnosperms. The selective syringyl degradation was also reflected in observed lower ratios of syringic acid to syringaldhyde found in these wetlands. These results were supported by gradually lowering S/G ratios found for humic acids extracted from these wetland soils along the salinity gradient from freshwater marsh to saline marsh (Dodla et al., 2012a). As shown in Table 2, these wetland soils contain substantial fractions of slowly releasing AHC, even though they were found to be dominated in passive pools of SOC (Dodla et al., 2012b). Furthermore, Bianchi et al. (2009) noted a rapid degradation of DOM in these wetlands, especially those in the estuary leading to the inner Louisiana shelf based on a decreasing trend of S/G ratios. Because of the short distance from the upper marshes to the inner shelf, they attributed this rapid decomposition of DOM to the photodegradation of dissolved lignin, as the microbial degradation of lignin was generally slower (Benner and Opsahl, 2001; Opsahl, 2005). It is expected that the photochemical process breaks down large organic molecules into smaller ones, increasing their lability for assimilation by aquatic bacteria. On the other hand, this photodegradation could also convert organic C to dissolved inorganic C as CO_2 and other C gases (Mopper and Kieber, 2002). A recent estimate has indicated that marsh organic C inputs to the Mississippi plume region accounted for 19 to 34% and 3 to 15% in spring and fall, respectively (Bianchi et al., 2011). Clearly the lability of SOC in these MRDP wetlands has a significant effect on oxygen demand requirement in the northern Gulf of Mexico.

Conclusions

Among different MRDP wetland ecosystems, freshwater and brackish marsh soils contain higher LOC than saline marsh and forest swamp soils, which is highly correlated to the total organic C contents of these soils and consistent with the soil–sediment C accumulation rates in these wetlands. Labile organic C as estimated by aerobic incubation as well as salt and cold water extractions is generally less than 0.5% of SOC. Hot water extraction and especially acid hydrolysis yield much higher LOC estimates (1.2–3.49% and 12.8–43% of SOC, respectively) than those by aerobic mineralization, cold water, and salt extractions. There are generally close correlations among AMC, CWEC, and SEC, which are supported by their similar relations to SOC functionality. While carboxyl C (mainly organic acids) was the primary contributor of LOC by cold water and salt extractions, aerobic mineralization includes the positive contribution of both carboxyl C and O/N-alkyl C (polysaccharides, amino acids, and proteins). On the other hand, LOC by hot water extraction captures carbonyl C (mainly ketonic and aldeyhyde compounds), whereas slowly releasing organic C by acid hydrolysis includes the contribution of alkyl C (aliphatics) besides O/N-alkyl C. The positive contribution of alkyl C in these MRDP soils to AHC is different from that reported for upland soils. In addition, higher LOC in freshwater marsh soils than in the saline and forest swamp soils facilitates higher denitrification potential of the former. The presence of phenolic aldehydes and ketones in these wetland systems could, however, inhibit denitrification. Mississippi River diversion and seawater intrusion likely exert an integrated effect on the lability of SOM. While clay sediment additions reduce LOC degradation, which is evident by lowered CO_2 emissions, both divalent Ca^{2+} and monovalent K^{+} enhance the retention of DOC, although the magnitude of these effects varies with specific wetland soils. Research evidence suggests selective degradation of syringyl structures of lignin structure over guaiacyl in both bulk SOM and DOM in the MRDP wetlands. In addition, the DOM degradation by photochemical process likely enhances its lability and contribution to the northern Gulf of Mexico hypoxia.

References

Amelung, W., S. Brodowski, A. Sandhage-Hofmann, and R. Bol. 2008. Combining biomarker with stable isotope analyses for assessing the transformation and turnover of soil organic matter. Adv. Agron. 100:155–250. doi:10.1016/S0065-2113(08)00606-8

Amthor, J.S., and M.A. Huston. 1998. Terrestrial ecosystem response to global change: A Research strategy, ORNL/TM-1998/27. Oak Ridge Natl. Lab., Oak Ridge, TN.

Balaria, A., Johnson, C.E., and Xu, Z. 2009. Molecular-scale characterization of hot-water-extractable organic matter in organic horizons of a forest soil. Soil Sci. Soc. Am. J. 73: 812–821. doi:10.2136/sssaj2008.0075

Baldock, J.A., J.M. Oades, P.N. Nelson, T.M. Skene, A. Golchin, and P. Clark. 1997. Assessing the extent of decomposition of natural organic materials using solid-state ^{13}C NMR spectroscopy. Aust. J. Soil Res. 35:1061–1084. doi:10.1071/S97004

Bano, N., M.A. Moran, and R.E. Hodson. 1998. Photochemical formation of labile organic matter from two components of dissolved organic carbon in a freshwater wetland. Aquat. Microb. Ecol. 16:95–102. doi:10.3354/ame016095

Barras, J.A. 2006. Land area change in coastal Louisiana after the 2005 hurricanes: A series of three maps. U.S. Geological Survey Open-File Report 06-1274.

Bengtsson, M.M., K. Wagner, E.R. Herberg, N.R. Burns, W. Wanek, and T.J. Battin. 2012. Interactions between recalcitrant and labile organic carbon in streams—Can stream biofilms mediate a priming effect? Geophys. Res. Abstr. 14:12713.

Benner, R., and S. Opsahl. 2001. Molecular indicators of the sources and transformation of dissolved organic matter in the Mississippi plume. Org. Geochem. 32:597–611. doi:10.1016/S0146-6380(00)00197-2

Bianchi, T.S., S.F. Dimarco, J.H. Cowan Jr., R.D. Hetland, P. Chapman, J.W. Day, and M.A. Allison. 2010. The science of hypoxia in the northern Gulf of Mexico: A review. Sci. Total Environ. 408:1471–1484. doi:10.1016/j.scitotenv.2009.11.047

Bianchi, T.S., S.F. DiMarco, R.W. Smith, and K.M. Schreiner. 2009. A gradient of dissolved organic carbon and lignin from Terrebonne-Timbalier Bay estuary to the Louisiana shelf (USA). Mar. Chem. 117:32–41. doi:10.1016/j.marchem.2009.07.010

Bianchi, T.S., L.A. Wysocki, K. M. Schreiner, T.R. Filley, D.R. Corbett, A.S. Kolker. 2011. Sources of terrestrial organic carbon in the Mississippi plume region: Evidence for the importance of coastal marsh Inputs. Aquat. Geochem. 17:431–456.

Boesch, D.F., and R.E. Turner. 1984. Dependence of fishery species on salt marshes: The role of food and refuge. Estuaries 7:460–468. doi:10.2307/1351627

Breulmann, M., E. Schulz, K. Weibhuhn, and F. Buscot. 2012. Impact of the plant community composition on labile soil organic carbon, soil microbial activity and community structure in semi-natural grassland ecosystems of different productivity. Plant Soil 352:253–265. doi:10.1007/s11104-011-0993-6

Bridgham, S.D., and C.J. Richardson. 2003. Endogenous versus exogenous nutrient control over decomposition and mineralization in North Carolina peatlands. Biogeochemistry 65:151–178. doi:10.1023/A:1026026212581

Burford, J.R., and J.M. Bremner. 1975. Relationships between the denitrification capacities of soils and total, water soluble and readily decomposable soil organic matter. Soil Biol. Biochem. 7:389–394. doi:10.1016/0038-0717(75)90055-3

Chabbi, A., and C. Rumpel. 2009. Organic matter dynamics in agro-ecosystems—The knowledge gaps. Eur. J. Soil Sci. 60:153–157. doi:10.1111/j.1365-2389.2008.01116.x

Cuevas, L.A., J.E. Egge, T.F. Thingstad, and B. Topper. 2011. Organic carbon and mineral nutrient limitation of oxygen consumption, bacterial growth and efficiency in the Norwegian Sea. Polar Biol. 34:871–882. doi:10.1007/s00300-010-0944-3

Curtin, D., C.E. Wright, M.H. Beare, and F.M. McCallum. 2006. Hot water-extractable nitrogen as an indicator of soil nitrogen availability. Soil Sci. Soc. Am. J. 70:1512–1521. doi:10.2136/sssaj2005.0338

Davidson, E.A., and I.A. Janssens. 2006. Temperature sensitivity of soil carbon decomposition and feedbacks to climate change. Nature 440:165–173. doi:10.1038/nature04514

DeLaune, R.D., A. Jugsujinda, J.L. West, C.B. Johnson, and M. Kongchum. 2005. A screening of the capacity of Louisiana freshwater wetlands to process nitrate in diverted Mississippi River water. Ecol. Eng. 25:315–321. doi:10.1016/j.ecoleng.2005.06.001

DeLaune, R.D., and S.R. Pezeshki. 2003. The role of soil organic carbon in maintaining surface elevation in rapidly subsiding U.S. Gulf of Mexico coastal marshes. Water Air Soil Pollut. 3:167–179. doi:10.1023/A:1022136328105

DeLaune, R.D., and J.R. White. 2011. Will coastal wetlands continue to sequester carbon in response to an increase in global sea level?: A case study of the rapidly subsiding Mississippi River deltaic plain. Clim. Change 10.1007/s10584–011-0089–6.

Dodla, S.K. 2009. Physical and chemical factors controlling carbon gas emissions and organic matter transformation in coastal wetlands. Ph.D. diss., Louisiana State Univ., Baton Rouge.

Dodla, S.K., J.J. Wang, and R.L. Cook. 2012a. molecular composition of humic acids from coastal wetland soils along a salinity gradient. Soil Sci. Soc. Am. J. 76:1592–1605. doi:10.2136/sssaj2011.0346

Dodla, S.K., J.J. Wang, and R.D. DeLaune. 2012b. Characterization of labile organic carbon in coastal wetland soils of the Mississippi River deltaic plain: Relationships to carbon functionalities. Sci. Total Environ. 435-436:151–158. doi:10.1016/j.scitotenv.2012.06.090

Dodla, S.K., J.J. Wang, R.D. DeLaune, and G. Breitenbeck. 2009. Carbon gas production under different electron acceptors in a freshwater marsh soil. Chemosphere 76:517–522. doi:10.1016/j.chemosphere.2009.03.013

Dodla, S.K., J.J. Wang, R.D. DeLaune, and R.L. Cook. 2008. Denitrification potential and its relation to organic carbon quality in three wetland soils. Sci. Total Environ. 407:471–480. doi:10.1016/j.scitotenv.2008.08.022

Dou, F., C.L. Ping, L. Guo, and T. Jorgenson. 2008. Estimating the impact of seawater on the production of soil water extractable organic carbon during coastal erosion. J. Environ. Qual. 37:2368–2374. doi:10.2134/jeq2007.0403

Fang, C., P. Smith, J.B. Moncrieff, and J.U. Smith. 2005. Similar response of labile and resistant soil organic matter pools to changes in temperature. Nature 433:57–59. doi:10.1038/nature03138

Fröberg, M., D. Berggren, B. Bergkvist, C. Bryant, and H. Knicker. 2003. Contribution of Oi, Oe, and Oa horizons to dissolved organic matter in forest floor leachates. Geoderma 113:311–322. doi:10.1016/S0016-7061(02)00367-1

Furukawa, Y. 2000. Energy-filtering transmission electron microscopy (EFTEM) and electron energy-loss spectroscopy (EELS) investigation of clay-organic matter aggregates in aquatic sediments. Org. Geochem. 31:735–744. doi:10.1016/S0146-6380(00)00043-7

Ghani, A., M. Dexter, and K.W. Perrott. 2003. Hot-water extractable carbon in soils: A sensitive measurement for determining impacts of fertilisation, grazing and cultivation. Soil Biol. Biochem. 35:1231–1243. doi:10.1016/S0038-0717(03)00186-X

Gondar, D., R. Lopez, S. Fiol, J.M. Antelo, and F. Arce. 2005. Characterization and acid–base properties of fulvic and humic acids isolated from two horizons of an ombrotrophic peat bog. Geoderma 126:367–374. doi:10.1016/j.geoderma.2004.10.006

González-Pérez, M., P.V. Torrado, L.A. Colnago, L. Martin-Neto, X.L. Otero, D.M.B.P. Milori, and F.H. Gomes. 2008. ^{13}C NMR and FTIR spectroscopy characterization of humic acids in spodosols under tropical rain forest in southeastern Brazil. Geoderma 146:425–433. doi:10.1016/j.geoderma.2008.06.018

Gregorich, E.G., M.H. Beare, U. Stoklas, and P. St-Georges. 2003. Biodegradability of soluble organic matter in maize-cropped soils. Geoderma 113:237–252. doi:10.1016/S0016-7061(02)00363-4

Greenan, C.M., T.B. Moorman, T.C. Kaspar, T.B. Parkin, and D.B. Jaynes. 2006. Comparing carbon substrates for denitrification of subsurface drainage water. J. Environ. Qual. 35:824–829.

Harvey, H.R., J.H. Tuttle, and J.T. Bell. 1995. Kinetics of phytoplankton decay during simulated sedimentation: Changes in biochemical composition and microbial activity under oxic and anoxic conditions. Geochim. Cosmochim. Acta 59:3367–3377. doi:10.1016/0016-7037(95)00217-N

Hatcher, P.G., R. Rowan, and M.A. Mattingly. 1980. ^{1}H and ^{13}C NMR of marine humic acids. Org. Geochem. 2:77–85. doi:10.1016/0146-6380(80)90023-6

Haynes, R.J. 2005. Labile organic matter fractions as central components of the quality of agricultural soils; An Overview. Adv. Agron. 85:221–268. doi:10.1016/S0065-2113(04)85005-3

Jokic, A., J.N. Cutler, D.W. Anderson, and F.L. Walley. 2004. Detection of heterocyclic N compounds in whole soils using N-XANES spectroscopy. Can. J. Soil Sci. 84: 291–293.

Kalbitz, K., J. Schmerwitz, D. Schwesig, and E. Matzner. 2003. Biodegradation of soil-derived dissolved organic matter as related to its properties. Geoderma 113:273–291. doi:10.1016/S0016-7061(02)00365-8

Keil, R.G., D.B. Montlucon, F.G. Prahl, and J.I. Hedges. 1994. Sorptive preservation of labile organic matter in marine sediments. Nature 370:549–552. doi:10.1038/370549a0

Kleber, M. 2010. What is recalcitrant soil organic matter? Environ. Chem. 7:320–332. doi:10.1071/EN10006

Kleber, M., P. Sollins, and R. Sutton. 2007. A conceptual model of organo-mineral interactions in soils: self-assembly of organic molecular fragments into zonal structures on mineral surfaces. Biogeochemistry 85:9–24.

Kögel-Knabner, I. 1997. ^{13}C and ^{15}N NMR spectroscopy as a tool in soil organic matter studies. Geoderma 80:243–270. doi:10.1016/S0016-7061(97)00055-4

Krull, E.E., J.A. Baldock, and J.O. Skjemstad. 2003. Importance of mechanisms and processes of the stabilization of soil organic matter for modeling carbon turnover. Funct. Plant Biol. 30:207–222. doi:10.1071/FP02085

Lane, R.L., J.W. Day, Jr., and B. Thibodeaux. 1999. Water quality analysis of freshwater diversion at Caernarvon, Louisiana Estuar. 22:327–336.

Leenheer, J.A., and J.P. Croue. 2003. Characterizing dissolved aquatic organic matter: Understanding the unknown structure is key to better treatment of drinking water. Environ. Sci. Technol. 37:18A–26A. doi:10.1021/es032333c

Mahieu, N., D.S. Powlson, and E.W. Randall. 1999. Statistical analysis of published carbon-13 CPMAS NMR spectra of soil organic matter. Soil Sci. Soc. Am. J. 63:307–319. doi:10.2136/sssaj1999.03615995006300020008x

Malcom, R.L. 1990. The uniqueness of humic substances in each of soil, stream and marine environments. Anal. Chim. Acta 232:19–30. doi:10.1016/S0003-2670(00)81222-2

Mao, J.-D., W.-G. Hu, K. Schmidt-Rohr, G. Davies, E.A. Ghabbour, and B. Xing. 2000. Quantitative characterization of humic substances by solid-state carbon-13 Nuclear Magnetic Resonance. Soil Sci. Soc. Am. J. 64:873–884. doi:10.2136/sssaj2000.643873x

Marschner, B., and K. Kalbitz. 2003. Controls of bioavailability and biodegradability of dissolved organic matter in soils. Geoderma 113:211–235. doi:10.1016/S0016-7061(02)00362-2

McLauchlan, K.K., and S.E. Hobbie. 2004. Comparison of labile organic matter fractionation techniques. Soil Sci. Soc. Am. J. 68:1616–1625. doi:10.2136/sssaj2004.1616

Mendonca, A., C.D. Aemando, and E.B.H. Santos. 2004. Spectroscopic properties of sedimentary humic acids from a salt marsh (Ria de Aveiro, Portugal): Comparison of sediments colonized by *Halimione portulacoides* (L.) Aellen and non-vegetated sediments. Biogeochemistry 69:159–174.

Mopper, K., and D.J. Kieber. 2002. Photochemistry and the cycling of carbon, sulfur, nitrogen and phosphous. In: D. Hansell and C. Carson, editors, Biogeochemistry of marine organic matter. Academic Press, Waltham, MA.

Nozawa, T., and Y. Maruyama. 1988. Denitrification by a soil bacterium with phthalate and other aromatic compounds as substrate. J. Bacteriol. 170:2501–2505.

Nyman, J.A., R.D. DeLaune, and W.H. Patrick, Jr. 1990. Wetland soil formation in the rapidly subsiding Mississippi River deltaic plain: Mineral and organic matter relationships. Estuar. Coast. Shelf Sci. 31:57–69. doi:10.1016/0272-7714(90)90028-P

Opsahl, S.P. 2005. Organic carbon composition and oxygen metabolism across a gradient of seasonally inundated limesink and riparian wetlands in the southeast Coastal Plain, USA. Biogeochemistry 76:47–68. doi:10.1007/s10533-005-2074-5

Otto, A., and M.J. Simpson. 2007. Analysis of soil organic matter biomarkers by sequential chemical degradation and gas chromatography-mass spectrometry. J. Sep. Sci. 30:272–282. doi:10.1002/jssc.200600243

Parton, W.J., D.S. Schimel, C.V. Cole, and D. Ojima. 1987. Analysis of factors controlling soil organic matter levels in the Great Plains grasslands. Soil Sci. Soc. Am. J. 51:1173–1179. doi:10.2136/sssaj1987.03615995005100050015x

Paul, E.A., S.J. Morris, R.T. Conant, and A.F. Plante. 2006. Does the acid hydrolysis-incubation method measure meaningful soil organic carbon pools? Soil Sci. Soc. Am. J. 70:1023–1035. doi:10.2136/sssaj2005.0103

Rabalais, N.N., W.J. Wiseman, R.E. Turner, B.K. SenGupta, and Q. Dortch. 1996. Nutrient changes in the Mississippi river and system responses on the adjacent continental shelf. Estuaries 19(2B):386–407. doi:10.2307/1352458

Ran, Y., K. Sun, Y. Yang, B. Xing, and E. Zeng. 2007. Strong sorption of phenanthrene by condensed organic matter in soils and sediments. Environ. Sci. Technol. 41:3952–3958. doi:10.1021/es062928i

Reddy, K.R., and R.D. DeLaune. 2008. Biogeochemistry of wetlands: Science and applications. CRC Press, Boca Raton, FL.

Rietz, D.N., and R.J. Haynes. 2003. Effects of irrigation–induced salinity and sodicity on soil microbial activity. Soil Biol. Biochem. 35:845–854. doi:10.1016/S0038-0717(03)00125-1

Rovira, P., and R. Vallejo. 2007. Labile, recalcitrant, and inert organic matter in Mediterranean forest soils. Soil Biol. Biochem. 39:202–215. doi:10.1016/j.soilbio.2006.07.021

Said-Pullicino, D., K. Kaiser, G. Guggenberger, and G. Gigliotti. 2007. Changes in chemical composition of water extractable organic matter during composting: Distribution between stable and labile organic matter pools. Chemosphere 66:2166–2176. doi:10.1016/j.chemosphere.2006.09.010

Saidy, A.R., R.J. Smernik, J.A. Baldock, K. Kaiser, J. Sanderman, and L.M. Macdonald. 2012. Effects of clay mineralogy and hydrous iron oxides on labile organic carbon stabilization. Geoderma 173–174:104–110. doi:10.1016/j.geoderma.2011.12.030

Sanderman, J., T. Maddern and J. Baldock. 2014. Similar composition but differential stability of mineral retained organic matter across four classes of clay minerals. Biogeochemistry 121:409–424. doi:10.1007/s10533-014-0009-8

Scheel, T., L. Haumaier, R.H. Ellerbrock, J. Rühlmann, and K. Kalbitz. 2008. Properties of organic matter precipitated from acidic forest soil solutions. Org. Geochem. 39:1439–1453. doi:10.1016/j.orggeochem.2008.06.007

Schellekens, J., P. Buueman, I. Fraga, and A. Martinez-Cortiza. 2011. Holocene vegetation and hydrologic changes inferred from molecular vegetation markers in peat, Penido Vello (Galicia, Spain). Palaeogeogr. Palaeoclimatol. Palaeoecol. 299:56–89. doi:10.1016/j.palaeo.2010.10.034

Schmidt, M.W.I., M.S. Torn, S. Abiven, T. Dittmar, G. Guggenberger, I.A. Janssens, M. Kleber, I. Kogel-Knabner, J. Lehmann, D.A.C. Manning, P. Nannipieri, D.P. Rasse, S. Weiner, and S.E. Trumbore. 2011. Persistence of soil organic matter as an ecosystem property. Nature 478:49–56. doi:10.1038/nature10386

Senesi, N., T.M. Miano, M.R. Provenzano, and G. Brunetti. 1989. Spectroscopic and compositional comparative characterization of I.H.S.S. reference and standard fulvic and humic acids of various origin. Sci. Total Environ. 81-82:143–156. doi:10.1016/0048-9697(89)90120-4

Senesi, N., C. Plaza, G. Brunetti, and A. Plolo. 2007. A comparative survey of recent results on humic-like fractions in organic amendments and effects on native soil humic substances. Soil Biol. Biochem. 39:1244–1262.

Silveira, M.L., N.M. Comerford, K.R. Reddy, W.T. Cooper, and H. El-Rifai. 2008. Characterization of soil organic carbon pools by acid hydrolysis. Geoderma 144:405–414. doi:10.1016/j.geoderma.2008.01.002

Sirivedhin, T., and K.A. Gray. 2006. Factors affecting denitrification rates in experimental wetlands: Field and laboratory studies. Ecol. Eng. 26:167–181. doi:10.1016/j.ecoleng.2005.09.001

Siskin, M., and A.R. Katritzky. 1991. Reactivity of organic compounds in hot water: Geochemical and technological implications. Science 254:231–237. doi:10.1126/science.254.5029.231

Six, J., P. Callewaert, S. Lenders, S. De Gryze, S.J. Morris, E.G. Gregorich, E.A. Paul, and K. Paustian. 2002. Measuring and understanding carbon storage in afforested soils by physical fractionation. Soil Sci. Soc. Am. J. 66:1981–1987. doi:10.2136/sssaj2002.1981

Søndergaard, M., and M. Middelboe. 1995. A cross-system analysis of labile dissolved organic carbon. Mar. Ecol. Prog. Ser. 118:283–294.

Stange, R.R., Jr., J. Raph, J. Peng, J.J. Sims, S.L. Midland, and R.E. McDonald. 2001. Acidolysis and hot water extraction provide new insight into the composition of the induced "lignin-like" material from squash fruit. Phytochemistry 57:1005–1011. doi:10.1016/S0031-9422(01)00096-6

Stevenson, F.J. 1994. Humus chemistry. 2nd ed. Wiley-Intersci., New York.

Stockmann, U., M.A. Adams, J.W. Crawford, D.J. Field, N. Henakaarchchi, M. Jenkins, B. Minasny, A.B. McBratney, V.R. Courcelles, K. Singh, I. Wheeler, L. Abbott, D.A. Angers, J. Baldockd, M. Birde, P.C. Brookes, C. Chenu, J.D. Jastrow, R. Lal, J. Lehmann, A.G. O'Donnell, W.J. Parton, D. Whitehead, and M. Zimmermann. 2013. The knowns, known unknowns and unknowns of sequestration of soil organic carbon. Agric. Ecosyst. Environ. 164:80–99. doi:10.1016/j.agee.2012.10.001

Sparks, D.L. 2003. Environmental soil chemistry. 2nd ed. Academic Press, New York.

Tan, Z.X., R. Lal, R.C. Izaurralde, and W.M. Post. 2004. Biochemically protected soil organic carbon at the North Appalachian experimental watershed. Soil Sci. 169:423–433. doi:10.1097/01.ss.0000131227.51226.68

Torn, M.S., C.W. Swanston, C. Castanha, and S.E. Trumbore. 2009. Storage and turnover of organic matter in soil. In: N. Senesi, B. Xing, and P.M. Huang, editors, Biophsico-chemical processes involving natural organic matter in environmental systems. John Wiley & Sons, New York.

Van Soest, P.J. 1994. The nutritional ecology of the ruminant. 2nd ed. Cornell Univ. Press. Ithaca, NY.

von Lützow, M., I. Kögel-Knabner, K. Ekschmitt, H. Flessa, G. Guggenberger, E. Matzner, and B. Marschner. 2007. Soil fractionation methods: Relevance to functional pools and stabilization mechanisms. Soil Biol. Biochem. 39:2183–2207. doi:10.1016/j.soilbio.2007.03.007

Wang, J.J., S.K. Dodla, R.D. DeLaune, W.H. Hudnall, and R.L. Cook. 2011. Soil carbon characteristics in two Mississippi river deltaic marshland soils. Wetlands 31:157–166. doi:10.1007/s13157-010-0130-y

Wang, J., C. Song, X. Wang, and Y. Song. 2012. Changes in labile soil organic carbon fractions in wetland ecosystems along a latitudinal gradient in Northeast China. Catena 96:83–89. doi:10.1016/j.catena.2012.03.009

Wang, Q.K., and S.L. Wang. 2007. Soil organic matter under different forest types in southern China. Geoderma 142:349–356. doi:10.1016/j.geoderma.2007.09.006

Wilson, C.A., and M.A. Allison. 2008. An equilibrium profile model for retreating marsh shorelines in southeast Louisiana. Estuar. Coast. Shelf Sci. 80:483–494. doi:10.1016/j.ecss.2008.09.004

Yavitt, J.B., C.J. Williams, and R.K. Weider. 2004. Soil chemistry versus environmental controls on production of CH_4 and CO_2 in northern peatlands. Eur. J. Soil Sci. 56:169–178. doi:10.1111/j.1365-2389.2004.00657.x

Zhang, J.B., C.C. Song, and W.Y. Yang. 2006. Land use effects on the distribution of labile organic carbon fractions through soil profiles. Soil Sci. Soc. Am. J. 70:660–667. doi:10.2136/sssaj2005.0007

Zou, X.M., H.H. Ruan, Y. Fu, X.D. Yang, and L.Q. Sha. 2005. Estimating soil labile organic carbon and potential turnover rates using a sequential fumigation–incubation procedure. Soil Biol. Biochem. 37:1923–1928. doi:10.1016/j.soilbio.2005.02.028

Sources, Early Diagenesis, and Structure of Organic Matter in the Pearl River Delta

Yong Ran* and Yulong Zhang

Abstract

Sediment organic matter (OM) plays important roles in biological, chemical, and physical processes within the aquatic ecosystem, and arguably plays a major role in maintaining long-term ecological stability in the changing world. Despite its importance, there is a great deal still unknown about sediment OM chemistry. In this chapter, the development of sophisticated analytical methods has been reviewed for investigating sediment OM compositions, constituents, structure, transformation, and fates. The methods have also recently been applied to study the sources, early diagenesis, and fates of sediment OM in the Pearl River Delta and the northern South China Sea (NSCS), provided unique insight with respect to OM structures and compositions. It is found that soil OM, old OM (such as kerogen), and polymethene are important during the transport of OM from terrestrial ecosystem to marine ecosystem. The OM preservation pattern in the NSCS seems tightly related to the key factors including primary production, sediment accumulation rate, remineralization rate, and water oxygen concentration. The longstanding query regarding the sediment OM stability and recalcitrance would be better understood when comprehensive molecular methods are used in tandem to get insights on the structure, source, age, degradation stage, and interactions of specific OM components.

Abbreviations: BC, black carbon; BHP, bacteriohopanepolyol; CP-MAS, cross polarization–magic angle spinning; CSRA, compound specific radiocarbon analysis; DAPA, diaminopimelic acid; DI, degradation index; GC-MS, gas chromatography–mass spectrometry; KC, kerogen carbon; LC-MS, liquid chromatography–mass spectrometry; NHC, nonhydrolyzable C; NMR, nuclear magnetic resonance; NR, North River; NSCS, northern South China Sea; OM, organic matter; POC, particulate organic carbon; SML, surface mixed layer; SPM, suspended particulate matter; THAA, total hydrolyzable amino acids; TOC, total organic carbon; WR, West River.

Yong Ran and Yulong Zhang, State Key Laboratory of Organic Geochemistry, Guangzhou Institute of Geochemistry, Chinese Academy of Sciences, 511 Kehuajie Tianhe, Guangzhou 510640, China (gammalong@163.com). *Corresponding author (yran@gig.ac.cn).

doi:10.2136/sssaspecpub62.2014.0047

Labile Organic Matter—Chemical Compositions, Function, and Significance in Soil and the Environment. SSSA Special Publication 62. Zhongqi He and Fengchang Wu, editors.

The major sources of OM in marine systems may be either marine (primary production) or terrestrial (eroded soil and plant material), with freshwater aquatic production generally contributing an insignificant component compared with the other two sources (Ittekkot, 1988; Lobbes et al., 2000; Meybeck, 1982; Onstad et al., 2000). Marine phytoplankton produces OM in the euphotic zone of the ocean in response to the controlling factors that modify nutrient supply and light. Terrigenous OM including humic substances is supplied to marine systems by both river input and coastal erosion, with supply by aeolian processes relatively minor in coastal (shelf) seas (Stein, 1991). In global carbon budgets (Eglinton and Repeta, 2004) and most local or regional budgets (cf. Durrieu de Madron et al., 2000; Johannessen et al., 2003; Macdonald et al., 1998; Thomas et al., 2005), marine primary production supplies a much larger quantity of OM than the combined terrigenous sources. Nevertheless, the terrestrial and marine components of the organic carbon cycle both need to be studied because their behaviors and fates are affected by intrinsic (compositional) differences and by the influence of different important processes (Hedges and Keil, 1995). Terrigenous and marine sources and the major active processes affecting each component may also respond differently to global climate change (Chen et al., 2003a, 2003b). Terrigenous OM generally includes a large fraction of biologically resistant (recalcitrant) compounds (e.g., humic substances) such as lignin, kerogen, and OM, which were already strongly degraded by soil microbes (Bergamaschi et al., 1997; Hedges and Oades, 1997). Marine OM is generally much more labile and, for example, grazing by zooplankton frequently limits phytoplankton biomass (Wassmann et al., 2004). Only about 20% of net primary production settles out of ocean surface waters and fluxes decrease exponentially down the water column, with less than 10% of surface production generally remaining at depths of hundreds of meters and only about 1% in deep ocean areas (Betzer et al., 1984; Hedges and Keil, 1995; Suess, 1980; Wassmann et al., 2004). Despite its intrinsic recalcitrance, terrigenous OM can also be recycled in the marine environment. Recent terrigenous OM is recycled quite efficiently in some places, for example, >65% on the Beaufort Sea shelf (Goñi et al., 2005). Ancient terrigenous OM fractions also comprise kerogen or old soil OM and humic substances, albeit to a lesser extent (<25% [Goñi et al., 2005]). Terrigenous OM supplied by rivers undergoes intensive reworking in estuaries and deltas (Eglinton and Repeta, 2004), with some estimates suggesting losses of as much as two-thirds of the global fluvial input of particulate OM (POM) in these environments (Keil et al., 1997). The fractions of terrigenous OM lost to leaching (conversion to dissolved OM) vs. completely mineralized (to CO_2) vs. exported or buried are probably highly variable, and the controlling mechanisms in most systems are not yet well known.

The small portion of marine OM and variable portion of terrigenous OM evading recycling in the water column and surface sediments is ultimately preserved and buried in seafloor sediments (Hedges and Keil, 1995). Although the detailed mechanisms of OM preservation in sediments are not yet clear, several principal factors contribute (cf. Eglinton and Repeta, 2004). In addition to the intrinsic properties (recalcitrance) of the compounds comprising OM, preservation and burial is enhanced by high sedimentation rates, stabilization of OM by sorption onto sedimentary mineral surfaces, and sustained anoxic conditions in the sediments (Hedges and Keil, 1995; Henrichs, 1995). Coastal marine sediments show large differences in their OM content and composition, which reflects variations in the supply of marine and terrigenous OM, the environmental conditions under which the OM was produced, and the transport and transformation processes controlling OM fates.

Geochemical Investigation of Marine Organic Matter

Sediments represent a repository for OM entering and produced within the ocean and, as such, form a basis from which the organic carbon cycle may be studied over various space and time scales (Altabet and Francois, 1994; Calvert et al., 1992; Hedges et al., 1982; Muzuka and Hillaire-Marcel, 1999; Schubert and Stein, 1996; Stein, 1991). Of principal importance in applying sediment core data to the study of organic carbon cycling is an understanding of the sedimentary context of the cores and, in particular, estimation of sedimentation rate. Sedimentation rates provide a basis for estimating dates associated with various sediment core sequences and thereby obtaining a record of changes in the OM within the system, and they also allow estimation of rates of other processes (e.g., organic carbon burial). The atmospherically delivered natural radionuclide ^{210}Pb (half-life 22.26 yr) has been widely used to estimate accumulation rates in recent sediments, including sediments from the Arctic Ocean (Baskaran and Naidu, 1995; Huh et al., 1997) and a few cores from Hudson Bay (d'Anglejan and Biksham, 1988; Lockhart et al., 1998). The possible influence of surface mixing processes (e.g., bioturbation) on the depth distribution of tracer substances is an important consideration in marine environments (Anderson et al., 1988). The influence of mixing processes may be addressed through ^{210}Pb modeling approaches that incorporate a surface mixed layer (SML) and a deeper layer where the sediments are relatively unmixed (Carpenter et al., 1985; Lavelle et al., 1986). Sedimentation rates are derived from the portions of the ^{210}Pb profiles below the SML. Additional sedimentary tracers improve confidence in ^{210}Pb-derived rates (Boer et al., 2006; Smith, 2001). Caesium-137 (half-life 30.0 yr), which has a well-known input history associated with fallout from atomic weapons testing (largely between 1952 and 1964), has been widely used as an independent time marker to verify accumulation parameters derived from ^{210}Pb (Baskaran and Naidu, 1995; Fuller et al., 1999). Contaminants with known input histories (e.g., Pb) provide possible additional tracers.

Suspended and sedimentary OM in coastal areas is generally extremely heterogeneous, containing various particles resulting from a wide range of inputs and numerous transport and transformation processes. Characterizing OM through the application of geochemical proxies provides a means of querying the sediment record for information of relevance to organic carbon cycling (Hedges and Oades, 1997; Meyers, 1994; Stein and Macdonald, 2004). Elemental (carbon and nitrogen) ratios and stable carbon and nitrogen isotope ratios (13-C and 15-N) are among the most widely used 'bulk' proxies providing information about OM origin (Meyers, 1994; Stein, 1991). Low carbon/nitrogen (C/N) ratios and high (or enriched) 13-C values in marine sediments both distinguish OM derived from marine sources (phytoplankton or ice algae) from contributions of terrigenous OM, which is characteristically nitrogen-poor and isotopically depleted (Fernandes and Sicre, 2000; Hedges et al., 1988b; Meyers, 1994; Ruttenberg and Goñi,

1997). Use of these proxies together helps confirm that differences are indeed related to the source and not postproduction processes (e.g., microbial degradation, zooplankton scavenging). Elemental and isotopic compositions of various source materials (e.g., terrestrial plant matter, marine phytoplankton) are well characterized in a general sense for Arctic waters (Gaye et al., 2007; Goñi and Montgomery, 2000; Lobbes et al., 2000; Naidu et al., 2000; Schubert and Calvert, 2001; Schubert and Stein, 1996), and additional site-specific samples may be analyzed to increase confidence in regional source signatures. Stable nitrogen isotope ratios (^{15}N) of sedimentary OM have also been used in some environments as a tracer of marine vs. terrigenous OM sources (enriched and depleted, respectively [Naidu et al., 2000]), but in other environments these values provide a proxy for surface water nutrient conditions when marine OM was being produced (Altabet, 2004; Altabet and Francois, 1994; Francois et al., 1992). Although geochemical bulk parameters such as C/N ratios and ^{13}C values are important proxies for identifying general origins of OM, specific biomarkers often provide more precise information about the specific sources contributing to complex mixtures of OM, like those commonly found in coastal sediments (Meyers, 1994; Stein, 1991; Stein and Macdonald, 2004). The detailed identification of terrigenous sources is clearly important for linking variation to changes in specific watershed sources (e.g., vegetation types) or processes (overland runoff vs. river bank erosion). One of the best proxies for terrestrial (vascular plant) carbon input is lignin because it is a major component of terrestrial vascular plants but essentially absent from marine organisms (Goñi and Hedges, 1995; Hedges and Mann, 1979). Lignin components differentiate angiosperm and gymnosperm sources, woody vs. nonwoody tissues, and fresh vs. degraded material. Nevertheless, biomarkers like lignin complement rather than replace bulk proxies because they generally only directly represent very small proportions of the total organic carbon (TOC) content of sediments; for example, lignin commonly composes only a few percent of the TOC content of the sediments (Goñi and Montgomery, 2000, 2005). Bulk proxies and specific biomarkers used in combination generally yield the most confident OM source apportionments and hence greater insight into the influence of transport and transformation processes.

The origins and diagenetic state of OM buried in marine sediments have been investigated by various techniques, including bulk parameters (e.g., C/N ratios and stable carbon isotopic compositions) and organic biomarkers (e.g., lignin phenols and alkanes). Specifically, stable carbon isotopic ratios and C/N ratios are frequently used to distinguish terrestrial and marine sources of sedimentary OM (Bianchi et al., 2002; Gordon and Goñi, 2003; Hu et al., 2006; Yu et al., 2010). Biomarkers derived from terrestrial plants, marine plankton, and bacteria can provide more specific and complementary data for tracing the origins of OM. For example, lignin phenols are very useful tracers of terrestrial OM because they are derived exclusively from vascular plants, and lignin-derived parameters, such as the ratio of syringyl-to-vanillyl phenols (S/V), cinnamyl-to-vanillyl phenols (C/V), and acid-to-aldehyde (Ad/Al) of vanillyl and syringyl phenols, are indicative of terrestrial plant sources and the extent of diagenetic alterations (e.g., Hedges and Mann, 1979; Opsahl and Benner, 1995). Total hydrolyzable amino acid content and composition can also provide insights about OM sources and diagenetic alterations (Cowie and Hedges, 1992; Dauwe and Middelburg, 1998; Kaiser and Benner, 2008; Keil et al., 2000). In addition, some amino acids, such as the D-enantiomers of alanine, glutamic, and aspartic acid, and diaminopimelic acid (DAPA), provide information regarding bacterial sources (Kaiser and Benner, 2008; Lomstein et al., 2009).

Solid-State Nuclear Magnetic Resonance Spectroscopy

Solid-state nuclear magnetic resonance (NMR) spectroscopy is commonly used to characterize complex humic substances (Preston, 1996; Conte and Piccolo, 1999; Mopper et al., 2007), including humic and fulvic acids (Thorn, 1987; Hatcher et al., 1989; Smernik et al., 2004) and humin (Preston and Newman, 1995). The primary technique used in these studies is the solid-state ^{13}C cross polarization–magic angle spinning (CP-MAS) technique. By using the solid-state ^{13}C CP-MAS spectroscopy, our understanding of chemical structures of humic substances has markedly improved over decades of extensive investigations (Thorn, 1987; Wilson, 1987). For example, it is well documented that algae-derived marine humic acids generally contain less aromatics than lignin-derived terrigenous ones (Mao et al., 2007; Mopper et al., 2007). Despite the knowledge gained by these NMR studies by using ^{13}C CP-MAS, the full potential of NMR for characterizing humic substances has yet to be realized (Mopper et al., 2007). Carbon-13 CP-MAS has drawbacks such as loss of structural information and poor quantification (Schmidt-Rohr and Spiess, 1994; Mao et al., 2000; Duer, 2004; Smernik et al., 2006). Moreover, because of their complex natures, ^{13}C CP-MAS or direct polarization-magic angle spinning NMR spectra of humic substances generally reveal broad, poorly resolved peaks pertaining to major functional groups. Therefore, ambiguous assignments can be caused by the overlapping chemical shifts of numerous different structural moieties (Mopper et al., 2007). Thus, other analytical methods have also been used in sediment OM characterization as reviewed in the sections below.

Biomarker Methods

Biomarker methods have vastly improved the fundamental understanding of OM composition and ecological responses to environmental change (Feng et al., 2010, Crow et al., 2009; Clemente et al., 2011). Biomarkers, which are analogous to OM fingerprints, are compounds that can be traced to a specific plant, microbial, or anthropogenic source because they retain their carbon backbones during abiotic and biotic degradation (Amelung et al., 1996). Biomarkers are used widely in marine organic geochemistry, but have also been used in soil OM studies to determine terrestrial OM sources as well as relative stages of decomposition (Hedges and Mann, 1979; Hedges et al., 1988a; Van Bergen et al., 1997; Bull et al., 2000). Many biomarkers are the products of chemolysis reactions on larger starting materials. For example, phospholipid fatty acids, lignin-derived phenols and base hydrolysis products (a variety of hydroxyalkanoic acids) are parts of larger constituents found in OM. Lignin phenol monomers and dimers are isolated after CuO oxidation, which targets ether bonds in polymeric lignin (Hedges and Mann, 1979; Goñi et al., 1993; Otto and Simpson, 2006a). Base hydrolysis is used to break ester bonds found in cutin and suberin biopolymers, releasing components that are amenable to analysis (Otto and Simpson, 2006b; Mendez-Millan et al., 2010). Biomarkers are predominantly measured by gas chromatography–mass spectrometry (GC-MS), a method that facilitates separation and accurate quantification of individual OM components (Simoneit 2005; Amelung et al., 1996). In most cases, biomarker compounds must be derivatized to make them more volatile before gas chromatography analysis. In some cases, larger intact biomarkers such as bacteriohopanepolyols (BHPs) (Talbot et al., 2001, 2003, 2007a,2007b; Xu et al., 2009) and branched glycerol dialkyl glycerol tetraethers (Weijers et al., 2006; Peterse et al., 2010; Zhu et al., 2011) are measured using liquid chromatography–mass spectrometry (LC-MS). With LC-MS, the side chain information for isolated BHPs provides an additional level of organism specificity, sometimes to the species level (Fox et al., 1998; Talbot et al., 2008; Xu et al., 2009). In contrast, analysis of the cleaved hopanoid by GC–MS is performed after removal of the side chain (Innes et al., 1997; Winkler et al., 2001; Shunthirasingham and Simpson, 2006). In addition to providing quantitative source information, biomarkers can also be used to study OM degradation and turnover. Organic matter can undergo a variety of processes in sediment, which include abiotic and biotic degradation as well as preservation of intact constituents. For example, common plant steroids include β-sitosterol, stigmasterol, and stigmastanol (Otto and Simpson, 2005). These steroids can be preserved or biodegraded. The ratio of the precursor biomarkers to the degradation products can be used to assess plant steroid fates in OM. Similarly, other biomarker ratios can be used to study a variety of processes related to OM dynamics, such as the fate of plant- or microbially-derived compounds and the impact of forest fires on OM composition. Acid-to-aldehyde ratios for lignin-derived phenols commonly are applied to study the stability of lignin in various ecosystems (Hedges et al., 1988a; Goñi et al., 1993; Crow et al., 2009; Thevenot et al., 2010; Clemente et al., 2011). For example, the acid-to-aldehyde ratio of vanillyl- and syringyl-type phenols increases with progressive lignin oxidation. Thus, it is a useful measure of lignin degradation. Other ratios relating to cutin and suberin also provide information regarding the relative degradation state of these resistant components of OM. Phospholipid fatty acid ratios can also be used to quantify microbial stress due to declining substrate availability (Pautler et al., 2010). Biomarker methods have several advantages over NMR techniques. Mass spectrometry is generally more sensitive than NMR and is able to detect low concentrations of biomarkers (nanogram range or lower depending on the instrument and compound of interest). Furthermore, separation via GC and LC enables the identification and quantification of individual components, whereas NMR spectra may have overlapping signals that prevent precise quantification of individual components. Biomarker methods can also be applied more routinely than NMR because of the relatively economical and low maintenance. Gas chromatography–mass spectrometry and LC-MS instruments are more widely available than more costly NMR spectrometers. Despite these advantages, biomarker methods only detected a small percentage of OM. For example, a sequential extraction method that released a large variety of biomarkers only accounted for 17% of the TOC in the sample (Otto and Simpson, 2007). This is in sharp contrast to the high percentage of OM that can be measured with solid-state ^{13}C NMR (practically all of the OM). It is noteworthy that biomarker and NMR methods are complementary (Otto and Simpson, 2007; Simpson et al., 2008). For example, degradation ratios for plant-derived steroids, cutin, suberin, and lignin phenols corresponded to degradation trends measured with alkyl/O-alkyl ratios by solid-state ^{13}C NMR for four prairie soils (Simpson et al., 2008). Finally, biomarker methods are targeted and selective methods, and one would employ a specific extraction procedure for a specific group (e.g., CuO oxidation to isolate lignin phenols [Otto and Simpson, 2007]). Nuclear magnetic resonance methods are nontargeted, and because NMR is nonselective, no a priori knowledge about the sample is needed. Thus, depending on the information required, NMR or biomarkers can be used.

Isotopic Methods

Radiocarbon isotope geochemistry can provide constraints on the quantity of kerogen in soils and sediments. Ancient sedimentary organic carbon is a significant component of the suspended load on active ocean margins (Blair et al., 2003; Goñi et al., 2005; Komada et al., 2005; Raymond et al., 2004). As the average kerogen C concentration of the suspended load is about 0.5%, it was estimated that kerogen carbon (KC) flux could be 80 Tg C/yr, accounting for about 30 to 50% of the organic carbon output of rivers to the world's oceans (Blair et al., 2003; Meybeck, 1993). Kerogen may contribute as much as 50 to 70% of the suspended particulate organic carbon (POC) in the Lanyang Hsi of Taiwan and the Santa Clara River of southern California (Kao and Liu, 1996; Masiello and Druffel, 2001). However, the radiocarbon method cannot directly isolate KC for property characterization (Blair et al., 2003; Goñi et al., 2005).

Sediment OM turnover and relative age can be assessed by using biomarkers in combination with isotopic methods (Amelung et al., 1996; Feng et al., 2011; Mendez-Millan et al., 2011) or radiocarbon techniques (Eglinton et al., 1996; Hou et al., 2010; Ingalls et al., 2010). For example, compound specific isotope analysis has been used to measure the turnover and dynamics of OM components in disturbed systems where new sediment carbon inputs carry a distinct ^{13}C isotopic signature as compared with that of native (older) sediment OM. Similar methods can be used when labeled compounds, such as when ^{13}C or ^{15}N enriched substrates, are introduced to the sediment (Kuzyakov and Bol, 2005; Providoli et al., 2006; Huygens et al., 2008; Dijkstra, 2009; Dijkstra et al., 2011; Glaser et al., 2012; Hatton et al., 2012). The variation in ^{12}C vs. ^{13}C isotopes (expressed as $\delta^{13}C$) can also be used to identify variations in C3 vs. C4 plant-derived OM as well as to investigate OM turnover mechanisms (Blagodatskaya et al., 2011; Christensen et al., 2011; Dijkstra et al., 2008, 2010; Mendez-Millan et al., 2012; Morgan et al., 2011; Torn et al., 2011). The use of radiocarbon to examine the fate of specific OM components in sediment over decadal and millennial timescales can be examined by simultaneously using $\Delta^{14}C$ to measure OM age and biomarkers to evaluate specific compounds. This technique, referred to as compound specific radiocarbon analysis (CSRA), was developed to study the fate of OM in marine systems (Eglinton et al., 1996). In the future, application of CSRA will develop our understanding of the difference between total sediment OM residence time and that of specific OM components.

Sources, Diagenesis, and Fate of Organic Matter in the Riverine and Marine Sediments

These methods have also recently been applied to study the sources, early diagenesis, and fates of sedimentary OM (Fig. 1 for the sampling sites) in the Pearl River Delta and the NSCS and have provided unique insight with respect to OM structure and composition. The Zhujiang (Pearl) River is one of the major rivers in China, with a length of 2214 km and a drainage area of 453,690 km^2. It flows to the South China Sea and consists of three main rivers: the Xijiang (West) River, the Beijing (North) River, and the Dongjiang (East) River. The West River basin is dominated by limestone, whereas in the East River basin granite and shale are the main rock types. The upper reaches of the West River drain the karstic limestone region in South China, whereas the East River drains the South China granite area. The North River basin is composed of different rock types, including limestone, shales, clastic rocks, and granite (Zhang et al., 2001). The climate of the Pearl River Delta is subtropical monsoon and humid, with an annual average temperature of about 22°C. In summer it is around 28°C and in winter about 14°C. The annual precipitation is more than 1500 mm. Biogeochemical processes and mechanical erosions are very intense in the Pearl River basin.

The South China Sea is the second largest marginal sea in the world. The deep central basin is bordered by two broad shelves, representing 52% of the South China Sea area (Higginson et al., 2003). The Pearl River discharges onto the northern shelf of the South China Sea (Fig. 1) (Hu et al., 2006). This river system is located in a tropical–subtropical region with warm temperatures, high rainfall, and elevated rates of biogeochemical processes (Ni et al., 2008; Ran et al., 2004, 2009). Agricultural land (C3 and C4 plants) and conifer-dominated forests (C3 plants) dominate in the lower drainage basin, and grasses (mostly C4 plants) dominate the upper drainage basin (Li et al., 2011; Sun et al., 1999). The Pearl River discharges ~80% of its annual water flux (~ 3.3×10^{11} m^3), and ~94% of its annual suspended sediment flux (~2.5×10^7 t) during flood periods between April and September (Higginson et al., 2003). The flux of particulate OC from the Pearl River is 9.2×10^5 t/yr and accounts for ~0.1 to 0.2% of the total OC exported by the worldwide rivers (Gao et al., 2002; Ni et al., 2008). The Pearl River estuary retains ~80% of the river's suspended sediments during the wet season, with the remainder being delivered to the adjacent shelf and slope (Hu et al., 2006; Zhang et al., 2014).

The NSCS is characterized by a relatively broad shelf that receives discharge from the Pearl River, the second largest river in China in terms of annual runoff flux (Fig. 1) (Higginson et al., 2003). Terrestrial OM is a notable component of sedimentary OM in the Pearl River estuary and its western coast (He et al., 2010; Yang et al., 2011; Yu et al., 2010; Zhang et al., 2009), but concentrations of terrestrial OM decline rapidly across the shelf (Hu et al., 2006, 2008). In the present study, we used multiple tracers of terrestrial and marine OM and a three end member mixing model to further investigate the sources and diagenetic transformations of sedimentary OM in the NSCS.

Sources and Diagenetic State of Suspended Particulate Organic Matter in the Pearl River

The average suspended particulate matter (SPM) concentration (21.4 ± 17.1 mg/L) in the Pearl River (Table 1) is significantly lower than in the Yangtze River (68 mg/L [Wu et al., 2007]) and Mississippi River (152 mg/L [Kendall et al., 2001]), whereas the average TOC concentration (3.7%) for SPM in the Pearl River (Zhang et al., 2014) is higher than in the Yangtze River (1.1% [Wu et al., 2007]) and Mississippi River (1.6% [Bianchi et al., 2007]). Soils and plankton can be important sources of SPM in rivers (Hedges et al., 1997), and the TOC concentration in soil

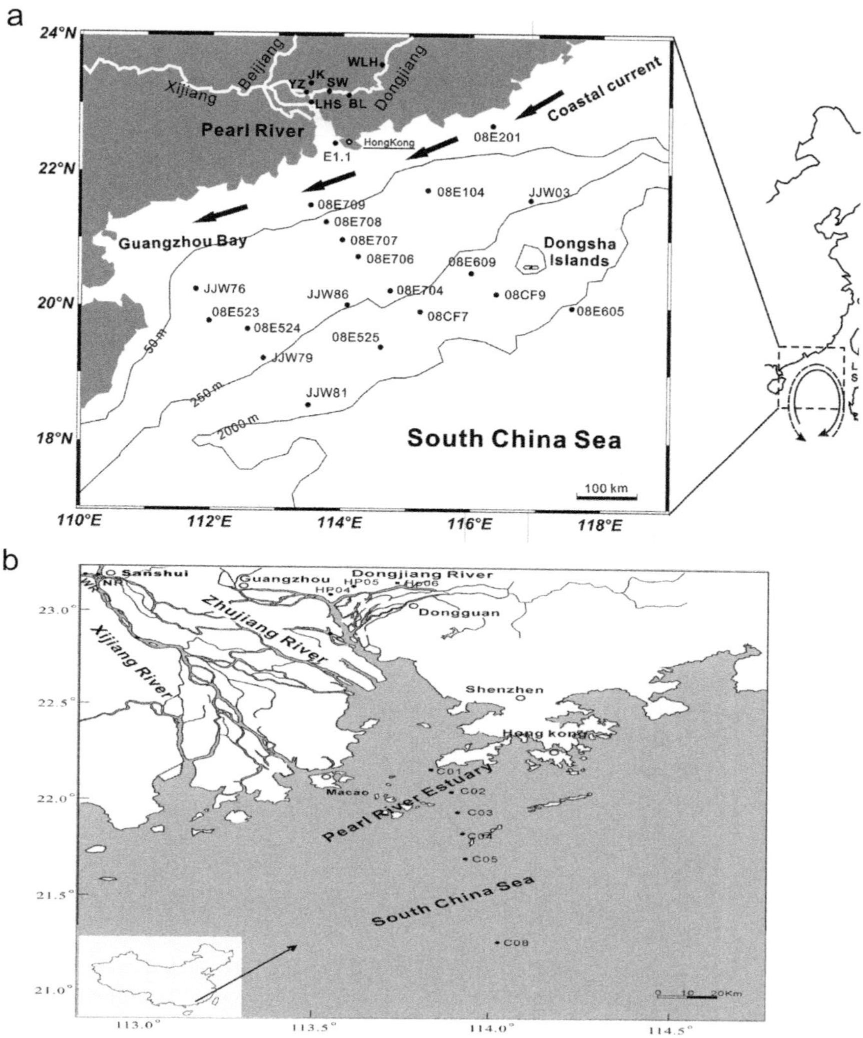

Fig. 1. (a) Sample locations in the Pearl River system and the South China Sea. The solid and the dashed curves with arrows represent the surface circulation in the summer and winter, respectively. (b) Sampling locations of the three soils and eight sediments in the Pearl River Delta and Estuary.

is much lower than in plankton (~30% [Cowie and Hedges, 1994]). It is noted that the TOC concentration in the SPM of the Pearl River is higher than in the soils (2.1% [Yu et al., 2010]) in the Pearl River catchment, probably indicating the occurrence of plankton-derived OM in SPM. The $\delta^{13}C$ values of the SPM in the Pearl River ranged from −36.9 to −26.3‰, with the lowest value observed in the JK reservoir sample (Table 1). The average $\delta^{13}C$ value (−28.3‰) is somewhat depleted in ^{13}C compared with several other large rivers, such as the Mississippi (−18.5‰

Table 1. Bulk properties of suspended particulate matter and sediments.†

Site	Date (Year/month)	Depth	SPM	TOC	TN	Δ¹³C	C/N
		m	mg/L	%	%	‰	
		SPM in the Pearl River					
Reservoir							
WLH	2009/11	0.5–1	2.12	5.76	0.67	−28.3	10.0
JK	2010/1	0.5–1	4.15	19.01	3.82	−36.9	5.8
River							
BL	2010/1	0.5–1	8.21	4.11	0.64	−30.9	7.5
LHS1	2010/3	0.5–1	34.4	1.35	0.13	−26.3	12.1
LHS2	2010/7	0.5–1	28.6	4.53	0.58	−29.0	9.1
SW	2010/7	0.5–1	16.7	3.43	0.50	−28.6	8.0
YZ1	2008/12	0.5–1	24.4	6.36	1.05	n.a.	7.1
YZ2	2009/4	0.5–1	52.4	4.36	0.43	−26.7	11.8
		Sediments in the South China Sea					
Inner shelf							
E1.1	2010/12	17		0.90	0.10	n.a.	11.0
08E201	2008/8	30		0.19	0.02	−21.8	10.3
08E709	2008/8	43		0.83	0.11	−21.6	8.8
Outer shelf							
08E708	2008/8	63		0.57	0.07	−21.5	9.1
08E707	2008/8	80		0.21	0.02	−21.4	10.8
JJW76	2010/4	82		0.19	0.03	−21.4	9.0
08E706	2008/8	84		0.51	0.07	−21.7	8.4
08E523	2008/8	93		n.a.	n.a.	n.a.	n.a.
08E104	2008/8	98		0.56	0.07	−21.1	8.8
08E524	2008/8	144		0.36	0.05	−21.7	7.9
08E704	2008/8	180		0.40	0.05	−21.3	9.1
JJW86	2010/4	190		0.25	0.03	−21.3	9.7
JJW79	2010/4	247		0.14	0.02	−21.6	10.3
JJW3	2010/4	260		0.15	0.02	−21.8	9.0
Slope							
08E609	2008/8	562		0.38	0.05	n.a.	8.6
08CF9	2008/8	937		0.55	0.09	−20.5	7.3
08CF7	2008/8	1094		0.54	0.07	−20.7	8.7
08E525	2008/8	1192		0.75	0.11	−20.9	7.9
JJW81	2010/4	1394		0.66	0.10	−20.6	8.1
08E605	2008/8	2379		n.a.	n.a.	n.a.	n.a.

† SPM, suspended particulate matter; n.a., not applicable; TOC, total organic carbon; TN, total nitrogen.

to −26.4‰ [Onstad et al., 2000]), Columbia (−27.9‰ [Kendall et al., 2001]), and Yangtze (−24.4‰ [Wu et al., 2007]). However, these $\delta^{13}C$ values are comparable with those reported previously for the Pearl River (−28.7‰ [Yu et al., 2010]). Yu et al. (2010) investigated the stable carbon isotopic compositions of vegetation in the Pearl River catchment and reported that C4 plants have enriched $\delta^{13}C$ (avg. −13.1‰) values compared with C3 plants (avg. −29.0‰) and SPM in the river. Riverine plankton are also depleted in $\delta^{13}C$ (avg. −30.0‰ [Kendall et al., 2001]) and have similar isotopic compositions as C3 vegetation in the Pearl River catchment. Overall, the relatively depleted $\delta^{13}C$ values of SPM in the Pearl River indicate a minor contribution of C4 plants. However, a relatively depleted $\delta^{13}C$ of −24.1‰ for the soils indicate that ~30% of soil OC in the Pearl River catchment could be derived from C4 plants, which is consistent with the fact that agricultural plants, including sugarcane and corn, are dominant in the downstream catchment (Yu et al., 2010). Furthermore, the C/N ratios (5.8–12.1) indicate particulate OM in the Pearl River is largely derived from both terrestrial OM (>8.0) and plankton (5.0–8.0) (Table 1). There is a significant correlation between $\delta^{13}C$ values and N/C ratios ($R^2 = 0.92$) of riverine particles, indicating soil OM and riverine plankton are the major sources of SPM (Zhang et al., 2014).

Proteins are more bioreactive than bulk carbon in natural OM, so the fraction of OC as amino acids decreases with increasing decomposition, and carbon-normalized yields of total hydrolyzable amino acids (THAA) (%OC) serve as indicators of the diagenetic state of OM (Cowie and Hedges, 1992; Cowie and Hedges, 1994; Davis et al., 2009). The THAA yields (OC%) of riverine SPM ranged from 22.6 to 37.4% and were only slightly lower than those (23.8–59.4%; avg. 38.1%) of fresh plankton (Table 2) (Cowie and Hedges, 1992; Kaiser and Benner, 2009). The high amino acid yields indicate riverine SPM is relatively fresh and has undergone minimal diagenetic alteration. The high THAA (OC%) yields indicate plankton are an important source of SPM in the lower reach of the Pearl River. The composition of amino acids also provides insights about the diagenetic state of OM, and a degradation index (DI) has been developed for this purpose (Dauwe and Middelburg, 1998). The DI values decrease from positive values (1–1.5) for fresh phytoplankton to negative values (less than −1) for diagenetically altered OM in deep sea sediments (Dauwe et al., 1999). The amino acid DI values for suspended particles in the Pearl River range from −0.5 to 0.8 (avg. 0.1) and are generally indicative of minor to moderate alteration (Table 2).

The carbon-normalized yields of lignin phenols (Λ_8) in riverine suspended particles ranged from 0.06 to 1.12 mg/100 mg OC (Table 3). The reservoir samples (WLH and JK) had low lignin yields, which is consistent with a substantial plankton component in these samples. Overall, the suspended particles in the Pearl River had lower lignin yields than in the Mississippi River (1.40 mg/100 mg OC [Bianchi et al., 2007]), the Atchafalaya River (2.82 mg/100 mg OC [Gordon and Goñi, 2003]), and several other North American rivers (1.33 mg/100 mg OC [Onstad et al., 2000]). The significant contribution of plankton to SPM in the Pearl River is responsible for the low C-normalized yields of lignin. The ratios of syringyl-to-vanillyl phenols (S/V) ranged from 0.72 to 1.22, reflecting the predominant contribution of angiosperm vegetation (Zhang et al., 2014).

The (Ad/Al)v and (Ad/Al)s ratios in suspended particles ranged from 0.37 to 0.75 and 0.35 to 0.59, respectively (Table 3). Elevated Ad/Al ratios (>0.4) are indicative of degraded OM material and are commonly observed in soil-derived OM

Table 2. Concentrations and yields of total hydrolyzable amino acids, degradation indices, and yields of diaminopimelic acid in the samples.†

Site	THAA		DI	DAPA
	μmol/g	%OC‡		nmol/mg OC
		SPM in the Pearl River		
Reservoir				
WLH	256.0	22.6	0.0	4.2
JK	1321.9	37.4	0.8	5.3
River				
BL	285.8	35.6	0.0	4.4
LHS1	61.6	23.3	0.1	n.a.
LHS2	247.6	28.2	0.1	4.2
SW	156.5	22.7	−0.5	6.9
YZ1	360.0	29.4	0.4	4.4
YZ2	196.4	23.2	0.2	5.4
		Sediments in the South China Sea		
Inner shelf				
E1.1	16.4	9.2	−0.3	10.9
08E201	5.2	13.9	−0.6	n.a.
08E709	12.9	7.8	−0.2	21.3
Outer shelf				
08E708	7.9	7.9	2.0	9.7
08E707	5.0	12.1	−0.4	13.1
JJW76	5.6	14.2	−0.7	46.4
08E706	5.1	4.9	−1.0	n.a.
08E523	7.5	n.a.	−0.8	n.a.
08E104	14.1	12.4	−0.8	10.7
08E524	7.1	9.6	−0.9	19.8
08E704	5.7	7.1	−0.3	19.9
JJW86	4.4	8.7	−0.6	29.2
JJW79	5.7	19.4	−0.2	140.9
JJW3	3.8	11.8	−1.2	41.1
Slope				
08E609	12.8	16.8	−0.7	18.7
08CF9	8.2	7.4	−1.0	10.8
08CF7	9.8	8.9	−0.9	13.5
08E525	13.1	8.3	−1.2	9.4
JJW81	13.1	9.4	−1.3	14.6
08E605	13.3	n.a.	−1.8	n.a.

† THAA, Total hydrolyzable amino acids; DI, degradation indices; DAPA, diaminopimelic acid.

‡ THAA (%OC) is percentage of OC presents as amino acids in sediment TOC.

Table 3. Yields of lignin phenols and phenol ratios in the samples.

Site	V	S	C	Λ_8†	C/V	S/V	(Ad/Al)v‡	(Ad/Al)s‡	3,5-BD/V
	——— mg/100 mg OC ———								
				SPM in the Pearl River					
Reservoir									
WLH	0.02	0.02	0.03	0.06	1.43	0.93	0.61	0.35	0.42
JK	0.02	0.03	0.03	0.07	1.12	1.19	0.75	0.57	1.12
River									
BL	0.12	0.15	0.10	0.38	0.84	1.22	0.74	0.59	0.19
LHS1	0.48	0.48	0.16	1.12	0.33	1.00	0.50	0.46	0.11
LHS2	0.40	0.32	0.11	0.84	0.27	0.80	0.42	0.40	0.06
SW	0.20	0.23	0.11	0.54	0.55	1.13	0.66	0.54	0.17
YZ1	0.39	0.28	0.10	0.77	0.25	0.72	0.48	0.47	0.11
YZ2	0.38	0.31	0.13	0.83	0.35	0.81	0.37	0.37	0.06
				Sediments in the South China Sea					
Inner shelf									
E1.1	0.52	0.47	0.14	1.12	0.27	0.90	0.44	0.39	0.08
08E201	0.17	0.12	0.03	0.32	0.17	0.69	0.44	0.38	0.13
08E709	0.09	0.07	0.03	0.19	0.37	0.79	0.45	0.41	0.12
Outer shelf									
08E708	0.09	0.06	0.04	0.19	0.40	0.70	0.59	0.53	0.14
08E707	0.03	0.02	0.02	0.07	0.52	0.55	0.61	0.45	0.58
JJW76	0.07	0.04	0.02	0.13	0.36	0.63	0.59	0.50	0.35
08E706	0.02	0.01	0.01	0.03	0.33	0.73	0.81	0.60	0.68
08E523	n.a.	n.a.	n.a.	n.a.	0.37	0.63	0.77	0.49	1.10
08E104	0.09	0.05	0.03	0.16	0.31	0.58	0.58	0.51	0.23
08E524	0.13	0.08	0.02	0.23	0.13	0.66	0.58	0.51	0.28
08E704	0.07	0.05	0.02	0.13	0.24	0.81	0.73	0.60	0.36
JJW86	0.02	0.02	0.01	0.04	0.36	0.73	1.04	0.54	1.38
JJW79	0.07	0.05	0.02	0.14	0.34	0.69	0.69	0.49	0.58
JJW3	0.08	0.05	0.01	0.14	0.14	0.70	0.49	0.41	0.28
Slope									
08E609	0.18	0.12	0.03	0.33	0.16	0.64	0.68	0.59	0.37
08CF9	0.04	0.03	0.01	0.08	0.38	0.90	0.76	0.54	0.65
08CF7	0.03	0.02	0.01	0.07	0.49	0.78	0.93	0.65	0.68
08E525	0.02	0.02	0.02	0.06	1.01	0.87	0.96	0.71	0.99
JJW81	0.02	0.02	0.02	0.05	0.96	0.86	1.17	0.61	1.23
08E605	n.a.	n.a.	n.a.	n.a.	0.58	0.94	1.01	0.66	0.70

† V, S, and C are yields of total vanillyl, total syringyl, and total cinnamyl phenols, respectively.

‡ Λ_8 = V + S + C; and (Ad/Al)v and (Ad/Al)s are the ratios of acid/aldehyde of vanillic and syringic phenols, respectively.

(Gordon and Goñi, 2003). Based on Ad/Al ratios, it appears soil OM was present in SPM from the Pearl River. In addition, the CuO oxidation product, 3,5-dihydroxybenozic acid (3,5-BD), accumulates in soils relative to vanillyl phenols, and elevated ratios of 3,5-BD/V (>0.1) provide additional evidence of soil-derived OM (Gordon and Goñi, 2003; Houel et al., 2006; Kuzyk et al., 2008; Louchouarn et al., 1997). The 3,5-BD/V ratios ranged from 0.06 to 1.12 in this investigation (Table 3), further indicating that soils were an important source of terrigenous OM in suspended particles (Zhang et al., 2014).

Sources of Sedimentary Organic Matter in the Northern South China Sea

Total organic carbon concentrations were <1.0% in sediments throughout the NSCS, with relatively elevated values occurring on the inner shelf and slope (Table 1). A linear regression of total nitrogen vs. TOC in sediments had an intercept of zero, indicating sedimentary nitrogen is predominantly of organic origin (Zhang et al., 2014). The C/N ratio in sedimentary OM ranged from 7.3 to 11.0 (Table 1). The C/N ratios decreased with increasing water depth, suggesting an increasing contribution of marine OM in outer shelf and slope sediments.

The stable carbon isotopic composition ($\delta^{13}C$) of OM ranged from −21.8 to −20.5‰ (avg. −21.3‰) (Table 1), indicating a predominantly marine origin (Zhang et al., 2014; Tesi et al., 2007). The $\delta^{13}C$ values gradually became more enriched in ^{13}C from the inner shelf (−21.7‰) to the outer shelf (−21.5‰) and slope (−20.7‰). This pattern was consistent with declining C/N ratios along this gradient, further supporting the increasing contributions of marine OM offshore. Goñi et al. (1998) and Gordon and Goñi (2003) observed a similar enrichment of $\delta^{13}C$ values in northern Gulf of Mexico sediments. They attributed this trend, in part, to an increasing contribution of C4 grass-derived soil OM from the Mississippi River drainage basin. Soils in the Pearl River drainage basin have an average $\delta^{13}C$ composition of −24.1‰ (Yu et al., 2010), indicating a potential contribution of C4-derived OM to sediments in the NSCS margin (Zhang et al., 2014).

Lignin yields ranged from 0.03 to 1.12 mg/100 mg OC and generally decreased from the inner shelf to the slope (Table 3). Yields in NSCS sediments were generally lower than those in other marginal sediments (e.g., Prahl et al., 1994; Requejo et al., 1991; Sánchez-García et al., 2009). The NSCS shelf is relatively broad (250 km), resulting in extensive remobilization and decomposition before burial (Blair and Aller, 2012; Burdige, 2005). Declining lignin yields across the shelf most likely result from biodegradation and hydrodynamic sorting. The latter process can lead to the retention of coarse plant debris in the delta and estuary, and the preferential delivery of fine soil OM to offshore sediments (Keil et al., 1998).

Diaminopimelic acid is a unique amino acid that is found in the cell wall polymer, peptidoglycan, of Gram negative bacteria (Jørgensen et al., 2003). The occurrence of DAPA in marine sediments is indicative of both living bacteria and remnants of bacterial cells (Lomstein et al., 2009). The DAPA yields in the majority of NSCS sediments fall in the same ranges as those of the investigated marine bacteria, indicating that bacteria are an important source of amino acids in these sediments (Table 2) (Zhang et al., 2014).

Diagenesis of Sedimentary Organic Matter in the Northern South China Sea

The amino acid DI values in NSCS sediments ranged from −1.8 to −0.2 (Table 2). The DI values decrease from the inner shelf to the slope, indicating an increasing degree of diagenetic alteration with water depth. Amino acids comprised a major fraction of SPM in the Pearl River, but it is unlikely they survive transport from the river to marine sediments because of their high reactivity (Zhang et al., 2014; Chen et al., 2004). In addition, marine OM appears to be the dominant source of amino acids in these NSCS sediments. Previous studies in the South China Sea reported that primary production was 406 mg C m^{-2} d^{-1} (Liu et al., 2007), and it appears that only 0.1 to 0.2% of the carbon is buried in sediments (Zhao et al., 2009).

The THAA (%OC) values ranged from 4.9 to 19.4% in NSCS sediments and were generally in the range reported for ocean margin sediments (Zhang et al., 2014; Keil et al., 2000). No consistent pattern of THAA (%OC) was observed from inner shelf to slope sediments. It appears the different sources of amino acids in sediment OM have variable THAA (%OC) yields and amino acid composition, leading to varying patterns in the yields and compositions of amino acids in NSCS sediments.

Ratios of (Ad/Al)v and 3,5-BD/V can be used to indicate the extent of diagenetic alterations and sources of terrestrial OM (Gordon and Goñi, 2003; Houel et al., 2006; Kuzyk et al., 2008). Ratios of (Ad/Al)v and 3,5-BD/V ranged from 0.44 to 1.17 and 0.08 to 1.38, respectively, in the NSCS sediments (Table 3). Most of the sediments had high ratios of Ad/Al (>0.40) and 3,5-BD/V (>0.10), and are consistent with a soil OM source (Kuzyk et al., 2008). Furthermore, the slope sediments exhibited 3,5-BD/V and (Ad/Al)v ratios (Table 3) which are similar to those of mineral-bound soil OM, whereas the inner shelf sediments have ratios indicating contributions from the surface litter layer of soils (Zhang et al., 2014).

Increasing ratios of (Ad/Al)v and 3,5-BD/V were observed from shelf to slope sediments (Table 3), indicating increasing diagenetic alterations of terrestrial OM with water depth (Zhang et al., 2014). Hydrodynamic sorting, which leads to the preferential delivery of soil OM to offshore sediments (Keil et al., 1998), could also contribute to the observed increase in (Ad/Al)v and 3,5-BD/V with increasing water depth. Several high (Ad/Al)v ratios in the outer shelf sediments indicated diagenetically altered OM, which was consistent with the elevated 3,5-BD/V diagenetic indicator (Table 3). Although ocean kelp is a possible source for 3,5-BD (Goñi et al., 2000), the strong correlation between 3,5-BD/V and (Ad/Al)v ratios indicated that most 3,5-BD was derived from the land.

Sources and Compositions of Nonhydrolyzable Carbon and/or Black Carbon

Nonhydrolyzable C (NHC) and black C (BC) in three soils and 22 bulk and size-fractionated sediments from the Pearl River Delta and Estuary, South China (Fig. 1b and Table 4), were measured on treatment by a HCl/HF/trifluoroacetic acid method, and by a combustion method at 375°C (Ran et al., 2007, 2009; Nguyen et al., 2004; Sun et al., 2008). They were also characterized by elemental analysis, radiocarbon accelerated mass spectroscopy, solid-state ^{13}C CP-MAS NMR. The OC contents of the investigated samples ranged from 0.076% to 1.20% in the Pearl River region except for the North River (NR) sample (Table 4). As the NR river

Table 4. Percentages of nonhydrolyzable organic carbon, black carbon, and total organic carbon in the soils and sediments.†

Site	OC_o	OC_t	NHC/OC	NHC	BC/OC	BC
	——— % (w/w) ———		%	% (w/w)	%	% (w/w)
			Marine sediments			
C01	1.02 ± 0.079‡	36.9	35.4	0.361	15.3	0.133
C02	0.336 ± 0.031	30.7	31.4	0.106	9.15	0.032
C03	0.618 ± 0.020	23.2	34.3	0.212	17.3	0.049
C04	0.680 ± 0.036	4.97	34.9	0.237	4.14	0.012
C05	0.536 ± 0.027	23.6	25.6	0.137	4.73	0.032
C08	0.568 ± 0.017	34.4	45.4	0.258	nd	nd
			Soils			
HP04	1.77 ± 0.140	50.7	70.5	1.248	12.5	0.633
HP05	3.35 ± 0.177	52.3	64.8	2.171	10.3	1.14
HP06	2.47 ± 0.047	55.8	55.3	1.349	16.2	0.753
			River sediments and their fractions			
WR	0.446 ± 0.026	27.3	48.2	0.215	nd	nd
WR-A§	0.336 ± 0.016	0.328	28.2	0.095	nd	nd
WR-1¶	0.089 ± 0.023	0.140	57.7	0.060	nd	nd
WR-2	0.160 ± 0.022	0.196	44.9	0.072	nd	nd
WR-3	0.422 ± 0.107	4.74	44.8	0.189	nd	nd
WR-4	0.520 ± 0.003	25.5	37.1	0.193	nd	nd
WR-5	0.801 ± 0.013	28.5	28.9	0.231	nd	nd
C08-2¶	0.320 ± 0.009	2.19	29.8	0.095	nd	nd
C08-3	0.420 ± 0.001	22.5	37.8	0.159	nd	nd
C08-4	0.220 ± 0.030	1.74	75.7	0.167	nd	nd
C08-5	1.20 ± 0.000	25.2	84.7	1.017	nd	nd
NR-1¶	3.67 ± 0.335	59.8	77.7	2.852	nd	nd
NR-2	1.78 ± 0.020	29.5	62.3	1.109	nd	nd
NR-3	1.22 ± 0.155	42.8	67.2	0.819	nd	nd
NR-4	0.708 ± 0.027	17.4	78.9	0.558	nd	nd
NR-5	1.50 ± 0.013	45.3	74.5	1.12	nd	nd

† NHC, nonhydrolyzable organic carbon; BC, black carbon; OC, organic carbon; nd, not detected.

‡ Represents averages and standard deviations of total organic carbon.

§ WR represents the residual fraction remained after humic acid extraction.

¶ 1, 2, 3, 4, 5 represent >180 μm, 180–61 μm, 61–30 μm, 30–2 μm, and 2 μm fractions, respectively.

sediment was collected near a busy harbor, the sediment size fractions generally contained relatively high OC contents, ranging from 0.708 to 3.67%. The three soils samples were collected from an industrial region, and their OC contents of 1.77 to 3.35% were comparable to those of the NR samples (Table 4).

The West River (WR) and NR sediments and one estuary sediment (C08) were fractionated into five size fractions (>180 μm, Fraction 1; 180–61 μm, Fraction 2; 61–30 μm, Fraction 3; 30–2 μm, Fraction 4; and 2 μm, Fraction 5), respectively (Fig. 1b and Table 4). The river sediments contained coarser sand particles (>180 μm and 180–61 μm fractions) than the estuary sediment (C08). The silt fraction (61–30 μm), fine silt fraction (30–2 μm), and clay (<2 μm) fraction in the C08 estuary sediment was slightly higher than in the NR river sediment and much higher than in the WR river sediment (Ran et al., 2007, 2009). The OC contents in the bulk sample and in the six size fractions were comparable (Table 4).

The NHC contents in respect to sediment and soil weights varied from 0.060 to 2.85%, accounting for 25.6 to 84.7% of the OC contents (Table 4). The two fine fractions in C08, and three soils, and NR size fractions with NHC/OC ranging from 55.3% to 84.7%, were generally richer in NHC than the other sediment samples with NHC/OC, ranging from 25.6 to 57.7% (Table 4). The average NHC accounted for 51.9% of the OC in the 25 samples with a standard deviation of 19.6%.

Black carbon accounted for 4.14 to 17.3% of the OC in the soil and sediment samples (Table 4). The BC contents on average accounted for 11.2% of the OC contents in the samples with a standard deviation of 5.03%. This value was close to the mean value of 9% reviewed from around 300 sediments globally (Cornelissen et al., 2005). In previous studies, charcoal was reported to account for 5 to 45% of OC for soils (Song et al., 2002), and BC contents were within ranges of 15 to 30% OC for marine sediments (Middelburg et al., 1999).

The O/C atomic ratios of the isolated NHC in the soil and sediment samples ranged from 0.11 to 0.34, and the H/C atomic ratios ranged from 0.56 to 1.06 (Table 5). The O/C and H/C atomic ratios, when plotted on the Van Krevelen diagram, were similar to the type of kerogen III derived mainly from high plant materials, which was rich in lignin with aromatic structure. The degree of maturation increased as the ratios decreased (Li et al., 2002; Ran et al., 2009). The O/C ratios for NHC were much lower than the average O/C ratios (0.53) for typical humic acids extracted from soils and sediments (Allen-King et al., 2002; Gunasekara and Xing, 2003; Kang and Xing, 2005). The BC samples measured by the combustion method at 375°C from the three soil samples had atomic O/C and H/C ratios of 0.078 to 0.13 and 0.09 to 0.12, respectively (Table 5), suggesting that BC had higher degree of maturation than KC, which is similar to NHC.

The C/N ratios in all the samples ranged from 21 to 208, suggesting that the NHC and BC fractions contained very low N chemicals (Table 5). For the surface sediments in the river and the sea, the C/N ratios generally ranged from 5 to 10 (Wang et al., 1998). However, the terrestrial and ancient OM had much higher C/N ratios (Hedges and Keil, 1995). Based on the C/N ratios, it is concluded that the acid hydrolysis method with trifluoroacetic acid and hydrochloride acid in this study could efficiently remove young OM such as proteins, peptides, and nuclei acids (Ran et al., 2007).

The $\Delta^{14}C$ values for the selected NHC samples ranged from −321 to −431‰ (Table 6). Comparable ^{14}C values for acid (6 M HCl) insoluble carbon were also reported for other coastal and marine sediments (Wang et al., 1998). These compositions translated to conventional ^{14}C ages of 3840 to 4530 yr before present in the samples (Ran et al., 2009). The old ages suggested significant contribution of ancient carbon sources to the sediments. This ancient OM included bitumen or kerogens found in sedimentary rocks that had been thermally matured, or the entrainment of terrigeneous OM pre-aged as a result of long residence times within the drainage basin. The latter OM also contained chemically recognizable biochemicals such as, for example, lignin, albeit in a more extensively degraded

Table 5. Elemental (C, H, N, O) compositions of the isolated kerogen carbon and black carbon, and their atomic H/C, O/C, C/N ratios.

Sample	C	H	N	O	H/C	O/C	C/N
	% (w/w)						
NHC							
C01	36.9 ± 0.21	2.49 ± 0.04	0.38 ± 0.00	9.260.67	0.81	0.19	113
C02	30.7 ± 0.14	2.41 ± 0.19	0.34 ± 0.13	10.8 ± 0.72	0.94	0.27	105
C03	23.2 ± 0.49	1.82 ± 0.04	0.23 ± 0.02	7.49 ± 0.06	0.94	0.24	118
C04	4.97 ± 0.02	0.378 ± 0.02	0.13 ± 0.04	2.24 ± 0.01	0.91	0.34	45
C05	23.6 ± 0.57	1.80 ± 0.04	0.28 ± 0.02	8.79 ± 0.66	0.92	0.28	98
C08	34.4 ± 0.49	2.71 ± 0.25	0.46 ± 0.15	8.8 ± 0.35	0.94	0.19	87
HP04	50.7 ± 0.14	2.45 ± 0.07	0.32 ± 0.05	13.6 ± 010.	0.58	0.2	185
HP05	52.3 ± 0.42	3.72 ± 0.05	0.56 ± 0.08	16.7 ± 1.3	0.85	0.24	109
HP06	55.9 ± 0.49	2.63 ± 0.01	0.36 ± 0.03	12.5 ± 0.07	0.56	0.17	181
WR	27.3 ± 0.00	2.42 ± 0.02	0.78 ± 0.03	11.2 ± 0.09	1.06	0.31	41
WR-A	0.328 ± 0.02	0.046 ± 0.00	0.18 ± 0.02	0.10 ± 0.01	1.69	0.22	2
WR-1	0.14 ± 0.01	0.117 ± 0.02	0.16 ± 0.00	0.03 ± 0.01	8.6	0.14	1
WR-2	0.20 ± 0.01	0.028 ± 0.01	0.34 ± 0.06	0.05 ± 0.00	1.71	0.18	1
WR-3	4.74 ± 0.09	0.158 ± 0.02	0.75 ± 0.24	nd	0.4	nd	7
WR-4	25.5 ± 0.35	1.76 ± 0.04	1.40 ± 0.09	5.41 ± 0.06	0.83	0.16	21
WR-5	28.5 ± 028.	2.22 ± 0.04	1.15 ± 0.12	nd	0.94	nd	29
C08–2	2.19 ± 0.03	1.19 ± 001	0.08 ± 0.00	nd	6.55	nd	32
C08–3	22.5 ± 0.57	1.46 ± 0.04	0.38 ± 0.14	5.11 ± 0.02	0.78	0.17	69
C08–4	1.74 ± 0.10	0.145 ± 0.02	0.04 ± 0.00	0.25 ± 0.03	1	0.11	51
C08–5	25.2 ± 0.78	2.02 ± 0.17	0.59 ± 0.32	4.90 ± 0.24	0.96	0.15	50
NR-1	59.8 ± 0.28	4.24 ± 0.27	0.430 ± .09	13.3 ± 0.36	0.85	0.17	162
NR-2	29.5 ± 0.64	1.44 ± 0.01	0.20 ± 0.01	3.83 ± 0.23	0.58	0.1	172
NR-3	42.8 ± 0.99	2.24 ± 0.06	0.24 ± 0.05	8.87 ± 0.35	0.63	0.16	208
NR-4	17.4 ± 0.92	1.01 ± 0.05	0.12 ± 0.02	5.20 ± 0.35	0.7	0.22	169
NR-5	45.30 ± .21	4.16 ± 0.38	0.35 ± 0.04	12.6 ± 0.47	1.1	0.21	151
BC							
HP04	16.4 ± 0.35	0.528 ± 0.06	0.118 ± 0.02	1.72 ± 0.08	0.09	0.078	162
HP05	17.3 ± 0.49	0.772 ± 0.19	0.177 ± 0.04	3.09 ± 0.09	0.12	0.13	114
HP06	34.1 ± 0.71	0.825 ± 0.09	0.262 ± 0.002	5.33 ± 0.55	0.09	0.12	152

and altered state (Goñi et al., 2005). However, quantitatively discriminating between the incorporation of fossil OC and the entrainment of terrigeneous OM from old soils or paleosols was difficult and beyond the scope of this work. For the purpose of the following discussion, we assumed that all of the ancient OC was fossil in origin (i.e., ^{14}C dead). By using isotopic mass balance, it was possible to quantitatively constrain ancient and modern carbon contributions in these samples (Blair et al., 2003; Goñi et al., 2005). Assuming simple two end member mixing, the fraction of modern OC (f_m) and ancient OC in each sample ($f_a = 1\text{-}f_m$) could be estimated from the $\Delta^{14}C$ values presented in Table 6 by the following equation:

$$\Delta^{14}C = f_m \Delta^{14}C_m + f_a \Delta^{14}C_a \qquad [1]$$

where $\Delta^{14}C_m$ and $\Delta^{14}C_a$ represented the $\Delta^{14}C$ signature of the modern and ancient end members, respectively. Values of −1000 and +100‰ were used for$\Delta^{14}C_a$ and-$\Delta^{14}C_m$ as suggested by Blair et al. (2003). The isotopic mass balance calculations quantitatively demonstrated the importance of ancient OM, which accounted for 39 to 49% of the OC in the NHC samples (Table 6). The average value of f_a in the five NHC samples was 44% of the OC (Table 6). The above result is generally consistent with other recent investigations.

Ancient sedimentary organic carbon was a significant component of the suspended load on active margins and globally (Blair et al., 2003). Kerogen may contribute as much as 50 to 70% of the suspended POC from the Lanyang of Taiwan and the Santa Clara River of southern California (Kao and Liu, 1996; Masiello and Druffel, 2001). The nonprotein alkyl carbon consisted of 20.4 to 35.8% of OC when OC abundance was higher than 1% in several ocean sediments (Gélinas et al., 2001). Schmidt et al. (1996) reported the influence of brown coal emissions from a briquette factory on soil OM. They found that the content of OC originating from brown coal contamination in bulk soil and size fractions contributed 51 to 97% of OC. Song et al. (2002) reported that the kerogen and BC fractions amounted to 57.8 to 80.6% of the OC contents in one soil and three sediments by using a 0.1 M NaOH extraction method. However, as these fractions had obvious oxygen-containing peaks in the ^{13}C NMR spectra (Song et al., 2002), they were possibly overestimated. Another oxidation method using sodium persulfate to remove amorphous OM also showed that condensed OM had considerable O-alkyl C and carboxyl C, and was likely to contain some amounts of labile organic components (carbohydrates, peptides, fatty acids) (Xing and Chen, 1999; Cuypers et al., 2002).

The ^{13}C NMR spectra of the NHC fractions revealed a large contribution from alkyl carbon (0–45 ppm) and aryl carbon (93–148 ppm), but a minor or

Table 6. Radiocarbon measurements for the selected nonhydrolyzable organic carbon samples.

Sample	OC	$\Delta^{14}C$	^{14}C apparent age	f_a
	%	‰		
C01	36.9	−390	3840 ± 40	0.45
C05	23.6	−422	4400 ± 40	0.48
C08	34.4	−431	4530 ± 40	0.49
C08–5	25.2	−321	3110 ± 40	0.39
WR	27.3	−401	4120 ± 40	0.46

negligible contribution from methoxyl carbon (45–63 ppm), carbohydrate (63–93 ppm), O-aryl (148–165 ppm), and carboxyl (165–187 ppm) carbon (Table 7). The aromaticity was calculated for each of the isolated OM based on the peak areas for aromatic carbon (93–148 ppm) and total carbon (0–165 ppm), and ranges from 28.5 to 55%.

The ^{13}C NMR spectra of the isolated NHC were very similar to that of the same OM fraction isolated from a marine sediment (Ran et al., 2009; Gélinas et al., 2001) and of the kerogen isolate from the Borden sand (Ran et al., 2003). The NHC fraction (Gélinas et al., 2001) was reportedly rich in polymethylene carbon and similar in abundance and chemical composition to nonprotein alkyl carbon, slow relaxing components of proton spin relaxation editing technique, and kerogen from type I oil shales in the OC-rich ocean sediment samples. Matured and unweathered kerogen often contained only alkyl and aromatic carbon, but less matured amorphous kerogen may have other functional groups indicative of its precursor materials such as humic substances. However, humic substances generally had relatively large contents of the methoxyl, carbohydrate, alkyl-substituted aromatic, oxygen-substituted aromatic/phenolic, and carboxylic carbon as indicated in previous studies (Cuypers et al., 2002; Gunasekara and Xing, 2003; Kang and Xing, 2005; Ran et al., 2002). It was obvious that the isolation method was effective in removing young OM materials such as lignin, carbohydrates, and humic substance. Finally, relatively high percentages of aromatic carbon in the isolated NHC from the soils and sediments indicated that they may originate from different thermally matured bitumen or kerogen, aged terrigeneous OM, BC, and high plants containing lignin. It was also consistent with the conclusion that the isolated NHC is similar to kerogen III according to the Van Krevelen diagram (Tissot and Welte, 1984).

The minor carboxylic peaks may be related to the less matured amorphous kerogen as indicated above and/or the oxidation of kerogen (Table 7) (Ran et al., 2007). As oxidation condition was prevailing in the soil samples and reduction condition was dominating in the sediment environment, the KC was more easily oxidized in the soil than in the sediment. Hence, the carbonyl and carboxylic peaks were higher in the soils than in the sediments. The result was consistent with a recent study on kerogen degradation during black shale weathering (Petsch et al., 2001). They reported that the highly aliphatic components remained largely unaltered, while the mixed aliphatic–aromatic component lost aliphatic carbon and accumulated carboxylic–carbonyl oxidation products.

Summary

Molecular-level techniques have facilitated our ability to identify key sediment OM composition, structure, transformation, and fates, and their changes with the changing environmental factors. The molecular methods highlighted within this short review have demonstrated sensitivity for identifying chemical changes in sediment OM. Marine coastal seas are important sites of organic carbon cycling. Projected changes in temperatures, river inflows, and environmental conditions will affect this cycling, with consequences for local ecosystems and global carbon cycles. In-depth understanding of OM sources and processes controlling production, transport, and burial is required to predict and measure the effects of changes. In the case study of this chapter, various geochemical tools were applied

Table 7. Cross polarization–magic angle spinning ^{13}C nuclear magnetic resonance spectra.

Sample	Alkyl, 0–45 ppm	Methoxyl, 45–63 ppm	Carbohydrate, 63–93 ppm	Aryl, 93–148 ppm	O-Aryl, 148–165 ppm	Carboxyl, 165–187 ppm	Carbonyl, 187–220 ppm	Aromaticity, %†
C01	49.46	4.44	0.94	37.03	0.98	0.89	6.23	40.92
C02	59.05	4.3	2.69	27.67	1.08	1.55	3.66	30.33
C03	56.49	6.45	0	28.33	1.05	2.23	5.45	31.82
C04	41.96	6.02	8.86	27.66	5.04	3.84	6.62	36.52
C05	56.51	3.33	2.82	28.22	1.07	2.73	5.33	31.86
C08	61.36	2.28	2.07	24.83	1.38	2.85	5.26	28.52
HP04	30.03	4.65	4.86	45.43	2.91	2.98	9.14	55.01
HP05	52.03	5.24	2.44	30.59	2.62	1.63	5.45	35.74
HP06	36.69	4.2	1.42	39.64	6.43	2.68	8.93	52.12

† Aromaticiy = 100 × aromatic C (93–165 ppm)/[aromatic C (93–165 ppm) + aliphatic C (0–93 ppm)].

to study the sources, structure, transformation, and fates of OM in sediments and suspended particulate samples in the Pearl River region of the southern China. The case study research identified the molecular-level indicators that can be directly applied to various aspects of diagenesis, transformation, and fates of OM. These indicators would in turn result in improved management and monitoring strategies for our changing world.

Acknowledgments

This study was supported by a key joint project of the National Natural Foundation of China (NNSFC), the Guangdong Province Natural Science Foundation (U1201235), and a general project of the NNSFC (41473103). This is contribution No. IS-2018 from GIGCAS.

References

Allen-King, R.M., P. Grathwohl, and W.P. Ball. 2002. New modeling paradigms for the sorption of hydrophobic organic chemicals to heterogeneous carbonaceous matter in soils, sediments, and rocks. Adv. Water Resour. 25:985–1016. doi:10.1016/S0309-1708(02)00045-3

Altabet, M.A. 2004. Oceanic nitrogen isotope biogeochemistry—A status report. Spring Meeting, American Geophysical Union, Portland, OR. 26–30 Jan. 2004

Altabet, M.A., and R. Francois. 1994. Sedimentary nitrogen isotopic ratio as a recorder for surface ocean nitrate utilization. Global Biogeochem. Cycles 8:103–116. doi:10.1029/93GB03396

Amelung, W., M.V. Cheshire, and G. Guggenberger. 1996. Determination of neutral and acidic sugars in soil by capillary gas-liquid chromatography after trifluoroacetic acid hydrolysis. Soil Biol. Biochem. 28:1631–1639.

Anderson, R.F., R.F. Bopp, K.O. Buesseler, and P.E. Biscaye. 1988. Mixing of particles and organic constituents in sediments from the continental-shelf and slope off Cape-Cod-seep. I. Results. Cont. Shelf Res. 8:925–946. doi:10.1016/0278-4343(88)90082-9

Baskaran, M., and A.S. Naidu. 1995. ^{210}Pb-derived chronology and the fluxes of ^{210}Pb and ^{137}Cs isotopes into continental shelf sediments, East Chukchi Sea, Alaskan Arctic. Geochim. Cosmochim. Acta 59:4435–4448. doi:10.1016/0016-7037(95)00248-X

Bergamaschi, B.A., E. Tsamakis, R.G. Keil, T.I. Eglinton, D.B. Montlucon, and J.I. Hedges. 1997. The effect of grain size and surface area on organic matter, lignin and carbohydrate concentration, and molecular compositions in Peru Margin sediments. Geochim. Cosmochim. Acta 61:1247–1260. doi:10.1016/S0016-7037(96)00394-8

Betzer, P.R., W.J. Showers, E.A. Laws, C.D. Winn, G.R. Ditullio, and P.M. Kroopnick. 1984. Primary productivity and particle fluxes on a transect of the equator at 153-degree-W in the Pacific-ocean. Deep Sea Res., Part A 31:1–11.

Bianchi, T.S., S. Mitra, and B.A. McKee. 2002. Sources of terrestrially-derived organic carbon in lower Mississippi River and Louisiana shelf sediments: Implications for differential sedimentation and transport at the coastal margin. Mar. Chem. 77:211–223. doi:10.1016/S0304-4203(01)00088-3

Bianchi, T.S., L.A. Wysocki, M. Stewart, T.R. Filley, and B.A. McKee. 2007. Temporal variability in terrestrially-derived sources of particulate organic carbon in the lower Mississippi River and its upper tributaries. Geochim. Cosmochim. Acta 71:4425–4437. doi:10.1016/j.gca.2007.07.011

Blagodatskaya, E., T. Yuyukina, S. Blagodatsky, and Y. Kuzyakov. 2011. Turnover of soil organic matter and of microbial biomass under C-3-C-4 vegetation change: Consideration of C-13 fractionation and preferential substrate utilization. Soil Biol. Biochem. 43:159–166. doi:10.1016/j.soilbio.2010.09.028

Blair, N. E. and Aller, R. C., 2012. The fate of terrestrial organic carbon in the marine environment. Annu. Rev. Mar. Sci. 4, 17.1–17.23.

Blair, N.E., E.L. Leithold, S.T. Ford, K.A. Peeler, J.C. Holmes, and D.W. Perkey. 2003. The persistence of memory: The fate of ancient sedimentary organic carbon in a modern sedimentary system. Geochim. Cosmochim. Acta 67:63–73. doi:10.1016/S0016-7037(02)01043-8

Boer, W., G.D. van den Bergh, H. de Haas, H.C. de Stigter, R. Gieles, and T.C.E. van Weering. 2006. Validation of accumulation rates in Teluk Banten (Indonesia) from commonly applied ^{210}Pb models, using the 1883 Krakatau tephra as time marker. Mar. Geol. 227:263–277. doi:10.1016/j.margeo.2005.12.002

Bull, I.D., P.F. van Bergen, C.J. Nott, P.R. Poulton, and R.P. Evershed. 2000. Organic geochemical studies of soils from the Rothamsted classical experiments—V. The fate of lipids in different long-term experiments. Org. Geochem. 31:389–408. doi:10.1016/S0146-6380(00)00008-5

Burdige, D. J., 2005. Burial of terrestrial organic matter in marine sediments: A re-assessment. Global Biogeochem.Cycles 19. doi:10.1029/2004GB002368

Calvert, S.E., R.M. Bustin, and T.F. Pedersen. 1992. Lack of evidence for enhanced preservation of sedimentary organic matter in the oxygen minimum if the Gulf of California. Geology 20:757–760. doi:10.1130/0091-7613(1992)020<0757:LOEFEP>2.3.CO;2

Carpenter, R., M. Peterson, and J. Bennett. 1985. ^{210}Pb-derived sediment accumulation and mixing rates for the greater Puget Sound region. Mar. Geol. 64:291–312. doi:10.1016/0025-3227(85)90109-4

Chen, C.T.A., K.K. Liu, and R.W. Macdonald. 2003a. Continental margin exchanges. In: M.F.R. Fasham, editor, Ocean biogeochemistry: The role of the ocean carbon cycle in global change. chap 3. Springer, Berlin. p. 53–97.

Chen, J., Y. Li, K. Yin, and H. Jin. 2004. Amino acids in the Pearl River Estuary and adjacent waters: Origins, transformation and degradation. Cont. Shelf Res. 24:1877–1894. doi:10.1016/j.csr.2004.06.013

Chen, Y.L., H.Y. Chen, and Y.H. Lin. 2003b. Distribution and downward flux of Trichodesmium in the South China Sea as influenced by the transport from the Kuroshio Current. Mar. Ecol. Prog. Ser. 259:47–57. doi:10.3354/meps259047

Christensen, B.T., J.E. Olesen, E.M. Hansen, and I.K. Thomsen. 2011. Annual variation in δ^{13}C values of maize and wheat: Effect on estimates of decadal scale soil carbon turnover. Soil Biol. Biochem. 43:1961–1967. doi:10.1016/j.soilbio.2011.06.008

Clemente, J.S., A.J. Simpson, and M.J. Simpson. 2011. Association of specific organic matter compounds in size fractions of soils under different environmental controls. Org. Geochem. 42:1169–1180. doi:10.1016/j.orggeochem.2011.08.010

Conte, P., and A. Piccolo. 1999. Conformational arrangement of dissolved humic substances. Influence of solution composition on association of humic molecules. Environ. Sci. Technol. 33:1682–1690. doi:10.1021/es9808604

Cornelissen, G., O. Gustafsson, T.D. Bucheli, M.T.O. Jonker, A.A. Koelmans, and P.C.M. Van Noort. 2005. Extensive sorption of organic compounds to black carbon, coal, and kerogen in sediments and soils: Mechanisms and consequences for distribution, bioaccumulation, and biodegradation. Environ. Sci. Technol. 39:6881–6895. doi:10.1021/es050191b

Cowie, G.L., and J.I. Hedges. 1992. Sources and reactivities of amino-acids in a coastal marine-environment. Limnol. Oceanogr. 37:703–724. doi:10.4319/lo.1992.37.4.0703

Cowie, G.L., and J.I. Hedges. 1994. Biochemical indicators of diagenetic alteration in natural organic matter mixtures. Nature 369:304–307. doi:10.1038/369304a0

Crow, S.E., K. Lajtha, T.R. Filley, C.W. Swanston, R.D. Bowden, and B.A. Caldwell. 2009. Sources of plant-derived carbon and stability of organic matter in soil: Implications for global change. Glob. Change Biol. 15:2003–2019. doi:10.1111/j.1365-2486.2009.01850.x

Cuypers, C., T. Grotenhuis, K.G.J. Nierop, E.M. Franco, A. de Jager, and W. Rulkens. 2002. Amorphous and condensed organic matter domains: The effect of persulfate oxidation on the composition of soil/sediment organic matter. Chemosphere 48:919–931. doi:10.1016/S0045-6535(02)00123-6

d'Anglejan, B., and G. Biksham. 1988. Late winter early spring sedimentation off the Great Whale River, Southern Hudson-Bay. Can. J. Earth Sci. 25:930–933. doi:10.1139/e88-091

Dauwe, B., and J.J. Middelburg. 1998. Amino acids and hexosamines as indicators of organic matter degradation state in North Sea sediments. Limnol. Oceanogr. 43:782–798. doi:10.4319/lo.1998.43.5.0782

Dauwe, B., J.J. Middelburg, P.M.J. Herman, and C.H.R. Heip. 1999. Linking diagenetic alteration of amino acids and bulk organic matter reactivity. Limnol. Oceanogr. 44:1809–1814. doi:10.4319/lo.1999.44.7.1809

Davis, J., K. Kaiser, and R. Benner. 2009. Amino acid and amino sugar yields and compositions as indicators of dissolved organic matter diagenesis. Org. Geochem. 40:343–352. doi:10.1016/j.orggeochem.2008.12.003

Dijkstra, F.A. 2009. Modeling the flow of ^{15}N after a ^{15}N pulse to study long-term N dynamics in a semiarid grassland. Ecology 90:2171–2182. doi:10.1890/08-1172.1

Dijkstra, F.A., D. Blumenthal, J.A. Morgan, D.R. LeCain, and R.F. Follett. 2010. Elevated CO_2 effects on semi-arid grassland plants in relation to water availability and competition. Funct. Ecol. 24:1152–1161. doi:10.1111/j.1365-2435.2010.01717.x

Dijkstra, F.A., E. Pendall, A.R. Mosier, J.Y. King, D.G. Milchunas, and J.A. Morgan. 2008. Long-term enhancement of N availability and plant growth under elevated CO_2 in a semi-arid grassland. Funct. Ecol. 22:975–982. doi:10.1111/j.1365-2435.2008.01398.x

Dijkstra, P., J.J. Dalder, P.C. Selmants, S.C. Hart, G.W. Koch, E. Schwartz, and B.A. Hungate. 2011. Modeling soil metabolic processes using isotopologue pairs of position-specific C-13-labeled glucose and pyruvate. Soil Biol. Biochem. 43:1848–1857. doi:10.1016/j.soilbio.2011.05.001

Duer, M.J. 2004. Introduction to solid-state NMR spectroscopy. Blackwell, Malden, MA.

Durrieu de Madron, X.D., A. Abassi, S. Heussner, A. Monaco, J.C. Aloisi, O. Radakovitch, P. Giresse, R. Buscail, and P. Kerherve. 2000. Particulate matter and organic carbon budgets for the Gulf of Lions (NW Mediterranean). Oceanol. Acta 23:717–730. doi:10.1016/S0399-1784(00)00119-5

Eglinton, T.I., L.I. Aluwihare, J.E. Bauer, E.R.M. Druffel, and A.P. McNichol. 1996. Gas chromatographic isolation of individual compounds from complex matrices for radiocarbon dating. Anal. Chem. 68:904–912. doi:10.1021/ac9508513

Eglinton, T.I., and D.J. Repeta. 2004. Organic matter in the contemporary ocean. In: H. Elderfield, editor, The oceans and marine geochemistry. Elsevier Pergamon, London.

Feng, X., K.M. Hills, A.J. Simpson, J.K. Whalen, and M.J. Simpson. 2011. The role of biodegradation and photo-oxidation in the transformation of terrigenous organic matter. Org. Geochem. 42:262–274. doi:10.1016/j.orggeochem.2011.01.002

Feng, X., A.J. Simpson, W.H. Schlesinger, and M.J. Simpson. 2010. Altered microbial community structure and organic matter composition under elevated CO_2 and N fertilization in the duke forest. Glob. Change Biol. 16:2104–2116. doi:10.1111/j.1365-2486.2009.02080.x

Fernandes, M.B., and M.A. Sicre. 2000. The importance of terrestrial organic carbon inputs on Kara Sea shelves as revealed by *n*-alkanes, OC and δ^{13}C values. Org. Geochem. 31:363–374. doi:10.1016/S0146-6380(00)00006-1

Fox, P.A., J. Carter, and P. Farrimond. 1998. Analysis of bacteriohopanepolyols in sediment and bacterial extracts by high performance liquid chromatography atmospheric pressure chemical ionization mass spectrometry. Rapid Commun. Mass Spectrom. 12:609–612. doi:10.1002/(SICI)1097-0231(19980529)12:10<609::AID-RCM210>3.0.CO;2-U

Francois, R., M.A. Altabet, and L.H. Burckle. 1992. Glacial to interglacial changes in surface nitrate utilization in the Indian sector of the Southern Ocean as recorded by sediment delta N-15. Paleoceanography 7:589–606. doi:10.1029/92PA01573

Fuller, C., A. Van Geen, M. Baskaran, and R. Anima. 1999. Sediment chronology in San Francisco Bay, California, defined by ^{210}Pb, ^{234}Th, ^{137}Cs, and ^{239,240} Pu. Mar. Chem. 64:7–27. doi:10.1016/S0304-4203(98)00081-4

Gao, Q., Z. Tao, C. Shen, Y. Sun, W. Yi, and C. Xing. 2002. Riverine organic carbon in the Xijiang River (South China): Seasonal variation in content and flux budget. Environ. Geol. 41:826–832. doi:10.1007/s00254-001-0460-4

Gaye, B., K. Fahl, L.A. Kodina, N. Lahajnar, B. Nagel, D. Unger, and A.C. Gebhardt. 2007. Particulate matter fluxes in the southern and central Kara Sea compared to sediments: Bulk fluxes, amino acids, stable carbon and nitrogen isotopes, sterols and fatty acids. Cont. Shelf Res. 27:2570–2594. doi:10.1016/j.csr.2007.07.003

Gélinas, Y., J.A. Baldock, and J.I. Hedges. 2001. Demineralization of marine and freshwater sediments for CP/MAS ^{13}C NMR analysis. Org. Geochem. 32:677–693. doi:10.1016/S0146-6380(01)00018-3

Glaser, B., M. Benesch, M. Dippold, and W. Zech. 2012. In situ ^{15}N and ^{13}C labelling of indigenous and plantation tree species in a tropical mountain forest (Munessa, Ethiopia) for subsequent litter and soil organic matter turnover studies. Org. Geochem. 42:1461–1469. doi:10.1016/j.orggeochem.2011.06.010

Goñi, M.A., and J.I. Hedges. 1995. Sources and reactivities of marine-derived organic matter in coastal sediments as determined by alkaline CuO oxidation. Geochim. Cosmochim. Acta 59:2965–2981. doi:10.1016/0016-7037(95)00188-3

Goñi, M.A., and S. Montgomery. 2000. Alkaline CuO oxidation with a microwave digestion system: Lignin analyses of geochemical samples. Anal. Chem. 72:3116–3121. doi:10.1021/ac991316w

Goñi, M.A., B. Nelson, R.A. Blanchette, and J.I. Hedges. 1993. Fungal degradation of wood lignins: Geochemical perspectives from CuO-derived phenolic dimers and monomers. Geochim. Cosmochim. Acta 57:3985–4002. doi:10.1016/0016-7037(93)90348-Z

Goñi, M.A., K.C. Ruttenberg, and T.I. Eglinton. 1998. A reassessment of the sources and importance of land-derived organic matter in surface sediments from the Gulf of Mexico. Geochim. Cosmochim. Acta 62:3055–3075. doi:10.1016/S0016-7037(98)00217-8

Goñi, M.A., M.B. Yunker, R.W. Macdonald, and T.I. Eglinton. 2000. Distribution and sources of organic biomarkers in arctic sediments from the Mackenzie River and Beaufort Shelf. Mar. Chem. 71:23–51.

Goñi, M.A., M.B. Yunker, R.W. Macdonald, and T.I. Eglinton. 2005. The supply and preservation of ancient and modem components of organic carbon in the Canadian Beaufort Shelf of the Arctic Ocean. Mar. Chem. 93:53–73. doi:10.1016/j.marchem.2004.08.001

Gordon, E.S., and M.A. Goñi. 2003. Sources and distribution of terrigenous organic matter delivered by the Atchafalaya River to sediments in the northern Gulf of Mexico. Geochim. Cosmochim. Acta 67:2359–2375. doi:10.1016/S0016-7037(02)01412-6

Gunasekara, A.S., and B.S. Xing. 2003. Sorption and desorption of naphthalene by soil organic matter: Importance of aromatic and aliphatic components. J. Environ. Qual. 32:240–246. doi:10.2134/jeq2003.2400

Hatcher, P.G., H.E. Lerch III, A.L. Bates, and T.V. Verheyen. 1989. Solid-state ^{13}C nuclear magnetic resonance studies of coalified gymnosperm xylem tissue from Australian brown coals. Org. Geochem. 14:145–155. doi:10.1016/0146-6380(89)90068-5

Hatton, P.-J., M. Kleber, B. Zeller, C. Moni, A.F. Plante, K. Townsend, L. Gelhaye, K. Lajtha, and D. Derrien. 2012. Transfer of litter-derived N to soil mineral-organic associations: Evidence from decadal ^{15}N tracer experiments. Org. Geochem. 42:1489–1501. doi:10.1016/j.orggeochem.2011.05.002

He, B., M. Dai, W. Zhai, L. Wang, K. Wang, J. Chen, J. Lin, A. Han, and Y. Xu. 2010. Distribution, degradation and dynamics of dissolved organic carbon and its major compound classes in the Pearl River estuary, China. Mar. Chem. 119:52–64. doi:10.1016/j.marchem.2009.12.006

Hedges, J.I., R.A. Blanchette, K. Weliky, and A.H. Devol. 1988a. Effects of fungal degradation on the CuO oxidation-products of lignin- A controlled laboratory study. Geochim. Cosmochim. Acta 52:2717–2726. doi:10.1016/0016-7037(88)90040-3

Hedges, J.I., W.A. Clark, and G.L. Cowie. 1988b. Fluxes and reactivitites of organic matter in a coastal marine bay. Limnol. Oceanogr. 33:1137–1152. doi:10.4319/lo.1988.33.5.1137

Hedges, J.I., J.R. Ertel, and E.S. Leopold. 1982. Lignin geochemistry of a late quaternary sediment core from lake Washington. Geochim. Cosmochim. Acta 46:1869–1877. doi:10.1016/0016-7037(82)90125-9

Hedges, J.I., and R.G. Keil. 1995. Sedimentary organic matter preservation: An assessment and speculative synthesis. Mar. Chem. 49:81–115. doi:10.1016/0304-4203(95)00008-F

Hedges, J.I., R.G. Keil, and R. Benner. 1997. What happens to terrestrial organic matter in the ocean? Org. Geochem. 27:195–212. doi:10.1016/S0146-6380(97)00066-1

Hedges, J.I., and D.C. Mann. 1979. Characterization of plant tissues by their lignin oxidation products. Geochim. Cosmochim. Acta 43:1803–1807. doi:10.1016/0016-7037(79)90028-0

Hedges, J.I., and J.M. Oades. 1997. Comparative organic geochemistries of soils and marine sediments. Org. Geochem. 27:319–361. doi:10.1016/S0146-6380(97)00056-9

Henrichs, S.M. 1995. Sedimentary organic matter preservation: An assessment and speculative synthesis—A comment. Mar. Chem. 49:127–136. doi:10.1016/0304-4203(95)00012-G

Higginson, M.J., J.R. Maxwell, and M.A. Altabet. 2003. Nitrogen isotope and chlorin paleoproductivity records from the Northern South China Sea: Remote vs. local forcing of millennial- and orbital-scale variability. Mar. Geol. 201:223–250. doi:10.1016/S0025-3227(03)00218-4

Hou, J., Y. Huang, C. Brodsky, M.R. Alexandre, A.P. McNichol, J.W. King, F.S. Hu, and J. Shen. 2010. Radiocarbon dating of individual lignin phenols: A new approach for establishing chronology of Late Quaternary lake sediments. Anal. Chem. 82:7119–7126. doi:10.1021/ac100494m

Houel, S., P. Louchouarn, M. Lucotte, R. Canuel, and B. Ghaleb. 2006. Translocation of soil organic matter following reservoir impoundment in boreal systems: Implications for in situ productivity. Limnol. Oceanogr. 51:1497–1513. doi:10.4319/lo.2006.51.3.1497

Hu, J., P.a. Peng, G. Jia, B. Mai, and G. Zhang. 2006. Distribution and sources of organic carbon, nitrogen and their isotopes in sediments of the subtropical Pearl River estuary and adjacent shelf, Southern China. Mar. Chem. 98:274–285. doi:10.1016/j.marchem.2005.03.008

Hu, J., G. Zhang, K. Li, P.a. Peng, and A.R. Chivas. 2008. Increased eutrophication offshore Hong Kong, China during the past 75 years: Evidence from high-resolution sedimentary records. Mar. Chem. 110:7–17. doi:10.1016/j.marchem.2008.02.001

Huh, C.-A., N.G. Pisias, J.M. Kelley, T.C. Maiti, and A. Grantz. 1997. Natural radionuclides and plutonium in sediments from the western Arctic Ocean: Sedimentation rates and pathways of radionuclides. Deep Sea Res. Part II Top. Stud. Oceanogr. 44:1725–1743. doi:10.1016/S0967-0645(97)00040-4

Huygens, D., P. Boeckx, P.H. Templer, L. Paulino, O. Van Cleemput, C. Oyarzun, C. Mueller, and R. Godoy. 2008. Mechanisms for retention of bioavailable nitrogen in volcanic rainforest soils. Nat. Geosci. 1:543–548. doi:10.1038/ngeo252

Ingalls, A.E., E.E. Ellis, G.M. Santos, K.E. McDuffee, L. Truxal, R.G. Keil, and E.R.M. Druffel. 2010. HPLC purification of higher plant-derived lignin phenols for compound specific radiocarbon analysis. Anal. Chem. 82:8931–8938. doi:10.1021/ac1016584

Innes, H.E., A.N. Bishop, I.M. Head, and P. Farrimond. 1997. Preservation and diagenesis of hopanoids in recent lacustrine sediments of Priest Pot, England. Org. Geochem. 26:565–576. doi:10.1016/S0146-6380(97)00017-X

Ittekkot, V. 1988. Global trends in the nature of organic matter in river suspensions. Nature 332:436–438. doi:10.1038/332436a0

Johannessen, S.C., R.W. Macdonald, and D.W. Paton. 2003. A sediment and organic carbon budget for the greater Strait of Georgia. Estuarine Coastal Shelf Sci. 56:845–860. doi:10.1016/S0272-7714(02)00303-7

Jørgensen, N.O.G., R. Stepanaukas, A.G.U. Pedersen, M. Hansen, and O. Nybroe. 2003. Occurrence and degradation of peptidoglycan in aquatic environments. FEMS Microbiol. Ecol. 46:269–280. doi:10.1016/S0168-6496(03)00194-6

Kaiser, K., and R. Benner. 2008. Major bacterial contribution to the ocean reservoir of detrital organic carbon and nitrogen. Limnol. Oceanogr. 53:99–112. doi:10.4319/lo.2008.53.1.0099

Kaiser, K., and R. Benner. 2009. Biochemical composition and size distribution of organic matter at the Pacific and Atlantic time-series stations. Mar. Chem. 113:63–77. doi:10.1016/j.marchem.2008.12.004

Kang, S.H., and B.S. Xing. 2005. Phenanthrene sorption to sequentially extracted soil humic acids and humins. Environ. Sci. Technol. 39:134–140. doi:10.1021/es0490828

Kao, S.J., and K.K. Liu. 1996. Particulate organic carbon export from a subtropical mountainous river (Lanyang Hsi) in Taiwan. Limnol. Oceanogr. 41:1749–1757. doi:10.4319/lo.1996.41.8.1749

Keil, R.G., L.M. Mayer, P.D. Quay, J.E. Richey, and J.I. Hedges. 1997. Loss of organic matter from riverine particles in deltas. Geochim. Cosmochim. Acta 61:1507–1511. doi:10.1016/S0016-7037(97)00044-6

Keil, R.G., E. Tsamakis, J.C. Giddings, and J.I. Hedges. 1998. Biochemical distributions (amino acids, neutral sugars, and lignin phenols) among size-classes of modern marine sediments from the Washington coast. Geochim. Cosmochim. Acta 62:1347–1364. doi:10.1016/S0016-7037(98)00080-5

Keil, R.G., E. Tsamakis, and J.I. Hedges. 2000. Early diagenesis of particulate amino acids in marine systems. In: G.A. Goodfriend et al., editors, Perspectives in amino acid and protein geochemistry. Oxford Univ. Press, Oxford, UK.

Kendall, C., S.R. Silva, and V.J. Kelly. 2001. Carbon and nitrogen isotopic compositions of particulate organic matter in four large river systems across the United States. Hydrol. Processes 15:1301–1346. doi:10.1002/hyp.216

Komada, T., Druffel, E. R. M., and Hwang, J., 2005. Sedimentary rocks as sources of ancient organic carbon to the ocean: An investigation through $\delta^{14}C$ and delta ^{13}C signatures of organic compound classes. Global Biogeochem.Cycles 19, GB2017. doi:10.1029/2004GB002347

Kuzyakov, Y., and R. Bol. 2005. Three sources of CO_2 efflux from soil partitioned by ^{13}C natural abundance in an incubation study. Rapid Commun. Mass Spectrom. 19:1417–1423. doi:10.1002/rcm.1938

Kuzyk, Z.Z.A., M.A. Goñi, G.A. Stern, and R.W. Macdonald. 2008. Sources, pathways and sinks of particulate organic matter in Hudson Bay: Evidence from lignin distributions. Mar. Chem. 112:215–229. doi:10.1016/j.marchem.2008.08.001

Lavelle, J.W., G.J. Massoth, and E.A. Crecelius. 1986. Accumulation rates of recent sediments in Puget Sound, Washington. Mar. Geol. 72:59–70. doi:10.1016/0025-3227(86)90099-X

Li, K., J. Yin, L. Huang, J. Zhang, S. Lian, and C. Liu. 2011. Distribution and abundance of thaliaceans in the northwest continental shelf of South China Sea, with response to environmental factors driven by monsoon. Cont. Shelf Res. 31:979–989. doi:10.1016/j.csr.2011.03.004

Li, L., Y. Ran, G.Y. Sheng, J.M. Fu, P.A. Peng, and W.L. Huang. 2002. Molecular structure characterization of humic acids progressively extracted from a peat soil. Chinese J. Anal. Chem. 30:1303–1307.

Liu, K.K., Y.J. Chen, C.M. Tseng, I.I. Lin, H.B. Liu, and A. Snidvongs. 2007. The significance of phytoplankton photo-adaptation and benthic–pelagic coupling to primary production in the South China Sea: Observations and numerical investigations. Deep Sea Res. Part II Top. Stud. Oceanogr. 54:1546–1574. doi:10.1016/j.dsr2.2007.05.009

Lobbes, J.M., H.P. Fitznar, and G. Kattner. 2000. Biogeochemical characteristics of dissolved and particulate organic matter in Russian rivers entering the Arctic Ocean. Geochim. Cosmochim. Acta 64:2973–2983. doi:10.1016/S0016-7037(00)00409-9

Lockhart, W.L., P. Wilkinson, B.N. Billeck, R.A. Danell, R.V. Hunt, G.J. Brunskill, J. Delaronde, and V. St Louis. 1998. Fluxes of mercury to lake sediments in central and northern Canada inferred from dated sediment cores. Biogeochemistry 40:163–173. doi:10.1023/A:1005923123637

Lomstein, B.A., J. Niggemann, B.B. Jørgensen, and A.T. Langerhuus. 2009. Accumulation of prokaryotic remains during organic matter diagenesis in surface sediments off Peru. Limnol. Oceanogr. 54:1139–1151. doi:10.4319/lo.2009.54.4.1139

Louchouarn, P., M. Lucotte, R. Canuel, J.-P. Gagné, and L.-F. Richard. 1997. Sources and early diagenesis of lignin and bulk organic matter in the sediments of the Lower St. Lawrence Estuary and the Saguenay Fjord. Mar. Chem. 58:3–26. doi:10.1016/S0304-4203(97)00022-4

Macdonald, R.W., S.M. Solomon, R.E. Cranston, H.E. Welch, M.B. Yunker, and C. Gobeil. 1998. A sediment and organic carbon budget for the Canadian Beaufort shelf. Mar. Geol. 144:255–273. doi:10.1016/S0025-3227(97)00106-0

Mao, J., R.M. Cory, D.M. McKnight, and K. Schmidt-Rohr. 2007. Characterization of a nitrogen-rich fulvic acid and its precursor algae from solid state NMR. Org. Geochem. 38:1277–1292. doi:10.1016/j.orggeochem.2007.04.005

Mao, J.D., W.G. Hu, K. Schmidt-Rohr, G. Davies, E.A. Ghabbour, and B.S. Xing. 2000. Quantitative characterization of humic substances by solid-state carbon-13 nuclear magnetic resonance. Soil Sci. Soc. Am. J. 64:873–884. doi:10.2136/sssaj2000.643873x

Masiello, C.A., and E.R. Druffel. 2001. Carbon isotope geochemistry of the Santa Clara River. Global Biogeochem. Cycles 15:407–416. doi:10.1029/2000GB001290

Mendez-Millan, M., M.-F. Dignac, C. Rumpel, and S. Derenne. 2011. Can cutin and suberin biomarkers be used to trace shoot and root-derived organic matter? A molecular and isotopic approach. Biogeochemistry 106:23–38. doi:10.1007/s10533-010-9407-8

Mendez-Millan, M., M.F. Dignac, C. Rumpel, and S. Derenne. 2010. Quantitative and qualitative analysis of cutin in maize and a maize-cropped soil: Comparison of CuO oxidation, transmethylation and saponification methods. Org. Geochem. 41:187–191. doi:10.1016/j.orggeochem.2009.10.008

Mendez-Millan, M., M.F. Dignac, C. Rumpel, D.P. Rasse, G. Bardoux, and S. Derenne. 2012. Contribution of maize root derived C to soil organic carbon throughout an agricultural soil profile assessed by compound specific C-13 analysis. Org. Geochem. 42:1502–1511. doi:10.1016/j.orggeochem.2011.02.008

Meybeck, M. 1982. Carbon, nitrogen, and phosphorous transport by world rivers. Am. J. Sci. 282:401–450. doi:10.2475/ajs.282.4.401

Meybeck, M. 1993. C, N, P and S in rivers: From sources to global inputs. In: R. Wollast et al., editors, Interactions of C, N, P and S Biogeochemical cycles and global change. Springer, Berlin. p. 163–193.

Meyers, P.A. 1994. Preservation of elemental and isotopic source identification of sedimentary organic matter. Chem. Geol. 114:289–302. doi:10.1016/0009-2541(94)90059-0

Middelburg, J.J., J. Nieuwenhuize, and P. van Breugel. 1999. Black carbon in marine sediments. Mar. Chem. 65:245–252. doi:10.1016/S0304-4203(99)00005-5

Mopper, K., A. Stubbins, J.D. Ritchie, H.M. Bialk, and P.G. Hatcher. 2007. Advanced instrumental approaches for characterization of marine dissolved organic matter: Extraction techniques, mass spectrometry, and nuclear magnetic resonance spectroscopy. Chem. Rev. 107:419–442. doi:10.1021/cr050359b

Morgan, J.A., D.R. LeCain, E. Pendall, D.M. Blumenthal, B.A. Kimball, Y. Carrillo, D.G. Williams, J. Heisler-White, F.A. Dijkstra, and M. West. 2011. C_4 grasses prosper as carbon dioxide eliminates desiccation in warmed semi-arid grassland. Nature 476:202–205. doi:10.1038/nature10274

Muzuka, A.N.N., and C. Hillaire-Marcel. 1999. Burial rates of organic matter along the eastern Canadian margin and stable isotope constraints on its origin and diagenetic evolution. Mar. Geol. 160:251–270. doi:10.1016/S0025-3227(99)00022-5

Naidu, A.S., L.W. Cooper, B.P. Finney, R.W. Macdonald, C. Alexander, and I.P. Semiletov. 2000. Organic carbon isotope ratios ($\delta^{13}C$) of Arctic Amerasian Continental shelf sediments. Int. J. Earth Sci. 89:522–532. doi:10.1007/s005310000121

Nguyen, T.H., I. Sabbah, and W.P. Ball. 2004. Sorption nonlinearity for organic contaminants with diesel soot: Method development and isotherm interpretation. Environ. Sci. Technol. 38:3595–3603. doi:10.1021/es0499748

Ni, H.G., F.H. Lu, X.L. Luo, H.Y. Tian, and E.Y. Zeng. 2008. Riverine inputs of total organic carbon and suspended particulate matter from the Pearl River Delta to the coastal ocean off South China. Mar. Pollut. Bull. 56:1150–1157. doi:10.1016/j.marpolbul.2008.02.030

Onstad, G.D., D.E. Canfield, P.D. Quay, and J.I. Hedges. 2000. Sources of particulate organic matter in rivers from the continental USA: Lignin phenol and stable carbon isotope compositions. Geochim. Cosmochim. Acta 64:3539–3546. doi:10.1016/S0016-7037(00)00451-8

Opsahl, S., and R. Benner. 1995. Early diagenesis of vascular plant tissues: Lignin and cutin decomposition and biogeochemical implications. Geochim. Cosmochim. Acta 59:4889–4904. doi:10.1016/0016-7037(95)00348-7

Otto, A., M.J. Simpson. 2005. Degradation and preservation of vascular plant-derived biomarkers in grassland and forest soils from Western Canada. Biogeochemistry 74:377–409.

Otto, A., and M.J. Simpson. 2006a. Evaluation of CuO oxidation parameters for determining the source and stage of lignin degradation in soil. Biogeochemistry 80:121–142. doi:10.1007/s10533-006-9014-x

Otto, A., and M.J. Simpson. 2006b. Sources and composition of hydrolysable aliphatic lipids and phenols in soils from western Canada. Org. Geochem. 37:385–407. doi:10.1016/j.orggeochem.2005.12.011

Otto, A., and M.J. Simpson. 2007. Analysis of soil organic matter biomarkers by sequential chemical degradation and gas chromatography–mass spectrometry. J. Sep. Sci. 30:272–282. doi:10.1002/jssc.200600243

Pautler, B.G., A.J. Simpson, D.J. McNally, S.F. Lamoureux, and M.J. Simpson. 2010. Arctic permafrost active layer detachments stimulate microbial activity and degradation of soil organic matter. Environ. Sci. Technol. 44:4076–4082. doi:10.1021/es903685j

Peterse, F., G.W. Nicol, S. Schouten, and J.S.S. Damste. 2010. Influence of soil pH on the abundance and distribution of core and intact polar lipid-derived branched GDGTs in soil. Org. Geochem. 41:1171–1175. doi:10.1016/j.orggeochem.2010.07.004

Petsch, S.T., R.J. Smernik, T. Eglinton, and J.M. Oades. 2001. A solid state ^{13}C-NMR study of kerogen degradation during black shale weathering. Geochim. Cosmochim. Acta 65:1867–1882. doi:10.1016/S0016-7037(01)00572-5

Prahl, F.G., J.R. Ertel, M.A. Goñi, M.A. Sparrow, and B. Eversmeyer. 1994. Terrestrial organic carbon contributions to sediments on the Washington margin. Geochim. Cosmochim. Acta 58:3035–3048. doi:10.1016/0016-7037(94)90177-5

Preston, C., and R. Newman. 1995. A long-term effect of N fertilization on the ^{13}C CPMAS NMR of de-ashed soil humin in a second-growth Douglas-fir stand of coastal British Columbia. Geoderma 68:229–241. doi:10.1016/0016-7061(95)00051-6

Preston, C.M. 1996. Applications of NMR to soil organic matter analysis: History and prospects. Soil Sci. 161:144–166. doi:10.1097/00010694-199603000-00002

Providoli, I., H. Bugmann, R. Siegwolf, N. Buchmann, and P. Schleppi. 2006. Pathways and dynamics of (15)NO_3^- and (15)NH_4^+ applied in a mountain Picea abies forest and in a nearby meadow in central Switzerland. Soil Biol. Biochem. 38:1645–1657. doi:10.1016/j.soilbio.2005.11.019

Ran, Y., W. Huang, P.S.C. Rao, G. Sheng, and J. Fu. 2002. The role of condensed organic matter in the nonlinear sorption of hydrophobic organic contaminants by a peat and sediments. J. Environ. Qual. 31:1953–1962. doi:10.2134/jeq2002.1953

Ran, Y., K. Sun, B.S. Xing, and C.D. Shen. 2009. Characterization of condensed organic matter in soils and sediments. Soil Sci. Soc. Am. J. 73:351–359. doi:10.2136/sssaj2007.0385

Ran, Y., K. Sun, Y. Yang, B. Xing, and E. Zeng. 2007. Strong sorption of phenanthrene by condensed organic matter in soils and sediments. Environ. Sci. Technol. 41:3952–3958.

Ran, Y., B. Xiao, W. Huang, P. Peng, D. Liu, J. Fu, and G. Sheng. 2003. Kerogen in aquifer material and its strong sorption for nonionic organic pollutants. J. Environ. Qual. 32:1701–1709. doi:10.2134/jeq2003.1701

Ran, Y., B.S. Xing, P.S.C. Rao, and J.M. Fu. 2004. Importance of adsorption (hole-filling) mechanism for hydrophobic organic contaminants on an aquifer kerogen isolate. Environ. Sci. Technol. 38:4340–4348. doi:10.1021/es035168+

Raymond, P.A., J.E. Bauer, N.F. Caraco, J.J. Cole, B. Longworth, and S.T. Petsch. 2004. Controls on the variability of organic matter and dissolved inorganic carbon ages in northeast US rivers. Mar. Chem. 92:353–366. doi:10.1016/j.marchem.2004.06.036

Requejo, A.G., J.S. Brown, P.D. Boehm, and T.C. Sauer. 1991. Lignin geochemistry of North American coastal and continental shelf sediments. Org. Geochem. 17:649–662. doi:10.1016/0146-6380(91)90007-7

Ruttenberg, K.C., and M.A. Goñi. 1997. Phosphorus distribution, C:N:P ratios, and $\delta^{13}C_{oc}$ in arctic, temperate, and tropical coastal sediments: Tools for characterizing bulk sedimentary organic matter. Mar. Geol. 139:123–145. doi:10.1016/S0025-3227(96)00107-7

Sánchez-García, L., J.R. de Andrés, J.A. Martín-Rubí, and P. Louchouarn. 2009. Diagenetic state and source characterization of marine sediments from the inner continental shelf of the Gulf of Cádiz (SW Spain), constrained by terrigenous biomarkers. Org. Geochem. 40:184–194. doi:10.1016/j.orggeochem.2008.11.001

Schmidt, M. W. I., H. Knicker, P. G. Hatcher, and I. KogelKnabner. 1996. Impact of brown coal dust on the organic matter in particle-size fractions of a Mollisol. Org. Geochem. 25:29–39. doi:10.1016/S0146-6380(96)00104-0

Schmidt-Rohr, K., and H.W. Spiess. 1994. Multidimensional solid-state NMR and polymers. Academic Press, San Diego, CA.

Schubert, C.J., and S.E. Calvert. 2001. Nitrogen and carbon isotopic composition of marine and terrestrial organic matter in Arctic Ocean sediments: Implications for nutrient utilization and organic matter composition. Deep Sea Res. Part I 48:789–810. doi:10.1016/S0967-0637(00)00069-8

Schubert, C.J., and R. Stein. 1996. Deposition of organic carbon in Arctic Ocean sediments: Terrigenous supply vs marine productivity. Org. Geochem. 24:421–436. doi:10.1016/0146-6380(96)00042-3

Shunthirasingham, C., and M.J. Simpson. 2006. Investigation of bacterial hopanoid inputs to soils from Western Canada. Appl. Geochem. 21:964–976. doi:10.1016/j.apgeochem.2006.03.007

Simoneit, B.R.T. 2005. A review of current applications of mass spectrometry for biomarker/ molecular tracer elucidations. Mass Spectrom. Rev. 24:719–765. doi:10.1002/mas.20036

Simpson, M.J., A. Otto, and X. Feng. 2008. Comparison of solid-state carbon-13 nuclear magnetic resonance and organic matter biomarkers for assessing soil organic matter degradation. Soil Sci. Soc. Am. J. 72:268–276. doi:10.2136/sssaj2007.0045

Smernik, R.J., R.S. Kookana, and J.O. Skjemstad. 2006. NMR characterization of ^{13}C-benzene sorbed to natural and prepared charcoals. Environ. Sci. Technol. 40:1764–1769. doi:10.1021/ es051895o

Smernik, R.J., D. Olk, and N. Mahieu. 2004. Quantitative solid-state ^{13}C NMR spectroscopy of organic matter fractions in lowland rice soils. Eur. J. Soil Sci. 55:367–379. doi:10.1111/j.1365-2389.2004.00600.x

Smith, J.N. 2001. Why should we believe ^{210}Pb Sediment geochronologies? J. Environ. Radioact. 55:121–123. doi:10.1016/S0265-931X(00)00152-1

Song, J.Z., P.A. Peng, and W.L. Huang. 2002. Black carbon and kerogen in soils and sediments. 1. Quantification and characterization. Environ. Sci. Technol. 36:3960–3967. doi:10.1021/ es025502m

Stein, R. 1991. Accumulation of organic carbon in marine sediments: Results from the Deep Sea Drilling Project/Ocean Drilling Program (DSDP/ODP). Lecture Notes in Earth Sciences, 34. Springer, Berlin, 217 pp.

Stein, R. and Macdonald, R. W., 2004. Organic carbon budget: Arctic ocean vs. global ocean. In: R. Stein and R.W. Macdonald, editors, Organic carbon cycle in the arctic ocean, Springer, Berlin. p. 315–322.

Suess, E. 1980. Particulate organic carbon flux in the oceans—Surface productivity and oxygen utilization. Nature 288:260–263. doi:10.1038/288260a0

Sun, K., Y. Ran, Y. Yang, and B. Xing. 2008. Sorption of phenanthrene by nonhydrolyzable organic matter from different size sediments. Environ. Sci. Technol. 42:1961–1966. doi:10.1021/ es7024627

Sun, X., X. Li, and H.-J. Beug. 1999. Pollen distribution in hemipelagic surface sediments of the South China Sea and its relation to modern vegetation distribution. Mar. Geol. 156:211–226. doi:10.1016/S0025-3227(98)00180-7

Talbot, H.M., M. Rohmer, and P. Farrimond. 2007a. Rapid structural elucidation of composite bacterial hopanoids by atmospheric pressure chemical ionisation liquid chromatography/ion trap mass spectrometry. Rapid Commun. Mass Spectrom. 21:880–892. doi:10.1002/rcm.2911

Talbot, H.M., M. Rohmer, and P. Farrimond. 2007b. Structural characterization of unsaturated bacterial hopanoids by atmospheric pressure chemical ionisation liquid chromatography/ion trap mass spectrometry. Rapid Commun. Mass Spectrom. 21:1613–1622. doi:10.1002/rcm.2997

Talbot, H.M., R.E. Summons, L.L. Jahnke, C.S. Cockell, M. Rohmer, and P. Farrimond. 2008. Cyanobacterial bacteriohopanepolyol signatures from cultures and natural environmental settings. Org. Geochem. 39:232–263. doi:10.1016/j.orggeochem.2007.08.006

Talbot, H.M., D.F. Watson, J.C. Murrell, J.F. Carter, and P. Farrimond. 2001. Analysis of intact bacteriohopanepolyols from methanotrophic bacteria by reversed-phase high-performance liquid chromatography-atmospheric pressure chemical ionisation mass spectrometry. J. Chromatogr. A 921:175–185. doi:10.1016/S0021-9673(01)00871-8

Talbot, H.M., D.F. Watson, E.J. Pearson, and P. Farrimond. 2003. Diverse biohopanoid compositions of non-marine sediments. Org. Geochem. 34:1353–1371. doi:10.1016/S0146-6380(03)00159-1

Tesi, T., S. Miserocchi, M.A. Goni, and L. Langone. 2007. Source, transport and fate of terrestrial organic carbon on the western Mediterranean Sea, Gulf of Lions, France. Mar. Chem. 105:101–117.

Thevenot, M., M.-F. Dignac, and C. Rumpel. 2010. Fate of lignins in soils: A review. Soil Biol. Biochem. 42:1200–1211. doi:10.1016/j.soilbio.2010.03.017

Thomas, H., Y. Bozec, H.J.W. de Baar, K. Elkalay, M. Frankignoulle, L.S. Schiettecatte, G. Kattner, and A.V. Borges. 2005. The carbon budget of the North Sea. Biogeosciences 2:87–96. doi:10.5194/bg-2-87-2005

Thorn, K.A. 1987. Structural characteristics of the IHSS Suwannee River fulvic and humic acids determined by solution state C-13 NMR spectroscopy. Sci. Total Environ. 62:175–183. doi:10.1016/0048-9697(87)90498-0

Tissot, B.P., and D.H. Welte. 1984. Petroleum Formation and Occurrence. Springer, New York.

Torn, M.S., S.C. Biraud, C.J. Still, W.J. Riley, and J.A. Berry. 2011. Seasonal and interannual variability in C-13 composition of ecosystem carbon fluxes in the U.S. Southern Great Plains. Tellus B Chem. Phys. Meterol. 63:181–195. doi:10.1111/j.1600-0889.2010.00519.x

VanBergen, P.F., I.D. Bull, P.R. Poulton, and R.P. Evershed. 1997. Organic geochemical studies of soils from the Rothamsted classical experiments—1. Total lipid extracts, solvent insoluble residues and humic acids from Broadbalk wilderness. Org. Geochem. 26:117–135. doi:10.1016/S0146-6380(96)00134-9

Wang, X.C., E.R.M. Druffel, S. Griffin, C. Lee, and M. Kashgarian. 1998. Radiocarbon studies of organic compound classes in plankton and sediment of the northeastern Pacific Ocean. Geochim. Cosmochim. Acta 62:1365–1378. doi:10.1016/S0016-7037(98)00074-X

Wassmann, P., Bauerfeind, E., Fortier, M., Fukuchi, M., Hargrave, B., Moran, B., Noji, T., Nothig, E. M., Olli, K., Peinert, R., Sasaki, H., and Shevchenko, V., 2004. Particulate organic carbon flux to the Arctic Ocean sea floor. In: R. Stein and R.W. MacDonald, editors, Organic carbon cycle in the Arctic Ocean. Springer, Berlin. p. 101–138.

Weijers, J.W.H., S. Schouten, O.C. Spaargaren, and J.S.S. Damste. 2006. Occurrence and distribution of tetraether membrane lipids in soils: Implications for the use of the TEX86 proxy and the BIT index. Org. Geochem. 37:1680–1693. doi:10.1016/j.orggeochem.2006.07.018

Wilson, M.A. 1987. NMR techniques and applications in geochemistry and soil chemistry. Pergamon Press, New York.

Winkler, A., L. Haumaier, and W. Zech. 2001. Variation in hopanoid composition and abundance in forest soils during litter decomposition and humification. Org. Geochem. 32:1375–1385. doi:10.1016/S0146-6380(01)00115-2

Wu, Y., T. Dittmar, K.-U. Ludwichowski, G. Kattner, J. Zhang, Z.Y. Zhu, and B.P. Koch. 2007. Tracing suspended organic nitrogen from the Yangtze River catchment into the East China Sea. Mar. Chem. 107:367–377. doi:10.1016/j.marchem.2007.01.022

Xing, B., and Z. Chen. 1999. Spectroscopic evidence for condensed domains in soil organic matter. Soil Sci. 164:40–47. doi:10.1097/00010694-199901000-00006

Xu, Y., M.P. Cooke, H.M. Talbot, and M.J. Simpson. 2009. Bacteriohopanepolyol signatures of bacterial populations in Western Canadian soils. Org. Geochem. 40:79–86. doi:10.1016/j.orggeochem.2008.09.003

Yang, S., M. Tang, W.W.S. Yim, Y. Zong, G. Huang, A.D. Switzer, and Y. Saito. 2011. Burial of organic carbon in Holocene sediments of the Zhujiang (Pearl River) and Changjiang (Yangtze River) estuaries. Mar. Chem. 123:1–10. doi:10.1016/j.marchem.2010.07.001

Yu, F., Y. Zong, J.M. Lloyd, G. Huang, M.J. Leng, C. Kendrick, A.L. Lamb, and W.W.S. Yim. 2010. Bulk organic d^{13}C and C/N as indicators for sediment sources in the Pearl River delta and estuary, southern China. Estuar. Coast. Shelf Sci. 87:618–630. doi:10.1016/j.ecss.2010.02.018

Zhang, L., K.D. Yin, L. Wang, F.R. Chen, D.R. Zhang, and Y.Q. Yang. 2009. The sources and accumulation rate of sedimentary organic matter in the Pearl River Estuary and adjacent coastal area, Southern China. Estuar. Coast. Shelf Sci. 85:190–196. doi:10.1016/j.ecss.2009.07.035

Zhang, Y.L., K. Kaiser, L. Li, D.N. Zhang, Y. Ran, and R. Benner. 2014. Sources, distributions, and early diagenesis of sedimentary organic matter in the Pearl River region of the South China Sea. Mar. Chem. 158:39–48. doi:10.1016/j.marchem.2013.11.003

Zhang, C.S., and L.J. Wang 2001. Multi-element geochemistry of sediments from the Pearl River system, China. Appl. Geochem. 16:1251–1259.

Zhao, M., P. Wang, J. Tian, and J. Li. 2009. Biogeochemistry and the carbon reservoir. In: P. Wang and Q. Li, editors, The South China Sea. Springer, Amsterdam, the Netherlands.

Zhu, C., J.W.H. Weijers, T. Wagner, J.-M. Pan, J.-F. Chen, and R.D. Pancost. 2011. Sources and distributions of tetraether lipids in surface sediments across a large river-dominated continental margin. Org. Geochem. 42:376–386. doi:10.1016/j.orggeochem.2011.02.002

Comparison of Distribution of Total Organic Matter and Its Key Fractions in Coastal Lagoon Sediments from Estonia, Finland, and Sweden for Use as Healing Muds

Monika Übner

Abstract

In medical spas, different natural sediments are used that contain a number of organic and inorganic compounds. Five certified Estonian healing muds originated from the sea and freshwater lakes. The curative effect of natural sediments is related to organic matter and mostly to humic substances. The aim of this chapter is to compare the content of total organic matter and its key fractions in certified muds, with coastal lagoon sediment samples analyzed to give a preliminary assessment of their suitability for use as healing mud. Sediment samples have been collected from Estonian, Finnish, and Swedish coastal lagoons around the Baltic Sea. Sandier sediments from Gotland had the lowest water content. Inre Fjärden and Lillträsket (Åland) coastal lagoon sediments had a high content of organic matter and lipids. Inre Fjärden (Åland) and Linnulaht Bay (Saaremaa) sediment samples are rich in humic substances. Therefore, the yield of organic compounds in the sediments from Åland and Saaremaa was higher. The next step will be to characterize these samples microbiologically and determine the content of heavy metals and radioactive elements.

Abbreviations: FA, fulvic acids; HA, humic acids.

Monika Übner, Pärnu College, University of Tartu, Pärnu 80010, Estonia (monika.ubner@ut.ee).
doi:10.2136/sssaspecpub62.2014.0048

Labile Organic Matter—Chemical Compositions, Function, and Significance in Soil and the Environment. SSSA Special Publication 62. Zhongqi He and Fengchang Wu, editors.

Since ancient times, local natural sediments have been used for healing purposes. These sediments are called peloids. Gomes et al. (2013) proposed a new definition of peloids: "Peloid is a maturated mud or muddy dispersion with healing and/or cosmetic properties, composed of a complex mixture of fine-grained natural materials of geologic and/or biologic origins, mineral water or sea water, and commonly organic compounds from biological metabolic activity." Maturation could take place either in the natural environment (natural maturation) or in tanks (artificial maturation) during a variable time (Gomes et al., 2013). Nowadays natural (Vreca and Dolenec, 2005; Tserenpil et al., 2010) and artificial peloids (Pozo et al., 2013; Fernández-González et al., 2013) are widely used in spas. Natural peloids are located in different geographical areas. Medical spas used marine and lake sediments, peats, clays, etc. All these peloids must have been tried and tested as having a positive effect on the human organism. In the case of environmentally relevant impact on peloid deposits, anthropogenically harmful substances like heavy metals must be checked. Also they must not contain any harmful microorganisms (Quality Criteria of the European Spas Association, 2006).

Natural sediments contain different inorganic and organic compounds, the impact of which on the human organism is not yet clear. Also, there are no good suggestions as to what is the ideal concentration of specific compounds to attain an optimal response to treatment (Klöcking and Helbig, 2005, Odabasi et al., 2007). It is believed that therapeutic sediments contain biologically active humic substances, which are formed on decomposing plants and the organisms that fed on them. Therefore, these humic substances also contain nitrogen-rich particles that originate from the phytoplankton of the water body (Abate and Masini, 2001). Labile nitrogen-containing proteinaceous material are incorporated mainly in humic substances, where they are protected from biodegradation (Zang et al., 2000). Fourier transform infrared spectra of marine sediment humic substances showed that the nitrogen present is mostly forming as part of amide groups (Giovanela et al., 2004). However, the composition and content of humic substances depends on the origin of the sediment (Esteves and Duarte, 2001; Übner et al., 2004). In the characterization of humic substances, the two fractions usually determined are humic acids (HA) soluble in alkali and fulvic acids (FA) soluble in water at all pH values. On the other hand, HA has the ability to exert antioxidant and antiradical effects, which are important for preventing the oxidative stress which is associated with a number of diseases (Efimova et al., 2013). In addition to these hydrophilic components of humic substances, lipophilic components may contribute to the healing effect of therapeutic sediments (healing mud) (Odabasi et al., 2007; Tserenpil et al., 2010).

Different coastal areas could have different environmental conditions, as many coastal areas have richer vegetation and animal life and larger sedimentation of organic material than the open sea. Transportation and accumulation of sediments vary at the bottom of coastal areas (Musielak et al., 2000). In the coastal lagoons connected with the sea, the mud is sandier and sometimes mixed with gravel after a storm. There are also differences in coastal lakes. In one side of the lake, the mud layer is absent and the lake bottom is sandy or rocky. No deposition of fine materials occurs, and the sediment water content is less than 50%. In the other side, there is a layer of clay under a thin mud layer. In a quiet part of the coastal lake, a relatively thick layer of mud can be found. These sediments have high water content (>75%) and also a high content of organic matter. These areas are usually situated in the innermost parts of the coastal area (Musielak et al., 2000; Kisand and Übner, 2012). The sedimentary organic matter in coastal areas is mainly derived from inputs of terrestrial material and bacterial production in the sediments themselves. The relative importance of those sources is determined by local climate, nutrient supply, hydrodynamic conditions, and water properties. Changes in these factors and human disturbance may be reflected in the sedimentary organic matter (Pinturier-Geiss et al., 2002). In the Baltic Sea countries, most coastal lagoons have been separated from the sea because of tectonic rising after the Ice Age. Some of them are still temporarily connected to the sea. Lagoon water contains different chlorides and sulfates, and the bottom is covered with a thick layer of mud. Abundance of plants on the upper sediment layer is related with the consistency of the sediment: either more mineral or organic (Ott and Tamre, 2012). Marine sediments also have anthropogenic influences which can impact the sediments and alter their therapeutic properties (Miko et al., 2007).

Currently, there are only five certified healing muds originating from the sea (Haapsalu, Käina, Mullutu) and freshwater lakes (Ermistu, Värska) in Estonia. Thus, we have collected sediment samples from numerous coastal lagoons situated in ecologically clean nature conservation areas for evaluating their potential as healing muds. In this chapter we compared the levels of total organic matter and their key fractions in the natural sediments from the coastal lagoons of Estonia (Saaremaa and the Pärnu County area), Finland (Åland), and Sweden (Gotland) with certified healing muds from Estonia to give a preliminary assessment of their suitability for use as healing mud. This preliminary study reveals the differences in the organic matter composition between different coastal lagoon sediments which may result in different effects as healing muds.

Materials and Methods

Sampling Areas

Fifty-two sediment samples were collected from different coastal lagoons in Estonia (Saaremaa and Pärnu County), Finland (Åland), and Sweden (Gotland) (Table 1).

Organic sediment in Estonian coastal lakes around Saaremaa Island is mostly located on bluish gray clay and its content is low. There is only a thin layer of lake mud (up to 20 cm) on the sediment surface, and the silt is thicker in the lower layer between lake mud and clay (Kisand and Übner, 2012). Sediment samples were collected from nine coastal lakes (Table 1). All of those lakes are parts of different nature conservation areas. Ott et al. (2012) studied the main characterizing data of Saaremaa coastal lakes. Põldealuse Bay is located far away from the sea. The southeast corner sediment was mainly sandy with large stones. In the other sites, there was 15 cm of greenish gray semi-liquid jelly-like mud in the top layer followed by dark mud and clayey silt. Aenga Bay had a connection to the sea, small water volume and very low depth. In the section close to the sea, the soft sediment layer was the highest, approximately 40 cm, dominated by greenish gray mud, clayey mud and clay. At some places, Aenga Bay had a stony bottom and gravel, sand, and silt under a thin layer of mud. In Oessaare Bay, near the river inflow, there was only 25 cm of sediment. The upper 20 cm of it was very fluid greenish mud, then 10 cm of gravelly mud, and under it, thick sandy sediment and large cobblestones. In other areas, the mud layer has reduced, and the sediment turns quickly into sandy clay (Ott et al., 2012). The sediment upper layer

Table 1. Sampling sites with their corresponding geographical coordinates.

Sample	Sample place	Geographic coordinates	
	Gotland (Sweden)		
1	G1-1	57°10′5.63″N	18°29′0.82″E
2	G1-2	57°10′4.84″N	18°28′55.99″E
3	G1-3	57°10′6.17″N	18°28′53.65″E
4	G2-1	57°6′50.98″N	18°25′27.37″E
5	G2-2	57°6′50.76″N	18°25′31.19″E
6	G3-1	57°0′47.23″N	18°11′59.14″E
7	G3-2	57°0′45.90″N	18°11′58.45″E
8	G3-3	57°0′43.88″N	18°11′54.02″E
9	Mölnorträsk-1	57°55′34.75″N	19°8′37.10″E
10	Mölnorträsk-2	57°55′36.23″N	19°8′38.62″E
11	Farnavik-1	57°56′9.10″N	19°6′33.62″E
12	Farnavik-2	57°56′2.94″N	19°6′6.73″E
13	Bondansträsk	57°58′17.65″N	19°8′26.99″E
14	Paviken-1	57°27′6.80″N	18°8′40.20″E
15	Paviken-2	57°27′15.44″N	18°8′36.71″E
	Pärnu County (Estonia)		
16	Võiste-1	58°12′35.35″N	24°27′58.00″E
17	Võiste-2	58°12′40.64″N	24°28′9.26″E
18	Võiste-3	58°12′39.20″N	24°28′10.49″E
19	Võiste-4	58°12′14.11″N	24°28′9.12″E
20	Võiste-5	58°12′11.23″N	24°28′5.38″E
21	Võiste-6	58°12′8.53″N	24°28′17.04″E
22	Võiste-7	58°11′56.58″N	24°27′53.32″E
23	Häädemeeste-1	58°5′15.18″N	24°28′47.42″E
24	Häädemeeste-2	58°5′25.22″N	24°28′48.90″E
25	Häädemeeste-3	58°5′39.88″N	24°28′25.10″E
26	Häädemeeste-4	58°5′36.78″N	24°28′39.11″E
27	Häädemeeste-5	58°10′26.11″N	24°27′57.96″E
	Saaremaa (Estonia)		
28	Põldealuse Bay	58°18′39.31″N	22°53′22.16″E
29	Aenga Bay	58°18′18.70″N	22°54′5.90″E
30	Oessaare Bay	58°19′42.62″N	22°52′26.15″E
31	Poka Bay	58°19′19.24″N	22°51′23.62″E
32	Laidevahe Bay-1	58°18′6.23″N	22°51′37.73″E
33	Laidevahe Bay-2	58°18′51.16″N	22°51′36.47″E
34	Vägara Bay	58°14′8.00″N	22°19′27.00″E
35	Linnulaht Bay	58°15′3.83″N	22°26′39.79″E
36	Mullutu Bay	58°14′39.48″N	22°21′30.84″E
37	Suurlaht Bay	58°15′2.52″N	22°24′26.94″E
	Åland (Finland)		
38	Gästgivars Fladan	60°4′51.80″N	20°13′10.40″E
39	Lillfjärd-1	60°19′22.30″N	19°55′3.04″E
40	Lillfjärd-2	60°19′17.20″N	19°55′14.30″E
41	Lillfjärd-3	60°19′27.60″N	19°55′9.90″E
42	Lillträsket-1	60°16′0.20″N	20°10′57.90″E
43	Lillträsket-2	60°16′5.20″N	20°10′58.20″E
44	Lillträsket-3	60°16′11.80″N	20°10′55.60″E
45	Korrvik-1	60°17′14.00″N	19°46′38.30″E
46	Korrvik-2	60°17′13.60″N	19°46′42.00″E
47	Inre Fjärden-1	60°16′39.01″N	19°35′54.84″E
48	Inre Fjärden-2	60°16′22.30″N	19°35′57.50″E
49	Inre Fjärden-3	60°16′13.20″N	19°35′56.70″E
50	Kråkskärsfjärden-1	60°17′23.00″N	19°36′11.50″E
51	Kråkskärsfjärden-2	60°17′24.50″N	19°36′22.00″E
52	Kråkskärsfjärden-3	60°17′22.70″N	19°36′1.30″E

contained mainly calcite (Murumets, 2012). Poka Bay is a small corner of Oessaare Bay, but their sediments had quite a different thickness. In the northwestern part, the thickness of softer and more viscous sediment was 50 cm, covering stones (Ott et al., 2012). Viscosity is related to the higher content of clay minerals (Murumets, 2012). Laidevahe Bay water body is formed by the sea. This bay is cornered and rich in little islands. Because of that the bottom sediments vary quite a lot in different areas of the lake. In the southwest section of the lake, a layer of up to 50 cm of greenish gray mud can be found, turning into silt and clay sediment below (Ott et al., 2012). The sediment's upper layer contained more clay minerals (Murumets, 2012). Vägara Bay is located amid forests and swamps and therefore the water contained much organic matter, mainly humic substances. The sediment surface had a greenish gray mud that turned into greenish gray clay. Linnulaht Bay is a relict coastal lake because the seawater does not flow directly into it. The lake has been under a long-term influence of people, and the water is muddy and contains many nutrients. In the southwestern part of the lake, the muddy sediment layer was approximately 1.5 m thick. Mullutu Bay has developed steadily for a long time and the sea had a relatively low impact on the water. The sediment upper layer was formed by light fluid mud (Ott et al., 2012), which contained mainly quartz and feldspar (Murumets, 2012). A layer of greyish brown clay started at the depth of approximately 50 cm. Suurlaht Bay was separated from the sea a relatively long time ago. A thicker mud layer was in the southwestern part of the lake (Ott et al., 2012), and the upper 10 cm contained mainly quartz (Murumets, 2012). Mud from Oessaare Bay, Poka Bay, and Suurlaht Bay is in the healing muds list but only the latter deposit is used actively in medical spas.

Coastal lagoons in Pärnu County are located in the territory of the Luitemaa Nature Reserve, where the coastal meadows have been managed for years. Because of this, some lagoons have been deepened. During the excavation, excessive sediment was raised to the shore. These lagoons are small and do not have names. Eleven samples were taken from this area (Table 1). Some lagoons are directly connected to the sea. The mud on the surface layer was light. The thickness of the sandy mud was up to 20 cm and the color varies from greenish brown to black. Mineralogical analyses showed that sediments from the Võiste area contained more clay minerals, but sediments from the Häädemeeste area contained more quartz-like sandy minerals (Murumets, 2012).

Sediment samples from Gotland (Sweden) were collected from seven coastal lagoons (Table 1). Some of them are small and do not have names. The mud layer in those lagoons was very thin and often located on limestone plateaus (Übner, 2012). The thickness of eastern coast sediments was up to 10 cm and in shaded areas up to 19 cm. There was dark gray sandy sediment in southern coastal lagoons. The sandy lagoon sediments from the Fårö Island in the north were white and contained small dark particles. The sediment thickness was up to 10 cm. The sampling site of the western coast was done on the territory of the Paviken nature reserve that is managed with cattle. A stream from the bog flows in the coastal lake, bringing along humic-rich sediments and water. The sediments were 20 cm deep, followed by a sandy layer. There was dark gray mud and under it slightly sandy mud. Gotland sediments contained mainly quartz, but sediment sample from Bondansträsk contained more carbonates than other samples (Murumets, 2012).

The lakes in the northern area of Åland are more oligotrophic and poor in calcium. They are located on the foot slopes of the hills or adjacent to clay areas at lower elevation. The lowland lakes are surrounded by deciduous forests, fields and pastures. These lakes are rich in vegetation and are eutrophic (Carlsson, 2001). Sediment samples were collected from six coastal lagoons (Table 1). Two or three samples were taken from bigger lagoons.

Methods

The sediment samples were collected using a gravitational Uwitec corer. The collected samples were cut in the field into 10-cm-thin sections which were stored in plastic bags in a refrigerator below 4°C until laboratory analysis. The samples of five certified Estonian healing muds came from spas that used them in mud therapy.

All sediment samples were homogenized. For sediment water content determination, the samples were dried at 105°C overnight. Content of organic matter was determined by weight loss on ignition at 550°C for 4 h.

The procedure for assessing the yield of HA and FA has previously been described in detail by Übner et al. (2004). Humic substances were extracted from nondried homogenized sediments to avoid possible changes caused by drying. The humid sediment sample (40 g) was stirred for 5 h with 100 mL of 0.2 M NaOH and then allowed to stand for 20 h at room temperature. The suspension was centrifuged at 5000 × *g* for 30 min. The alkaline extract contained HA and FA. The alkaline extract was acidified to pH 2 by addition of 6 M HCl. After 20 h, this solution was centrifuged at 5000 × *g* for 20 min, and the soluble FA and insoluble HA were separated. The precipitate was repeatedly washed with distilled water until free of Cl^- (control with $AgNO_3$) and dried.

The FA solution was cleaned through the Amberlite XAD-7 resin column for desalination. After removal of salts with 0.01 M HCl, the adsorbed FA was recovered by elution with 0.1 M NaOH solution directly onto the strong cation exchanger Amberlite IR-120. The gathered FA fraction was dried.

Lipids were separated from dry sediment by treating it repeatedly with the mixture of chloroform and ethanol (2:1), and the separated soluble lipid fraction was dried.

Since the depth of the sediment layer in Gotland was in almost all cases not more than 10 cm, the properties of the top layers at all sampling points were compared. This study gives a preliminary assessment of the suitability of the different sediments as healing mud. For statistical analyses, the Pearson correlation coefficients and Student's *t*-test were calculated.

Results and Discussion

Sediment Water Content

Sediments from the Saaremaa and Åland coastal lagoons had a high water content (Fig. 1). The water content was the highest in the sediments from Inre Fjärden (Åland) (Samples 47, 48, and 49) being 97 to 98% in the surface sediment layer. Inre Fjärden is Finland's westernmost lake. Sediments from Mullutu Bay and Suurlaht Bay (Saaremaa) contained more dry matter, and the water content was only 37 and 59%, respectively. These samples consisted of quartz-rich sand minerals up to 73% of dry matter (Murumets, 2012).

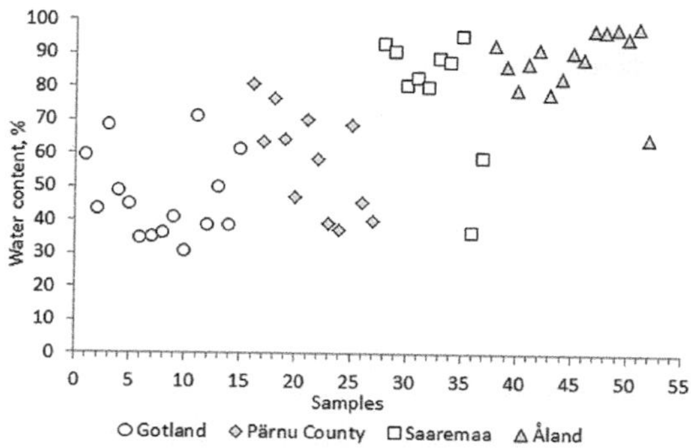

Fig. 1. Water content (%) of the sediment samples in a depth of 0 to 10 cm.

The water content of sediments from the lagoons in Pärnu County was not very high, being in the range of 38 to 81%. The water content was higher in some dredged lagoons (Samples 16 and 18) in the Võiste area where the content of clay minerals was 61% of the dry matter (Murumets, 2012). The lowest water content was in some coastal lagoons, Samples 23 and 24, in the Häädemeeste area, being 39 and 38%, respectively. These samples contained more quartz up to 77% of dry matter (Murumets, 2012).

The samples from Gotland contained less water, being in the range of 31 to 71% (Fig. 1). These samples contained more quartz in the range of 59 to 77% (Murumets, 2012). Recently proposed classification of peloids (Gomes et al., 2013) which have a potential therapeutic application are based on the origin and the content of inorganic mineral and organic components. The content of phyllosilicates is mainly determined from sediment mineral composition. Phyllosilicates are crystalline clay minerals (Konta, 2009). When comparing mineralogical data (Murumets, 2012) with sediment water content, a good negative correlation with silicates (quartz and feldspar) and positive correlation with the phyllosilicates (mica, chlorite, illite, kaolinite, and talc) were found (Table 2). Therefore, a higher water content of natural sediment samples indicated a low content of quartz-rich sand minerals and a high content of clay minerals in these sediments. The mud samples, which have a low content of clay particles and a high content of quartz and feldspars, were difficult to handle as the mud had a thick consistency and a low content of water (Karakaya et al., 2010). Also, high water content is an important factor for high heat capacity of the peloid when applied to the patient's body at 40 to 45°C for 20 to 30 min (Veniale et al., 2007).

Table 2. Pearson's correlation matrix for some sediment mineralogical characteristics for Samples 1, 2, 5 through 7, 9, 13, 14, 17, 18, 23, 24, 30–32, 36, and 37 ($P < 0.001$) (adapted from Murumets, 2012).

	Water content	Silicates	Phyllosilicates
Water content	1		
Silicates	−0.9007	1	
Phyllosilicates	0.6438	−0.7354	1

Table 3. Chemical properties of Estonian certified healing muds.†

	Mullutu	Haapsalu	Käina	Ermistu	Värska
Water content, %	91.2	76.7	56.3	90.4	89.1
Organic matter, % in DM	24.3	11.4	5.3	56.4	36.8
Lipids, % in DM	0.56	0.47	0.11	0.64	0.90
HA, mg g^{-1}, in DM	27.2	12.4	4.8	87.9	89.7
FA, mg g^{-1} in DM	3.5	1.8	1.2	6.2	11.7

† DM, dry matter; HA, humic acid; FA, fulvic acid.

There were five certified healing muds of Estonia for comparison. Three of them are marine muds, Haapsalu, Mullutu, and Käina and two freshwater lake muds, Ermistu and Värska. The water content of those muds was in the range of 56 to 91% (Table 3). The lowest value was in the Käina mud (56.3%) and the highest in the Mullutu mud (91.2%). Water content of most coastal lagoon sediment samples from Saaremaa and Åland were in that range. Most sediment samples from Gotland (Samples 2, 4–10, and 12–14) and some from the Pärnu County (Samples 20, 23, 24, 26, and 27) contained too much dry matter due to their higher content of sand minerals. These sediments have too thick a consistency and are not suitable for mud therapy for the difficulties in handling.

Yield of Organic Matter in Sediment Samples

The organic matter content differed among the four sampling regions (Fig. 2). Åland coastal lakes samples were characterized by a higher organic matter content. In some cases it was up to 68%. The highest values were in the sediments from Inre Fjärden (Samples 47–49). The highest values among Saaremaa sediment samples came from Linnulaht Bay (39%). This coastal lake is considered to be one of the most nutrient-rich water bodies (Ott et al., 2012), which was also reflected in the higher organic matter content of the sediments. The high content of organic matter in Lake Inre Fjärden may reflect a higher level of nutrients in lake water, but in this study the nitrogen and phosphorous content was not determined. Coastal lagoons in the Pärnu County contained less organic matter, reaching 10% in most cases. And the sediments from Gotland contained less organic matter, mostly less than 5% of the dry matter.

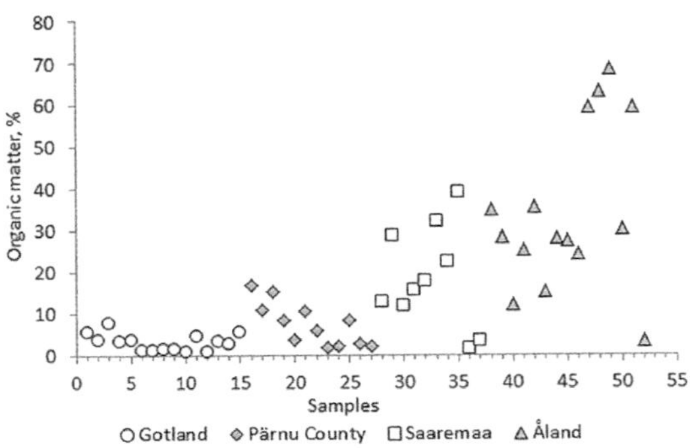

Fig. 2. Yield of organic matter (% in dry matter) in the sediment samples.

Table 4. Pearson's correlation matrix for sediment chemical characteristics in all samples.

	Water content	Organic matter	Lipids	HA	FA
Water content	1				
Organic matter	0.8171***	1			
Lipids	0.7595***	0.6581***	1		
HA	0.6720***	0.9028**	0.5807***	1	
FA	0.7101***	0.9537***	0.5805***	0.9678***	1

** $P < 0.01$.

*** $P < 0.001$.

Among Estonian healing muds, the organic matter content of marine muds was in the range of 5.4 to 24.3% and that of lake muds in the range of 36.9 to 56.4% (Table 3). The lowest value was in the Käina mud and the highest in the Ermistu mud. Yield of organic matter in the sediment samples analyzed from Gotland and Pärnu County was more similar to the healing mud samples of marine origin, but most of them remained below the lowest value of certified healing muds. Suitable samples in the case of organic matter are 1, 3, 15 to 19, 21, 22, and 25. Therefore, these samples may not be suitable for mud therapy in terms of organic matter content. Some samples from Saaremaa and Åland were more similar to the type of lake mud of healing mud samples. Samples from Inre Fjärden had more organic matter than Estonian healing muds.

In samples with high organic matter content, the water content of natural sediment was too high. Organic matter content was in good positive correlation with water content (Table 4). Because organic matter can contain more water, the high content of organic components usually led to higher water content in sediment. A similar pattern was also seen in peat which consisted mostly of organic matter (Fukue et al., 2006). If we consider the high water content and calculate the organic matter content based on the wet weight of natural sediment, the organic matter content in all samples was below 4.7% of wet weight. Higher levels were calculated in the Lillträsket (Åland) and Laidevahe Bay (Saaremaa) coastal lakes and some Võiste area sampling points of Pärnu County. Lower values were in some of the Gotland's and Saaremaa's sandier sediment samples. Estonian certified marine muds had 2.1 to 2.7% and lake muds 4.0 to 5.4% of organic matter in sediment wet weight. Most Gotland sediment samples (Samples 4–14) and some samples from the Häädemeeste area of Pärnu County (Samples 20, 23, 24, and 26) remained below these levels. The coastal lakes Põldealuse Bay, Linnulaht Bay, Mullutu Bay, Suurlaht Bay from Saaremaa and Inre Fjärden and Kråkskärsfjärden from Åland had lower values, in the range of 0.9 to 1.9%. These samples may not have enough organic compounds in natural humid samples, and the therapeutic effect may be too small. However, there are no limit values for the content of organic matter in healing mud, and these suggestions are based on the data of Estonian certified healing muds.

Yield of Total Lipids

The lipid class includes a number of different compounds, mostly hydrocarbons and phospholipids. The levels of the latter are related to biomass, and their concentration decreased with increasing depth of sediments. Therefore, the lipid content shows the presence of fresh organic matter (Pinturier-Geiss et al., 2002).

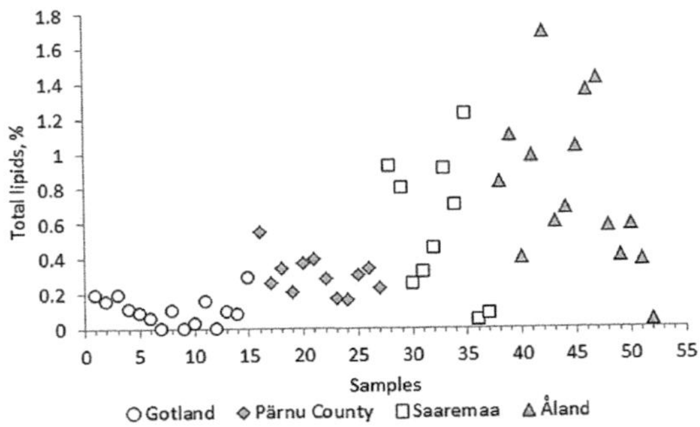

Fig. 3. Total lipid content (% in dry matter) in the sediment samples.

In Gotland sediment samples the lipid content was the lowest, not more than 0.29% of dry matter (Fig. 3). The highest lipid concentration was detected in the Lillträsket (Åland) sediment, up to 1.7% of the sediment dry matter. Among Saaremaa coastal lakes sediment samples, Linnulaht Bay (1.2%) had more lipids than others. Sediment samples from Pärnu County contained not more than 0.4% of lipids.

Estonian certified healing muds contained lipids in the range of 0.1 to 0.9% of dry matter, marine muds up to 0.6%, and lake muds more than 0.6% (Table 3). Among the sediments studied, the lipid content of some samples from Saaremaa (Mullutu Bay and Suurlaht Bay) and Gotland (Mölnorträsk and Bondansträsk) coastal lakes remained under 0.1%. The content of total organic matter in those humid sediment samples was also too low.

There was a strong positive correlation between the organic matter and lipid content in sediment samples (Table 4). Odabasi et al. (2007) concluded that the chemical structure of therapeutic mud does not only contain hydrophilic organic compounds such as HA and FA, but also lipophilic organic substances, especially low-molecular-weight fatty acid–structured compounds and sulphoglycolipids. Marine sediments contain acyclic and cyclic hydrocarbons, which indicate their genetic relation to algae and marine organisms (Voronkov et al., 2009). Lipids from hydrosulfide peloids contain more fatty esters and steroids (Tserenpil et al., 2010). However, there is no consensus about substances responsible for the chemical effect of mud therapy either (Odabasi et al., 2007).

Humic Substances in Sediment Samples

Humic substances are organic macromolecules of microbiological, vegetative, and animal origin. Humic materials in sediments have been observed to be closely connected with balneotherapy efficiency (Peña-Méndez et al., 2005). The soluble humic substances content (HA and FA) of marine sediments was low, amounting for only a few percent of dry matter. There were more HA than FA in sediment samples (Fig. 4 and 5). The highest values were in some Åland's eutrophic lakes (Inre Fjärden: 138.8 mg g^{-1} HA, 24.8 mg g^{-1} FA). Sediment samples from Saaremaa coastal lakes contained up to 33.6 mg g^{-1} HA and up to 10.4 mg g^{-1} FA. The lowest values were in the sediments from Mullutu Bay (2.1 mg g^{-1} HA, 0.3 mg g^{-1} FA) and Suurlaht Bay (2.3 mg g^{-1} HA, 0.8 mg g^{-1} FA), and the highest values in Linnulaht

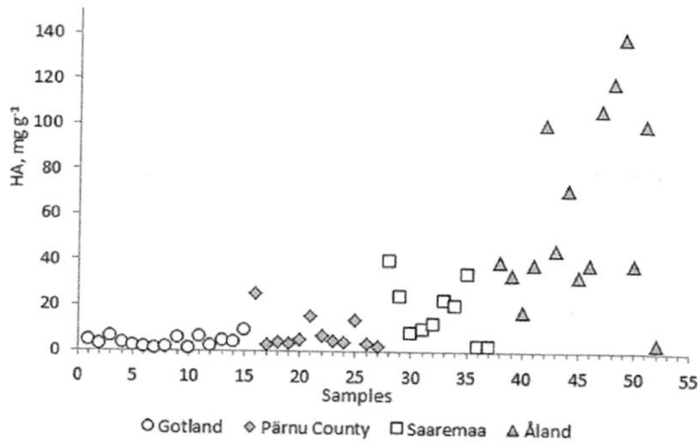

Fig. 4. Yield of humic acid (HA) (mg g^{-1} in dry matter) in the sediment samples.

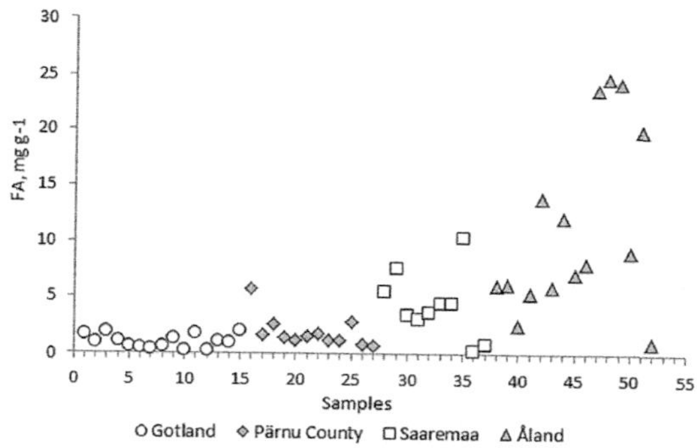

Fig. 5. Yield of fulvic acid (FA) (mg g^{-1} in dry matter) in the sediment samples.

Bay. In Pärnu County's coastal lagoon sediments, the yield of humic substances was also lower than in Saaremaa and Åland's lagoon sediment samples. Values of HA were in the range of 2.3 to 25.2 mg g^{-1} and FA 0.7 to 5.7 mg g^{-1}. The highest values were in small lagoons of active herding areas. In Gotland's coastal lagoons, the content of humic substances was in the range of 1.0 to 8.9 mg g^{-1} HA and 0.2 to 2.0 mg g^{-1} FA. The highest value was in the Paviken coastal lagoon sediment.

Certified healing mud samples had HA in marine muds in the range of 0.5 to 2.7% and lake muds 8.8 to 9.0% of sediment dry matter. The content of FA was lower than HA content. In marine muds it was 0.1 to 0.3% and in lake muds 0.6 to 1.2% of sediment dry matter. Comparison of the sediment samples analyzed with certified healing muds revealed that the humic substances content was suitable in most samples from Saaremaa (except samples from Mullutu Bay, Laidevahe Bay, and Suurlaht Bay) and Åland coastal lakes (except Sample 52 from Kråkskärsfjärden). Gotland's coastal lagoon sediments did not contain enough humic substances. Only Sample 3 (Ålarve) and Sample 15 (Paviken) were suitable. Most samples from Pärnu County lacked the HA fraction, but Samples 16, 21, 22, and 25 had a higher content of humic substances.

The HA fraction contains high molecular mass compounds, which are also bioactive and have different positive effects on the human body (Klöcking and Helbig, 2005). However, it is not known how high the level of the humic substances should be to classify the sediment as healing mud. HA is the most important component of humic substances. This fraction helps break up clay, assists in transferring micronutrients, and enhance water retention (Peña-Méndez et al., 2005). Fulvic acid is soluble at all pH values and has low molecular mass. They are more mobile in the environment than HA, and FA contain more oxygen and less carbon than HA (Übner et al., 2004).

The yields of humic substances in all sediment samples studied were in excellent positive correlation with HA and FA. Organic matter also had a very good positive correlation with HA and FA (Table 4). The same good correlations were found in a previous study (Calace et al., 2006). Humic acid and FA had good positive correlation with sediment water content. This characteristic is related to the humic substances capability to hold water (Peña-Méndez et al., 2005). Lipids and humic substances also had positive correlation because lipids may bind to humic substances (Lehtonen et al., 2001).

Since there are no standard values for different bioactive compounds in the composition of healing mud, only the yield of humic substances, organic matter, and total lipids were determined in this study. When comparing those data with certified Estonian healing muds, it was found that all sediment samples from Åland (except Sample 52) and most from Saaremaa were suitable. Only the analyzed Mullutu Bay, Laidevahe Bay, and Suurlaht Bay samples contained an insufficient level of bioactive organic compounds. The coastal lagoons in the Pärnu County are small and only a few samples were rich in those active compounds, but they are located in the area of active herding and, therefore, the sediments might contain pathogenic microorganisms. In Gotland the content of organic matter, humic substances, and lipids sediments is low. Only sediments from Ålarve and Paviken contain enough of these substances. For the next step, it will be necessary to clarify if those sediments are microbiologically safe and unpolluted with radioactive elements and heavy metals.

Conclusions

The content of humic substances, organic matter, and lipids of sediments from different coastal lagoons of Estonia, Sweden, and Finland were studied. An analysis to determine the amount of those bioactive compounds was chosen for the preliminary assessment of the suitability of the sediments as healing mud. Natural sediments which are used as healing mud must be obtained from an ecologically clean area. This requirement was fulfilled for most sediments of Saaremaa Island. Here it is not allowed to discharge untreated sewage into the coastal lagoon, and mining of natural resources (except healing mud) is prohibited around the lake. Åland coastal lagoon sediments were rich in humic substances and lipids, and for the next step it will be necessary to determine the presence of pathogenic microorganisms and harmful elements. The samples from Pärnu County are not suitable for pelotherapy in most cases because of active herding in that area and lack of organic compounds. In Gotland's sediment samples, the yield of bioactive compounds was too low and the content of quartz-rich sand minerals was too high. Some samples had higher values of humic substances, but the sufficient safety of these sediments for human health has to be determined.

Acknowledgments

This work was supported by the Central Baltic INTERREG IVA Program and by the Regional Competence Centre Development program, European Regional Development Fund.

References

Abate, M., and J.C. Masini. 2001. Acid-basic and complexation properties of a sedimentary humic acid. A study on the Barra Bonita Reservoir of Tietê River, São Paulo State, Brazil. J. Braz. Chem. Soc. 12:109–116. doi:10.1590/S0103-50532001000100015

Calace, N., N. Cardellicchio, B.M. Petronio, M. Pietrantonio, and M. Pietroletti. 2006. Sedimentary humic substances in the northern Adriatic Sea (Mediterranean Sea). Mar. Environ. Res. 61:40–58. doi:10.1016/j.marenvres.2005.05.002

Carlsson, R. 2001. Freshwater snail communities and lake classification. An example from the Aland Islands, southwestern Finland. Limnologica 31:129–138. doi:10.1016/S0075-9511(01)80007-4

Efimova, I.V., S.L. Khil'ko, O.V. Smirnova, V.S. Berezhnoi, and V.I. Rybachenko. 2013. Antioxidant properties of hymatomelanic acids from brown coal. Solid Fuel Chem. 47:193–196. doi:10.3103/S0361521913040058

Esteves, V.I., and A.C. Duarte. 2001. Differences between humic substances from riverine, estuarine, and marine environments observed by fluorescence spectroscopy. Acta Hydrochim. Hydrobiol. 28:359–363. doi:10.1002/1521-401X(20017)28:7<359::AID-AHEH359>3.0.CO;2-6

Fernández-González, M.V., J.M. Martin-Garcia, G. Delgado, J. Párraga, and R. Delgado. 2013. A study of the chemical, mineralogical and physicochemical properties of peloids prepared with two medicinal mineral waters from Lanjarón Spa (Granada, Spain). Appl. Clay Sci. 80–81:107–116. doi:10.1016/j.clay.2013.06.011

Fukue, M., Y. Sato, K. Uehara, Y. Kato, and Y. Furukawa. 2006. Contamination of sediments and proposed containment technique in a wood pool in Shimizu. In: M. Fukue et al., editors, Contaminated sediments: Evaluation and remediation techniques. ASTM, Japan. p. 32–43.

Giovanela, M., E. Parlanti, E.J. Soriano-Sierra, M.S. Soldi, and M.M.D. Sierra. 2004. Elemental compositions, FT-IR spectra and thermal behavior of sedimentary fulvic and humic acids from aquatic and terrestrial environments. Geochem. J. 38:255–264. doi:10.2343/geochemj.38.255

Gomes, C., M.I. Carretero, M. Pozo, F. Maraver, P. Cantista, F. Armijo, J.L. Legido, F. Teixeira, M. Rautureau, and R. Delgado. 2013. Peloids and pelotherapy: Historical evolution, classification and glossary. Appl. Clay Sci. 75–76:28–38. doi:10.1016/j.clay.2013.02.008

Karakaya, M.Ç., N. Karakaya, S. Sarioglan, and M. Koral. 2010. Some properties of thermal muds of some spas in Turkey. Appl. Clay Sci. 48:531–537. doi:10.1016/j.clay.2010.02.005

Kisand, A., and M. Übner. 2012. Sediments of coastal lagoons and their biochemical characterisation. In: M. Kose et al., editors, Coastal lagoons in Estonia and in the central Baltic Sea region: Development history, geology and hydrology, biodiversity and nature conservation value. Tartu Univ. Press, Tartu, Estonia. p. 28–31.

Klöcking, R., and B. Helbig. 2005. Medical aspects and applications of humic substances. In: A. Steinbüchel and R.H. Marchessault, editors, Biopolymers for medical and pharmaceutical applications. Wiley, Weinheim, Germany. p. 3–16.

Konta, J. 2009. Phyllosilicates in the sediment-forming process: Weathering, erosion, transportation, and deposition. Acta Geodyn. Geomater. 153:13–43.

Lehtonen, K., K. Hänninen, and M. Ketola. 2001. Structurally bound lipids in peat humic acids. Org. Geochem. 32:33–43. doi:10.1016/S0146-6380(00)00158-3

Miko, S., G. Koch, S. Mesić, M.Š. Miko, M. Šparica, and P. Vreča. 2007. Influence of land use in small karst watersheds on the chemical status of peloid sediments on the eastern Adriatic coast. J. Soils Sediments 7:303–312. doi:10.1065/jss2007.10.254

Murumets, S. 2012. Chemical characterisation of coastal region lagoons in western Estonia and Gotland Island. M.Sc. thesis, Tartu University, Tartu, Estonia.

Musielak, S., K. Furmanczyk, L. Håkanson, P. Jonsson, and J. Persson. 2000. The Baltic Sea and its coasts. In: L.-C. Lundin, editor, Sustainable water management in the Baltic Sea basin. Book 1: The waterscape. Uppsala Univ. Programme, Uppsala, Sweden. p. 28–38.

Odabasi, E., H. Gul, E. Macit, M. Turan, and O. Yildiz. 2007. Lipophilic components of different therapeutic mud species. J. Altern. Complement. Med. 13:1115–1118. doi:10.1089/acm.2007.0504

Ott, I., K. Maileht, T. Krause, A. Palm, K. Karus, T. Feldmann, A. Kisand, A. Rakko, H. Timm, K. Kübar, K. Palm, and M. Kose. 2012. Results of studies of coastal lagoons. In: M. Kose et al., editors, Coastal lagoons in Estonia and in the central Baltic Sea region: Development history, geology and hydrology, biodiversity and nature conservation value. Tartu Univ. Press, Tartu, Estonia. p. 57–80.

Ott, I., and R. Tamre. 2012. Formation, development and typology of coastal lagoons. In: M. Kose et al., editors, Coastal lagoons in Estonia and in the central Baltic Sea region: Development history, geology and hydrology, biodiversity and nature conservation value. Tartu Univ. Press, Tartu, Estonia. p. 15–21.

Peña-Méndez, E.M., J. Havel, and J. Patocka. 2005. Humic substances compounds of still unknown structure: Applications in agriculture, industry, environment, and biomedicine. J. Appl. Biomed. 3:13–24.

Pinturier-Geiss, L., L. Méjanelle, B. Dale, and D.A. Karlsen. 2002. Lipids as indicators of eutrophication in marine coastal sediments. J. Microbiol. Methods 48:239–257. doi:10.1016/S0167-7012(01)00326-8

Pozo, M., M.I. Carretero, F. Maraver, E. Pozo, I. Gómez, F. Armijo, and J.A.M. Rubi. 2013. Composition and physico-chemical properties of peloids used Spanish spas: A comparative study. Appl. Clay Sci. 83–84:270–279. doi:10.1016/j.clay.2013.08.034

Quality Criteria of the European Spas Association. 2006. European Spas Association. http://dev.sm.vectweb.com/media/66/File/quality_criteria.pdf (accessed 15 Feb. 2012).

Tserenpil, Sh., G. Dolman, and M.G. Voronkov. 2010. Organic matters in healing muds from Mongolia. Appl. Clay Sci. 49:55–63. doi:10.1016/j.clay.2010.04.002

Übner, M. 2012. Comparison of sediments from Estonian and Gotland coastal lagoons. In: M. Kose, et al., editors, Coastal lagoons in Estonia and in the central Baltic Sea region: Development history, geology and hydrology, biodiversity and nature conservation value. Tartu Univ. Press, Tartu, Estonia. p. 98–101.

Übner, M., M. Treuman, A. Viitak, and M. Lopp. 2004. Properties of humic substances from the Baltic Sea and Lake Ermistu mud. J. Soils Sediments 4:24–29. doi:10.1007/BF02990825

Veniale, F., A. Bettero, P.G. Jobstraibizer, and M. Setti. 2007. Thermal muds: Perspectives of innovations. Appl. Clay Sci. 36:141–147. doi:10.1016/j.clay.2006.04.013

Voronkov, M.G., G.M. Dolmaa, Sh. Tserenpil, O. Ugtahbayar, G. Ganzaya, and K.A. Abzaeva. 2009. Chemical composition of peloids from Gurvan Nuur Middle Lake. Dokl. Chem. 426:115–117. doi:10.1134/S0012500809060019

Vreca, P., and T. Dolenec. 2005. Geochemical estimation of copper contamination in the healing mud from Makirina Bay, central Adriatic. Environ. Int. 31:53–61. doi:10.1016/j.envint.2004.06.009

Zang, X., J.D.H. van Heemst, K.J. Dria, and P.G. Hatcher. 2000. Encapsulation of protein in humic acid from a histosol as an explanation for the occurrence of organic nitrogen in soil and sediment. Org. Geochem. 31:679–695. doi:10.1016/S0146-6380(00)00040-1

Printed and bound by CPI Group (UK) Ltd, Croydon, CR0 4YY
16/09/2026
14963421-0003

AMS SHORT COURSE LECTURE NOTES
Introductory Survey Lectures
A Subseries of Proceedings of Symposia in Applied Mathematics

Volume 40 MATRIX THEORY AND APPLICATIONS
Edited by Charles R. Johnson (Phoenix, Arizona, January 1989)

Volume 39 CHAOS AND FRACTALS: THE MATHEMATICS BEHIND THE COMPUTER GRAPHICS
Edited by Robert L. Devaney and Linda Keen (Providence, Rhode Island, August 1988)

Volume 38 COMPUTATIONAL COMPLEXITY THEORY
Edited by Juris Hartmanis (Atlanta, Georgia, January 1988)

Volume 37 MOMENTS IN MATHEMATICS
Edited by Henry J. Landau (San Antonio, Texas, January 1987)

Volume 36 APPROXIMATION THEORY
Edited by Carl de Boor (New Orleans, Louisiana, January 1986)

Volume 35 ACTUARIAL MATHEMATICS
Edited by Harry H. Panjer (Laramie, Wyoming, August 1985)

Volume 34 MATHEMATICS OF INFORMATION PROCESSING
Edited by Michael Anshel and William Gewirtz (Louisville, Kentucky, January 1984)

Volume 33 FAIR ALLOCATION
Edited by H. Peyton Young (Anaheim, California, January 1985)

Volume 32 ENVIRONMENTAL AND NATURAL RESOURCE MATHEMATICS
Edited by R. W. McKelvey (Eugene, Oregon, August 1984)

Volume 31 COMPUTER COMMUNICATIONS
Edited by B. Gopinath (Denver, Colorado, January 1983)

Volume 30 POPULATION BIOLOGY
Edited by Simon A. Levin (Albany, New York, August 1983)

Volume 29 APPLIED CRYPTOLOGY, CRYPTOGRAPHIC PROTOCOLS, AND COMPUTER SECURITY MODELS
By R. A. DeMillo, G. I. Davida, D. P. Dobkin, M. A. Harrison, and R. J. Lipton (San Francisco, California, January 1981)

Volume 28 STATISTICAL DATA ANALYSIS
Edited by R. Gnanadesikan (Toronto, Ontario, August 1982)

Volume 27 COMPUTED TOMOGRAPHY
Edited by L. A. Shepp (Cincinnati, Ohio, January 1982)

Volume 26 THE MATHEMATICS OF NETWORKS
Edited by S. A. Burr (Pittsburgh, Pennsylvania, August 1981)

Volume 25 OPERATIONS RESEARCH: MATHEMATICS AND MODELS
Edited by S. I. Gass (Duluth, Minnesota, August 1979)

Volume 24 GAME THEORY AND ITS APPLICATIONS
Edited by W. F. Lucas (Biloxi, Mississippi, January 1979)

Volume 23 MODERN STATISTICS: METHODS AND APPLICATIONS
Edited by R. V. Hogg (San Antonio, Texas, January 1980)

Volume 22 NUMERICAL ANALYSIS
Edited by G. H. Golub and J. Oliger (Atlanta, Georgia, January 1978)

Volume 21 MATHEMATICAL ASPECTS OF PRODUCTION AND DISTRIBUTION OF ENERGY
Edited by P. D. Lax (San Antonio, Texas, January 1976)

PROCEEDINGS OF SYMPOSIA IN APPLIED MATHEMATICS

AMS SHORT COURSE LECTURE NOTES

Introductory Survey Lectures

published as a subseries of

Proceedings of Symposia in Applied Mathematics

PROCEEDINGS OF SYMPOSIA
IN APPLIED MATHEMATICS
Volume 40

MATRIX THEORY AND APPLICATIONS

Charles R. Johnson, Editor

American Mathematical Society
Providence, Rhode Island

LECTURE NOTES PREPARED FOR THE
AMERICAN MATHEMATICAL SOCIETY SHORT COURSE

MATRIX THEORY AND APPLICATIONS

HELD IN PHOENIX, ARIZONA
JANUARY 10–11, 1989

The AMS Short Course Series is sponsored by the Society's Committee on Employment and Educational Policy (CEEP). The series is under the direction of the Short Course Advisory Subcommittee of CEEP.

Library of Congress Cataloging-in-Publication Data

Matrix theory and applications/Charles R. Johnson, editor.
p. cm. – (Proceedings of symposia in applied mathematics, ISSN 0160-7634; v. 40. AMS short course lecture notes)
"Held in Phoenix, Arizona, January 10–11, 1989"–T.p. verso.
Includes bibliographical references.
ISBN 0-8218-0154-6 (alk. paper)
1. Matrices congresses. I. Johnson, Charles R. II. American Mathematical Society. III. Series: Proceedings of symposia in applied mathematics; v. 40. IV. Series: Proceedings of symposia in applied mathematics. AMS short course lecture notes.
QA188.M395 1990 90-30-584
512.9′434–dc20 CIP

1980 *Mathematics Subject Classification*. (1985 *Revision*).
Primary 15-02.

Printed in the United States of America.
The paper used in this book is acid-free and falls within the guidelines
established to ensure permanence and durability. ∞
Portions of this publication were typeset using AMS-TeX,
the American Mathematical Society's TeX macro system.

10 9 8 7 6 5 4 3 2 1 95 94 93 92 91 90

Table of Contents

Preface

As a subfield of mathematics, matrix theory continues to enjoy a renaissance that has accelerated during the past decade, though its roots may be traced much further back. This is due in part to stimulation from a variety of applications and to the considerable interplay with other parts of mathematics, but also to a great increase in the number and vitality of specialists in the field. As a result, the once popular misconception that the subject has been fully researched has been largely dispelled. The interest on the part of the American Mathematical Society and the approximately 140 participants in the Short Course (at the January 1989 Phoenix Meeting) on which this volume is based is a reflection of this change. The steady growth in quality and volume of the subject's three principal journals, Linear Algebra and its Applications, Linear and Multilinear Algebra, and the SIAM Journal on Matrix Analysis and Applications is another. Approximately 500 different authors have published in one of these three journals in the last two years. Geographically, strong research centers in matrix theory have developed recently in Portugal and Spain, Israel, the Netherlands, Belgium, and Hong Kong.

The purpose of the Short Course was to present a sample of the ways in which modern matrix theory is stimulated by its interplay with other subjects. Though the course was limited to seven speakers, the "other subjects" represented included combinatorics, probability theory, statistics, operator theory and control theory, algebraic coding theory, partial differential equations, and analytic function theory. Among other important examples, numerical analysis, optimization, physics and economics are, unfortunately, at most lightly touched. There is no limit to the specific examples that might be cited.

One of the ingredients in the recent vitality of matrix theory is the variety of points of view and tools brought to the subject by researchers in different areas. This is responsible for a number of important trends in current research. For example, the notion of majorization (mentioned in the talk by Olkin) has become pervasive in a historically brief period of time. The trend away from the "basis-free" point of view is illustrated by work in combinatorial matrix theory (Brualdi, Johnson), the Hadamard product

(Horn) and nonnegative matrices (Mukherjea). There are many quite worthy issues that are at least excruciatingly difficult to view in a basis-free way, and freedom from the basis-free view has opened many exciting avenues of research. On the other hand, recognition of the "right" problem dependant symmetries can provide vital insight (Diaconis). The synergy between matrix theory and systems theory has had a tremendous impact on both, and on the now highly mathematically driven field of electrical engineering (Gohberg). The immense variety of tools and problems illustrates a reason for use of the term "matrix theory" or "matrix analysis" in place of "linear algebra". A large portion of current work is neither primarily linear nor primarily algebraic in nature. No point of view on what the subject is or where it is going could, or should, be without substantial disagreement. This only reflects the remarkable breadth of interest enjoyed by the subject. For an historical perspective on the nature and role of the subject the reader might enjoy the prefaces to each of the following: *Recent Advances in Matrix Theory* (Schneider 1964); *A Survey of Matrix Theory and Matrix Inequalities* (Marcus and Minc, 1964); *Linear Algebra and its Applications*, volume 1 (Alan Hoffman, 1968); and *Matrix Analysis* (Horn and Johnson, 1985). A glimpse of the contagious appeal of the subject is communicated by Olga Taussky in her November 1988 *Monthly* article "How I Became a Torchbearer for Matrix Theory".

As organizer, I would like to again thank each of the speakers for a contribution that will advance both the subject and the general understanding of it. The significant time necessary to prepare both a talk and then subsequent paper is much appreciated. The community would also like to thank the American Mathematical Society for recognizing, and providing a forum for, the subject.

Charles R. Johnson
College of William and Mary

Proceedings of Symposia in Applied Mathematics
Volume 40, 1990

THE MANY FACETS OF COMBINATORIAL MATRIX THEORY

Richard A. Brualdi

ABSTRACT. I take a very broad view of combinatorial matrix theory. Combinatorial matrix theory is concerned with the use of matrix theory and linear algebra in proving combinatorial theorems and in describing and classifying combinatorial constructions, and it is also concerned with the use of combinatorial ideas and reasoning in the finer analysis of matrices and with intrinsic combinatorial properties of matrix arrays.

1. THE DETERMINANT. It should come as no surprise that combinatorial theory and matrix theory could interact to form a subject called "combinatorial matrix theory". One of the first matrix functions one studies is the determinant, and the determinant has a very combinatorial definition: The determinant of an n by n matrix

(1.1) $$A = (a_{ij} : i = 1,...,n; j = 1,...,n)$$

is given by

(1.2) $$\det A = \sum_{\pi} (\text{sign } \pi) \prod_{i=1}^{n} a_{i\pi(i)}$$

where the summation is over all permutations π of $\{1,...,n\}$, and where sign π is $+1$ or -1 according as π is an even or odd permutation. We may impart even more combinatorial spirit to the determinant through the use of directed graphs (or digraphs).

A digraph Γ of order n has a set of n vertices, usually taken to be the set $\{1,...,n\}$, and a set $\mathcal{A}$ of arcs where each arc is an ordered pair (i,j) of not necessarily distinct vertices. The arcs of Γ can be regarded as a subset of the positions of the matrix (1.1). The entry a_{ij} at the position (i,j) of A is the weight assigned by A to the arc (i,j) of Γ. If $a_{ij} = 1$ or 0 according as (i,j) is or is not an arc of Γ (i,j = 1,...,n), then A is the adjacency matrix of Γ. We may use these arc weights in order to assign weights to other objects associated with Γ. However this may be done, we define the weight of a set of objects to be the sum of the weights of the objects in the set.

This paper was written during a period in which the author was partially supported by National Science Foundation Grant No. DMS–8521521.

Let k be a positive integer. A <u>cycle</u> γ <u>of length</u> k is a sequence

(1.3) $$(i_1,\ldots,i_k\, ,\, i_1)$$

of vertices such that $i_1,\ldots,i_k$ are distinct and $(i_1,i_2),\ldots,(i_{k-1},i_k),(i_k,i_1)$ are arcs of Γ (referred to as the <u>arcs</u> of γ). (If $i_1,\ldots,i_k$ are not assumed to be distinct, then (1.3) is called a <u>circuit</u>.) We define the <u>weight of a cycle</u> (1.3) to be

(1.4) $$-a_{i_1 i_2} \cdots a_{i_{k-1} i_k} a_{i_k i_1} ,$$

the negative of the products of the weights of its arcs. When we refer to <u>disjoint cycles</u>, we mean cycles with no vertex in common. The <u>weight of a pairwise disjoint union of cycles</u> is the product of the weights of the individual cycles.

(1.5) <u>Example</u>: Let Γ be the digraph of order 6 pictured below. Then

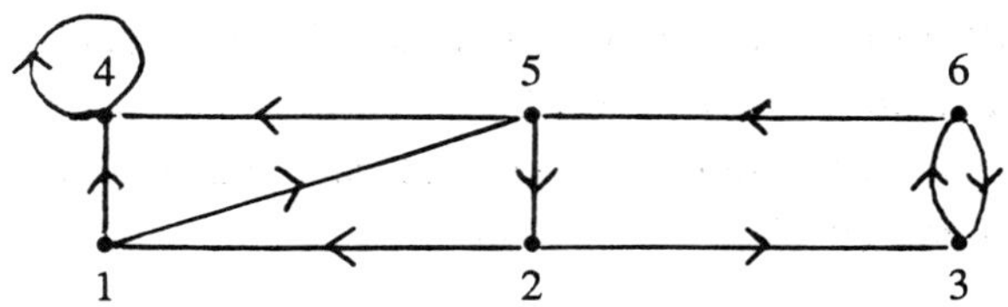

(1.6) $$\{(1,5,2,1),\ (3,6,3),\ (4,4)\}$$

is a pairwise disjoint union of three cycles of Γ having weight

$$(-a_{15}a_{52}a_{21})(-a_{36}a_{63})(-a_{44}) = -a_{15}a_{21}a_{36}a_{44}a_{52}a_{63}\ . \qquad \square$$

To a permutation π of $\{1,\ldots,n\}$ we associate a <u>permutation digraph</u> $\Gamma(\pi)$ of order n whose set of arcs is $\{(i,\pi(i)):\ i = 1,\ldots,n\}$. There is a one–to–one correspondence between the cycles of the permutation π and the cycles of the digraph $\Gamma(\pi)$. For example, let π be the permutation of $\{1,\ldots,6\}$ defined by

$$\pi(1) = 5,\ \pi(2) = 1,\ \pi(3) = 6,\ \pi(4) = 4,\ \pi(5) = 2,\ \pi(6) = 3.$$

Then $\Gamma(\pi)$ is pictured as

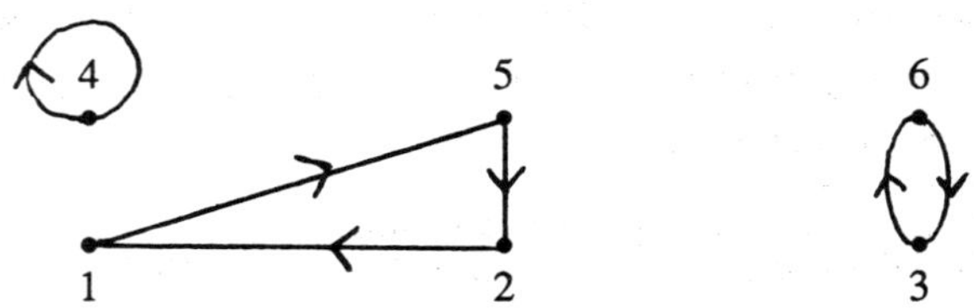

with the three cycles (1.6). Moreover

$$\pi = (1\ 5\ 2)(4)(3\ 6)$$

is a representation of π as a product of pairwise disjoint permutation cycles of lengths 3,1, and 2 respectively.

Each permutation digraph $\Gamma(\pi)$ is the pairwise disjoint union of cycles encompassing all vertices and its weight is

$$(1.7) \qquad (-1)^{\#\pi} \prod_{i=1}^{n} a_{i\pi(i)}$$

where $\#\pi$ is the number of cycles of π. Suppose the cycles of π have lengths $\ell_1,\ldots,\ell_k$, respectively, where $k = \#\pi$. Then

$$\text{sign } \pi = (-1)^{(\ell_1-1)+\cdots+(\ell_k-1)} = (-1)^{n-k} = (-1)^n(-1)^k .$$

Hence the weight (1.7) of $\Gamma(\pi)$ equals

$$(1.8) \qquad (\text{sign } \pi) \prod_{i=1}^{n} (-a_{i\pi(i)}) .$$

Let $\mathcal{P}(n)$ denote the set of all permutation digraphs of order n. From the above discussion we conclude that

$$(1.9) \qquad \det(-A) = \text{weight}(\mathcal{P}(n)) ,$$

a very combinatorial interpretation of the determinant.

Let I_n denote the identity matrix of order n. Then $\det(I_n-A)$ is the sum of the determinants of all principal submatrices of $-A$. It now follows that

$$(1.10) \qquad \det(I_n-A) = \text{weight}(\mathcal{P}^*(n))$$

where $\mathcal{P}^*(n)$ denotes the set of all permutation digraphs whose vertices form a subset of $\{1,\ldots,n\}$.

While the above combinatorial interpretation of the determinant has been known for some time, we have relied on the particular description in [Ze].

It was Jurkat and Ryser [JuRy1] who first showed that the determinant of an n by n matrix can be represented as the product of n matrices. Since the determinant is a scalar, the first matrix in the product must have one row and the last matrix must have one column. A very short and very combinatorial derivation of this factorization was given in [BrSh] within the general framework of a graded poset. We consider here only the graded poset P of all subsets of $\{1,\ldots,n\}$ partially ordered by inclusion and graded by cardinality.

Let P_i denote the collection of all i–element subsets of $\{1,\ldots,n\}$ $(i = 0,1,\ldots,n)$, and let $\alpha_1^{(i)}, \alpha_2^{(i)}, \ldots,\alpha_{\binom{n}{i}}^{(i)}$ be a listing, say in lexicographic order, of the sets in P_i. The chains of length n (the maximum possible length) in P are of the form

$$(1.11) \qquad \phi = \alpha_{i_0=1}^{(0)} \subset \alpha_{i_1}^{(1)} \subset \alpha_{i_2}^{(2)} \subset \cdots \subset \alpha_{i_n=1}^{(n)} = \{1,\ldots,n\} .$$

Let j_k be the unique element of $\alpha^{(k)}_{i_k}$ not belonging to $\alpha^{(k-1)}_{i_{k-1}}$ $(k = 1,\dots,n)$. Then

(1.12) $$j_1 j_2,\dots,j_n$$

is a permutation of $\{1,\dots,n\}$. Conversely, given a permutation (1.12) we obtain a chain (1.11) by defining $\alpha^{(0)}_{i_0} = \phi$ and $\alpha^{(k)}_{i_k} = \{j_1,\dots,j_k\}$ $(k = 1,\dots,n)$.

We use the n by n matrix (1.1) to assign weights to pairs consisting of an $(i-1)$–element set $\alpha^{(i-1)}_k$ in P_{i-1} and an i–element set $\alpha^{(i)}_\ell$ in P_i :

(1.13) $$w^{(i)}_{k\ell} = \begin{cases} 0, & \text{if } \alpha^{(i-1)}_k \not\subset \alpha^{(i)}_\ell \\ (-1)^{c_j} a_{ij}, & \text{if } \alpha^{(i-1)}_k \cup \{j\} = \alpha^{(i)}_\ell \\ & \text{and } c_j \text{ elements of } \alpha^{(i-1)}_k \\ & \text{are greater than } j. \end{cases}$$

The $\binom{n}{i-1}$ by $\binom{n}{i}$ matrix

$$W^{(i)} = (w^{(i)}_{k\ell} : k = 1,\dots,\tbinom{n}{i-1};\ \ell = 1,\dots,\tbinom{n}{i})$$

is the <u>weighted incidence matrix</u> for P_{i-1} and P_i $(i = 1,\dots,n)$. We observe that the entries of $W^{(i)}$ depend only on the entries in row i of A.

(1.14) <u>Example</u>: Let $n = 3$. Then the partially ordered set P with weights assigned as above is pictured below. The (formally) zero weights do not appear in this diagram which is the usual Hasse diagram of P with labels (the weights) on its edges.

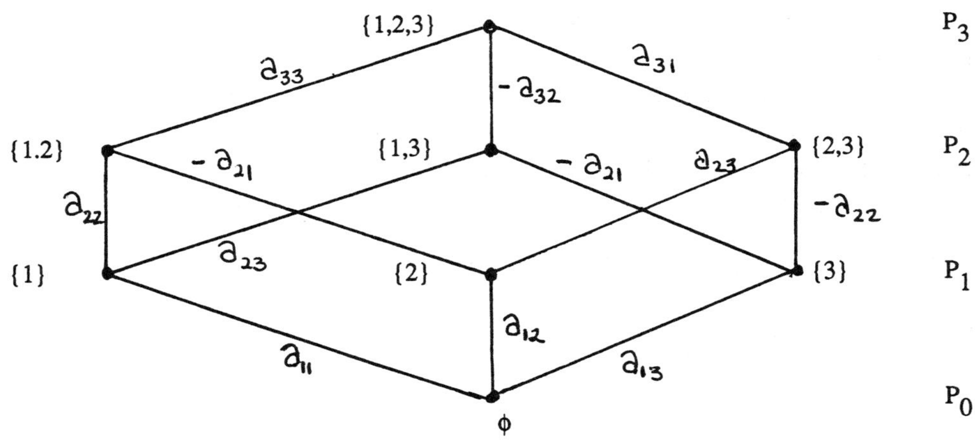

The chain $\phi \subset \{3\} \subset \{1,3\} \subset \{1,2,3\}$ corresponds to the permutation 3,1,2. □

We now define the <u>weight of the chain</u> (1.11) to be

(1.15) $$w^{(1)}_{i_0 i_1} w^{(2)}_{i_1 i_2} \cdots w^{(n)}_{i_{n-1} i_n} .$$

If (1.11) corresponds to the permutation π in (1.12) then denoting the number of inversions of π by $\text{inv}(\pi)$, we see that (1.15) equals

$$(-1)^{\text{inv}(\pi)} a_{1j_1} a_{2j_2} \cdots a_{nj_n} = (\text{sign } \pi) \prod_{i=1}^{n} a_{i\pi(i)} .$$

Let $\mathcal{C}_n$ denote the set of all chains in P having length n. It now follows that

(1.16) $$\det A = \text{weight}(\mathcal{C}_n) .$$

On the other hand the matrix

$$W^{(1)} W^{(2)} \cdots W^{(n)}$$

is a 1 by 1 matrix whose unique entry equals

(1.17) $$\sum_{i_{n-1}} \cdots \sum_{i_2} \sum_{i_1} w^{(1)}_{1i_1} w^{(2)}_{i_1 i_2} \cdots w^{(n)}_{i_{n-1}1} .$$

Since the product $w^{(1)}_{1i_1} w^{(2)}_{i_1 i_2} \cdots w^{(n)}_{i_{n-1}1}$ is zero if $\phi = \alpha^{(0)}_1, \alpha^{(1)}_{i_1}, \ldots, \alpha^{(n-1)}_{i_{n-1}}, \alpha^{(n)}_1 = \{1,\ldots,n\}$ is not a chain, we conclude that

(1.18) $$\text{weight}(\mathcal{C}_n) = W^{(1)} W^{(2)} \cdots W^{(n)} .$$

Combining (1.16) and (1.18) we obtain

(1.19) $$\det A = W^{(1)} W^{(2)} \cdots W^{(n)} ,$$

a <u>matrix factorization of the determinant</u>.

(1.20) <u>Example</u>: Let n = 3. Then the factorization (1.19) is $W^{(1)}W^{(2)}W^{(3)}$ where

$$W^{(1)} = \begin{array}{c|ccc} & \{1\} & \{2\} & \{3\} \\ \hline \phi & a_{11} & a_{12} & a_{13} \end{array} ,$$

$$W^{(2)} = \begin{array}{c|ccc} & \{1,2\} & \{1,3\} & \{2,3\} \\ \hline \{1\} & a_{22} & a_{23} & 0 \\ \{2\} & -a_{21} & 0 & a_{23} \\ \{3\} & 0 & -a_{21} & -a_{22} \end{array} ,$$

$$W^{(3)} = \begin{array}{c} \{1,2\} \\ \{1,3\} \\ \{2,3\} \end{array} \overset{\{1,2,3\}}{\begin{bmatrix} a_{33} \\ -a_{32} \\ a_{31} \end{bmatrix}} . \qquad \square$$

If we define the weights $w_{k\ell}^{(i)}$ by

$$w_{k\ell}^{(i)} = \begin{cases} 0, & \text{if} \quad \alpha_k^{(i-1)} \not\subset \alpha_\ell^{(i)} \\ a_{ij}, & \text{if} \quad \alpha_k^{(i-1)} \cup \{j\} = \alpha_\ell^{(i)} \end{cases} ,$$

then we obtain as above matrices $W^{(1)},W^{(2)},...,W^{(n)}$ where now

$$W^{(1)}W^{(2)} \cdots W^{(n)} = \text{per } A ,$$

the permanent of the matrix A.

2. Matrix–theoretic Proofs of Combinatorial Theorems.

In this section we illustrate how matrix theory has been used to prove theorems in combinatorics. These theorems have formulations which appear to be "purely combinatorial", but the proofs given show that the theorems have a hidden matrix–theoretic meaning, which is provided by the incidence matrix.

Let $X = \{x_1,...,x_v\}$ be a nonempty set of v elements, and let $X_1,...,X_b$ be b not necessarily distinct subsets of X. The incidence matrix of the subsets $X_1,...,X_b$ of X is the b by v (0,1)–matrix

$$A = (a_{ij}: 1 \le i \le b; \quad 1 \le j \le v)$$

where $a_{ij} = 1$ if $x_j \in X_i$ and $a_{ij} = 0$ if $x_j \notin X_i$. Row i of the incidence matrix displays the elements of the set X_i while column j displays the sets containing the element x_j. The incidence matrix gives a complete description of the subsets $X_1,...,X_b$ of the set X. The importance of the incidence matrix in combinatorial theory rests with the observation that combinatorial statements concerning the subsets $X_1,...,X_b$ of X can often be formulated as algebraic statements about the matrix A.

We now describe a class of configurations which are central in combinatorial theory. Let k and t be nonnegative integers and let λ be a positive integer. Then the subsets $X_1,...,X_b$ of $X = \{x_1,...,x_v\}$ form a t–design provided:

(2.1) Each X_i contains k elements;

(2.2) For each subset T of t elements of X, there are exactly λ of the sets $X_1,\ldots,X_b$ which contain T.

A t–design is usually denoted as $S_\lambda(t,k,v)$, and the sets $X_1,\ldots,X_v$ are called blocks. A t–design with $\lambda = 1$ is called a Steiner system. A 2–design is commonly known as a balanced incomplete block design (BIBD). Trivial examples of t–designs are obtained by taking $X_1,\ldots,X_b$ to be the family of all the distinct k–element subsets of X (thus $b = \binom{v}{k}$) and $\lambda = \binom{v-t}{k-t}$). The following are some basic facts about t–designs [Ha,Wil1,Wil2]

(2.3) $b = \lambda\binom{v}{t}/\binom{k}{t}$.

(2.4) A t–design $S_\lambda(t,k,v)$ is also an i–design $S_{\lambda_i}(i,k,v)$ where $\lambda_i = \lambda\binom{v-i}{t-i}/\binom{k-i}{t-i}$ $(i = 0,1,\ldots,t)$. In particular, the λ_i are integers.

According to (2.3) the integer b is determined by $v,k,\lambda,$ and t. A fundamental problem in combinatorial theory is to determine if a t–design $S_\lambda(t,k,v)$ exists for given v,k,λ and t. A 1–design $S_\lambda(1,k,v)$ is a very general combinatorial configuration consisting of b subset $X_1,\ldots,X_b$ of a v–element set $X = \{x_1,\ldots x_v\}$ where each X_i contains exactly k elements and each x_j is contained in exactly λ blocks. By (2.4) a t–design $S_\lambda(t,k,v)$ with $t \geq 2$ is both a 1–design $S_{\lambda_1}(1,k,v)$ and a 2–design $S_{\lambda_2}(2,k,v)$. We now consider 2–designs $S_\lambda(2,k,v)$ and denote the number λ_1 of blocks containing a specified element by r. By (2.3) and (2.4) $bk = rv$ and $r(k-1) = \lambda(v-1)$.

Let $A = (a_{ij})$ be the b by v incidence matrix of an $S_\lambda(2,k,v)$, and let A^T denote the transpose of A. The (i,j)–entry of A^TA is

$$\sum_{u=1}^{b} a_{ui}a_{uj},$$

which equals the number λ of blocks containing x_i and x_j if $i \neq j$ and equals the number r of blocks containing x_i if $i = j$. Let I_m denote the m by m identity matrix, and let J_{mn} denote the m by n matrix each of whose entries equals 1 (abbreviated to J_m if $m = n$). Then the properties (2.1) and (2.2) of a 2–design are entirely equivalent to the matrix equations

(2.5) $$AJ_{vb} = kJ_b\,,\quad A^TA = (r-\lambda)I_v + \lambda J_v\,.$$

Since each element of X is in exactly r blocks, A also satisfies

(2.6) $$J_{v,b}A = rJ_v\,.$$

We now assume that $v > k$, equivalently that $r > \lambda$. The matrix $(r-\lambda)I_v + \lambda J_v$ has eigenvalues $r+(v-1)\lambda$ and $\lambda-r$ ($v-1$ times), and it is easily checked that its inverse satisfies

(2.7) $$((r-\lambda)I_v + \lambda J_v)^{-1} = \frac{1}{r-\lambda}(I_v - \frac{\lambda}{rk}J_v)\,.$$

It follows from (2.5) that the v by v matrix A^TA is nonsingular. Since A is a b by v matrix we obtain

(2.8) <u>Fisher's inequality</u>: Let $v > k$. Then the number b of blocks and the number v of points in a 2–design $S_\lambda(2,k,v)$ satisfy $b \geq v$. □

We recall the following basic fact from linear algebra:

(2.9) Let C be an m by n matrix with rank equal to n. Then the <u>orthogonal projection</u> from the m–dimensional real vector space m onto the column space W of C is given by the m by m matrix

(2.9a) $$P = C(C^TC)^{-1}C^T\,.$$

The matrix P is symmetric and idempotent with eigenvalues 0 ($m-n$ times) and 1 (n times). The orthogonal projection of $\mathbb{R}^m$ onto the orthogonal complement of W is given by the m by m matrix $Q = I_m - P$. The matrix Q is symmetric and idempotent with eigenvalues 0 (n times) and 1 ($m-n$ times). In particular, both P and Q are positive semidefinite. □

We apply (2.9) to the incidence matrix A of an $S_2(\lambda,k,v)$. The matrix A^TA is nonsingular, and using (2.5), (2.6), (2.7), and (2.9a) we calculate the orthogonal projections P and Q:

$$\begin{aligned}P &= A(A^TA)^{-1}A^T\\ &= A((r-\lambda)I_v + \lambda J_v)^{-1}A^T\\ &= A\frac{1}{r-\lambda}(I_v - \frac{\lambda}{rk}J_v)A^T;\end{aligned}$$

hence

(2.10) $$P = \frac{1}{r-\lambda}(AA^T - \frac{\lambda k}{r}J_b)\,,$$

and

(2.11) $$Q = I_b - P = I_b - \frac{1}{r-\lambda}(AA^T - \frac{\lambda k}{r} J_b) .$$

Because Q is positive semidefinite each of its principal submatrices has a nonnegative determinant. An m by m principal submatrix of Q is obtained as follows. Let $X_1',\dots,X_m'$ be m of the blocks of a 2–design $S_2(\lambda,k,v)$, and let X_i' and X_j' have μ_{ij} elements in common $(i,j = 1,\dots,m)$. Let $U = (\mu_{ij}: i,j = 1,\dots,m)$. Then

$$Q_m = I_m - \frac{1}{r-\lambda}(U - \frac{\lambda k}{r} J_m)$$

is a principal submatrix of Q, and hence

(2.12) $$\det Q_m \geq 0 .$$

The 2^b-1 inequalities of the form (2.12) are Connor's inequalities [Co,Hal,Wil2]. The diagonal entries of Q_m are equal to $(r-k)/r$. For $i \neq j$ the (i,j)–entry of Q_m equals $(\lambda k - r\mu_{ij})/r(r-\lambda)$. The case $m = 1$ of (2.12) asserts that $r \geq k$; because $bk = rv$, this is equivalent to Fisher's inequality (2.8). The case $m = 2$ of (2.12) is equivalent to the statement that the number μ of elements common to two different blocks satisfies

(2.13) $$(r-k)(r-\lambda) \geq |\lambda k - r\mu| .$$ □

A symmetric 2–design is an $S_2(\lambda,k,v)$ with $b=v$ (and hence $r=k$). For a symmetric design (2.13) is equivalent to $\mu = \lambda$. Thus in a symmetric design two distinct blocks have exactly λ elements in common. The incidence matrix A of a symmetric design thus satisfies the matrix equations

(2.14) $$J_vA = AJ_v = kJ_v , \quad AA^T = A^TA = (k-\lambda)I_v + \lambda J_v .$$

The second set of equations in (2.14) implies that if there is a symmetric 2–design $S_2(\lambda,k,v)$, then the matrices I_v and $(k-\lambda)I_v + \lambda J_v$ are rationally congruent. This observation leads to the following necessary conditions of Bruck, Ryser, and Chowla [Hal,Ry2] for the existence of a symmetric $S_2(\lambda,k,v)$:

(2.15) If v is odd, then the equation

$$x^2 = (k-\lambda)y^2 + (-1)^{\frac{v-1}{2}} \lambda z^2$$

has a solution in integers x,y, and z not all equal to 0. □

A pairwise balanced design is obtained by removing the condition (2.1) in the definition of a 2–design. Thus subsets $X_1,\dots,X_b$ of $X = \{x_1,\dots,x_v\}$ form a pairwise balanced design of index $\lambda \geq 1$ if and only if the incidence matrix A satisfies

(2.16) $$A^TA = \mathrm{diag}(r_1-\lambda,\dots,r_v-\lambda) + \lambda J_v$$

where r_i is the number of elements in the block X_i $(i = 1,\dots,v)$. To avoid trivial situations we make the assumptions that each block contains at least two elements and no block contains all the elements of X (this implies that $r_i > \lambda$ $(i = 1,\dots,v)$). The matrix A^TA in (2.16) is then the sum of a positive definite and positive semidefinite matrix and hence is nonsingular. We thus obtain Majindar's result [Ma] that the number b of blocks and the number v of points in a pairwise balanced design satisfy $b \geq v$. Using matrix–theoretic methods Ryser [Ry3] and independently Woodall [Wo] proved:

(2.17) A pairwise balanced design of index λ for which $b = v$ is either a symmetric 2–design or has exactly two block sizes k_1 and k_2 where $k_1+k_2 = v+1$. (In the latter case the pairwise balanced design is called a λ–design). □

For each $\lambda \geq 1$ there are only finitely many λ–designs [Wo] but their classification except when λ is a prime or $1 \leq \lambda \leq 10$ has not been completed [Se].

Using higher order incidence matrices Wilson [Wil1,Wil2] has generalized Connor's inequalities to arbitrary t–designs.

We close this section by applying matrix methods in a striking way in order to obtain a theorem of Graham and Pollak [GrPo]. A graph G of order n has a set of n vertices, usually taken to be the set $\{1,2,\dots,n\}$, and a set E of edges where each edge is an unordered pair $\{i,j\}$ of distinct vertices. One may think of a graph as a digraph Γ with the property that if (i,j) is an arc, then so is (j,i). Thus G has an n by n symmetric incidence matrix $A = (a_{ij})$ where $a_{ii} = 0$ $(i = 1,\dots,n)$. Of particular interest is the complete graph K_n for which $\{i,j\}$ is an arc for each pair of distinct vertices i and j. The adjacency matrix of K_n is J_n-I_n. A graph G is called bipartite provided its set of vertices can be partitioned into two sets U_1 and U_2 and each edge contains one vertex from U_1 and one vertex from U_2. In case $\{i,j\}$ is an edge for all $i \in U_1$ and $j \in U_2$, then G is a complete bipartite graph. If $|U_1| = a$ and $|U_2| = b$, then the complete bipartite graph G is denoted by $K_{a,b}$. Let $G_1,\dots,G_r$ be complete bipartite graphs whose vertex sets are subsets of $\{1,\dots,n\}$. Then $\{G_1,\dots,G_r\}$ is called a bipartite decomposition of K_n provided each edge of K_n is an edge of exactly one of $G_1,\dots,G_r$. It is easy to obtain bipartite decompositions $\{G_1,\dots,G_{n-1}\}$ of K_n. Let G_i have vertices $\{i,\dots,n\}$ and edges $\{(i,i+1),(i,i+2),\dots,(i,n)\}$. Thus G_i is a complete bipartite graph $K_{1,n-i}$, and we have a bipartite decomposition $\{K_{1,n-1},K_{1,n-2},\dots,K_{1,2}\}$ of K_n. The proof of the following result of Graham and Pollak

[GrPo] is due to Peck [Pe]:

(2.18) The smallest integer r for which the complete graph K_n has a bipartite decomposition $\{G_1,\dots,G_r\}$ equals $n-1$.

Proof. Let $\{G_1,\dots,G_r\}$ be a bipartite decomposition of K_n where G_i is a K_{a_i,b_i}. Let G_i' be the bipartite graph with vertex set $\{1,\dots,n\}$ which has the same set of edges as G_i' ($i = 1,\dots,n$). The n by n adjacency matrix A_i of G_i', after relabeling its vertices, has the form

$$\begin{bmatrix} O & J_{a_ib_i} & O \\ J_{b_ia_i} & O & O \\ O & O & O \end{bmatrix}. \tag{2.19}$$

More precisely, there is a permutation matrix P_i such that $P_iA_iP_i^T$ equals (2.19). The statement that $\{G_1,\dots,G_r\}$ is a bipartite decomposition of K_n is equivalent to the matrix equation

$$J_n-I_n = A_1 + A_2 + \cdots + A_r. \tag{2.20}$$

Let B_i be the rank 1 matrix obtained from A_i by replacing either $J_{a_ib_i}$ or $J_{b_ia_i}$ in (2.19) by a zero matrix of the same size, and let

$$Q_i = A_i - 2B_i \quad (i = 1,\dots,r). \tag{2.21}$$

Then $P_iQ_iP_i^T$ has the form

$$\begin{bmatrix} O & J_{a_ib_i} & O \\ -J_{a_ib_i} & O & O \\ O & O & O \end{bmatrix},$$

and hence Q_i is a real skew symmetric matrix. Using (2.20) and (2.21) we now obtain

$$J_n-I_n = (Q_1 + \cdots + Q_r) + 2(B_1 + \cdots + B_r). \tag{2.22}$$

Let Q be the real skew symmetric matrix $Q_1 + \cdots + Q_r$. Then (2.22) implies that

$$I_n + Q = J_n - 2(B_1 + \cdots + B_r). \tag{2.23}$$

Now Q being a real skew symmetric matrix has pure imaginary eigenvalues, and so I_n+Q is nonsingular. In (2.23) we have an n by n nonsingular matrix written as the sum of $r+1$ matrices of rank 1. Since the rank of a sum of matrices never exceeds the sum of their ranks, we conclude $r+1 \geq n$. □

The original proof of (2.18) by Graham and Pollak is also matrix–theoretic. In addition there is a proof by Tverberg [Tv] which uses the Cauchy Schwarz inequality.

3. Combinatorial Matrix Analysis

As used by Johnson [Jo] the term combinatorial matrix analysis refers to "the use of combinatorial methodology or thought to better understand matrix structure or a particular matrix–theoretical problem". One of the earliest examples of combinatorial matrix analysis occurs in the work of Frobenius [Va,HoJo1] on spectral properties of (entrywise) nonnegative matrices. This work is now classical and we shall only discuss it briefly.

Let $A = (a_{ij})$ be an n by n matrix. The digraph of A, denoted by $\Gamma(A)$, is the digraph of order n with vertices $1,\ldots,n$ in which there is an arc (i,j) if and only if $a_{ij} \neq 0$ $(i,j = 1,\ldots,n)$. Thus the arcs of $\Gamma(A)$ correspond to the nonzero entries of A. (If, in the terminology of section 1 we use A to assign weights to $\Gamma(A)$, then an arc (i,j) of $\Gamma(A)$ would be assigned weight a_{ij}). The adjacency matrix of $\Gamma(A)$ as defined in section 1, is the n by n (0,1)–matrix whose 1's are in exactly those positions in which A has a nonzero entry. The matrix A is called reducible provided there is a permutation matrix P such that

$$PAP^T = \begin{bmatrix} A_1 & O \\ A_{21} & A_2 \end{bmatrix} \tag{3.1}$$

where A_1 and A_2 are square (non–vacuous) matrices. Reducibility of the matrix A is equivalent to the following cut–set property of the digraph $\Gamma(A)$: The vertices of the digraph can be partitioned into two nonempty sets V_1 and V_2 such that there is no arc (i,j) in the digraph with $i \in V_1$ and $j \in V_2$ (see below).

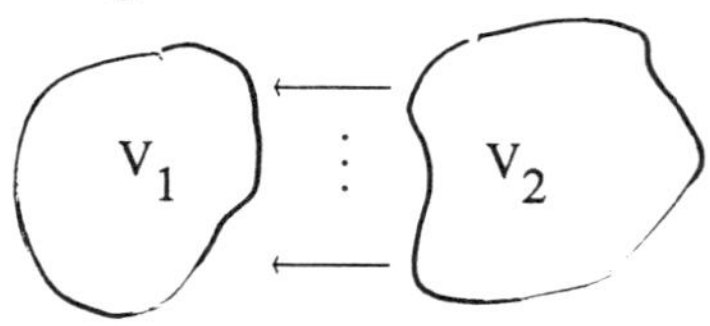

The set of arcs from V_2 to V_1 is a cut–set. The matrix A is irreducible provided it is not reducible. A digraph is called strongly connected provided it does not have the cut–set property. We thus have:

(3.2) The matrix A is irreducible if and only if the digraph $\Gamma(A)$ is strongly connected. □

A path from vertex i to vertex j in a digraph is a sequence $i = i_0, i_1,\ldots,i_k = j$ of $k \geq 0$ not necessarily distinct vertices such that $(i_0,i_1),\ldots,(i_{k-1},i_k)$ are arcs of the digraph (these arcs are called the arcs of the path). The property of strong connectivity of a digraph is equivalent to the following property:

(3.3) For each ordered pair i,j of vertices there is a path from i to j.

Generalizing results on positive matrices of Perron, Frobenius showed how the digraph of an irreducible n by n nonnegative matrix A determines certain spectral properties of A, which are thus independent of the weight assigned by A to the arcs of $\Gamma(A)$. (Frobenius did not use the terminology of digraphs and expressed his ideas differently.) Let k be the greatest common divisor of the lengths of the cycles of the irreducible n by n nonnegative matrix A. One of Frobenius' contributions is:

(3.4) There is a positive number r such that r is a simple eigenvalue of the irreducible matrix A, and no eigenvalue of A has a modulus which exceeds r. There exists an (entrywise) positive eigenvector of A corresponding to r. The eigenvalues of A of modulus r are the k numbers $e^{(2\pi i)t/k}$ $(t = 0,1,\dots,k-1)$, and indeed the spectrum of A, regarded as a point set in the complex plane, is invariant under the rotation about the origin through the angle $(2\pi i)/k$. □

From (3.4) we conclude that the number k of eigenvalues of maximum modulus r of the irreducible nonnegative matrix A depends only on the digraph $\Gamma(A)$. Suppose $k = 1$, so that the lengths of the cycles of $\Gamma(A)$ are relatively prime. The matrix A then has a unique eigenvalue of maximum modulus r, and A is called <u>primitive</u>. It can be shown that A is primitive if and only if A^p is an (entrywise) positive matrix for some positive integer p. Now the (i,j)–entry of A^p equals

$$\sum_{i_{p-1}=1}^{n} \cdots \sum_{i_1=1}^{n} a_{ii_1} a_{i_1 i_2} \cdots a_{i_{p-1}j},$$

which is nonzero if and only if there exists a path $i,i_1,i_2,\dots,i_{p-1},j$ of length p from i to j. We thus have:

(3.5) A nonnegative matrix is primitive if and only if there exists a positive integer p such that for each ordered pair i,j of vertices of $\Gamma(A)$ there is a path of length p from i to j. □

(It follows readily that a nonnegative matrix A for which A^p is a positive matrix for some positive integer p is irreducible.) The smallest positive integer p such that the $p^{\underline{th}}$ power of the primitive matrix A is a positive matrix ($\Gamma(A)$ has property (3.5)) is the <u>exponent</u> e(A) of A. The largest exponent of an n by n primitive matrix is n^2-2n+2. With the recent work of Lewin and Vitek {LeVi], Shao [Sh], and Zhang [Zh], the determination of those integers between 1 and n^2-2n+2 which are exponents of an n by n primitive matrix is now settled.

One of the most useful theorems in matrix theory is Geršgorin's theorem [HoJo1] for the location of the eigenvalues of a matrix. Let $A = (a_{ij})$ be an n by n complex matrix with eigenvalues $\lambda_1,\ldots,\lambda_n$ and let

$$r_i = \sum_{\substack{j=1 \\ j\neq i}}^{n} |a_{ij}|$$

be the sum of the moduli of the off–diagonal elements in row i of A $(i = 1,\ldots,n)$. Then Geršgorin's theorem asserts

(3.6) The union of the disks

$$D_i = \{z: |z-a_{ii}| \leq r_i\} \qquad (i = 1,\ldots,n)$$

contains all of the eigenvalues $\lambda_1,\ldots,\lambda_n$ of A.

Geršgorin's theorem implies that if A is a diagonally dominant matrix, that is

$$|a_{ii}| > r_i \qquad (i = 1,\ldots,n) ,$$

then 0 is not an eigenvalue of A and hence A is nonsingular. Conversely, knowing that diagonally dominant matrices are always nonsingular one immediately obtains Geršgorin's theorem. This is so because $\lambda_i I-A$ is singular for each eigenvalue λ_i of A and hence cannot be diagonally dominant.

We can better understand and improve Geršgorin's theorem by considering the digraph $\Gamma(A)$ of the n by n matrix $A = (a_{ij})$. For simplicity we assume that $n > 1$ and A is irreducible. Thus $\Gamma(A)$ is strongly connected from which it follows that every arc of $\Gamma(A)$ is an arc of some cycle. Let $\mathcal{C}(A)$ denote the set of cycles of $\Gamma(A)$ having length at least 2. We then have the following theorem of Brualdi [Br4]:

(3.7) Let $A = (a_{ij})$ be an n by n irreducible matrix with $n > 1$. If

(3.7a) $$\prod_{\gamma} |a_{ii}| > \prod_{\gamma} r_i \quad \text{for all } \gamma \in \mathcal{C}(A),$$

then A is nonsingular.

The notation $\prod_{\gamma}$ above means to take the product over all vertices i which appear in the cycle γ. Since A is assumed to be irreducible in (3.7), the strict inequalities (3.7a) can be replaced by non–strict inequalities as long as one remains strict. We sketch the proof of

(3.7).

Suppose A is singular and satisfies (3.7a). There exists a nonzero vector $x = (x_1,\ldots,x_n)$ such that $Ax^T = 0$. Because every arc of $\Gamma(A)$ is an arc of some cycle, (3.7a) implies that $a_{ii} \neq 0$ $(i = 1,\ldots,n)$. Consider the digraph Γ^* obtained from $\Gamma(A)$ by deleting those vertices i for which $x_i = 0$ and all arcs which contain at least one of these vertices. (Γ^* is the digraph of a principal submatrix of A.) Since no diagonal entry of A is zero, for each vertex i of Γ^* there exists a vertex j of Γ^* different from i such that (i,j) is an arc of Γ^*. It now follows fairly easily that there is a cycle $\gamma^* = (i_1,i_2,\ldots,i_k,i_{k+1} = i_1)$ of Γ^* (and hence of Γ) such that $|x_{i_{j+1}}| \geq |x_p|$ for all p such that (i_j,p) is an arc of Γ^* $(j = 1,\ldots,k)$. For $1 \leq i \leq n$, let $S_i = \{j \neq i\colon\ x_j \neq 0 \text{ and } a_{ij} \neq 0\}$ We now compute for $j = 1,\ldots,k$:

$$|a_{i_j i_j} x_{i_j} = -\sum_{\substack{p=1\\ p\neq i_j}}^{n} a_{i_j p}\, x_p = -\sum_{p\in S_{i_j}} a_{i_j p}\, x_p$$

and

$$|a_{i_j i_j}|\ |x_{i_j}| \leq \sum_{p\in S_{i_j}} |a_{i_j p}|\ |x_p| \leq \sum_{p\in S_{i_j}} |a_{i_j p}|\ |x_{i_{j+1}}|$$

$$\leq r_{i_j} |x_{i_{j+1}}| .$$

Multiplying we obtain

$$\prod_{\gamma^*} |a_{ii}| \prod_{\gamma^*} |x_i| \leq \prod_{\gamma^*} r_i \prod_{\gamma^*} |x_i| .$$

Since $x_i \neq 0$ for all vertices in γ^*, we conclude that

$$\prod_{\gamma^*} |a_{ii}| \leq \prod_{\gamma^*} r_i ,$$

contradicting (3.7a). □

If A is diagonally dominant, then (3.7a) surely holds and hence by (3.7) A is nonsingular, and Geršgorin's theorem follows. From (3.7) we obtain the more general eigenvalue inclusion region [Br4].

(3.8) Let $A = (a_{ij})$ be an n by n irreducible matrix. Then the union of the regions

$$D_\gamma = \{z: \prod_\gamma |z-a_{ii}| \le \prod_\gamma r_i\}$$

contains all the eigenvalues of A. □

We now describe some investigations of Johnson, Grone, Sá, and Wolkowicz [JoGrSaWo] of a much different kind (see also the article by Johnson in this volume). Suppose one wishes to construct a positive semi–definite hermitian n by n matrix $A = [a_{ij}]$ by filling in the entries one at a time. Since A is to be hermitian, each time an entry a_{ij} with $i \neq j$ is specified we should next specify a_{ji} to be $\bar{a}_{ji}$. So we will think of entries as being specified in pairs except when a diagonal entry is specified. But diagonal entries of hermitian matrices are real and those of positive semi–definite matrices are nonnegative. Thus we should be cautious to put only nonnegative real numbers on the main diagonal. We need to be even more cautious. An hermitian matrix is positive semi–definite if and only if each principal submatrix of A has a nonnegative determinant. Thus whenever we have specified all of the entries of a principal submatrix, we should have been cautious to make its determinant nonnegative. Even exercising all this caution, there is no guarantee that we will be able to complete the construction of the positive semi–definite hermitian matrix A. We may have progressed to a point (with caution!) in which not all entries have been specified and it is not possible to specify another entry without obtaining a principal submatrix with a negative determinant.

(3.9) Example: Suppose we have specified the entries of a 4 by 4 matrix as follows:

$$\begin{bmatrix} 1 & 1 & & -1 \\ 1 & 1 & 1 & \\ & 1 & 1 & 1 \\ -1 & & 1 & 1 \end{bmatrix}.$$

The entries on the main diagonal are positive and insofar as the matrix has been specified the matrix is hermitian. No principal submatrix of order 3 has been fully specified, and those of order 2 that have been specified have determinant equal to zero. Suppose we now specify the (1,3)–entry with the complex number a and thus the (3,1)–entry with its conjugate $\bar{a}$. Then two principal submatrices of order 3 have been specified, and their determinants are, respectively,

$$2\,\mathrm{Re}(a) - (1+|a|^2) \quad \text{and} \quad -2\,\mathrm{Re}(a) - (1+|a|^2)\,.$$

But for no complex number a are both of these determinants nonnegative. We thus have reached an impasse in this case. □

It is in general a difficult multivariable problem to decide whether a partially specified matrix can be completely specified so that the result is a positive semidefinite matrix. We can think of specifying the entries of a matrix as described above as specifying the edges of a graph with vertices 1,2,...,n: when we specify a_{ij} (and thus a_{ji}) we join vertices i and j by an edge. Specified diagonal entries correspond to loops (edges joining a vertex to itself) in the graph. Thus each partially specified matrix A has an associated graph (with loops) G(A). Rather than requiring one partially specified matrix to be completeable to a positive semidefinite hermitian matrix, we now require a large infinite set to be completeable making the problem more tractable.

Let G be a graph with vertices 1,...,n. Let $\mathcal{C}(G)$ denote the set of all partially specified n by n matrices $A = (a_{ij})$ having the following properties:

(3.10) If a_{ij} is specified, then so is a_{ji} and $a_{ji} = \bar{a}_{ij}$ (i,j = 1,...,n);

(3.11) A fully specified principal submatrix of A has a nonnegative determinant;

(3.12) G(A) = G.

We note that (3.12) implies that each matrix in $\mathcal{C}(G)$ has the same specified entries. A graph G is chordal (or triangulated) provided each of its cycles of length at least four has the property that there are two distinct vertices of the cycle which are joined by an edge in G but not joined by an edge of the cycle. The theorem of [JoGrSaWo] on completions is the following:

(3.13) Let G be a graph with vertices 1,...,n. Each partially specified matrix in $\mathcal{C}(G)$ can be completed to a positive semidefinite matrix if and only if the graph G is chordal. □

We close this section by showing how combinatorial analysis arises in the problem of diagonal similarity of matrices. Let $A = (a_{ij})$ and $B = (b_{ij})$ be two n by n matrices. Although our implicit assumption in this section has been that our matrices are complex matrices, the entries of A and B can be from any field F. The matrices A and B are diagonally similar provided there exists a nonsingular diagonal matrix D such that

(3.14) $$DAD^{-1} = B.$$

Let the diagonal entries of D be $d_1,\ldots,d_n$ where $d_i \neq 0$ (i = 1,...,n) and write $D = \mathrm{diag}(d_1,\ldots,d_n)$. The relation (3.14) implies that A and B have identical digraphs, but $\Gamma(A) = \Gamma(B)$ does not in general imply that A and B are diagonally similar.

Suppose $\Gamma(A) = \Gamma(B)$, and let us write Γ for the common digraph of A and B. Let $\gamma = (i_1,\ldots,i_k,i_1)$ be a cycle of Γ. Then as described in section 1, each of A and B assigns a weight to γ and we write $wt_A(\gamma) = -a_{i_1 i_2} \cdots a_{i_{k-1} i_k} a_{i_k i_1}$ and

$wt_B(\gamma) = -b_{i_1 i_2} \cdots b_{i_{k-1} i_k} b_{i_k i_1}$. (The negative signs are of no consequence in what follows but in order to conform with our earlier definition we retain them.) Suppose (3.14) holds. Then $b_{ij} = d_i a_{ij} d_j^{-1}$ (i,j = 1,...,n), and it follows that $wt_A(\gamma) = wt_B(\gamma)$ for each cycle γ of Γ. For irreducible matrices the converse holds as well and this fact was first proved by Fiedler and Ptak [FiPa,SaSc].

(3.15) Let $A = (a_{ij})$ and $B = (b_{ij})$ be n by n matrices, and assume that A is irreducible. Then A and B are diagonally similar if and only if $\Gamma(A) = \Gamma(B)$ and

(3.15a) $wt_A(\gamma) = wt_B(\gamma)$ for every cycle γ of $\Gamma(A)$.

Because of the discussion preceding (3.15) we have only to show the converse. Suppose that $\Gamma(A) = \Gamma(B)$ and (3.15a) holds. Then B is also irreducible. As above let Γ be the common digraph of A and B. Because A is irreducible, Γ is strongly connected. We proceed by induction on the number of arcs of Γ. Now a strongly connected digraph can always be obtained by starting with a cycle and successively adding paths between two vertices:

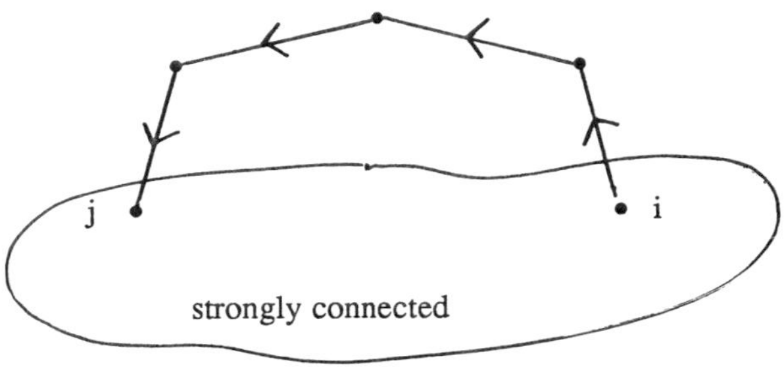

First suppose that Γ is a cycle. Without loss of generality, Γ consists of the vertices and arcs of the cycle (1,2,...,n,1). Then (3.15a) implies that $a_{12} \ldots a_{n-1,n} a_{n1} = b_{12} \ldots b_{n-1,n} b_{n1}$. Let $d_1 = 1$ and define, recursively, $d_{i+1} = \frac{d_i a_{i,i+1}}{b_{i,i+1}}$ (i = 1,...,n−1). Then $b_{i,i+1} = d_i a_{i,i+1} d_{i+1}^{-1}$ (i = 1,...,n−1), and we have

$$a_{12} \ldots a_{n-1,n} a_{n1} = d_1 a_{12} \ldots a_{n-1,n} d_n^{-1} b_{n1} .$$

Then $b_{n1} = d_n a_{n1} d_1^{-1}$, and we can conclude $DAD^{-1} = B$. Now suppose that Γ is obtained from a strongly connected digraph Γ' by adjoining a path from vertex i to vertex j of Γ'. For ease of notation, let us assume that the path is (3,2,1,n) (the same kind of proof works in general). Thus

$$A=\left[\begin{array}{cc|c} 0 & 0 & a_{1n} \\ a_{21} & 0 & \\ \hline & a_{32} & \\ & & A' \\ & & \end{array}\right], \quad B=\left[\begin{array}{cc|c} 0 & 0 & b_{1n} \\ b_{21} & 0 & \\ \hline & b_{32} & \\ & & B' \\ & & \end{array}\right]$$

where A' and B' are irreducible, and unspecified entries equal 0. By the inductive hypothesis it follows that there is a nonsingular diagonal matrix $D' = (d_3,\dots,d_n)$ such that $D'A'D'^{-1} = B'$. Define $d_2 = \dfrac{d_3 a_{32}}{b_{32}}$ and $d_1 = \dfrac{d_2 a_{21}}{b_{21}}$. Then $d_3 a_{32} d_2^{-1} = b_{32}$ and $d_2 a_{21} d_1^{-1} = b_{21}$. Since there is a path in Γ' from vertex n to vertex 3, it follows as in the first case that $d_1 a_{1n} d_n^{-1} = b_{1n}$ as well. Hence letting $D = \mathrm{diag}(d_1,\dots,d_n)$, we see that $DAD^{-1} = B$. □

The analysis of matrices from a combinatorial point of view is an active and exciting topic in matrix theory. We have touched on only a small part of its progress.

4. Combinatorial Matrix Algebra

Let $A = (a_{ij})$ be an n by n matrix and let, as in section 3, $\Gamma(A)$ denote the digraph of A. Each arc of $\Gamma(A)$ is assigned a weight by A. For each positive integer p and each pair of integers i and j with $1 \le i,j \le n$, the (i,j)–entry of A^p equals the sum of the weights of all the paths from i to j of length p in $\Gamma(A)$. This observation and others like it form the basis for a combinatorial approach to matrix identities which has produced elegant proofs of some classical results. One of the first combinatorial proofs of a matrix algebraic identity is Swan's proof [SW1,SW2,Bo] of the Amitsur–Levitzki Theorem [AmLe,Pi].

The standard polynomial of degree n is the polynomial.

(4.1) $$[y_1,y_2,\dots,y_n] = \sum_{\pi} (\mathrm{sign}\ \pi) y_{\pi(1)} y_{\pi(2)} \cdots y_{\pi(n)}$$

where the summation extends over all permutations π of $\{1,\dots,n\}$. We have $[y_1] = y_1$ and

(4.2) $$[y_1,y_2] = y_1y_2 - y_2y_1 .$$

The equation

$$[y_1,y_2,\dots,y_n] = 0$$

is the standard identity of degree n, and in view of (4.2), the standard identities can be regarded as a sequence of progressively weaker versions of the commutative law. Let R be a commutative ring with identity 1. The Amitsur–Levitzki theorem asserts the following:

(4.3) The ring $M_n(R)$ of n by n matrices over R satisfies the standard identity of order 2n; that is, if $A_1,\ldots,A_{2n}$ are n by n matrices over R, then

(4.3a) $$[A_1,A_2,\ldots,A_{2n}] = O .$$

The ring $M_n(R)$ is actually an R–module, and since the standard polynomial is an R–multilinear function of its arguments, the validity of (4.3a) for all $A_1,\ldots,A_{2n}$ follows from its validity for all choices of $A_1,\ldots,A_{2n}$ from a basis of the R–module $M_n(R)$. Let E_{ij} denote the n by n matrix all of whose entries are zero except for the (i,j)–entry which equals 1 (i,j = 1,...,n). If follows that it suffices to know (4.3a) when each of the matrices $A_1,\ldots,A_{2n}$ is contained in the set $\{E_{ij}: i,j = 1,\ldots,n\}$. Let $A_k = E_{i_kj_k}$ $(k = 1,\ldots,2n)$. Let π be a permutation of $\{1,2,\ldots,2n\}$. Then:

(4.4) The product $A_{\pi(1)}A_{\pi(2)}\cdots A_{\pi(2n)}$ equals $E_{i_{\pi(1)}j_{\pi(2n)}}$ provided $j_{\pi(1)} = i_{\pi(2)}$, $j_{\pi(2)} = i_{\pi(3)},\ldots,j_{\pi(2n-1)} = i_{\pi(2n)}$, and equals a zero matrix otherwise. □

We now construct a <u>directed multigraph</u> Γ^* of order n. (That Γ^* is a directed multigraph means that each arc has a positive integral multiplicity and thus the arcs determine a multiset. We could regard these multiplicities as weights of the arcs, but for our present purposes we need to think of the multiplicity as the number of times the arc occurs in Γ^*. Thus an arc (i,j) has multiplicity 3 means that three 'copies' of (i,j) occur in Γ^*.) The vertices of Γ^* are $\{1,2,\ldots,n\}$ and the multiset of arcs of Γ^* is the multiset of ordered pairs (i_k,j_k) for $k = 1,\ldots,2n$. An <u>Eulerian path</u> from vertex i to vertex j in Γ^* is a path in Γ^* from i to j which includes each arc of Γ^* exactly once. In particular it follows that an Eulerian path has length 2n, the number of arcs in Γ^*. We can now restate (4.4) in the following terms:

(4.5) The product $A_{\pi(1)}A_{\pi(2)}\cdots A_{\pi(2n)}$ equals $E_{i_{\pi(1)}j_{\pi(2n)}}$ if $i_{\pi(1)},j_{\pi(1)},j_{\pi(2)},\ldots,j_{\pi(2n-1)},j_{\pi(2n)}$ is an Eulerian path in Γ^* and equals a zero matrix otherwise. □

Each Eulerian path γ in Γ^* corresponds to an ordered arrangement of the arcs (i_k,j_k) $(k = 1,\ldots,2n)$ of Γ^* and hence to a permutation π of $\{1,2,\ldots,2n\}$. We define the <u>sign of the Eulerian path</u> γ to be the sign of the corresponding permutation π. We now have

(4.6) The (i,j)–entry of $[A_1,A_2,\dots,A_{2n}]$ equals $\sum_{\gamma_{ij}} \text{sign}(\gamma_{ij})$ where the summation extends over all Eulerian paths γ_{ij} in Γ from i to j (i,j = 1,...,2n).

Hence the Amitsur–Levitzki Theorem (4.3) is equivalent to the following combinatorial result:

(4.7) If Γ^* is a directed multigraph of order n with (at least) 2n arcs, then

$$\sum_{\gamma_{ij}} \text{sign}(\gamma_{ij}) = 0 . \qquad \square$$

We refer to [Sw1,Sw2,Bo] for a proof of (4.7). We have shown here how a very algebraic–looking identity about matrices over a commutative ring R with identity can be reduced to a purely combinatorial identity which is independent of the ring R.

Another recent instance of the preceding phenomenon is the proof by Laue [La] of a fundamental trace identity. Let π be a permutation of {1,...,k}. As noted in section 1, there is a one–to–one correspondence between the cycles of π and the cycles of the permutation digraph $\Gamma(\pi)$. We denote the cycles of π by $\pi_1,\pi_2,\dots,$ and we write $j \in \pi_i$ to mean that j occurs in the cycle π_i. As above R denotes a commutative ring with identity. The following trace identity goes back to Frobenius [Fr,Le,Pr]:

(4.8) Let k and n be integers with $k > n$, and let $A_1,\dots,A_k$ be n by n matrices over R. Then

$$\text{(4.8a)} \qquad \sum_{\pi} \text{sign}(\pi) \prod_{\pi_i} \text{Tr}\Big(\prod_{p \in \pi_i} A_p\Big) = 0 .$$

In (4.8a) the summation is over all permutations π of {1,...,k}, the first product is over all the cycles π_i of π, and the second product is over all $p \in \pi_i$ taken in the cyclical order of π_i. Thus if $\pi_i = (3,7,5)$, then $\prod_{p\in\pi_i} = A_3A_7A_5$. Since $\text{Tr}(XY) = \text{Tr}(YX)$, $\text{Tr}(A_3A_7A_5) = \text{Tr}(A_7A_5A_3) = \text{Tr}(A_5A_3A_7)$.

The function on the left side of (4.8a) is an R–multilinear function of its arguments $A_1,\dots,A_k$. It thus suffices to show that (4.8a) holds when $A_t = E_{i_tj_t}$ where $1 \le i_t,j_t \le n$ (t = 1,...,k). Let Γ^* be the directed multigraph of order n constructed as in our discussion of the Amitsur–Levitzki theorem. In particular, Γ^* has $k > n$ edges. Let π be a permutation of {1,...,k}, and let π_i be a cycle of π. Then $\text{Tr}(\prod_{p\in\pi_i} A_p)$ equals 1 if the arcs (i_p,j_p) for

$p \in \pi_i$ (taken in the cyclical order of π_i) are the arcs of a circuit (see section 1) of Γ^*, and equals 0 otherwise. Hence $\prod_{\pi_i} Tr(\prod_{p\in\pi_i} A_p)$ equals 1 if the cycles $\pi_1, \pi_2, \ldots$ of π correspond to a partitioning of the multiset of arcs of Γ^* into circuits of Γ^*. While each permutation π of $\{1,2,\ldots,k\}$ corresponds to a permutation of the k arcs $(i_1,j_1),(i_2,j_2),\ldots,(i_k,j_k)$ of Γ^*, not all correspond to a partitioning of the arcs of Γ^* into circuits. The result (4.8) is equivalent to the following combinatorial theorem:

(4.9) If Γ^* is a directed multigraph of order n with $k > n$ arcs, then the number of the partitions of Γ^* into circuits which correspond to an even permutation of $\{1,2,\ldots,k\}$ equals the number of partitions corresponding to an odd permutation of $\{1,2,\ldots,k\}$.

The proof of (4.9) is easy. Since Γ^* has more arcs then it has vertices, there are two arcs α and β which have the same initial vertex (it is permissible that $\alpha = \beta = (i,j)$ but then (i,j) must have multiplicity at least equal to 2). There is a one–to–one sign reversing correspondence (indeed an involution) between the circuit partitions of Γ^* in which α and β belong to the same circuit and those in which α and β belong to different circuits. This correspondence is pictured below:

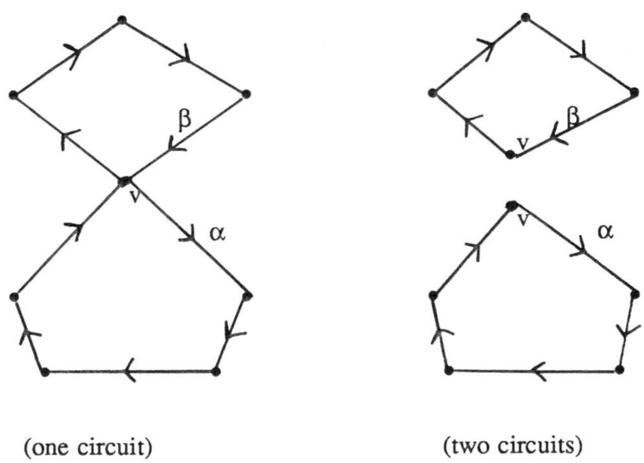

□

We have discussed above two identities which are satisfied by matrices over a commutative ring R with identity. While these identities may be new to some, the following identity is surely known to all. Let A be an n by n matrix over a commutative ring R with identity, and let

(4.10) $$\chi_A(\lambda) = \det(\lambda I - A)$$

be the characteristic polynomial of A. Then the Cayley–Hamilton Theorem asserts:

(4.11) $$\chi_A(A) = 0\,.$$

Straubing [St,Ze] recently gave an elegant combinatorial proof of (4.11) which we now outline. The characteristic polynomial (4.10) is the polynomial

$$\chi_A(\lambda) = \lambda^n + \sigma_1\lambda^{n-1} + \cdots + \sigma_k\lambda^{n-k} + \cdots + \sigma_n$$

where σ_i is the sum of the determinants of all the i by i principal submatrices of $-A$ $(i = 1,\ldots,n)$. Thus (4.11) is equivalent to the identity

(4.12) $$A^n + \sigma_1 A^{n-1} + \cdots + \sigma_k A^{n-k} + \cdots + \sigma_n I_n = O\,.$$

To prove (4.12) one needs to show that each entry of the matrix on the left hand side of (4.12) is 0. As in section 1 we use A to assign weights to the arcs of a digraph of order n. It follows from (1.9) (see also (1.10)) that σ_k is the sum of the weights of all permutation digraphs whose vertices form a subset of cardinality k of $\{1,\ldots,n\}$. Moreover the (i,j)–entry of A^{n-k} is the sum of the weights of all paths from i to j of length $n-k$. Let X_{ij} be the set of pairs (γ,π) where γ is a path from i to j of length of at most n, π is a collection of cycles which have no vertex in common, and the number of arcs of γ plus the number of arcs of π equals n. Define the weight of the pair (γ,π) to be

$$\mathrm{wt}(\gamma)\mathrm{wt}(\pi) = (-1)^{\#\pi}(\text{product of all arc weighs of } \gamma \text{ and } \pi)$$

where $\#\pi$ is the number of cycles of π. Then the (i,j)–entry of the matrix on the left hand side of (4.12) equals the weight of the set X_{ij}. Thus (4.12) is equivalent to:

(4.13) $$\text{weight } (X_{ij}) = 0 \qquad (i,j = 1,\ldots,n)\,.$$

We can establish (4.13) by showing that there is a one–to–one correspondence (indeed an involution) from X_{ij} to itself in which corresponding pairs (γ,π) have weights of opposite sign. This correspondence is defined as follows. Let (γ,π) be in X_{ij}. Since the sum of the number of arcs of γ and the number of arcs of π equals n (the number of vertices), either there is a vertex of γ which is also a vertex of one of the cycles of π, or the path γ contains a repeated vertex and hence 'contains' a cycle (possibly both). We traverse γ until we first arrive at a vertex k which has been previously visited or is a vertex of a cycle of π. In the first instance, γ contains a cycle γ_0 through vertex k whose vertices are disjoint from those in π; we then 'remove' γ_0 from γ and include it in π giving a $(\gamma',\pi') \in X_{ij}$. In the second instance, π contains a cycle π_0 one of whose vertices is k; we remove π_0 from π and 'join' it to γ giving a $(\gamma',\pi') \in X_{ij}$. In either instance $\mathrm{wt}(\gamma',\pi') = -\mathrm{wt}(\gamma,\pi)$ and (4.13) is proved. □

We conclude this section by mentioning that combinatorial considerations arise also in the determination of the elementary divisors of a matrix and hence in the determination of the Jordan canonical form [Br5,Br6].

5. Matrix Theory as a Tool in Combinatorics

In section 2 we have illustrated how matrix theory and linear algebra have been used to prove combinatorial theorems. Matrix theory and linear algebra have also been used extensively to classify combinatorial objects and in general to extract combinatorial information from algebraic information. The vast subject of spectral graph theory [CvDoSa,CvDoGuTo] is concerned with the relationships between the (linear) algebraic properties of the spectra of matrices associated with a graph (e.g. the adjacency matrix of a graph) and the combinatorial properties of the graph. A graph with n vertices is called strongly regular provided there are integers k, λ and μ such that each vertex is adjacent to k other vertices (the graph is k–regular), any two adjacent vertices are mutually adjacent to λ other vertices, and any two non–adjacent vertices are mutually adjacent to μ other vertices. It is customary to exclude empty (i.e. no edges) and complete graphs as strongly regular graphs. The integers n, k, λ, and μ are the parameters of a strongly regular graph. Examples of strongly regular graphs are the pentagon $(n=5,k=2,\lambda=0,\mu=1)$, the famous Petersen graph $(n=10,k=3,\lambda=0,\mu=1)$, the graph of the regular octahedron $(n=6,k=4,\lambda=2,\mu=4)$, and the cocktail party graphs $CP(m)$, obtained by removing m pairwise vertex disjoint edges from the complete graph K_{2m} $(n = 2m, k = 2(m-1), \lambda = 2(m-2), \mu = 2(m-1))$. The adjacency matrix A of a strongly regular graph satisfies

$$A^2 = (\lambda-\mu)A + \mu J_n + (r-\mu)I_n \tag{5.1}$$

where J_n is the n by n matrix of all 1's. It follows from (5.1) that A has three distinct eigenvalues k, λ_1, and λ_2. Conversely, a k–regular graph whose adjacency matrix has three distinct eigenvalues is a strongly regular graph.

Let G be a graph of order n with m edges and let B be the m by n incidence matrix of the edges of G as defined in section 2. Since each edge of G is a subset of two vertices, B is a (0,1)–matrix with exactly two 1's in every row. Since edges of G are distinct (as subsets of vertices), we have

$$BB^T = A + 2I_m \tag{5.2}$$

where A is a symmetric (0,1)–matrix with zero trace. The matrix A is the adjacency matrix of a graph $L(G)$ of order m, called the line graph of G. The vertices of $L(G)$ are the edges of G and two vertices $\{i,j\}$ and $\{k,\ell\}$ of $L(G)$ are adjacent if and only if the edges $\{i,j\}$ and $\{k,\ell\}$ of G have a vertex in common. Since BB^T is a positive semidefinite matrix, it follows from (5.2) that the smallest eigenvalue of the adjacency matrix A of $L(G)$ is at least -2. Cameron, Goethals, Seidel, and Shult [CaGoSeSh] have characterized graphs for which the smallest eigenvalue of the adjacency matrix is at least -2. We briefly describe their characterization.

Let G be a graph of order n, and let $a_1,\ldots,a_n$ be nonnegative integers. The generalized line graph $L(G;a_1,\ldots,a_n)$ is obtained as follows. Begin with (vertex disjoint) graphs $L(G)$ and $CP(a_1),\ldots,CP(a_n)$ and then include extra edges: the vertex $\{i,j\}$ of $L(G)$ (edge of G) is joined by an edge to each vertex in $CP(a_i)$ and $CP(a_j)$ $(i,j=1,\ldots,n)$. The graph $L(G;0,\ldots,0)$ is the same as the line graph $L(G)$. The following characterization [CaGoSeSh] holds:

(5.3) The smallest eigenvalue of the adjacency matrix of a graph G is at least -2 if and only if

(5.3a) G is (isomorphic to) a generalized line graph, or

(5.3b) G can be represented (in a way we do not describe here) by one of the classical real root systems E_6, E_7 and E_8. □

For his solution of Shannon's problem on the capacity of a pentagon, Lovász [Lo1] introduced a new matrix theoretic parameter for the study of graphs which has turned out to have other implications as well. Shannon's problem is a problem in communication theory. Consider an alphabet of n letters $x_1,x_2,\ldots,x_n$ which are used to encode information being sent through a noisy communications channel. Because of channel noise the received letter may be confounded with the letter sent. Let us define a graph G of order n with vertices $x_1,\ldots,x_n$ and join two distinct vertices x_i and x_j by an edge $\{x_i,x_j\}$ if and only if x_i is confoundable with x_j. Two k–letter words are confoundable provided corresponding letters are either equal or confoundable. Thus two k–letter words are not confoundable provided they are not confoundable in at least one position. Define $\alpha_k(G)$ to be the maximum number of k–letter words such that no two distinct words are confoundable $(k=1,2,\ldots)$. The number of $\alpha(G) = \alpha_1(G)$ is the maximum number of letters in our alphabet no two of which are confoundable, that is, the maximum number of vertices of G no pair of which form an edge. Thus $\alpha(G)$ is the stability number of G. Let H_1 and H_2 be two graphs with vertex sets V_1 and V_2 respectively. The strong product $H_1 * H_2$ is the graph with vertex set $V_1 \times V_2$ in which vertices (u_1,u_2) and (w_1,w_2) are joined by an edge if and only if $u_1 = w_1$ and $\{u_2,w_2\}$ is an edge of H_2, $u_2 = w_2$ and $\{u_1,w_1\}$ is an edge of H_2, or $\{u_1,w_1\}$ is an edge of H_1 and $\{u_2,w_2\}$ is an edge of H_2. Define the kth power $G^{(k)}$ of G to be

$G * \cdots * G$ (k factors). Thus two k–tuples of vertices of G are joined by an edge in $G^{(k)}$ provided vertices in corresponding positions are either equal or joined by an edge in G. Then it follows that $\alpha_k(G)$ equals the stability number $\alpha(G^{(k)})$ of the kth power of G. The number

(5.4) $$\Theta(G) = \sup\{\sqrt[k]{\alpha(G^{(k)})}:\ k = 1,2,\ldots\}$$

is called the Shannon capacity of the graph. Shannon asked for the determination of $\Theta(G)$ for G equal to the pentagon.

Let A be the adjacency matrix of G, and let $\mathcal{A}$ denote the set of all n by n real symmetric matrices $B = [b_{ij}]$ such that $b_{ii} = 1$ $(i=1,\ldots,n)$ and $b_{ij} = 1$, if a_i and a_j are joined by an edge in G $(1 \le i < j \le n)$. Let $\lambda_{max}(B)$ denote the largest eigenvalue of B. Define

(5.5) $$\mathcal{V}(G) = \min\{\lambda_{max}(B) : B \in \mathcal{A}\}\ .$$

We then have [Lo1]

(5.6) $$\Theta(G) \le \mathcal{V}(G)\ .$$ □

In case of a regular graph the number $\mathcal{V}(G)$ in (5.5) can be bounded in terms of the smallest eigenvalue of the adjacency matrix of G [Lo1,Lo2]:

(5.7) If G is a k–regular graph of order n and λ_{min} is the smallest eigenvalue of the adjacency matrix of G, then

(5.7a) $$\mathcal{V}(G) \le \frac{-n\ \lambda_{min}}{k - \lambda_{min}}\ .$$ □

Now let G be the pentagon. Then $\alpha(G) = 2$ and $\alpha(G^{(2)}) = 5$, and we obtain $\Theta(G) \ge \sqrt{5}$ by (5.4). The adjacency matrix of G is a circulant matrix with smallest eigenvalue equal to $-2 \operatorname{Cos} \pi/5 = -2\sqrt{5}/(5-\sqrt{5})$. Using (5.7a) we conclude that $\mathcal{V}(G) = \sqrt{5}$. Now (5.6) implies:

(5.8) The Shannon capacity of the pentagon is $\sqrt{5}$. □

The quantity $\mathcal{V}(G)$ associated with a graph G of order n is related to two fundamental parameters of G. These are the chromatic number $\chi(G)$ which equals the fewest number of colors required to color the vertices of G so that no edge contains two vertices of the same color, and the clique number $\omega(G)$ which equals the largest number of vertices every distinct pair of which form an edge. Clearly $\omega(G) \le \chi(G)$. Grötschel, Lovász and Schrijver [GrLoSc1,GrLoSc2,Lo2] proved that $\mathcal{V}(G)$ falls in between these parameters:

(5.9) $$\omega(G) \le \mathcal{V}(G) \le \chi(G)\ .$$ □

For a graph G the determination of $\omega(G)$ and $\chi(G)$ is believed to be a computationally difficult problem, but the quantity $\mathcal{V}(G)$ (more precisely, a rational approximation to any specified degree of accuracy) can be computed using an algorithm the number of steps of which is bounded by a polynomial in the number n of vertices. It follows that a graph G for which $\omega(G) = \chi(G)$ has the property that its chromatic number $\chi(G)$ can be computed in a polynomial number of steps. Now there is a class of graphs G called perfect graphs for which $\omega(H) = \chi(H)$ for every induced subgraph H. The chordal graphs discussed in section 3 are examples of perfect graphs. We conclude that the chromatic number of chordal graphs can be computed in a polynomial number of steps.

In this section we have barely scratched the surface of a large and fascinating body of knowledge.

6. The rest of the story

Of course, we cannot hope to tell the rest of the story in the remaining few pages, nor do we pretend to have told the full story up to this point. A more complete exposition of combinatorial matrix theory will be found in the forthcoming book by Brualdi and Ryser [BrRy]. We want in this final section to touch on some aspects of combinatorial matrix theory which are of a different nature from those considered thus far.

One of the earliest 'purely' combinatorial theorems is the elementary but nonetheless remarkable theorem of König concerning (0,1)–matrices. Let A be an m by n (0,1)–matrix and define the term rank $\rho(A)$ of A to be the maximum number of 1's in A with no two of the 1's on the same row or column. König's theorem [Ry2] asserts:

(4.1) $\rho(A)$ equals the minimum number of rows and columns in A which contain all the 1's in A. □

König's theorem is now the foundation for the extensive theory of matchings in graphs [LoPl]. It is also a special instance of the Ford and Fulkerson theorem on maximum flows in a network [FoFu]. Although König's theorem above is described in terms of (0,1)–matrices, it applies to all matrices in which the entries are partitioned into two sets (those corresponding to 0 and those corresponding to 1).

One approach to the study of (0,1)–matrices is to study properties of a class of (0,1)–matrices which is combinatorially defined. One such class is the class studied in detail by Ryser and others [Ry2,Br2] consisting of all m by n (0,1)–matrices with a prescribed number of 1's in each row and column. Let $R = (r_1,\dots,r_m)$ and $S = (s_1,\dots,s_n)$ be nonnegative integral vectors, and assume without loss of generality that $r_1 \geq \cdots \geq r_m$ and $s_1 \geq \cdots \geq s_n$. Then $\mathfrak{A}(R,S)$ is the class of the m by n (0,1)–matrices with row sum vector R and column sum vector S. Let

$$\bar{\rho} = \max\{\rho(A): A \in \mathfrak{A}(R,S)\}$$

and

$$\tilde{\rho} = \min\{\rho(A): A \in \mathfrak{A}(R,S)\}$$

be, respectively, the maximum and minimum term rank that can be achieved by a (0,1)–matrix with row sum vector R and column sum vector S. Knowing $\bar{\rho}$ and $\tilde{\rho}$ allows us to conclude that

(6.1) $$\tilde{\rho} \le \rho(A) \le \bar{\rho} \quad (A \in \mathfrak{A}(R,S)),$$

and (6.1) gives the best possible inequalities for the term rank of a matrix based only on its row and column sum vector. The value of $\bar{\rho}$ was determined by Ryser [Ry2,BrRo]:

(6.2) $$\bar{\rho} = \min\{k\ell + \sum_{i>k} r_i - \sum_{j<\ell} s_j + k + \ell:\ k = 0,1,\ldots,m \text{ and } \ell = 0,1,\ldots,n\}.$$ □

A formula for $\tilde{\rho}$ is more complicated. A formula was derived by Haber [Ha] and was later simplified by Brualdi [Br3]. Much more is known about the class $\mathfrak{A}(R,S)$. A survey of the results up to about 1980 appears in [Br2].

Another class of matrices which is of combinatorial significance is the class Ω_n of n by n doubly stochastic matrices. An n by n matrix A is a doubly stochastic matrix provided each of its entries is a nonnegative real number and each row and column sum of A equals 1. The class Ω_n is a convex polytope in real n^2–dimensional space and has dimension $(n-1)^2$: Ω_n lies on the flat of dimension $(n-1)^2$ determined by the row and column sum constraints and on this flat it has a nonempty interior. Combinatorial considerations arise as a result of the classical theorem of Birkhoff [Ry2] which asserts:

(6.3) The extreme points of the convex polytope Ω_n are the n by n permutation matrices.

□

Combinatorial considerations also arise as follows. A face of Ω_n consists of all those doubly stochastic matrices which have 0's in certain prescribed positions. Let A be an n by n (0,1)–matrix and let

(6.4) $$\mathcal{F}(A) = \{X: X \in \Omega_n, X \le A\}.$$

In (6.4) the inequality $X \le A$ is to be interpreted as entrywise inequality. Since no entry of a doubly stochastic matrix can exceed 1, (6.4) defines the face consisting of those matrices in Ω_n which have 0's where A has 0's (and possibly elsewhere). Thus every face of Ω_n is of the form $\mathcal{F}(A)$ for some n by n (0,1)–matrix A. This correspondence between faces and matrices is not one–to–one due to the row and column constraints. A (0,1)–matrix A has total support provided for each 1 of A there is some permutation matrix P containing that 1 with $P \le A$. The correspondence $A \longrightarrow \mathcal{F}(A)$ between n by n (0,1)–matrices with total support and faces of Ω_n is a one–to–one correspondence. It follows from the above that (0,1)–matrices have geometrical significance as faces of the polytope Ω_n. A study of (0,1)–matrices from this point of view was initiated by Brualdi and Gibson [BrGi1,BrGi2,BrGi3]. There are many other convex sets of matrices whose extreme points have combinatorial significance. These include the transportation polytopes $\mathcal{T}(R,S)$ of nonnegative matrices with prescribed row and column sum vectors (see e.g. [JuRy2,Br1]) and the convex hull $\mathfrak{A}(R,S)^c$ of the matrices in the class $\mathfrak{A}(R,S)$ [BrHaHw].

Now let $H = (h_{ij})$ denote an n by n matrix with $h_{ij} = \pm 1$ for $i,j = 1,\ldots,n$. We call H a (1,–1)–matrix. From the classical Hadamard determinant theorem [Had] we obtain

(6.5)
$$|\det H| \le \prod_{i=1}^{n} \sqrt{\sum_{j=1}^{n} h_{ij}^2} = n^{n/2}$$

with equality if and only if

(6.5a)
$$HH^T = nI_n .$$
□

If H satisfies (6.5a), then also $HH^T = nI_n$. An n by n (1,–1)–matrix H which satisfies (6.5a) is called a <u>Hadamard matrix</u>. It is straightforward to prove that if there is an n by n Hadamard matrix, then n is a multiple of 4, and it has been conjectured that the converse holds. While the conjecture has not been proved, many constructions are known [StSt]. Hadamard matrices are equivalent to the symmetric 2–designs $S_2(\lambda,v,k)$ discussed in section 2 with $\lambda = n-1$, $v = 4n-1$ and $k = 2n - 1$[Ry1]. They have also been used to construct good error correcting codes [MacSl]. Let

(6.6)
$$\alpha_n = \max\{\det B\colon B \text{ an } n \times n \ (1,-1)\text{–matrix}\}$$

be the maximum determinant of an n by n (1,–1)–matrix. (Since interchanging two rows of a matrix changes the sign of the determinant, this is the same as the maximum absolute value of the determinants.) Thus by (6.5),

(6.7)
$$\alpha_n \le n^{n/2}$$

with equality if and only if an n by n Hadamard matrix exists. Thus if n is not a multiple of 4, we have strict inequality in (6.7). Ehlich [Eh1,Eh2] has obtained bounds for α_n in case n is not a multiple of 4. Let

(6.8)
$$\beta_n = \max\{\det A\colon A \text{ an } n \text{ by } n \ (0,1)\text{–matrix}\}$$

be the maximum determinant (or absolute value thereof) of an n by n (0,1)–matrix.

Williamson [Wi] showed that each of $\{\alpha_n: n = 1,2,...\}$ and $\{\beta_n: n = 1,2,...\}$ determines the other by the following argument.

Let B be an n by n (1,–1)–matrix. We multiply certain rows and certain columns of B by –1 and obtain a matrix C with $|\det C| = |\det B|$ such that the first row of C contains only 1's and the first column of C below the first row contains only –1's. Adding row 1 of C to each of row 2,...,n, and taking the determinant, we see that

$$\det C = \det D$$

where D is an n–1 by n–1 matrix of 0's and 2's. Let D = 2E where E is a matrix of 0's and 1's. Then

$$|\det B| = 2^{n-1}|\det E| \ . \tag{6.9}$$

Conversely, starting with an n–1 by n–1 (0,1)–matrix E we may reverse the above steps and obtain an n by n (1,–1)–matrix B satisfying (6.9). It follows that

$$\alpha_n = 2^{n-1}\beta_{n-1} \qquad (n \geq 2) \ . \tag{6.10}$$ □

Thus the problem of determining the maximum determinant of an n by n (1,–1)–matrix is equivalent to that of determining the maximum determinant of an n–1 by n–1 (0,1)–matrix.

One approach to the problem of the maximum determinant of an n by n (0,1)–matrix is to try to determine the maximum determinant over a combinatorially interesting class of (0,1)–matrices. For example, Ryser [Ry1] obtained an upper bound for the determinant of a v by v (0,1)–matrix with k 1's in every row and column and showed that equality hold if any only if there exists a certain symmetric 2–design $S_2(\lambda,v,k)$. We know very little about the maximum (absolute value of the) determinant of a matrix in a square class $\mathfrak{A}(R,S)$. Brualdi and Solheid [BrSo] investigated the class of n by n (0,1)–matrices whose 0's form an acyclic pattern, determined the maximum determinant, and characterized those matrices with maximum determinant.

A class of matrices related to the class of Hadamard matrices is the class of conference matrices [MacSl]. An n by n matrix C of 1's, –1's, and 0's is a conference matrix provided its diagonal entries are 0, each non–diagonal entry is 1 or –1, and $CC^T = (n-1)I_n$. In order for an n by n conference matrix to exist n must be even.

We hope that this short exposition may have persuaded the listener and reader of the fascinating and multifaceted nature of combinatorial matrix theory.

References

[AmLe] S. A. Amitsur and J. Levitzki, Minimal identities for algebras, Proc. Amer. Math. Soc. 1 (1950), 449–463.

[Bo] B. Bollobás, Graph Theory An Introductory Course, Springer–Verlag (New York), 1979.

[Br1] R. A. Brualdi, Convex sets of non–negative matrices, Canadian J. Math. 20 (1968), 144–157.

[Br2] R. A. Brualdi, Matrices of zeros and ones with fixed row and column sum vectors, Lin. Alg. Applics. 33 (1980), 159–231.

[Br3] R. A. Brualdi, On Haber's minimum term rank formula, Europ. J. Combinatorics 2 (1981), 17–20.

[Br4] R. A. Brualdi, Matrices, eigenvalues, and directed graphs, Lin. Multilin. Alg. 11 (1982), 143–165.

[Br5] R. A. Brualdi, Combinatorial verification of the elementary divisors of tensor products, Lin. Alg. Applics. 71 (1985), 31–47.

[Br6] R. A. Brualdi, The Jordan canonical form: an old proof, Amer. Math. Monthly 94 (1987), 257–267.

[BrGi1] R. A. Brualdi and P. M. Gibson, The convex polyhedron of doubly stochastic matrices: I. Applications of the permanent function, J. Combinatorial Theory (A) 22 (1977), 194–230.

[BrGi2] R. A. Brualdi and P. M. Gibson, The convex polyhedron of doubly stochastic matrices: II. The graph of Ω_n, J. Combinatorial Theory (B) 22 (1979), 175–198.

[BrGi3] R. A. Brualdi and P. M. Gibson, The convex polyhedron of doubly stochastic matrices: III. Affine and combinatorial properties of Ω_n, J. Combinatorial Theory (A), 22 (1977), 338–351.

[BrHaHw] R. A. Brualdi, D. J. Hartfiel, and S. –G. Hwang, On assignment functions, Lin. Multilin. Alg. 19 (1986), 203–219.

[BrRo] R. A. Brualdi and J. A. Ross, On Ryser's maximum term rank formula, Lin. Alg. Applics. 29 (1980), 33–38.

[BrRy] R. A. Brualdi and H. J. Ryser, Combinatorial Matrix Theory, in preparation.

[BrSh] R. A. Brualdi and B. L. Shader, Matrix factorizations of determinants and permanents, J. Combinatorial Theory (A), to appear.

[BrSo] R. A. Brualdi and E. S. Solheid, Maximum determinants of complementary acyclic matrices of zeros and ones, Discrete Math. 61 (1986), 1–19.

[CaGoSeSh] P. Cameron, J. M. Goethals, J. J. Seidel, and E. E. Shult, Line graphs, root systems, and elliptic geometry, J. Algebra 43 (1976), 305–327.

[Co] W. S. Connor, Jr., On the structure of balanced incomplete block designs, Ann. Math. Statist. 23 (1952), 57–71.

[CvDoSa] D. Cvetković, M. Doob, and H. Sachs, Spectra of graphs 2nd ed., Academic Press, (New York), 1982.

[CvDoGuTo] D. Cvetković, M. Doob, I. Gutman, and A. Torĝasev, Recent results in the theory of graph spectra, North Holland (Amsterdam), 1988.

[Eh1] H. Ehlich, Determinantenabschätzungen für binäre Matrizen, Math. Zeitschr. 83 (1964), 123–132.

[Eh2] H. Ehlich, Determinantenabschätzungen für binäre Matrizen mit $n \equiv 3 \bmod 4$, Math. Zeitschr. 84 (1964), 438–447.

[FiPt] M. Fiedler and V. Ptak, Cyclic products and an inequality for determinants, Czech. Math. J. 19 (1969), 428–450.

[FoFu] L. R. Ford, Jr. and D. R. Fulkerson, Flows in Networks, Princeton Univ. Press (Princeton), 1962.

[Fr] F. G. Frobenius, Über die Primfactoren der Gruppendeterminante, Sitzungsberichte der Königlich Preussischen Akodemie der Wissenschaften zu Berlin (1896), 1343–1382.

[GraPo] R. L. Graham and H. O. Pollak, On the addressing problem for loop switching, Bell Syst. Tech. J. 50 (1971), 2495–2519.

[GrLoSc] M. Grötschel, L. Lovász, and A. Schrijver, The ellipsoid method and its consequences in combinatorial optimization, Combinatorica 1 (1981), 169–197.

[GrLoSc] M. Grötschel, L. Lovász, and A. Schrijver, Polynomial algorithms for perfect graphs, Topics on perfect graphs, C. Berge and V. Chvátal, eds., North–Holland (Amsterdam), 1984, 326–356.

[Ha] R. M. Haber, Minimal term rank of a class of (0,1)–matrices, Canad. J. Math. 12 (1960), 462–475.

[Had] J. Hadamard, Résolution d'une question relative and déterminants, Bull. Sci. Math. 17 (1893), 240–246.

[Hal] M. Hall, Jr., Combinatorial Theory 2nd ed., Wiley (New York), 1986.

[HoJo1] R. Horn and C. R. Johnson, Matrix Analysis, Cambridge Univ. Press (Cambridge), 1985.

[HoJo2] R. Horn and C. R. Johnson, Topics in Matrix Analysis, Cambridge Univ. Press (Cambridge), 1988.

[Jo] C. R. Johnson, Combinatorial matrix analysis: an overview, Lin. Alg. Applics. 107 (1988), 3–15.

[JoGrSaWo] C. R. Johnson, R. Grone, E. Marques de Sá, and H. Wolkowicz, Positive definite completions of partial hermitian matrices, Lin. Alg. Applics. 58 (1984), 109–124.

[JuRy1] W. B. Jurkat and H. J. Ryser, Matrix factorizations of determinants and permanents, J. Algebra 3 (1966), 1–27.

[JuRy2] W. B. Jurkat and H. J. Ryser, Term ranks and permanents of nonnegative matrices, J. Algebra 5 (1967), 342–357.

[La] H. Laue, A graph theoretic proof of the fundamental trace identity, Discrete Math. 69 (1988), 197–198.

[Le] J. S. Lew, The generalized Cayley–Hamilton theorem in n dimensions, Z. Angew. Math. Phys. 17 (1966), 650–653.

[LeVi] M. Lewin and Y. Vitek, A system of gaps in the exponent set of primitive matrices, Illinois J. Math. 25 (1981), 87–98.

[Lo1] L. Lovász, On the Shannon capacity of a graph, IEEE Trans. Inform. Theory IT–25 (1979), 1–7.

[Lo2] L. Lovász, An algorithmic theory of numbers, graphs and convexity, Soc. for Indust. Applied Math. (Philadelphia), 1986.

[LoPl] L. Lovász and M. D. Plummer, Matching theory, North–Holland (Amsterdam), 1986.

[MacSl] F. J. MacWilliams and N. J. A. Sloane, The theory of error–correcting codes, North–Holland (Amsterdam), 1977.

[Ma] K. N. Majindar, On the parameters and intersections of blocks of balanced incomplete block designs, Ann. Math. Statist. 33 (1962), 1200–1205.

[Pe] G. W. Peck, A new proof of a theorem of Graham and Pollak, Discrete Math. 49 (1984), 327–328.

[Pi] R. S. Pierce, Associative Algebras, Springer–Verlag (New York), 1982.

[Pr] C. Procesi, The invariant theory of $n \times n$ matrices, Adv. in Math. 19 (1976), 306–381.

[Ry1] H. J. Ryser, Maximal determinants in combinatorial investigations, Canad. J. Math. 8 (1956), 245–249.

[Ry2] H. J. Ryser, Combinatorial Mathematics, Carus Math. Mon. No. 14, Math. Assoc. of America (Washington), 1963.

[Ry3] H. J. Ryser, An extension of a theorem of de Bruijn and Erdös on combinatorial designs, J. Algebra 10 (1968), 246–261.

[SaSc] B. D. Saunders and H. Schneider, Flows on graphs applied to diagonal similarity and diagonal equivalence for matrices, Discrete Math. 24 (1978), 205–220.

[Se] A. Seress, On λ–designs with $\lambda = 2p$, preprint.

[Sh] J. Shao, On a conjecture about the exponent set of primitive matrices, Lin. Alg. Applics. 65 (1985), 91–123.

[ShSi] S. S. Shrikhande and N. M. Singhi, On the λ–design conjecture, Utilitas Math. 9 (1976), 301–318.

[St] H. Straubing, A combinatorial proof of the Cayley–Hamilton theorem, Discrete Math. 43 (1983), 273–279.

[StSt] A. P. Street and D. J. Street, Combinatorics of Experimental Design, Clarendon Press (Oxford), 1987.

[Sw1] R. G. Swan, An application of graph theory to algebra, Proc. Amer. Math. Soc. 14 (1963), 367–373.

[Sw2] R. G. Swan, Correction to "An application of graph theory to algebra", Proc. Amer. Math. Soc. 21 (1969), 349–380.

[Tv] H. Tverberg, On the decomposition of K_n into complete bipartite graphs, J. Graph Theory 6 (1982), 493–494.

[Va] R. S. Varga, Matrix iterative analysis, Prentice–Hall (Englewood Cliffs), 1962.

[Wi] J. Williamson, Determinants whose elements are 0 and 1, Amer. Math. Monthly 53 (1946), 427–434.

[Wil1] R. M. Wilson, Incidence matrices of t–designs, Lin. Alg. Applics. 46 (1982), 73–82.]

[Wil2] R. M. Wilson, The theory of t–designs, Enumeration and Design, D. M. Jackson and S. A. Vanstone, eds., Academic Press (New York), 1984.

[Wo] D. R. Woodall, Square λ–linked designs, Proc. London Math. Soc. 20 (1970), 669–687.

[Ze] D. Zeilberger, A combinatorial approach to matrix algebra, Discrete Math. 56 (1985), 61–72.

[Zh] K. M. Zhang, On Lewin and Vitek's conjecture about the exponent set of primitive matrices, Linear Alg. Applics. 96 (1987), 101–108.

DEPARTMENT OF MATHEMATICS
UNIVERSITY OF WISCONSIN
MADISON, WISCONSIN 53706

Proceedings of Symposia in Applied Mathematics
Volume **40**, 1990

Patterned Matrices

PERSI DIACONIS

ABSTRACT. If the pattern in a matrix can be captured as invariance under a group, the spectral properties of the matrix can be written down using Fourier analysis. Such "generalized circulants" are closed under sum, product, transpose, conjugation. They all commute and are explicitly simultaneously diagonalizable. Applications to Markov Chains, Radon transforms, and coding theory are presented. The ideas extend to give natural classes of matrices whose powers are circulants.

1. Introduction

This is a study of patterned matrices with spectral properties deducible from their group of symmetries. The easiest examples are circulants. These are matrices of form

$$\begin{matrix} a & b & c & d \\ d & a & b & c \\ c & d & a & b \\ b & c & d & a \end{matrix}$$

where each row is derived from the row above by shifting right cyclically.

Section 2 lists the well known properties of circulants. In brief, the class of circulants is closed under sum, product, inverse and transpose. All circulants commute. They are simultaneously diagonalized by the Fourier matrix and thus have explicitly given eigenvalues and eigenvectors. If you run into a circulant in the course of a problem you are happy to make its acquaintance.

Circulants are invariant under the cyclic group in the sense that

$$M_{ij} = M_{\pi(i)\pi(j)}$$

for $\pi(i) = i + 1 \cdot (\mathrm{mod}\, n)$. It is natural to define the symmetry group of a matrix as the set of all permutations π leaving it invariant. One can also begin with a group G of permutations and consider all invariant matrices as G-circulants. The well known properties of circulants all have analogs which now involve the representation theory of the group. These topics are developed in section 3.

1980 *Mathematics Subject Classification* (1985 *Revision*). Primary 15-02; Secondary 20C15.

Section 3 then specializes to matrices of the form

$$M_{st} = f(ts^{-1})$$

for $s, t \in G$. These *convolution* matrices are invariant, with G acting in the usual way by left multiplication, if and only if the function f is constant on conjugacy classes: $f(sts^{-1}) = f(t)$: $M_{as,at} = f(ats^{-1}a^{-1}) = f(ts^{-1}) = M_{st}$. An example from S_3 is

$$\begin{matrix} a & b & b & b & c & c \\ b & a & c & c & b & b \\ b & c & a & c & b & b \\ b & c & c & a & b & b \\ c & b & b & b & a & c \\ c & b & b & b & c & a \end{matrix} \tag{1-1}$$

Matrices with this pattern commute. Their eigenstructure can be explicitly given using the character theory of the symmetric group. Similarly, every finite group gives rise to a simple collection of matrices by $M_{st} = f(ts^{-1})$ where f is constant on conjugacy classes.

I encountered these generalized circulants in joint work with Mehrdad Shahshahani analyzing repeated shuffling of a deck of cards. They proved useful in the analysis of Radon transform in joint work with Ron Graham. They had earlier been applied in work of Rothaus and Thompson (1966) to show the non-existence of perfect codes. More recently Chilling (1987b) and others have applied the spectral decomposition to finite group theory. Andersson and Perlman (1988) use these ideas to get generalizations of Hadamard's inequality. These applications are discussed in sections 3C and 4E.

Section 4 outlines the most broadly useful generalization: If the matrices are indexed by a finite set X on which G acts transitively then X can be identified with elements of G and matrices of the form $M_{xy} = f(x^{-1}y)$ can be defined. For these to be well defined, f must satisfy $f(k_1xk_2) = f(x)$, for all k in $K \subset G$ the isotropy subgroup. All such patterned matrices commute if and only if (G, X) form what is called a Gelfand pair. The eigenvalues are given as explicit spherical transforms of the matrix entries. Everything can be computed. This includes the constant on conjugacy set up but there are hundreds of other examples that occur naturally, including matrices indexed by the vertices of distance transitive graphs. An example, analyzed further in section 4 is

$$\begin{matrix} a & b & b & b & b & c \\ b & a & b & c & b & b \\ b & b & a & b & c & b. \\ b & c & b & a & b & b \\ b & b & c & b & a & b \\ c & b & b & b & b & a \end{matrix}$$

Section 5 presents examples such as the following

$$\begin{matrix} a & b & c & d & e \\ c & d & e & a & b \\ e & a & b & c & d. \\ b & c & d & e & a \\ d & e & a & b & c \end{matrix}$$

Here each row is shifted by 2. These are matrices whose powers are circulants.

A generalization to arbitrary groups, natural applications, and a nice theory are presented.

2. Circulants

A. Basic Properties. An $n \times n$ matrix with complex entries of form

$$\begin{matrix} a_0 & a_1 & \cdots & a_{n-1} \\ a_{n-1} & a_0 & \cdots & a_{n-2} \\ a_{n-2} & a_{n-1} & \cdots & a_{n-3} \\ \vdots & & & \vdots \\ a_1 & a_2 & \cdots & a_0 \end{matrix}$$

is called a circulant. These matrices have a long history carefully surveyed by Muir (1890–1923). They are still a topic of research interest as the reader can discover by browsing through recent years of the journal *Linear Algebra and its Applications.*

A lovely introduction to circulants is given by Davis (1979). This book discusses numerous applications. It seems like a worthwhile project to see in what way the properties described by Davis extend to the generalizations suggested here.

One way to organize and understand properties of circulants recognizes them as the matrix form of convolution: Treat the index set $0, 1, 2, \cdots n-1$ as the integers mod n. Then a circulant has the form

$$M_{ij} = f(j-i) \quad \text{with} \quad f(k) = a_k, \quad \text{subscripts} \quad \text{mod } n.$$

Multiplying M on the left by the row vector $(g(0), g(1) \cdots, g(n-1))$ yields

$$\sum_{i=0}^{n-1} g(i) f(j-i) \quad \text{as} \quad j\text{th entry.} \tag{2-1}$$

It is natural to interpret (2-1) in the language of probability. Suppose $g(0), g(1) \cdots g(n-1)$ are the respective chances of a random walk taking initial step $0, 1, \ldots, n-1$, and $f(0), \ldots, f(n-1)$ the chance of the second step in a walk. Then, the chance that after two steps the walk winds up at j is given by (2-1)—the convolution of g and f. This will be denoted $f * g(j)$ in what follows.

The Fourier transform of f at frequency j is

$$\hat{f}(j) = \sum_{k=0}^{n-1} e^{2\pi i jk/n} f(k). \tag{2-2}$$

As usual, $\widehat{f * g}(j) = \hat{f}(j) \cdot \hat{g}(j)$ and f can be recovered from $\hat{f}$ by

$$f(k) = \frac{1}{n} \sum_{j=0}^{n-1} e^{-2\pi ijk/n} \hat{f}(j). \tag{2-3}$$

Fix a function f and define a circulant by $M_{ij} = f(j-i)$. Take the function $g_k(j) = e^{2\pi ijk/n}$ as a row vector. Then

$$(g_k M)_j = \sum_i g_k(i) f(j-i) = \sum_i g_k(j-i) f(i) = g_k(j) \sum_i g_k(-i) f(i).$$

Thus the complex exponentials $g_k(\cdot)$ are left eigenvectors of M with eigenvalue $\hat{f}(-k)$. The usual summing of the complex exponentials to 0 implies that the Fourier matrix

$$F_{jk} = \frac{e^{2\pi ijk/n}}{\sqrt{n}} \tag{2-4}$$

is unitary and the spectral decomposition becomes the following: M is a circulant with first row $f(0), f(1) \cdots f(n-1)$ if and only if

$$M = F^* \mathrm{Diag}(\hat{f}(0), \hat{f}(-1) \cdots \hat{f}(-(n-1)))F. \tag{2-5}$$

Here $\hat{f}$ is defined at (2-2) and F is the Fourier matrix defined at (2-4).

All properties of circulants follow from this spectral representation:

(2-6) The class of circulants is closed under sum, product, transpose, conjugation and inverse.

Davis (1979) discusses generalized inverses of circulants.

(2-7) All circulants commute. They are simultaneously diagonalized by the Fourier matrix F of (2-4).

B. Some Applications. Circulants appear in many mathematical problems. Davis (1979) gives several examples. The present section lists some applied problems where circulants make a natural appearance.

EXAMPLE 2.1. RANDOM WALK. Consider a particle constrained to hop about on n points arranged in a circle. At each time the particle hops left or right with probability 1/2. This is a cyclic version of the classical drunkard's walk. Index the points as $0, 1, 2, \ldots, n-1$. The chance of moving from i to j is thus

$$M_{ij} = \begin{cases} \frac{1}{2} & \text{if } |i-j| = 1 \\ 0 & \text{otherwise.} \end{cases}$$

The matrix M is a circulant with first row $(0, \frac{1}{2}, 0, \ldots, 0, \frac{1}{2})$. Using the explicit diagonalization one can answer a question such as "What happens after many steps?" Answer: the walk becomes uniformly distributed. More specifically, the chance that a walk starting at i winds up at j after k steps is $\frac{1}{n}$ plus an error term that becomes exponentially small provided $k \gg n^2$. This assumes n is odd (otherwise the walk is at an even position after an even number of steps).

One may ask other questions such as "How long on average before the walk returns to where it started?" Answer: about n^2. "How long before the walk hits every point?" Answer: about n^2. These and other classical problems about random walks are discussed in Feller (1968). Diaconis (1988) gives details using the tools developed here.

If X_k is the position of the walk at time k, the walk can be described as

$$X_k = X_{k-1} + \varepsilon_k \pmod{n}, \tag{2-8}$$

where ε_k are random variables taking value ± 1 with probability $\frac{1}{2}$. Using formula (2-8) the disturbance terms can be generalized to have an arbitrary distribution $P\{\varepsilon_k = j\} = f(j)$, say. The associated Markov chain is still a circulant with first row $(f(0), \ldots, f(n-1))$.

EXAMPLE 2.2. SMOOTHING DATA. Consider $x_0, x_1, x_2, \ldots, x_{n-1}$ a sequence of observations collected in time. A plot of i versus x_i may be smoothed replacing x_i by e.g., $(x_{i-1} + x_i + x_{i+1})/3$ or $(x_{i-1} + 2x_i + x_{i+1})/4$. Such linear smoothers can be represented as linear maps on sequences of length n. The matrices of linear smoothers are circulants. In this context, the eigenvalues of the associated circulants are important in studies of stability and invertibility—a "smooth" is a 1-1 encoding of data if and only if no eigenvalue vanishes. Buja, Hastie, and Tibshirini (1989) give a comprehensive survey of this topic.

EXAMPLE 2.3. COVARIANCE MATRICES. If $X_0, X_1, \ldots, X_{N-1}$ is a series of random variables, the covariance between X_i and X_j is defined by

$$R_{ij} = E(X_i - \mu_i)(X_j - \mu_j) \quad \text{with} \quad \mu_i = E(X_i).$$

The collection of covariances are often collected in a covariance motion Σ. This has (i, j) entry R_{ij}. In applied work Σ is to be estimated from data. This is a hard job and a variety of simplifying assumptions are commonly employed. Stationarity implies that R_{ij} is a function of $i - j$ only. If this is interpreted cyclically, Σ is a circulant. The cyclic assumption is sometimes valid from physical considerations—for example, X_i could be readings of seismic recorders placed in a circle around Mt. St. Helens. Anderson (1984) is a classical reference to circularly symmetric covariance matrices. Perlman (1987) gives a modern discussion of this and its generalizations.

EXAMPLE 2.4. CYCLIC CODES. A binary code is a collection of binary vectors of length n. It is called *cyclic* if it is closed under cyclic rearrangements. For example, $\{000, 001, 010, 100\}$ is a cyclic code. A code is *quasi-cyclic* if there is some integer s such that every cyclic shift of a code word by s is again a code word. Such codes have an elegant theory. They include the important examples of the BCH codes.

The assignment of a given input message to a code word is often a linear operation. The matrix of this linear map is called the generator matrix of the code. By convention the generator matrix is of the form $[I|A]$, where I is an identity matrix and A is a matrix. There is a rich class of naturally occurring codes whose generator matrix has A a circulant. Chapter (16, §§6,7) of Macwilliams and Sloane (1977) discusses these results. In particular, they show that such "double circulant" codes are automatically quasi-cyclic with $s = 2$.

EXAMPLE 2.5. CYCLIC DESIGNS. Consider an experiment designed to help a manufacturer decide how much vanilla to put into ice cream. Subjects will rate the ice cream on a scale from 0 to 100. Suppose 7 levels of vanilla are being considered and each subject will be asked to taste 3 different levels (after several tastings, ice cream tastes like colored sugar water). A design for such an experiment consists in a determination of which levels each subject will taste. A classically studied example is

$$\{1\ 2\ 4\}\ \{2\ 3\ 5\}\ \{3\ 4\ 6\}\ \{4\ 5\ 7\}\ \{1\ 5\ 6\}\ \{2\ 6\ 7\}\ \{1\ 3\ 7\}$$

Thus the first person will be given levels 1, 2, and 4 to taste, and so on, there being 7 subjects. In all, 21 numbers will be reported—the ratings of all subjects on all samples tasted.

Classical analysis of such designed experiments tries to disentangle things into 'people effects' and 'treatment effects'. After all some people love ice cream and may give all levels quite a high rating. Others may be uniformly low. Taking differences between the ratings of a single taster helps to eliminate subject effects (also called block effects). One then combines information across subjects to get estimates of the difference between any pair of treatment levels.

The example above was formed by taking a single starting block $\{1, 2, 4\}$ and translating by repeatedly adding 1 (mod 7). These are called *cyclic* block designs. An important ingredient in the analysis of cyclic designs is the concurrence matrix. This is a matrix indexed by treatment levels with (i, j) entry the number of times levels i and j are in the same block. Returning to the central theme, the concurrence matrix of a cyclic design is a circulant.

Statistical models can be introduced as a formal way of defining treatment effects. The accuracy with which such effects can be estimated is intimately related to the eigenvalues of the concurrence matrix. For example if d_{ij} is the best unbiased estimate of the difference between the ith and jth treatment, $\sum_{i,j}$ variance (d_{ij}) is proportional to the harmonic mean of the eigenvalues of the concurrance matrix. Designs that minimize this harmonic mean are called "A optimal".

The closed form expressions for the eigenvalues plays an important part in comparing and evaluating designs. The eigenvectors also have natural interpretations as fundamental contrasts in treatment levels.

Bailey (1989) and John (1987) are good references on these matters. In particular, Bailey shows how everything generalizes naturaly to Abelian group designs which give rise to generalized circulants to be defined in section 3.

Further examples. Ablow and Brenner (1963) show how circulants arise naturally in a class of mechanical problems involving a system of springs.

I believe there are dozens of other examples and would be pleased to hear from others on this matter.

3. Circulants indexed by groups

A. One natural generalization of circulants replaces the integers mod n by a finite group G. Let f be a real or complex function defined on G. Define a $|G| \times |G|$ convolution matrix M via

$$M_{st} = f(ts^{-1}) \quad \text{for } s, t \in G. \tag{3-1}$$

These matrices appear at the beginning of the theory of group characters with Frobenius' work generalizing the expression of the determinant of a circulant as a polynomial in the entries. See Hawkins (1978). An elegant simplification results from taking f to be constant on conjugacy classes:

$$f(sts^{-1}) = f(t) \quad \text{for all } s, t \in G. \tag{3-2}$$

Such functions are called *class functions*. The associated matrices will be called *class circulants*.

REMARKS. (1) For the integers mod n (or any Abelian group) $sts^{-1} = t$ for every $s, t \in G$. Thus all functions are class functions.

(2) For non-Abelian groups, elements are lumped together in conjugacy classes. For example, S_3–the 6-element symmetric group–has 3 conjugacy classes $\{id\}, \{(1,2),(1,3),(2,3)\}, \{(1,2,3),(1,3,2)\}$. Here, permutations are being written in cycle notation, so $(1,2,3)$ takes 1 to 2 and 2 to 3 and 3 to 1. In S_3 a class function only takes on 3 values. Call them a, b, c. If the group is written in the order above, the 6×6 matrix is shown in (1-1) above.

The main result of this section shows that the class circulants are closed under sum, product, transpose, conjugation and inverse. All class circulants commute and are explicitly simultaneously diagonalizable. These results are proved in section B below which gives a crash course in elementary group representation theory. Section C gives examples and applications. Section D relates the definition above to work of Chillag (1987a,b, 1988) who gives applications to group theory.

B. A crash course in group representations. Let G be a finite group. A *representation* assigns matrices to group elements in such a way that the product of group elements is assigned to the product of their matrices. Formally, a representation is a homomorphism $\rho\colon G \to \mathrm{GL}(V)$. Thus $\rho(st) = \rho(s)\rho(t)$. Here V is a vector space over the real or complex numbers of dimension $d_\rho < \infty$.

Let f be a function from G into the complex numbers. Define the *Fourier transform* at ρ by

$$\hat{f}(\rho) = \sum_{t \in G} f(t)\rho(t).$$

If g is another function define *convolution* by

$$f * g(s) = \sum_{t \in G} f(st^{-1})g(t).$$

As usual, Fourier transform turns convolution into multiplication

$$\widehat{f * g}(\rho) = \hat{f}(\rho)\hat{g}(\rho).$$

This is easily checked by multiplying the matrices on the right hand side.

A representation is *irreducible* if 0 and V are the only invariant subspaces of V. Thus if W is a subspace of V such that $\rho(s)W \subset W$ for each $s \in G$ then $W = 0$ or V.

EXAMPLE. Take $G = S_n$ the symmetric group. Every mathematician is familiar with 3 representations:

(i) The *trivial representation*: V is 1-dimensional, $\rho(s)v = v$.
(ii) The *alternating representation*: V is 1-dimensional, $\rho(s)v = \mathrm{sgn}(s)v$.
(iii) The *n-dimensional representation*: V is n-dimensional, generated by the standard basis vectors $e_1, e_2, \ldots, e_n$. If s is a permutation, define $\rho(s)e_i = e_{s(i)}$ and extend by linearity. The matrix of the linear map $\rho(\cdot)$ is the usual permutation matrix in this basis.

The n-dimensional representation is not irreducible: If W is all multiples of $e_1 + \cdots + e_n$, then W is invariant under permuting coordinates. Similarly $W^\perp = \{x \in \mathbf{C}^n \colon x_1 + \cdots + x_n = 0\}$ is invariant and can be shown to be irreducible. There are other irreducible representations of S_n.

The *Fourier inversion theorem* gives a recipe for reconstructing f from $\hat{f}(\rho)$ at all irreducible representations of G:

$$f(s) = \frac{1}{|G|}\sum_{\rho} d_\rho \mathrm{tr}(\rho(s^{-1})\hat{f}(\rho)).$$

EXAMPLE. Take $G = \mathbf{Z}(n)$ the integers mod(n). There are n irreducible representations; all are 1-dimensional. A 1-dimensional linear map is given by multiplying by a complex number. Here the jth representation

$$\rho_j(k) = e^{2\pi ijk/n}.$$

The Fourier transform at the jth representation becomes the usual discrete Fourier transform:

$$\hat{f}(\rho_j) = \sum_{k=0}^{n-1} e^{2\pi ijk/n} f(k).$$

The Fourier inversion theorem becomes

$$f(k) = \frac{1}{n}\sum_{j=0}^{n-1} e^{-2\pi ijk/n}\hat{f}(\rho_j).$$

The kind of elementary representation theory needed to follow this paper is developed in the first 30 pages of Serre (1977). Other fine references are Lederman (1988), Naimark-Stern (1982), or the encyclopedic Curtis-Reiner (1962).

To describe the diagonalization of convolution matrices enumerate the group as $G = \{s_1, s_2, \ldots, s_N\}$. Let $M_{ij} = f(s_j s_i^{-1})$ with f an arbitrary function.

Let $\rho_1, \rho_2, \ldots, \rho_k$ be the irreducible representations, with ρ_i of dimension d_i. These may be taken as unitary, so $\rho(s)$ is a unitary matrix for every $s \in G$.

The diagonalization is built up in stages. For each representation ρ_j, define a $d_j^2 \times d_j^2$ block diagonal matrix:

$$D_j = \begin{pmatrix} \hat{f}(\rho_j) & & 0 \\ & \ddots & \\ 0 & & \hat{f}(\rho_j) \end{pmatrix}.$$

Let

$$D = \begin{pmatrix} D_1 & & 0 \\ & \ddots & \\ 0 & & D_k \end{pmatrix}. \tag{3-3}$$

This D is $N \times N$ because $d_1^2 + d_2^2 + \cdots + d_k^2 = N$ (Serre (1978, page 18)). Let

$$\psi_j(s) = \sqrt{\frac{d_j}{N}}(\rho_j(s)_{11} \cdots \rho_j(s)_{d_j,1}, \rho_j(s)_{21} \cdots, \rho_j(s)_{d_j d_j})$$
$$\varphi(s) = (\psi_1(s), \psi_2(s), \ldots, \psi_k(s))^t.$$

Thus φ is an $N \times 1$ column. Finally let Φ be the $N \times N$ matrix with ith column $\varphi(s_i)$

$$\Phi = (\varphi(s_1), \varphi(s_2), \cdots, \varphi(s_N)). \tag{3-4}$$

THEOREM 1. *Let f be a function from a finite group G into $\mathbf{C}$. Let $M_{st} = f(ts^{-1})$ be an $N \times N$ matrix with $N = |G|$. Then*

$$M = \Phi^* D \Phi \tag{3-5}$$

with $\Phi^*\Phi = I$, a unitary matrix defined in (3-4) *above and D as in* (3-3).

PROOF. The decomposition follows from the Fourier inversion theorem:

$$\begin{aligned} M_{ab} &= \frac{1}{N}\sum_{j=1}^{K} d_j \mathrm{tr}[\hat{f}(\rho_j)\rho_j(s_b)\rho_j(s_a^{-1})] \\ &= \frac{1}{N}\sum_{j=1}^{K} d_j \mathrm{tr}[\rho_j(s_a^{-1})\hat{f}(\rho_j)\rho_j(s_b)] \\ &= \sum_{j=1}^{K} \psi_j(s_a)^* D_j \psi_j(s_b). \end{aligned}$$

The last equality follows by expanding the trace. The property $\Phi^*\Phi = I$ follows from the orthogonality of the matrix entries of the irreducible representations. This is one of the standard corollaries of Schur's lemma (see, e.g., Serre (1978, pg. 14)). □

REMARK. The decomposition of M is not a diagonalization, indeed the matrices $\hat{f}(\rho_j)$ that make up D_j can be arbitrary: just choose them and define f through the Fourier inversion theorem. In the following corollary, class functions are shown to have diagonal $\hat{f}(\rho_j)$ and in section 4 Gelfand pairs are shown to have diagonal tansforms.

Corollary 1. *Let* $M_{st} = f(ts^{-1})$ *with* f *a class function (i.e.,* $f(sts^{-1}) = f(t)$*). Then* M *can be unitarily diagonalized as*

$$M = \Phi^* D \Phi$$

with Φ *unitary as defined in* (3-4) *above, and* D *a diagonal matrix made of blocks* $\Delta_1, \Delta_2, \ldots, \Delta_k$ *with*

$$\Delta_i = c_i I_{d_i^2 \times d_i^2}$$

and

$$c_i = \frac{1}{d_i} \sum_{t \in G} f(t) \overline{\chi}_i(t); \qquad \chi_i(t) = \mathrm{Tr}(\rho_i(t))\text{–}\textit{the character of } \rho \textit{ at } t.$$

Proof. Let ρ be a unitary representation and observe that $\hat{f}(p)$ commutes with $\rho(s)$:

$$\rho(s)\hat{f}(\rho)\rho(s^{-1}) = \sum_{t \in G} f(t)\rho(sts^{-1}) = \hat{f}(\rho).$$

Schur's lemma now implies that $\hat{f}(\rho) = cI$. Calculate the constant by taking traces: $d_\rho c = \Sigma f(s)\chi_\rho(s)$. Thus in Theorem 1, $D_i = c_i I_{d_i^2 \times d_i^2}$.

Remarks. The explicit diagonalization was used in Rothaus and Thompson (1966) for the symmetric group. Diaconis and Shahshahani (1981) use it in general. It was written out in this explicit form in Matthews (1985). It has frequently been rediscovered, see, e.g., Zieschang (1988). I presume it is folklore going back to the beginnings of the subject.

Corollary 2. *Let* G *be a finite group with* $|G| = N$. *Let* f *be a class function with associated matrix* $M_f = f(ts^{-1})$. *The class of such matrices is closed under sum, product, transpose, conjugation, and inverse. All such* M_f *commute and are simultaneously diagonalizable by the matrix* Φ *of* (3-4).

C. Applications. The explicit decomposition has been applied to problems of random walk, perfect codes, and inversion of Radon transforms. These applications are detailed in Chapter 3 of Diaconis (1988) and will only be treated briefly here.

Consider first the problem of random transpositions studied by Diaconis and Shahshahani (1981). Begin with n cards face down on a table. Card 1 is at the left, card 2 next, and card n at the right of the row. Cards are repeatedly mixed by random transpositions. Thus the left hand chooses a card at random, the right hand chooses a card at random (so left = right with probability $\frac{1}{n}$) and then the two cards are transposed. This corresponds to the probability

$$f(\pi) = \begin{cases} \frac{1}{n} & \text{if } \pi = id \\ \frac{2}{n^2} & \text{if } \pi \text{ is a transposition} \\ 0 & \text{otherwise.} \end{cases}$$

The identity and the transpositions form conjugacy classes and so $f(\pi)$ is a class function. If this problem is analyzed as a Markov chain, the transition matrix is M_f and the explicit diagonalization allows one to show $\frac{1}{2}n \log n + cn$ transpositions are necessary and suffice to mix up n cards.

Rothaus and Thompson (1966) used exactly the same ideas to study the existence of perfect codes in the symmetric group. Here the group is treated as a metric space using the minimum number of transpositions required to bring π to σ as a metric. One seeks a code–that is, a set of permutations–such that balls of a fixed radius about each code-word are disjoint and cover S_n. Such a code is called *perfect*. Taking the unit ball gives as indicator function

$$f(\pi) = \begin{cases} 1 & \text{if } \pi = id \text{ or a transportation} \\ 0 & \text{otherwise.} \end{cases}$$

If g is the indicator function of the code, then perfectness translates into $f * g(\pi) \equiv 1$. Fourier transforming this relation, gives strong restrictions to $\hat{g}(\rho)$. Rothaus and Thompson used these results to show that perfect codes on S_n do not exist for any n such that a prime p divides $1 + \binom{n}{2}$ with $p > \sqrt{n} + 2$. Laura Chihara (1987) has used these ideas to show that perfect codes do not exist for other classical finite groups.

A third application of the explicit diagonalizatoin appears in work of Diaconis and Graham (1985) on Radon transforms. Very roughly, the Radon transform replaces a function f with an average and then one asks about uniqueness and inversion. If f is defined on a group and

$$Rf(s) = \sum_{t \in B} f(st)$$

where B is a Ball in a metric on G, then the Radon transform can be represented as a convolution matrix. If the metric is bi-invariant, $d(a, b) = d(sa, sb) = d(as, as)$, the matrix will be a class circulant. The transform is invertible if and only if the numbers c_i of Corollary 1 are non-zero.

This brief sketch omits many rich mathematical details. Morrison (1986) and Fill (1989) are nice examples of how far the simple ideas outlined above can be pushed.

D. A Dual Theory. In a series of papers, Eiichi Bannai (see Banni and Ito (1984, p. 119, 182) and David Chillag (1987a,b, 1988) have developed a parallel theory. With present notation, it is easy to relate their results to the development of section B. Let A be the space of class functions on a finite group G. This is a linear space over the complex numbers. It can be considered as a commutative algebra in two ways. The classical way to form an algebra is by convolution since the convolution of two class functions is a class function.

A second natural way to form a commutative algebra uses the pointwise product of two functions. This leads to Chillag's generalized circulants. To explain, take the characters of the irreducible representations as a basis for the algebra A under pointwise product. These are orthonormal under the usual inner product

$$\langle f|g\rangle = \frac{1}{|G|}\sum_{s\in G} f(s)g^*(s).$$

Fix a class function f and treat it as a linear map on A by

$$L_f(g)(x) = f(x)g(x).$$

In the basis given by the characters L_f has matrix with i, j entry

$$M_{ij} = \langle f\chi_i|\chi_j\rangle.$$

Chillag calls these matrices generalized circulants. They are simultaneously diagonalizable as

$$M = X^{-1}DX.$$

With X the character matrix having i, j entry $\chi_i(c_j)$ the value of the ith character at the jth conjugacy class. The matrix D is diagonal with (i, i) element $f(c_i)$. Note that the matrix X is not unitary in non-Abelian examples.

From this representation one sees that all the usual properties of circulants hold. One way to understand these circulants is to look back at Theorem 1. The matrix D_k has a lot of redundancy, with repeated blocks down the diagonal. For class functions, one can reduce this redundancy by looking only at one block for each representation.

One can also consider f acting on A by convolution. This gives a linear map which also leads to simultaneously diagonalizable matrices. This is essentially what was done in section B. It is important to note that the two algebra structures on A are isomorphic (both are commutative semi-simple algebras of the same dimension). The Fourier transform takes one algebra to the other. Giving the isomorphism explicitly is a classical unsolved problem of identifying characters and conjugacy classes in a natural way. See Mackey (1975).

Having two ways of writing the same thing can be enormously useful. Bannai uses these ideas to explain a starteling observation of McKay (1980) who found that the eigenvectors of certain Dynkin diagrams, thought of as graphs, are the character table of a finite subgroup of SU(2). Chillag applies has ideas to group theory in Chillag (1987b). The results are refreshing and thought-provoking.

4. A General Theory

A. Homogeneous spaces and Gelfand pairs. The examples above give one generalization of circulants for any group. There are other possible generalizations. In this section a natural general theory is developed.

Let X be a finite set. Let M_{xy} be a matrix with complex entries indexed by X. Let G be a group of permutations of X preserving M. Thus $s \in G$ implies $M_{sxsy} = M_{xy}$ for all x, y. For example, the set of all permutations preserving M always forms a group.

To start a theory, suppose G acts transitively on X so that for each x, y in X there is $s \in G$ such that $sx = y$. Fix $x_0 \in X$. Let $K = \{s\colon sx_0 = x_0\}$. Identify X with cosets of K: choose $id, x_1, x_2, \cdots$ in G such that $G = K \cup x_1K \cup x_2K \cup \cdots$ as a disjoint union. Identify x_0 with the identity and X with $\{x_0, x_1, \cdots\}$. There is no essential dependence on the choice of x_0. Any $s \in G$ can be uniquely written $s = x_ik$ for some x_i.

Define a function f on G by

$$f(xk) = f(x) = M_{x,x_0}.$$

By definition f is right invariant under K: $f(sk) = f(s)$. The definition also yields that f is K left invariant: $f(ks) = f(s)$.

Going backward; if f is given as a K-bi-invariant function on G. Let $L(X)$ be the vector space of complex valued functions on X. Define a linear mapping $T_f\colon L(X) \to L(X)$ by

$$T_f g(x) = \sum_{y \in X} g(y) f(y^{-1}x).$$

Observe that T is well defined. If $L(X)$ is given its usual basis of delta functions $\delta_z(x) = 1$ if $x = z$, zero otherwise, the matrix of T_f for this basis is $M_{xy} = f(y^{-1}x)$. If g above is K left invariant so is $T_f g$.

This gives a 1-1 correspondence between matrices indexed by $X = G/K$ and invariant under G and K-bi-invariant functions on G. In this correspondence, matrix product corresponds to convolution. Let the algebra of G-invariant matrices be denoted $H(G, K)$. These objects are often called Hecke algebras. Curtis and Reiner (1962) (1986 II.D) contains a clear elementary treatment. Takacs (1986) contains an extensive bibliography. Hecke algebras are an important part of modern group theory and are still in a very active state of development.

Example. (Class circulants). Let X itself be a group, whose elements are denoted $x, y, \cdots$. Let G be the product group $X \times X$. This acts on X by $(s,t)x = sxt^{-1}$. Taking $x_0 = id$. The isotropy subgroup K is seen to be the diagonal subgroup $\{(s,s)\}$. A matrix M_{xy} is G invariant if and only if $M_{xy} = f(y^{-1}x)$ with f constant on conjugacy classes.

The group G operates on X and so on $L(X)$. Thus $L(X)$ is a representation of G. By Mashkes' Theorem (Serre 1977, p. 7) $L(X)$ decomposes into a direct sum of invariant irreducible subspaces.

$$L(X) = V_0 \oplus V_1 \oplus \cdots \oplus V_j. \tag{4-1}$$

Define the pair (G, K) to be a *Gelfand pair* if no V_i is isomorphic to a V_j for $i \neq j$. We say the direct sum decomposition is multiplicity free. Gelfand pairs have an extensive modern development. Recent surveys are given by Conway and Sloane (1987) who give applications to packing and covering. Letac (1981) and Diaconis (1988) are surveys aimed at probabilists and statisticians.

The connection with circulants comes in the following theorem proved, for example, in Diaconis (1988, Chapter 3F).

THEOREM 2. *Let G be a finite group acting transitively on a finite set X with isotropy subgroup K. The following are equivalent*

(a) *The algebra of G invariant matrices is commutative.*
(b) *G/K is a Gelfand pair.*
(c) *There exists a unitary matrix U such that for every invariant M*

$$M = U^* DU$$

with D a block diagonal matrix having j blocks (j appears in (4-1) above) the ith block is itself diagonal with only the $(1, 1)$ entry nonzero. This entry is $\sum_x s_i(x) M_{x,x_0}$, where the spherical function s_i is the unique K-bi-invariant function in V_i having $s_i(x_0) = 1$.

REMARKS. 1. There are literally hundreds of classes of Gelfand pairs known. Diaconis (1988; 3G) gives an annotated bibliography. The class functions form a commutative algebra and so a Gelfand pair. Distance transitive graphs generate an interesting class of examples where the eigenvalues and eigenvectors can be explicitly given. See Biggs (1974) and Bannai-Ito (1984, 1986). Stanton (1984) gives an extremely clear discussion of Gelfand pairs arising from Coxeter groups where the spherical functions are classically studied orthogonal polynomials.

2. Invariant matrices can be block diagonalized without the Gelfand pair condition. Start with a matrix M_{xy}. Let G be the largest group such that $M_{sxsy} = M_{xy}$. Take K as the isotropy subgroup. Suppose

$$L(X) = V_0 \oplus V_1 \cdots \oplus V_j = W_0 \oplus W_1 \oplus \cdots \oplus W_\ell,$$

where V_i are irreducible representations and W_i is a direct sum of isomorphic irreducible representations. Suppose that W_i lumps together ℓ_i of the V's. Then, there is a unitary U such that every invariant M can be written

$$M = U^* DU$$

with D block diagonal, the ith block of form $\left(\begin{smallmatrix} A & 0 \\ 0 & 0 \end{smallmatrix}\right)$ with A $\ell_i \times \ell_i$ arbitrary. There are $\ell + 1$ blocks. The ith block can be given reasonably explicitly in terms of Fourier analysis. See, for example, Dieudonné (1978). An example is presented in the last section of this paper. The result follows from Theorem 1 in section 3 and Frobenius reciprocity.

3. It is an interesting algorithmic question to calculate the symmetry group of a given matrix. If the group is to be transitive, all rows (and columns) must be permutations of the first row. Fix a row (say, the first). The set of permutations arising from the rows is one set of candidates (if all entries in the first row are distinct, these must all work).

B. An Example. Fix integers k, n, $0 < k \leq n/2$. Let the set X consist of the k-element subsets of $\{1, 2, \ldots, n\}$. Thus $|X| = \binom{n}{k}$. The symmetric group S_n acts transitively on X. The subgroup fixing $\{1, 2, \ldots, k\}$ is $S_k \times S_{n-k}$. Thus X can be identified with $S_n/S_k \times S_{n-k}$.

When $n = 4, k = 2$, X can be identified with the vertices of an octahedron

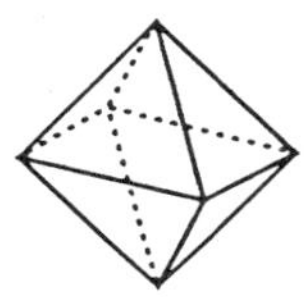

FIGURE 1

Random walk on this graph gives rise to a matrix of form

$$\begin{pmatrix} a & b & b & b & b & c \\ b & a & b & c & b & b \\ b & b & a & b & c & b \\ b & c & b & a & b & b \\ b & b & c & b & a & b \\ c & b & b & b & b & a \end{pmatrix}.$$

This is invariant under S_4 and can be explicitly diagonalized using Theorem 2.

For general n, k, this example arose in joint work with Shahshahani in analyzing a model of diffusion proposed by Bernoulli and Laplace. Consider two urns. The left contains k red balls, the right contains $n - k$ black balls. At each time, a ball is chosen at random from each urn and the two balls are switched. It is intuitively clear that after many switches the urns are all mixed up, about half red and half black. Using the machinery above, Shahshahani and I showed that order $n \log n$ switches are required. The paper gives an explicit diagonalization. Here the spherical functions are "dual Hahn polynomials".

C. Greenhalgebras. One way to interpolate between conjugacy invariant and bi-invariance under a subgroup considers a finite group G and subgroups $H \supset K$ with K normal in H. Let $H(G; H, K)$ be the set of complex functions on G conjugacy invariant under H (so $f(s) = f(hsh^{-1})$) and bi-invariant under K (so $f(s) = f(k_1 s k_2)$). These form an algebra under convolution.

Taking $H = G, K = id$ gives conjugacy invariance. Taking $K = H$ gives bi-invariance. Shahshahani and I discovered that this algebra is commutative for the example $S_n \supset S_k \times S_{n-k} \supset S_k$. See also Travis (1974). Andy Greenhalgh (1989) and Hirschman (1972) give necessary and sufficient conditions on G, H, K to insure commutativity in general. Any irreducible representation of H that restricts to an isotypic direct sum of 1-dimensional representations of K must induce up to a multiplicity free representation of G.

Charles Dunkl (1972) has pointed out that this can be recast in the language of Gelfand pairs: form the product group $G \times K$ with subgroup $H \times K$. The algebra of bi-invariant functions in this product set up is isomorphic to $H(G; H, K)$. Thus $H(G; H, K)$ is commutative if and only if $G \times K | H \times K$ is a Gelfand pair.

Greenhalgh carried out the details of the following random walk example: consider n balls labelled $1, 2, \ldots, n$. Place k of them in a rack in order $1, 2, \ldots, k$. Place the remaining $n-k$ in a bag. Each time pick a ball from the bag and one from the rack and switch. All of the spherical functions underlying the diagonalization of this Markov chain are computable from character theory. Greenhalgh gives a sharp form to the following rough summary–$n \log n + cn$ steps are necessary and suffice to mix things up.

D. Some matrix theory. Working in a different setting, David Chillag (1988) has pointed out that many of the computations can be usefully regarded as change of basis formulas. Let A be a commutative algebra of dimension k over $\mathbf{C}$ (assumed semisimple) with $\Sigma = \{\sigma_1, \ldots, \sigma_k\}$ and $\Delta = \{\delta_1, \ldots, \delta_k\}$ a pair of bases of A. For $a \in A$ define a $k \times k$ matrix $M(a, \Sigma, \Delta)$ by $a\sigma_i = \Sigma m_{ij}(a; \Sigma, \Delta)\delta_j$.

A standard result (e.g., Suprunenko and Tyshkevich (1968)) is that A has a basis $\mathscr{E} = \{\varepsilon_1 \cdots \varepsilon_k\}$ such that $a\varepsilon_i = a(i)\varepsilon_i$ for all a (so $\mathscr{E}$ is a basis of eigenvectors of all a simultaneously), $\varepsilon_i \varepsilon_j = 0$ for $i \neq j$, and $\varepsilon_i^2 = \varepsilon_i$. It is convenient to take one of the bases above as this $\mathscr{E}$. If a $k \times k$ matrix X is defined by $\sigma_i = \sum_{j=1}^{k} x_{ij} \cdot \varepsilon_j$, Chillag shows that $a \mapsto M(a, \Sigma, \Sigma)$ is an isomorphism of algebras. Hence, for any basis Σ, the $M(a, \Sigma, \Sigma)$ may be regarded as "circulants"–they are closed under sum, product, they all commute and are simultaneously diagonalized by the matrix X:

$$M(a, \Sigma, \Sigma) = X \operatorname{Diag}(a(1), a(2), \ldots, a(k)) X^{-1}.$$

In the setting of Gelfand pairs, A may be taken as the algebra of bi-invariant functions, $\mathscr{E}$ the spherical functions, and Σ the indicator functions of double cosets. This gives the usual Fourier inversion theorem.

The diagonalizations that have been natural in my applications have involved matrices indexed by groups or homogeneous spaces. This brief section shows that (as Letac (1981) has noted) one can also work with matrices of size the dimension of the algebra. This throws away some redundancy, but often also some ease of interpretation. Chillag (1988) gives many variants of these change-of-basis of formulas which can be carried over to the present setting.

E. A final example. One way to produce matrices invariant under a group is to start with a fixed matrix and symmetrize it. An elegant collection of applications of this idea appears in Andersson and Perlman (1988). They work with Hermitian matrices M. If G is a finite group acting as permutations, define permutation matrices $\pi(s)$ for $s \in G$ and

$$\widehat{M} = \frac{1}{|G|} \cdot \sum_{s \in G} \pi(s) M \pi^*(s).$$

Thus $\pi(s)\widehat{M}\pi(s)^* = \widehat{M}$ so $\widehat{M}$ can be explicitly diagonalized (if G acts transitively).

If $\lambda(M)$ is the vector of eigenvalues of M, it is a standard inequality that $\lambda(M)$ majorizes $\lambda(\widehat{M})$. Since the eigenvalues of $\widehat{M}$ can be calculated in closed form (at least in principle) this gives a collection of inequalities for the original eigenvalues.

Andersson and Perlman adapted this idea from an idea of Marshall and Olkin. These last authors worked with $\pi(s)$ ranging over the group of all $n!2^n$ sign change matrices. Then $\widehat{M} = \text{Diag}(M_{11}, M_{12}, \cdots M_{nn})$. Thus any symmetric convex function of the eigenvalues of M is smaller than the same function applied to the diagonal entries. In particular, for positive definite symmetric matrices one gets Hadamard's inequality

$$\det(M) \leq \prod M_{jj}.$$

In this last application the sign change group acts in a more general way than permutations. Most of the theory developed here goes over to such generalized symmetry. It seems like a worthwhile project to work this out carefully.

5. Matrices whose Powers are Circulants

A. Garden variety circulants indexed by $\mathbf{Z}(n)$ have each row shifted right by 1. Consider a matrix which has rows shifted by 2:

$$\begin{matrix} a & b & c & d & e \\ d & e & a & b & c \\ b & c & d & e & a. \\ e & a & b & c & d \\ c & d & e & a & b \end{matrix}$$

These matrices can be explicitly Jordanized. They have the property that an appropriate power is a circulant. For $\mathbf{Z}(P)$, these results were derived in Friedman (1958), (1961) and Ablow-Brenner (1963). They are nicely described by Davis (1979, Chapter 6). As explained below the Jordanization also follows from standard representation theory.

These results can be generalized to the kinds of generalized circulants introduced in previous sections. The results are easiest to describe for circulants derived from convolutions in section 3.

Let G be a finite group. Let $A: G \to G$ be an automorphism. Thus $A(st) = A(s)A(t)$. Let $f: G \to \mathbf{C}$ be any function.

Define a $|G| \times |G|$ matrix M indexed by G as

$$M_{st} = f(tA(s^{-1})). \tag{5-1}$$

EXAMPLE. If $G = \mathbf{Z}(n)$, and $A(j) = aj$. Then the matrix M has each row shifted right by a. More general groups and homomorphisms have rows a more general permutation of the first row.

LEMMA. *Let M be defined as in* (5-1). *If $A^k = Id$ then the kth power of M is a convolution matrix which can be unitarily block diagonalized as in Theorem* 1 *of Section* 3.

PROOF. A motivated proof can be given using the language of Markov chains. Define a chain taking values on G by $X_0 = id, X_n = A(X_{n-1}) * \varepsilon_n$ with t_i independent and identically distributed random variables $P\{\varepsilon_i = s\} = f(s)$. This assumes f is a probability but the identity proved will hold for all probabilities and so for all f. Express the X_n in terms of ε_i:

$$X_0 = id, \qquad X_1 = \varepsilon_1, \qquad X_2 = A(\varepsilon_1) \cdot \varepsilon_2, \qquad X_3 = A^2(\varepsilon_1) \cdot A(\varepsilon_2) \cdot \varepsilon_3, \cdots$$

$$X_n = A^n(\varepsilon_1) \cdot A^{n-1}(\varepsilon_2) \cdots \varepsilon_n. \tag{5-2}$$

Now $A^k = id$, so

$$X_{kj} = \underbrace{A^{k-1}(\varepsilon_1) \cdot A^{k-2}(\varepsilon_2) \cdots \varepsilon_k}_{\delta_1} \cdot \underbrace{A^{k-1}(\varepsilon_{k+1}) \cdots \varepsilon_{2k}}_{\delta_2} \cdots \underbrace{A^{k-1}(\varepsilon_{(j-1)k+1}) \cdots \varepsilon_{jk}}_{\delta_j} \cdot$$

This represents X_{kj} as the j-fold convolution of a basic step δ, where δ_i are independent and identically distributed since t_i are. □

REMARKS. 1. If f is a class function, so is the image law of f under A^j. It follows that the law of δ_i above is a class function and so has a diagonal Fourier transform. Thus for class functions the matrix (5-1) has its kth power unitarily diagonalizable. The eigenvalues of the kth power are explicitly available using character theory. The eigenvalues of the original matrix are thus kth roots of these values. For applications this may be all the information needed. A bit more generally, (5-2) expresses the law of X_n as a convolution of class functions for any n.

2. As a special case, if G is Abelian, all functions are class functions so there is no restriction. In this case it is easy to give the explicit Jordanization of the original matrix. Briefly, the matrix is the matrix of a linear transformation of convolution by f on the vector space of all functions on the group with basis of delta functions at the group elements. Make the unitary

change of basis to a basis of the characters. The automorphism A acts on these, splitting them into orbits. Here $A\chi(s) = \chi(As)$. These orbits span subspaces invariant under M. On such a subspace $M\chi = \hat{f}(\chi) \cdot A\chi$, so the matrix of M has $\hat{f}(x)$ below the diagonal in the upper right hand corner as x ranges through the characters in the orbit. Such matrices are of the form πD with π a basic circulant and D diagonal. Davis (1979) gives their explicit Jordanization.

3. The shifted circulants of this section can be put into a more group theoretic framework. The powers of A generate a cyclic group $c(k)$ which acts on G. Form the semi-direct product of $c(k)$ and G. This acts on $L(G)$ to give a representation. The discussion above amounts to an explicit decomposition of this representation along the lines of the Mackey-Wigner method of "little subgroups". See Serre (1977, Chapter 8).

4. Let G be the integers mod(n) with n even. Then multiplying by 2 is no longer an automorphism. Ablow and Brenner give an explicit Jordanization for this case as well. It seems worthwhile trying to generalize their results to the present setting when A is merely a homomorphism.

5. In joint work with Rosemary Bailey and Chris Rowley a natural generalization of the ideas of this section to matrices indexed by a Homogeneous space has been developed. Let X be a finite set on which a group G acts transitively with isotropy subgroup K. Identify X with coset representations of K in G. A convolution matrix can be defined for each K bi-invariant function f by $M_{xy} = f(y^{-1}x)$. Let π be a permutation of X and consists the matrix M^π obtained by permuting the rows of M by π. For a convolution matrix M, when does M^π have the property that some power of it is again a convolution matrix? A sufficient condition can be given in terms of $\mathrm{End}_G(L(X))$ the linear maps from $L(X)$ into itself that commute with the action of G. Let $\overline{G}$ be the permutations of X that commute with $\mathrm{End}_G(L(X))$ so $G \subseteq \overline{G}$. If π is in the normalizer of $\overline{G}$ and π has order k then $(M^\pi)^k$ is a convolution matrix if M is.

If G, K in Gelfand, M^π can be explicitly Jordanized for such π's. Specialize to $\pi \in G$ and to the random walk set up, $X_n = \pi((X_{n-1} * \varepsilon_n)$ can be explicitly Jordanized provided the law of ε is K bi-invariant.

B. An example. On the integers mod p, consider the family of processes

$$X_n = a_n X_{n-1} + b_n \pmod p$$

where (a_n, b_n) are independent. Such processes arise in analyzing random number generators with varying multipliers. For example, the first congruential generator employed by Lehmer was $X_n = aX_{n-1} \pmod p$. This worked at a constant rate but was called upon at varying times (depending on the time required to complete other loops in the program). This results in an effectively random a.

A second familiar specialization has $a_n \equiv 1$. This is then a random walk (mod p). Suppose p is odd and $b_n = \pm 1$ with probability $\frac{1}{2}$. Then standard techniques show that the walk tends to a uniform distribution. More quantitatively:

$$\sum \left| P\{X_n - j\} - \frac{1}{p} \right| \leq \alpha e^{-\beta n/p^2} \quad \text{for universal } \alpha, \beta.$$

This estimate shows n has to be larger than p^2 for the walk to effectively forget where it started. This is proved in detail in Chung-Diaconis-Graham (1987) or in Diaconis (1988).

Varying slightly, consider $X_n = 2X_{n-1} + b_n \pmod p$, with $b_n = 0, \pm 1$ with probability $\frac{1}{3}$. Chung-Diaconis-Graham (1987) show this deterministic doubling dramatically speeds up the walk: order $\log p \log\log p$ steps are necessary and suffice to have the walk become random.

The matrix of this last Markov chain has the form described in this section. If $2^k = 1 \pmod p$ its kth power is a circulant. This forms the basis for the calculations in Chung-Diaconis-Graham (1987).

The argument of lemma 1 above allows random values of a_n to be dealt with: In a basis given by the characters, the matrix of the Markov chain is a band-diagonal matrix. For example, $X_n = a^{t_n} X_{n-1} + b_n$ with $t_n = \pm 1$ with probability $\frac{1}{2}$ and b_n arbitrary leads to Jacobi matrices. The extensive theory of orthogonal polynomials can now be called upon to give estimates for eigenvalues. This theory is at present in active development by myself and my students.

ACKNOWLEDGEMENT. I thank Dan Rockmore, Rosemary Bailey, Chris Rowley, and John McKay for their help.

REFERENCES

1. C. Ablow and J. Brenner, *Roots and canonical forms for circulant matrices*, Trans. Amer. Math. Soc. **107** (1963), 360–376.
2. T. W. Anderson, *Multivariate Analysis*, 2nd ed., Wiley, New York, 1984.
3. S. A. Andersson and M. Perlman, *Group-invariant analogues of Hadamard's inequality*, Linear Alg. Appl. **110** (1988), 91–116.
4. R. Bailey, *Cyclic designs and factorial designs*, R. C. Bose Symposium on Probability, Statistics, and Design of Experiments (S. Delhi (K. Sen ed.), 1989.
5. E. Bannai and T. Ito, *Algebraic Combinatorics*, Benjamin, Menlo Park, CA, 1984.
6. ———, *Current research on algebraic combinatorics*, Graphs Combin. **2** (1986), 287–308.
7. N. Biggs, *Algebraic Graph Theory*, Cambridge University Press, London, 1974.
8. A. Buja, T. Hastie and R. Tibshirini, *Linear smoothers and additive models*, Ann. Statist. **17** (1989), 453–555.
9. L. Chihara, *On the zeros of the Askey-Wilson polynomials, with application to coding theory*, SIAM J. Math. Anal. **18** (1987), 183–207.
10. D. Chillag, *Generalized circulants and class functions of finite groups*, Linear Alg. Appl. **93** (1987), 191–208.
11. ———, *Characters, nonnegative matrices and generalized circulants*, Proc. Sympos. Pure Math. **47** (1987), 33–40.
12. ———, *Generalized circulants and class functions of finite groups. II*, Linear Alg. Appl. **108** (1988), 199–212.

13. F. Chung, P. Diaconis and R. L. Graham, *A random walk problem arising in random number generation*, Ann. Probab. **15** (1987), 1148–1165.
14. J. Conway and N. Sloane, *Sphere Packings, Lattice and Groups*, Springer Verlag, New York, 1987.
15. C. W. Curtis and I. Reiner, *Representation of Finite Groups and Associative Algebras*, Wiley-Interscience, New York, 1962.
16. ——, *Methods of Representation Theory, Vol. I*, Wiley-Interscience, New York, 1981.
17. ——, *Methods of Representation Theory, Vol. II*, Wiley-Interscience, New York, 1987.
18. P. Davis, *Circulant Matrices*, Wiley, New York, 1979.
19. P. Diaconis, *Group Representations in Probability and Statistics*, Institute of Mathematical Statistics, Hayward, CA, 1988.
20. P. Diaconis and R. Graham, *The Radon transform on Z_2^k*, Pacific J. Math. **118** (1985), 323–345.
21. P. Diaconis and M. Shahshahani, *Generating a random permutation with random transpositions*, Z. Wahrsch. Verw. Gebiete **57** (1981), 159–179.
22. ——, *Time to reach stationarity in the Bernoulli-Laplace diffusion model*, SIAM J. Math. Anal. **18** (1987), 208–218.
23. J. Dieudonné, *Treatise on Analysis, VI*, Academic Press, New York, 1978.
24. C. Dunkl, *Structures isomorphic to compact homogeneous spaces*, Technical report, Dept. Math., University of Virginia, 1972.
25. J. Fill, *The Radon transform on $\mathbf{Z}_n$*, Discrete Math. (1989), 262–283.
26. B. Friedman, *n-commutative matrices*, Math. Ann. **136** (1958), 343–347.
27. ——, *Eigenvalues of composite matrices*, Proc. Camb. Phil. Soc. **57** (1961).
28. A. Greenhalgh, *Measures on Groups with Subgroup Invariance Properties*, Ph.D. Thesis, Dept. of Math., Stanford University, 1989.
29. T. Hawkins, *The Creation of the Theory of Group Characters*, Rice University Studies 64, 1978, pp. 57–71.
30. I. Hirschman, *Integral equations on certain compact homogeneous spaces*, SIAM J. Math. Anal. **3** (1972), 314–343.
31. J. John, *Cyclic Designs*, London, Chapman Hall, 1987.
32. N. W. Lederman, *The Theory of Group Characters*, Cambridge University Press, Cambridge, 1988.
33. G. Letac, *Problèmes classiques sur un couple de Gelfand*, Lecture Notes in Math, **861**, Springer-Verlag, New York, 1981.
34. G. Mackey, *On the structure of the set of conjugate classes in certain locally compact groups*, Symposia Mathematica **16** (1975), 433–467.
35. J. MacWilliams and W. Sloane, *The Theory of Error Correcting Codes*, North-Holland, Amsterdam, 1977.
36. P. Matthews, *Covering problems for random walks on spheres and finite groups*, Ph.D. Thesis, Dept. of Statistics, Stanford University, 1985.
37. J. McKay, *Graphs, singularities, and finite groups*, Proc. Symp. Pure Appl. Math. (37), Amer. Math. Soc., Providence, Rhode Island, 1980, pp. 183–186.
38. J. A. Morrison, *Weighted averages of Radon transforms on $\mathbf{Z}_2^k$*, SIAM J. Algebraic Discrete Methods **7** (1986), 404–413.
39. T. Muir, *The Theory of Determinants in the Historical Order of Development*, Vols. 1–4, London, 1890–1923.
40. M. A. Naimark and D. I. Stern, *Theory of Group Representations*, Springer-Verlag, New York, 1982.
41. M. Perlman, *Group symmetry covariance models*, Statist. Sci **2** (1987), 421–425.
42. D. Rothaus and J. G. Thompson, *A combinatorial problem in the symmetric group*, Pacific J. Math. **18** (1966), 175–178.
43. J. P. Serre, *Linear Representations of Finite Groups*, Springer-Verlag, New York, 1977.
44. D. Stanton, *Orthogonal Polynomials and Chevalley Groups*, R. Askey et al (EDS), Special Functions: Group Theoretical Aspects and Applications, Dordrecht, Boston, 1984, pp. 87–92.
45. D. Supruneko and R. Tyshkevich, *Commutative Matrices*, Academic Press, New York.

46. L. Takacs, *Harmonic analysis on Schur algebras and its application in the theory of probability*, J. W. Chao, W. Wayczynski, (eds.), Probability Theory and Harmonic Analysis, Marcel Dekker, New York, 1986, pp. 227–283.
47. B. Travis, *Spherical functions on finite groups*, J. Alg. **29** (1974), 65–76.
48. P. H. Zieshang, *Cayley graphs of finite groups*, J. Alg. **118** (1988), 447–454.

DEPARTMENT OF MATHEMATICS, HARVARD UNIVERSITY, CAMBRIDGE, MASSACHUSETTS 02138

Proceedings of Symposia in Applied Mathematics
Volume **40**, 1990

Tangential Interpolation Problems For Rational Matrix Functions

JOSEPH A. BALL, ISRAEL GOHBERG, AND LEIBA RODMAN

ABSTRACT. We generalize to matrix functions a number of basic interpolation problems, namely, those of Lagrange and of Nevanlinna and Pick together with zero and pole interpolation for rational functions. Here only the simplest cases of the matrix interpolation problems are presented to keep the exposition elementary and self-contained.

Table of Contents

1. Introduction. In the last couple of decades there has been a surge of renewed interest in generalizations to matrix and operator valued functions of a variety of classical interpolation problems. Partly this renewed interest is due to recently found manifold applications in circuit and systems theory; see [F,H2] and the references there. This modern work in interpolation theory has developed along several lines; for further details and references to the work of other authors, we refer the reader to [BGR1].

The authors have recently developed an approach to handle elaborate bi-tangential high order interpolation problems in an elementary self-contained way (see [BGR1, BGR2, BGR3]). The main influences on this work are the projective geometry approach in [BH], the analysis of zero and pole structure

1980 *Mathematics Subject Classification* (1985 *Revision*). Primary 30E05.

Research of these authors was partially supported by the National Science Foundation.
Research of this author was partially supported by the Air Force Office of Scientific Research (Grant AFOSR-87-0287).

for a rational matrix function and the connection of this with a realization of the function as the transfer function of a linear system in [BGK, GKLR], and the preliminary synthesis of these ideas in [BR]. A full account will appear in the forthcoming monograph [BGR4].

In this article we illustrate the approach by solving (with complete proofs) the simplest tangential interpolation problems for rational matrix functions. For a slightly different account, see [BGR1] where a more general exposition of zero and pole structure for a rational matrix function is also included.

In Section 2 we state the simplest tangential Nevanlinna-Pick interpolation problem and reduce the solution to the construction of a rational matrix function Θ having a certain zero and pole structure together with additional symmetry properties; the ideas for this section come primarily from [BH] and [H1]. In Section 3 we find a necessary and sufficient condition for such a function $\Theta(z)$ to exist, and when this condition is satisfied, an explicit formula for the unique solution $\Theta(z)$. The ideas for this section come primarily from [BR,AG] and the earlier work of [GKLR]. Section 4 completes the proof of the reduction presented in Section 2 and Section 5 indicates how the approach also leads to a solution of a matrix analogue of the classical Lagrange interpolation problem. Finally Section 6 sketches the connections with systems theory and how the same approach can handle more general problems.

Notation used throughout the paper is: $\mathbf{C}^{m\times n}$ is the set of $m \times n$ matrices (with complex entries); $i\mathbf{R}$ stands for the imaginary axis and Π^+ for the open right half-plane. For hermitian matrices A and B we use $A > B$ (resp. $A \geq B$) to designate that $A - B$ is positive definite (resp. positive semidefinite); the notation $A < B$ and $A \leq B$ is used analogously.

2. The tangential Nevanlinna-Pick interpolation problem. We consider here the simplest case of the tangential Nevanlinna-Pick interpolation problem over the right half plane $\Pi^+ = \{z\colon \operatorname{Re} z > 0\}$. In this problem we are given n distinct points $z_1, z_2, \ldots, z_n$ in Π^+, n nonzero row vectors $x_1, \ldots, x_n \in \mathbf{C}^{1\times M}$ and n row vectors $y_1, \ldots, y_n \in \mathbf{C}^{1\times N}$. The problem is to find a rational $M \times N$ matrix function $F(z)$ such that

(i) $F(z)$ is analytic on Π^+

(ii) $\sup_{z\in\Pi^+} \|F(z)\| < 1$

and

(iii) $x_i F(z_i) = y_i$ for $1 \leq i \leq n$.

From the classical case [N] we expect that, when solutions exist, the set of all solutions is described by a linear fractional parametrization

$$F = (\Theta_{11}G + \Theta_{12})(\Theta_{21}G + \Theta_{22})^{-1} \tag{2.1}$$

where

$$\Theta = \begin{bmatrix} \Theta_{11} & \Theta_{12} \\ \Theta_{21} & \Theta_{22} \end{bmatrix}$$

is a rational $(M+N)\times(M+N)$ matrix function which is determined by the data of the problem and G is an arbitrary rational matrix function satisfying conditions (i) and (ii) only (with G in place of F).

With the assumption that the solution of the problem has this form, we now proceed to analyze what properties the linear fractional parametrizer Θ must have. In order that F satisfy (i) for *any* choice of G satisfying (i) and (ii), a natural property to impose on Θ is

(I) $\quad\Theta(z)$ *is analytic in* $\Pi^+ \cup i\mathbf{R}$, *including at infinity.*

Indeed, if Θ satisfies (I) then the analycity of F on Π^+ (F as in (2.1)) holds for any G satisfying (i) and (ii) for which we can show $\det(\Theta_{21}G+\Theta_{22})$ does not vanish on Π^+; we will return to this side issue later.

The linear fractional formula (2.1) itself can be rewritten in the more linear form as

$$\begin{bmatrix} F \\ I \end{bmatrix} = \begin{bmatrix} \Theta_{11} & \Theta_{12} \\ \Theta_{21} & \Theta_{21} \end{bmatrix} \begin{bmatrix} G \\ I \end{bmatrix} (\Theta_{21}G+\Theta_{22})^{-1}. \tag{2.2}$$

From this we get the identity

$$\begin{aligned} F^*F - I &= [F^*I]\begin{bmatrix} I & 0 \\ 0 & -I \end{bmatrix}\begin{bmatrix} F \\ I \end{bmatrix} \\ &= X^*[G^*I]\Theta^* J\Theta \begin{bmatrix} G \\ I \end{bmatrix} X \end{aligned} \tag{2.3}$$

where $X=(\Theta_{21}G+\Theta_{22})^{-1}$ and $J=I_M\oplus -I_N$. Next, if we assume that

$$\Theta(z)^*J\Theta(z) = J \tag{2.4}$$

then we may simplify (2.3) further to

$$F(z)^*F(z) - I = X(z)^*(G(z)^*G(z)-I)X(z).$$

If we assume that $X(z)$ is invertible for all z under consideration, this leads to

$$F(z)^*F(z) - I < 0 \Leftrightarrow G(z)^*G(z) - I < 0$$

or equivalently

$$\|F(z)\| < 1 \Leftrightarrow \|G(z)\| < 1.$$

If instead of (2.4) we have only the weaker condition

$$\Theta(z)^*J\Theta(z) \le J \tag{2.5}$$

then from (2.3) we are still able to conclude

$$F(z)^*F(z) - I \le X(z)^*(G(z)^*G(z)-I)X(z)$$

so we have the implication in one direction

$$\|G(z)\| < 1 \Rightarrow \|F(z)\| < 1. \tag{2.6}$$

One cannot expect a nonconstant matrix function to satisfy (2.4) on all of Π^+ and maintain analycity. Indeed, if (2.4) holds for all $z \in \Pi^+$, then in particular $\Theta(z)$ is invertible and

$$\overline{(\Theta(z))} = (\Theta(z))^{*T} = J\Theta(z)^{-1T}J$$

is analytic in Π^+ as well. However, an analytic function whose complex conjugate is analytic as well, must be a constant. As a reasonable compromise we shall impose

$$\text{(II)} \qquad \begin{aligned} \Theta(z)^* J\Theta(z) &= J \text{ for } z \in i\mathbf{R} \\ \Theta(z)^* J\Theta(z) &\leq J \text{ for } z \in \Pi^+. \end{aligned}$$

In any case we shall attempt to reverse the implication (2.6) only when F also satisfies the interpolation conditions (iii).

We next seek additional conditions to impose on Θ so that F as in (2.1) in addition satisfies (iii) when G satisfies (i) and (ii). We rewrite (iii) in the form of homogeneous interpolation conditions on the block column function $\begin{bmatrix} F(z) \\ I \end{bmatrix}$:

$$(2.7) \qquad [x_i, -y_i]\begin{bmatrix} F(z_i) \\ I \end{bmatrix} = 0.$$

By (2.2) this becomes

$$(2.8) \qquad [x_i, -y_i]\Theta(z_i)\begin{bmatrix} G(z_i) \\ I \end{bmatrix} X(z_i) = 0 \quad \text{for } 1 \leq i \leq n.$$

As $G(z)$ may be chosen to have any value at z_i and we eventually hope to show that the associated function $X(z) = (\Theta_{21}(z)G(z)+\Theta_{22}(z))^{-1}$ is analytic at z_i, to guarantee (2.5) it is natural to impose

$$\text{(III)} \qquad [x_i, -y_i]\Theta(z_i) = 0 \text{ for } 1 \leq i \leq n.$$

(recall that $x_i \neq 0$).

If Θ satisfies additional constraints of the form in (III) at additional points z_i in Π^+ then any F arising as in (2.1) from a G satisfying (i) and (ii) we would expect to solve additional interpolation constraints of the form (iii). Also if $u(z)$ is a row vector function analytic at z_i with $u(z_i) = [x_i, -y_i]$ and if the zero of $u(z)\Theta(z)$ at z_i is not simple, then the associated $F(z)$ will satisfy higher order interpolation conditions at z_i in addition to the simple one in (iii). Similarly, if $[u_i, -v_i]$ is linearly independent of $[x_i, -y_i]$ and satisfies a condition of the form in (III), then the $F(z)$ generated by (2.1) necessarily satisfies independent interpolation conditions in addition to those listed in (iii). As we want to construct Θ so that (2.1) parameterizes *all* solutions F of (i), (ii) and (iii) we impose:

(IV) $\det\Theta(z)$ *has only simple zeros in* Π^+ *occurring at* $z_1, \ldots, z_n$.

together with the strengthened form of (III).

The left kernel $\operatorname{Ker}_\ell \Theta(z_i)$ defined by

(III) $$\operatorname{Ker}_\ell \Theta(z_i) = \{u \in \mathbf{C}^{1\times(M+N)} : u\Theta(z_i) = 0\}$$

is the one dimensional subspace spanned by $[x_i, -y_i]$ *for* $1 \le i \le n$.

This completes a heuristic argument to show that if Θ satisfies (I)–(IV) and F and G are related as in (2.1), then F satisfies (i), (ii), and (iii) if and only if G satisfies (i) and (ii). Gaps in the argument concern the analyticity of $X(z) = (\Theta_{21}(z)G(z) + \Theta_{22}(z))^{-1}$ in Π^+ for any G satisfying (i) and (ii) and an analysis as to when the implication (2.6) can be reversed. In fact these gaps can be plugged as the following theorem shows. We postpone the formal proof to Section 4.

THEOREM 2.1. *Suppose there exists a rational* $(M+N)\times(M+N)$ *matrix function* $\Theta(z)$ *which satisfies* (I)–(IV):

(I) $\Theta(z)$ *is analytic on* $\Pi^+ \cup i\mathbf{R}$, *including infinity.*

$$\Theta(z)^* J\Theta(z) = J \text{ for } z \in i\mathbf{R}$$

(II) $$\Theta(z)^* J\Theta(z) \le J \quad \text{for } z \in \Pi^+$$

(III) $$\operatorname{Ker}_\ell \Theta(z_i) = \operatorname{span}\{[x_i, -y_i]\} \text{ for } 1 \le i \le n$$

(IV) $\det\Theta(z)$ *has only simple zeros in* Π^+ *at* $z_1, \ldots, z_n$.

Then there exists $M \times N$ *matrix functions* $F(z)$ *which satisfy* (i)–(iii):

(i) $F(z)$ *is analytic on* $\Pi^+ \cup i\mathbf{R}$, *including at infinity.*

(ii) $\sup_{z\in\Pi^+} \|F(z)\| < 1$

(iii) $x_iF(z_i) = y_i$ *for* $1 \le i \le n$.

Moreover, given $\Theta(z)$ *satisfying* (I)–(IV), $F(z)$ *is a solution of* (i)–(iii) *if and only if* $F(z)$ *is given by* (2.1) *for a* $G(z)$ *which satisfies* (i) *and* (ii) (*with* G *in place of* F).

The question of existence of a rational matrix function satisfying conditions (I)–(IV) is settled by the following theorem which will be proved in Section 3.

THEOREM 3.2. *There exists a rational matrix function satisfying conditions* (I)–(IV) *in Theorem* 2.1 *if and only if the Hermitian matrix*

(2.9) $$\Lambda = \left[\frac{x_ix_j^* - y_iy_j^*}{z_i + \bar{z}_j}\right]_{1\le i,j\le n}$$

is positive definite. In this case the unique such matrix function $\Theta(z)$ *with value* I *at infinity is given by*

(2.10)

$$\Theta(z) = \begin{bmatrix} I_M & 0 \\ 0 & I_N \end{bmatrix} + \begin{bmatrix} -x_1^* & \cdots & -x_n^* \\ y_1^* & \cdots & y_n^* \end{bmatrix} \begin{bmatrix} (z+\bar{z}_1)^{-1} & & 0 \\ & \ddots & \\ 0 & & (z+\bar{z}_n)^{-1} \end{bmatrix} \Lambda^{-1} \begin{bmatrix} x_1 & -y_1 \\ \vdots & \vdots \\ x_n & -y_n \end{bmatrix}$$

By combining Theorems 2.1 and 2.2 we see that positive definiteness of Λ in (2.9) is sufficient for the existence of many solutions of the tangential interpolation problem (i)–(iii). It turns out that it is also necessary.

THEOREM 2.3. *A necessary and sufficient condition for the tangential Nevanlinna-Pick interpolation problem* (i)–(iii) *to have a solution is that the matrix* Λ *in* (2.9) *be positive definite. In this case one such solution* $F(z)$ *is given by*

$$F(z) = [X^*(zI + A_\zeta^*)^{-1}\Lambda^{-1}Y][I + Y^*(zI + A_\zeta^* - \Lambda^{-1}YY^*)^{-1}\Lambda^{-1}Y]$$

where

$$X = \begin{bmatrix} x_1 \\ \vdots \\ x_n \end{bmatrix}, \qquad Y = \begin{bmatrix} y_1 \\ \vdots \\ y_n \end{bmatrix}$$

and

$$A_\zeta = \begin{bmatrix} z_1 & & 0 \\ & \ddots & \\ 0 & & z_n \end{bmatrix}.$$

PROOF. If Λ is positive definite, then we may construct Θ with properties (I)–(IV) by (2.10) as in Theorem 2.2. Then Theorem 2.1 gives a solution F of the tangential Nevanlinna-Pick interpolation problem (i)–(iii) corresponding to each G satisfying (i) and (ii) via the linear fractional formula (2.1). The particular F in (2.11) arises by inserting $G = 0$ and using Lemma 3.1.2 to compute $\Theta_{22}(z)^{-1}$. The necessity of the positive definiteness of Λ for solutions of the Nevanlinna-Pick problem to exist is classical; we do not prove this here. For an elementary approach to the proof of this necessity condition which fits in with the ideas here, see [BGR1]. A straightforward operator-theoretic proof (with the disk in place of the right-half plane) appears in [S] and Proposition V.8.1 of [NF]. □

3. Construction of Θ. In this section we discuss how one might go about constructing a rational matrix function which satisfies the conditions (I)–(IV) arrived at in the previous section.

By way of motivation we first discuss the scalar version of (I)–(IV) ($M + N = 1$) with $J = 1$. Without loss of generality we may take all vectors $[x_i, -y_i]$ to be the number 1. Then the problem is simply to construct a scalar rational function with no poles in the closed right half plane with values of modulus 1 on the imaginary axis and having only simple zeros in the right half plane which occur at the points $z_1, \ldots, z_n$. The solution by elementary classical complex analysis is well known to be given by the Blaschke product

$$\Theta(z) = \Pi_{i=1}^{n} \frac{z - z_i}{z + \bar{z}_i}.$$

Note however that this case is useless for the tangential Nevanlinna-Pick interpolation problem; even for the scalar Nevanlinna-Pick problem ($M = N = 1$), we must construct a 2×2 matrix function $\Theta(z)$ satisfying (I)–(IV) with $J = \begin{bmatrix} 1 & 0 \\ 0 & -1 \end{bmatrix}$.

To handle the matrix version of (I)–(IV) we show that the derived matrix function $\Theta(z)$ has a simple partial fraction expansion. Simple direct calculations then yield necessary and sufficient conditions for such a $\Theta(z)$ to exist. The following simple result on uniqueness of partial fraction representations for rational matrix functions will be useful in the sequel.

LEMMA 3.0.1. *Suppose $w_1, \ldots, w_n$ are distinct complex numbers, $\psi_1, \ldots, \psi_n, \tilde{\psi}_1, \ldots, \tilde{\psi}_n$ are nonzero $p \times 1$ column vectors and $\gamma_1, \ldots, \gamma_n, \tilde{\gamma}_1, \ldots, \tilde{\gamma}_n$ are $2n$ nonzero $1xp$ row vectors.*

Define rational $p \times p$ matrix functions $R(z)$ and $\widetilde{R}(z)$ by

$$R(z) = I + \sum_{j=1}^{n} (z - w_j)^{-1} \psi_j \gamma_j$$

and

$$\widetilde{R}(z) = I + \sum_{j=1}^{n} (z - w_j)^{-1} \tilde{\psi}_j \tilde{\gamma}_j.$$

Then $R(z) = \widetilde{R}(z)$ if and only if there exist nonzero complex numbers $c_1, \ldots, c_n$ for which

$$\tilde{\psi}_j = c_j \psi_j, \qquad \tilde{\gamma}_j = c_j^{-1} \gamma_j.$$

PROOF. Suppose $R(z) = \widetilde{R}(z)$. Then

$$\psi_j \gamma_j = \lim_{z \to w_j} (z - w_j) R(z) = \tilde{\psi}_j \tilde{\gamma}_j.$$

Then the rank 1 matrix $\psi_j \gamma_j = \tilde{\psi}_j \tilde{\gamma}_j$ has image space $\operatorname{span}\{\psi_j\} = \operatorname{span}\{\tilde{\psi}_j\}$. Hence $\tilde{\psi}_j = c_j \psi_j$ for a nonzero complex number c_j. As ψ_j is one-to-one as a linear transformation from $\mathbf{C}^1$ into $\mathbf{C}^p$, it follows also that $\tilde{\gamma}_j = c_j^{-1} \gamma_j$. □

3.1 Zero-pole interpolation. In this subsection we consider a basic problem of interpolation of prescribed null and pole structure for a rational matrix function. The solution of this problem is an intermediate step in the construction of a rational matrix function satisfying (I)–(IV) to be presented in Section 3.3.

We first consider the following basic lemma which relates null and pole structure at a given point to the Laurent expansion of the function or of its inverse at that point.

LEMMA 3.1.1. *Suppose* $W(z)$ *is a rational* $p \times p$ *matrix function analytic at the point* z_0 *in* $\mathbf{C}$, φ_0 *is a nonzero* $1 \times p$ *row vector and* ψ_0 *is a nonzero* $p \times 1$ *column vector.*

(a) *Then* $\det W(z)$ *has a simple zero at* z_0 *and*

$$\operatorname{Ker}_\ell W(z_0) = \operatorname{span}\{\varphi_0\}$$

if and only if the Laurent expansion for $W(z)^{-1}$ *at* z_0 *has the form*

$$W(z)^{-1} = \frac{\chi_0 \varphi_0}{z - z_0} + [\textit{analytic at } z_0]$$

for some nonzero $p \times 1$ *column vector* χ_0.

(b) *Then* $\det W(z)$ *has a simple zero at* z_0 *and*

$$\operatorname{Ker} W(z_0) = \operatorname{span}\{\psi_0\}$$

if and only if the Laurent expansion for $W(z)^{-1}$ *at* z_0 *has the form*

$$W(z)^{-1} = \frac{\psi_0 \gamma_0}{z - z_0} + [\textit{analytic at } z_0]$$

for some nonzero $1 \times p$ *row vector* ψ_0.

PROOF. Note that (b) is simply the transposed version of (a) so we need prove only (a). Suppose first that $\det W(z)$ has a simple zero at z_0 and $\operatorname{Ker}_\ell W(z_0) = \operatorname{span}\{\varphi_0\}$. Let $W(z)^{-1}$ have Laurent expansion

$$W(z)^{-1} = \sum_{j=-q}^{\infty} (z - z_0)^j \widetilde{W}_j$$

where $\widetilde{W}_{-q} \neq 0$; since $W(z_0)$ is not invertible, certainly $q \geq 1$. As $\widetilde{W}_{-q} \neq 0$ we can choose a row vector α_0 so that $\alpha_0 \widetilde{W}_{-q} \neq 0$. Then we have

$$\alpha_0 W(z)^{-1} = (z - z_0)^{-q} \gamma_0(z) \tag{3.1.1}$$

where γ_0 is analytic at z_0 and $\gamma_0(z_0) = \alpha_0 \widetilde{W}_{-q} \neq 0$. Rewrite (3.1.1) as

$$(z - z_0)^q \alpha_0 = \gamma_0(z) W(z). \tag{3.1.2}$$

In particular $\gamma_0(z_0) W(z_0) = 0$. Since by assumption $\operatorname{Ker}_\ell W(z_0) = \operatorname{span}\{\varphi_0\}$ we conclude that

$$\alpha_0 \widetilde{W}_{-q} = \gamma_0(z_0) = c\varphi_0 \tag{3.1.3}$$

for some nonzero complex number c. Since $\gamma_0(z_0) \neq 0$ we may choose $p - 1$ row vectors $\gamma_2, \gamma_3, \ldots, \gamma_p$ so that

$$\det(\operatorname{col.}(\gamma_0(z_0), \gamma_2, \ldots, \gamma_p)) \neq 0.$$

If the zero of $\gamma_0(z) W(z)$ at z_0 were not simple, then

$$\det\{[\operatorname{col.}(\gamma_0(z), \gamma_2, \ldots, \gamma_p)] \cdot W(z)\} = \det[\operatorname{col.}(\gamma_0(z), \gamma_2, \ldots, \gamma_p)] \cdot \det W(z)$$

would have a higher order zero at z_0. Since by assumption $\det W(z)$ has a simple zero at z_0, we conclude that the zero of $\gamma_0(z)W(z)$ at z_0 must be simple. Thus $q = 1$ from (3.1.2). Finally, since α_0 was chosen to be any row vector for which $\alpha_0 \widetilde{W}_{-1} \neq 0$, we conclude from (3.1.3) that

$$\text{Im}_\ell \, \widetilde{W}_{-1} = \text{span}\{\varphi_0\}.$$

Here and elsewhere we denote by $\text{Im}_\ell \, X$ the left image of an $m \times n$ matrix X

$$\text{Im}_\ell \, X = \{xX \colon x \text{ } is \text{ } a \text{ } 1 \times \text{ } m \text{ } row\}.$$

From this it is elementary that there is a nonzero $p \times 1$ column vector χ_0 such that $\widetilde{W}_{-1} = \chi_0 \varphi_0$ as needed.

Conversely suppose $W(z)^{-1}$ has a Laurent expansion at z_0 of the form

$$W(z)^{-1} = \frac{\chi_0 \varphi_0}{(z - z_0)} + [analytic \text{ } at \text{ } z_0]$$

for a nonzero column vector χ_0. Multiply both sides on the right by $(z - z_0)W(z)$ to deduce

$$(z - z_0)I = \chi_0 \varphi_0 W(z) + (z - z_0)[analytic \text{ } at \text{ } z_0]W(z). \tag{3.1.4}$$

By assumption $W(z)$ itself is analytic at z_0. Thus we may evaluate (3.1.4) at z_0 to deduce that $\chi_0 \varphi_0 W(z_0) = 0$. Since χ_0 is a nonzero column vector, this in turn gives $\varphi_0 W(z_0) = 0$. If $\tilde{\varphi}_0$ is any other row vector for which $\tilde{\varphi}_0 W(z_0) = 0$, the first part of the proof gives that $\tilde{\varphi}_0 \in \text{Im}_\ell \, Res_{z=z_0} W(z)^{-1} = \text{span}\{\varphi_0\}$ as asserted. Next rewrite (3.1.4) as

$$\begin{aligned}(z - z_0)I &= (\chi_0 \varphi_0 + (z - z_0)[analytic \text{ } at \text{ } z_0]) \cdot W(z) \\ &=: V(z)W(z).\end{aligned}$$

Taking determinants of both sides gives

$$(z - z_0)^p = \det V(z) \cdot \det W(z). \tag{3.1.5}$$

Since $\chi_0 \varphi_0$ is a rank 1 matrix, it is easy to see that $\det V(z)$ has a zero at z_0 of order at least $p - 1$. Thus from (3.1.5) we deduce that $\det W(z)$ has a zero at z_0 of order at most 1. Since $\det W(z_0) = 0$, we get that $\det W(z)$ has a zero at z_0 of order exactly equal to 1. □

We now state the interpolation problem which we wish to analyze in this subsection. We are given distinct points $z_1, z_2, \ldots, z_n \in \mathbf{C}$ and nonzero $1 \times p$ row vectors $\varphi_1, \varphi_2, \ldots, \varphi_n$, as well as additional distinct points $w_1, w_2, \ldots, w_n \in \mathbf{C}$ and nonzero $p \times 1$ column vectors $\psi_1, \psi_2, \ldots, \psi_n$. To avoid unwanted complications we assume that the set of points $\{z_1, \ldots, z_n\}$ is disjoint from the set $\{w_1, \ldots, w_n\}$. The problem is to construct a rational $p \times p$ matrix function $W(z)$ with value I at infinity such that

(3.1.6i) *For* $i = 1, \ldots, n$, $W(z)$ *is analytic at* z_i, $\det W(z)$ *has a simple zero at* z_i, *and* $\text{Ker}_\ell \, W(z_i) = \text{span}\{\varphi_i\}$.

(3.1.6ii) *For* $j = 1, \ldots, n$ $W(z)^{-1}$ *is analytic at* w_j, $\det W(z)^{-1}$ *has a simple zero at* w_j, *and* $\operatorname{Ker} W(w_j)^{-1} = \operatorname{span}\{\psi_j\}$

(3.1.6iii) *The poles of* $W(z)$ *occur only at* $w_1, w_2, \ldots, w_n$. *The poles of* $W(z)^{-1}$ *occur only at* $z_1, z_2, \ldots, z_n$.

If $W(z)$ is to be a solution of (3.1.6), from (3.1.6i) and Lemma 3.1.1(a) we know that $W(z)^{-1}$ has a simple pole at z_i with residue at z_i of the form $\chi_i \varphi_i$ for some nonzero column vector χ_i. By (3.1.6iii) we are requiring that $W(z)^{-1}$ has no other poles, and we also require that $W(\infty) = I$. Thus by Liouville's Theorem we know that a solution W of (3.1.6) must be such that its inverse $W(z)^{-1}$ has the partial fraction expansion

$$W(z)^{-1} = I - \sum_{i=1}^{n} (z - z_i)^{-1} \chi_i \varphi_i \tag{3.1.7}$$

where now the nonzero column vectors $\chi_i, \ldots, \chi_n$ are the unknowns. Similarly, from (3.1.6ii), (3.1.6iii) and Lemma 3.1.1(b) (applied to W^{-1} rather that to W), we deduce that a solution W must itself have a partial fraction expansion

$$W(z) = I + \sum_{j=1}^{n} (z - w_j)^{-1} \psi_j \gamma_j \tag{3.1.8}$$

for some nonzero row vectors $\gamma_1, \gamma_2, \ldots, \gamma_n$ to be determined. To examine the consequence of (3.1.7) and (3.1.8) holding for the same rational matrix function $W(z)$, we need to be able to compute the inverse of a rational matrix function given in the form of a partial fraction expansion with simple poles. For this we need the following lemma.

LEMMA 3.1.2. *Suppose* $z_1, \ldots, z_n$ *are distinct points in* $\mathbf{C}$, $\chi_1, \ldots, \chi_n$ *are nonzero* $p \times 1$ *column vectors,* $\varphi_1, \ldots, \varphi_n$ *are nonzero* $1 \times p$ *row vectors, and* $V(z)$ *is the rational* $p \times p$ *matrix function given by*

$$V(z) = I - \sum_{i-1}^{n} (z - z_i)^{-1} \chi_i \varphi_i$$

Then

$$V(z)^{-1} = I + C_\chi (zI - (A_\zeta + B_\varphi C_\chi))^{-1} B_\varphi \tag{3.1.9}$$

where

$$C_\chi = [\chi_1, \chi_2, \ldots, \chi_n]$$

$$A_\zeta = \begin{bmatrix} z_1 & & 0 \\ & \ddots & \\ 0 & & z_n \end{bmatrix}, \quad B_\varphi = \begin{bmatrix} \varphi_1 \\ \varphi_2 \\ \vdots \\ \varphi_n \end{bmatrix} \tag{3.1.10}$$

PROOF. Since $V(z)$ is a rational matrix function with value I at infinity it is clear from the adjoint formula for the inverse of a matrix that $V(z)^{-1}$ exists as a rational matrix function. Using the formulas for C_χ, B_φ and A_ζ as in (3.1.10) and defining $\Omega_\zeta(z)$ by

$$\Omega_\zeta(z) = (zI - A_\zeta)^{-1} \tag{3.1.11}$$

we may write $V(z)$ in the more compact matrix form

$$V(z) = I - C_\chi \Omega_\zeta(z) B_\varphi \tag{3.1.12}$$

Computation of $V(z)^{-1}$ amounts to solving the equation

$$V(z)u = v \tag{3.1.13}$$

for v in terms of u.

Substituting (3.1.12) into (3.1.13) gives

$$u - C_\chi \Omega_\zeta(z) B_\varphi u = v \tag{3.1.14}$$

If we set

$$y = B_\varphi u \tag{3.1.15}$$

(3.1.14) becomes

$$u = C_\chi \Omega_\zeta(z) y + v$$

Multiplication by B_φ now gives

$$y - B_\varphi C_\chi \Omega_\zeta(z) y = B_\varphi v$$

Thus

$$y = (I - B_\varphi C_\chi \Omega_\zeta(z))^{-1} B_\varphi v.$$

Plug this into (3.1.14) to get

$$\begin{aligned} u &= [C_\chi \Omega_\zeta(z)(I - B_\varphi C_\chi \Omega_\zeta(z))^{-1} B_\varphi + I]v \\ &= [I + C_\chi(\Omega_\zeta(z)^{-1} - B_\varphi C_\chi)^{-1} B_\varphi]v. \end{aligned}$$

Recalling the definition (3.1.11) of $\Omega_\zeta(z)$ now gives the desired formula (3.1.9) for $V(z)^{-1}$. □

From Lemma 3.1.2 we conclude that $W(z) = V(z)^{-1}$ is given by (3.1.9) and (3.1.10) if $W(z)^{-1}$ is given by (3.1.7). Thus for the interpolation problem (3.1.6i)–(3.1.6iii) to have a solution, there must be a choice of nonzero column vectors $\chi_i, \dots, \chi_n$ and nonzero row vectors $\gamma_i, \dots, \gamma_n$ for which $V(z)^{-1}$ given by (3.1.9) and (3.1.10) is the same as $W(z)$ given by (3.1.8). In particular the n poles of $W(z)$ in (3.1.8), namely $w_1, w_2, \dots, w_n$, must be the same as the poles of $V(z)^{-1}$ in (3.1.9). But the poles of $V(z)^{-1}$ clearly occur at the n eigenvalues of the $n \times n$ matrix $A_\zeta + B_\varphi C_\chi$. Thus the eigenvalues of $A_\zeta + B_\varphi C_\chi$ must consist precisely of the n complex numbers $w_1, \dots, w_n$, i.e., $A_\zeta + B_\varphi C_\chi$ must be similar to the diagonal matrix

$$A_\pi = \begin{bmatrix} w_1 & & & 0 \\ & w_2 & & \\ & & \ddots & \\ 0 & & & w_n \end{bmatrix}.$$

Hence there must exist an invertible $n \times n$ matrix S such that

$$A_\zeta + B_\varphi C_\chi = S A_\pi S^{-1}. \tag{3.1.16}$$

Moreover, by using this similarity it is easy to convert the expression (3.1.9) for $V(z)^{-1}$ to a partial fraction form to make comparison with (3.1.8) more convenient.

Specifically, plug (3.1.16) into (3.1.9) to get

$$\begin{aligned} V(z)^{-1} &= I + C_\chi (zI - SA_\pi S^{-1})^{-1} B_\varphi \\ &= I + C_\chi S (zI - A_\pi)^{-1} S^{-1} B_\varphi \\ &= I + \sum_{j=1}^{n} (z - w_j)^{-1} \tilde{\psi}_j \tilde{\gamma}_j \end{aligned} \tag{3.1.17}$$

where $\tilde{\psi}_j$ is the j-th column of $C_\chi S$ and $\tilde{\gamma}_j$ is the j-th row of $S^{-1}B_\varphi$.

By Lemma 3.0.1 we deduce that equality of $V(z)^{-1}$ given by (3.1.9) with $W(z)$ given by (3.1.8) forces $\tilde{\psi}_j = c_j \psi_j$ and $\tilde{\gamma}_j = c_j^{-1} \gamma_j$ for nonzero complex numbers $c_1 \dots, c_n$. As the γ_j's are unknown, we may rescale γ_j if necessary so that we have the equality $\tilde{\psi}_j = \psi_j$, that is

$$C_\chi S = C_\psi. \tag{3.1.18}$$

When we plug (3.1.18) into (3.1.16) and rearrange, we get

$$S A_\pi - A_\zeta S = B_\varphi C_\psi \tag{3.1.19}$$

By computing the (i, j)-th entry of each side of (3.1.19), we see that the matrix $S = [s_{ij}]_{1 \le i,j \le n}$ satisfies (3.1.19) if and only if $s_{ij} (1 \le i, j \le n)$ satisfies the system of scalar equations

$$s_{ij} w_j - z_i s_{ij} = \varphi_i \psi_j.$$

Trivially the unique solution s_{ij} of this system is given by

$$s_{ij} = \varphi_i \psi_j / (w_j - z_i)$$

We have thus arrived at a good piece of the following result.

THEOREM 3.1.3. *There exists a rational matrix function $W(z)$ with value I at infinity satisfying* (3.1.6i)–(3.1.6iii) *if and only if the matrix S given by*

$$S = \left[\frac{\varphi_i \psi_j}{w_j - z_i} \right]_{1 \le i,j \le n} \tag{3.1.20}$$

is invertible. In this case the unique such $W(z)$ s given by

$$W(z) = I + \sum_{j=1}^{n} (z - w_j)^{-1} \psi_j \gamma_j \tag{3.1.21}$$

with inverse given by

$$W(z)^{-1} = I - \sum_{i=1}^{n} (z - z_i)^{-1} \chi_i \varphi_i \tag{3.1.22}$$

where $\begin{bmatrix} \gamma_1 \\ \vdots \\ \gamma_n \end{bmatrix} = S^{-1} \begin{bmatrix} \varphi_1 \\ \vdots \\ \varphi_n \end{bmatrix}$ *and*

$$[\chi_1, \ldots, \chi_n] = [\psi_1, \ldots, \psi_n] \cdot S^{-1}$$

PROOF. The argument preceding the statement of the theorem gives that necessarily the unique solution S (3.1.20) of (3.1.19) is invertible if a solution $W(z)$ of (3.1.6i)–(3.1.6iii) with value I at infinity exists. Moreover, from (3.1.17) we have $W(z) = V(z)^{-1} = I + C_\chi S(zI - A_\pi)^{-1} S^{-1} B_\varphi$. Combining this with (3.1.18) gives the desired formula (3.1.21) for $W(z)$. Also from (3.1.7) we see that $W(z)^{-1} = I - C_\chi (zI - A_\zeta)^{-1} B_\varphi$. When this is combined with (3.1.18) we get the indicated formula (3.1.22) for $W(z)^{-1}$.

Conversely suppose that the matrix S given by (3.1.20) is invertible. Define a matrix function $W(z)$ by (3.1.21). To see that $W(z)^{-1}$ is given by (3.1.22) we simply observe that (3.1.21) and (3.1.22) can be rewritten in the form

$$W(z) = I + C_\psi (zI - A_\pi)^{-1} S^{-1} B_\varphi$$

and

$$W(z)^{-1} = I - C_\psi S^{-1} (zI - A_\zeta)^{-1} B_\varphi,$$

where

$$A_\pi = \begin{bmatrix} w_1 & & 0 \\ & \ddots & \\ 0 & & w_n \end{bmatrix}, \qquad A_\zeta = \begin{bmatrix} z_1 & & 0 \\ & \ddots & \\ 0 & & z_n \end{bmatrix},$$

$$B_\varphi = \begin{bmatrix} \varphi_1 \\ \vdots \\ \varphi_n \end{bmatrix}, \qquad C_\psi = [\psi_1, \ldots, \psi_n]$$

Now

$$\begin{aligned} &[I - C_\psi S^{-1}(zI - A_\zeta)^{-1} B_\varphi][I + C_\psi (zI - A_\pi)^{-1} S^{-1} B_\varphi] \\ &= I + C_\psi S^{-1} (zI - A_\zeta)^{-1} \cdot [-S(zI - A_\pi) + (zI - A_\zeta)S - B_\varphi C_\psi] \\ &\cdot (zI - A_\pi)^{-1} S^{-1} B_\varphi \\ &= I + C_\psi S^{-1} (zI - A_\zeta)^{-1} \cdot [SA_\pi - A_\zeta S - B_\varphi C_\psi] \cdot \\ &(zI - A_\pi)^{-1} S^{-1} B_\varphi \\ &= I \end{aligned}$$

where we use that S satisfies the matrix equation (3.1.19). To show that $W(z)$ satisfies (3.1.6i)–(3.1.6iii), by Lemma 3.1.1 and the partial fraction representations (3.1.21) and (3.1.22) for $W(z)$ and $W(z)^{-1}$ it remains only to show that the rows of $S^{-1}B_\varphi$ and the columns of $C_\psi S^{-1}$ are nonzero. Thus suppose that the i-th row of $S^{-1}B_\varphi$ were zero. Then also the i-th row of $S^{-1}B_\varphi C_\psi$ would be zero. But from (3.1.19) we have

$$S^{-1} B_\varphi C_\psi = S^{-1}(SA_\pi - A_\zeta S)$$

$$= A_\pi - S^{-1}A_\zeta S.$$

Since the i-th row of A_π is $w_i e_i^T$ (where e_i is the i-th standard basis vector $[0,\dots,0,1,0,\dots,0]^T$ in $\mathbf{C}^n$), our hypothesis forces the i-th row of $S^{-1}A_\zeta S$ to be $w_i e_i^T$. But this property is enough to imply

$$e_i^T \cdot S^{-1}A_\zeta S = w_i e_i^T,$$

and hence w_i is an eigenvalue of $S^{-1}A_\zeta S$. As the eigenvalues of $S^{-1}A_\zeta S$ are the same as the eigenvalues of A_ζ, namely $z_i, z_2, \dots, z_n$, this leads to a contradiction. Thus necessarily the rows of $S^{-1}B_\varphi$ are nonzero as needed. By an analogous argument one can show that the columns of $C_\psi S^{-1}$ are all nonzero. This completes the proof of Theorem 3.1.2. □

3.2 J-unitary and J-inner matrix functions.

We are now ready to tackle the problem of constructing a rational matrix function $\Theta(z)$ satisfying the conditions (I)–(IV) needed to complete the solution of the tangential Nevanlinna-Pick interpolation problem discussed in Section 2.

We first consider the following variation of problem (3.1.6i)–(3.1.6iii). We are given n nonzero $1 \times p$ row vectors $\varphi_1, \dots, \varphi_n$ and n distinct points $z_1, \dots, z_n$ in the open right half plane Π^+. We denote by $\overline{\Pi^+}$ the closure of Π^+ in the extended complex plane ($\overline{\Pi^+} = \{z \in \mathbf{C}\colon \operatorname{Re} z \geq 0\} \cup \{\infty\}$). The problem is to find a rational $p \times p$ matrix function $\Theta(z)$ with value I at infinity such that

(3.2.1i) *For $i = 1, \dots, n$ $\Theta(z)$ is analytic at z_i, $\det\Theta(z)$ has a simple zero at z_i and $\operatorname{Ker}_l \Theta(z_i) = \operatorname{span}\{\varphi_i\}$*

(3.2.1ii) *$\Theta(z)$ is analytic in the closed right half plane $\overline{\Pi^+}$ and the only zeros of $\det\Theta(z)$ in $\overline{\Pi^+}$ are the points $z_1, z_2, \dots, z_n$*

(3.2.1iii) *At all points z and $-\bar{z}$ of analyticity of $\Theta(z)$ the identity $\Theta(-\bar{z})^* J(\theta)(z) = J$ holds*

Here $J = J^* = J^{-1}$ is a $p \times p$ signature matrix.

Suppose $\Theta(z)$ is a solution of (3.2.1i)–(3.2.1iii). From (3.2.1i) together with (3.2.1ii) we see that the poles of $\Theta(z)^{-1}$ in $\overline{\Pi^+}$ occur precisely at the points $z_1, z_2, \dots, z_n$. Moreover, from the identity $\Theta(z)^{-1} = J\Theta(-\bar{z})^* J$ implied by (3.2.1iii), the analyticity of Θ in the closed right half plane implies the analyticity of $\Theta(z)^{-1}$ in the closed left half plane. Hence in fact $z_1, \dots, z_n$ account for all the poles of $\Theta(z)^{-1}$ in the entire complex plane including infinity. From condition (3.2.1i) together with Lemma 3.1.1 we know that the pole of $\Theta(z)^{-1}$ at z_i is simple with residue of the form $\chi_i\varphi_i$ for some nonzero column vector $\chi_i (1 \leq i \leq n)$. Then by Liouville's Theorem, a solution $\Theta(z)$ of (3.2.1i)–(3.2.1iii) must have the partial fraction form

$$\Theta(z)^{-1} = I + \sum_{i=1}^{n} (z - z_i)^{-1} \chi_i \varphi_i \tag{3.2.2}$$

From (3.2.2) we compute

$$J\Theta(-\bar{z})^{*-1} J = I - \sum_{i=1}^{n} (z + \bar{z}_i)^{-1} J \varphi_i^* \gamma_i$$

(where γ_i is the nonzero row vector $\chi_i^* J$). But from (3.2.2iii)

$$\Theta(z) = J\Theta(-\bar{z})^{*-1} J.$$

Thus we arrive at

$$\Theta(z) = I - \sum_{i=1}^{n} (z + \bar{z}_i)^{-1} J \varphi_i^* \gamma_i \tag{3.2.3}$$

where $\gamma_i \neq 0$ for $1 \leq i \leq n$. Now by Lemma 3.0.1 and the discussion above, we conclude that $\Theta(z)$ is a solution of a problem of the form (3.1.6i)–(3.1.6iii) (with Θ in place of W), where we set $w_i = -\bar{z}_i$ and $\psi_i = -J\varphi_i^*$ for $1 \leq i \leq n$. Then by Theorem 3.1.3 the matrix H given by

$$H = \left[\frac{\varphi_i J \varphi_j^*}{z_i + \bar{z}_j} \right]_{1 \leq i,j \leq n}$$

is necessarily invertible if Θ is a solution of (3.2.1i)–(3.2.1iii). Conversely, if the matrix H is invertible, by Theorem 3.1.3 we may construct a solution Θ of (3.1.6i)–(3.1.6iii) with the data having the special form $w_i = -\bar{z}_i$ and $\psi_i = -J\varphi_i^*$; it remains only to check that this function indeed satisfies all the conditions (3.2.1i)–(3.2.1iii). We have established most of the following.

THEOREM 3.2.1. *Let the points $z_1, \dots, z_n$ in Π^+ and nonzero row vectors $\varphi_1, \dots, \varphi_n$ be given. Then there exists a rational matrix function $\Theta(z)$ with value I at infinity which satisfies conditions* (3.2.1i)–(3.2.1iii) *if and only if the Hermitian matrix*

$$H = \left[\frac{\varphi_i J \varphi_j^*}{z_i + \bar{z}_j} \right]_{1 \leq i,j \leq n} \tag{3.2.4}$$

is invertible. In this case the unique such matrix function $\Theta(z)$ is given by

$$\Theta(z) = I - \sum_{j=1}^{n} (z + \bar{z}_j)^{-1} J \varphi_j^* \gamma_j \tag{3.2.5}$$

with

$$\Theta(z)^{-1} = I - \sum_{i=1}^{n} (z - z_i)^{-1} \chi_i \varphi_i \tag{3.2.6}$$

where

$$\begin{bmatrix} \gamma_i \\ \vdots \\ \gamma_n \end{bmatrix} = H^{-1} \begin{bmatrix} \varphi_1 \\ \vdots \\ \varphi_n \end{bmatrix}$$

and

$$[\chi_1, \ldots, \chi_n] = [-J\varphi_1^*, \ldots, -J\varphi_n^*] \cdot H^{-1}.$$

PROOF. It remains only to prove sufficiency. If H is invertible, then Theorem 3.1.3 implies that the formulas (3.2.5) and (3.2.6) give a rational matrix function $\Theta(z)$ which satisfies conditions (3.1.6i)–(3.1.6iii) for the case where the data has the extra symmetry property $w_i = -\bar{z}_i$ and $\psi_i = -J\varphi_i^*$. From (3.1.5i) we see that $\Theta(z)$ satisfies (3.2.1i). From the formulas (3.2.5) and (3.2.6) we see that (3.2.1ii) is satisfied. Finally, to see that $\Theta(z)$ satisfies (3.2.1iii), note that when we compute $J\Theta(-\bar{z})^*J$ from (3.2.5) we recover (3.2.6). □

To solve the tangential Nevanlinna-Pick interpolation problem, in addition to conditions of the type (3.2.1i)–(3.2.1iii) for $\Theta(z)$ we require

(3.2.1iv) $$\text{for } z \in \Pi^+, \Theta(z)^* J\Theta(z) \leq J$$

Since the solution to (3.2.1i)–(3.2.1iii) is already unique (when it exists), it remains only to analyze when this unique solution (3.2.5) also satisfies (3.2.1iv). To do this we write (3.2.5) in a more matricial form

$$\Theta(z) = I - JB^*(zI + A^*)^{-1}H^{-1}B$$

where

$$B = \begin{bmatrix} \varphi_1 \\ \vdots \\ \varphi_n \end{bmatrix}, A = \begin{bmatrix} z_1 & & 0 \\ & \ddots & \\ 0 & & z_n \end{bmatrix}.$$

We now compute

$$\begin{aligned} \Theta(z)^* J\Theta(z) &= [I - B^*H^{-1}(\bar{z}I + A)^{-1}BJ]J[I - JB^*(zI + A^*)^{-1}H^{-1}B] \\ &= J - B^*H^{-1}(\bar{z}I + A)^{-1}B - B^*(zI + A^*)^{-1}H^{-1}B \\ &+ B^*H^{-1}(\bar{z}I + A)^{-1}BJB^*(zI + A^*)^{-1}H^{-1}B \\ &= J - B^*H^{-1}(\bar{z}I + A)^{-1}[H(zI + A^*) \\ &\quad + (\bar{z}I + A)H - BJB^*](zI + A^*)^{-1}H^{-1}B. \end{aligned}$$

Noting that H satisfies the Lyapunov equation

$$HA^* + AH = BJB^*$$

we get

$$H(zI + A^*) + (\bar{z}I + A)H - BJB^* = (z + \bar{z})H.$$

Thus

$$\Theta(z)^* J\Theta(z) = J - B^*H^{-1}(\bar{z}I + A)^{-1}(z + \bar{z})H(zI + A^*)^{-1}H^{-1}B$$

so

$$\frac{J-\Theta(z)^*J\Theta(z)}{z+\bar{z}} = B^*H^{-1}(\bar{z}I+A)^{-1}H(zI+A^*)^{-1}H^{-1}B. \tag{3.2.7}$$

From this we see that $\Theta(z)$ is J-contractive in $\Pi^+ = \{z \in \mathbf{C}\colon z+\bar{z} > 0\}$ if the Hermitian matrix H is positive definite. This proves one direction of the following result.

THEOREM 3.2.2. *There exists a matrix function $\Theta(z)$ with value I at infinity which satisfies the conditions* (3.2.1i)–(3.2.1iv) *if and only if the Hermitian matrix*

$$H = \left[\frac{\varphi_i J \varphi_j^*}{z_i + \bar{z}_j}\right]_{1\le i,j\le n}$$

is positive definite. In this case such a $\Theta(z)$ is unique and is given by (3.2.5) *with inverse given by* (3.2.6).

PROOF. It remains only to prove necessity. Thus, suppose $\Theta(z)$ is J-contractive for z in Π^+. From (3.2.7) we see that to show that H is positive definite it suffices to show that

$$\operatorname{span}\{(zI+A^*)^{-1}H^{-1}\,By\colon\ z\in\Pi^+,\ y\in\mathbf{C}^p\} = \mathbf{C}^n. \tag{3.2.8}$$

By duality (3.2.8) follows if we show that the only vector $x \in \mathbf{C}^n$ which satisfies

$$x^T(zI+A^*)^{-1}H^{-1}\,By = 0 \tag{3.2.9}$$

for all $z \in \Pi^+$ and $y \in \mathbf{C}^p$ is $x = 0$. Recalling that

$$A = \begin{bmatrix} z_1 & & 0 \\ & \ddots & \\ 0 & & z_n \end{bmatrix}, \qquad H^{-1}B = \begin{bmatrix} \gamma_1 \\ \vdots \\ \gamma_n \end{bmatrix}$$

and writing out the components of x as $x^T = [x_1, \ldots, x_n]$, (3.2.9) can be written out more explicitly as

$$R(z) = \sum_{i=1}^{n}(z+\bar{z}_i)^{-1}x_i^T\gamma_i y = 0 \tag{3.2.10}$$

for all $z \in \Pi^+$ and $y \in \mathbf{C}^p$. By analytic continuation (3.2.10) continues to hold for all $z \in \mathbf{C}\setminus\{-\bar{z}_1, \ldots, -\bar{z}_n\}$, and hence

$$0 = \lim_{z\to-\bar{z}_i}(z+\bar{z}_i)R(z) = x_i^T\gamma_i y \tag{3.2.10}$$

for $i = 1, \ldots, n$ for all $y \in \mathbf{C}^p$. As in the proof of Theorem 3.1.3, we know that the rows γ_i of $H^{-1}B$ are nonzero. Thus for each i there is a choice of y in $\mathbf{C}^p$ so that $\gamma_i y \neq 0$. Thus (3.2.10) forces the scalar x_i to be 0 for each i, and hence $x^T = [x_1, \ldots, x_n]$ as required. □

We are now in a position to complete the proof of Theorem 2.2.

PROOF OF THEOREM 2.2. Note first that if a rational matrix function $\Theta(z)$ satisfies $\Theta(z)^* J\Theta(z) = J$ for $z \in i\mathbf{R}$ then since $z = -\bar{z}$ for $z \in i\mathbf{R}$ we have

$$\Theta(-\bar{z})^* J\Theta(z) = J \tag{3.2.11}$$

for $z \in i\mathbf{R}$. As $\Theta(-\bar{z})^*$ is analytic at any z_0 such that $\Theta(z)$ is analytic at $-\bar{z}_0$, (3.2.11) continues to hold at all points z of analycity of $\Theta(z)$. With this observation we see that the problem of constructing a matrix function $\Theta(z)$ which satisfies (I)–(IV) is equivalent to producing a $\Theta(z)$ satisfying (3.2.1i)–(3.2.1iv) with the special choice of data

$$\varphi_i = [x_i, -y_i], \qquad J = I_M \oplus -I_N.$$

$(p = M + N)$. When this form of the data is plugged into Theorem 3.2.1 the matrix H (3.2.4) is the same as the matrix Λ (2.9) and Theorem 2.2 is immediate from Theorem 3.2.2. □

4. Proof of Theorem 2.1.

The goal of this section is to give a complete proof of Theorem 2.1. We begin with a preliminary lemma.

LEMMA 4.1. *Suppose that the* $(M+N) \times (M+N)$ *matrix* $\begin{bmatrix} A & B \\ C & D \end{bmatrix}$ *is a* J*-contractive matrix*

$$\begin{bmatrix} A^* & C^* \\ B^* & D^* \end{bmatrix} \begin{bmatrix} I_M & 0 \\ 0 & -I_N \end{bmatrix} \begin{bmatrix} A & B \\ C & D \end{bmatrix} \le \begin{bmatrix} I_M & 0 \\ 0 & -I_N \end{bmatrix}.$$

Then D is invertible and

$$\|D^{-1}C\| < 1.$$

PROOF. We first show that $\begin{bmatrix} A & B \\ C & D \end{bmatrix}$ is J-contractive if and only if its adjoint $\begin{bmatrix} A & B \\ C & D \end{bmatrix}^* = \begin{bmatrix} A^* & C^* \\ B^* & D^* \end{bmatrix}$ is also J-contractive. Indeed, if $\begin{bmatrix} A & B \\ C & D \end{bmatrix}$ is J-contractive, then in particular

$$B^*B - D^*D = [0, I] \begin{bmatrix} A^* & C^* \\ B^* & D^* \end{bmatrix} \begin{bmatrix} I & 0 \\ 0 & -I \end{bmatrix} \begin{bmatrix} A & B \\ C & D \end{bmatrix} \begin{bmatrix} 0 \\ I \end{bmatrix} \le -I.$$

Thus

$$B^*B + I \le D^*D$$

and since D is square, necessarily D is invertible. This means then in the system of equations

$$\begin{bmatrix} A & B \\ C & D \end{bmatrix} \begin{bmatrix} x_1 \\ x_2 \end{bmatrix} = \begin{bmatrix} y_1 \\ y_2 \end{bmatrix} \tag{4.1}$$

we can solve for y_1, x_2 in terms of y_2, x_1. The result is

$$\begin{bmatrix} BD^{-1} & A - BD^{-1}C \\ D^{-1} & -D^{-1}C \end{bmatrix} \begin{bmatrix} y_2 \\ x_1 \end{bmatrix} = \begin{bmatrix} y_1 \\ x_2 \end{bmatrix}. \tag{4.2}$$

That $\begin{bmatrix} A & B \\ C & D \end{bmatrix}$ is J-contractive means

$$\|y_1\|^2 - \|y_2\|^2 \le \|x_1\|^2 - \|x_2\|^2$$

or equivalently

$$\|y_1\|^2 + \|x_2\|^2 \le \|y_2\|^2 + \|x_1\|^2,$$

whenever x_1, x_2, y_1, y_2 is a solution of (4.1). Since (4.1) and (4.2) are equivalent we get in general:

$$(4.3) \qquad \begin{bmatrix} A & B \\ C & D \end{bmatrix} \text{ is J-contractive if and only if } \begin{bmatrix} BD^{-1} & A - BD^{-1}C \\ D^{-1} & -D^{-1}C \end{bmatrix} \text{ is contractive.}$$

But a matrix is contractive in the usual sense if and only if its adjoint is also contractive. Hence the J-contractiveness of $\begin{bmatrix} A & B \\ C & D \end{bmatrix}$ implies the contractiveness of

$$\begin{bmatrix} BD^{-1} & A - BD^{-1}C \\ D^{-1} & -D^{-1}C \end{bmatrix}^* = \begin{bmatrix} D^{*-1}B^* & D^{*-1} \\ A^* - C^*D^{*-1}B^* & -C^*D^{*-1} \end{bmatrix}.$$

Then also the matrix

$$\begin{bmatrix} 0 & -I \\ I & 0 \end{bmatrix} \begin{bmatrix} D^{*-1}B^* & D^{*-1} \\ A^* - C^*D^{*-1}B^* & -C^*D^{*-1} \end{bmatrix} \begin{bmatrix} 0 & I \\ -I & 0 \end{bmatrix} = \begin{bmatrix} C^*D^{*-1} & A^* - C^*D^{*-1}B^* \\ D^{*-1} & -D^{*-1}B^* \end{bmatrix}$$

is contractive. Now by the general principle (4.3) we get that

$$\begin{bmatrix} A^* & C^* \\ B^* & D^* \end{bmatrix} = \begin{bmatrix} A & B \\ C & D \end{bmatrix}^*$$

is J-contractive as asserted.

Next, from the J-contractiveness of $\begin{bmatrix} A^* & C^* \\ B^* & D^* \end{bmatrix}$ we get

$$CC^* - DD^* = [0\ I] \begin{bmatrix} A & B \\ C & D \end{bmatrix} \begin{bmatrix} I & 0 \\ 0 & -I \end{bmatrix} \begin{bmatrix} A^* & C^* \\ B^* & D^* \end{bmatrix} \begin{bmatrix} 0 \\ I \end{bmatrix} \le -I.$$

Therefore

$$D^{-1}CC^*D^{*-1} \le I - D^{-1}D^{*-1}$$

so $\|C^*D^{*-1}x\| < \|x\|$ for all $x \neq 0$. Thus $\|D^{-1}C\| < 1$ as asserted. □

PROOF OF THEOREM 2.1. We suppose Θ is a matrix function which satisfies (I)–(IV). Let G be an $M \times N$ matrix function which satisfies (i) and (ii). Since Θ satisfies (I) and (II), $\Theta_{22}(z)$ is analytic and invertible on $\Pi^+ \cup i\mathbf{R}$ (including at infinity) and

$$\sup_{z \in \Pi^+} \|\Theta_{22}(z)^{-1}\Theta_{21}(z)\| < 1$$

by Lemma 4.1. Thus $\sup_{z\in\Pi^+} \|\Theta_{22}(z)^{-1}\Theta_{21}(z)G(z)\| < 1$ and hence $\Theta_{21}G+\Theta_{22} = \Theta_{22}(\Theta_{22}^{-1}\Theta_{21}G + I)$ is analytic and invertible on Π^+. We conclude that $F = (\Theta_{11}G+\Theta_{12})(\Theta_{21}G+\Theta_{22})^{-1}$ is analytic on $\Pi^+ \cup i\mathbf{R}$ (including infinity). Once this is established the implication (2.6) is rigorous, so we have established that F satisfies (i) and (ii). To show that F satisfies (iii) is equivalent to showing that F satisfies (2.7), or equivalently that (2.8) holds for $1 \leq i \leq n$. As X and G are analytic at each z_i this is an immediate consequence of (III).

The converse is somewhat more delicate. Suppose F is a solution of (i), (ii), and (iii). Define rational matrix functions G_1 and G_2 by

$$\begin{bmatrix} F \\ I \end{bmatrix} = \Theta \begin{bmatrix} G_1 \\ G_2 \end{bmatrix}. \tag{4.4}$$

We claim G_1 and G_2 are analytic on Π^+, i.e., that $\Theta^{-1}\begin{bmatrix} F \\ I \end{bmatrix}$ is analytic on Π^+.

By (IV) the only poles of Θ^{-1} in Π^+ occur at $z_1, \dots, z_n$. If z_i is one of these poles by (III) and (IV) together with Lemma 3.1.1, $\Theta(z)^{-1}$ has a Laurent expansion near z_i of the form

$$\Theta(z)^{-1} = \sum_{j=-1}^{\infty} \Theta_j^{(i)}(z - z_i)^j$$

where

$$\operatorname{Im}_\ell \Theta_{-1}^{(i)} = \operatorname{span}\{[x_i, -y_i]\}.$$

Here the *left image* of $\Theta_{-1}^{(i)}$ is defined by

$$\operatorname{Im}_\ell \Theta_{-1}^{(i)} = \{u\Theta_{-1}^{(i)} : u \in \mathbf{C}^{1\times(M+N)}\}.$$

From the interpolation condition (iii) we conclude that $\Theta^{-1}\begin{bmatrix} F \\ I \end{bmatrix}$ is analytic at z_i, for each $i = 1, 2, \dots, n$. Thus $\begin{bmatrix} G_1 \\ G_2 \end{bmatrix} = \Theta^{-1}\begin{bmatrix} F \\ I \end{bmatrix}$ is analytic on Π^+ as asserted.

We next claim that $\det G_2$ does not vanish on Π^+. Since $\Theta(z)$ is J-unitary for $z \in i\mathbf{R}$, we have

$$\begin{aligned} 0 > F(z)^*F(z) - I &= [F(z)^* \; I]J\begin{bmatrix} F(z) \\ I \end{bmatrix} \\ &= [G_1(z)^* \; G_2(z)^*]\Theta(z)^*J\Theta(z)\begin{bmatrix} G_1(z) \\ G_2(z) \end{bmatrix} \\ &= [G_1(z)^* \; G_2(z)^*]J\begin{bmatrix} G_1(z) \\ G_2(z) \end{bmatrix} \\ &= G_1(z)^* G_1(z) - G_2(z)^*G_2(z) \end{aligned} \tag{4.5}$$

for $z \in i\mathbf{R}$. Thus if $G_2(z)x = 0$ and $x \neq 0$ for some $z \in i\mathbf{R}$ we have

$$0 > \|G_1(z)x\|^2 \geq 0$$

a contradiction. Thus $\det G_2(z) \neq 0$ for $z \in i\mathbf{R}$, and the above argument implies

$$\|G_1(z)G_2(z)^{-1}\| < 1$$

for $z \in i\mathbf{R}$. As by Lemma 2.2 $\|\Theta_{22}(z)^{-1}\Theta_{21}(z)\| < 1$ for $z \in i\mathbf{R}$, we obtain that

$$I + t\Theta_{22}(z)^{-1}\Theta_{21}(z)G_1(z)G_2(z)^{-1}$$

is invertible for $z \in i\mathbf{R}$ and $0 \leq t \leq 1$. So by a standard homotopy argument

$$\text{wno}\,\det(I + \Theta_{22}(z)^{-1}\Theta_{21}(z)G_1(z)G_2(z)^{-1}) = \text{wno}\,I = 0,$$

where wno $f(z)$ is the winding number or net change in the argument of $f(z)$ along $i\mathbf{R}$. From (4.4) we have $I = \Theta_{21}G_1 + \Theta_{22}G_2$, so

$$\begin{aligned}(4.6)\qquad 0 &= \text{wno}\,\det I = \text{wno}\,\det(\Theta_{21}G_1 + \Theta_{22}G_2)\\ &= \text{wno}\,\det(\Theta_{22}[\Theta_{22}^{-1}\Theta_{21}G_1G_2^{-1} + I]G_2)\\ &= \text{wno}\,\det\Theta_{22} + \text{wno}\,\det(\Theta_{22}^{-1}\Theta_{21}G_1G_2^{-1} + I) + \text{wno}\,\det G_2\\ &= \text{wno}\,\det G_2.\end{aligned}$$

Here we used the fact established in the first part of the proof that Θ_{22} is analytic and invertible on Π^+, so wno $\det\Theta_{22} = 0$ by the Argument Principle. As we already established that G_2 is analytic on Π^+, again by the Argument Principle (4.6) implies that $\det G_2$ has no zeros in Π^+ as claimed.

We now define $G = G_1G_2^{-1}$. As G_1 and G_2^{-1} are each analytic on Π^+ we see that G satisfies (i). By (4.5), $\sup_{z\in i\mathbf{R}}\|G(z)\| < 1$. Then by the maximum modulus theorem G satisfies (ii). Moreover, from (4.4) we see that

$$F = \Theta_{11}G_1 + \Theta_{12}G_2 = (\Theta_{11}G + \Theta_{12})G_2$$

and

$$I = \Theta_{21}G_1 + \Theta_{22}G_2 = (\Theta_{21}G + \Theta_{22})G_2.$$

Combining these two gives that $F = (\Theta_{11}G + \Theta_{12})(\Theta_{21}G + \Theta_{22})^{-1}$. Thus any solution F of (i), (ii), (iii) has the form (2.1) with G satisfying (i) and (ii) as asserted. □

5. Tangential Lagrange interpolation problem.

In this section we pause to show how the tools developed here can be used to solve a less structured more affine version of the tangential Nevanlinna-Pick problem. Specifically we drop the norm constraint demanded before for a solution and replace the right half plane Π^+ by a general subset σ of the complex plane.

We consider here only the simplest case of the *tangential Lagrange interpolation problem*; for the extension of the ideas presented here to a much more general setting together with complete details, we refer the reader to [BGR2]. We are given a subset $\sigma \subset \mathbf{C}$, n distinct points $z_1, \ldots, z_n$ in σ, n nonzero $1 \times M$ row vectors $x_1, \ldots, x_n$, and n $1 \times N$ row vectors $y_1, \ldots, y_n$.

The problem is to describe the set of rational $M \times N$ matrix functions $F(z)$ for which

$$F(z) \text{ is analytic on } \sigma \tag{5.1}$$

and

$$x_i F(z_i) = y_i \text{ for } 1 \leq i \leq n. \tag{5.2}$$

Much of the analysis done in the preceding sections for the Nevanlinna-Pick problem is relevant and applies equally well to this tangential Lagrange interpolation problem where the norm constraint is removed. To give more flexibility, we shall seek to parametrize solutions F in the more general linear fractional form

$$F = (\Theta_{11}G_1 + \Theta_{12}G_2)(\Theta_{21}G_1 + \Theta_{22}G_2)^{-1} \tag{5.3}$$

where (G_1, G_2) is a pair of rational matrix functions (of sizes $M \times N$ and $N \times N$ respectively) sweeping some appropriate parameter set, and $\Theta(z) = \begin{bmatrix} \Theta_{11}(z) & \Theta_{12}(z) \\ \Theta_{21}(z) & \Theta_{22}(z) \end{bmatrix}$ is a rational $(M+N) \times (M+N)$ matrix function to be built from the data of the problem. The arguments presented in Sections 2-4 can easily be adapted to produce the following result.

THEOREM 5.1. *Suppose that σ is a subset of $\mathbf{C}$, $z_1, \ldots, z_n$ are distinct points in σ, $x_1, \ldots, x_n$ are nonzero vectors in $\mathbf{C}^{1\times M}$ and $y_1, \ldots, y_n$ are vectors in $\mathbf{C}^{1\times N}$. Let $\Theta(z) = [\Theta_{ij}(z)]_{1\leq i,j\leq 2}$ be any rational $(M+N)\times(M+N)$ matrix function such that*

$$\Theta(z) \text{ is analytic on } \sigma \tag{5.4}$$

$$\text{For } 1 \leq i \leq n,\ \operatorname{Ker}_\ell \Theta(z_i) = \operatorname{span}\{[x_i, -y_i]\} \tag{5.5}$$

$$\det \Theta(z) \text{ has only simple zeros in } \sigma \text{ occurring at } z_1, \ldots, z_n \tag{5.6}$$

Then the matrix function $F(z)$ satisfies (5.1) and (5.2) if and only if

$$F = (\Theta_{11}G_1 + \Theta_{12}G_2)(\Theta_{21}G_1 + \Theta_{22}G_2)^{-1}$$

for rational $M \times N$ and $N \times N$ matrix functions G_1 and G_2 such that

$$G_1 \text{ and } G_2 \text{ are analytic on } \sigma \tag{5.7}$$

and

$$\det(\Theta_{21}G_1 + \Theta_{22}G_2) \text{ has no zeros in } \sigma. \tag{5.8}$$

Once the existence of such a rational matrix function $\Theta(z)$ is established, then it is easy to see that there are many pairs of functions (G_1, G_2) which satisfy (5.7) and (5.8). Indeed from the conditions (5.4)–(5.6) and the fact that each of the vectors $x_1, \ldots, x_n$ is nonzero, it follows that

$$\operatorname{Im} \Theta_{21}(z) + \operatorname{Im} \Theta_{22}(z) = \mathbf{C}^N$$

for each z in σ, and hence a generic pair of functions (G_1, G_2) satisfying (5.7) also satisfies (5.8). Theorem 3.3.1 thus reduces the solution of the tangential Lagrange interpolation problem (5.1)–(5.2) to the construction of a matrix function $\Theta(z)$ which satisfies (5.4)–(5.6).

To construct such a function $\Theta(z)$ we apply the ideas from Section 2. Note that conditions (5.5)–(5.6) are of the same form as (3.1.5i), and thus by Lemma 3.1.1 can be expressed in the equivalent form

(5.9) *For* $i = 1, \ldots, n$, $\Theta(z)^{-1}$ *has a simple pole at* z_i

such that $\operatorname{Im}_\ell \operatorname{Res}_{z=z_i} \Theta(z)^{-1} = \operatorname{span}\{[x_i, -y_i]\}$.

Condition (5.6) requires that $\Theta(z)^{-1}$ have no poles in σ other than $z_1, \ldots, z_n$. As a matter of economy, let us try to find $\Theta(z)$ so that $z_1, \ldots, z_n$ are the only poles of $\Theta(z)^{-1}$ in all of $\mathbf{C}$. If we insist that $\Theta(\infty) = I$, then by Lemma 3.1.1 and Liouville's Theorem we must have

$$\Theta(z)^{-1} = I + \widetilde{C}(zI - A_\zeta)^{-1} B_\zeta \tag{5.10}$$

where

$$A_\zeta = \begin{bmatrix} z_1 & & 0 \\ & \ddots & \\ 0 & & z_n \end{bmatrix}, \quad B_\zeta = \begin{bmatrix} x_1 & -y_1 \\ \vdots & \vdots \\ x_n & -y_n \end{bmatrix} \tag{5.11}$$

and where

$$\widetilde{C} = [u_1, \ldots, u_n] \text{ where } u_j \neq 0 \text{ for } 1 \leq j \leq n \tag{5.12}$$

is an $(M + N) \times n$ matrix. We can use Lemma 3.1.2 to deduce from (5.10)

$$\Theta(z) = I - \widetilde{C}(zI - (A_\zeta - B_\zeta \widetilde{C}))^{-1} B_\zeta. \tag{5.13}$$

Then under the conditions specified above, any $\Theta(z)$ given by (5.11)–(5.13) satisfies (5.5)–(5.6). For the remaining condition (5.4) all that is required is

(5.14) *The eigenvalues of* $A_\zeta - B_\zeta \widetilde{C}$ *are disjoint from* σ.

The problem of choosing a $\widetilde{C}$ so that (5.14) is satisfied for a given pair (A_ζ, B_ζ) is well-known in systems theory as the "pole placement problem" (see [K]). Since in our case the vectors $\{x_i, y_i]$ are nonzero, (A_ζ, B_ζ) is a full range pair, i.e. $\operatorname{span}\{\operatorname{Im} A_\zeta^n B_\zeta \colon n \geq 0\} = \mathbf{C}^n$. Then the existence of such a $\widetilde{C}$ (and moreover one with all columns nonzero) is known, as long as σ is a proper subset of $\mathbf{C}$. We thus have obtained the following result.

THEOREM 5.2. *Suppose σ is a proper subset of $\mathbf{C}$. Let $z_1, \ldots, z_n$, $x_1, \ldots, x_n$, $y_1, \ldots, y_n$ be as in Theorem 5.1 and define $\Theta(z)$ by*

$$\Theta(z) = I - \widetilde{C}(zI - (A_\zeta - B_\zeta \widetilde{C}))^{-1} B_\zeta$$

where (A_ζ, B_ζ) is defined by (5.11) *and where $\widetilde{C}$ satisfies* (5.12) *and* (5.14). *Then $\Theta(z)$ satisfies* (5.4)–(5.6) *and thus provides a linear fractional parametrization of all solutions of the tangential Lagrange interpolation problem as in Theorem* 5.1.

We remark that once one particular solution K of the interpolation problem (5.1)–(5.2) is constructed, then it is possible to construct a Θ satisfying all the conditions (5.4)–(5.6) which has the block triangular form

$$\Theta = \begin{bmatrix} \psi & K \\ 0 & I \end{bmatrix}.$$

With this choice of Θ the linear fractional formula (5.3) assumes the simpler affine form

$$F = K + \psi G$$

where the requirement on $G = G_1 G_2^{-1}$ corresponding to F satisfying (5.1)–(5.2) is simply analyticity on σ. For complete details, we refer to [BGR2].

Note also that the assumption that σ is a proper subset of $\mathbf{C}$ precludes the guarantee of finding polynomial interpolants by this procedure. To handle the case $\sigma = \mathbf{C}$, we must build a *matrix polynomial* $\Theta(z)$ (i.e. a rational matrix function whose only pole is at infinity) which satisfies (5.5)–(5.6). This is also possible but we must use more general realization formulas.

6. Connections with systems theory and generalizations.

Here we have obtained with a complete proof the linear fractional parametrization of all solutions of the simplest tangential Nevanlinna-Pick interpolation problem in a direct manner by using only elementary ideas from linear algebra and complex analysis. However this development belies the manner in which this proof was discovered, where the general framework and construction of systems theory had a great influence on the authors. The connection with systems theory is not merely a source of inspiration; it is also essential for the development of a systematic more general machinery needed to handle matrix interpolation problems much more complicated than the simplest case presented here.

In this section we sketch the basic ideas of systems theory relevant to the approach to interpolation problems presented here; for more complete details we refer the reader to [BGR1,GLR,K].

A useful tool from systems theory for the study of rational matrix functions is the ideas of realizing a rational matrix function as the *transfer function* of a linear system. We consider only the case of a square rational matrix function with value I at infinity. The collection of matrices (A, B, C) is said to be a *realization* for the rational $p \times p$ matrix function $W(z)$ if $W(z)$ is given by

$$W(z) = I + C(zI - A)^{-1}B. \tag{6.1}$$

Here A, B, C have sizes $q \times q$, $q \times p$ and $p \times q$; in general q (the size of the *state space* on which A acts) is much larger than p. Let us consider multiplication by $W(z)$ as mapping an input rational vector function $\hat{u}(z)$ to an output rational vector function $\hat{y}(z)$

$$\hat{y}(z) = W(z)\hat{u}(z). \tag{6.2}$$

If $W(z)$ is given by (6.2) and we introduce an auxiliary variable $\hat{x}(z)$ by

$$\hat{x}(z) = (zI - A)^{-1}B\hat{u}(z)$$

then (6.2) can alternatively be expressed as

$$z\hat{x}(z) = A\hat{x}(z) + B\hat{u}(z)$$
$$\hat{y}(z) = C\hat{x}(z) + \hat{u}(z).$$

If we now apply inverse Laplace transform we get

$$\begin{aligned} \dot{x}(t) &= Ax(t) + Bu(t), \quad x(0) = 0 \\ y(t) &= Cx(t) + u(t). \end{aligned} \tag{6.3}$$

The relation (6.2) then expresses multiplication by $W(z)$ as the input-output map in the frequency domain for the linear system (6.3); for this reason the formula (6.1) is said to *realize* $W(z)$ as the *transfer function* of the linear system (6.3).

Note that for the case where A is diagonal

$$A = \begin{bmatrix} w_1 & & 0 \\ & \ddots & \\ 0 & & w_q \end{bmatrix} \tag{6.4a}$$

and C has columns $u_1, \ldots, u_q$

$$C = [u_1, \ldots, u_q] \tag{6.4b}$$

and B has rows $v_1, \ldots, v_q$

$$B = \begin{bmatrix} v_1 \\ \vdots \\ v_q \end{bmatrix}, \tag{6.4c}$$

then the formula (3.1.1) for $W(z)$ is equivalent to a matrix partial fraction expansion for $W(z)$

$$W(z) = I + \sum_{\alpha=1}^{q} u_\alpha v_\alpha (z - w_\alpha)^{-1}. \tag{6.5}$$

which played a key role in our analysis in Section 3.

A realization (6.1) is said to be *minimal* if

(i) (C, A) is a *null-kernel pair*, i.e.

$$\bigcap_{j=0}^{\infty} \operatorname{Ker} CA^j = (0)$$

and

(ii) (A, B) is a *full-range pair*, i.e.

$$\operatorname{span}\{\operatorname{Im} A^j B \colon j \geq 0\} = \mathbf{C}^q.$$

For the special case where (A, B, C) are as in (6.4a)–(6.4c), (C, A) is a null-kernel pair exactly when each column u_i of C is nonzero, and (A, B) is a null range pair exactly when each row v_j of B is nonzero.

In the system theory literature for physical reasons a null-kernel pair is called an *observable* pair and a full range pair is called a *controllable* pair (see e.g. [K]). It turns out that a realization (A, B, C) is minimal if and only if the size q of the state space is minimal among all possible realizations for the same function $W(z)$. If (A, B, C) is a realization for $W(z)$ and S is a $q \times q$ invertible matrix, then one can easily see that $(S^{-1}AS,\ S^{-1}B,\ CS)$ is also a realization; moreover (A, B, C) is minimal if and only if $(S^{-1}AS,\ S^{-1}B,\ CS)$ is minimal. The following theorem (known as the *state space isomorphism theorem*) gives that this is the extent of nonuniqueness of a minimal realization for a given function $W(z)$; for a proof the reader can check any one of a number of sources (e.g. [BGK, BGR1, GLR, K]). Lemma 3.0.1 is a crude form of this more general result.

THEOREM 6.1. *If (A_1, B_1, C_1) and (A_2, B_2, C_2) are two minimal realizations for the same rational matrix function $W(z)$, then (A_1, B_1, C_1) and (A_2, B_2, C_2) are similar, i.e. there exists an invertible matrix S for which*

$$\begin{aligned} A_2 &= S^{-1}A_1S \\ B_2 &= S^{-1}B_1 \\ C_2 &= C_1S. \end{aligned}$$

If $W(z)$ is a rational matrix function with value I at infinity, then $W(z)^{-1}$ exists as a rational matrix function. The next result describes how to compute a realization for $W(z)^{-1}$ in terms of a realization for $W(z)$, and can be seen as a generalization of Lemma 3.1.2. The formula can be proved either by direct check or can be derived by inverting the system (i.e. interchanging the roles of $u(t)$ and $y(t)$) in (6.3); for complete details, we refer to [BGK,BGR1,GLR].

THEOREM 6.2. *If (A, B, C) is a realization for $W(z)$ (as in (6.1)), then $(A^\times, B, -C)$, where*

$$A^\times = A - BC,$$

is a realization for $W(z)^{-1}$. Moreover, (A, B, C) is minimal if and only if $(A^\times, B, -C)$ is minimal.

Finally, we remark that the zero-pole interpolation problem (3.1.6i)–(3.1.6iii) can be formulated in the following way:

Given the null-kernel pair (C, A_π) and the full range pair (A_ζ, B) find a rational matrix function $W(z)$ so that $W(z)$ has a minimal realization of the form (6.6)

$$W(z) = I + C(zI - A_\pi)^{-1}\widetilde{B}$$

for some $\widetilde{B}$, and $W(z)^{-1}$ has a minimal realization of the form

$$W(z)^{-1} = I - \widetilde{C}(zI - A_\zeta)^{-1}B$$

for some $\widetilde{C}$.

For the problem (3.1.6i)–(3.1.6iii), the given matrices C, A_π, A_ζ, B have the special form

$$C = [\psi_1, \dots, \psi_n],\ A_\pi = \begin{bmatrix} w_1 & & 0 \\ & \ddots & \\ 0 & & w_n \end{bmatrix} \quad \textit{with } \psi_i \neq 0$$

$$A_\zeta = \begin{bmatrix} z_1 & & 0 \\ & \ddots & \\ 0 & & z_n \end{bmatrix},\ B = \begin{bmatrix} \varphi_1 \\ \vdots \\ \varphi_n \end{bmatrix} \quad \textit{with } \varphi_j \neq 0.$$

For more general null-kernel pairs (C, A_π) and full-range pairs (A_ζ, B) the problem (6.6) can still be formulated, and indeed still can be interpreted as a zero-pole interpolation problem. The following result is an exact generalization of Theorem 3.1.3. The proof follows exactly as the proof of Theorem 3.1.3, where one now uses the state space isomorphism theorem (Theorem 6.1) and the realization formula for the inverse (Theorem 6.2); for details we refer the reader to [GKLR] or [BGR1].

THEOREM 6.3. *Let the null kernel pair (C, A_π) and the full range pair (A_ζ, B) be given. Then there exists a rational matrix function $W(z)$ which is a solution of the problem* (6.6) *if and only if there exists an invertible solution S of the Sylvester equation*

$$SA_\pi - A_\zeta S = BC \tag{6.7}$$

In this case $W(z)$ may be taken to be

$$W(z) = I + C(zI - A_\pi)^{-1}S^{-1}B$$

with inverse given by

$$W(z)^{-1} = I - CS^{-1}(zI - A_\zeta)^{-1}B.$$

This theorem can also be adapted to handle the case where one wants a solution $\Theta(z)$ with J-unitary values on the imaginary line and J-contractive values on the right half plane (i.e., generalizations of Theorems 3.2.1 and 3.2.2). This leads to solutions of more complicated interpolation problems; a complete systematic development will appear in the forthcoming monograph [BGR4].

BIBLIOGRAPHY

[AG] D. Alpay and I. Gohberg, *Unitary rational matrix functions*, Topics in Interpolation Theory of Rational Matrix Functions (ed. I. Gohberg), OT33 Birkhauser, Basel, 1988, pp. 175–222.

[BGR1] J. A. Ball, I. Gohberg, and L. Rodman, *Realization and interpolation of rational matrix functions*, Topics in Interpolation Theory of Rational Matrix Functions (ed. I. Gohberg), OT33 Birkhauser, Basel, 1988, pp. 1–72.

[BGR2] ——, *Two sided Lagrange-Sylvester interpolation problems for rational matrix functions*, submitted to Proceedings of AMS Summer Institute on Operator Theory/Operator Algebras and Applications, Durham, N.H.

[BGR3] ———, *Two-sided Nudelman interpolation problem for rational matrix functions*, (to appear in Marcel-Dekker volume dedicated to M. Cotlar).

[BGR4] ———, *Interpolation for Rational Matrix Functions*, monograph in preparation.

[BH] J. A. Ball and J. W. Helton, *A Beurling-Lax theorem for the Lie group* $U(m, n)$ *which contains most classical interpolation theory*, J. Operator Theory **9** (1983), 107–142.

[BR] J. A. Ball and A. C. M. Ran, *Local inverse spectral problems for rational matrix functions*, Integral Equations and Operator Theory **10** (1987), 349–415.

[BGK] H. Bart, I. Gohberg, and M. A. Kaashoek, *Minimal Factorization of Matrix and Operator Functions*, OT1 Birkhauser, Basel, 1979.

[F] B. Francis, *A Course in* H_∞ *Control Theory*, Lecture Notes in Control and Information Sciences No. 88, Springer-Verlag, Berlin, 1987.

[GKLR] I. Gohberg, M. A. Kaashoek, L. Lerer, and L. Rodman, *Minimal divisors of rational matrix functions with prescribed zero and pole structure*, in Topics in Operator Theory, Systems and Networks (ed. H. Dym, I. Gohberg), OT12 Birkhauser, Basel, 1984, pp. 241–275.

[GLR] I. Gohberg, P. Lancaster and L. Rodman, *Invariant Subspaces of Matrices and Applications*, Wiley, New York, 1986.

[H1] J. W. Helton, *The distance of a function to* H^∞ *in the Poincare metric, electrical power transfer*, J. Functional Anal. **38** (1980), 273–314.

[H2] J. W. Helton, *Operator Theory, Analytic Functions, Matrices, and Electrical Engineering*, CBMS No. 68, American Math. Soc., Providence, 1987.

[K] T. Kailath, *Linear Systems*, Prentice Hall, Englewood Cliffs, 1980.

[NF] B. Sz.-Nagy and C. Foias, *Harmonic Analysis of Operators on Hilbert Space*, Amer. Elsevier, New York, 1970.

[N] R. Nevanlinna, *Ueber beschraenkte analytische Funktionen*, Ann. Acad. Sci. Fenn. **32** (1929), No. 7.

[S] D. Sarason, *Generalized interpolation in* H^∞, Trans. Amer. Math. Soc. **127** (1967), 179–203.

Department of Mathematics
Virginia Tech
Blacksburg, VA 24061

School of Mathematical Sciences
The Raymond and Beverly Sackler Faculty of Exact Sciences
Tel-Aviv University
Tel-Aviv, Ramat-Aviv 69978
ISRAEL

and

Department of Mathematics
College of William and Mary
Williamsburg, VA 23185

Proceedings of Symposia in Applied Mathematics
Volume **40**, 1990

THE HADAMARD PRODUCT

ROGER A. HORN

TABLE OF CONTENTS

1980 *Mathematics Subject Classification.* Primary 15A99; Secondary 15-02, 15A42, 15A45, 15A69.

Key words and phrases. Hadamard product, Schur product, positive definite matrix, singular values.

1. Introduction, Definitions, and Basic Facts. The *Hadamard product* of two matrices $A = [a_{ij}]$ and $B = [b_{ij}]$ with the same dimensions (not necessarily square) with entries in a given ring is the entry-wise product $A \circ B \equiv [a_{ij}b_{ij}]$, which has the same dimensions as A and B. For example, if

$$A = \begin{bmatrix} 1 & 2 & 3 \\ 4 & 5 & 6 \end{bmatrix} \text{ and } B = \begin{bmatrix} \pi & e & 0 \\ i & 0 & 1 \end{bmatrix}$$

then

$$A \circ B = \begin{bmatrix} \pi & 2e & 0 \\ 4i & 0 & 6 \end{bmatrix}.$$

This seemingly naive product arises naturally in analysis and in many diverse areas of mathematics, but it is rarely even mentioned in linear algebra texts, perhaps because it is explicitly basis-dependent. Though some might reject it as "unnatural" for this reason alone, that would be unfortunate, as there are nice theorems to be found! The beautiful Perron-Frobenius theory of matrices with positive or nonnegative entries is, of course, another example of a basis-dependent topic that has a rich and useful development. Like the Perron-Frobenius theory, the Hadamard product should perhaps be thought of not as a topic from *linear algebra*, but rather one from *matrix analysis*.

If A and B are matrices with the same dimensions but not square, then the ordinary matrix product AB is not defined, but $A \circ B$ is defined; in the square case both products are defined. Like the ordinary product, the Hadamard product is associative and distributes over matrix addition. The identity element with respect to Hadamard multiplication is the matrix with all entries equal to 1, which we denote by J; the Hadamard-invertible matrices are those with no zero entries. If the underlying ring is commutative, then the Hadamard product is commutative, an important difference between it and the ordinary matrix product.

For the rest of our discussion, we shall consider only matrices with complex entries; we denote the set of m-by-n matrices with complex entries by $M_{m,n}$ and we set $M_n \equiv M_{n,n}$. Thus, M_n is an algebra in two ways, both with ordinary matrix addition, but one with the ordinary matrix product and one with the Hadamard product. Many interesting questions about the Hadamard product arise from the interplay between these two algebras; sometimes there are analogous results, but sometimes there are striking differences.

We denote the *transpose* of a matrix $A = [a_{ij}]$ by $A^T \equiv [a_{ji}]$; its *adjoint* is the conjugate transpose $A^* \equiv \overline{A}^T$; if $A \in M_n$, its *trace* is $\text{tr}(A) = a_{11} + \cdots + a_{nn}$, the sum of its main diagonal entries; for all $A, B \in M_n$, $\text{tr}(AB) = \text{tr}(BA)$. A nonzero vector $x \in \mathbb{C}^n$ is an *eigenvector* of $A \in M_n$ if there is a scalar λ such that $Ax = \lambda x$; λ is an *eigenvalue* of A corresponding to the eigenvector x. If λ is a given eigenvalue of $A \in M_n$, then $\{x \in \mathbb{C}^n\colon\ Ax = \lambda x\}$ is a subspace of $\mathbb{C}^n$, called the *eigenspace of A corresponding to the eigenvalue* λ. The *spectral radius* of a given matrix $A \in M_n$ is $\rho(A) \equiv \max\{|\lambda|: \lambda \text{ is an eigenvalue of } A\}$. The ordinary *Euclidean norm* of a vector $x \in \mathbb{C}^n$ is $\|x\|_2 = +(x^*x)^{\frac{1}{2}}$. For a given vector $x \in \mathbb{C}^n$, the matrix $\text{diag}(x) \in M_n$ is a diagonal matrix whose main diagonal entries are the respective entries of x; for a given $A = [a_{ij}] \in M_n$, diag(A) is the vector of main diagonal entries $[a_{11}, \ldots a_{nn}]^T \in \mathbb{C}^n$. The usual n-by-n identity matrix is denoted by $I \equiv \text{diag}(e)$ where $e \equiv [1 \ldots 1]^T \in M_n$. A matrix $U \in M_n$ is *unitary* if and only if $U^* U = I$; U is unitary if and only if $\| Ux \|_2 = \|x\|_2$ for all $x \in \mathbb{C}^n$.

A matrix $A \in M_n$ is *Hermitian* if and only if $A = A^*$. A Hermitian matrix A is *positive definite* if $x^* Ax > 0$ for all nonzero $x \in \mathbb{C}^n$; A is *positive semidefinite* if $x^* Ax \geq 0$ for all $x \in \mathbb{C}^n$; A is *negative definite* or *semidefinite*, respectively, if $-A$ is positive definite or semidefinite, respectively. A Hermitian matrix is positive semidefinite (respectively, positive definite) if and only if all its eigenvalues are nonnegative (respectively, positive). A positive semidefinite matrix is positive definite if and only if it is nonsingular. Since the determinant of a square matrix is the product of its eigenvalues and its trace is the sum of its eigenvalues, a positive semidefinite matrix has a nonnegative determinant that is positive if and only if the matrix is positive definite; a positive semidefinite matrix has a nonnegative trace that is zero if and only if the matrix is zero. If $A = C^* C$ for some $C \in M_{m,n}$, then A is positive semidefinite since $x^* Ax = x^* C^* Cx = (Cx)^*(Cx) \geq 0$. If $A \in M_n$ is Hermitian and $C \in M_n$ is nonsingular, then the *congruence* $C^* A\, C$ is a Hermitian matrix with the same number of positive, negative, and zero eigenvalues as A; this is *Sylvester's Inertia Theorem.*

If $A \in M_n$ is Hermitian, then there is a unitary U and a real diagonal matrix $\Lambda = \text{diag}([\lambda_1, \ldots, \lambda_n]^T)$ such that $A = U \Lambda U^*$; this factorization is known as as a *spectral decomposition.* If $A \in M_n$ is positive semidefinite and $A = U \Lambda U^*$ is a spectral decomposition, then $A^{\frac{1}{2}} \equiv U \Lambda^{\frac{1}{2}} U^*$ is the unique positive semidefinite matrix such that $(A^{\frac{1}{2}})^2 = A$; of course, $\Lambda^{\frac{1}{2}} \equiv \text{diag}(+\sqrt{\lambda_1}, \ldots, +\sqrt{\lambda_n})$; similarly, $A^{\alpha} \equiv U \Lambda^{\alpha} U^*$ for any $\alpha > 0$. Although the square root is the most frequently-used positive non-integer power of a positive semidefinite matrix, arbitrary positive powers are defined in the same way: $A^p \equiv U \Lambda^p U^*$. If A is positive semidefinite and $A^{\frac{1}{2}} = [a_1 \ \ldots \ a_n]$ is partitioned according to its columns, then

$$A = A^{\frac{1}{2}} A^{\frac{1}{2}} = A^{\frac{1}{2}*} A^{\frac{1}{2}} = [a_1 \ \ldots \ a_n]^* \, [a_1 \ \ldots \ a_n] = [a_i^* a_j],$$

so a positive semidefinite matrix is a *Gram matrix* of inner products of a set of vectors $\{a_i\}$ (and conversely); A is positive definite if and only the set of vectors $\{a_i\}$ is *independent*; A is positive semidefinite with positive main diagonal entries if and only if $\{a_i\}$ does not contain the zero vector; A is a *correlation matrix* if and only if $\{a_i\}$ is a set of *unit vectors.*

If A is positive semidefinite, $x \neq 0$, and $x^* Ax = 0$, then x is an eigenvector of A and $Ax = 0$. If $A \in M_n$ has only real eigenvalues (in particular, if A is Hermitian), we denote the algebraically smallest and largest eigenvalues by $\lambda_{min}(A)$ and $\lambda_{max}(A)$, respectively. If A is positive definite, then $\lambda_{max}(A) = 1/\lambda_{min}(A^{-1})$. If $A \in M_n$ is Hermitian, then $A + |\lambda_{min}(A)| I$ is positive semidefinite; if A is positive semidefinite, then $A - \lambda_{min}(A)I$ and $\lambda_{max}(A)I - A$ are both positive semidefinite. *The Rayleigh–Ritz Theorem* gives a variational characterization for the smallest and largest eigenvalues of a Hermitian matrix A: $\lambda_{min}(A) = \min \{x^* Ax\colon \ x \in \mathbb{C}^n$ and $\|x\|_2 = 1\}$ and $\lambda_{max}(A) = \max \{x^* Ax\colon x \in \mathbb{C}^n$ and $\|x\|_2 = 1\}$. The variational characterization in the Rayleigh-Ritz Theorem is of great importance in matrix analysis for at least two reasons:

1. It converts a highly *nonlinear* operation—computing an eigenvalue, which is a root of a complicated polynomial function of the entries of the matrix—into a *quasi-linear* operation—finding an extremum with respect to x of x^*Ax, which is a *linear* function of A. For example, if $A, B \in M_n$ are Hermitian, if $x \in \mathbb{C}^n$ is an arbitrary unit vector, and if $t \in [0,1]$ is given, then $x^*[tA + (1-t)B]x = t(x^*Ax) + (1-t)x^*Bx$, and hence

$$\min\{\lambda_{min}(A), \lambda_{min}(B)\} \le t\lambda_{min}(A) + (1-t)\lambda_{min}(B) \le t(x^*Ax) + (1-t)(x^*Bx)$$

$$\le t\lambda_{max}(A) + (1-t)\lambda_{max}(B) \le \max\{\lambda_{max}(A), \lambda_{max}(B)\}$$

The Rayleigh-Ritz Theorem then gives the useful bounds

$$\alpha(A, B) \equiv \min\{\lambda_{min}(A), \lambda_{min}(B)\} \le \lambda_{min}[tA + (1-t)B]$$

$$\le \lambda_{max}[tA + (1-t)B] \le \max\{\lambda_{max}(A), \lambda_{max}(B)\} \equiv \beta(A, B), \text{ for all } t \in [0,1] \quad (1.1)$$

for the eigenvalues of a convex combination of Hermitian matrices. As another example of how the quasi-linearization in the Rayleigh-Ritz Theorem can be useful, suppose $A, B \in M_n$ with A Hermitian and B positive semidefinite. Then $x^*(A \pm B)x = x^*Ax \pm x^*Bx$ and $x^*Bx \ge 0$ for every unit vector $x \in \mathbb{C}^n$. Thus, $\lambda_{min}(A + B) = \min\{x^*(A+B)x\colon x \in \mathbb{C}^n \text{ and } \|x\|_2 = 1\} \ge \min\{x^*Ax\colon x \in \mathbb{C}^n \text{ and } \|x\|_2 = 1\} = \lambda_{min}(A)$. The same argument shows that $\lambda_{max}(A + B) \ge \lambda_{max}(A)$; these observations are known as *the monotonicity theorem.*

2. It also gives a powerful tool for dealing with properties of submatrices and leads to useful bounds and interlacing inequalities. A matrix $\hat{A} \in M_k$ is a *principal submatrix* of $A = [a_{ij}] \in M_n$ if there are indices $1 \le m_1 < \cdots < m_k \le n$ for which $\hat{A} = [a_{m_i, m_j}]$, $i, j = 1, \ldots, k$. If A is Hermitian, then $\lambda_{max}(\hat{A}) = \max\{x^*Ax : x \in \mathbb{C}^n, \|x\|_2 = 1, \text{ and } x_i = 0 \text{ if } i \ne m_1, \ldots, m_k\} \le \max\{x^*Ax\colon x \in \mathbb{C}^n \text{ and } \|x\|_2 = 1\} = \lambda_{max}(A)$ because the second maximization is over a set of vectors that includes all of those that were used in the first maximization. Using a similar argument for the minimum eigenvalue, we have the important bounds

$$\lambda_{min}(A) \le \lambda_{min}(\hat{A}) \le \lambda_{max}(\hat{A}) \le \lambda_{max}(A) \quad (1.2)$$

for the eigenvalues of any principal submatrix $\hat{A}$ of a Hermitian matrix A. Taking the principal submatrix to be 1-by-1 shows that all the main diagonal entries of a Hermitian matrix lie in the interval $[\lambda_{min}(A), \lambda_{max}(A)]$. There is a generalization of (1.2) to eigenvalues of A and $\hat{A}$ other than just the smallest and largest: The *interlacing eigenvalues theorem* says that

$$\lambda_{n-k+i}(A) \le \lambda_i(\hat{A}) \le \lambda_i(A) \qquad \text{for } \hat{A} \in M_k \text{ a principal submatrix of a Hermitian } A \in M_n \text{ and } i = 1,\dots, k, \tag{1.3}$$

where the eigenvalues $\lambda_1 \ge \lambda_2 \ge \cdots$ are arranged in algebraically decreasing order. These bounds give (1.2) for $i = 1$ and $i = k$. In particular, if $\hat{A}$ has m positive (respectively, negative) eigenvalues, then A must have at least m positive (respectively, negative) eigenvalues.

The lower bound in (1.2) makes it clear that if $A = [a_{ij}]$ is positive semidefinite, then any principal submatrix of A is positive semidefinite, and that if A is positive definite, then any principal submatrix is positive definite; taking the principal submatrix to be 2-by-2 gives the useful bound

$$\det \begin{bmatrix} a_{ii} & a_{ij} \\ a_{ji} & a_{jj} \end{bmatrix} = a_{ii}a_{jj} - |a_{ij}|^2 \ge 0, \text{ or } a_{ii}a_{jj} \ge |a_{ij}|^2 \tag{1.4}$$

for the entries of a positive semidefinite matrix. In particular, if $a_{ii} = 0$, then *every* entry in row i and column i is zero and the matrix is singular; this useful fact is also evident from the representation of A as a Gram matrix $A = [a_i^* a_j]$, since $a_{ii} = 0$ if and only if the vector $a_i = 0$.

An analog of the spectral theorem that is valid for all $A \in M_{m,n}$ is the *singular value decomposition* $A = V\Sigma W^*$, where $V \in M_m$ and $U \in M_n$ are unitary, $\Sigma = [\sigma_{ij}] \in M_{m,n}$ with $\sigma_{ij} = 0$ for all $i \ne j$, and $\sigma_1 \equiv \sigma_{11} \ge \cdots \ge \sigma_q \equiv \sigma_{qq} \ge 0$, $q \equiv \min\{m,n\}$; $\sigma_1,\dots, \sigma_q$ are the *singular values* of A. The columns of V and W are the left and right *singular vectors* of A, respectively. The squared singular values of A are the eigenvalues of AA^* or, equivalently, of A^*A. The number of positive singular values of a matrix is equal to its *rank*. The largest singular value of $A \in M_{m,n}$ is equal to the *spectral norm* of A: $\sigma_1 = |||A|||_2 \equiv \max\{\|Ax\|_2\colon x \in \mathbb{C}^n \text{ and } \|x\|_2 = 1\}$. If $A \in M_n$ is Hermitian, then $|||A|||_2 = \max\{|\lambda_{min}(A)|, |\lambda_{max}(A)|\}$; if A is positive semidefinite, then $|||A|||_2 = \lambda_{max}(A)$. It follows immediately from the definition that $|||AB|||_2 \le |||A|||_2\,|||B|||_2$ for all $A, B \in M_n$, that is, the spectral norm is *submultiplicative* with respect to the ordinary matrix product. Another variational characterization of the spectral norm is $|||A|||_2 \equiv \max\{|x^*Ay|\colon \|x\|_2 = \|y\|_2 = 1\}$, and if we use it in an argument similar to the one leading to (1.2), we find that

$$|||\hat{A}|||_2 \le |||A|||_2 \qquad \text{for } \hat{A} \in M_{r,s} \text{ a submatrix of } A \in M_{m,n}; \tag{1.5}$$

the submatrix $\hat{A}$ need not be a principal submatrix of A. If $U \in M_n$ is unitary, then $|||U|||_2 = 1$.

Another useful factorization, available for all $A \in M_n$, is the *Schur unitary triangularization* $A = U\Delta U^*$, where $U, \Delta \in M_n$, U is unitary, Δ is upper triangular, and, necessarily, the main diagonal entries of Δ are exactly the eigenvalues of A. The upper triangular factor Δ in the Schur unitary triangularization of A is diagonal if and only if A is *normal*, that is, A commutes with its adjoint, $AA^* = A^*A$.

A function $N(\cdot)\colon M_{m,n} \to \mathbb{R}_+$ is a *norm* if for every $c \in \mathbb{C}$ and every $A, B \in M_{m,n}$ we have

$N(A) \geq 0$ with $N(A) = 0$ if and only if $A = 0$, $N(cA) = |c| N(A)$, and $N(A + B) \leq N(A) + N(B)$. A norm $N(\cdot)$ on $M_{m,n}$ is *unitarily invariant* if $N(UAV) = N(A)$ for every $A \in M_{m,n}$ and all unitary $U \in M_m$, $V \in M_n$; the spectral norm is unitarily invariant, as is the *Frobenius norm* (or *Euclidean norm*) defined by $\|A\|_2 \equiv [\mathrm{tr}(A^*A)]^{\frac{1}{2}}$. The spectral norm and the Frobenius norm are easily seen to be *self adjoint* norms: $N(A^*) = N(A)$ for all A; in fact, *every* unitary invariant norm is self-adjoint.

See [Horn & Johnson, 1987a] and [Horn & Johnson, 1989] for more information about these and other basic notions from matrix analysis.

2. Some Historical Remarks. Two areas of current research on Hadamard products that have a classical tradition are questions about *quadratic forms* and various *norm inequalities*, many of which arise by analogy with known inequalities for the ordinary matrix product. Seminal results of both types are in [Schur, 1911], which seems to be the first systematic study of algebraic and analytic properties of the entry-wise product. In this memorable paper, Schur proves that if A and B are positive semidefinite matrices of the same size, then $A \circ B$ is also positive semidefinite—a *product theorem* that now usually bears Schur's name. Schur comments that his product theorem is a result that "despite its simplicity seems not to be known". Schur also proves the submultiplicativity inequality $|||A \circ B|||_2 \leq |||A|||_2 \, |||B|||_2$ for the spectral norm and gives some improvements in this bound when both matrices are positive semidefinite.

Because of the originality and clarity of Schur's 1911 paper, there is ample justification to call the entry-wise matrix product the *Schur product*, and some writers do so, e.g., [Bellman, 1970, p. 30], [Davis, 1962], [Bapat & Sunder, 1985], and [Ong, 1984]. The name *Hadamard product* seems to be more common, however, perhaps because it was used—exactly once and apparently for the first time in print—in P. Halmos' influential text *Finite Dimensional Vector Spaces* [Halmos, 1948, p. 144; 1958, p. 174]. Some authors, e.g., [Marcus & Thompson, 1963], cite Halmos as their authority for use of the term, while others have found Halmos' use of the term puzzling [Styan, 1973, p. 217]. In [Bennett, 1977] we read that "the term 'Hadamard product' [was] coined by von Neumann and introduced in the literature by Halmos." Professor Halmos, when asked why he associated Hadamard with a matrix product that, prior to his 1948 book, seems to have borne no name in print, attributed the appellation to von Neumann in his lectures at the Institute for Advanced Study in the early 1940's [private communication, December 5, 1988].

Von Neumann's usage may have been the result of the long-lasting influence of a famous 1899 paper in which Hadamard studied two Maclaurin series $f(z) = \Sigma a_n z^n$ and $g(z) = \Sigma b_n z^n$ with positive radii of convergence and their composition $h(z) = \Sigma a_n b_n z^n$, obtained as a coefficient-wise product [Hadamard, 1899]. He showed that the function $h(\cdot)$ can be obtained from $f(\cdot)$ and $g(\cdot)$ by an integral convolution, and he proved that any singularity z_1 of $h(\cdot)$ must be of the form $z_1 = z_2 z_3$, where z_2 and z_3 are singularities of $f(\cdot)$ and $g(\cdot)$, respectively. Even though Hadamard never mentioned entry-wise products of matrices in this paper, the importance of his product theorem for power series and his mathematical eminence (he had proved the Prime Number Theorem in 1896) gave analysts a reason to associate Hadamard's name with term-by-term products of all kinds. For three examples, see: [Loewner & Netanyahu, 1959] for theorem on products of power series of Hadamard's type; [Ruscheweyh & Sheil-Small, 1974] for the solution of a long-outstanding conjecture of Pólya and Schoenberg on the 'Hadamard product' of convex conformal maps of the unit disc; and [Shields & Wallen, 1971], where the authors define "the Hadamard product of the matrices A and B, by analogy with the Hadamard product of two power series (the name seems to have been suggested by von Neumann)."

With studied impartiality, some authors refer to the entry-wise product as the *Schur-Hadamard product* [Marshall & Olkin, 1979, pp. 258] or the *Hadamard-Schur (entrywise) product* [Friedland, 1989]. Others call it the *pointwise product* [Biggs, 1976], *pointwise multiplication* [Varopoulos, 1974], or *element-product* [Fiedler, 1957] (but *Hadamard product* is used in [Fiedler, 1961] and subsequent papers). In [Stout, 1981b], however, the author eschews ambivalence and takes a stand in his first paragraph that the entry-wise product "will herein be called Schur multiplication, as opposed to the more common, but falsely crediting, 'Hadamard multiplication'." Finally, there are authors who have had a public change of heart on this issue: A note added in proof at the end of [Bennett, 1976] refers to "a forthcoming paper, *Hadamard multipliers*", but when the paper appeared as [Bennett, 1977], it was entitled *Schur Multipliers.*

As in so many other instances in the history of science, both Hadamard and Schur were anticipated by even earlier work, which we describe briefly here as the first of several examples of how the Hadamard product arises in various applications.

2.1 Example. A Uniqueness Theorem for Elliptic Partial Differential Equations. In 1894, Th. Moutard considered two real n-by-n symmetric semidefinite matrices $A = [a_{ij}]$ and $B = [b_{ij}]$ and the sum of the entries of their entry-wise product $\Theta \equiv \Sigma a_{ij} b_{ij}$. He knew that if A is positive definite then there is some real n-by-n $C = [c_{ij}]$ such that $A = C^T C$, and he showed that the sum Θ could be represented as

$$\Theta = \sum_{i,\,j=1}^{n} a_{ij} b_{ij} = \sum_{i,\,j=1}^{n} \sum_{k=1}^{n} c_{ki} c_{kj} b_{ij} = \sum_{k=1}^{n} \left[\sum_{i,\,j=1}^{n} c_{ki} b_{ij} c_{kj} \right].$$

Moutard concluded that if A is positive definite and B is positive semidefinite, then Θ cannot be negative, and he used his result to obtain a uniqueness theorem for boundary value problems for real linear elliptic partial differential equations [Moutard, 1894, Note II, pp.57–60].

A linear partial differential operator of the form

$$Lu \equiv \sum_{i,\,j=1}^{n} a_{ij}(x) \frac{\partial^2 u}{\partial x_i \partial x_j} + \sum_{i=1}^{n} b_i(x) \frac{\partial u}{\partial x_i} - c(x)u, \quad x \in \mathcal{D} \subset \mathbb{R}^n, \quad u \in C^2(\mathcal{D}) \tag{2.1}$$

is said to be *elliptic* in a domain $\mathcal{D}$ if the matrix $[a_{ij}(x)]$ is positive definite at every point $x \in \mathcal{D}$. From now on, assume that L is elliptic and that the coefficient function $c(x)$ is positive in the interior of $\mathcal{D}$. Suppose there is a real-valued twice-continuously-differentiable solution $u(x)$ to $Lu = 0$ in $\mathcal{D}$. If u has a local minimum at an interior point $\xi \in \mathcal{D}$, then all its first derivatives vanish at ξ and its Hessian matrix $[\partial^2 u/\partial x_i \partial x_j]$ is positive semidefinite there, so the first-order sum in (2.1) vanishes and the second-order sum—necessarily nonnegative—is of the form considered by Moutard. Since

$$\sum_{i,j=1}^{n} a_{ij}(\xi) \left.\frac{\partial^2 u}{\partial x_i \partial x_j}\right|_{x=\xi} = c(\xi)u(\xi),$$

the left-hand side is nonnegative by Moutard's result, and $c(\xi) > 0$ by assumption, it follows that *u cannot be negative at a local minimum in the interior of* $\mathcal{D}$. But the Hessian of u is negative semidefinite at an interior local maximum, so it follows from the same argument that *u cannot be positive at a local interior maximum.* Thus, subject to suitable continuity and smoothness assumptions, it follows that if $u = 0$ everywhere on the boundary of $\mathcal{D}$, then $u \equiv 0$ in $\mathcal{D}$. This gives a fundamental uniqueness theorem for real linear elliptic partial differential equations: If u_1 and u_2 are two solutions of $Lu = f$ in $\mathcal{D}$ that both satisfy the same boundary condition $u_1 = g = u_2$ on the boundary of $\mathcal{D}$, for given functions f and g, then their difference $u \equiv u_1 - u_2$ satisfies $Lu = 0$ in $\mathcal{D}$ and vanishes on the boundary, and hence must be identically zero. A familiar classical example of such an equation is $\nabla^2 u - \lambda u = f$, where λ is a given positive scalar. Notice that the key idea in this approach to an important problem from analysis is a simple algebraic fact about Hadamard products of real symmetric positive semidefinite matrices.

The assumption that the coefficient $c(x)$ in (2.1) be *positive* in the interior of $\mathcal{D}$ was essential for the algebra in the preceding argument, but a careful analytic argument allows one to derive the same conclusion from the weaker assumption that $c(x)$ is merely *nonnegative* in the interior of $\mathcal{D}$, a result often referred to as the (first) *Hopf lemma.* In particular, Moutard's approach leads to a uniqueness theorem for the boundary value problem for Laplace's equation $\nabla^2 u = 0$.

Not to give Moutard too much credit for his insight, we should note that he (unlike Schur) did not consider the algebraic quadratic form $x^*(A \circ B)x$ but looked only at the particular sum that is its value for $x = e$. Hadamard, without any reference to Moutard, considered sums of the same form in his influential book on the partial differential equations of wave propagation and hydrodynamics [Hadamard, 1903]. However, if one is in the right frame of mind, an understanding of even this special case leads easily to a proof of Schur's apparently more general product theorem; L. Fejér devotes the entire first section of a 1918 paper on partial differential equations to a review of Moutard's result and a proof that it is equivalent to what he termed the "remarkable theorem of J. [Issai] Schur" [Fejér, 1918], but Moutard's contribution seems thereafter to have sunk into obscurity.

Finally, a comment on notation. Apparently neither Hadamard nor Schur felt the need for any notation for the entry-wise product, so there is no historical precedent for later authors to follow. Our notation $A \circ B$ for the Hadamard product is common but not universal. Other notations seen in the literature are $A * B$ [Styan, 1973] and [Marcus & Minc, 1964], $A \odot B$ [Djoković, 1965], $A \times B$ [Mirsky, 1955, p. 421], $A \boxdot B$ [Rao, 1970], and $A \bullet B$ [Cowen, 1990].

The Hadamard product seems to be one of those useful mathematical ideas that does not readily find a niche in the standard textbook literature, but appears again and again in the specialized journals as facts about it are repeatedly re-discovered by various authors who need them for some other purpose.

3. Schur's Theorems and Recognition of the Hadamard Product in Various Ways. In his 1911 paper, Schur proved that the entry-wise product of two positive semidefinite matrices is positive semidefinite, and he gave quantitative bounds on the eigenvalues of this product in terms of the eigenvalues of one factor and the main diagonal entries of the other. He also considered general matrices, not necessarily positive semidefinite or even Hermitian, and gave three bounds on the spectral norm of the entry-wise product.

3.1 Schur's Theorems. Let $A = [a_{ij}]$, $B \in M_n$ be given. Then

(a) If A and B are positive semidefinite, then so is $A \circ B$.

(b) If A and B are positive semidefinite, then

$$(\min a_{ii})\, \lambda_{min}(B) \leq \lambda_{min}(A \circ B) \leq \lambda_{max}(A \circ B) \leq (\max a_{ii})\, \lambda_{max}(B)$$

(c) If B is positive definite and if A is positive semidefinite with all its main diagonal entries positive, then $A \circ B$ is positive definite.

(d) If A is positive semidefinite, then $|||\, A \circ B \,|||_2 \leq (\max a_{ii})\, |||\, B \,|||_2$.

(e) If $A = X^* Y$ with $X, Y \in M_{r,n}$, then $|||\, A \circ B \,|||_2 \leq c_1(X) c_1(Y)\, |||\, B \,|||_2$, where $c_1(Z)$ denotes the maximum Euclidean length of the columns of Z.

(f) $|||\, A \circ B \,|||_2 \leq |||\, A \,|||_2\, |||\, B \,|||_2$.

By *The Schur Product Theorem*, one usually means the simple assertion (a). The assertion in (c) follows directly from the lower bound in (b), which is one of many possible quantitative forms of the qualitative assertion (a); for a discussion of some others, see [Horn & Johnson, 1989, Section 5.3]. The upper bounds in (b) and (d) are, for positive semidefinite A, an improvement on the general upper bound in (f) since $\max a_{ii} \leq \lambda_{max}(A) = |||\, A \,|||_2$ in this case. The bound in (e) implies the bounds in (d) and (f): for (f), take $r = n$, $X = I$, and $Y = A$ and notice that $c_1(A) \leq |||\, A \,|||_2$; for (d), take $X = Y = A^{\frac{1}{2}}$ and notice that the main diagonal entries of A are the respective Euclidean column lengths of $A^{\frac{1}{2}}$. For a simple direct proof of 3.1(e), see [Ando, Horn, & Johnson, 1987, p. 364]; these authors, like [Walter, 1986] and probably others, were unaware of Schur's 1911 discovery of this useful general inequality.

The bounds in (b) follow easily from (a) and the monotonicity theorem. Let $A, B \in M_n$ be positive semidefinite and observe that $A \circ B = A \circ [B - \lambda_{min}(B)I] + \lambda_{min}(B)(A \circ I)$. Since $B - \lambda_{min}(B)I$ is positive semidefinite, the product $A \circ [B - \lambda_{min}(B)I]$ is positive semidefinite by (a), and $\lambda_{min}(A \circ B) \geq \lambda_{min}[\lambda_{min}(B)\,(A \circ I)] = \lambda_{min}(B)\,\lambda_{min}(A \circ I) = \min a_{ii}\,\lambda_{min}(B)$ by the monotonicity theorem. The upper bound follows from the same argument by considering $B - \lambda_{max}(B)I$, which is negative semidefinite.

In what follows, we provide proofs (usually, several different proofs) of each of Schur's theorems, and discuss various generalizations and variations.

In proving theorems, it can be helpful to recognize that the Hadamard product can be obtained from familiar processes of matrix theory in various ways. We now discuss a variety of techniques for dealing with the Hadamard product, and illustrate their utility by giving several more proofs of Schur's Theorem as well as other results about the Hadamard product.

3.2 The Hadamard Product as a Submatrix of the Kronecker Product. If $A = [a_{ij}] \in M_{m,n}$ and $B = [b_{ij}] \in M_{p,q}$ are given, then their *Kronecker product* $A \otimes B$ (also called the *tensor product*) is the mp-by-nq matrix $[a_{ij}B]$. If $A, B \in M_{m,n}$ and one looks at the entries of $A \otimes B$ that lie in the intersections of columns $1, n+2, 2n+3, 3n+4, \ldots, (n-1)n+n = n^2$ with rows $1, m+2, 2m+3, 3m+4, \ldots, (m-1)m+m = m^2$, one finds exactly the entries of the Hadamard product $A \circ B$! Thus, *$A \circ B$ is a submatrix of $A \otimes B$* when A and B are the same size, and it is a *principal submatrix* if, in addition, A and B are both square. Thus, recognition of the Hadamard product as a submatrix of the Kronecker product is likely to give useful information about concepts for which one knows inequalities or inclusion relations when passing from the full matrix to a submatrix. Examples of such concepts are: singular values and rank of a general complex rectangular matrix, the field of values of a square complex matrix, eigenvalues of a Hermitian matrix, and the spectral radius of a square matrix with nonnegative entries.

A handy property of the Kronecker product that is easily checked by direct computation is the *mixed product rule:*

$$(X_1 Y_1) \otimes (X_2 Y_2) = (X_1 \otimes X_2)(Y_1 \otimes Y_2). \tag{3.2.1}$$

One also checks easily that the Kronecker product of two matrices that are diagonal, symmetric, normal, Hermitian, or unitary, respectively, is a matrix with the same property.

Let $A, B \in M_{m,n}$ be given and have singular value decompositions $A = V_1 \Sigma_1 W_1^*$ and $B = V_2 \Sigma_2 W_2^*$. Then the mixed product rule permits us to write

$$A \otimes B = (V_1 \otimes V_2)(\Sigma_1 \otimes \Sigma_2)(W_1 \otimes W_2)^*, \tag{3.2.2}$$

which is evidently a singular value decomposition of $A \otimes B$. In particular, the singular values of $A \otimes B$ are the pairwise products of the respective singular values of A and B, so the spectral norm of $A \otimes B$ (which is its largest singular value) is the product of the spectral norms of A and B, that is, $|||A \otimes B|||_2 = |||A|||_2\, |||B|||_2$. Since $A \circ B$ is a submatrix of $A \otimes B$, it follows from (1.5) that

$$|||A\circ B|||_2 \le |||A\otimes B|||_2 = |||A|||_2\,|||B|||_2 \text{ for all } A,\, B\in M_{m,n}. \tag{3.2.2a}$$

This inequality proves 3.1(f). See [Varopoulos, 1974, p. 90] for a discovery, proof, and application of the inequality (3.2.2a); one may infer from comments in [Varopoulos, 1975, pp. 181-2] that the author was unaware of Schur's theorem 3.1(f).

Since the number of positive singular values of a matrix equals its rank, and since the rank of a submatrix is not greater than the rank of the full matrix, the preceding discussion permits us to conclude that *rank is Hadamard submultiplicative:*

$$\operatorname{rank}(A\circ B) \le \operatorname{rank}(A\otimes B) = (\operatorname{rank} A)(\operatorname{rank} B).$$

If $A\in M_n$, $B\in M_m$ have Schur unitary triangularizations $A = U_1\Delta_1U_1^*$ and $B = U_2\Delta_2U_2^*$, then

$$A\otimes B = (U_1\otimes U_2)(\Delta_1\otimes\Delta_2)(U_1\otimes U_2)^*, \tag{3.2.3}$$

which is evidently a Schur unitary triangularization of $A\otimes B$. Since the main diagonal entries of $\Delta_1\otimes\Delta_2$ are the pairwise products of the main diagonal entries of Δ_1 and Δ_2, it is clear that the eigenvalues of $A\otimes B$ are the pairwise products of the respective eigenvalues of A and B. In particular, the spectral radius of $A\otimes B$ is the product of the spectral radii of A and B, that is, $\rho(A\otimes B) = \rho(A)\rho(B)$ for all $A\in M_n$, $B\in M_m$.

Now suppose that $A, B\in M_n$ are *Hermitian*. By the interlacing eigenvalues theorem, each of the eigenvalues of the principal submatrix $A\circ B$ is bounded from above and below by eigenvalues of $A\otimes B$—which are just pairwise products of the respective eigenvalues of A and B—according to the inequalities (1.3). The simplest bounds are for the algebraically largest and smallest eigenvalues of $A\circ B$, which are bounded from above and below by the algebraically largest and smallest eigenvalues of $A\otimes B$. These two extremal eigenvalues are among the four points in the set

$$\mathcal{E}(A,B) \equiv \{\lambda_{min}(A)\lambda_{min}(B),\ \lambda_{min}(A)\lambda_{max}(B),\ \lambda_{max}(A)\lambda_{min}(B),\ \lambda_{max}(A)\lambda_{max}(B)\}$$

$$\text{for Hermitian } A,\, B\in M_n. \tag{3.2.4}$$

Without further assumptions on the signs of the extremal eigenvalues of A and B, all we can say is that

$$\min\{\lambda\in\mathcal{E}(A,B)\} \le \lambda_{min}(A\circ B) \le \lambda_{max}(A\circ B) \le \max\{\lambda\in\mathcal{E}(A,B)\}$$

$$\text{for Hermitian } A,\, B\in M_n. \tag{3.2.5}$$

If A and B are both *positive semidefinite*, the Hermitian matrix $A \otimes B$ must be positive semidefinite because its eigenvalues, the products of the eigenvalues of A and B, are all nonnegative. Since every principal submatrix of a positive semidefinite matrix is positive semidefinite, and the special principal submatrix $A \circ B$ must therefore be positive semidefinite, this gives a simple conceptual proof of the Schur Product Theorem—the qualitative part 3.1(a) of Schur's Theorem. Moreover, since $\lambda_{min}(A \otimes B) = \min\{\lambda \in \mathcal{E}(A, B)\} = \lambda_{min}(A)\lambda_{min}(B)$ and $\lambda_{max}(A \otimes B) = \max\{\lambda \in \mathcal{E}(A, B)\} = \lambda_{max}(A)\lambda_{max}(B)$ in this case, (3.2.5) gives the bounds

$$\lambda_{min}(A)\lambda_{min}(B) \leq \lambda_{min}(A \circ B) \leq \lambda_{max}(A \circ B) \leq \lambda_{max}(A)\lambda_{max}(B)$$

$$\text{for positive semidefinite } A, B \in M_n. \qquad (3.2.6)$$

This upper bound is just 3.1(f) in the positive semidefinite case, but the lower bound here is not as good as the lower bound in 3.1(b).

If $A, B \in M_n$ are matrices with nonnegative entries, then so are $A \circ B$ and $A \otimes B$. The spectral radius $\rho(A)$ is an important quantity in the study of nonnegative matrices, and it is known that if $\hat{C}$ is a principal submatrix of a matrix $C \in M_n$ with nonnegative entries, then $\rho(\hat{C}) \leq \rho(C)$ [Horn & Johnson, 1987a, 8.1.20]. Thus, we have the upper bound

$$\rho(A \circ B) \leq \rho(A \otimes B) = \rho(A)\rho(B) \qquad \text{for } A, B \in M_n \text{ with nonnegative entries,} \qquad (3.2.7)$$

that is, the spectral radius is Hadamard submultiplicative on square matrices with nonnegative entries. For many more results about Hadamard products of nonnegative matrices, see [Horn & Johnson, 1989, 5.7].

The *field of values* of $A \in M_n$ (also called the *numerical range*) is the set

$$F(A) \equiv \{x^*Ax\colon\ x \in \mathbb{C}^n, \|x\|_2 = 1\} \qquad (3.2.8)$$

in the complex plane, a compact set that contains all the eigenvalues of A and is known to be convex. It follows from the quasi-linearity of the definition (3.2.8) that $F(A+B)$ is contained in the set sum of $F(A)$ and $F(B)$, that is,

$$F(A+B) \subset F(A) + F(B) \qquad \text{for all } A, B \in M_n. \qquad (3.2.9)$$

It is known that $F(A \otimes B)$ always contains the convex hull of the set product of $F(A)$ and $F(B)$, that is,

$$F(A \otimes B) \supset \text{Co}[F(A)F(B)] \qquad \text{for all } A, B \in M_n, \qquad (3.2.10)$$

and that this containment is an equality if at least one of the factors is normal:

$$F(A \otimes B) = \text{Co}[F(A)F(B)] \quad \text{for } A, B \in M_n, \text{ at least one of } A, B \text{ is normal.} \qquad (3.2.11)$$

Finally, it is evident from the definition (3.2.8) that there is an inclusion relation for the field of values of a principal submatrix:

$$F(\hat{A}) \subset F(A) \qquad \text{if } \hat{A} \text{ is a principal submatrix of } A \in M_n. \qquad (3.2.12)$$

Since $A \circ B$ is a principal submatrix of $A \otimes B$, we can now use (3.2.12) and (3.2.11) to conclude that

$$F(A \circ B) \subset \text{Co}[F(A)F(A)] \text{ for } A, B \in M_n \text{ if at least one of } A, B \text{ is normal.} \qquad (3.2.13)$$

The example $A = B = \begin{bmatrix} 0 & 2 \\ 0 & 0 \end{bmatrix}$ shows that the inclusion in (3.2.13) cannot hold in general; in this case $F(A) = F(B) = F(A)F(B)$ is the unit disc, but $F(A \circ B)$ is a disc with radius 2 centered at the origin.

Since the field of values of a normal matrix is exactly the convex hull of its eigenvalues, if *both* A and B are normal and have respective eigenvalues $\{\alpha_i\}$ and $\{\beta_i\}$, it follows from (3.2.13) that $F(A \circ B)$ is contained in the convex hull (a convex polygon) of the points $\{\alpha_i \beta_j : i, j = 1, \ldots, n\}$ [Marcus & Thompson, 1963, Theorem 1].

An important scalar function associated with the field of values is the *numerical radius* $r(A)$, which is the radius of the smallest disc centered at the origin that contains the field of values of A:

$$r(A) \equiv \max\{|x^* A x| : x \in \mathbb{C}^n \text{ and } \|x\|_2 = 1\}. \qquad (3.2.14)$$

It follows from (3.2.9) that the nonnegative function $r(\cdot)$ is subadditive, that is, satisfies the triangle inequality

$$r(A + B) \leq r(A) + r(A) \qquad \text{for all } A, B \in M_n, \qquad (3.2.15)$$

and, in fact, $r(\cdot)$ is actually a norm on M_n. However, the numerical radius is not submultiplicative with respect to ordinary matrix multiplication, that is, $r(AB) \leq r(A)r(B)$ is *not* generally true. It is known that $r(AB) \leq 4\, r(A)r(B)$ for all $A, B \in M_n$, and that $r(AB) \leq 2\, r(A)r(B)$ if at least one factor is normal; both bounds are sharp.

For the Hadamard product, if A is normal, we have

$$r(A \circ B) = \max\{|z| : z \in F(A \circ B)\} \leq \max\{|z| : z \in F(A \otimes B)\}$$

$$= \max\{|z| : z \in \text{Co}[F(A)F(B)]\} = \max\{|z| : z \in F(A)F(B)\}$$

$$= \max\{|z_1 z_2|: z_1 \in F(A), z_2 \in F(B)\}$$

$$= \max\{|z_1|: z_1 \in F(A)\} \max\{|z_2|: z_2 \in F(B)\} = r(A)r(B),$$

so the numerical radius is Hadamard submultiplicative if at least one factor is normal:

$$r(A \circ B) \le r(A)r(B) \qquad \text{for } A, B \in M_n, \text{ at least one of } A, B \text{ is normal.} \qquad (3.2.16)$$

Every $A \in M_n$ can be written (uniquely) as a sum of a Hermitian and skew-Hermitian matrix, $A = \frac{1}{2}(A + A^*) + \frac{1}{2}(A - A^*) \equiv H(A) + S(A)$, and both summands are normal, so for any $A, B \in M_n$ we can use (3.5.15) to compute

$$r(A \circ B) = r([H(A) + S(A)] \circ B) \le r(H(A) \circ B) + r(S(A) \circ B)$$

$$\le r(H(A))r(B) + r(S(A))r(B) = \tfrac{1}{2}r(A + A^*)r(B) + \tfrac{1}{2}r(A - A^*)r(B)$$

$$\le \tfrac{1}{2}[r(A) + r(A^*)]r(B) + \tfrac{1}{2}[r(A) + r(A^*)]r(B).$$

But $F(C^*) = \overline{F(C)}$ (which is the reflection of $F(C)$ across the real axis) for any $C \in M_n$, so $r(A^*) = r(A)$ and hence

$$r(A \circ B) \le 2\, r(A)r(B) \qquad \text{for all } A, B \in M_n. \qquad (3.2.17)$$

The example following (3.2.13) shows that this bound is sharp. For more information about the field of values and the numerical radius, and proofs of all the preceding assertions, see [Horn & Johnson, 1987a, 5.7] and [Horn & Johnson, 1989, Chapters 1 (especially 1.7.22 *ff.*) and 4 (especially 4.2.16 *ff.*)]

Although the connection between Hadamard and Kronecker products is well worth knowing, it does not seem to be a sufficiently powerful tool to prove all the theorems one would like about Hadamard products. Some remarkable Hadamard product inequalities are proved in [Ando, 1979] with very clever use of Kronecker products. Explicit recognition of $A \circ B$ as a principal submatrix of $A \otimes B$ seems to have occurred first in the literature in the 1950's [Toyama, 1952, p. 199] and [Marcus & Khan, 1959].

3.3 The Hadamard Product as the Trace of an Ordinary Product; The Frobenius Norm. In using variational arguments to calculate the eigenvalues or singular values of a matrix A, one often looks at the values of x^*Ax or x^*Ay as x and y range over a set of unit vectors in $\mathbb{C}^n$. In this regard, it can be helpful to know the following easily verified identity for the sesquilinear form generated by the Hadamard product:

$$\text{For } A, B \in M_n \text{ and } x, y \in \mathbb{C}^n,\ y^*(A \circ B)x = \operatorname{tr}[\operatorname{diag}(y)^* A \operatorname{diag}(x)\, B^T]. \qquad (3.3.1)$$

As a demonstration of the utility of this observation, suppose A, $B \in M_n$ are positive semidefinite and let $x \in \mathbb{C}^n$ be given. Then

$$x^*(A \circ B)x = \operatorname{tr}[\operatorname{diag}(x)^* A \operatorname{diag}(x) B^T] = \operatorname{tr}[B^{\frac{1}{2}T} \operatorname{diag}(x)^* A^{\frac{1}{2}} A^{\frac{1}{2}} \operatorname{diag}(x) B^{\frac{1}{2}T}]$$

$$= \operatorname{tr}[(A^{\frac{1}{2}} \operatorname{diag}(x) B^{\frac{1}{2}T})^* (A^{\frac{1}{2}} \operatorname{diag}(x) B^{\frac{1}{2}T})] \geq 0 \tag{3.3.2}$$

because this is the trace of a matrix that is of the form $C^* C$ and hence is positive semidefinite. This is a second proof of the Schur Product Theorem.

The stronger assertion 3.1(c) also follows from the identity (3.3.1): If $A = [a_{ij}]$, $B \in M_n$ are positive semidefinite, if $x^*(A \circ B)x = 0$, and if we set $C = [c_{ij}] \equiv A^{\frac{1}{2}} \operatorname{diag}(x) B^{\frac{1}{2}T}$, then (3.3.2) says that $0 = \operatorname{tr} C^* C = \Sigma |c_{ij}|^2$, which means that $C = 0$. If B is positive definite (and hence is nonsingular), then so is $B^{\frac{1}{2}}$ and we have $0 = CB^{-\frac{1}{2}T} = A^{\frac{1}{2}} \operatorname{diag}(x)$, which means that every column of $A^{\frac{1}{2}}$ corresponding to a nonzero entry of x must be zero. Since the main diagonal entries of A are the squares of the Euclidean lengths of the columns of $A^{\frac{1}{2}}$, assuming that all $a_{ii} > 0$ forces us to conclude that $x = 0$, that is, $A \circ B$ must be positive *definite* if B is positive definite, A is positive semidefinite, and all $a_{ii} > 0$.

The trace identities (3.3.1) and (3.3.2) can also be used to derive quantitative bounds. For this purpose it is useful to know that the function $\langle A, B\rangle_F \equiv \operatorname{tr}(B^* A)$ is an *inner product* on $M_{m,n}$, often called the *Frobenius inner product*. The associated norm $\|A\|_2 \equiv (\langle A, A\rangle_F)^{\frac{1}{2}} = [\operatorname{tr}(A^* A)]^{\frac{1}{2}}$ is the *Frobenius* or ℓ_2 *norm* on $M_{m,n}$. It is evident from the definition that

$$\text{The Frobenius norm is unitarily invariant} \tag{3.3.3}$$

since $\| UAV \|_2 = \operatorname{tr}[(UAV)^*(UAV)] = \operatorname{tr}(V^* A^* U^* UAV) = \operatorname{tr}(V^* A^* A V) = \operatorname{tr}(A^* A V V^*) = \operatorname{tr}(A^* A) = \| A \|_2^2$ for any unitary $U \in M_m$, $V \in M_n$. The singular value decomposition $A = V \Sigma W^*$ can be used to derive an inequality between the Frobenius norm and the spectral norm: $\| A \|_2 = \| V \Sigma W^* \|_2 = \| \Sigma \|_2 = (\sigma_1^2 + \cdots + \sigma_q^2)^{\frac{1}{2}} \geq \sigma_1 = |\!|\!| A |\!|\!|_2$, where $q \equiv \min\{m,n\}$. If $A \in M_{m,n}$ and $B = [b_1 \ \ldots \ b_k] \in M_{n,k}$, then $\| AB \|_2^2 = \| Ab_1 \|_2^2 + \cdots + \| Ab_k \|_2^2 \leq |\!|\!| A |\!|\!|_2^2 \| b_1 \|_2^2 + \cdots + |\!|\!| A |\!|\!|_2^2 \| b_k \|_2^2 = |\!|\!| A |\!|\!|_2^2 \| B \|_2^2$. Thus, we have the upper bounds

$$\| AB \|_2 \leq |\!|\!| A |\!|\!|_2 \| B \|_2 \leq \| A \|_2 \| B \|_2 \quad \text{for } A \in M_{m,n},\ B \in M_{n,k}. \tag{3.3.4}$$

In particular, the Frobenius norm is submultiplicative with respect to the ordinary product. To show that it is also submultiplicative with respect to the Hadamard product, notice that for any x, $y \in \mathbb{C}^n$ we have

$$\| x \circ y \|_2^2 = \sum_{i=1}^{n} |x_i|^2 |y_i|^2 \leq \left[\sum_{i=1}^{n} |x_i|^2\right] \left[\sum_{i=1}^{n} |y_i|^2\right] = \| x \|_2^2 \| y \|_2^2$$

and hence if $A = [\, a_1 \; \cdots \; a_n \,]$, $B = [\, b_1 \; \cdots \; b_n \,] \in M_{m,n}$ then

$$\| A \circ B \|_2^2 = \sum_{i=1}^{n} \| a_i \circ b_i \|_2^2 \le \sum_{i=1}^{n} \| a_i \|_2^2 \, \| b_i \|_2^2 \le \left[\sum_{i=1}^{n} \| a_i \|_2^2 \right] \left[\sum_{i=1}^{n} \| b_i \|_2^2 \right],$$

which gives the desired analog of 3.1(f) for the Frobenius norm:

$$\| A \circ B \|_2 \le \| A \|_2 \, \| B \|_2 \quad \text{for } A, B \in M_{m,n} \tag{3.3.5}$$

There is also a lower bound corresponding to (3.3.4): If $A \in M_n$ is nonsingular and $B \in M_{n,k}$, then $\| B \|_2 = \| A^{-1}AB \|_2 \le ||| A^{-1} |||_2 \, \| AB \|_2$ and hence

$$\| AB \|_2 \ge \left[||| A^{-1} |||_2 \right]^{-1} \| B \|_2 \quad \text{for } B \in M_{n,k} \text{ and nonsingular } A \in M_n. \tag{3.3.6}$$

For $A \in M_n$, let $A = U \Delta U^*$ be a Schur unitary triangularization, where $U \in M_n$ is unitary and $\Delta \in M_n$ is upper triangular and the entries of $\text{diag}(\Delta) = [\lambda_1 \; \cdots \; \lambda_n]^T$ are the eigenvalues of A. From this it follows that $\| A \|_2^2 = \| U \Delta U^* \|_2^2 = \| \Delta \|_2^2 \ge \| \text{diag}(\Delta) \|_2^2$, that is,

$$\| A \|_2 \ge \| [\lambda_1 \; \cdots \; \lambda_n]^T \|_2, \text{ where } A \in M_n \text{ has eigenvalues } \lambda_1, \ldots, \lambda_n. \tag{3.3.7}$$

Now let $A, B \in M_n$ be positive definite and let $x \in \mathbb{C}^n$ be a given unit vector. Starting with (3.3.2), we have

$$x^*(A \circ B)x = \| A^{\frac{1}{2}} \text{diag}(x) B^{\frac{1}{2}T} \|_2^2 \tag{3.3.8}$$

$$= \| (A^{\frac{1}{2}} B^{\frac{1}{2}T})(B^{-\frac{1}{2}})^T \text{diag}(x) B^{\frac{1}{2}T} \|_2^2$$

$$\ge ||| (A^{\frac{1}{2}} B^{\frac{1}{2}T})^{-1} |||_2^{-2} \, \| B^{-\frac{1}{2}T} \text{diag}(x) B^{\frac{1}{2}T} \|_2^2 \qquad \text{[by (3.3.6)]}$$

$$\ge ||| (A^{\frac{1}{2}} B^{\frac{1}{2}T})^{-1} |||_2^{-2} \, \| x \|_2^2 \qquad \text{[by (3.3.7)]}$$

$$= ||| B^{-\frac{1}{2}T} A^{-\frac{1}{2}} |||_2^{-2} \cdot 1 = \left[\lambda_{max} \left[(B^{-\frac{1}{2}T} A^{-\frac{1}{2}})^* (B^{-\frac{1}{2}T} A^{-\frac{1}{2}}) \right] \right]^{-1}$$

$$= \left[\lambda_{max} [A^{-\frac{1}{2}} (B^{-1})^T A^{-\frac{1}{2}}] \right]^{-1} = \lambda_{min}([A^{-\frac{1}{2}} (B^{-1})^T A^{-\frac{1}{2}}]^{-1})$$

$$= \lambda_{min}(A^{\frac{1}{2}} B^T A^{\frac{1}{2}})$$

$$= \lambda_{min}(A^{-\frac{1}{2}} [AB^T] A^{\frac{1}{2}}) = \lambda_{min}(AB^T).$$

The not-necessarily Hermitian matrix AB^T is similar to the positive definite matrix $A^{\frac{1}{2}} B^T A^{\frac{1}{2}}$, so all of its eigenvalues are positive. Notice that the inequality $x^*(A\circ B)x \geq \lambda_{min}(AB^T)$ is trivial if either A or B is singular. Since our inequality holds for all unit vectors $x \in \mathbb{C}^n$, the Rayleigh-Ritz Theorem permits us to conclude that

$$\lambda_{min}(A\circ B) \geq \lambda_{min}(AB^T) \quad \text{for positive semidefinite } A,\, B \in M_n. \tag{3.3.9a}$$

Fiedler has shown that equality holds in (3.3.9a) if and only if $AB^T = \alpha I$ for some $\alpha > 0$ [Fiedler, 1983]. A similar argument gives the companion bound

$$\lambda_{min}(A\circ B) \geq \lambda_{min}(AB) \quad \text{for positive semidefinite } A,\, B \in M_n. \tag{3.3.9b}$$

The two lower bounds (3.3.9a,b) have a very different character from the lower bound in 3.1(b). For example, if $A = J$ then JB and JB^T are both singular, so (3.3.9a,b) both give the trivial lower bound $\lambda_{min}(B) = \lambda_{min}(J\circ B) \geq 0$; 3.1(b) gives $\lambda_{min}(B) = \lambda_{min}(J\circ B) \geq 1\cdot\lambda_{min}(B)$. On the other hand, if A is positive definite and $B = A^{-1}$, then (3.3.9b) gives $\lambda_{min}(A\circ A^{-1}) \geq \lambda_{min}(AA^{-1}) = \lambda_{min}(I) = 1$, while 3.1(b) gives $\lambda_{min}(A\circ A^{-1}) \geq \min a_{ii}\, \lambda_{min}(A^{-1}) = \min a_{ii} / \lambda_{max}(A)$, a lower bound that is at most one and is strictly less than one if A is not a scalar multiple of the identity since $\min a_{ii} \leq \max a_{ii} \leq \lambda_{max}(A)$.

Both of the lower bounds (3.3.9a,b) are discussed in [Horn & Johnson, 1989, Sections 5.3,4]; (3.3.9a) is in [Fiedler, 1983] and (3.3.9b) is in [Johnson & Elsner, 1987]. For another proof of (3.3.9a,b), see the remarks following (3.7.19a,b).

Suppose we order the eigenvalues of a positive semidefinite matrix $X \in M_n$ from largest to smallest as $\lambda_{max}(X) = \lambda_1(X) \geq \lambda_2(X) \geq \cdots \geq \lambda_n(X) = \lambda_{min}(X)$. The inequality (3.3.9b) says that $\lambda_n(A\circ B) \geq \lambda_n(AB)$, and Oppenheim's inequality (6.1) says that

$$\lambda_1(A\circ B)\cdots\lambda_n(A\circ B) = \det(A\circ B) \geq (\det A)(\det B) = \det(AB) = \lambda_1(AB)\cdots\lambda_n(AB)$$

for positive semidefinite A, $B \in M_n$. These two known results are the cases $k = 1$ and n of the conjectured family of inequalities

$$\prod_{i=k}^{n} \lambda_i(A\circ B) \geq \prod_{i=k}^{n} \lambda_i(AB) \quad \text{for positive semidefinite } A,\, B \in M_n,\ k = 1,\ldots, n, \tag{3.3.10}$$

that were proposed in [Johnson & Bapat, 1987] as an open problem.

For square matrices, the derivation of (3.3.5) given is straightforward but unnecessary, given (3.3.4); actually, the derivations of both (3.3.4) and (3.3.5) are unnecessary, given that $\|A\|_2 \geq |||A|||_2$, which follows immediately from the singular value decomposition. The reason is that something much more general is true:

Let $N(\cdot)$ be a given unitarily invariant norm on M_n. The following are equivalent:

(3.3.11)

(a) $N(\cdot)$ is submultiplicative with respect to the ordinary product.
(b) $N(\cdot)$ is submultiplicative with respect to the Hadamard product.
(c) $N(A) \geq |||A|||_2$ for every $A \in M_n$.
(d) $N(A) \geq \rho(A)$ for every $A \in M_n$.

An immediate consequence of (c) is that *the set of submultiplicative unitarily invariant norms on M_n is a convex set.* The hypothesis of unitary invariance is essential in (3.3.11), for when it is dropped the only implications that are generally true are (a) ⇒ (d) and (c) ⇒ (d) [Horn & Johnson, 1987b, pp. 102-5].

The bound $\|AB\|_2 \leq |||A|||_2 \|B\|_2$ in (3.3.4) has an analog for the Hadamard product, and both have sweeping generalizations [Ando, Horn, & Johnson, 1987, pp. 361-2]:

For any unitarily invariant invariant norm $N(\cdot)$ on M_n, $N(AB) \leq |||A|||_2 N(B)$ and $N(A \circ B) \leq |||A|||_2 N(B)$ for all $A, B \in M_n$.

(3.3.12)

3.4 Hadamard Product with a Rank One Factor. If $B = [b_{ij}] \in M_n$ and $x = [x_i]$, $y = [y_i] \in \mathbb{C}^n$, then the Hadamard product of the rank one matrix xy^* with B is readily recognized as an ordinary product:

$$(xy^*) \circ B = [x_i \bar{y}_j] \circ [b_{ij}] = [x_i b_{ij} \bar{y}_j] = \operatorname{diag}(x)\, B \operatorname{diag}(y)^*.$$

If $y = x$ then this identity becomes $(xx^*) \circ B = \operatorname{diag}(x)\, B \operatorname{diag}(x)^*$, which is obviously positive semidefinite if B is. This simple observation leads to a straightforward proof of 3.1(a) of Schur's Theorem, and to proofs of 3.1(f) and 3.1(b) as well.

If $A \in M_n$ is positive semidefinite, it has a spectral decomposition $A = U \Lambda U^*$ with a unitary $U = [u_{ij}] = [u_1 \ldots u_n]$ and a nonnegative diagonal $\Lambda = \operatorname{diag}(\lambda_1, \ldots, \lambda_n)$. The spectral decomposition also can be written in the equivalent form

$$A = \lambda_1 u_1 u_1^* + \cdots + \lambda_n u_n u_n^* \tag{3.4.1}$$

as a nonnegative linear combination of rank one Hermitian matrices. Then

$$A \circ B = \sum_{i=1}^{n} \lambda_i (u_i u_i^*) \circ B = \sum_{i=1}^{n} \lambda_i \operatorname{diag}(u_i)\, B \operatorname{diag}(u_i)^*,$$

which is evidently positive semidefinite if B is, since it is a nonnegative linear combination of positive semidefinite matrices. This third proof of the Schur Product Theorem is essentially the proof given in [Bellman, 1970, p. 95].

The same kind of calculation also gives another proof of the quantitative part 3.1(b) of Schur's Theorem. Let $A = [a_{ij}]$, $B \in M_n$ be positive semidefinite. Notice that one of the consequences of the Rayleigh-Ritz theorem is that $z^*Bz \le \lambda_{max}(B)\, z^*z$ for any vector $z \in \mathbb{C}^n$. Now let x be any given unit vector and use the spectral decomposition (3.4.1) to compute

$$x^*(A \circ B)x = x^*\left[\sum_{i=1}^{n} \lambda_i (u_i u_i^*) \circ B\right]x = \sum_{i=1}^{n} \lambda_i\, x^* \operatorname{diag}(u_i)\, B \operatorname{diag}(u_i)^* x$$

$$= \sum_{i=1}^{n} \lambda_i (x \circ \bar{u}_i)^*\, B\, (x \circ \bar{u}_i)$$

$$\le \sum_{i=1}^{n} \lambda_i \lambda_{max}(B)(x \circ \bar{u}_i)^*\, (x \circ \bar{u}_i) \tag{3.4.2}$$

$$= \lambda_{max}(B) \sum_{i=1}^{n} \lambda_i (x \circ \bar{u}_i)^*\, I\, (x \circ \bar{u}_i) = \lambda_{max}(B) \sum_{i=1}^{n} \lambda_i\, x^* \operatorname{diag}(u_i)\, I \operatorname{diag}(u_i)^* x$$

$$= \lambda_{max}(B)\, x^*\left[\sum_{i=1}^{n} \lambda_i (u_i u_i^*) \circ I\right]x = \lambda_{max}(B)\, x^*(A \circ I)x$$

$$= \lambda_{max}(B) \sum_{k=1}^{n} |x_k|^2 a_{kk} \le \lambda_{max}(B) \max_i a_{ii}. \tag{3.4.3}$$

Since the unit vector x is arbitrary, another reference to the Rayleigh-Ritz Theorem concludes this proof of the upper bound in (c) of Schur's Theorem. The lower bound in 3.1(b) follows from a similar argument: use the lower bound $(x \circ \bar{u}_i)^* B (x \circ \bar{u}_i) \ge \lambda_{min}(B)\, (x \circ \bar{u}_i)^*\, (x \circ \bar{u}_i)$ in (3.4.2) and the lower bound $a_{kk} \ge \min a_{ii}$ in (3.4.3).

Finally, these same ideas also can be used to prove the norm inequality 3.1(f). Let A, $B \in M_n$ be given, not necessarily positive semidefinite or even Hermitian, and invoke the singular value decomposition $A = V \Sigma W^*$, where $V = [v_{ij}] = [v_1 \dots v_n]$ and $W = [w_{ij}] = [w_1 \dots w_n]$ are unitary and $\Sigma = \operatorname{diag}(\sigma_1, \dots, \sigma_n)$ with $|||\, A\, |||_2 = \sigma_1 \ge \cdots \ge \sigma_n \ge 0$. The singular value decomposition has the equivalent form

$$A = \sigma_1 v_1 w_1^* + \cdots + \sigma_n v_n w_n^*, \tag{3.4.4}$$

which gives A as a nonnegative linear combination of rank one matrices. Recall that $|x^*By| \le |||\, B\, |||_2\, \|x\|_2\, \|y\|_2$ for any x, $y \in \mathbb{C}^n$, with equality when x and y are chosen to be appropriate left and right singular vectors of B. Now let x and y be a given unit vectors and compute

$$|x^*(A\circ B)y| = \left| x^*\left[\sum_{i=1}^{n}\sigma_i(v_iw_i^*)\circ B\right]y\right| \le \sum_{i=1}^{n}\sigma_i\,|x^*\mathrm{diag}(v_i)\,B\,\mathrm{diag}(w_i)^*y|$$

$$= \sum_{i=1}^{n}\sigma_i\,|(x\circ\bar{v}_i)^*\,B\,(y\circ\bar{w}_i)| \le \sum_{i=1}^{n}\sigma_i|||\,B\,|||_2\|x\circ\bar{v}_i\|_2\|y\circ\bar{w}_i\|$$

$$\le |||\,B\,|||_2\sum_{i=1}^{n}|||\,A\,|||_2\,\|x\circ\bar{v}_i\|_2\|y\circ\bar{w}_i\|_2 = |||\,A\,|||_2\,|||\,B\,|||_2\sum_{i=1}^{n}\|x\circ v_i\|_2\|y\circ w_i\|_2$$

$$\le |||\,A\,|||_2\,|||\,B\,|||_2\left[\sum_{i=1}^{n}\|x\circ v_i\|_2^2\right]^{\frac{1}{2}}\left[\sum_{i=1}^{n}\|y\circ w_i\|_2^2\right]^{\frac{1}{2}}$$

$$= |||\,A\,|||_2\,|||\,B\,|||_2\left[\sum_{i=1}^{n}\sum_{k=1}^{n}|x_kv_{ki}|^2\right]^{\frac{1}{2}}\left[\sum_{i=1}^{n}\sum_{k=1}^{n}|y_kw_{ki}|^2\right]^{\frac{1}{2}}$$

$$= |||\,A\,|||_2\,|||\,B\,|||_2\left[\sum_{k=1}^{n}|x_k|^2\sum_{i=1}^{n}|v_{ki}|^2\right]^{\frac{1}{2}}\left[\sum_{k=1}^{n}|y_k|^2\sum_{i=1}^{n}|w_{ki}|^2\right]^{\frac{1}{2}}$$

$$= |||\,A\,|||_2\,|||\,B\,|||_2$$

since every row of the unitary matrices V and W has unit length.

If $x, y, z, w \in \mathbb{C}^n$, then $(xy^*)\circ(zw^*) = (x\circ z)(y\circ w)^*$, so the Hadamard product of two rank one matrices has rank one. Furthermore, the sum of k rank one matrices has rank at most k, and a matrix of rank k can be written as a sum of exactly k rank one matrices (one such expansion is given by (3.4.4)). If $A, B \in M_n$ are given with

$$A = \sum_{i=1}^{\rho_1} x_iy_i^*,\quad \rho_1 = \mathrm{rank}\,A,\quad \text{and}\quad B = \sum_{j=1}^{\rho_2} z_jw_j^*,\quad \rho_2 = \mathrm{rank}\,B,$$

then

$$A\circ B = \sum_{i=1}^{\rho_1}\sum_{j=1}^{\rho_2}(x_i\circ z_j)(y_i\circ w_j)^*$$

has rank at most $\rho_1\rho_2$, so the Hadamard product is submultiplicative with respect to rank:

$$\operatorname{rank}(A \circ B) \leq (\operatorname{rank} A)(\operatorname{rank} B) \tag{3.4.5}$$

A stronger statement can be made in the positive semidefinite case. Since any zero main diagonal entry of a positive semidefinite matrix must lie in a zero row and column, a positive semidefinite matrix with exactly k positive main diagonal entries has rank at most k. Suppose $B \in M_n$ is positive definite. If $A \in M_n$ is positive semidefinite and has exactly k positive main diagonal entries, then the same is true of the positive semidefinite matrix $A \circ B$, which therefore has rank at most k. On the other hand, denote by $\hat{A}$ the k-by-k principal submatrix of A whose entries are the intersection of the rows and columns of A corresponding to its nonzero main diagonal entries. Let $\hat{B}$ denote the corresponding principal submatrix of B, which must be positive semidefinite. Then $\hat{A}$ is positive semidefinite and has positive main diagonal entries, so $\hat{A} \circ \hat{B}$ is positive definite (and hence is nonsingular) by 3.1(c). Since $A \circ B$ contains the k-by-k nonsingular principal submatrix $\hat{A} \circ \hat{B}$, it has rank at least k. Thus, we have shown [Ballantine, 1968]:

> If $A \in M_n$ is positive semidefinite and has exactly k positive main diagonal entries, and if $B \in M_n$ is positive definite, then $A \circ B$ is positive semidefinite and has rank k. (3.4.6)

An immediate application of this result is to the following Hadamard factorization problem [Ballantine, 1968]:

> A given positive semidefinite matrix $C \in M_n$ can be factored as $C = A \circ B$ with A positive semidefinite and B positive definite if and only if the number of positive main diagonal entries of C is equal to rank C. (3.4.7)

If C can be factored as asserted, we have just shown that the rank of C equals the number of positive main diagonal entries of A, which is evidently equal to the number of positive main diagonal entries of $A \circ B = C$. Conversely, if the rank of C is equal to the number of its positive main diagonal entries, then the principal submatrix $\hat{C}$ corresponding to those positive main diagonal entries is positive definite. Define $B \equiv [b_{ij}]$, where $b_{ij} \equiv c_{ij}$ except if $i = j$ and $c_{ii} = 0$, in which case $b_{ii} \equiv 1$. Define $A = [a_{ij}]$, where $a_{ij} \equiv 1$ if $c_{ii}c_{jj} \neq 0$ (*i.e.*, if c_{ij} is an entry of $\hat{C}$), otherwise $a_{ij} \equiv 0$. Then A is positive semidefinite, B is positive definite, and $A \circ B = C$.

In this same spirit, it is easy to show that every positive definite matrix $C \in M_n$ has a Hadamard factorization $C = A \circ B$ with both A and B positive definite. For $\alpha \geq 1$ define $J_\alpha, K_\alpha \in M_n$ by $J_\alpha \equiv (1 - \alpha)I + \alpha J$, $K_\alpha \equiv (1 - \alpha^{-1})I + \alpha^{-1}J$, and note that K_α is positive definite when $\alpha > 1$. Since $C = C \circ J_1$ is positive definite, it is clear from a continuity argument that $C \circ J_\alpha$ is positive definite for some $\alpha_0 > 1$, in which case $C = (C \circ J_{\alpha_0}) \circ K_\alpha$. This argument is in [Djokovic, 1965]; an explicit computable value for α_0 is given in [Majindar, 1963].

Since the number of positive main diagonal entries of a positive semidefinite matrix is an upper bound on its rank, an immediate consequence of (3.4.6) is the weaker but still useful statement [Djoković, 1965]:

> If $A \in M_n$ is positive semidefinite and $B \in M_n$ is positive definite, then $\operatorname{rank}(A \circ B) \geq \operatorname{rank} A$. (3.4.8)

Of course, $\operatorname{rank}(A \circ B) > \operatorname{rank} A$ is possible: just choose $A = J$, which has rank 1.

The principal submatrix idea used to prove (3.4.6) can also be used to prove a variant of this result [Ballantine, 1968]:

> If $A, B \in M_n$ are positive semidefinite and B has positive main diagonal entries, then $\operatorname{rank}(A \circ B) \geq \operatorname{rank} A$. If, in addition, B has rank 1, then $\operatorname{rank}(A \circ B) = \operatorname{rank} A$. (3.4.9)

In this case, let $k = \operatorname{rank} A$, so there is a k-by-k principal submatrix $\hat{A}$ of A that is nonsingular and therefore is positive definite. Let $\hat{B}$ be the corresponding principal submatrix of B. Then $\hat{B}$ is positive semidefinite and has positive main diagonal entries, so $\hat{A} \circ \hat{B}$ is positive definite by 3.1(c). Since $\hat{A} \circ \hat{B}$ is a nonsingular principal submatrix of $A \circ B$, it follows that $\operatorname{rank}(A \circ B) \geq k = \operatorname{rank} A$. The final assertion is immediate from (3.4.1), since B must have the form $B = xx^*$, where $x \in \mathbb{C}^n$ has no zero entries. Then $\operatorname{rank}(A \circ B) = \operatorname{rank}[A \circ (xx^*)] = \operatorname{rank}[\operatorname{diag}(x)\ A\ \operatorname{diag}(x)^*] = \operatorname{rank} A$, since $\operatorname{diag}(x)$ has full rank.

It is sometimes convenient to know the following characterizations of positive definiteness.

> $A \in M_n$ is positive semidefinite if and only if $A \circ B$ is positive semidefinite for every positive semidefinite $B \in M_n$. (3.4.10)

and

> $A \in M_n$ is positive semidefinite if and only if
>
> $$e^T(A \circ B)e = \sum_{i,j=1}^{n} a_{ij} b_{ij} \geq 0 \qquad (3.4.11)$$
>
> for every positive semidefinite $B \in M_n$.

In both cases, the necessity of the stated condition follows from 3.1(a). Sufficiency in (3.4.10) follows from choosing $B = J$, while sufficiency in (3.4.11) follows from choosing a rank one $B = xx^*$ and noting that $e^T(A \circ xx^*)e = x^*Ax$.

3.5 The Hadamard Product as an Ordinary Product of Stacked Matrices. There is another approach to proving Schur's norm inequality 3.1(f) that is much simpler and more direct than the preceding argument, and it does not require the singular value decomposition. Let $A, B \in M_n$ be given, partition $A = [a_1 \ldots a_n]$ and $B = [b_1 \ldots b_n]$ according to their columns, and write the identity matrix as $I = [e_1 \ldots e_n]$ (so e_i is the ith standard unit basis vector). Define

$$\mathcal{A} \equiv [\text{diag}(a_1) \;\dots\; \text{diag}(a_n)] \in M_{n,n^2} \tag{3.5.1}$$

and

$$\mathcal{B} \equiv [e_1 \otimes b_1 \;\dots\; e_n \otimes b_n] \in M_{n^2,n}. \tag{3.5.2}$$

The matrix $\mathcal{A}$ is constructed from n diagonal blocks of the same size as A, stacked side-by-side; the kth block is formed from the kth column of A, $k = 1,\dots n$. The matrix $\mathcal{B}$ has n columns, each of which is a vertical stack of n n-vectors, all but one of which are zero; the nonzero vector in the kth stack is the kth column of B, $k = 1,\dots, n$. In each successive column of $\mathcal{B}$ the corresponding column of B appears one position further down in the stack. One checks easily that the Hadamard product of A and B is equal to the ordinary product of $\mathcal{A}$ and $\mathcal{B}$:

$$A \circ B = \mathcal{A}\mathcal{B}. \tag{3.5.3}$$

Recognizing that the Hadamard product can be written in this form does not seem to be very helpful in questions involving positive definite matrices, since Hermicity and positive definiteness of A and B are not reflected in any obvious properties of $\mathcal{A}$ and $\mathcal{B}$. However, since the spectral norm is submultiplicative with respect to the ordinary product, we have the immediate inequality

$$|||\, A \circ B \,|||_2 = |||\, \mathcal{A}\mathcal{B} \,|||_2 \le |||\, \mathcal{A} \,|||_2 \, |||\, \mathcal{B} \,|||_2. \tag{3.5.4}$$

A calculation reveals that $\mathcal{B}^*\mathcal{B} = \text{diag}(\|b_1\|_2^2,\dots, \|b_n\|_2^2)$, so the spectral norm of $\mathcal{B}$ is the largest Euclidean column length of B, which we denote by $c_1(B)$. Similarly, if we write $A^T = [\alpha_1 \dots \alpha_n]$ (so that the row vectors $\alpha_1^T,\dots, \alpha_n^T$ are the rows of A), then $\mathcal{A}\mathcal{A}^* = \text{diag}(\|\alpha_1\|_2^2,\dots, \|\alpha_n\|_2^2)$ and the spectral norm of $\mathcal{A}$ is the largest Euclidean row length of A, which we denote by $r_1(A)$. Since any row or column length of a matrix is a lower bound for its spectral norm (by (1.5)), this simple argument proves and strengthens 3.1(f):

$$|||\, A \circ B \,|||_2 \le r_1(A)\; c_1(B) \le r_1(A)\, |||\, B \,|||_2 \le |||\, A \,|||_2 \, |||\, B \,|||_2. \tag{3.5.5}$$

Since $|||\, (A \circ B)^T \,|||_2 = |||\, A \circ B \,|||_2$, the role of the maximum row and column Euclidean lengths can be reversed in these inequalities, and the resulting inequalities can be combined with (3.5.5) to give inequalities in which the maximum row and column lengths appear symmetrically:

$$|||\, A \circ B \,|||_2 \le [r_1(A)\; c_1(A)]^{\frac{1}{2}}\, [r_1(B)\; c_1(B)]^{\frac{1}{2}} \le [r_1(A)\; c_1(A)]^{\frac{1}{2}}\, |||\, B \,|||_2 \le |||\, A \,|||_2 \, |||\, B \,|||_2 \tag{3.5.6}$$

Schur gave a different proof for the first of these inequalities in [Schur, 1911; Satz II, p. 8] and then derived 3.1(f) from it [Schur, 1911; Satz III, pp. 8-9]; see (5.7.5).

3.6 The Hadamard Product as a Sum of Rank One Matrices. Let $A, B \in M_{m,n}$ be given and partition them according to their rows and columns as

$$A = [\, a_1 \;\cdots\; a_n \,],\quad B = [\, b_1 \;\cdots\; b_n \,],\quad a_i, b_i \in \mathbb{C}^m,\ i = 1, \ldots, n$$

$$A^T = [\, \alpha_1 \;\cdots\; \alpha_m \,],\quad B^T = [\, \beta_1 \;\cdots\; \beta_m \,],\quad \alpha_j, \beta_j \in \mathbb{C}^n,\ j = 1, \ldots, m$$

Let $E_{ii} \in M_m$ and $\mathcal{E}_{ii} \in M_n$ denote square matrices with a single 1 entry in position i,i and zeroes elsewhere. It is straightforward to check that

$$A \circ B = \sum_{i=1}^{m} E_{ii} B \operatorname{diag}(\alpha_i) \tag{3.6.1a}$$

and

$$A \circ B = \sum_{j=1}^{n} \operatorname{diag}(a_j)\, B \mathcal{E}_{jj} \tag{3.6.1b}$$

express $A \circ B$ as a sum of products of a rank one matrix and a diagonal matrix. For $X \in M_{m,n}$, let $c_1(X)$ [respectively, $r_1(X)$] denote the maximum Euclidean length of the columns (respectively, the rows) of X, and let $\| X \|_\infty$ denote the maximum absolute value of the entries of X.

These simple representations lead to a surprisingly good bound for the spectral norm of $A \circ B$. Let $x = [x_i] \in \mathbb{C}^n$ be an arbitrary unit vector and use (3.6.1a) to compute

$$x^*[(A\circ B)^*(A\circ B)]x = \sum_{i,\,j=1}^{m} x^* \operatorname{diag}(\alpha_i)^*\, B^* E_{ii} E_{jj} B \operatorname{diag}(\alpha_j) x$$

$$= \sum_{i=1}^{m} (x \circ \alpha_i)^*\, [B^*(E_{ii}B)]\, (x \circ \alpha_i) \le \sum_{i=1}^{m} |||\, B^*(E_{ii}B)\, |||_2\, \| x \circ \alpha_i \|_2^2$$

$$\le \sum_{i=1}^{m} |||\, B\, |||_2\, |||\, E_{ii}B\, |||_2 \sum_{j=1}^{n} | x_j a_{ij} |^2 = |||\, B\, |||_2 \sum_{j=1}^{n} | x_j |^2 \sum_{i=1}^{m} | a_{ij} |^2 \| \beta_j \|$$

$$\le |||\, B\, |||_2 \sum_{j=1}^{n} | x_j |^2\, \| a_j \|^2\, r_1(B) \le |||\, B\, |||_2\, c_1(A)^2\, r_1(B).$$

Since $r_1(B) \le |||\, B\, |||_2$ and $c_1(A) \le m^{\frac{1}{2}} \| A \|_\infty$, we have

$$|||\, A \circ B\, |||_2 \le c_1(A)\, [r_1(B)\, |||\, B\, |||_2\,]^{\frac{1}{2}} \le \tfrac{1}{2} c_1(A)\, [r_1(B) + |||\, B\, |||_2\,]$$

$$\le c_1(A)\, |||\, B\, |||_2 \le m^{\frac{1}{2}} \| A \|_\infty\, |||\, B\, |||_2 \quad \text{for } A, B \in M_{m,n}. \tag{3.6.2a}$$

A similar calculation with (3.6.1b) [or apply (3.6.2a) to $(A \circ B)^T$] gives the companion bounds

$$|||\, A \circ B\, |||_2 \le r_1(A)\, [c_1(B)\, |||\, B\, |||_2\,]^{\frac{1}{2}} \le \tfrac{1}{2} r_1(A)\, [c_1(B) + |||\, B\, |||_2\,]$$

$$\le r_1(A)\, |||\, B\, |||_2 \le n^{\frac{1}{2}} \|A\|_\infty\, |||\, B\, |||_2 \quad \text{for } A,\, B \in M_{m,n}. \tag{3.6.2b}$$

Multiplying these two bounds gives the combined bound

$$|||\, A \circ B\, |||_2 \le [r_1(A)\; c_1(A)]^{\frac{1}{2}}\, [r_1(B)\; c_1(B)]^{\frac{1}{4}}\, [\, |||\, B\, |||_2\,]^{\frac{1}{2}}$$

$$\le [r_1(A)\; c_1(A)]^{\frac{1}{2}}\, |||\, B\, |||_2 \le [mn]^{\frac{1}{2}} \|A\|_\infty\, |||\, B\, |||_2 \quad \text{for } A,\, B \in M_{m,n}. \tag{3.6.3}$$

If $r_1(A)$ and $c_1(A)$ are replaced by the upper bound $|||\, A\, |||_2$ in these inequalities, one recovers Schur's bound 3.1(f). The bounds $|||\, A \circ B\, |||_2 \le [r_1(A)$ or $c_1(A)]\, |||\, B\, |||_2$ in (3.6.2a,b) were obtained with a different argument in [Ong, 1984]. See [Johnsen, 1977] for a different proof of the outer inequalities in (3.6.2a,b) when $m = n$, that is, $|||\, A \circ B\, |||_2 \le n^{\frac{1}{2}}\, \|A\|_\infty\, |||\, B\, |||_2$ for $A,\, B \in M_n$; Johnsen also identifies the cases of equality in these bounds.

3.7 The Hadamard Product as a Submatrix of a Positive Semidefinite Block Matrix. The Schur Product Theorem, which applies only to positive semidefinite matrices, can be combined with various transformations of a given matrix into a larger positive semidefinite matrix to give useful information about Hadamard products, *vid.* [Horn & Johnson, 1985; 7.3.7, 7.7.6-8] for other uses of block Hermitian matrices.

The basic observation is that for any given $X \in M_{m,n}$ the $(m+n)$-by-$(m+n)$ block matrix

$$\mathcal{X}_{I,I} \equiv \begin{bmatrix} I & X \\ X^* & I \end{bmatrix}$$

is positive semidefinite if and only if X is a *contraction*, that is, $|||\, X\, |||_2 \le 1$. Moreover, $\mathcal{X}_{I,I}$ is positive *definite* if and only if X is a *proper* contraction, that is, $|||\, X\, |||_2 < 1$. The reason is clear from a simple calculation: Let $z \equiv \begin{bmatrix} x \\ y \end{bmatrix} \in \mathbb{C}^{m+n}$ with $x \in \mathbb{C}^m$ and $y \in \mathbb{C}^n$ and note that $z^* \mathcal{X}_{I,I}\, z = \|x\|_2^2 + \|y\|_2^2 + 2\,\mathrm{Re}(x^* X y) \ge 0$ for all x and y if and only if $\|x\|_2^2 + \|y\|_2^2 \ge 2\,|x^* X y|$ for all x and y; these equivalent inequalities are strict for all x and y not both zero if and only if $\mathcal{X}_{I,I}$ is positive definite. The bound $|||\, X\, |||_2 \le 1$ in the positive semidefinite case and the bound $|||\, X\, |||_2 < 1$ in the positive definite case now follow from choosing x and y to be suitable unit left and right singular vectors of X, while the reverse implications follow from the basic inequality $|x^* X y| \le |||\, X\, |||_2\, \|x\|_2\, \|y\|_2$ and the arithmetic-geometric mean inequality.

A modest generalization of the preceding observation is that

> If $L \in M_m$ and $M \in M_n$ are given positive semidefinite matrices and $X \in M_{m,n}$, then the Hermitian block matrix

$$\mathcal{X}_{L,M} \equiv \begin{bmatrix} L & X \\ X^* & M \end{bmatrix} \tag{3.7.1}$$

is positive semidefinite if and only if there is a contraction $C \in M_{m,n}$ such that $X = L^{\frac{1}{2}} C M^{\frac{1}{2}}$. If L and M are positive definite, then this *contraction criterion* is equivalent to having $|||\, L^{-\frac{1}{2}} X M^{-\frac{1}{2}} \,|||_2 \le 1$. Moreover, $\mathcal{X}_{L,M}$ is positive definite if and only if L and M are positive definite and $|||\, L^{-\frac{1}{2}} X M^{-\frac{1}{2}} \,|||_2 < 1$.

If L and M are positive definite, this assertion follows from the observation that the congruence

$$\begin{bmatrix} L^{-\frac{1}{2}} & 0 \\ 0 & M^{-\frac{1}{2}} \end{bmatrix} \begin{bmatrix} L & X \\ X^* & M \end{bmatrix} \begin{bmatrix} L^{-\frac{1}{2}} & 0 \\ 0 & M^{-\frac{1}{2}} \end{bmatrix} = \begin{bmatrix} I & L^{-\frac{1}{2}} X M^{-\frac{1}{2}} \\ M^{-\frac{1}{2}} X^* L^{-\frac{1}{2}} & I \end{bmatrix}$$

preserves positive semidefiniteness, and hence $\mathcal{X}_{L,M}$ is positive semidefinite (respectively, positive definite) if and only if the matrix $C \equiv L^{-\frac{1}{2}} X M^{-\frac{1}{2}}$ is a contraction (respectively, a proper contraction). The case of singular L or M (which can happen only if $\mathcal{X}_{L,M}$ is singular) follows from a limiting argument.

If the block matrix $\mathcal{X}_{L,M}$ in (3.7.1) is positive semidefinite, then $X = L^{\frac{1}{2}} C M^{\frac{1}{2}}$ for some contraction C, and hence $|||\, X \,|||_2 = |||\, L^{\frac{1}{2}} C M^{\frac{1}{2}} \,|||_2 \le |||\, L^{\frac{1}{2}} \,|||_2 \, |||\, C \,|||_2 \, |||\, M^{\frac{1}{2}} \,|||_2 \le |||\, L^{\frac{1}{2}} \,|||_2 \, |||\, M^{\frac{1}{2}} \,|||_2$. This gives the useful bound

$$|||\, X \,|||_2^2 \le \lambda_{max}(L)\lambda_{max}(M) = |||\, L \,|||_2 \, |||\, M \,|||_2 \tag{3.7.2}$$

if the block matrix $\mathcal{X}_{L,M}$ in (3.7.1) is positive semidefinite. More generally, if we set $|X| \equiv (X^*X)^{\frac{1}{2}}$, it is known [Horn & Mathias, 1990b, Theorem 2.3] that

$$\|\, |X|^p \,\|^2 \le \|\, L^p \,\| \, \|\, M^p \,\| \text{ for all } p > 0 \text{ and any unitarily invariant norm } \|\cdot\|. \tag{3.7.3}$$

Since $|X|$ has the same singular values as X, the inequality (3.7.3) is $\|\, X \,\| \le \|\, L \,\| \, \|\, M \,\|$ when $p = 1$. In addition to these unitarily invariant norm inequalities, there are inequalities of similar form for the entry-wise ℓ_p norms on matrices $A = [a_{ij}] \in M_{m,n}$ given by

$$\begin{aligned} \|\, A \,\|_p &\equiv \Big[\sum_{i=1}^{m} \sum_{j=1}^{n} |a_{ij}|^p \Big]^{1/p} \quad \text{for } 1 \le p < \infty \\ \|\, A \,\|_\infty &\equiv \max\{\, |a_{ij}| : 1 \le i \le m,\ 1 \le j \le n \} \end{aligned} \tag{3.7.4}$$

It is shown in [Horn & Mathias, 1990b, Theorem 4.2] that

$\|\, X \,\|_{2p}^2 \le \|\, L \,\|_{2p} \, \|\, M \,\|_{2p}$ for all positive integers p and for all real $p \ge$

$m+n-2$ if the block matrix $\mathcal{X}_{L,M}$ in (3.7.1) is positive semidefinite. One may replace $2p$ by p in this bound if X has real nonnegative entries and if all the entries of L and M have nonnegative real part.

(3.7.5)

To apply these results to the study of Hadamard products, one takes two choices of L, M, and X that make the corresponding Hermitian block matrices $\mathcal{X}_{L,M}$ positive semidefinite and then forms their Hadamard product, which will then be positive semidefinite. The contraction criterion (3.7.1) and the inequalities (3.7.2,3,5) then give some necessary conditions. In the following discussion, we give several illustrations of this process.

First, let $A, B \in M_n$ be given and let $A = X^* Y$ for some $X, Y \in M_{r,n}$. Then the two $2n$-by-$2n$ block matrices

$$\begin{bmatrix} X^*X & A \\ A^* & Y^*Y \end{bmatrix} = \begin{bmatrix} X^*X & X^*Y \\ Y^*X & Y^*Y \end{bmatrix} = \begin{bmatrix} X \\ Y \end{bmatrix}^* \begin{bmatrix} X \\ Y \end{bmatrix} \tag{3.7.6}$$

and

$$\begin{bmatrix} |||B|||_2 I & B \\ B^* & |||B|||_2 I \end{bmatrix} \tag{3.7.7}$$

are both positive semidefinite. Their Hadamard product

$$\begin{bmatrix} |||B|||_2[I \circ (X^*X)] & (A \circ B) \\ (A \circ B)^* & |||B|||_2[I \circ (Y^*Y)] \end{bmatrix} \tag{3.7.8}$$

is positive semidefinite, and hence (3.7.2) gives the bound

$$|||A \circ B|||_2^2 \le |||B|||_2^2 \, |||I \circ (X^*X)|||_2 \, |||I \circ (Y^*Y)|||_2.$$

But $I \circ X^*X$ and $I \circ Y^*Y$ are diagonal matrices whose main diagonal entries are the squared Euclidean lengths of the columns of X and Y, respectively, so $|||I \circ (X^*X)|||_2 = c_1(X)^2$ and $|||I \circ (Y^*Y)|||_2 = c_1(Y)^2$. This gives Schur's bound 3.1(e):

$$|||A \circ B|||_2 \le c_1(X)\, c_1(Y)\, |||B|||_2. \tag{3.7.9}$$

If we apply (3.7.3) with $p = 1$ to (3.7.8), we obtain the more general bound

$$\| A \circ B \| \le [\, \| I \circ (X^*X) \| \, \| I \circ (Y^*Y) \| \,]^{\frac{1}{2}} \, |||B|||_2 \le c_1(X) c_1(Y) \, \| I \| \, |||B|||_2, \tag{3.7.10}$$

valid for any unitarily invariant norm $\| \cdot \|$.

The bound (3.7.9) and 3.1(e) can be considerably better than 3.1(f). For example, if $A = J \in M_n$, then $|||\, J\,|||_2 = n$ and hence Schur's bound 3.1(f) gives

$$|||\, B\,|||_2 = |||\, J \circ B\,|||_2 \leq |||\, J\,|||_2\, |||\, B\,|||_2 = n\, |||\, B\,|||_2$$

for any $B \in M_n$. However, $J = ee^T$, so we may write $J = X^* Y$ with $X = Y = e^T$ and all of the column lengths of X and Y are 1. The bound

$$|||\, B\,|||_2 = |||\, J \circ B\,|||_2 \leq c_1(e^T)\; c_1(e^T)\; |||\, B\,|||_2 = |||\, B\,|||_2$$

given by (3.7.9) is clearly more satisfactory.

One universally valid factorization $A = X^* Y$ in 3.1(e) is given by the singular value decomposition $A = V \Sigma W^*$, where V, $W \in M_n$ are unitary and $\Sigma = \operatorname{diag}(\sigma_1, \ldots, \sigma_n)$ with $\sigma_1 \geq \cdots \geq \sigma_n \geq 0$. Set $X \equiv \Sigma^{\frac{1}{2}} V^*$, $Y \equiv \Sigma^{\frac{1}{2}} W^*$, so $A = X^* Y$, $X^* X = V \Sigma V^* = (AA^*)^{\frac{1}{2}} \equiv P$, and $Y^* Y = W \Sigma W^* = (A^* A)^{\frac{1}{2}} \equiv Q$. Denote the largest main diagonal entry of P by $p_1(A)$, and that of Q by $q_1(A)$, so $p_1(A) = c_1(X)^2$ and $q_1(Y) = c_1(Y)^2$. Then 3.1(e) gives

$$|||\, A \circ B\,|||_2 \leq [p_1(A) q_1(A)]^{\frac{1}{2}}\, |||\, B\,|||_2 \qquad \text{for } A,\, B \in M_n, \tag{3.7.11}$$

a bound that was obtained in a very different way by [Walter, 1986], who was evidently unaware of Schur's theorem 3.1(e). A one-parameter family of bounds that includes (3.7.11) can be obtained by factoring $A = X^*_\alpha Y_{1-\alpha}$, where $X_\alpha \equiv \Sigma^\alpha V^*$, $Y_{1-\alpha} \equiv \Sigma^{1-\alpha} W^*$, and $\alpha \in [0, 1]$. If A is Hermitian, then $P = Q = (A^2)^{\frac{1}{2}} = U\, |\Lambda|\, U^* \equiv |\, A\, |$, where $A = U \Lambda U^*$ is a spectral decomposition. In this special case, (3.7.11) becomes

$$|||\, A \circ B\,|||_2 \leq p_1(A)\, |||\, B\,|||_2 \qquad \text{for } A,\, B \in M_n \text{ and } A \text{ Hermitian.} \tag{3.7.12}$$

Finally, if A is positive semidefinite, then $|\, A\, | = A$ and (3.7.12) becomes

$$|||\, A \circ B\,|||_2 \leq \max a_{ii}\, |||\, B\,|||_2 \quad \text{for } A,\, B \in M_n \text{ and } A \text{ positive semidefinite,} \tag{3.7.13}$$

which is 3.1(d). If B is also positive semidefinite, then $|||\, B\,|||_2 = \lambda_{max}(B)$, $|||\, A \circ B\,|||_2 = \lambda_{max}(A \circ B)$, and (3.7.13) becomes the upper bound in 3.1(b). Thus, the bound 3.1(e) for general matrices may be thought of not only as an improvement to 3.1(f), but also as a generalization to all matrices of the upper bound in 3.1(b), which applies only to positive semidefinite matrices.

For a given $A \in M_n$, consider the Hadamard product mapping $\mathcal{H}_A \colon M_n \rightarrow M_n$ given by $\mathcal{H}_A(X) = A \circ X$. The bound 3.1(e) gives a useful and versatile upper bound on the operator norm of $\mathcal{H}_A$ induced by the spectral norm:

$$\max\, \{|||\, A \circ B\,|||_2 \colon B \in M_n \text{ and } |||\, B\,|||_2 \leq 1\}$$

$$\leq \inf\{c_1(X)c_1(Y): X, Y \in M_{r,n}, X^*Y = A, r \geq 1\} \tag{3.7.14}$$

It is known that *equality* holds in (3.7.14), a result attributed in [Paulsen, 1986, 7.7] to U. Haagerup. Moreover, it is shown in [Ando, Horn, & Johnson, 1987, p. 363] that (3.7.14) can be generalized from the spectral norm to any unitarily invariant norm $\|\cdot\|$ on M_n:

$$\| A\circ B \| \leq c_1(X)c_1(Y) \| B \| \quad \text{for } A, B \in M_n, X, Y \in M_{r,n}, X^*Y = A, r \geq 1. \tag{3.7.15}$$

One can investigate the operator norm of the Hadamard product operator $\mathcal{H}_A$ induced by other norms on M_n. For example, if the spectral norm is replaced by the numerical radius norm $r(\cdot)$ defined in (3.2.14), it is known [Ando & Okubo, 1991] that for any given $A \in M_n$,

$$\max\{ r(A\circ B) : B \in M_n \text{ and } r(B) \leq 1\}$$

$$= \inf\{c_1(X)c_1(Y): X, Y \in M_{r,n}, W \in M_r \text{ and } |||W|||_2 \leq 1, X^*WY = A, r \geq 1\} \tag{3.7.16}$$

A second application of the Hermitian block technique (3.7.1), but with a focus on criteria that restrict the main diagonal blocks rather than the off-diagonal blocks, is to the special matrix $A\circ A^{-1}$. Let $A \in M_n$ be positive definite. Then the Hermitian block matrices

$$\begin{bmatrix} A & I \\ I & A^{-1} \end{bmatrix} \text{ and } \begin{bmatrix} A^{-1} & I \\ I & A \end{bmatrix} \tag{3.7.17}$$

are positive semidefinite since $C_1 \equiv A^{-\frac{1}{2}} I A^{\frac{1}{2}} = I$ and $C_2 \equiv A^{\frac{1}{2}} I A^{-\frac{1}{2}} = I$ are both contractions. Taking the Hadamard product, we find that

$$\begin{bmatrix} A\circ A^{-1} & I \\ I & A^{-1}\circ A \end{bmatrix} = \begin{bmatrix} A\circ A^{-1} & I \\ I & A\circ A^{-1} \end{bmatrix} \tag{3.7.18}$$

is positive semidefinite, and hence $C \equiv (A\circ A^{-1})^{-\frac{1}{2}} I (A\circ A^{-1})^{-\frac{1}{2}} = (A\circ A^{-1})^{-1}$ is a contraction. But $1 \geq |||C|||_2 = |||(A\circ A^{-1})^{-1}|||_2 = \lambda_{max}[(A\circ A^{-1})^{-1}] = 1/\lambda_{min}(A\circ A^{-1})$ means that

$$\lambda_{min}(A\circ A^{-1}) \geq 1 \quad \text{for } A \text{ positive definite,} \tag{3.7.19a}$$

a bound that also follows from (3.3.9b) with $B = A^{-1}$. See [Bapat & Kwong, 1987] and [Horn & Johnson, 1989, Theorem 5.4.6] for a family of block generalizations of (3.7.19a) and the companion inequality

$$\lambda_{min}(A\circ A^{-T}) \geq 1 \quad \text{for } A \text{ positive definite,} \tag{3.7.19b}$$

which appeared first in [Fiedler, 1957]. Although (3.7.19a,b) are implied by (and are apparently weaker than) (3.3.9b,a), it is interesting to note that (3.7.19a,b) actually imply (3.3.9b,a) [Horn & Johnson, 1989, Problem 1 in Section 5.4].

As a third application of the block matrix technique (3.7.1), let $A, B \in M_{m,n}$ and consider the positive semidefinite block matrices

$$\begin{bmatrix} A \\ I \end{bmatrix} \begin{bmatrix} A \\ I \end{bmatrix}^* = \begin{bmatrix} AA^* & A \\ A^* & I \end{bmatrix} \text{ and } \begin{bmatrix} B \\ I \end{bmatrix} \begin{bmatrix} B \\ I \end{bmatrix}^* = \begin{bmatrix} BB^* & B \\ B^* & I \end{bmatrix}, \tag{3.7.20}$$

which are evidently positive semidefinite, and their Hadamard product

$$\begin{bmatrix} (A\,A^*)\circ(BB^*) & A\circ B \\ (A\circ B)^* & I \end{bmatrix}. \tag{3.7.21}$$

The contraction criterion says that $A\circ B = [(AA^*)\circ(BB^*)]^{\frac{1}{2}}C$ for some contraction C, and hence $(A\circ B)(A\circ B)^* = [(AA^*)\circ(BB^*)]^{\frac{1}{2}}C\,C^*[(AA^*)\circ(BB^*)]^{\frac{1}{2}}$. But then for any unit vector $x \in \mathbb{C}^n$ we have

$$\begin{aligned} x^*(A\circ B)(A\circ B)^*x &= x^*[(AA^*)\circ(BB^*)]^{\frac{1}{2}}C\,C^*[(AA)^*\circ(BB^*)]^{\frac{1}{2}}x \\ &= \|C^*[(AA^*)\circ(BB^*)]^{\frac{1}{2}}x\|_2^2 \le |||C^*|||_2\, \|[(AA)^*\circ(BB^*)]^{\frac{1}{2}}x\|_2^2 \\ &\le 1\, \|(AA^*)\circ(BB^*)]^{\frac{1}{2}}x\|_2^2 = x^*[(AA^*)\circ(BB^*)]x \end{aligned}$$

and hence

$$(AA^*)\circ(BB^*) - (A\circ B)(A\circ B)^* \text{ is positive semidefinite for all } A, B \in M_{m,n}. \tag{3.7.22}$$

If $A, B \in M_n$ are Hermitian, this says that

$$A^2\circ B^2 - (A\circ B)^2 \text{ is positive semidefinite.} \tag{3.7.23}$$

See (4.1.16a,b) for a consequence of (3.7.23) when A, B are positive semidefinite. In the positive definite case, there is a generalization that extends (3.7.22) to a one-parameter family of inequalities:

$$A^p\circ B^p - (A\circ B)^p \text{ is positive semidefinite for all } p \in [1, 2] \text{ and all positive definite } A, B. \tag{3.7.24}$$

For a proof of this and many other related inequalities, and extensions to products of more than two matrices, see [Ando, 1979, Theorem 11(i)].

Recognizing that

$$\begin{bmatrix} I \\ B^* \end{bmatrix} \begin{bmatrix} I \\ B^* \end{bmatrix}^* = \begin{bmatrix} I & B \\ B^* & B^*B \end{bmatrix} \tag{3.7.25}$$

is also positive semidefinite, we can form the positive semidefinite matrix

$$\begin{bmatrix} AA^* & A \\ A^* & I \end{bmatrix} \circ \begin{bmatrix} I & B \\ B^* & B^*B \end{bmatrix} = \begin{bmatrix} I\circ(AA^*) & A\circ B \\ (A\circ B)^* & I\circ(B^*B) \end{bmatrix} \tag{3.7.26}$$

for $A, B \in M_{m,n}$. Using (3.7.3) with $p = 1$ and any unitarily invariant norm $\|\cdot\|$, we obtain

$$\| A\circ B \|^2 \le \| I\circ(AA^*) \| \, \| I\circ(B^*B) \| \le \| AA^* \| \, \| B^*B \|, \tag{3.7.27}$$

where the second inequality follows from (3.7.15) with $X = Y = I$. For another proof and the corresponding inequality for the usual matrix product, see [Horn & Mathias, 1990a, Theorem 3.1].

As a fourth application of (3.7.1), let $A, B \in M_n$, write $|X| \equiv (X^*X)^{\frac{1}{2}}$, and suppose A and B are normal. A direct calculation using the spectral theorem shows that the factors in the Hadamard product

$$\begin{bmatrix} |A| & A \\ A^* & |A| \end{bmatrix} \circ \begin{bmatrix} |B| & B \\ B^* & |B| \end{bmatrix} = \begin{bmatrix} |A|\circ|B| & A\circ B \\ (A\circ B)^* & |A|\circ|B| \end{bmatrix} \tag{3.7.28}$$

are positive semidefinite, and hence (3.7.3) gives

$$\| A\circ B \| \le \| \, |A|\circ|B| \, \| \quad \text{for normal } A, B \in M_n \tag{3.7.29}$$

for any unitarily invariant norm $\|\cdot\|$. The hypothesis of normality cannot be omitted in (3.7.29); see [Horn & Mathias, 1990b, Example 3.7] for a 3-by-3 counterexample with A positive definite and B not normal.

Any of the preceding types of positive semidefinite block matrices can be combined to apply (3.7.1-3). For example, if $A = [a_{ij}] \in M_n$ is positive semidefinite and $B \in M_n$, (3.7.28) and (3.7.7) give the positive semidefinite block matrix

$$\begin{bmatrix} A & A \\ A & A \end{bmatrix} \circ \begin{bmatrix} |||B|||_2 I & B \\ B^* & |||B|||_2 I \end{bmatrix} = \begin{bmatrix} |||B|||_2(I\circ A) & A\circ B \\ (A\circ B)^* & |||B|||_2(I\circ A) \end{bmatrix}$$

Thus,

$$\| A\circ B \| \le \| A\circ I \| \, \sigma_1(B) \le \max_i a_{ii} \| I \| \, \sigma_1(B) \tag{3.7.30}$$

for every unitarily invariant norm $\|\cdot\|$, $A, B \in M_n$, A positive semidefinite. This generalizes

3.1(d), to which it reduces when $\|\cdot\|$ is the spectral norm.

The inequalities (3.7.5) can be applied to all of the preceding positive semidefinite Hadamard products of block matrices to obtain ℓ_p norm inequalities analogous to the unitarily invariant norm inequalities we have been discussing.

4. Further Examples Involving the Hadamard Product. In Example 2.1 we have already discussed the Moutard uniqueness theorem as an illustration of how the Hadamard product arises naturally in an application. To further illustrate that Hadamard products arise in interesting problems from diverse areas of mathematics, we offer a selection of examples.

4.1 Example. The Derivative of a Hermitian Family and Monotone Matrix Functions. A matrix-valued function of a real variable $C(t) = [c_{ij}(t)]$ is said to be *continuous, differentiable, integrable,* etc. if each of the entry functions $c_{ij}(t)$ has the property of interest, and the derivative or integral of $C(t)$ is the matrix obtained by entry-wise differentiation or integration. For example, if $A, B \in M_n$ are given and $C(t) \equiv (1-t)B + tA$ for $0 \le t \le 1$, then the derivative $C'(t) = A - B$ is easily computed. Sometimes, particularly in perturbation theory, one wants to differentiate a function of a matrix that depends on a parameter. For example, if A and B are both positive definite, and if the function is the square root function, then $f(C(t)) \equiv [(1-t)B + tA]^{\frac{1}{2}}$ is well-defined and positive definite for $0 \le t \le 1$, but since it is not clear how to write its entries as explicit functions of t and the entries of A and B, it is also not clear how to use the definition to compute the derivative of $f(C(t))$.

Fortunately, Daleckiĭ and Kreĭn have a theorem that permits us to calculate the derivative of a function—differentiable in the ordinary scalar sense—of a differentiably parametrized family of Hermitian matrices, and it involves the Hadamard product in a very interesting way [Daleckiĭ & Kreĭn, 1951; 1965], [Horn & Johnson, 1989, Section 6.6]. Let $H(t) \equiv [h_{ij}(t)] \in M_n$ be a given continuously differentiable Hermitian matrix-valued function of a real variable t. For each t, let $H(t)$ have a spectral decomposition $H(t) = U(t)\Lambda(t)U^*(t)$, where $U(t) = [u_1(t) \dots u_n(t)] \in M_n$ is unitary and $\Lambda(t) = \mathrm{diag}(\lambda_1(t),\dots,\lambda_n(t))$ is real diagonal. Let $f(\cdot)$ be a given real-valued function of a real variable that is continuously differentiable in a neighborhood of all of the points $\lambda_i(t)$. If we write $E_i(t) \equiv u_i(t)u_i^*(t)$ for the rank one Hermitian matrix that projects onto the ith eigenvector of $H(t)$, then the spectral decomposition of $H(t)$ can be re-written as

$$H(t) = \sum_{i=1}^{n} \lambda_i(t)E_i(t) = U(t)\Lambda(t)U^*(t)$$

and $f(H(t))$ is defined by

$$f(H(t)) \equiv \sum_{i=1}^{n} f(\lambda_i(t))E_i(t) = U(t)\begin{bmatrix} f(\lambda_1(t)) & & 0 \\ & \ddots & \\ 0 & & f(\lambda_n(t)) \end{bmatrix} U^*(t)$$

$$= U(t)\, f(\Lambda(t))\, U^*(t). \tag{4.1.0}$$

Since $f(\cdot)$ is real-valued, $f(H(t))$ is necessarily Hermitian. This definition has the important property that for any unitary $U \in M_n$, $U^* f(H(t))\, U = f(U^* H(t)\, U)$, and that $H(t)$ and $f(H(t))$ have the same eigenvectors, that is, $H(t)$ and $f(H(t))$ commute.

The Daleckiĭ-Kreĭn formula for the derivative is

$$\frac{d}{dt} f(H(t)) = \sum_{i,j=1}^{n} \frac{f(\lambda_i(t)) - f(\lambda_j(t))}{\lambda_i(t) - \lambda_j(t)} E_i(t) \frac{d(H(t))}{dt} E_j(t), \tag{4.1.1}$$

where the difference quotient is to be replaced by the derivative $f'(\lambda_i(t))$ if $\lambda_i(t) = \lambda_j(t)$. It is convenient to re-write this formula as

$$\frac{d}{dt} f(H(t)) = \sum_{i,j=1}^{n} \frac{f(\lambda_i(t)) - f(\lambda_j(t))}{\lambda_i(t) - \lambda_j(t)} u_i(t)u_i^*(t)\; H'(t)\; u_j(t)u_j^*(t)$$

$$= \sum_{i,j=1}^{n} \frac{f(\lambda_i(t)) - f(\lambda_j(t))}{\lambda_i(t) - \lambda_j(t)} u_i(t)[u_i^*(t)\; H'(t)\; u_j(t)]u_j^*(t)$$

$$= \sum_{i,j=1}^{n} \frac{f(\lambda_i(t)) - f(\lambda_j(t))}{\lambda_i(t) - \lambda_j(t)} [u_i^*(t)\; H'(t)\; u_j(t)]\; u_i(t)u_j^*(t) \tag{4.1.2}$$

Although the preceding transformations make it evident that the Daleckiĭ-Kreĭn formula has something to do with Hadamard products, some further manipulations will clarify the point. Write $\eta_{ij}(t) \equiv u_i^*(t)H'(t)u_j(t)$ for $i, j = 1,\ldots, n$, and notice that $[\eta_{ij}(t)] = U^*(t)\; H'(t)\; U(t)$. For given points $\xi_1,\ldots, \xi_n$ in the domain of $f(\cdot)$, define the *Loewner matrix* or *difference quotient matrix*

$$K_f(\{\xi_i\}) \equiv \left[\frac{f(\xi_i) - f(\xi_j)}{\xi_i - \xi_j} \right] \in M_n \tag{4.1.3}$$

with the continuing understanding that entries for which $\xi_i = \xi_j$ are taken to be the derivative $f'(\xi_i)$. If we write $\kappa_{ij}(t) \equiv [f(\lambda_i(t)) - f(\lambda_j(t))]/[\lambda_i(t) - \lambda_j(t)]$ for the difference quotients that appear in (4.1.2), then $[\kappa_{ij}(t)] = K_f(\{\lambda_i(t)\})$. Finally, notice that for any matrix $G = [\gamma_{ij}] \in M_n$ we have the algebraic identity

$$\sum_{i,j=1}^{n} \gamma_{ij} u_i(t) u_j^*(t) = U\, G\, U^*.$$

The sum (4.1.2) can now be written as

$$\sum_{i,j=1}^{n} \kappa_{ij}(t)\, \eta_{ij}(t)\, u_i(t) u_j^*(t) = U(t)\, [\kappa_{ij}(t) \eta_{ij}(t)]\, U^*(t)$$

and hence the Daleckiĭ-Kreĭn formula is

$$\frac{d}{dt} f(H(t)) = U(t) \Big[K_f(\{\lambda_i(t)\}) \circ [U^*(t) H'(t)\, U(t)] \Big]\, U^*(t), \tag{4.1.4}$$

where $H(t) = U(t)\, \Lambda(t)\, U^*(t)$ is a spectral decomposition and $\Lambda(t) = \mathrm{diag}(\lambda_1(t), \ldots, \lambda_n(t))$. Thus, the derivative of $f(H(t))$ is unitarily similar to a Hadamard product of the Loewner matrix $K_f(\{\lambda_i(t)\})$ and a matrix that is unitarily similar to $H'(t)$. Notice that the main diagonal entries of $K_f(\{\lambda_i(t)\})$ are $\kappa_{ii}(t) = f'(\lambda_i(t))$.

Although the explicit form of the representation (4.1.4) leaves open the unpleasant possibility that $\frac{d}{dt} f(H(t))$ might depend on the choice of the diagonalizing unitary similarity $U(t)$, it is an essential part of the Daleckiĭ-Kreĭn Theorem that it does not. Thus, (4.1.4) gives a well-defined formula for $\frac{d}{dt} f(H(t))$, in which we are free to utilize any convenient unitary $U(t)$ that diagonalizes $H(t)$.

Since the eigenvalues $\lambda_i(t)$ of $H(t)$ are continuous in t [Horn & Johnson, 1987a, Appendix D], the Loewner matrix $K_f(\{\lambda_i(t)\})$ is continuous in t. Unfortunately, under the given conditions (and even if they are strengthened to require that $H(t)$ be infinitely differentiable in t) there may be no continuous choice of diagonalizing unitary similarities $U(t)$; for a 2-by-2 example originally due to Rellich, see [Kato, 1982, p. 128]). If $H(t)$ is an *analytic* function of the real parameter t, however, there is always an *analytic* choice of $U(t)$ [Kato, 1982, Theorem II.6.2]. Even though our hypotheses of continuous differentiability of $H(t)$ and $f(t)$ are insufficient to ensure that $U(t)$ can be chosen to be continuous, we are fortunate that, as a whole, the expression for $\frac{d}{dt} f(H(t))$ given by (4.1.4) is always a continuous function of t under these conditions, even if $U(t)$ is not [Horn & Johnson, 1989, Section 6.6].

The formula (4.1.4) has an important application to the theory of monotone matrix functions, which arise in scattering theory, electrical engineering, and quantum mechanics. If $A, B \in M_n$ are Hermitian, we write $A \succeq B$ if $A - B$ is positive semidefinite and we write $A \succ B$ if $A - B$ is positive definite; this is often called *the Loewner partial order* on the n-by-n Hermitian matrices. With this notation, to say that A is positive semidefinite is the same as saying $A \succeq 0$. A real-valued differentiable function $f(\cdot)$ on a real interval (a, b) is said to be *a monotone matrix function of order n on* (a, b) if $f(A) \succeq f(B)$ for every pair of n-by-n Hermitian matrices $A \succeq B$ whose eigenvalues lie in (a, b).

If $f(\cdot)$ is a monotone matrix function of order n on (a, b) and if $a < s < t < b$, then $tI \succ sI$ and hence $f(t)I = f(tI) \succeq f(sI) = f(s)I$. Thus, $f(t) - f(s) \geq 0$, that is, $f(\cdot)$ is a monotone (increasing) scalar function on (a, b) in the ordinary sense, and hence $f'(t) \geq 0$ for all $t \in (a, b)$. However, there are differentiable ordinary monotone functions on (a, b) that are not matrix monotone.

Let A and B be given n-by-n Hermitian matrices whose eigenvalues lie in a real interval (a, b), and let $f(\cdot)$ be a given differentiable real valued function on (a, b). Let $H(t) \equiv B + t(A - B) = tA + (1 - t)B$, so $H(0) = B$, $H(1) = A$, and $H'(t) = A - B$. We know from (1.1) that all the eigenvalues of $H(t)$ are in the interval (a, b) for $0 \leq t \leq 1$, so $f(H(t))$ is defined and has a derivative given by (4.1.4) for $0 \leq t \leq 1$. To say that $A \succeq B$, is, of course, to say that $x^*Ax \geq x^*Bx$ for every $x \in \mathbb{C}^n$, so for any given unit vector x it is natural to consider the real-valued function $\varphi_x(t) \equiv x^* f(H(t))x$. The mean value theorem gives

$$x^*f(A)x - x^*f(B)x = \varphi_x(1) - \varphi_x(0) = \varphi_x'(c) = x^*\left[\frac{d}{dt} f(H(t))\bigg|_{t=c}\right] x \tag{4.1.5}$$

for some $c \in \mathbb{C}^n$, so if we want to conclude that $f(A) \succeq f(B)$, it is sufficient to have $\varphi_x'(t) \geq 0$ for all $t \in [0, 1]$ and all $x \in \mathbb{C}^n$. Thus, it is sufficient that the matrix $df(H(t))/dt$ given by (4.1.4) be positive semidefinite for every $t \in [0, 1]$. Now compute

$$\begin{aligned} \varphi_x'(t) = x^*\left[\frac{d}{dt} f(H(t))\right] x &= x^* U(t) \left[K_f(\{\lambda_i(t)\}) \circ [U^*(t)H'(t)U(t)]\right] U^*(t)x \\ &= [U^*(t)x]^*\left[K_f(\{\lambda_i(t)\}) \circ \left[U^*(t)[A - B]U(t)\right]\right][U^*(t)x]. \end{aligned} \tag{4.1.6}$$

If we now assume that $A - B$ is positive semidefinite, then the congruence $U^*(t)[A - B]U(t)$ is also positive semidefinite for all $t \in [0, 1]$. If we further assume that the matrix $K_f(\{\lambda_i(t)\})$ is positive semidefinite for all $t \in [0, 1]$, we can conclude from the Schur Product Theorem that $\varphi'(t) \geq 0$ and hence that $f(A) \succeq f(B)$. This proves half of the following basic characterization of monotone matrix functions.

Loewner's Theorem. Let $f(\cdot)$ be a given continuously differentiable real-valued function on the real interval (a, b) and let n be a given positive integer. Then $f(\cdot)$ is a monotone matrix function of order n on (a, b) if and only if for every choice of points $\xi_1, \ldots, \xi_n \in (a, b)$ the Loewner matrix

$$K_f(\{\xi_i\}) \equiv \left[\frac{f(\xi_i) - f(\xi_j)}{\xi_i - \xi_j}\right] \tag{4.1.7}$$

is positive semidefinite. If $\xi_i = \xi_j$, the i, j entry is taken to be $f'(\xi_i)$.

Proof: The sufficiency of the stated condition has already been established. To show that it is also necessary, suppose the given function $f(\cdot)$ is a monotone matrix function of order n on (a, b) and let A and B be given n-by-n Hermitian matrices whose eigenvalues lie in the interval (a, b). Let $H(t) \equiv B + t(A - B)$, so $H(t) \succeq H(0) = B$ for all $t \in [0, 1]$ and hence $f(H(t)) \succeq f(H(0))$ for all $t \in [0, 1]$. Now let x be any given unit vector, let $t \in (0, 1]$, and observe that

$$x^* \left[\frac{f(H(t)) - f(H(0))}{t} \right] x \geq 0,$$

so if we let $t \downarrow 0$ we have

$$x^* \left[\frac{d}{dt} f(H(t)) \bigg|_{t=0} \right] x = [U^*(0)x]^* \Big[K_f(\{\lambda_i(0)\}) \circ \Big[U^*(0)[A - B] U(0) \Big] \Big] [U^*(0)x]$$

$$= [U^*x]^* \Big[K_f(\{\lambda_i\}) \circ \Big[U^*[A - B] U \Big] \Big] [U^*x] \geq 0,$$

where $H(0) = B = U \Lambda U^*$ and $\Lambda = \operatorname{diag}(\lambda_1, \ldots, \lambda_n)$. Since the unit vector x is arbitrary, this means that the matrix $K_f(\{\lambda_i\}) \circ \Big[U^*[A - B] U \Big]$ is positive semidefinite for every choice of the Hermitian matrices A and $B = U \Lambda U^*$ whose eigenvalues lie in (a, b) and are such that $A \succeq B$. Now choose $A = A_\epsilon \equiv B + \epsilon U J U^*$, where $\epsilon > 0$ is small enough so that all of the eigenvalues of A_ϵ lie in the open interval (a, b). Then $A_\epsilon - B = \epsilon U^* J U$ is positive semidefinite and the positive semidefiniteness of

$$K_f(\{\lambda_i\}) \circ \Big[U^*[A_\epsilon - B] U \Big] = K_f(\{\lambda_i\}) \circ \Big[U^*[\epsilon U J U^*] U \Big]$$

$$= \epsilon K_f(\{\lambda_i\}) \circ J = \epsilon K_f(\{\lambda_i\})$$

implies that $K_f(\{\lambda_i\})$ is positive semidefinite. This concludes the argument since we are free to choose $B = \operatorname{diag}(\lambda_1, \ldots, \lambda_n)$ for arbitrary points $\lambda_1, \ldots, \lambda_n \in (a, b)$. □

Positive definiteness of the Loewner matrix is a very strong condition, and valuable information can be obtained from drawing the consequences of the fact that its low-dimensional principal submatrices must all be positive semidefinite. Let $f(\cdot)$ be a given monotone matrix function of some given order $n \geq 2$ on the real interval (a, b). The 1-by-1 principal submatrices of $K_f(\{\xi_i\})$ are just the main diagonal entries $\kappa_{ii} = f'(\xi_i)$, so we have confirmation of the fact that $f'(t) \geq 0$ on (a, b). Positive semidefiniteness of all 2-by-2 principal submatrices is equivalent to having

$$\begin{bmatrix} f'(s) & \dfrac{f(s) - f(t)}{s - t} \\ \dfrac{f(s) - f(t)}{s - t} & f'(t) \end{bmatrix} \tag{4.1.8}$$

positive semidefinite for all $s, t \in (a, b)$. Using the fact that the determinant of this matrix must be nonnegative, we obtain the inequality

$$f'(s)\, f'(t)\, (s-t)^2 \geq [f(s) - f(t)]^2 \text{ for all } s, t \in (a, b). \tag{4.1.9}$$

If $f'(s) = 0$ for some $s \in (a, b)$, it follows that $f(t) = f(s)$ for all $t \in (a, b)$:

> A monotone matrix function $f(s)$ of order $n \geq 2$ on (a, b) is either constant or is strictly increasing on all of (a, b), that is, either $f'(t) \equiv 0$ for all $t \in (a, b)$, or $f'(t) > 0$ for all $t \in (a, b)$. (4.1.10)

In particular, every entry of the Loewner matrix of a nonconstant monotone matrix function is *positive.*

Now let $t \in (a, b)$ be given, take $s = t + h$, with $h \neq 0$ but sufficiently small so that $t + h \in (a, b)$, write $\Delta \equiv [f(t+h) - f(t)]/h$, and compute the following congruence of (4.1.8):

$$\begin{bmatrix} 1 & \frac{1}{h} \\ 0 & -\frac{1}{h} \end{bmatrix}^T \begin{bmatrix} f'(t+h) & \Delta \\ \Delta & f'(t) \end{bmatrix} \begin{bmatrix} 1 & \frac{1}{h} \\ 0 & -\frac{1}{h} \end{bmatrix}$$

$$= \begin{bmatrix} f'(t+h) & \dfrac{f'(t+h) - \Delta}{h} \\ \dfrac{f'(t+h) - \Delta}{h} & \dfrac{f'(t+h) - 2\Delta + f'(t)}{h^2} \end{bmatrix} \tag{4.1.11}$$

This matrix must also be positive semidefinite for all $t, t + h \in (a, b)$, $h \neq 0$. Now use the mean value theorem to write $\Delta = f'(c)$ for some c between t and $t + h$, and use the fact that the main diagonal entries of (4.1.11) must be nonnegative to obtain the inequality

$$f'(t+h) - 2\Delta + f'(t) = f'(t+h) - 2f'(c) + f'(t) \geq 0.$$

It follows from continuity of $f'(\cdot)$ that

> If $f(\cdot)$ is a monotone matrix function of order $n \geq 2$ on (a, b), then $f'(\cdot)$ is a *convex* function on (a, b). (4.1.12)

To draw a final consequence from (4.1.8), it is convenient to make the additional assumption that $f(\cdot)$ is *three* times continuously differentiable. With this assumption, we can use Taylor's Theorem to expand the entries of (4.1.11) and obtain:

$$\begin{bmatrix} f'(t) & \frac{1}{2!}f''(t) \\ \frac{1}{2!}f''(t) & \frac{1}{3!}f'''(t) \end{bmatrix} + O(h) \text{ as } h \to 0,$$

which again must be positive semidefinite. Now let $h \to 0$ and conclude that

> If $f(\cdot)$ is a three times continuously differentiable monotone matrix function of order $n \geq 2$ on (a, b), then the matrix
>
> $$\begin{bmatrix} f'(t) & \frac{1}{2!}f''(t) \\ \frac{1}{2!}f''(t) & \frac{1}{3!}f'''(t) \end{bmatrix} \tag{4.1.13}$$
>
> is positive semidefinite for every $t \in (a, b)$. In particular, if $f(\cdot)$ is not constant, then $[f'(t)]^{-\frac{1}{2}}$ is a concave function on (a, b).

The last assertion is equivalent to the differential inequality $2f'f''' - 3(f'')^2 \geq 0$, which is exactly what one obtains by requiring that the determinant of (4.1.13) be nonnegative.

All of these conditions have followed from looking at 2-by-2 principal submatrices of the Loewner matrix. If one looks at larger principal submatrices, similar manipulations lead to the conclusion that

> If $f(\cdot)$ is a monotone matrix function of order n on (a, b) with $2n-1$ continuous derivatives, then the matrix
>
> $$\left[\frac{1}{(i+j-1)!}\, f^{(i+j-1)}(t)\right] \quad i, j = 1,\ldots, n \tag{4.1.14}$$
>
> is positive semidefinite for every $t \in (a,b)$. In particular, if $f(\cdot)$ is not constant, then all the odd order derivatives $f', f''',\ldots, f^{(2n-1)}$ are positive on (a, b).

In fact, the smoothness assumption can be dropped—careful attention to the analysis shows that if $f(\cdot)$ is merely continuously differentiable, then positive semidefiniteness of the Loewner matrix of order n *implies* that $f(\cdot)$ has $2n-3$ continuous derivatives and that $f^{(2n-3)}(\cdot)$ is convex, and hence has a finite derivative except for at most a countable number of points. Since that derivative is monotone increasing, it is differentiable almost everywhere. Thus, if the extra smoothness assumption is dropped in (4.1.14), one still has the same conclusion for almost all $t \in (a, b)$. Conversely, it is known that positive semidefiniteness of the matrix (4.1.14) on (a,b) is sufficient for a sufficiently smooth function $f(\cdot)$ to be a monotone matrix function of order n on (a,b) [Donoghue, 1974, Theorem V in Chapter VIII].

The identity function $f(t) = t$ is a monotone matrix function on every real interval since its Loewner matrix $K_f = J$ is positive semidefinite (and rank one, hence singular for $n > 1$). The function $f(t) = -1/t$ is a monotone matrix function on the positive real line since its Loewner matrix $K_f(\{\xi_i\}) = [1/\xi_i\xi_j]$ is also rank one positive semidefinite. A less trivial example of a monotone matrix function is the square root function $f(t) = t^{\frac{1}{2}}$ on the positive real line. If $\xi_1,\ldots, \xi_n$ are given positive real numbers, then the Loewner matrix for the square root function is

$$K_f(\{\xi_i\}) = \left[\frac{\xi_i^{\frac{1}{2}} - \xi_j^{\frac{1}{2}}}{\xi_i - \xi_j}\right] = \left[\frac{1}{\xi_i^{\frac{1}{2}} + \xi_j^{\frac{1}{2}}}\right] = [\kappa_{ij}],$$

and the matrix entries have the integral representation

$$\frac{1}{\xi_i^{\frac{1}{2}} + \xi_j^{\frac{1}{2}}} = \int_0^\infty e^{-(\xi_i^{\frac{1}{2}} + \xi_j^{\frac{1}{2}})s}\, ds.$$

For any $n = 1, 2,\ldots$ and any given vector $x = [x_i] \in \mathbb{C}^n$ we have

$$x^* K_f(\{\xi_i\})x = \int_0^\infty \sum_{i,\,j=1}^{n} e^{-(\xi_i^{\frac{1}{2}} + \xi_j^{\frac{1}{2}})s}\, \bar{x}_i x_j\, ds = \int_0^\infty \left|\sum_{i=1}^{n} e^{-\xi_i^{\frac{1}{2}} s}\, \bar{x}_i\right|^2 ds \geq 0, \qquad (4.1.15)$$

with strict inequality for all $x \neq 0$ if and only if the n positive points $\xi_1,\ldots, \xi_n$ are *distinct*. Thus, the Loewner matrix $K_f(\{\xi_i\})$ for the square root function is always positive semidefinite and is positive definite whenever all the points $\xi_1,\ldots, \xi_n$ are distinct. Loewner's theorem now guarantees that the square root function is a monotone matrix function of every order $n = 1, 2,\ldots$ on the positive real line: If A and B are given positive definite matrices of the same size and if $A \succeq B$, then $A^{\frac{1}{2}} \succeq B^{\frac{1}{2}}$. As an application of this fact, re-write (3.7.23) as $A^2 \circ B^2 \succeq (A \circ B)^2$ and apply the square root function to both sides:

$$(A^2 \circ B^2)^{\frac{1}{2}} \succeq A \circ B \quad \text{whenever } A, B \in M_n \text{ are positive semidefinite.} \qquad (4.1.16a)$$

Replacing each matrix by its square root gives the equivalent inequality

$$(A \circ B)^{\frac{1}{2}} \succeq A^{\frac{1}{2}} \circ B^{\frac{1}{2}} \quad \text{whenever } A, B \in M_n \text{ are positive semidefinite.} \qquad (4.1.16b)$$

Although it is evident from the definition that the functional composition of two monotone matrix functions with compatible range and domain must be a monotone matrix function, it is interesting to see how this useful observation follows from Loewner's Theorem and the Schur Product Theorem. Suppose f is a monotone matrix function on the interval (a, b) whose range is contained in the interval (c, d), and suppose g is a monotone matrix function on

(c, d). Then the composition $(f \circ g)(t) \equiv f(g(t))$ is a real-valued differentiable function on (a, b). Its Loewner matrix is

$$K_{f \circ g}(\{\xi_i\}) = \left[\frac{f(g(\xi_i)) - f(g(\xi_j))}{\xi_i - \xi_j}\right] = \left[\frac{f(g(\xi_i)) - f(g(\xi_j))}{g(\xi_i) - g(\xi_j)} \; \frac{g(\xi_i) - g(\xi_j)}{\xi_i - \xi_j}\right]$$

$$= K_f(\{g(\xi_i)\}) \circ K_g(\{\xi_i\}),$$

and since this is the Hadamard product of two positive semidefinite matrices it must be positive semidefinite by Schur's Theorem. As a consequence, for example, we know that the function $f(t) = -t^{-\frac{1}{2}} = -(t^{\frac{1}{2}})^{-1}$ must be a monotone matrix function on the positive real line.

So far in our discussion of monotone matrix functions we have used only the qualitative part 3.1(a) of Schur's Theorem, but the quantitative bounds 3.1(b,d,f) have important applications as well. Let A and B be given n-by-n Hermitian matrices whose eigenvalues lie in a given real interval (a, b), and let $f(\cdot)$ be a given differentiable real-valued function on (a, b). Let x be a given unit vector, let $H(t) \equiv B - t(A - B) = U(t)\,\Lambda(t)\,U^*(t)$ for $0 \le t \le 1$, let $\varphi_x(t) \equiv x^* f(H(t))x$, and recall from (4.1.5) and (4.1.6) that

$$x^*[f(A) - f(B)]x = \varphi_x(1) - \varphi_x(0) = \varphi_x'(c) = x^*\left[\frac{d}{dt} f(H(t))\bigg|_{t=c}\right]x \tag{4.1.17}$$

$$= [U^*(c)x]^*\left[K_f(\{\lambda_i(c)\}) \circ \left[U^*(c)[A - B]U(c)\right]\right][U^*(c)x]$$

for some $c \in [0, 1]$. We are interested in applying these ideas to the problem of estimating the modulus of continuity of $f(\cdot)$ with respect to the spectral norm and to the question of strict monotonicity of a monotone matrix function.

First consider the modulus of continuity of a not-necessarily-monotone matrix function $f(\cdot)$. Start with (4.1.17) and compute

$$|x^*[f(A) - f(B)]x| = |\varphi_x'(c)| = |x^*\left[\frac{d}{dt} f(H(t))\bigg|_{t=c}\right]x|$$

$$= |[U^*(c)x]^*\left[K_f(\{\lambda_i(c)\}) \circ \left[U^*(c)[A - B]U(c)\right]\right][U^*(c)x]|$$

$$\le |||\, K_f(\{\lambda_i(c)\}) \circ \left[U^*(c)[A - B]U(c)\right] |||_2$$

$$\le \max_t |||\, K_f(\{\lambda_i(t)\}) \circ \left[U^*(t)[A - B]U(t)\right] |||_2, \quad 0 \le t \le 1.$$

The Rayleigh-Ritz Theorem guarantees that there is some unit vector x for which

$$|x^*[f(A) - f(B)]x| = |||\, f(A) - f(B) \,|||_2,$$

so we have the bound

$$||| f(A) - f(B) |||_2 \leq \max_t ||| K_f(\{\lambda_i(t)\}) \circ \left[U^*(t)[A - B] U(t) \right] |||_2, \quad 0 \leq t \leq 1. \tag{4.1.18}$$

If we don't assume that $f(\cdot)$ is a monotone matrix function, we can use 3.1(f) to simplify this bound and obtain

$$||| f(A) - f(B) |||_2 \leq \max_t ||| K_f(\{\lambda_i(t)\}) |||_2 \, ||| U^*(t)[A - B] U(t) |||_2$$

$$= \max_t ||| K_f(\{\lambda_i(t)\}) |||_2 \, ||| A - B |||_2, \quad 0 \leq t \leq 1 \tag{4.1.19a}$$

$$\leq \max_{\xi_i} ||| K_f(\{\xi_i\}) |||_2 \, ||| A - B |||_2, \;\; \xi_1, \ldots, \xi_n \in [\alpha(A,B), \beta(A, B)], \tag{4.1.19b}$$

where the endpoints $\alpha(A,B)$ and $\beta(A,B)$ are defined in (1.1):

$$[\alpha(A,B), \beta(A,B)] \equiv [\min \{\lambda_{min}(A), \lambda_{min}(B)\}, \max \{\lambda_{max}(A), \lambda_{max}(B)\}]. \tag{4.1.20}$$

The maximum in (4.1.19b) is finite for each compact interval $[\alpha, \beta] \subset (a, b)$, and hence the maximum in (4.1.19b) is finite for each fixed pair of matrices A and B, because the entries of the Loewner matrix are all continuous functions of $\xi_1, \ldots, \xi_n$ on the compact interval $[\alpha, \beta]$. In particular, the bound (4.1.19b) shows that every continuously differentiable real-valued function $f(\cdot)$ on (a, b) generates a matrix-valued function of a matrix—according to the definition (4.1.0)—that is a norm-continuous function of its matrix argument:

> If $f(\cdot)$ is a differentiable real-valued function on (a, b), then the matrix-valued function on n-by-n Hermitian matrices defined by (4.1.0) is Lipschitz continuous with respect to the spectral norm (and hence with respect to every norm). Moreover, the matrix function $f(\cdot)$ is uniformly Lipschitz norm-continuous on the set of n-by-n Hermitian matrices whose eigenvalues lie in a given compact interval $[\alpha, \beta] \subset (a, b)$: If $[\alpha, \beta] \subset (a, b)$, then there exists a finite $L = L(\alpha, \beta, f, n)$ such that $||| f(A) - f(B) |||_2 \leq L ||| A - B |||_2$ for all Hermitian matrices $A, B \in M_n$ whose eigenvalues all lie in $[\alpha, \beta]$.

(4.1.21)

An explicit expression for a Lipschitz constant L is in (4.1.19b). But the bounds (4.1.19a,b) cannot be expected to be very good and can grow arbitrarily large as n increases. For example, the Loewner matrix for the identity function $f(t) = t$ is $K_f(\{\xi_i\}) = J$, whose spectral norm is n. Thus, (4.1.19a,b) gives the unsatisfactory bound $||| A - B |||_2 \leq n ||| A - B |||_2$.

If we now assume that the function $f(\cdot)$ is a *monotone* matrix function of order n, then the difference quotient matrix $K_f(\{\lambda_i(t)\}) = [\kappa_{ij}(t)]$ is positive semidefinite and we can apply 3.1(d) to (4.1.18) and obtain the more satisfactory bound

$$|||f(A) - f(B)|||_2 \le \max_t \max_i \kappa_{ii}(t) \, |||A - B|||_2 \qquad 0 \le t \le 1, \quad i = 1, \ldots, n$$

$$= \max_t \max_i f'(\lambda_i(t)) \, |||A - B|||_2 \qquad 0 \le t \le 1, \quad i = 1, \ldots, n$$

$$\le \max_s f'(s) \, |||A - B|||_2 \qquad s \in [\alpha(A,B), \beta(A,B)], \tag{4.1.22}$$

where the endpoints $\alpha(A,B)$ and $\beta(A,B)$ are defined in (4.1.20). As in (4.1.19b), this bound implies uniform Lipschitz continuity of the matrix function $f(\cdot)$ of order n on the set of n-by-n Hermitian matrices whose eigenvalues lie in a compact interval $[\alpha, \beta] \subset (a, b)$:

$$|||f(A) - f(B)|||_2 \le \max_s f'(s) \, |||A - B|||_2 \qquad s \in [\alpha, \beta]. \tag{4.1.23}$$

For the identity function, this bound at least gives a sensible result. For the square root function, $f'(t) = 1/2t^{\frac{1}{2}}$, so (4.1.22) gives the bound

$$|||A^{\frac{1}{2}} - B^{\frac{1}{2}}|||_2 \le \frac{1}{2\,[\min\{\lambda_{min}(A), \lambda_{min}(B)\}]^{\frac{1}{2}}} \, |||A - B|||_2 \quad \text{for positive definite } A, B \tag{4.1.24}$$

We now consider how (4.1.17) and the lower bound in 3.1(b) can be used to ensure that a monotone matrix function is strictly monotone. Let $A \succeq B \in M_n$ be Hermitian matrices whose eigenvalues lie in (a, b) and let $f(\cdot)$ be a monotone matrix function of order n on (a, b). Then for any unit vector x, we can start with (4.1.17) to write

$$x^*[f(A) - f(B)]x = [U^*(c)x]^* \Big[K_f(\{\lambda_i(c)\}) \circ \big[U^*(c)[A - B]\,U(c)\big]\Big][U^*(c)x]$$

$$\ge \lambda_{min}\Big[K_f(\{\lambda_i(c)\}) \circ \big[U^*(c)[A - B]\,U(c)\big]\Big]$$

$$\ge \min_i \kappa_{ii}(c)\, \lambda_{min}\big[U^*(c)[A - B]\,U(c)\big] = \min_i f'(\lambda_i(c))\, \lambda_{min}(A - B)$$

$$\ge \min_s f'(s)\, \lambda_{min}(A - B) \qquad s \in [\alpha(A,B), \beta(A,B)] = [\lambda_{min}(B), \lambda_{max}(A)],$$

where $[\alpha(A), \beta(B)]$ is given by (4.1.20) and $\alpha(A, B) = \lambda_{min}(B)$, $\beta(A, B) = \lambda_{max}(A)$ because we have assumed that $A \succeq B$. By the Rayleigh-Ritz Theorem we have

$$\lambda_{min}[f(A) - f(B)] \ge \min_s f'(s)\, \lambda_{min}(A - B) \qquad s \in [\lambda_{min}(B), \lambda_{max}(A)] \tag{4.1.25}$$

if $A \succeq B \in M_n$ are Hermitian and if $f(\cdot)$ is a monotone matrix function of order n on an open interval containing $[\lambda_{min}(B), \lambda_{max}(A)]$. Recall from (4.1.10) that a monotone matrix function of order $n \geq 2$ is either constant $[f'(s) = 0$ for all $s \in (a, b)]$ or strictly increasing $[f'(s) > 0$ for all $s \in (a, b)]$. Thus, the lower bound in (4.1.25) is *positive* if we have strict domination $A \succ B$ and if the monotone matrix function $f(\cdot)$ is not constant:

> Every nonconstant monotone matrix function of order $n \geq 2$ on (a, b) is *a strictly matrix monotone function of order n on* (a,b): $f(A) \succ f(B)$ whenever $A, B \in M_n$ are Hermitian matrices with all their eigenvalues in (a, b) and $A \succ B$.

(4.1.26)

The square root function $f(t) = t^{\frac{1}{2}}$ is therefore a strictly monotone matrix function on $(0, \infty)$, and for any two positive definite matrices $A \succ B$ the bound (4.1.25) gives

$$\lambda_{min}(A^{\frac{1}{2}} - B^{\frac{1}{2}}) \geq \tfrac{1}{2}[\lambda_{max}(A)]^{-\frac{1}{2}} \lambda_{min}(A - B) = \tfrac{1}{2}[\,|||\, A\, |||_2]^{-\frac{1}{2}} \lambda_{min}(A - B). \qquad (4.1.27)$$

In fact, we shall see that the square root function enjoys an even stronger matrix monotonicity property.

In the case of a scalar function, (4.1.26) is all there is to say about strict monotonicity since the real numbers are totally ordered. But the Loewner partial order $\succeq$ on Hermitian matrices is only a *partial* order for $n \geq 2$ and it can happen that $A \succeq B$ but not $A \succ B$—that is, $y^*(A - B)y \geq 0$ for all $y \in \mathbb{C}^n$, but $x^*(A - B)x = 0$ for some nonzero vector x—which happens for $A \succeq B$ if and only if $A - B$ is singular and $(A - B)x = 0$. Such a vector x need not be a common eigenvector of both A and B—all we know is that $Ax = Bx$. In such a case it could still be that $f(A) \succ f(B)$, but there is one situation in which it is futile to hope for this: Suppose a unit vector x is an eigenvector for both A and B associated with the same eigenvalue λ, so $x^*(A - B)x = x^*(\lambda x - \lambda x) = 0$. The definition (4.1.0) ensures that x is also an eigenvector for both $f(A)$ and $f(B)$ associated with the same eigenvalue $f(\lambda)$, so $x^*[f(A) - f(B)]x = x^*[f(\lambda)x - f(\lambda)x] = 0$ and we cannot have $f(A) \succ f(B)$. However, under a natural assumption about $f(\cdot)$, this is the only exceptional case, and the property 3.1(c) of the Hadamard product plays a crucial role in the argument. Before giving the theorem on strong monotonicity, we isolate the two key ideas in its proof. The first lemma gives the result that we want in the special case when B has distinct eigenvalues.

Lemma A. Let $f(\cdot)$ be a given monotone matrix function of order n on (a, b), and let $A, B \in M_n$ be given Hermitian matrices with all their eigenvalues in (a, b) and $A \succeq B$, so $f(A) \succeq f(B)$. Suppose the eigenvalues $\lambda_1(B) > \cdots > \lambda_n(B)$ of B are distinct and that the Loewner matrix $K_f(\{\lambda_i(B)\})$ is positive definite. If it is *not* the case that $f(A) \succ f(B)$, that is, if $x^*[f(A) - f(B)]x = 0$ for some unit vector x, then A and B have a common eigenvector for which they both have the same eigenvalue.

Proof: With the given hypotheses and notation, let $H(t) \equiv B + t(A - B)$ for $0 \le t \le 1$, and let $H(t) = U(t)\,\Lambda(t)\,U^*(t)$ be a spectral decomposition, with $H(0) = B \equiv U\Lambda\,U^*$ for a unitary U and a real diagonal $\Lambda = \mathrm{diag}(\lambda_1(B),\ldots,\lambda_n(B))$. Let $\varphi_x(t) \equiv x^* H(t)x$, so $\varphi_x(t)$ is a continuously differentiable monotone increasing function on [0, 1], which must be constant since $\varphi_x(1) - \varphi_x(0) = 0$. Use (4.1.17) to compute

$$0 = \varphi_x'(0) = [U^*(0)x]^*\Big[K_f(\{\lambda_i(0)\})\circ(U^*(0)[A-B]U(0))\Big][U^*(0)x]$$

$$= [U^*x]^*\Big[K_f(\{\lambda_i(B)\})\circ(U^*[A-B]U)\Big][U^*x]$$

$$= [U^*x]^*\Big[K_f(\{\lambda_i(B)\})\circ(U^*A\,U - U^*B\,U)\Big][U^*x]$$

$$= [U^*x]^*\Big[K_f(\{\lambda_i(B)\}\circ(U^*A\,U - \Lambda)\Big][U^*x].$$

Thus, the matrix $K_f(\{\lambda_i(B)\})\circ(U^*A\,U-\Lambda)$ must be singular, and since $K_f(\{\lambda_i(B)\})$ is positive definite and $U^*A\,U-\Lambda$ is positive semidefinite, 3.1.(d) implies that some main diagonal entry of $U^*A\,U-\Lambda$, say its k, k entry, is zero. But then the entire k^{th} row and column of $U^*A\,U-\Lambda$ is zero by (1.4), so all the off-diagonal entries in the kth row and column of $U^*A\,U$ are zero and the k, k entry of $U^*A\,U$ is $\lambda_k(B)$. This means that the kth unit basis vector e_k is a common eigenvector for both $U^*A\,U$ and Λ, for which they have the same eigenvalue $\lambda_k(B)$, and hence the kth column of U is a common eigenvector for A and B with common eigenvalue $\lambda_k(B)$. □

The second lemma gives a perturbation theorem that permits us to handle the case in which B does not have distinct eigenvalues.

Lemma B. Let $B \in M_n$ be a given Hermitian matrix with eigenvalues $\lambda_1 \ge \cdots \ge \lambda_n$, let $x \in \mathbb{C}^n$ be a given unit vector, and let $\epsilon > 0$ be given. Then there exists a Hermitian matrix $\hat{B}_{x,\epsilon} \in M_n$ such that

(a) $\hat{B}_{x,\epsilon}$ has distinct eigenvalues $\hat{\lambda}_1 > \cdots > \hat{\lambda}_n$,

(b) Every eigenvector of $\hat{B}_{x,\epsilon}$ is also an eigenvector of B,

(c) $B \succeq \hat{B}_{x,\epsilon}$,

(d) $\lambda_i \ge \hat{\lambda}_i \ge \lambda_i - \epsilon$ for $i = 1,\ldots, n$, and

(e) Using the definition (4.1.0), $f(B)x = f(\hat{B}_{x,\epsilon})x$ for every real-valued function $f(\cdot)$ whose domain includes the eigenvalues of both B and $\hat{B}_{x,\epsilon}$.

Proof: If B has n distinct eigenvalues, take $\hat{B}_{x,\epsilon} \equiv B$. If not, then B has a spectral decomposition $B = U\Lambda\,U^*$ in which $U = [u_1\ \ldots\ u_n]$ is unitary, $\Lambda = \mu_1 I_{n_1} \oplus \cdots \oplus \mu_k I_{n_k}$ with each $I_{n_i} \in M_{n_i}$, $n_1 + \cdots + n_k = n$, $k < n$, and $\mu_1 > \cdots > \mu_k$. For each $i = 1,\ldots, k$, the eigenspace S_i of B

corresponding to the eigenvalue μ_i is spanned by the n_i columns of U in positions $(n_1 + \cdots + n_{i-1}) + 1$, $(n_1 + \cdots + n_{i-1}) + 2, \ldots, (n_1 + \cdots + n_{i-1}) + n_i$, $i = 1, \ldots, k$. By projecting the given unit vector x onto the eigenspaces S_i we can write $x = x_1 + \cdots + x_k$, where each $x_i \in \mathbb{C}^n$, $x_i^* x_j = 0$ for all $i \neq j$, and $Bx_i = \mu_i x_i$ for $i = 1, \ldots, k$. For each eigenspace S_i in which $x_i \neq 0$, construct an ordered orthonormal basis $\hat{u}_1^{(i)} \equiv x_i / \|x_i\|_2, \hat{u}_2^{(i)}, \ldots, \hat{u}_{n_i}^{(i)}$ whose first vector is in the direction x_i; for each eigenspace S_i in which $x_i = 0$, take the ordered orthonormal basis $\hat{u}_1^{(i)}, \hat{u}_2^{(i)}, \ldots, \hat{u}_{n_i}^{(i)}$ to be the corresponding n_i columns of U. In either event, x is orthogonal to *all* of the basis vectors $\hat{u}_2^{(i)}, \ldots, \hat{u}_{n_i}^{(i)}$, and is *not* orthogonal to *some* of the basis vectors $\hat{u}_1^{(i)}$, $i = 1, \ldots, k$. Now let

$$\hat{U} \equiv [\hat{u}_1^{(1)} \ \hat{u}_2^{(1)} \ \ldots \ \hat{u}_{n_1}^{(1)} \ \ldots \ \hat{u}_1^{(k)} \ \hat{u}_2^{(k)} \ \ldots \ \hat{u}_{n_k}^{(k)}]$$

be the unitary matrix whose columns are the ordered basis vectors we have just constructed for the k eigenspaces of B and notice that $B = \hat{U} \Lambda \hat{U}^*$. Let $\delta \equiv \min \{\epsilon, \mu_i - \mu_{i+1}: i = 1, \ldots, k-1\}/n$, let $\hat{\lambda}_1 > \cdots > \hat{\lambda}_n$ be the respective distinct values $\mu_1, \mu_1 - \delta, \mu_1 - 2\delta, \ldots, \mu_1 - (n_1 - 1)\delta, \mu_2, \mu_2 - \delta, \mu_2 - 2\delta, \ldots, \mu_2 - (n_2 - 1)\delta, \ldots, \mu_k, \mu_k - \delta, \mu_k - 2\delta, \ldots, \mu_k - (n_k - 1)\delta$, let $\hat{\Lambda} \equiv \operatorname{diag}(\hat{\lambda}_1, \ldots, \hat{\lambda}_n)$, and take $\hat{B}_{x,\epsilon} \equiv \hat{U} \hat{\Lambda} \hat{U}^*$. The assertions (a) - (d) are immediately verified; (e) follows from observing that every entry of the vector $\hat{U}^* x$ corresponding to the perturbed eigenvalues is zero, and hence $\operatorname{diag}(f(\hat{\lambda}_1), \ldots, f(\hat{\lambda}_n)) \hat{U}^* x = \operatorname{diag}(f(\lambda_1), \ldots, f(\lambda_n)) \hat{U}^* x$, from which it follows that $f(\hat{B})x = \hat{U} \operatorname{diag}((f(\hat{\lambda}_1), \ldots, f(\hat{\lambda}_n)) \hat{U}^* x = f(B)x$. □

We say that $f(\cdot)$ is *strongly matrix monotone of order n on (a,b)* if $f(A) \succ f(B)$ for all Hermitian $A, B \in M_n$ whose eigenvalues lie in (a, b), satisfy $A \succeq B$, and are such that $x^*(A - B)x > 0$ for every common eigenvector x of A and B, that is, $A \succeq B$ and there is no common eigenvector for which A and B have the same eigenvalue. Notice that if $A \succ B$, or if $A \succeq B$ and they have no common eigenvectors, then they certainly satisfy the latter condition. A simple limiting argument shows that a strongly monotone matrix function must be a monotone matrix function.

An ordinary differentiable real valued function on a real interval is monotone increasing if and only if its derivative is nonnegative, and it is *strictly* (which is the same as *strongly*) monotone increasing if and only if its derivative is *positive*. We have seen that a function $f(\cdot)$ is a monotone matrix function of order n if and only if its Loewner matrix is positive semidefinite. The theorem is that $f(\cdot)$ is *strongly* matrix monotone if its Loewner matrix $K_f(\{\xi_i\})$ is positive *definite* whenever this is structurally possible—there is no hope of having it nonsingular if $\xi_i = \xi_j$ for some $i \neq j$, for in this case two rows of $K_f(\{\xi_i\})$ are identical.

Strong Monotonicity Theorem. Let $f(\cdot)$ be a given continuously differentiable real-valued function on the real interval (a, b), and let n be a given positive integer. If the Loewner matrix $K_f(\{\xi_i\})$ defined by (4.1.7) is positive definite for every choice of n distinct points $\xi_1, \ldots, \xi_n \in (a, b)$, then $f(\cdot)$ is strongly matrix monotone of order n on (a,b).

Proof: The hypotheses ensure that $f(\cdot)$ is a monotone matrix function of order n on (a, b). Let $A, B \in M_n$ be Hermitian matrices whose eigenvalues all lie in (a, b); suppose $A \succeq B$ and suppose there is no common eigenvector for which A and B have the same eigenvalue. If it is *not* the case that $f(A) \succ f(B)$, let x be a unit vector such that $x^*[f(A) - f(B)]x = 0$. For this vector x, use Lemma B to construct a Hermitian matrix $\hat{B} \equiv \hat{B}_{x,\epsilon}$ such that: the eigenvalues of $\hat{B}$ all lie in (a, b) (take $\epsilon > 0$ small enough to achieve this), $\hat{B}$ has distinct eigenvalues, any eigenvector of $\hat{B}$ is also an eigenvector of B, and $f(B)x = f(\hat{B})x$. Then $0 = x^*[f(A) - f(B)]x = x^*[f(A) - f(\hat{B})]x$, so Lemma A says there is a unit vector ξ and a real scalar λ such that $A\xi = \lambda\xi$ and $\hat{B}\xi = \lambda\xi$. Since every eigenvector of $\hat{B}$ is an eigenvector of B, it follows that $B\xi = \mu\xi$ for some real scalar μ, but since $A \succeq B \succeq \hat{B}$, we have $\lambda = \xi^* A\xi \geq \xi^* B\xi = \mu \geq \xi^*\hat{B}\xi = \lambda$, and hence $\mu = \lambda$, which contradicts our initial assumption. The conclusion is that there exists no unit vector x with the asserted property, and hence $f(A) \succ f(B)$. □

Since we have seen in (4.1.15) that the Loewner matrix of the square root function satisfies the positive definiteness condition of the theorem, it follows that the square root function is a strongly monotone matrix function of every order on $(0,\infty)$. The notion of strong matrix monotonicity seems to have appeared first in [Mathias, 1988], where there are interesting applications to the problem of determining when two partial orders on a convex cone of positive semidefinite matrices are equivalent.

There is an extensive and beautiful theory of monotone matrix functions, and there are still many open questions in the subject. Although the three examples of monotone matrix functions that we have discussed [$f(t) = t$, $f(t) = -1/t$, and $f(t) = t^{\frac{1}{2}}$] are monotone of order n for every $n = 1, 2, \ldots$, this is not always the case; for example, there are monotone matrix functions of order 2 that are not monotone of any order $n > 2$.

Functions that are matrix monotone of order n for *every* $n = 1, 2, \ldots$ on a real interval (a, b) are of special importance, and are called *monotone operator functions on* (a,b). It is known that a function $f(\cdot)$ is a monotone operator function on (a,b) if and only if it is real analytic on the real interval (a,b) and possesses an analytic continuation into the upper half plane that maps the upper half-plane into itself. Since this property holds on the interval $(0, \infty)$ for $f(t) = t^p$ for any $p \in (0, 1]$ (but not for any $p > 1$), and for $f(t) = \log t$, these functions are monotone operator functions on $(0,\infty)$. It is also known that a function $f(\cdot)$ is a monotone operator function on (a, b) if and only if it can be represented as

$$f(t) = \alpha t + \beta + \int_{-\infty}^{\infty} \left[\frac{1}{s-t} - \frac{s}{s^2+1}\right] d\mu(s) \tag{4.1.28}$$

where $\alpha \geq 0$, $\beta \in \mathbb{R}$, and $d\mu$ is a nonnegative Borel measure on $\mathbb{R}$ that has no mass in the interval (a, b) and for which the integral $\int (1 + s^2)^{-1}\, d\mu(s)$ on $(-\infty, \infty)$ is finite. For example, if $\alpha = 0$, $\beta = -1/\sqrt{2}$, and $d\mu = \sqrt{-s}\, ds/\pi$ on $(-\infty, 0)$ and $d\mu \equiv 0$ on $[0, \infty)$, then the resulting monotone operator function is $f(t) = \sqrt{t}$. The integral representation (4.1.28) provides the analytic continuation of $f(t)$ to the upper half-plane and verifies the mapping property

$$\operatorname{Im}[f(z)] = \frac{f(z) - \overline{f(z)}}{2i} = \alpha \operatorname{Im}(z) + \int_{-\infty}^{\infty} \frac{\operatorname{Im}(z)}{|s - z|^2}\, d\mu(s) > 0$$

whenever $\operatorname{Im}(z) > 0$.

Of course, one can check whether a given function $f(\cdot)$ is a monotone operator function without any recourse to analytic continuation or conformal mapping—just check whether the Loewner matrices $K_f(\{\xi_i\})$ of every order are positive semidefinite. A limiting argument shows that the n-by-n matrix $K_f(\{\xi_i\})$ is positive semidefinite for every order $n = 1, 2, \ldots$ if and only if the *Loewner kernel*

$$K_f(s,t) \equiv \frac{f(s) - f(t)}{s - t} \qquad\qquad K_f(s,s) \equiv f'(s)$$

is a *positive semidefinite kernel on* (a, b) in the following sense:

$$\int_a^b \int_a^b K_f(s, t)\, g(s) g(t)\, ds dt \geq 0 \tag{4.1.29}$$

for every continuous real-valued function $g(t)$ with compact support in (a, b). Conversely, if $f(\cdot)$ can be represented in the form (4.1.28), then direct substitution makes it clear that (4.1.29) is satisfied. It is curious that this real condition on a real interval is sufficient (and necessary) to guarantee that $f(\cdot)$ has an analytic continuation into the upper half-plane and, moreover, that this continuation has the important geometrical property that it maps the upper half-plane into itself.

If $f(\cdot)$ is a monotone operator function on (a,b), it is known that the sufficient condition in the Strong Monotonicity Theorem is also necessary, that is, a monotone operator function on (a,b) is strongly monotone of a given order n if and only if the Loewner matrix $K_f(\{\xi_i\})$ is positive definite for every choice of n distinct points $\xi_1, \ldots, \xi_n \in (a,b)$. Moreover, for a monotone operator function $f(\cdot)$, this condition is equivalent to $f(\cdot)$ *not* being a rational function of degree less than n. In particular, it follows that the monotone operator functions $f(t) = t^p$ for $p \in (0,1)$ and $f(t) = \log t$ are strongly monotone of all orders on $(0,\infty)$ [Mathias, 1988].

The classic paper on monotone matrix functions is [Loewner, 1934]. For an extensive bibliography and a survey of monotone matrix functions with an emphasis on the analytic theory, see [Donoghue, 1974]. The Daleckiĭ-Kreĭn formula (4.1.0) that makes the connection with Hadamard products and is the basis for our entire discussion was extended by Daleckiĭ to the non-Hermitian case [Daleckiĭ, 1965]; the formalism is similar for the diagonalizable non-Hermitian case, but is more complicated if the Jordan structure is not diagonal [Horn & Johnson, 1989, Section 6.6].

4.2 Example. Hadamard Products and the Lyapunov Equation $GA + A^*G = H$. In the study of stability of local equilibria for systems of ordinary differential equations, one is naturally led to consider matrices whose eigenvalues all lie in a half-plane; a square complex matrix is said to be *positive stable* if all its eigenvalues have positive real part. A positive definite matrix is positive stable, of course, but there is a more subtle connection between the positive definite matrices and the positive stable matrices. A celebrated theorem of Lyapunov says that a given matrix $A \in M_n$ is positive stable if and only if there is a positive definite $G \in M_n$ such that the Hermitian matrix $(GA) + (GA)^* = GA + A^*G$ is positive definite.

This observation suggests that it is worthwhile to study the *Lyapunov equation*

$$GA + A^*G = H \tag{4.2.1}$$

in which $A \in M_n$ is given and one wants to solve for G for various right-hand sides H. It is known that (4.2.1) has a unique solution $G = G_A(H)$ for every right-hand side $H \in M_n$ if and only if the eigenvalues $\{\lambda_1,\ldots,\lambda_n\}$ of A satisfy

$$\bar{\lambda}_i + \lambda_j \neq 0 \text{ for all } i, j = 1,\ldots, n \tag{4.2.2}$$

[Horn & Johnson, 1989, Section 4.4]. Let us make this assumption about the eigenvalues of A; in particular, this condition is clearly satisfied if A is positive stable. If, in addition, A is diagonalizable with $A = S\Lambda S^{-1}$ for some nonsingular $S \in M_n$ and $\Lambda = \operatorname{diag}(\lambda_1,\ldots, \lambda_n)$, then there is a simple formula for the solution $G_A(H)$ of (4.2.1) that involves the Hadamard product. Compute

$$GA + A^*G = GS\Lambda S^{-1} + (S^{-1})^*\bar{\Lambda}S^*G = H,$$

which is equivalent to

$$(S^*GS)\Lambda + \bar{\Lambda}(S^*GS) = S^*HS.$$

For any matrix $B = [b_{ij}] \in M_n$, $B\Lambda + \bar{\Lambda}B = [b_{ij}\lambda_j] + [\bar{\lambda}_i b_{ij}] = [(\bar{\lambda}_i + \lambda_j)b_{ij}]$, which is a Hadamard product. Now invoke the assumption (4.2.2) to define the Hadamard inverse matrix

$$L(A) \equiv [(\bar{\lambda}_i + \lambda_j)^{-1}], \quad i, j = 1,\ldots, n, \tag{4.2.3}$$

sometimes called the *Cauchy matrix* associated with the vector $\operatorname{diag}(\Lambda)$. Then $S^*GS = L(A) \circ (S^*HS)$, which yields the explicit formula

$$G_A(H) = (S^{-1})^*\{L(A) \circ [S^*HS]\}S^{-1} \qquad A = S\Lambda S^{-1} \tag{4.2.4}$$

for the solution to the Lyapunov equation $GA + A^*G = H$ when A is diagonalizable and its eigenvalues satisfy (4.2.2). Notice that the formula (4.2.4) automatically produces a Hermitian solution matrix $G_A(H)$ whenever the right-hand side H is Hermitian.

If A is positive stable, then Re $\lambda_i > 0$ for all $i = 1,\ldots, n$, $\bar{\lambda}_i + \lambda_j \neq 0$ for all $i, j = 1,\ldots, n$, the Cauchy matrix $L(A)$ has positive main diagonal entries, and the integral representation

$$\frac{1}{\bar{\lambda}_i + \lambda_j} = \int_0^{\infty} e^{-[\bar{\lambda}_i + \lambda_j]s}\, ds \tag{4.2.5}$$

is convergent for all $i, j = 1,\ldots, n$. In this case, for any $x \in \mathbb{C}^n$ we have

$$x^* L(A)x = \sum_{i,j=1}^{n} \int_0^{\infty} e^{-[\bar{\lambda}_i + \lambda_j]s}\, \bar{x}_i x_j\, ds = \int_0^{\infty} \Big|\sum_{i=1}^{n} e^{-\lambda_i s}\, x_i\Big|^2 ds \geq 0, \tag{4.2.6}$$

with strict inequality for all $x \neq 0$ if and only if the n points $\lambda_1,\ldots, \lambda_n$ are distinct. The conclusion is that if A is positive stable with eigenvalues $\lambda_1,\ldots, \lambda_n$, then the Cauchy matrix $L(A)$ defined in (4.2.3) is positive semidefinite and has positive main diagonal entries; moreover, $L(A)$ is positive definite if and only if the n points $\lambda_1,\ldots, \lambda_n$ are *distinct*. Thus, if A is diagonalizable and positive stable, (4.2.4) and Schur's Theorem 3.1(b, d) tell us that the unique solution $G_A(H)$ to (4.2.1) is positive *semi*definite whenever the right-hand side H is positive *semi*definite; the solution is positive definite if *either* H is positive definite, *or* the eigenvalues $\lambda_1,\ldots, \lambda_n$ are distinct, H is positive semidefinite, and S^*HS has positive main diagonal entries.

If A is not positive stable, but satisfies (4.2.2), there is no loss of generality to assume that its eigenvalues are arranged so that $\lambda_1,\ldots, \lambda_k$ have positive real part and $\lambda_{k+1},\ldots, \lambda_n$ have negative real part. It follows from the representation (4.2.6) that if the Cauchy matrix $L(A)$ associated with $\lambda_1,\ldots, \lambda_k, \lambda_{k+1},\ldots, \lambda_n$ is partitioned as

$$L(A) = \begin{bmatrix} L_{11} & L_{12} \\ L_{21} & L_{22} \end{bmatrix} \tag{4.2.7}$$

with $L_{11} \in M_k$ and $L_{22} \in M_{n-k}$, then L_{11} and $-L_{22}$ are both positive semidefinite and have positive main diagonal entries, since each is a Cauchy matrix of a set of complex numbers in the right half-plane: $\lambda_1,\ldots, \lambda_k$ for L_{11} and $-\lambda_{k+1},\ldots, -\lambda_n$ for L_{22}. If the points $\lambda_1,\ldots, \lambda_n$ are *distinct*, then L_{11} and $-L_{22}$ are actually positive definite.

Suppose the right-hand side H of (4.2.1) is positive definite, and partition the positive definite matrix $K \equiv S^*HS$ conformally with $L(A)$ as

$$K = S^*HS = \begin{bmatrix} K_{11} & K_{12} \\ K_{21} & K_{22} \end{bmatrix},$$

so that both principal submatrices K_{11} and K_{22} are positive definite. Then 3.1(c) guarantees that the Hadamard product

$$L(A)\circ[S^*HS] = L(A)\circ K = \begin{bmatrix} L_{11}\circ K_{11} & L_{12}\circ K_{12} \\ L_{21}\circ K_{21} & L_{22}\circ K_{22} \end{bmatrix} \tag{4.2.8}$$

has a k-by-k principal submatrix—$L_{11}\circ K_{11}$—that is positive definite and an $(n-k)$-by-$(n-k)$ principal submatrix—$L_{22}\circ K_{22}$—that is negative definite. If follows from the interlacing eigenvalues theorem that $L(A)\circ[S^*HS]$ has exactly k positive and $n-k$ negative eigenvalues. Since *Sylvester's Inertia Theorem* guarantees that this distribution of signs of the eigenvalues is preserved under nonsingular congruence, we conclude that

> If $A\in M_n$ is diagonalizable with eigenvalues $\lambda_1,\ldots,\lambda_n$ such that $\bar{\lambda}_i+\lambda_j\neq 0$ for all $i, j=1,\ldots, n$, and if $H\in M_n$ is positive definite, then the unique solution $G=G_A(H)$ to the Lyapunov equation $GA+A^*G=H$ has the same number of positive (respectively, negative) eigenvalues as A has eigenvalues in the open right (respectively, open left) half-plane.

(4.2.9)

This result, which we have derived using properties of Hadamard products under the assumption that A is diagonalizable, is a special case of the *General Inertia Theorem*, which asserts the same conclusion without the diagonalizability assumption [Horn & Johnson, 1989, Section 2.4]. The use of Hadamard products to approach the Lyapunov equation in this way is discussed in [Marcus & Thompson, 1963].

A key point in the preceding argument was the formulation of hypotheses on A and K to guarantee that the Hadamard product $L(A)\circ K$ in (4.2.8) has the same number of positive and negative eigenvalues as A has eigenvalues in the open right half-plane and open left half-plane, respectively. In the context of the Lyapunov equation it was very natural to assume that K was positive definite, in which case no assumption on the eigenvalues $\lambda_1,\ldots,\lambda_n$ of A other than (4.2.2) was necessary. However, exactly the same conclusion holds under a weaker hypotheses on K if one add the assumption that the eigenvalues are distinct:

> If $\lambda_1,\ldots,\lambda_n\in\mathbb{C}$ are distinct points such that $\bar{\lambda}_i+\lambda_j\neq 0$ for all $i, j=1,\ldots, n$, with exactly k of these points in the open right half-plane and exactly $n-k$ in the open left half-plane, and if $L\equiv[(\bar{\lambda}_i+\lambda_j)^{-1}]$, then $L\circ K$ is a nonsingular Hermitian matrix with exactly k positive and $n-k$ negative eigenvalues whenever $K\in M_n$ is positive semidefinite and has positive main diagonal entries.

(4.2.10)

The explanation is basically the same as for (4.2.9), but we interchange the factors to which we apply the two conditions in 3.1(c). If $\lambda_1,\ldots, \lambda_n$ are distinct, we have already noted that the two principal submatrices L_{11} and L_{22} in (4.2.7) are nonsingular and definite and hence the principal submatrices $L_{11}\circ K_{11}$ and $L_{22}\circ K_{22}$ in (4.2.8) are both nonsingular and definite since K_{11} and K_{22} are both positive semidefinite with positive main diagonal entries. See Section 5.4 for a discussion and characterization of the matrices—like the Cauchy matrix L associated with distinct points described in (4.2.10)—whose inertia is preserved under Hadamard product with any positive semidefinite matrix with positive main diagonal entries.

The formula (4.2.4), together with known inequalities for norms of Hadamard products, permits us to investigate how well-conditioned the problem (4.2.1) is. Assuming only that $A = S\Lambda S^{-1}$ is diagonalizable with eigenvalues satisfying (4.2.2), we can use 3.1(f) to obtain the bounds

$$\begin{aligned}
|||\, G_A(H)\,|||_2 &\le |||\,(S^{-1})^*\,|||_2\, |||\, L(A)\circ[S^*HS]\,|||_2\, |||\, S^{-1}\,|||_2 \\
&= |||\, S^{-1}\,|||_2^2\, |||\, L(A)\circ[S^*HS]\,|||_2 \qquad (4.2.11)\\
&\le |||\, S^{-1}\,|||_2^2\, |||\, L(A)\,|||_2\, |||\, S^*HS\,|||_2 \le |||\, S^{-1}\,|||_2^2\, |||\, S\,|||_2^2\, |||\, L(A)\,|||_2\, |||\, H\,|||_2 \\
&= \kappa(S)^2\, |||\, L(A)\,|||_2\, |||\, H\,|||_2,
\end{aligned}$$

for $A, H \in M_n$, with A satisfying (4.2.2). (4.2.12)

The quantity $\kappa(S) \equiv |||\, S\,|||_2\, |||\, S^{-1}\,|||_2 = \sigma_1(S)/\sigma_n(S) \ge 1$ is the *condition number* of the nonsingular matrix S with respect to the spectral norm, which can be very large if A has two eigenspaces that are nearly parallel. To be precise, if $S \in M_n$ is a given nonsingular matrix, then $\kappa(S) = \cot(\frac{1}{2}\theta)$, where θ is the minimum angle between Sx and Sy ($0 < \theta \le \pi/2$) over all pairs of orthonormal vectors $x, y \in \mathbb{C}^n$. Thus, the spectral condition number is large if and only if the angle between Sx and Sy is small for some pair of orthonormal vectors x and y [Horn & Johnson, 1987a, 7.4.32].

The reason for concern about the size of the condition number $\kappa(S)$ is that if the right-hand side H is subjected to a perturbation $H \to H + E$—perhaps the data for H come from experimental observations subject to measurement uncertainties, or perhaps we wish to assess the consequences of computer roundoff error—then the solution will be perturbed, $G_A(H) \to G_A(H+E) = G_A(H) + G_A(E)$, and the size of the perturbation in the solution is

$$|||\, G_A(H+E) - G_A(H)\,|||_2 = |||\, G_A(E)\,|||_2 \le \kappa(S)^2\, |||\, L(A)\,|||_2\, |||\, E\,|||_2.$$

If $\kappa(S)$ is large, then $\kappa(S)^2$ can be enormous and the upper bound we have on the perturbation in the solution can be large even if $|||\, E\,|||_2$ is small. This suggests that numerical solution of the problem (4.2.1) should be undertaken cautiously, as it could be very ill-conditioned.

If A is normal, however, then S can be taken to be unitary and then $\kappa(S) = 1$, so (4.1.1) is always well-conditioned in this case. But even for normal A, the bound (4.2.12) need not be very good. For example, when $A = I$ we have $L(A) = \frac{1}{2}J$, which gives the bound $|||\, G_A(H)\,|||_2 \le \frac{1}{2}n|||\, H|||_2$. This is not very satisfactory since we know that the solution is $G_A(H) = \frac{1}{2}H$.

Now assume that $A \in M_n$ is positive stable, so $L(A)$ is positive definite and we can apply the better bound (3.7.13) to (4.2.11) and obtain

$$|||\, G_A(H)\,||| \le \kappa(S)^2 \max_i (\bar{\lambda}_i + \lambda_i)^{-1} |||\, H|||_2$$

$$= \tfrac{1}{2}\, \kappa(S)^2 (\min_i \operatorname{Re} \lambda_i)^{-1} |||\, H|||_2 \quad \text{for positive stable } A. \tag{4.2.13}$$

If A is not only positive stable but normal (in particular, if A is positive definite), then $S = U$ can be taken to be unitary, $\kappa(S) = 1$, and we have the upper bound

$$|||\, G_A(H)\,|||_2 \le \tfrac{1}{2}(\min_i \operatorname{Re} \lambda_i)^{-1} |||\, H|||_2 \qquad \text{for } A \text{ positive stable and normal.} \tag{4.2.14}$$

Finally, if A is positive stable and normal, and if H is positive semidefinite, then we can use 3.1(b) to obtain a *lower* bound from (4.2.4):

$$\lambda_{min}[G_A(H)] = \lambda_{min}\Big[U\{L(A)\circ[U^*H\,U]\}U^*\Big] = \lambda_{min}\Big[L(A)\circ[U^*H\,U]\Big]$$

$$\ge \min_i (\bar{\lambda}_i + \lambda_i)^{-1} \lambda_{min}(U^*H\,U)$$

$$= \tfrac{1}{2}(\max_i \operatorname{Re} \lambda_i)^{-1} \lambda_{min}(H)$$

$$\text{for } A \text{ positive stable and normal and } H \succeq 0. \tag{4.2.15}$$

As an application of the solution formula (4.2.4) and the upper and lower bounds (4.2.13,15), suppose X, $Y \in M_n$ are given Hermitian matrices with X positive definite and Y positive semidefinite. Then

$$(X^{\frac{1}{2}} - Y^{\frac{1}{2}})(X^{\frac{1}{2}} + Y^{\frac{1}{2}}) + (X^{\frac{1}{2}} + Y^{\frac{1}{2}})(X^{\frac{1}{2}} - Y^{\frac{1}{2}}) = 2\,(X - Y),$$

so if we set $H \equiv 2\,(X - Y)$, $A \equiv X^{\frac{1}{2}} + Y^{\frac{1}{2}}$, and $G \equiv X^{\frac{1}{2}} - Y^{\frac{1}{2}}$, we have H Hermitian and A positive definite (and hence positive stable and normal), and G satisfies the Lyapunov equation $GA + A^*G = H$, as noted in [Chan & Kwong, 1985, pp. 535-6]. Let $A = U\Lambda U^*$ be a spectral decomposition and use (4.2.4) to write

$$X^{\frac{1}{2}} - Y^{\frac{1}{2}} = G_A(H) = U\{L(A)\circ[U^*H\,U]\}U^*$$

$$= 2\, U\{L(A)\circ[U^*(X-Y)\, U]\}U^*. \tag{4.2.16}$$

Thus, (4.2.14) gives the norm bound

$$|||\, X^{\frac{1}{2}} - Y^{\frac{1}{2}}\, |||_2 \le [\lambda_{min}(X^{\frac{1}{2}} + Y^{\frac{1}{2}})]^{-\frac{1}{2}} |||\, X - Y|||_2 \le [\lambda_{min}(X)]^{-\frac{1}{2}} |||\, X - Y|||_2,$$

$$\text{for } X, Y \in M_n,\ X \succ 0,\ Y \succeq 0, \tag{4.2.17}$$

which shows that the square root function is a norm-continuous function on the positive definite matrices. The intermediate bound in (4.2.17) corresponds to the identity $\sqrt{x} - \sqrt{y} = (x-y)/(\sqrt{x} + \sqrt{y})$ in the scalar case.

Now make the additional assumption that $X \succeq Y \succeq 0$, with $X \succ 0$, as before. Then $X - Y = H \succeq 0$ and it is evident from (4.2.16) and 3.1(a) that $G_A(H) \succeq 0$, that is, $X^{\frac{1}{2}} \succeq Y^{\frac{1}{2}}$! If $X \succeq Y \succeq 0$ and X is singular, consider $X_\epsilon \equiv X + \epsilon I$ for any given $\epsilon > 0$. Then $X_\epsilon \succeq Y \succeq 0$ and $X_\epsilon \succ 0$, so $X_\epsilon^{\frac{1}{2}} \succeq Y^{\frac{1}{2}}$. Now let $\epsilon \downarrow 0$ and conclude that $X^{\frac{1}{2}} \succeq Y^{\frac{1}{2}}$. This simple argument, which is completely independent of the theory of monotone matrix functions, shows that the square root function is a monotone operator function on the positive semidefinite matrices. Moreover, the lower bound (4.2.15) can be used with (4.2.16) to obtain:

$$\lambda_{min}(X^{\frac{1}{2}} - Y^{\frac{1}{2}}) \ge [\lambda_{max}(X^{\frac{1}{2}} + Y^{\frac{1}{2}})]^{-1}\lambda_{min}(X-Y) = |||\, X^{\frac{1}{2}} + Y^{\frac{1}{2}}\, |||_2^{-1}\, \lambda_{min}(X-Y)$$

$$\ge [|||\, X^{\frac{1}{2}}\, |||_2 + |||\, Y^{\frac{1}{2}}\, |||_2]^{-1}\lambda_{min}(X-Y)$$

$$\ge \tfrac{1}{2}[\lambda_{max}(X)]^{-\frac{1}{2}}\lambda_{min}(X-Y)$$

$$\text{for } X, Y \in M_n,\ X \succeq Y \succeq 0,\ X \succ 0. \tag{4.2.18}$$

Using the preceding limiting argument, it is clear that the condition $X \succ 0$ in (4.2.18) may be replaced with the weaker hypothesis $X \succeq 0$ but $X \neq 0$. Finally, (4.2.17) can be written as an eigenvalue inequality companion to (4.2.18):

$$\lambda_{max}(X^{\frac{1}{2}} - Y^{\frac{1}{2}}) \le [\lambda_{min}(X^{\frac{1}{2}} + Y^{\frac{1}{2}})]^{-\frac{1}{2}}\lambda_{max}(X-Y)$$

$$\le [\lambda_{min}(X)]^{-\frac{1}{2}}\lambda_{max}(X-Y)$$

$$\text{for } X, Y \in M_n,\ X \succeq Y \succeq 0,\ X \succ 0. \tag{4.2.19}$$

4.3 Example. Diagonal Entries and Eigenvalues of a Diagonalizable Matrix. Suppose $A = [a_{ij}] \in M_n$ is diagonalizable and has eigenvalues $\lambda_1, \ldots, \lambda_n$. Then there is a nonsingular $S \in M_n$ such that $A = S \Lambda S^{-1}$, where $\Lambda = \text{diag}(\lambda_1, \ldots, \lambda_n)$. One checks with a calculation that the vector of main diagonal entries of A and the vector of eigenvalues of A are related by

$$\begin{bmatrix} a_{11} \\ \vdots \\ a_{nn} \end{bmatrix} = [S \circ (S^{-1})^T] \begin{bmatrix} \lambda_1 \\ \vdots \\ \lambda_n \end{bmatrix}. \tag{4.3.1}$$

Thus, the vector of eigenvalues of A is transformed to the vector of its main diagonal entries by the coefficient matrix $S \circ (S^{-1})^T$, that is, $\text{diag}(A) = [S \circ (S^{-1})^T]\ \text{diag}(\Lambda)$.

One special case of this transformation is of particular interest. If A is normal, then $S = U = [u_{ij}]$ is unitary and the coefficient matrix in (4.3.1) has the special form $U \circ (U^{-1})^T = U \circ (U^*)^T = U \circ \overline{U} = [|u_{ij}|^2]$, a matrix that is *orthostochastic*, a special case of a *doubly stochastic* matrix (entry-wise nonnegative with all row and column sums one). In particular, when A is Hermitian then both its main diagonal and eigenvalues are real and are related by a doubly stochastic matrix. Since every doubly stochastic matrix is a convex combination of permutation matrices (a famous theorem of Birkhoff, see [Horn & Johnson, 1987a, Section 8.7]), it follows that the diagonal entries and eigenvalues of a Hermitian matrix are related by *majorization*, an observation with many useful consequences [Horn & Johnson, 1987a, Section 4.3], [Marshall & Olkin, 1979].

Using 3.1(f), we can obtain a bound for the norm of the transformation (4.3.1):

$$\|\text{diag}(A)\|_2 \le |||\, S \circ (S^{-1})^T |||_2\ \|\text{diag}(\Lambda)\|_2$$

$$\le |||\, S |||_2\ |||\, S^{-1} |||_2\ \|\text{diag}(\Lambda)\|_2 = \kappa(S)\ \|\text{diag}(\Lambda)\|_2$$

$$\text{for diagonalizable } A \in M_n \text{ with } A = S\Lambda S^{-1}, \tag{4.3.2}$$

where $\kappa(S)$ is the condition number of S with respect to the spectral norm, as discussed in Section 4.2.

4.4 Example. The Relative Gain Array. The special Hadamard product matrix that arose in the preceding example also plays a role in system engineering. In chemical engineering, a real square nonsingular *gain matrix* A describes the relation between inputs and outputs in a chemical process, and the matrix $A \circ (A^{-1})^T$ is called the *relative gain array.* The systems engineer inspects the relative gain array with the goal of determining a pairing of inputs and outputs to be used in facility design. See [McAvoy, 1983] for a discussion of the motivation for this approach and several case studies of its implementation; see also [Nett & Manousiouthakis, 1987]. For more information about mathematical properties of the mapping $A \to A \circ (A^{-1})^T$ and the related mapping $A \to A \circ A^{-1}$ see [Horn & Johnson, 1989, Section 5.4] and [Johnson & Shapiro, 1986].

4.5 Example. Covariance Matrices and Hadamard Products. Let $X = [X_1 \dots X_n]^T$ and $Y = [Y_1 \dots Y_n]^T$ be real or complex fourth order vector random variables with zero means, that is, $E(X) = [E(X_i)] = 0$ and $E(Y) = 0$, where $E(\cdot)$ denotes the expectation operator. The *covari-*

ance matrix of X has the general form

$$\mathrm{Cov}(X) \equiv E([X - E(X)][X - E(X)]^*),$$

which takes the special form $\mathrm{Cov}(X) = E(XX^*) = [E(X_i\overline{X}_j)]$ in our case since $E(X) = 0$. A common assumption (achievable for non-constant random variables by a multiplicative scaling) is that each component of X has unit variance, that is, $\mathrm{Var}(X) = E(|X|^2) = 1$ for $i = 1,\ldots, n$. In this case, the main diagonal entries of $\mathrm{Cov}(X)$ are all 1; such a matrix is called a *correlation matrix.*

If X and Y are independent, the second order vector random variable $Z \equiv X \circ Y$ has zero mean and has a covariance matrix that is obtained from the covariance matrices of X and Y as a Hadamard product:

$$\mathrm{Cov}(Z) = E(ZZ^*) = [E(X_iY_j\overline{X}_j\overline{Y}_j)] = [E(X_i\overline{X}_jY_j\overline{Y}_j)]$$

$$= [E(X_i\overline{X}_j)E(Y_i\overline{Y}_j)] = \mathrm{Cov}(X)\circ\mathrm{Cov}(Y).$$

One pleasant property of a covariance matrix is that it is easily shown to be positive semidefinite. For any $\xi = [\xi_i] \in \mathbb{C}^n$ we have

$$\xi^*\mathrm{Cov}(X)\xi = \sum_{i,\,j=1}^{n} E(X_i\overline{X}_j)\overline{\xi}_i\xi_i = \sum_{i,\,j=1}^{n} E(\overline{\xi}_iX_i\xi_j\overline{X}_j) = E\Big[\sum_{i,\,j=1}^{n} \overline{\xi}_iX_i\xi_j\overline{X}_j\Big]$$

$$= E\Big[\Big|\sum_{i=1}^{n}\overline{\xi}_iX_i\Big|^2\Big] \geq 0.$$

In this calculation we have used the facts that the expectation operator is linear and positive. Thus, since the Hadamard product of two covariance matrices is itself a covariance matrix, we may immediately infer that the Hadamard product of two covariance matrices is positive semidefinite. Since every positive semidefinite matrix is the covariance matrix of a Gaussian random vector, these observations give a probabilistic proof of the Schur product theorem.

If X and Y have covariance matrices A and B, respectively, then for any vector $c = [c_i] \in \mathbb{C}^n$ we have

$$\mathrm{Var}(c^*X) = E[(c^*X)(c^*X)^*] = \sum_{i,\,j=1}^{n} c_ia_{ij}\overline{c}_j = c^*Ac.$$

Then $\mathrm{Var}(c^*X) \geq \mathrm{Var}(c^*Y)$ if and only if $c^*Ac \geq c^*Bc$, an observation that gives a statistical interpretation of the Loewner partial order: $A \succeq B$ if and only if the linear combination $\overline{c}_1Y_1 + \cdots + \overline{c}_nY_n = c^*Y$ has a smaller variance than $\overline{c}_1X_1 + \cdots + \overline{c}_nX_n = c^*X$ for all $c \in \mathbb{C}^n$. If we now assume that X and Y are independent and set $Z \equiv X \circ Y$, then $\mathrm{Var}(c^* Z) = c^*(A \circ B)c$, so the bounds in 3.1(b) give

$$\|c\|_2^2 \min_i a_{ii} \lambda_{\min}(B) \leq \mathrm{Var}(c^* Z) = c^*(A \circ B)c \leq \|c\|_2^2 \max_i a_{ii} \lambda_{max}(B).$$

In particular, if X is normalized so that its covariance matrix is a correlation matrix, then

$$\|c\|_2^2 \lambda_{\min}(B) \leq \mathrm{Var}(c^* Z) \leq \|c\|_2^2 \lambda_{max}(B),$$

that is, for any X whose entries have unit variance, linear combinations of the entries of $Z = X \circ Y$ exhibit the same range of variances as linear combinations of the entries of Y. If both X and Y are normalized to have entries with unit variances, it follows that

$$\|c\|_2^2 \max\{\lambda_{\min}(A), \lambda_{\min}(B)\} \leq \mathrm{Var}(c^* Z) \leq \|c\|_2^2 \max\{\lambda_{max}(A), \lambda_{max}(B)\}.$$

For a survey of Hadamard product results useful in multivariate statistics, see [Styan, 1973].

4.6 Example. Index of Primitivity of a Nonnegative Matrix. Let $A \in M_n$ be a given nonnegative matrix. One says that A is *primitive* if $A^k > 0$ for some $k > 0$. If A is primitive, there is a least positive exponent, denoted $\gamma(A)$ and called the *index of primitivity* of A, for which $A^{\gamma(A)} > 0$ but $A^k \ngtr 0$ for all $k = 1, \ldots, \gamma(A) - 1$. Wielandt showed that $\gamma(A) \leq n^2 - 2n + 2$ and that this bound is best possible [Horn & Johnson, 1987a, Section 8.5].

M. Lewin studied the Hadamard product $A^q \circ A^{q+1}$ [Lewin, 1971], which is evidently a positive matrix if A is primitive and $q \geq \gamma(A)$. He showed that if $n > 2$ and $A \in M_n$ is nonnegative and irreducible, then $A \circ A^2 = 0$ if A is not primitive. Under these circumstances, therefore, a *sufficient* condition for A to be primitive is that $A \circ A^2 \neq 0$. Although the latter condition is not *necessary* for primitivity (see [Minc, 1988, p. 82] for a counterexample with $n = 5$), Lewin showed that a nonnegative irreducible $A \in M_n$ is primitive if and only if there is some positive integer q such that $A^q \circ A^{q+1} \neq 0$. If we denote by $q_0(A)$ the least positive integer for which the latter condition holds, then clearly $q_0(A) \leq \gamma(A) \leq n^2 - 2n + 2$. The potential value of these ideas as a criterion for primitivity lies in the fact that $q_0(A)$ can be substantially smaller than $\gamma(A)$. For example, Lewin shows that $q_0(A) = 1$ whenever $\gamma(A)$ achieves its maximum value of $n^2 - 2n + 2$, and that $q_0(A) = 2$ whenever $n > 3$ and $\gamma(A) = n^2 - 2n + 1$. A sharp upper bound on $q_0(A)$ analogous to Wielandt's bound on $\gamma(A)$ seems not to be known.

4.7 Example. Hadamard Products in Discrete Mathematics. The concept of association schemes in algebraic combinatorics is a fundamental framework within which coding theory and design theory can be given a unified development. The set of adjacency matrices of an association scheme is an algebra under both the usual matrix product and the Hadamard product (the *Bose–Messner algebra* of the scheme), and there is a natural and useful duality between these two algebraic structures. There is a set of real structure constants of a Bose-

Messner algebra called the *Krein parameters*. An important property of the Krein parameters is that they lie between 0 and 1 (the so-called *Krein conditions*), which follows from 3.1(a,f) since they can be recognized as eigenvalues of Hadamard products of real symmetric idempotent matrices. For further discussion see [Biggs, 1976], [Bannai & Ito, 1984, pp. 60–69], and [Cvetković, Doob, Gutman, and Torgašev, 1988, pp. 22–24].

A *coherent algebra* is a subalgebra $\mathcal{A}$ of M_n (under the usual matrix product) that is self-adjoint ($A \in \mathcal{A}$ if and only if $A^* \in \mathcal{A}$), closed under Hadamard products, and contains the identities I and J for both the usual and Hadamard products. Coherent algebras are precisely the adjacency algebras of coherent configurations. For a discussion and further references see [Higman, 1975], [Higman, 1987], and [Friedland, 1989].

The *formal incidence matrix* is a Hadamard product of matrices that arise in the study of combinatorial problems related to transversal theory. For a discussion and references to the literature, see [Ryser, 1973].

4.8 Example. Positive Definite Completions. A matrix $A \in M_n$ is said to be *partially defined* if only some of its entries are specified, with the unspecified entries treated as free variables. A *completion* of a partially defined matrix within a given class is simply a specification of the unspecified entries such that the whole matrix is in the given class. Interesting classes for which completion problems can be studied include: contractions, normal, positive semidefinite, and positive semidefinite with a special structure such as Toeplitz or Hankel.

If a partially defined matrix $A \in M_n$ has a positive semidefinite completion, it is clear that any defined principal submatrices of A must be positive semidefinite. More generally, $A \circ B$ must be positive semidefinite for any positive semidefinite $B \in M_n$ that has zero entries corresponding to the unspecified entries of A. In [Paulsen, Power, & Smith, 1989], this necessary condition is shown to be sufficient as well, and this characterization of partially defined matrices with positive semidefinite completions is shown to give a powerful unifying framework for a variety of previous results. For example, since a given $A \in M_n$ is a contraction if and only if the block matrix $\mathcal{A} \equiv \begin{bmatrix} I & A \\ A^* & I \end{bmatrix} \in M_{2n}$ is positive semidefinite (see Section 3.7), it is clear that a given partially defined $A \in M_n$ has a completion that is a contraction if and only if $\mathcal{A} \circ \mathcal{B}$ is positive semidefinite for every positive semidefinite $\mathcal{B} \equiv \begin{bmatrix} X & Y \\ Y^* & Z \end{bmatrix} \in M_{2n}$ such that Y has zero entries corresponding to the unspecified entries of A.

5. Generalizations of Schur's Theorem. The following sections show how virtually every aspect of Schur's theorems on Hadamard products can be generalized. They have been a font of inspiration for a great many mathematicians since 1911.

5.1 Fractional Hadamard Powers, Infinite Divisibility, and Hadamard Functions. Let $A = [a_{ij}] \in M_n$ be positive semidefinite. The Schur Product Theorem guarantees that $A \circ A = [(a_{ij})^2]$ and $A \circ (A \circ A) = [(a_{ij})^3]$ are positive semidefinite, as are all of the positive integer Hadamard powers $A^{\circ k} \equiv [(a_{ij})^k]$, $k = 1, 2, \ldots$. If the entries of A are all nonnegative, then one can also consider the non-integer Hadamard powers $A^{\circ\alpha} \equiv [(a_{ij})^\alpha]$, $\alpha > 0$, which are all real symmetric matrices and are positive semidefinite for $\alpha = 1, 2, \ldots$. Two natural questions arise at once: Are there *any* non-integer values of α for which $A^{\circ\alpha}$ is positive semidefinite? What can we say about a matrix A for which $A^{\circ\alpha}$ is positive semidefinite *for all* $\alpha > 0$?

The answer to the first question is: *$A^{\circ\alpha}$ is positive semidefinite for all $\alpha \geq n - 2$*, and *this inequality is sharp in the sense that if $\alpha < n - 2$ is* **not** *an integer, then there is a positive semidefinite $A \in M_n$ with positive entries such that $A^{\circ\alpha}$ is not positive semidefinite.* [FitzGerald & Horn, 1977a].

A matrix $B = [b_{ij}] \in M_n$ is said to be *conditionally positive definite* if $x^*Bx \geq 0$ for all $x \in \mathbb{C}^n$ whose entries sum to zero, that is, for all x such that $e^Tx = 0$. If all the entries of B are nonnegative, we say that B is *infinitely divisible* if $B^{\circ\alpha}$ is positive semidefinite for all $\alpha > 0$; if all the entries of B are positive, the entry-wise logarithm of B is denoted by $\log^\circ(B) \equiv [\log b_{ij}]$. A given matrix $B \in M_n$ is conditionally positive definite if and only if the entry-wise exponential $\exp^\circ(tB) \equiv [\exp(t\, b_{ij})]$ is positive semidefinite for all $t \geq 0$ [Horn, 1969a], [Loewner, 1966].

The answer to the second question is: If all the entries of A are *positive*—as they are for the Loewner matrix of any non-constant monotone matrix function—then *A is infinitely divisible if and only if $\log^\circ(A) \equiv [\log a_{ij}]$ is conditionally positive definite.* A non-constant monotone matrix function $f(\cdot)$ on a real interval (a, b) has a unique analytic continuation to the upper half-plane that maps the upper half-plane into itself; this mapping is *univalent* (classical function theorists would call such a mapping *schlicht*) if and only if its Loewner matrix $K_f(\{\xi_i\})$ of order n is infinitely divisible for all choices of n points $\xi_1, \ldots, \xi_n \in (a, b)$ and all $n = 1, 2, \ldots$ [FitzGerald, 1967], [Horn, 1967a], and [Horn, 1971].

We have just encountered several natural examples of *Hadamard functions of a matrix*—that is, the functions are applied entry-wise—and some of them have the pleasant property that they map the convex cone of positive semidefinite matrices of a given size into itself. It is easy to generate more such functions. Since all the integer Hadamard powers of positive semidefinite matrix are positive semidefinite, it follows that if $f(z) = a_0 + a_1 z + a_2 z^2 + \cdots$ is an analytic function whose Maclaurin series has nonnegative coefficients and radius of convergence $R > 0$, then the Hadamard function $f^\circ(A) \equiv [f(a_{ij})]$ is positive semidefinite whenever A is positive semidefinite and all $|a_{ij}| < R$.

There are also non-analytic Hadamard functions that map the cone of positive semidefinite matrices into itself; one example is the square of the absolute value function. Since $\overline{A}$ is positive semidefinite whenever A is, the Hadamard product $A \circ \overline{A} = [|a_{ij}|^2]$ is positive semidefinite whenever A is. However, the absolute value function itself does *not* have this property; see [Horn & Johnson, 1987a, 7.5] for a counterexample. No characterization is known for the class of all Hadamard functions that leave invariant the cone of positive semidefinite matrices. However, it is known that if a continuous real-valued function $f(\cdot)$ on $(0, \infty)$ is such that $f^\circ(A) \equiv [f(a_{ij})]$ is positive semidefinite whenever $A = [a_{ij}] \in M_n$ is positive semidefinite and all $a_{ij} > 0$, then $f(\cdot)$ is actually $n - 3$ times continuously differentiable, $f(\cdot)$ and all its derivatives through order $n-3$ are nonnegative, and $f^{(n-3)}(\cdot)$ is monotone increasing and convex [Horn, 1969a]. In particular, this necessary condition implies the assertion made previously that if $0 < \alpha < n-2$ and α is not an integer, then there must be some positive semidefinite $A \in M_n$ with positive entries for which $A^{\circ\alpha} \equiv [(a_{ij})^\alpha]$ is not positive semidefinite.

Under various special conditions on $A = [a_{ij}] \in M_n$, there are useful general classes of functions $f(\cdot)$ such that every Hadamard function $f^\circ(A) \equiv [f(a_{ij})]$ in the class is positive semidefinite. For example, if A is positive semidefinite and has nonnegative entries with $0 \leq a_{ij} \leq K$, then $f^\circ(A)$ is positive semidefinite for every function of the form

$$f(t) = \alpha_0 + \alpha_1 t + \cdots + \alpha_{n-2} t^{n-2} + \int_{n-2}^{\infty} t^\alpha \, d\mu(\alpha), \tag{5.1.1}$$

where all $\alpha_i \geq 0$, the measure $d\mu$ is nonnegative, and the integral converges on the interval $[0,K]$. As another example, if Re $A \equiv [\text{Re } a_{ij}] > 0$ and $-A$ is conditionally positive definite, then $f^\circ(A)$ is positive semidefinite for every Laplace transform

$$f(t) = \int_0^\infty e^{-\alpha t} \, d\mu(\alpha) \tag{5.1.2}$$

where the measure $d\mu$ is nonnegative and the integral converges for all $t > 0$. Such functions $f(\cdot)$ are often called *completely monotone* and are characterized by the property that $f(t)$ is infinitely differentiable on $(0,\infty)$ and $f^{(k)}(t) \geq 0$ for all $k = 1, 2, \ldots$. It is known that a given function $f(t)$ has the property that $f^\alpha(t)$ is completely monotone for all $\alpha > 0$ (these are the *infinitely divisible completely monotone* functions) if and only if either $f(t) \equiv 0$ or $f(t)$ is positive and continuously differentiable on $(0,\infty)$ and $-f'(t)/f(t)$ is completely monotone [Horn, 1967b, Theorem 4.4]; notice that $f(t) = 1/t$ satisfies the latter condition and is generated by (5.1.2) with $d\mu(\alpha) = d\alpha$. Thus, the Hadamard reciprocal $A^{\circ(-1)} = [1/a_{ij}]$ is positive semidefinite if Re $A > 0$ and $-A$ is conditionally positive definite; it is even infinitely divisible if A is real. In [Bapat, 1988a] it is shown that $-A$ is conditionally positive definite whenever A is a real symmetric matrix with positive entries and exactly one positive eigenvalue; it follows that the Hadamard reciprocal $A^{\circ(-1)} = [1/a_{ij}]$ of such a matrix is infinitely divisible.

5.2 Hermitian-Symmetric Inequalities. The Schur Product Theorem says that if $x^* A_i x \geq 0$ for all $x \in \mathbb{C}^n$, $i = 1, 2$, then $x^*(A_1 \circ A_2)x \geq 0$ for all $x \in \mathbb{C}^n$. What if the lower bound is not zero, but is a quadratic function of x? The *Grunsky Inequalities* from the theory of univalent analytic functions are

$$\sum_{i,j=1}^{n} \bar{x}_i x_j \left[\log \frac{1}{1 - z_i \bar{z}_j}\right] \geq \left| \sum_{i,j=1}^{n} x_i x_j \log\left[\frac{z_i z_j}{f(z_i) f(z_j)} \, \frac{f(z_i) - f(z_j)}{z_i - z_j}\right] \right|, \tag{5.2.1}$$

where $f(\cdot)$ is a regular analytic function in the unit disc $\{z: z \in \mathbb{C}, |z| < 1\}$, normalized so that $f(0) = 0$ and $f'(0) = 1$; the difference quotient is taken to be $f'(z_i)$ if $z_i = z_j$. It is known that $f(\cdot)$ satisfies these inequalities for all $x_1,\ldots, x_n \in \mathbb{C}$, all $z_1,\ldots, z_n$ in the unit disc, and all $n = 1, 2,\ldots$ if and only $f(\cdot)$ is *univalent*, that is, $f(z) = f(\zeta)$ if and only if $z = \zeta$. These inequalities are of the form

$$x^* A\, x \geq |x^T B\, x| \text{ for all } x \in \mathbb{C}^n \qquad A, B \in M_n, A = A^*, B = B^T, \tag{5.2.2}$$

where A is a positive semidefinite matrix whose entries $a_{ij} = -\log(1 - z_i\bar{z}_j)$ are related to the *Bergman kernel* for the unit disc and are independent of $f(\cdot)$, and B is a complex symmetric matrix whose entries are the log of a normalized difference quotient for $f(\cdot)$. If $A, B \in M_n$ satisfy the conditions in (5.2.2), we write $A \succeq_S B$ (the subscript S is to remind us that the lower bound is a *symmetric* function of x). The generalization of 3.1(a) to cover the inequalities (5.2.2) is that if $A_1 \succeq_S B_1$ and $A_2 \succeq_S B_2$, then $A_1 \circ A_2 \succeq_S B_1 \circ B_2$; more generally, this implication also holds for the Kronecker product [Horn, 1975a], [Friedland, 1975], [FitzGerald & Horn, 1977b]. Just as in the preceding section, if $h(z)$ is an analytic function whose Maclaurin series has nonnegative coefficients and radius of convergence R, if $A = [a_{ij}]$ and $B = [b_{ij}]$ satisfy $A \succeq_S B$, and if all $|a_{ij}| < R$, $|b_{ij}| < R$, it follows that $f^\circ(A) \succeq_S f^\circ(B)$. If we choose $f(z) = e^z$, this observation shows that the Grunsky inequalities (5.2.1) can be "exponentiated", that is, they remain true when the logs are deleted. This fact played an important role in attempts to solve the Bieberbach Conjecture [FitzGerald, 1972a,b].

There are similar results for the case in which the right-hand side of (5.2.2) is the absolute value of a Hermitian form, that is, $A, B \in M_n$ are Hermitian and $x^*Ax \geq |x^*Bx|$ for all $x \in \mathbb{C}^n$, denoted by $A \succeq_H B$ [Horn, 1975a].

5.3 Inequalities for Ky Fan k-Norms of Hadamard Products. Inequalities involving the spectral norm, such as 3.1(d,e,f), may be thought of as inequalities for the *one largest* singular value. With this perspective, one may seek analogous inequalities involving the *k largest* singular values in the form of the *Ky Fan k-norms* $N_k(\cdot)$, defined by

$$N_k(A) \equiv \sum_{i=1}^{k} \sigma_i(A), \quad A \in M_n, \quad k = 1,\ldots, n, \tag{5.3.1}$$

where $\sigma_1 \geq \cdots \geq \sigma_n \geq 0$ are the singular values of A. One can show that

$$N_k(A \circ B) \leq \sum_{i=1}^{k} \sigma_i(A)\sigma_i(B) \quad \text{for all } A, B \in M_n, k = 1,\ldots, n, \tag{5.3.2}$$

[Horn & Johnson, 1987b], [Okubo, 1987], [Bapat, 1987a], [Zhang, 1988] and, more generally, that

$$N_k(A \circ B) \le \sum_{i=1}^{k} c_i(X) c_i(Y) \sigma_i(B), \quad k = 1, \ldots, n,$$

$$\text{for all } A, B \in M_n \text{ and any } X \in M_{r,m}, \; Y \in M_{r,n} \text{ with } X^* Y = A, \tag{5.3.3}$$

where $c_1(Z) \ge \cdots \ge c_n(Z)$ denote the ordered Euclidean lengths of the columns of $Z \in M_{m,n}$. The cases $k = 1$ of (5.3.2, 3) are Schur's inequalities 3.1(f,e). One is interested in inequalities of this type involving sums of the k largest singular values because they are essential in establishing inequalities that hold for *all* unitarily invariant norms [Ando, Horn, & Johnson, 1987], [Horn & Johnson, 1987a, 7.4.47]; for another proof of (5.3.3), see [Bapat, 1989].

If we use the singular value decomposition to factor $A = V \Sigma W^* = X^* Y$ with $X = \Sigma^{\frac{1}{2}} V^*$ and $Y = \Sigma^{\frac{1}{2}} W^*$, then the ordered squared Euclidean column lengths of X are $c_i(X)^2 = p_i(A)$, where $p_i(A) \ge \cdots \ge p_n(A)$ are the ordered main diagonal entries of $P \equiv (AA^*)^{\frac{1}{2}}$; similarly, $c_i(Y)^2 = q_i(A)$, where $q_1(A) \ge \cdots \ge q_n(A)$ are the ordered main diagonal entries of $Q \equiv (A^* A)^{\frac{1}{2}}$. For this factorization, (5.3.3) becomes

$$N_k(A \circ B) \le \sum_{i=1}^{k} [p_i(A) q_i(A)]^{\frac{1}{2}} \sigma_i(B) \quad \text{for all } A, B \in M_n, \; k = 1, \ldots, n, \tag{5.3.4}$$

which specializes to Walter's inequality (3.7.11) for $k = 1$.

We can now combine the arithmetic-geometric mean inequality with Schur's classical majorization between the eigenvalues of P and Q (which are the singular values of A) and their main diagonal entries [Horn & Johnson, 1987a, Theorem 4.3.26] to obtain

$$\sum_{i=1}^{k} [p_i(A) q_i(A)]^{\frac{1}{2}} \le \sum_{i=1}^{k} \tfrac{1}{2}[p_i(A) + q_i(A)] \le \sum_{i=1}^{k} \tfrac{1}{2}[\sigma_i(A) + \sigma_i(A)] = \sum_{i=1}^{k} \sigma_i(A), \quad k = 1, \ldots, n.$$

Using these inequalities and a standard summation-by-parts argument [Horn & Johnson, 1989, Lemma 5.6.18], one may replace the terms $[p_i(A) q_i(A)]^{\frac{1}{2}}$ in (5.3.4) by $\sigma_i(A)$, which shows that (5.3.2) follows from (5.3.3) via (5.3.4).

Now suppose that $A, B \in M_n$ and that $A = [a_{ij}]$ is positive semidefinite, so the choice $X = Y = A^{\frac{1}{2}}$ gives the factorization $A = X^* Y = A^{\frac{1}{2}*} A^{\frac{1}{2}}$. Since the squared column lengths of $A^{\frac{1}{2}}$ are just the main diagonal entries of A, if we order them as $\{a_{11}, \ldots, a_{nn}\} \equiv \{\alpha_1(A), \ldots, \alpha_n(A)\}$ with $\alpha_1(A) \ge \cdots \ge \alpha_n(A) \ge 0$, then (5.3.3) becomes

$$N_k(A \circ B) \le \sum_{i=1}^{k} \alpha_i(A) \sigma_i(B) \quad \text{for } A, B \in M_n, \; A \succeq 0, \; k = 1, \ldots, n, \tag{5.3.5}$$

which is a generalization of 3.1(d). Since the eigenvalues of A, $\lambda_1(A) \ge \cdots \ge \lambda_n(A) \ge 0$, majorize its main diagonal entries, the weaker inequalities

$$N_k(A\circ B) \le \sum_{i=1}^{k} \lambda_i(A)\sigma_i(B) \quad \text{for } A, B \in M_n,\ A \succeq 0,\ k = 1,\dots, n \tag{5.3.6}$$

follow from (5.3.5) by standard manipulations from the theory of majorization [Marshall & Olkin, 1979]. If A is not only positive semidefinite but is a *correlation matrix* (all main diagonal entries equal to 1), then (5.3.5) becomes

$$N_k(A\circ B) \le \sum_{i=1}^{k} \sigma_i(B) \quad \text{for } A, B \in M_n,\ A \succeq 0 \text{ a correlation matrix},\ k = 1,\dots, n. \tag{5.3.7}$$

Finally, suppose that *both* A and B are positive semidefinite. Then so is $A\circ B$ and the singular values of $A\circ B$ and B are just their decreasingly ordered eigenvalues; in this event, (5.3.5) becomes

$$\sum_{i=1}^{k} \lambda_i(A\circ B) \le \sum_{i=1}^{k} \alpha_i(A)\lambda_i(B) \quad \text{for } A, B \in M_n,\ A, B \succeq 0,\ k = 1,\dots, n, \tag{5.3.8}$$

an inequality that appears in [Bapat & Sunder, 1985]. The weaker inequality (5.3.6) becomes

$$\sum_{i=1}^{k} \lambda_i(A\circ B) \le \sum_{i=1}^{k} \lambda_i(A)\lambda_i(B) \quad \text{for } A, B \in M_n,\ A, B \succeq 0,\ k = 1,\dots, n. \tag{5.3.9}$$

If A is a correlation matrix, (5.3.7) becomes

$$\sum_{i=1}^{k} \lambda_i(A\circ B) \le \sum_{i=1}^{k} \lambda_i(B) \quad \text{for } A, B \in M_n,\ A, B \succeq 0,\ A \text{ correlation},\ k = 1,\dots, n, \tag{5.3.10}$$

an inequality that has been applied to important problems in mathematical statistics [Koyak, 1987, pp. 1224–6].

Returning to the master inequalities (5.3.3), if we now take $X = I$ and $Y = A$ we obtain the inequalities

$$N_k(A\circ B) \le \sum_{i=1}^{k} c_i(A)\sigma_i(B) \quad \text{for all } A, B \in M_n,\ k = 1,\dots, n, \tag{5.3.11a}$$

which gives one of the bounds in (3.6.2a) when $k = 1$. Similarly, choosing $X = A$ and $Y = I$ in

(5.3.3) gives

$$N_k(A \circ B) \le \sum_{i=1}^{k} r_i(A)\sigma_i(B) \text{ for all } A, B \in M_n, k = 1,\dots, n, \tag{5.3.11b}$$

which gives one of the bounds in (3.6.2b) for $k = 1$.

A generalization of the combined bound (3.6.3)

$$N_k(A \circ B) \le \sum_{i=1}^{k} [r_i(A)\, c_i(A)]^{\frac{1}{2}}\sigma_i(B) \text{ for all } A, B \in M_n, k = 1,\dots, n, \tag{5.3.12}$$

has been obtained in [Ando, Horn, & Johnson, 1987, p. 350].

5.4 The Hadamard Product as an Inertia-Preserver. The *inertia* of a Hermitian matrix $A \in M_n$ is the integer row vector $i(A) \equiv (i_+(A), i_-(A), i_0(A))$, where $i_+(A)$, the *positive inertia*, is the number of positive eigenvalues of A, $i_-(A)$, the *negative inertia*, is the number of negative eigenvalues of A, and $i_0(A)$, the *zero inertia* or *nullity*, is the number of zero eigenvalues of A; of course, $i_+(A) + i_-(A) + i_0(A) = n$. Part 3.1(c) of Schur's Theorem says that if A is positive semidefinite and has positive main diagonal entries, then every matrix B with inertia $(n, 0, 0)$ has its inertia preserved by the Hadamard product operator $\mathcal{H}_A$: $B \to A \circ B$.

A natural question is: For a given inertia (n_1,n_2,n_3), how can we characterize the class of Hermitian matrices $B \in M_n$ such that $(n_1,n_2,n_3) = i(B) = i(A \circ B)$ for *every* positive semidefinite A with positive main diagonal entries? The answer is that this is exactly the class of *diagonally signed matrices* with the given inertia, that is, all Hermitian matrices B with the given inertia such that the positive, negative, and zero inertia of any principal submatrix $\hat{B}$ of B is equal to the number of positive, negative, and zero main diagonal entries of $\hat{B}$, respectively.

The interlacing eigenvalues principle for Hermitian matrices gives one obvious class of diagonally signed matrices that has a given inertia (n_1,n_2,n_3): all Hermitian matrices $B \in M_n$ that have the *inertia-explicit* block form

$$B = \begin{bmatrix} B_{11} & B_{12} & 0 \\ B_{21} & B_{22} & 0 \\ 0 & 0 & B_{33} \end{bmatrix}, B_{11} \in M_{n_1}, B_{22} \in M_{n_2}, 0 = B_{33} \in M_{n_3}, B_{11} \succ 0, -B_{22} \succ 0. \tag{5.4.1}$$

In fact, the class of diagonally signed matrices with inertia (n_1,n_2,n_3) is exactly the set of permutation similarities of matrices of the form (5.4.1). It is clear from 3.1(c) that the Hadamard product of a matrix of the form (5.4.1) with any positive semidefinite matrix with positive main diagonal entries is another matrix of the form (5.4.1), so the inertia-preserving property of these matrices is evident.

In Section 4.2 we have already encountered an important class of matrices that are permutation-similar to the form (5.4.1) in a natural way—Cauchy matrices $C = [1/(\overline{\xi}_i + \xi_j)]$ of a set of n distinct complex numbers $\{\xi_1,\dots, \xi_n\}$ with $\overline{\xi}_i + \xi_j \neq 0$ for all $i, j = 1,\dots, n$, of which n_1 have positive real part and $n_2 = n - n_1$ have negative real part. We saw in our discussion of (4.2.10) that the Hadamard product of such a Cauchy matrix C with any positive semidefinite matrix with positive main diagonal entries always has the same inertia as C. Of course, any positive semidefinite matrix with positive main diagonal entries is the Gram matrix of a set of nonzero vectors: If $A = [v_i^* v_j]$ and all $v_i \in \mathbb{C}^n$ are nonzero, then A is positive semidefinite and has positive main diagonal entries, and the Hadamard product $A \circ C = [a_{ij}/(\overline{\xi}_i + \xi_j)]$ has the same inertia as C, which can be read off from the signs of its main diagonal entries. This motivating example is in [Graves-Morris & Saff, 1988, pp. 80–5], where they needed to show only that the matrix $A \circ C$ is nonsingular; basic results about diagonally signed matrices are in [Graves-Morris & Johnson, 1988].

5.5 Hadamard Product Inequalities for Norms on Matrices Induced by ℓ_p Vector Norms. Schur's original inequalities 3.1(e,f) are for the spectral norm. Since the spectral norm is unitarily invariant, it is natural to seek generalizations of Schur's inequalities to the class of unitarily invariant norm, and we have done so in Sections 3.7 and 5.3. The spectral norm is also an induced norm, induced by the ℓ_2 (Euclidean) vector norm, so it is also natural to investigate analogs of Schur's inequalities for norms on matrices induced by the *Hölder p-norms* on $\mathbb{C}^n$:

$$\begin{aligned} \|x\|_p &\equiv \Big[\sum_{i=1}^{n} |x_i|^p\Big]^{1/p} \qquad p \geq 1 \\ \|x\|_\infty &\equiv \max\{|x_i| : i = 1,\dots, n\} \end{aligned} \tag{5.5.1}$$

The corresponding induced *p-q norms* are defined for $A \in M_n$ by

$$\|A\|_{p,q} \equiv \max\{\|Ax\|_q : x \in \mathbb{C}^n, \|x\|_p \leq 1\} \quad 1 \leq p, q \leq \infty. \tag{5.5.2}$$

The ordinary spectral norm is obtained for $p = q = 2$: $|||A|||_2 = \|A\|_{2,2}$.

G. Bennett has given a complete set of analogs of Schur's inequalities 3.1(f) for the p-q norms for $A, B \in M_n$ [Bennett, 1977, Proposition 2.1]:

$$\|A \circ B\|_{p,q} \leq \|\,|A \circ B|\,\|_{p,q} \leq \begin{cases} \|A\|_{p,\infty} \|B\|_{1,q} & \text{if } 1 \leq p \leq q \leq \infty & (5.5.3\text{a}) \\ \|A\|_{p,2} \|B\|_{2,q} & \text{if } \infty \geq p \geq 2 \geq q \geq 1 & (5.5.3\text{b}) \\ \|A\|_{p,q} \|B\|_{2,\infty} & \text{if } \infty \geq p \geq q \geq 2 & (5.5.3\text{c}) \\ \|A\|_{p,q} \|B\|_{1,2} & \text{if } 2 \geq p \geq q \geq 1 & (5.5.3\text{d}) \end{cases}$$

where $|C| \equiv [\,|c_{ij}|\,]$ for $C = [c_{ij}] \in M_n$.

It can be helpful to recognize that there are some special values of p and q for which these norms can be computed readily as absolute row or column sums. For $A = [a_{ij}] \in M_n$ we have

$$\| A \|_{1,q} = \max \{ \Big[\sum_{i=1}^{n} |a_{ij}|^q \Big]^{1/q} : j = 1,\ldots, n\}, \quad 1 \le q \le \infty \tag{5.5.4}$$

and

$$\| A \|_{p,\infty} = \max \{ \Big[\sum_{j=1}^{n} |a_{ij}|^{p^*} \Big]^{1/p^*} : i = 1,\ldots, n\}, \quad 1 \le p \le \infty, \; p^* \equiv p/(p-1). \tag{5.5.5}$$

In particular, for $p = q = 2$, inequality (5.5.3b) gives Schur's bound 3.1(f), while (5.5.3a,c,d) provide the bounds on $|||A \circ B|||_2$ given in (3.5.5). Since other values of p and q are less tractable, it can be helpful to know that $\| A \|_{p,q}$ is increasing in p, decreasing in q, and $\| A \|_{p,q} \le \| \,|A|\, \|_{p,q}$.

G. Bennett also studied analogs of the Kronecker product identity (3.2.2a), and found that for any $A, B \in M_n$

$$\| A \otimes B \|_{p,q} \ge \| A \|_{p,q} \| B \|_{p,q} \quad \text{for } 1 \le p, q \le \infty \tag{5.5.6}$$

and

$$\| A \otimes B \|_{p,q} = \| A \|_{p,q} \| B \|_{p,q} \quad \text{for } 1 \le p \le q \le \infty, \tag{5.5.7}$$

a result that generalizes (3.2.2a) [Bennett, 1977, Section 10]. In the complementary cases, he showed that if $1 \le q \le 2 \le p \le \infty$, then for all $A, B \in M_n$ and all $n = 1, 2, \ldots$ there exists a finite positive constant $K(p,q)$ (independent of n, A, and B) such that

$$\| A \otimes B \|_{p,q} \le K(p,q) \, \| A \|_{p,q} \| B \|_{p,q}. \tag{5.5.8}$$

However, if $1 \le q < p < 2$ and $K > 0$ is given, there exists some positive integer n and some $A \in M_n$ such that

$$\|A \otimes A \|_{p,q} > K \| A \|_{p,q}^2. \tag{5.5.9}$$

By duality, the same conclusion holds for $2 < q < p \le \infty$.

Implicit in Bennett's proof of (5.5.8) is a remarkable theorem of Grothendieck that gives a universal bound on the absolute value of the sum of the entries of a Hadamard product [Grothendieck, 1956]. Using the definition (5.5.2) and the fact that the Hölder vector norms $\|\cdot\|_1$ and $\|\cdot\|_\infty$ are duals, notice that

$$\| B \|_{\infty,1} = \max \{\| Bx \|_1 : x \in \mathbb{C}^n, \|x\|_\infty \le 1\}$$

$$= \max \{ \max \{ |y^* Bx| : y \in \mathbb{C}^n, \|y\|_\infty \le 1\} : x \in \mathbb{C}^n, \|x\|_\infty \le 1\}$$

$$= \max \{ |y^* Bx| : x, y \in \mathbb{C}^n, \|y\|_\infty \le 1, \|x\|_\infty \le 1\} \le \sum_{i,j=1}^{n} |b_{ij}|.$$

In general, the last inequality is strict; it is an equality if B is the Hadamard product of a nonnegative matrix and a rank 1 matrix. Grothendieck's theorem is that there exists a finite positive universal constant K_G such that

$$|e^T(A \circ B)e| = |\sum_{i,j=1}^{n} a_{ij} b_{ij}| \le K_G \, c_1(X) c_1(Y) \| B \|_{\infty,1}$$

for all $A, B \in M_n$, all $X, Y \in M_{r,n}$ such that $X^* Y = A$, all $r = 1, 2, \ldots$, and all $n = 1, 2, \ldots$. (5.5.10)

The precise value of K_G seems not to be known, but Grothendieck showed that

$$\tfrac{1}{2}\pi \le K_G \le 2 \sinh(\tfrac{1}{2}\pi);$$

the upper bound can be improved to $\sinh(\frac{1}{2}\pi)$ if all the matrices in (5.5.10) are restricted to have real entries. This inequality suggests analogies with Schur's bound 3.1(e) as well as with Moutard's and Fejér's studies of the quantity $e^T(A \circ B)e$ described in Example 2.1. For an accessible exposition of Grothendieck's inequality, see [Lindenstrauss & Pełczński, 1968].

5.6 Norm Inequalities for Other Products. The observation that there are basic inequalities such as 3.1(f) and (5.3.2) that are valid for both the ordinary product and the Hadamard product suggests that these two products may be members of a whole family of bilinear matrix products that satisfy these and other inequalities. From this point of view, [Horn, Mathias, & Nakamura, 1989, Theorem 3.1] characterized the bilinear matrix products that satisfy the basic singular value majorization inequality (5.3.2); both the ordinary and Hadamard products are easily shown to satisfy the conditions in this characterization, so one has, in this way, a unified proof of inequalities of the form (5.3.2) for both products. They also considered the partitioned products

$$A \square B \equiv [A_{ij} B_{ij}] \in M_n \tag{5.6.1}$$

where $A = [A_{ij}]_{i,j=1,\ldots,p}$, each $A_{ij} \in M_q$, $n = pq$, and similarly for B. The Hadamard product is obtained for $q = 1$, while the ordinary product is the case $p = 1$. It is easy to use the method in Section 3.5 to show that

$$|||A \square B|||_2 \le |||A|||_2 \, |||B|||_2 \tag{5.6.2}$$

for all choices of p and q and all $A, B \in M_n$, $n = pq$; more generally, under the same conditions it is shown that

$$\| A \,\Box\, B \| \leq \| A^*A \| \, \| B^*B \| \tag{5.6.3}$$

for all unitarily invariant norms $\|\cdot\|$, which includes the inequality (3.7.27) for $p = 1$ and its analog for the usual matrix product. Furthermore, one may add to the equivalent conditions in (3.3.11) the additional condition $N(\cdot)$ *is submultiplicative with respect to the* $\Box$ *product.* However, the basic majorization inequalities

$$\sum_{i=1}^{k} \sigma_i(A \,\Box\, B) \leq \sum_{i=1}^{k} \sigma_i(A)\sigma_i(B), \quad k = 1,\dots, n \tag{5.6.4}$$

hold for all $A, B \in M_{pq}$ only for $p = 1$ or $q = 1$, that is, only when $\Box$ is either the usual matrix product or the Hadamard product.

C. R. Johnson and P. Nylen have shown that within a large and natural class of bilinear matrix products that includes the usual product, the Hadamard product, and the $\Box$ products (5.6.1), essentially only the Hadamard product, the usual matrix product, and two exceptional nonstandard products obey singular value majorization inequalities of the form (5.6.4).

5.7 Bounds for the Spectral Norm of a Hadamard Product Involving Absolute Row and Column Norms. One way to prove 3.1(e) is to observe that for any $C = [\, c_{1\cdot} \;\dots\; c_{n\cdot} \,] \in M_n$ and any matrix norm $|||\cdot|||$ on M_n,

$$||| C |||_2^2 = \rho(CC^*) \leq ||| CC^* ||| \leq ||| C ||| \, ||| C^* |||; \tag{5.7.1}$$

see [Horn & Johnson, 1987a, Theorem 5.6.9]. If we take the matrix norm in (5.7.1) to be the maximum absolute column sum matrix norm $||| C |||_1 = \max\{\| c_{i\cdot} \|_1 : 1 \leq i \leq n\}$, where $\| c_{i\cdot} \|_1$ denotes the ℓ_1 vector norm on $\mathbb{C}^n$, then we have the bound

$$||| C |||_2^2 \leq ||| C |||_1 \, ||| C^* |||_1 = ||| C |||_1 \, ||| C |||_\infty, \tag{5.7.2}$$

where $||| C |||_\infty$ denotes the maximum absolute row sum matrix norm. Applying this inequality to $C = A \circ B$ for $A = [a_{ij}]$, $B = [b_{ij}]$ gives

$$\begin{aligned} ||| A \circ B |||_2^2 &\leq ||| A \circ B |||_1 \, ||| A \circ B |||_\infty \leq \max_i \sum_k | a_{ik} b_{ik} | \; \max_j \sum_k | a_{kj} b_{kj} | \\ &\leq \max_i \| a_{i\cdot} \| \, \| b_{i\cdot} \|^D \; \max_j \| a_{\cdot j} \| \, \| b_{\cdot j} \|^D, \end{aligned} \tag{5.7.3}$$

where $\|\cdot\|$ is any given absolute vector norm on $\mathbb{C}^n$ and $\|\cdot\|^D$ denotes its dual norm. We have written $A = [\, a_{1\cdot} \;\cdots\; a_{n\cdot} \,] = [\, a^T_{\cdot 1} \;\cdots\; a^T_{\cdot n} \,]$ to partition A according to its columns and rows, respectively. Carrying these inequalities one step further gives the general bound

$$|||\, A \circ B \,|||_2 \le [\max_i \|a_{i\cdot}\| \max_j \|a_{\cdot j}\|\,]^{\frac{1}{2}} \, [\max_i \|b_{i\cdot}\|^D \max_j \|b_{\cdot j}\|^D\,]^{\frac{1}{2}} \tag{5.7.4}$$

for any absolute vector norm $\|\cdot\|$ on $\mathbb{C}^n$ and its dual $\|\cdot\|^D$.

If we choose $\|\cdot\|$ to be the Euclidean vector norm, (5.7.4) gives

$$|||\, A \circ B \,|||_2 \le [r_1(A)c_1(A)]^{\frac{1}{2}} \, [r_1(B)c_1(B)]^{\frac{1}{2}}, \tag{5.7.5}$$

an inequality that we have obtained by different means in (3.5.6). This implies 3.1(e), and is essentially the original proof given in [Schur, 1911, Satz II].

If we choose $\|\cdot\|$ to be the ℓ_1 vector norm $\|\cdot\|_1$ in (5.7.4), we obtain

$$|||\, A \circ B \,|||_2 \le [\, |||\, A \,|||_1 \, |||\, A \,|||_\infty \,]^{\frac{1}{2}} \, \|\, B \,\|_\infty \le [n^{\frac{1}{2}} \, c_1(A) \, n^{\frac{1}{2}} \, r_1(A)]^{\frac{1}{2}} \, \|\, B \,\|_\infty$$

$$\le n^{\frac{1}{2}} \, [r_1(A)c_1(A)]^{\frac{1}{2}} \, \|\, B \,\|_\infty \le n^{\frac{1}{2}} \, |||\, A \,|||_2 \, \|\, B \,\|_\infty, \tag{5.7.6}$$

where $\|\, B\|_\infty$ is the maximum absolute value entry of B; see (3.6.3) for another version of these inequalities.

Many other inequalities can be obtained from (5.7.4) with other absolute norms, for example, other ℓ_p norms.

6. Hadamard Products, Determinants, and Permanents of Positive Semidefinite Matrices. The best-known inequality involving Hadamard products and determinants is surely Oppenheim's inequality for positive semidefinite $A = [a_{ij}]$, $B = [b_{ij}] \in M_n$ [Oppenheim, 1930]:

$$\det(A \circ B) \ge (a_{11} \cdots a_{nn}) \det B \ge (\det A)(\det B). \tag{6.1}$$

In explaining how he was led to this result, Oppenheim observes that the famous Hadamard determinant inequality $\det A \le a_{11} \cdots a_{nn}$ for positive definite A ([Hadamard, 1893], [Horn & Johnson, 1987a, Theorem 7.8.1]) can be written as $(\det A)(\det I) \le \det(A \circ I)$, which suggests the outer inequality in (6.1). In his 1930 paper, Oppenheim also proves the stronger inequality

$$\det(A \circ B) + (\det A)(\det B) \ge (b_{11} \cdots b_{nn}) \det A + a_{11} \cdots a_{nn} \det B, \tag{6.2}$$

whose discovery he credits to Schur. Oppenheim deduced all of these inequalities from Schur's qualitative theorem 3.1(c); his proofs are essentially the same as those in [Mirsky, 1955, Theorem 13.5.6] and [Horn & Johnson, 1987a, Theorem 7.8.6]. Schur did not publish his own, quite different, proof of (6.2) until 1937 [Schur, 1937].

R. B. Bapat has used simple properties of Schur complements of positive definite matrices to give block generalizations of Oppenheim's determinant inequalities (6.1) [Bapat, 1987b]. Suppose a given matrix $X \in M_n$, $n \geq 2$, is partitioned as

$$X = \begin{bmatrix} X_{11} & X_{12} \\ X_{21} & X_{22} \end{bmatrix},$$

where $X_{11} \in M_k$, $X_{22} \in M_{n-k}$, and $1 \leq k \leq n-1$. If X_{11} is nonsingular, define the *Schur complement of* X_{11} *in* X to be the matrix

$$\tilde{X}_{11} \equiv X_{22} - X_{21} X_{11}^{-1} X_{12} \in M_{n-k} \tag{6.3}$$

and compute

$$\begin{bmatrix} X_{11} & X_{12} \\ X_{21} & X_{22} \end{bmatrix} \begin{bmatrix} I & -X_{11}^{-1} X_{12} \\ 0 & I \end{bmatrix} = \begin{bmatrix} X_{11} & 0 \\ X_{21} & \tilde{X}_{11} \end{bmatrix}.$$

This identity shows that

$$\det X = (\det X_{11})(\det \tilde{X}_{11}). \tag{6.4}$$

In particular, the Schur complement $\tilde{X}_{11}$ must be nonsingular if X and X_{11} are nonsingular. Under suitable assumptions about nonsingularity, one checks [Horn & Johnson, 1987a, (7.7.5)] that X^{-1} can be written in conformally partitioned form (we need only the 2,2 block here) as

$$\begin{bmatrix} X_{11} & X_{12} \\ X_{21} & X_{22} \end{bmatrix}^{-1} = \begin{bmatrix} * & * \\ * & \tilde{X}_{11}^{-1} \end{bmatrix}.$$

We are interested here only in the case in which X is positive semidefinite and X_{11} is positive definite, which is certainly the case if X is positive definite; in these cases we also have $X_{21} = X_{12}^*$. Using a limiting argument in the singular case, we see that

> The Schur complement $\tilde{X}_{11} = X_{22} - X_{12}^* X_{11}^{-1} X_{12}$ of X_{11} in X is positive (semi)definite whenever X is positive (semi)definite and X_{11} is positive definite.

(6.5)

Now let given positive definite matrices $A, B \in M_n$ be partitioned conformally as

$$A = \begin{bmatrix} A_{11} & A_{12} \\ A_{12}^* & A_{22} \end{bmatrix} \text{ and } B = \begin{bmatrix} B_{11} & B_{12} \\ B_{12}^* & B_{22} \end{bmatrix},$$

and notice that

$$\hat{B} \equiv \begin{bmatrix} B_{11} & B_{12} \\ B_{12}^* & B_{22} - \tilde{B}_{11} \end{bmatrix} = \begin{bmatrix} B_{11} & B_{12} \\ B_{12}^* & B_{12}^* B_{11}^{-1} B_{12} \end{bmatrix}$$

$$= [(B_{11})^{\frac{1}{2}} \quad (B_{11})^{-\frac{1}{2}} B_{12}]^* \, [(B_{11})^{\frac{1}{2}} \quad (B_{11})^{-\frac{1}{2}} B_{12}]$$

is positive semidefinite. Thus, 3.1(a) and 3.1(b) guarantee that

$$A \circ \hat{B} = \begin{bmatrix} A_{11} \circ B_{11} & A_{12} \circ B_{12} \\ (A_{12} \circ B_{12})^* & A_{22} \circ (B_{22} - \tilde{B}_{11}) \end{bmatrix}$$

is positive semidefinite and $A_{11} \circ B_{22}$ is positive definite. In particular, (6.5) ensures that the Schur complement of $A_{11} \circ B_{11}$ in $A \circ \hat{B}$

$$A_{22} \circ (B_{22} - \tilde{B}_{11}) - (A_{12} \circ B_{12})^* (A_{11} \circ B_{11})^{-1} (A_{12} \circ B_{12})$$

is positive semidefinite, and hence

$$(A_{22} \circ B_{22}) - (A_{12} \circ B_{12})^* (A_{11} \circ B_{11})^{-1} (A_{12} \circ B_{12}) \succeq A_{22} \circ \tilde{B}_{11},$$

in which the left-hand side is the Schur complement of $A_{11} \circ B_{11}$ in $A \circ B$. This matrix inequality implies [Horn & Johnson, 1987a, Corollary 7.7.4] the determinant inequality

$$\det[(A_{22} \circ B_{22}) - (A_{12} \circ B_{12})^* (A_{11} \circ B_{11})^{-1} (A_{12} \circ B_{12})] \geq \det(A_{22} \circ \tilde{B}_{11}).$$

Finally, multiply both sides of the preceding inequality by $\det(A_{11} \circ B_{11})$, use (6.4) to recognize the left-hand side as $\det(A \circ B)$, and conclude from a limiting argument that

$$\det(A \circ B) \geq \det(A_{11} \circ B_{11}) \det(A_{22} \circ \tilde{B}_{11}) \text{ for all positive semidefinite } A, B \in M_n, \ n \geq 2. \quad (6.6)$$

Bapat's inequality (6.6) easily implies the stronger half of Oppenheim's inequality (6.1) with an induction argument: The implication is immediate for $n = 2$; if the implication is assumed for matrices of all sizes less than n and if $A_{11}, B_{11} \in M_k$ with $1 \leq k \leq n-1$, just multiply the inequalities

$$\det(A_{11}\circ B_{11}) \geq a_{11} \cdots a_{kk} (\det B_{11})$$

and

$$\det(A_{22}\circ \tilde{B}_{11}) \geq a_{k+1,k+1} \cdots a_{nn} (\det \tilde{B}_{11})$$

and use (6.4) to conclude that $\det(A\circ B) \geq a_{11} \cdots a_{nn} (\det B)$.

Bapat also showed that (6.6) implies

$$\det(A\circ A^{-1}) \geq \det[A_{11}\circ(A^{-1})_{11}] \det[A_{22}\circ(A_{22})^{-1}]$$

$$\geq \det[A_{11}\circ(A_{11})^{-1}] \det[A_{22}\circ(A_{22})^{-1}], \tag{6.7}$$

an inequality that is somewhat surprising since it is not always the case that $A\circ A^{-1} \succeq [A_{11}\circ(A_{11})^{-1}] \oplus [A_{22}\circ(A_{22})^{-1}]$. For some related inequalities, see [Horn & Johnson, 1989, Section 5.4].

Bergstrom's determinant inequalities for positive definite A, $B \in M_n$ are

$$\frac{\det(A + B)}{\det(A_i + B_i)} \geq \frac{\det A}{\det A_i} + \frac{\det B}{\det B_i}, \quad i = 1,\ldots, n, \tag{6.8}$$

where A_i, $B_i \in M_{n-1}$ denote the principal submatrices obtained by deleting the ith row and column; for a conventional proof using a quasilinearization, see [Horn & Johnson, 1987a, Problem 12 in Section 7.4]. In [Berenstein & Veinstein, 1987] it is shown that the determinant inequalities (6.8) are equivalent to the Hermitian form inequalities

$$(x+y)^*(A+B)^{-1}(x+y) \leq x^*A^{-1}x + y^*B^{-1}y \text{ for all } x, y \in \mathbb{C}^n; \tag{6.9}$$

[Marcus, 1967] showed earlier that (6.9) implies (6.8). This equivalent form of Bergstrom's inequalities is of interest to us here because Berenstein and Veinstein also show that

$$(x\circ y)^*(A\circ B)^{-1}(x\circ y) \leq (x^*A^{-1}x)(y^*B^{-1}y) \text{ for all } x, y \in \mathbb{C}^n, \tag{6.10}$$

which may therefore be thought of as a multiplicative analog of the Bergstrom inequalities. They go on to show that the inequalities (6.10) directly imply both of the determinant inequalities (6.1) and (6.2), so (6.10) may also be regarded as a generalization of Oppenheim's inequalities.

If $A = [a_{ij}] \in M_n$ is a correlation matrix, W. Watkins has shown [Watkins, 1988] that

$$\lambda_{min}(A\circ A^T) = \lambda_{min}(A\circ \bar{A}) \geq \det A, \tag{6.11}$$

which verifies half of a conjecture of R. Grone and R. Merris that the closed interval [det A,per A] contains the spectrum of $A \circ A^T$ [Grone & Merris, 1989]. If one considers the representation of the determinant of A as the signed sum of all $n!$ diagonal products in A, the *permanent*, denoted by per A, is the same sum in which all the signs are taken to be +. Since any positive semidefinite matrix with positive main diagonal entries may be transformed into a correlation matrix by a positive diagonal congruence, (6.11) leads to inequalities for general positive semidefinite matrices.

The permanental analog of Hadamard's determinant inequality, proved by M. Marcus ([Marcus, 1963] and [Marcus, 1964]), may also be regarded as a theorem about Hadamard products:

$$\text{per}\, A \geq \text{per}(A \circ I) = a_{11} \cdots a_{nn} \quad \text{for positive semidefinite } A = [a_{ij}] \in M_n. \qquad (6.12)$$

It is natural to conjecture, as Oppenheim did for the determinant, that

$$(\text{per}\, A)\,(\text{per}\, B) \geq \text{per}(A \circ B), \qquad (6.13)$$

and this is still an open problem. It is known that to prove (6.13) it suffices to prove it only for correlation matrices [Zhang, 1989], and it suffices to prove only the special case

$$(\text{per}\, A)^2 \geq \text{per}(A \circ A^T) = \text{per}(A \circ \bar{A}) \quad \text{for all positive semidefinite } A \in M_n \qquad (6.14)$$

[Chollet, 1982]; the latter inequality has been verified for $n = 2$ and 3 in [Gregorac & Hentzel, 1987]. For further discussion of permanental inequalities see [Merris, 1987] and [Bapat & Sunder, 1986]. A basic reference for the theory of permanents (with numerous results about Hadamard products and permanents) is [Minc, 1978].

Acknowledgements. It is a pleasure to acknowledge the help of Roy Mathias in the preparation of this survey. Many of the ideas discussed here are discussed in the two books [Horn & Johnson, 1987] and [Horn & Johnson, 1989], which are co-authored by Charles R. Johnson. Special thanks are due to T. Ando, R. B. Bapat, Rajendra Bhatia, Richard Brualdi, Carl Cowen, Chandler Davis, Michael Doob, Miroslav Fiedler, Shmuel Friedland, P. R. Graves-Morris, Paul Halmos, Robert Koyak, David Peterson, and Fuzhen Zhang, who contributed references, examples, or comments.

BIBLIOGRAPHY

G. T. ADAMS

1986 The bilateral Bergman shift, *Memoirs of the American Mathematical Society*, Vol. 63, No. 355.

T. ANDO

1979 Concavity of certain maps on positive definite matrices and applications to Hadamard products, *Linear Algebra Appl.* 26, 203–41.

1980 Inequalities for *M*-matrices, *Linear and Multilinear Algebra* 8, 291-316.

T. ANDO, R.A. HORN, and C.R. JOHNSON

1987 The singular values of a Hadamard product: A basic inequality, *Linear and Multilinear Algebra* 21, 345–65.

T. ANDO and K. OKUBO

1991? Norm of Schur multiplication, to appear in *Linear Algebra Appl.*

J. ARAZY

1982 Certain Schur-Hadamard multipliers in the Spaces C_p, *Proc. Amer. Math. Soc.* 86, 59–64.

C.S. BALLANTINE

1968 On the Hadamard product, *Math. Zeit.* 105, 365–6.

E. BANNAI and T. ITO

1984 *Algebraic Combinatorics I: Association Schemes*, Benjamin/Cummings, Menlo Park, California.

R. B. BAPAT

1987a Majorization and singular values, *Linear Multilinear Algebra* 21, 211-14.

1987b A refinement of Oppenheim's inequality, in *Current Trends in Matrix Theory*, edited by F. Uhlig and R. Grone, North Holland (Elsevier), New York, pp. 29-32.

1988a Multinomial probabilities, permanents and a conjecture of Karlin and Rinott, *Proc. Amer. Math. Soc.* 102, 467-72.

1988b The multinomial distribution and permanents, *Linear Algebra Appl.* 104, 201-5.

1989 Majorization and singular values II, *SIAM J. Matrix Analysis Appl.* 10, 429-34.

R. B. BAPAT and M. K. KWONG

1987 A generalization of $A \circ A^{-1} \geq I$, *Linear Algebra Appl.* 93, 107-12.

R. B. BAPAT and V. S. SUNDER

1985 On majorization and Schur products, *Linear Algebra Appl.* 72, 107-17.

1986 An extremal property of the permanent and the determinant, *Linear Algebra Appl.* 76, 153–63.

R. BELLMAN

1970 *Introduction to Matrix Analysis*, McGraw-Hill, New York, Second Edition.

G. BENNETT

1976 Unconditional convergence and almost everywhere convergence, *Z. Wahrscheinlich-*

keitstheorie verw. Gebiete 34, 135–55.

1977 Schur multipliers, *Duke Math. J.* 44, 603–39.

A. D. BERENSTEIN and A. D. VEINSTEIN

1987 Multiplicative analog of the Bergstrom inequality for Hadamard products, *Uspekhi Mat. Nauk* 42, No. 6 (258), 181-2 (in Russian).

N. BIGGS

1976 Automorphic graphs and the Krein condition, *Geometriae Dedicata* 5, 117-27.

N. N. CHAN and M. K. KWONG

1985 Hermitian matrix inequalities and a conjecture, *Amer. Math. Monthly* 92, 533–41.

M.-D. CHOI

1983 Tricks or treats with the Hilbert matrix, *Amer. Math. Monthly* 90, 301-12.

J. CHOLLET

1982 Is there a permanental analogue to Oppenheim's inequality?, *Amer. Math. Monthly* 89, 57–8.

D. CVETKOVIĆ, M. DOOB, I. GUTMAN, and A. TORGAŠEV

1988 *Recent Results in the Theory of Graph Spectra*, Vol. 36, Annals of Discrete Mathematics, North-Holland, Amsterdam.

C. C. COWEN

1990? An application of Hadamard multiplication to operators on weighted Hardy spaces, *Linear Algebra Appl.*, to appear.

J. L. DALECKIĬ

1965 Differentiation of non-Hermitian matrix functions depending on a parameter, *American Mathematical Society Translations, Series 2*, Vol. 47, American Mathematical Society, Providence, 73–87. Russian original in *Izv. Vyss. Ucebn. Zaved. Mat.* 1962, No. 2 (27), 52–64.

J. L. DALECKIĬ and S.G. KREĬN

1951 Formulas of differentiation according to a parameter of functions of Hermitian operators (Russian), *Dokl. Akad. Nauk SSSR* 76, 13-16.

1965 Integration and differentiation of functions of Hermitian operators and applications to the theory of perturbations, *American Mathematical Society Translations, Series 2*, Vol. 47, American Mathematical Society, Providence, 1-30. Russian original in *Voronez. Gos. Univ. Trudy Sem. Funkcional. Anal.* No. 1 (1956), 81-105.

C. DAVIS

1962 The norm of the Schur product operation, *Numerische Math.* 4, 343-344.

1963 Notions generalizing convexity for functions defined on spaces of matrices, 187-201, in *Convexity*, Vol. VII of Proceedings of Symposia in Pure Mathematics, edited by

V. Klee, American Mathematical Society, Providence.

T. E. DAVIS

1970 A note on Hadamard products, *Amer. Math. Monthly* 77, 1087.

W. F. DONOGHUE

1974 *Monotone Matrix Functions and Analytic Continuation*, Springer-Verlag, Berlin.

D. Ž. DJOKOVIĆ

1965 On the Hadamard product of matrices, *Math. Zeit.* 86, 395.

L. ELSNER, C. R. JOHNSON, and J. A. DIAS DA SILVA

1988 The Perron root of a weighted geometric mean of nonnegative matrices, *Linear Multilinear Algebra* 24, 1-13.

K. FAN

1964 Inequalities for *M*-matrices, *Proc. Koninkl. Nederl. Akademie van Wetenschappen, Series A*, 67, and *Indag. Math.* 26, 602-10.

1966 Some matrix inequalities, *Abhandlungen aus dem mathematischen Seminar der Universität Hamburg* 29, 185-96.

L. FEJÉR

1918 Über die Eindeutigkeit der Lösung der Linearen Partiellen Differentialgleichung zweiter Ordnung, *Math. Zeit.* 1, 70-9.

M. FIEDLER

1957 On some properties of Hermitian matrices, *Math.-Fyz. Casopis SAV*, VII, 3, 168-176 (in Czech with Russian and English summaries).

1961 Über eine Ungleichung für positiv definite Matrizen, *Math. Nachrichten* 23, 197-9.

1962 Relations between the diagonal elements of an *M*-matrix and its inverse matrix, *Math.-Fyz. Casopis SAV*, 12, 2,123-8.

1983 A note on the Hadamard product of matrices, *Linear Algebra Appl.* 49, 233-5.

M. FIEDLER and T. MARKHAM

1988a An inequality for the Hadamard product of an *M*-matrix and an inverse *M*-matrix, *Linear Algebra Appl.* 101, 1-8.

1988b On the range of the Hadamard product of a positive definite matrix and its inverse, *SIAM J. Matrix Analysis Applic.* 9, 343-7.

C. H. FITZGERALD

1967 On analytic continuation to a schlicht function, *Proc. Amer. Math. Soc.* 18, 788-92.

1972a Quadratic inequalities and coefficient estimates for schlicht functions, *Arch. Rational Mech. Anal.* 46, 356-68.

1972b Exponentiation of certain quadratic inequalities for schlicht functions, *Bull. Amer. Math. Soc.* 78, 209-10.

1977 Quadratic inequalities and analytic continuation, *J. D'Analyse Math.* 31, 19-47.

C. H. FITZGERALD and R. A. HORN

1977a On fractional Hadamard powers of positive definite matrices, *J. Math. Anal. Applic.* 61, 633-42.

1977b On the structure of Hermitian-symmetric inequalities, *J. London Math. Soc.* (2) 15, 419-30.

1982 On quadratic and bilinear forms in function theory, *Proc. London Math. Soc., Third Series* 44, 554-76.

C. K. FONG, H. RADJAVI, and P. ROSENTHAL

1987 Norms for matrices and operators, *J. Operator Theory* 18, 99-113.

S. FRIEDLAND

1975 Generalized Hadamard inequality and its applications, *Linear Multilinear Algebra* 2, 327-33.

1986 Limit eigenvalues of nonnegative matrices, *Linear Algebra Applic.* 74, 173-8.

1989 Coherent algebras and the graph isomorphism problem, *Discrete Applied Math.* 25, 73-98.

E. GEKELER

1981 On the pointwise matrix product and the mean value theorem, *Linear Algebra Appl.* 35, 183-91.

G. M. GOLUSIN

1969 *Geometric Theory of Functions of a Complex Variable*, Translations of Mathematical Monographs, Vol. 26, American Mathematical Society, Providence; original Russian language first edition published 1952, second Russian language edition, Moscow, 1966.

P. R. GRAVES-MORRIS and E. B. SAFF

1988 Row convergence theorems for generalized inverse, vector-valued Padé approximants, *J. Computational Appl. Math.* 23, 63-85.

P. R. GRAVES-MORRIS and C. R. JOHNSON

1989 A theorem on Hermitian matrices of indefinite signature, in *Applications of Matrix Theory*, edited by S. Barnett and M. J. C. Gysver, Oxford University Press, Oxford.

R. J. GREGORAC and I. R. HENTZEL

1987 A note on the analogue of Oppenheim's inequality for permanents, *Linear Algebra Applic.* 94, 109-12.

R. GRONE and R. MERRIS

1989? Conjectures on permanents, to appear in *Linear Multilinear Algebra.*

A. GROTHENDIECK

1956 Résume de la théorie métrique des produits tensoriels topologiques, *Boletim da Sociedade de Mathematica de Sao Paulo* 8, 1-79

J. HADAMARD

1893 Résolution d'une question relative aux déterminants, *Bulletin des Sciences Math.* (2) 17, 240-8.

1899 Théorème sur les séries entières, *Acta Math.* 22, 55-63.

1903 *Lecons sur la Propagation des Ondes et les Equations de l'Hydrodynamique*, Hermann, Paris; reprinted by Chelsea, New York, 1949.

P. R. HALMOS

1948 *Finite-Dimensional Vector Spaces,* Princeton University Press, Princeton.

1958 *Finite-Dimensional Vector Spaces*, Van Nostrand, Princeton, Second Edition.

P. R. HALMOS and V. S. SUNDER

1978 *Bounded Integral Operators on L^2 Spaces*, Ergebnisse der Mathematik und ihrer Grenzgebiete, Vol. 96, Springer-Verlag, Berlin.

D. G. HIGMAN

1975 Coherent configurations, *Geometriae Dedicata* 4, 1-32.

1987 Coherent algebras, *Linear Algebra Applic.* 93, 209-39.

R. A. HORN

1967a On boundary values of a schlicht mapping, *Proc. Amer Math. Soc.* 18, 782-7.

1967b On infinitely divisible matrices, kernels, and functions, *Z. Wahrscheinlichkeitstheorie verw. Geb.* 8, 219-30.

1969a The theory of infinitely divisible matrices and kernels, *Trans. Amer. Math. Soc.* 136, 269-86.

1969b Infinitely divisible positive definite sequences, *Trans. Amer. Math. Soc.* 136, 287-303.

1970a On certain power series and sequences, *J. London Math. Soc.* (2) 2, 160-2.

1970b On moment sequences and renewal sequences, *J. Math. Analysis Appl.* 31, 130-5.

1971 Schlicht mappings and infinitely divisible kernels, *Pacific J. Math.* 38, 423-30.

1975a On inequalities between Hermitian and symmetric forms, *Linear Algebra Applic.* 11, 189-218.

1975b Quadratic forms in harmonic analysis and the Bochner-Eberlein theorem, *Proc. Amer. Math. Soc.* 52, 263-70.

1977 On the moments of complex measures, *Math. Z.* 156, 1-11.

R. A. HORN and C. R. JOHNSON

1987a *Matrix Analysis*, Cambridge University Press, New York, second printing; first printing, 1985.

1987b Hadamard and conventional submultiplicativity for unitarily invariant norms on matrices, *Linear Multilinear Algebra* 20, 91-106.

1989 *Topics in Matrix Analysis*, Cambridge University Press, New York.

R. A. HORN and R. MATHIAS

1990a? An analog of the Cauchy-Schwarz inequality for Hadamard products and unitarily invariant norms, to appear in *SIAM J. Matrix Analysis Appl.*

1990b? Cauchy-Schwarz inequalities associated with positive definite matrices, to appear in *Linear Algebra Appl.*

R. A. HORN, R. MATHIAS, and Y. NAKAMURA

1989 Inequalities for unitarily invariant norms and bilinear matrix products, Technical Report No. 516, Department of Mathematical Sciences, The Johns Hopkins University, Baltimore, Maryland 21218.

E. C. JOHNSEN

1977 An inequality on the matrix spectral norm specifying Hadamard matrices, *Linear Multilinear Algebra*, 4, 277-9.

B. E. JOHNSON and J. P. WILLIAMS

1975 The range of a normal derivation, *Pacific J. Math.* 58, 105-122.

C. R. JOHNSON

1974a The Hadamard product of A and A^*, *Pacific J. Math.* 51, 477-81.

1974b Hadamard products of matrices, *Linear Multilinear Algebra* 1, 295-307.

1975 Notes on nonnegative Hadamard functions, *Linear Multilinear Algebra* 3, 187-192.

1977 A Hadamard product involving M-matrices, *Linear Multilinear Algebra* 4, 261-4.

1978 Partitioned and Hadamard product matrix inequalities, *J. Res. Nat. Bureau Standards* 83, 585-91.

1987 Closure properties of certain positivity classes of matrices under various algebraic operations, *Linear Algebra Applic.* 97, 243-7.

1988 Singular values of a Hadamard product: A master inequality, *Linear Algebra Appl.* 104, 198-201.

C. R. JOHNSON and R. B. BAPAT

1988 A weak multiplicative majorization conjecture for Hadamard products (problem proposed at the 1987 Utah State University matrix theory conference), *Linear Algebra Appl.* 104, 246-7.

C. R. JOHNSON and L. ELSNER

1987 The relationship between Hadamard and conventional multiplication for positive definite matrices, *Linear Algebra Appl.* 92, 231-40.

C. R. JOHNSON and T. MARKHAM

1985 Compression and Hadamard product inequalities, *Linear Multilinear Algebra* 18, 23-34.

C. R. JOHNSON, D.D. OLESKY, and P. VAN DEN DRIESSCHE

1985 Stability of *M*-matrix products, *Linear Multilinear Algebra* 18, 67-76.

C. R. JOHNSON and P. NYLEN

1989 Converse to a singular value inequality common to the Hadamard and conventional product, preprint.

C. R. JOHNSON and H. SHAPIRO

1986 Mathematical aspects of the relative gain array $A \circ A^{-T}$, *SIAM J. Algebraic Discrete Meth.* 7, 627-44.

C. R. JOHNSON and P. VAN DEN DRIESSCHE

1988 Interpolation of *D*-stability and sign stability, *Linear Multilinear Algebra* 23, 363-8.

S. KARLIN and F. OST

1985 Some monotonicity properties of Schur powers of matrices and related inequalities, *Linear Algebra Applic.* 68, 47-65.

T. KATO

1982 *A Short Introduction to Perturbation Theory for Linear Operators*, Springer-Verlag, New York.

R. KOYAK

1987 On measuring internal dependence in a set of random variables, *Annals of Statistics* 15, 1215-28.

S. KWAPIEŃ and A. PEŁCZYŃSKI

1970 The main triangle projection in matrix spaces and its applications, *Studia Math.* 34, 43-68.

M. LEWIN

1971 On exponents of primitive matrices, *Numer. Math.* 18, 154-61.

J. LINDENSTRAUSS and A. PEŁCZYŃSKI

1968 Absolutely summing operators in L_p-spaces and their applications, *Studia Math.* 29, 275-326.

K. LÖWNER (C. LOEWNER)

1934 Über monotone Matrixfunktionen, *Math. Zeit.* 38, 177-216.

1966a On schlicht-monotonic functions of higher order, *J. Math. Anal. Appl.* 14, 320-5.

1966b Determination of the critical exponent of the Green's function, in *Contemporary Problems in the Theory of Analytic Functions* (*International Conference on the Theory of Analytic Functions, Yerevan, 1965*), edited by M. A. Lavrent'ev, Nauka, Moscow, 1966, pp. 184-7. Also in *Charles Loewner, Collected Papers*, edited by L. Bers, Birkhäuser, Boston-Basel, 1988, pp. 480-4.

C. LOEWNER and E. NETANYAHU

1959 On some compositions of Hadamard type in classes of analytic functions, *Bull. Amer. Math. Soc.* 65, 284-6.

M. S. LYNN

1964 On the Schur product of *H*-matrices and non-negative matrices, and related inequalities, *Proc. Camb. Phil. Soc.* 60, 425-31.

K. N. MAJINDAR

1963 On a factorisation of positive definite matrices, *Canad. Math. Bull.* 6, 405-7.

M. MARCUS

1963 The permanent analogue of the Hadamard determinant theorem, *Bull. Amer. Math. Soc.* 69, 494-6.

1964 The Hadamard theorem for permanents, *Proc. Amer. Math. Soc.* 15, 967-73.

1967 An inequality for linear transformations, *Proc. Amer. Math. Soc.* 18, 793-7.

M. MARCUS and N. A. KHAN

1959 A note on the Hadamard product, *Canad. Math. Bull.* 2, 81-3.

M. MARCUS, K. KIDMAN, and M. SANDY

1984 Unitarily invariant generalized matrix norms and Hadamard products, *Linear Multilinear Algebra* 16, 197-213.

M. MARCUS and H. MINC

1964 *A Survey of Matrix Theory and Matrix Inequalities,* Allyn & Bacon, Boston.

M. MARCUS and M. SANDY

1988 Bessel's inequality in tensor space, *Linear Multilinear Algebra* 23, 233-49.

1990? Hadamard square roots, to appear in *SIAM J. Matrix Analysis Appl.*

M. MARCUS and R. C. THOMPSON

1963 The field of values of the Hadamard product, *Archiv Math.* 14, 283-8.

T. MARKHAM

1976 An application of theorems of Schur and Albert, *Proc. Amer. Math. Soc.* 59, 205-10.

A. W. MARSHALL and I. OLKIN

1979 *Inequalities: Theory of Majorization and its Applications*, Academic Press, New York.

R. MATHIAS

1988 Equivalence of two partial orders on a convex cone of positive semidefinite matrices, to appear in *Linear Algebra Appl.*

T. J. McAVOY

1983 *Interaction Analysis*, Instrument Society of America, Research Triangle Park.

R. MERRIS

1987 The permanental dominance conjecture, in *Current Trends in Matrix Theory*, edited by F. Uhlig and R. Grone, Elsevier (North-Holland), New York, pp. 213-23.

C. MICCHELLI

1986 Interpolation of scattered data: Distance matrices and conditionally positive definite functions, *Constr. Approx.* 2, 11-22.

H. MINC

1978 *Permanents*, Addison-Wesley, Reading, Massachusetts.

1988 *Nonnegative Matrices*, Wiley, New York.

L. MIRSKY

1955 *An Introduction to Linear Algebra*, Clarendon Press, Oxford. Reprinted in 1982 by Dover Publications, New York.

Th. MOUTARD

1894 Notes sur les équations derivées partielles, *J. de L'Ecole Polytechnique* 64, 55-69.

C. N. NETT and V. MANOUSIOUTHAKIS

1987 Euclidean condition and block relative gain: connections, conjectures, and clarifications, *IEEE Trans. Automatic Control* AC-32, 405-7.

K. OKUBO

1987 Hölder-type norm inequalities for Schur products of matrices, *Linear Algebra Applic.* 91, 13-28.

I. OLKIN

1985 A probabilistic proof of a theorem of Schur, *Amer. Math. Monthly* 92, 50-51.

S.-C. ONG

1984 On the Schur multiplier norm of matrices, *Linear Algebra Appl.* 56, 45-55.

A. OPPENHEIM

1930 Inequalities connected with definite Hermitian forms, *J. London Math. Soc.* 5, 114-9.

V. I. PAULSEN

1986 *Completely Bounded Maps and Dilations*, Pitman Research Notes in Mathematics No. 146, Wiley, New York.

1989 Schur products and matrix completions, *J. Functional Analysis* 85, 151-78.

Ch. POMMERENKE

1975 *Univalent Functions*, Vandenhoeck & Ruprecht, Göttingen.

C. R. RAO

1970 Estimation of heteroscedastic variances in linear models, *J. Amer. Stat. Assoc.* 65, 161-72.

W. H. RUCKLE

1975 Hadamard multipliers of infinite matrices, *Indiana Univ. Math. J.* 24, 949-57.

1985 Coordinatewise multipliers of infinite matrices, *Proc. Royal Irish Acad.* 85A, 119-29.

S. RUSCHEWEYH and T. SHEIL-SMALL

1973 Hadamard products of schlicht functions and the Pólya-Schoenberg conjecture, *Comment. Math. Helv.* 48, 119-35.

H. J. RYSER

1973 Indeterminates and incidence matrices, *Linear Multilinear Algebra* 1, 1949-57.

I. SCHUR

1911 Bemerkungen zur Theorie der beschränkten Bilinearformen mit unendlich vielen Veränderlichen, *J. Reine Angew. Math.* 140, 1-28.

1937 Über einige Ungleichungen im Matrizenkalkül, *Prace Matematyczno-Fizyczne* 44, 353-70.

H. S. SHAPIRO and A. L. SHIELDS

1961 On some interpolation problems for analytic functions, *Amer. J. Math.* 83, 513-32.

1962 On the zeros of functions with finite Dirichlet integral and some related function spaces, *Math. Z.* 80, 217-29.

A. L. SHIELDS and L. J. WALLEN

1971 The commutants of certain Hilbert space operators, *Indiana Univ. Math. J.* 20, 777-88.

Q. F. STOUT

1981a Schur multiplication on $B(\ell_p,\ell_2)$, *J. Operator Theory* 5, 231-43.

1981b Schur products of operators and the essential numerical range, *Trans. Amer. Math. Soc.* 264, 39-47.

J. STEWART

1976 Positive definite functions and generalizations: An historical survey, *Rocky Mountain J. Math.* 6, 409-34.

G. STYAN

1973 Hadamard products and multivariate statistical analysis, *Linear Algebra Appl.* 6, 217-40.

R. C. THOMPSON

1988 The Hadamard product (problem posed at the 1987 Utah State University matrix theory conference), *Linear Algebra Appl.* 104, 248-9.

H. TOYAMA

1952 *Theory of Matrices* (in Japanese), Kyoritsu-Shippan, Tokyo.

N. Th. VAROPOULOS

1974 On an inequality of von Neumann and an application of the metric theory of tensor products to operators theory, *J. Functional Analysis* 16, 83-100.

1975 A theorem on operator algebras, *Math. Scand.* 37, 173–82.

M. E. WALTER

1986 On the norm of a Schur product, *Linear Algebra Applic.* 79, 209–13.

W. WATKINS

1988 A determinantal inequality for correlation matrices, *Linear Algebra Applic.* 104, 59–63.

E. P. WIGNER and J. VON NEUMANN

1954 Significance of Loewner's theorem in the quantum theory of collisions, *Annals Math.* 59, 418–33.

F.-Z. ZHANG

1988 Another proof of a singular value inequality concerning Hadamard products of matrices, *Linear Multilinear Algebra* 22, 307–11.

1989 Notes on Hadamard products of matrices, *Linear Multilinear Algebra* 25, 237–42.

Proceedings of Symposia in Applied Mathematics
Volume **40**, 1990

Matrix Completion Problems: A Survey

CHARLES R. JOHNSON

I. Introduction.

a) *The Setting.* Let S denote a set of items, perhaps a field, such as the real or complex numbers, or a more arbitrary set. By a *partial matrix* we mean a rectangular array, the entries in certain locations of which are particular, known elements of S, while the remaining entries are unspecified and free to be chosen from S. We refer to the known entries as "specified" in contrast to the remaining unspecified entries, which are conventionally and suggestively denoted by ?'s. For example, if S is the field of rational numbers,

$$\begin{bmatrix} -1 & ? & \frac{1}{2} \\ ? & -\frac{2}{3} & 3 \end{bmatrix}$$

is an example of a partial matrix. A conventional matrix (all entries specified) is, of course, a special case. Often it will be convenient to refer to a part (submatrix, minor, etc.) of a partial matrix, each of whose entries is specified (or unspecified). In this event the adjective will be extended to the part. For example,

$$[-\tfrac{2}{3} \quad 3]$$

is a specified submatrix of the above example of a partial matrix. Also, modifiers applied to the term "partial matrix" will have the meaning naturally suggested by their use relative to an ordinary matrix. For example, an "Hermitian partial matrix", or "partial Hermitian matrix" is a square partial matrix that is Hermitian to the extent of its specified entries, i.e., the set S is the complex field and if the i, j entry is specified, then the j, i entry is also and the two are complex conjugates.

By a *completion* of a partial matrix $A = (a_{ij})$, we mean a conventional matrix $B = (b_{ij})$, of the same dimensions as A and with all entries from the set S, such that if a_{ij} is specified, then $b_{ij} = a_{ij}$. Thus, a completion is just a specification (from S) of all the unspecified entries, resulting in a conventional

---

1980 *Mathematics Subject Classification* (1985 *Revision*). Primary 15-02, 15A57, 15A15, 15A60, 15A03.

This work was supported in part by National Science Foundation grant DMS 88 02836.

matrix (over S). A *matrix completion problem*, then, is just the question as to whether a partial matrix has a completion in a certain class or (equivalently) with a certain property of interest. For example, we might ask if a partial Hermitian matrix has a positive definite completion, or if a given rectangular partial matrix over the complex numbers has a completion of rank 7. Many such problems carry with them an important inheritance structure. Often, having a completion with a certain property requires "obvious" necessary conditions of like kind upon the data (specified entries). For example, since every principal submatrix of a positive definite matrix is positive definite, if a partial Hermitian matrix is to have a positive definite completion, every specified principal submatrix must be positive definite, and, thus, every specified principal minor must be positive. Similarly, for a rectangular, complex partial matrix to have a completion of rank 7, no specified submatrix may have rank greater than 7.

The occurrence of such structure suggests matrix completion problems at another level. On the one hand, a matrix completion problem, as we have described it, in its most general form could be viewed as a request for a decision procedure, which, for each partial matrix, tells whether there is a completion of the desired type. Of course, for the answer to be "yes", any "obvious" necessary conditions must be met. What if they all are? Will the answer then be "yes"? (We distinguish here between obvious necessary conditions, of the sort mentioned above, and possibly more subtle conditions that would characterize completability and yield a complete decision procedure.) Generally, no, but is there a little more, and perhaps natural, structure that might be required of the partial matrices that would yield such a nice situation? A question of this type is the following. Which *patterns* for the specified entries insure that if certain "obvious" necessary conditions are met, then there will be a completion of desired type? This introduces a combinatorial (as opposed to magnitudinal or analytic) aspect into matrix completion problems, which turns out to be rather natural, but we will discuss it further in the context of the specific matrix completion problems that we survey here. Call this the *combinatorial matrix completion problem*; such study has become one major component in emerging work in "combinatorial matrix theory".

b) *Remarks.* The setting just described permits an immerse variety of specific problems, a nontrivial fraction (but not all) of which are mathematically interesting. Simply pick your favorite set S and a property of interest. Such problems have a variety of applied motivations including seismic reconstruction problems, entropy methods in statistical/physical problems, electrical engineering/systems theory, etc. They also provide a natural vehicle for mathematical understanding of structure; a good example is determinantal inequalities for positive definite matrices.

The general setting we have described is a good deal broader than that which we shall discuss. It includes at least two long standing (and important) lines of research that we mention here but discuss in no further detail. First, completion problems involving familiar combinatorial structures have been studied by combinatorists for some time as a means of better understanding these structures. For example, given a matrix of ± 1's, what is the smallest size Hadamard matrix into which it may be embedded; or, given a partial matrix whose specified entries are ± 1's may it be completed to an Hadamard matrix? Similarly for Latin squares—which n-by-n partial matrices over the set $\{1, 2, \ldots, n\}$ may be completed to Latin squares? Second, algebraically oriented matrix theorists have been studying for some time the possible invariants that may occur among completions of partial matrices whose specified entries consist of exactly one or two blocks with known invariants. For example, given the Jordan structure of the k-by-k complex matrix A, what are the possible Jordan structures of the n-by-n $(n > k)$ matrix

$$\begin{bmatrix} A & B \\ C & D \end{bmatrix}?$$

We survey here a selection of emerging work on a class of completion problems of a different sort. These involve positive definite completions, contraction completions and ranks of completions. Studied at the level of formality and generality we have described, this segment of completion problems is fairly new, but as often happens in mathematics, it encompasses some seemingly isolated classical results as special cases. For example, the Caratheodory-Fejer theorem of function theory may be viewed as a special positive definite completion result, and the Frobenius-König theorem of early combinatorial matrix theory is a very special rank completion result. Both are very special cases of recent results in these respective areas. All of the problems we survey enjoy some form of the inheritance structure mentioned earlier, as well as other aspects of commonality. One goal of research in this area, which we hope this survey furthers, is to better understand and unify existing work so as to support more general future work and easier solution of other completion problems.

II. Positive Definite Completions. This is the most heavily studied matrix completion problem of the sort we discuss, and we shall not be able even to mention all the interesting work upon it. The basic problem is to decide whether a given partial Hermitian matrix has a positive definite completion. There is a corresponding, and essentially equivalent, positive semidefinite completion problem. One may then go on and attempt to describe all positive definite (semidefinite) completions (they form a convex set), or to focus upon positive definite completions with additional important features. The obvious necessary condition on the data (positive definite completion case) is that every specified principal minor of the partial Hermitian matrix must be positive; equivalently, every specified principal submatrix must be positive definite. We refer to a partial Hermitian matrix meeting this minimal necessary condition as *partial positive definite.* The resulting combinatorial positive definite completion problem asks which patterns for the specified entries ensure that a partial positive definite matrix has a positive definite completion?

Call a square partial matrix *banded* if all entries within some band-width (symmetric w.r.t. the main diagonal) are specified and all entries outside this band-width are unspecified. An early positive definite completion result is the Caratheodory-Fejer theorem from function theory. In our language, it states that a (banded) Toeplitz (equal entries along bands) partial positive definite matrix always has a Toeplitz positive definite completion. This result was notably generalized in the stimulating paper **[DG1]**. Also translated to our language it included the following three major conclusions.

THEOREM DG. *Let H be a complex banded partial positive definite matrix. Then,* (1) *H has a (complex) positive definite completion.* (2) *Among the positive definite completions of H, there is a unique one with maximum determinant.* (3) *Among positive definite completions, the determinant maximizer is the unique one whose inverse is banded* (*at most same band-width*) *in the usual sense* (*all entries equal to zero outside some bands*).

This result, then, says that at least the banded patterns are among those sought in the combinatorial version of the positive definite completion problem. It should be noted that Theorem DG remains valid when S is the real, or even rational, field as well as the complex field, and that the authors also prove the more general result in which A is "block banded" (obvious interpretation). This could, though, just be viewed as the banded case with the set S consisting of matrices of various sizes.

The special case of the positive definite completion problem in which there is just one (two symmetrically placed) unspecified entry is worth noting. Because the general problem is unchanged under permutation similarity, this is the special case (of the case studied in **[DG1]**) in which the missing entry lies in the upper right (and lower left) corner. This case exhibits, in a simple way, many features of the problem, and is a substantial piece of the solution of more general problems. Consider the n-by-n partitioned partial Hermitian matrix

$$H = \begin{bmatrix} a & b^* & x \\ b & A & c \\ \overline{x} & c^* & d \end{bmatrix}$$

in which the maximal specified principal submatrices

$$B = \begin{bmatrix} a & b^* \\ b & A \end{bmatrix} \text{ and } C = \begin{bmatrix} A & c \\ c^* & d \end{bmatrix}$$

are positive definite, and $x \in \mathbf{C}$ is the unspecified entry. Using Sylvester's determinantal identity **[HJ]**, for example, we then have

$$\det H = \frac{\det B \det C - \left|\det \begin{bmatrix} b^* & x \\ A & c \end{bmatrix}\right|^2}{\det A}$$

$$= \frac{\det B \det C - \left|\det \begin{bmatrix} b^* & 0 \\ A & c \end{bmatrix} + (-1)^n x \det A\right|^2}{\det A}.$$

Since an Hermitian matrix is positive definite if and only if its leading principal minors are positive **[HJ]** and since B is positive definite, the matrix H is positive definite if and only if $\det H > 0$. Algebraic manipulation of this expression for $\det H$ reveals that $\det H > 0$ if and only if

$$\left| x - \frac{(-1)^{n-1} \det \begin{bmatrix} b^* & 0 \\ A & c \end{bmatrix}}{\det A} \right|^2 < \frac{\det B \det C}{(\det A)^2}.$$

Using Schur complements **[HJ]**, we find that

$$\det \begin{bmatrix} b^* & 0 \\ A & c \end{bmatrix} = (-1)^{n-1} b^* A^{-1} c \det A.$$

Substitution into the previous inequality then gives that $\det H > 0$ if and only if

$$|x - b^* A^{-1} c|^2 < \frac{\det B \det C}{(\det A)^2}.$$

Furthermore, $\det H$ is maximized exactly when

$$x = b^* A^{-1} c.$$

Since $\det B$, $\det C$ and $\det A > 0$ (the latter because it is a principal submatrix of the positive definite matrix B **[HJ]**), we may draw the following conclusions.

(1) Positive definite completions of H always exist under our assumption that B and C are positive definite.

(2) The set of all x that constitute a positive definite completion of H is the open disc with center:

$$\frac{(-1)^{n-1} \det \begin{bmatrix} b^* & 0 \\ A & c \end{bmatrix}}{\det A} = b^* A^{-1} c$$

and radius:

$$\frac{(\det B)^{1/2} (\det C)^{1/2}}{\det A}.$$

(3) The center of this disc uniquely maximizes $\det H$ among the positive completions of H, the maximum determinant being:

$$\frac{\det B \det C}{\det A}.$$

Since, by co-factors,

$$\frac{(-1)^{n+1}\det\begin{bmatrix} b^* & x \\ A & c \end{bmatrix}}{\det H}$$

is the $1, n$ entry of H^{-1} (if it exists), we also note that

(4) the corner entries ($1, n$ and $n, 1$) of H^{-1} are 0 exactly for the x that provides the unique determinant maximizing completion of H among the positive definite completions of H. These corner entries are in just the positions of our unspecified entry x in H.

It should also be noted if H is a partial positive definite *symmetric* matrix over the real field **R** (respectively, the rational field **Q**), then there are positive definite completions over **R** (respectively **Q**) and the determinant maximizing completion is from **R** (respectively **Q**). In the case of the real field **R**, the set of all positive definite completions is just a line segment with the same center and half-width as in the case of **C**.

Of course, conclusion (1) of Theorem DG may be proven by sequential application (down the bands one-at-a-time, going out) of the one-variable case, and conclusions (3) and (4) in the one variable case are indicative special cases of Theorem DG. Sequential application of the one variable case will continue to be important in the general problem.

In further considering the positive definite completion problem, we henceforth assume that all diagonal entries of any partial positive definite matrix under study are specified. For existence of positive definite completions this is no loss of generality, for, if not, any unspecified diagonal entries may be chosen arbitrarily large, leaving the question of a positive definite completion to rest upon that maximal partial principal submatrix all of whose diagonal entries are specified. For determinant maximization this is an obviously necessary regularity condition because the determinant of a positive definite matrix is monotone increasing in its diagonal entries.

Conclusions (2) and (3) of Theorem DG may be generalized to *any* partial positive definite matrix (independent of pattern) and proven in a simple way using optimization. This was first noted in **[GJSW]**.

THEOREM 1. *Let H be a partial positive definite matrix over* **C** *all of whose diagonal entries are specified, and suppose that H has a positive definite completion. Then, among all positive definite completions of H there is a unique one with maximum determinant. Among positive definite completions the determinant maximizer is characterized by having a zero entry in the inverse in every posititin in which H has an unspecified entry.*

PROOF OUTLINE. Consider the set of all positive *semi*definite completions of H. This set is nonempty (because of the assumption that H has a positive definite completion), closed and bounded (no off-diagonal entry can be too large because the diagonal entries are specified and every 2-by-2 principal submatrix must be positive semidefinite). In fact it is also convex. Since det is a continuous function on this compact set, the maximum of det is attained on it by Weierstrass' theorem. Any maximizing matrix must be in the (relative) interior of this set, because the boundary consists of positive semidefinite matrices that are not positive definite and therefore have determinant 0 (and there are matrices with positive determinant, in the interior, per assumption of the existence of positive definite completions).

The key to both uniqueness and the inverse 0-pattern is the not sufficiently well known fact that the determinant function is strictly log concave over the positive definite matrices. This may be proven via simultaneous diagonalization by congruence of two positive definite matrices, as in **[HJ]**, which reduces the question to the diagonal case. Since strictly log concave functions behave as strictly concave functions with respect to optimization, we conclude that there must be a *unique* determinant maximizing positive definite completion, which must occur at a (*unique*) critical point (because it lies in the interior of the constraint set). To see what it means to be a critical point we differentiate the determinant with respect to an unspecified entry h_{ij} of H:

$$\frac{\partial}{\partial h_{ij}} \det H = (-1)^{i+j} 2 \det H(i/j),$$

in which $H(i/j)$ denotes the submatrix of H resulting from deleting row i and column j. Thus,

$$\det H(i/j) = 0$$

for the determinant maximizing completion if the i, j entry is unspecified. But in view of the cofactor version for the inverse and the symmetry of H, we must have

$$H_{ij}^{-1} = H_{ji}^{-1} = 0,$$

as $\det H(i, j) = 0$ is the numerator of H_{ji}^{-1} (and H^{-1} exists for any positive definite completion). This explains the unique *inverse* zero pattern of the unique determinant maximizing completion. Strictly speaking, this argument is valid only in the case of real partial positive definite matrices (in which case it produces a real solution), but it can be adjusted, as in **[GJSW]**, to cover the complex case. □

Before discussing existence of positive definite completions, we mention an example that shows that not all patterns for the specified entries of a partial positive definite matrix insure a positive definite completion. Thus, the answer to the combinatorial positive definite completion question lies somewhere between the banded (or block banded) patterns of **[DG1]** and all patterns.

Consider the partial positive *semi*definite matrix

$$H = \begin{bmatrix} 1 & 1 & ? & -1 \\ 1 & 1 & 1 & x \\ ? & 1 & 1 & 1 \\ -1 & \overline{x} & 1 & 1 \end{bmatrix}$$

with an unspecified entry in the 1, 3 and 3, 1 positions and another (denoted by x) in the 2, 4 and 4, 2 positions. It is a simple exercise that the patterns that insure completability to a positive semidefinite matrix of a partial positive semidefinite matrix are the same as those that insure completability of a partial positive definite matrix to a positive definite. In any event we consider this partial positive semidefinite example for simplicity, but it could be converted to a partial positive definite one with the same point by adding small positive ε's to the diagonal entries.

In order to complete H to a positive semidefinite matrix, x must be chosen so that every principal submatrix in which it lies is positive semidefinite. Notice that x completes two different maximal 3-by-3 principal submatrices of H, the ones lying in rows and columns 2, 3, 4 and 1, 2, 4:

$$\underset{\{2,3,4\}}{\begin{bmatrix} 1 & 1 & x \\ 1 & 1 & 1 \\ \overline{x} & 1 & 1 \end{bmatrix}} \quad \text{and} \quad \underset{\{1,2,4\}}{\begin{bmatrix} 1 & 1 & -1 \\ 1 & 1 & x \\ -1 & \overline{x} & 1 \end{bmatrix}}.$$

A simple calculation, using the one variable case, shows that for the former to be positive semidefinite we must have $x = 1$ and for the latter we must have $x = -1$. Not both of these objectives may be achieved simultaneously. The problem, of course, is that x completes more than one different maximal principal submatrix and this happens to impose conflicting requirements upon x. It would not have helped to focus upon the other unspecified entry.

In order to discuss the existence of positive definite completions and describe a complete solution to the combinatorial positive definite completion problem we need a brief digression into relevant graph theory. First of all, it is natural to describe the *pattern* of the specified (and, therefore, unspecified) entries of an n-by-n partial Hermitian matrix $H = (h_{ij})$ with an undirected (because h_{ij} is specified when h_{ji} is) graph on n vertices. In this graph we place an edge joining vertex i and vertex j if and only if h_{ij} is specified. Because the diagonal entries are all assumed specified, we ignore loops at the vertices. For example, the graph associated with the partial Hermitian matrix

$$\begin{bmatrix} 5 & 2 & ? & 1 & -1 \\ 2 & 6 & ? & 2 & ? \\ ? & ? & 3 & i & ? \\ 1 & 2 & -i & 4 & ? \\ -1 & ? & ? & ? & 6 \end{bmatrix}$$

is

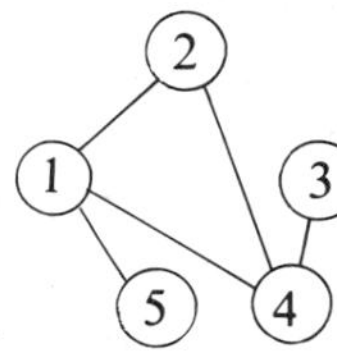

In any undirected graph, we say that 2 vertices are *adjacent* if they are joined by an edge. A *path* just a sequence of edges concatenated at vertices (i.e. a sequence of adjacent vertices);

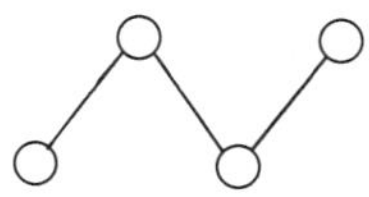

a *circuit* is a path with an "initial" and "terminal" vertex that coincide;

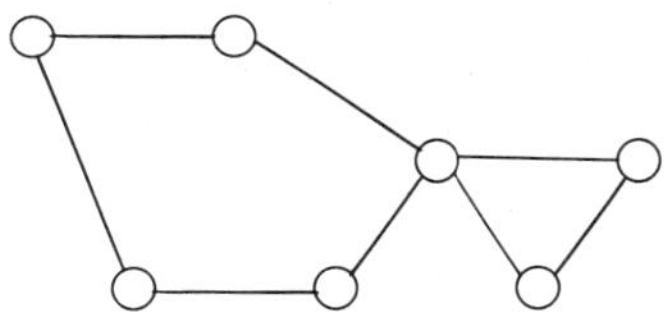

and a *simple circuit* is one that has no other self intersections

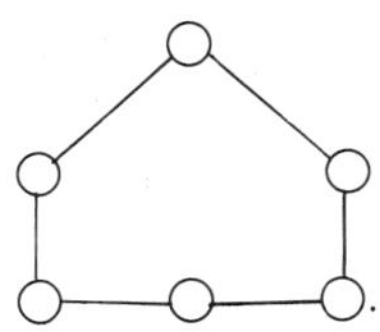

A *chord* of a circuit is just an edge joining two verticies (of the circuit) that are not adjacent in the circuit. A simple circuit is *minimal* if no proper subset of its vertices are themselves the vertices of a circuit. (Equivalently, the graph contains no chords for the circuit.) Note that an edge by itself does not consistute a circuit. The key notion is that of a chordal graph.

Definition: An undirected graph is called *chordal* if it has no minimal simple circuits of length (measured in number of edges) 4 or more. Equivalently, any simple circuit of length at least 4 must have a chord.

Chordal graphs turn out to be a very nicely structured, heavily studied class of graphs that arise in many ways. For example, they have previously arisen in numerical linear algebra in the analysis of Gaussian elimination; the usual undirected graph of a (combinatorially symmetric) "sparse" matrix is chordal if and only if the calculations in Gaussian elimination may be ordered so as to avoid any intermediate fill-in (independent of the numerical values of the nonzero entries). A standard reference is [**G**]. Chordal graphs are also nice from the point of view of computation. Most tasks may be carried out algorithmically in low polynomial order for chordal graphs, including many that are known to be difficult or *np*-complete for general graphs. Chordal graphs also form quite a large class that includes the "trees" and the graphs from banded (and block banded) patterns but also much more. The simplest example of a chordal graph that is not associated with a (block) banded pattern for any numbering of the vertices is the tree

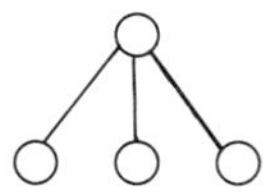

See [**JR3**] for a detailed discussion of where the block banded patterns sit within the chordal graphs.

We return to a few remaining graph theoretic concepts. An (undirected) graph is called *complete* if every pair of vertices is adjacent. An induced *subgraph* of a given graph is simply that graph resulting from retaining all edges both of whose vertices are contained in some selected subset of the vertices. A *clique* of a graph is an induced subgraph that is complete, and a clique is called *maximal* if its vertices do not constitute a proper subset of a clique. A (vertex) *separator* of two vertices i, j of a given graph is an induced subgraph whose removal leaves i and j but no path joining them; an i, j-separator is called *minimal* if it does not properly contain an i, j-separator.

It is now informative to return to the example partial positive semidefinite matrix that did not have a positive semidefinite completion:

$$H = \begin{bmatrix} 1 & 1 & ? & -1 \\ 1 & 1 & 1 & ? \\ ? & 1 & 1 & 1 \\ -1 & ? & 1 & 1 \end{bmatrix}.$$

The graph of the specified entries of H is

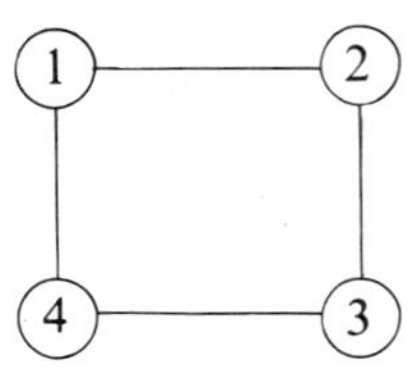

which is the simplest example of a nonchordal graph. When we tried to add the 2, 4 entry (edge) and remain partial positive semidefinite the two

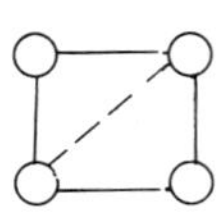

conflicting 3-by-3 principal submatrices corresponded to the two triangles

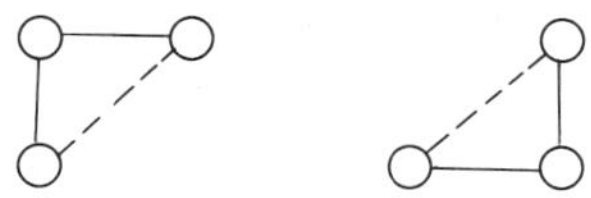

with the common (dotted) edge being a chord of the simple circuit of length 4. This is correctly suggestive of the general result proven in [**GJSW**].

THEOREM 2. *Every partial positive definite matrix with undirected graph G for its specified entries has a positive definite completion if and only if G is chordal.*

PROOF OUTLINE. Since any principal submatrix of a positive definite matrix is positive definite, any partial principal submatrix of a partial positive definite matrix that has a positive definite completion must itself have a positive definite completion. Since *non*chordal graphs are just those that have induced subgraphs consisting of just one "long" simple circuit, it suffices for necessity to give an example for each simple circuit of length $k \geq 4$ of a partial positive definite matrix that has no positive definite completion. This may be done in the same way as the 4-by-4 example we have discussed.

The key to sufficiency is to reduce the problem of finding a completion to a sequence of one-variable problems (already discussed). This may be done using a "chordal ordering" of the edges *missing* from G. It may be shown that for any chordal graph, the missing edges may be ordered so that their sequential addition results in a sequence of chordal graphs. Such an ordering is not unique (and there may be quite a few), but not every ordering will necessarily work. The addition of each such edge completes exactly one new maximal clique (principal submatrix) and gives a single one-variable problem to be solved. Its solution gives a new partial positive definite matrix with one more entry. Since the new pattern is chordal, this process may be repeated until it terminates in a positive definite completion of the original partial positive definite matrix. It should be noted that this (nonunique) one-step-at-a-time procedure, which is so closely linked to chordality, is both conceptually and computationally simple. It is also a natural paradigm for other completion problems and has further uses in the positive definite case. □

We conclude this section on the positive definite problem by mentioning some related results and making several remarks.

In [**JR1**] the possible inertias (distribution of positive, negative and zero eigenvalues) of Hermitian completions of partial Hermitian matrices are studied. Again the inertias of the maximal specified principal submatrices impose a constraint, this time through the interlacing inequalities. Under a necessary technical assumption about nonvanishing of certain key principal minors (which was automatic in the partial positive definite case), all inertias allowed by interlacing can be achieved in the chordal case. Additional, inertia-oriented results in the *banded* partial Hermitian (but not necessarily partial positive definite) case may be found in [**EGL2**, **EGL3**], and some of these may be generalized to the chordal case [**L**]. It follows from the positive definite results that in the chordal case there is a completion whose minimum eigenvalue is equal to the smallest of the minimum eigenvalues of all specified principal submatrices (this is, of course, the most it could be). The inertia results may be viewed as a generalization, but how large the minimum eigenvalue can be is an open question in the general (not necessarily chordal) case. Results somehow relating it to the graph structure and the smallest eigenvalue of any specified principal submatrix would be most welcome.

In [**JR3**] the case of partial matrices in which the set S is a ring of functions on a set X is considered. In the chordal case, if the partial matrix is partial positive definite for each element of X, there is, of course, a positive definite completion point-wise. Conditions on the ring that allow a point-wise positive definite completion with functions from the ring are studied. It is likely that stronger results of this type could be obtained.

Several authors have sought "nice" descriptions of the set of all positive (semi)definite completions of a partial positive definite matrix with given pattern. A natural strategy is to try to generalize the disc that occurs in the one variable case. Some examples include [**GKW**, **W2**, **W4**], among others. The approach has been to give a global parameterization of all positive semidefinite completions of a *banded* partial positive definite matrix via contractions. In the more general chordal case, a simple local parameterization may be given by describing the position in the disc for each successive one-variable problem along a chordal ordering of the missing edges (as in the proof of Theorem 2). A more global description of all positive semidefinite completions in the chordal (or less general) case is a subject of active research.

It is also natural to ask what can be said about existence of positive definite completions beyond the chordal case. It is clear that if only partial positive definite is assumed, the question becomes one of numerical values if the graph is not chordal. Two natural approaches may be suggested but neither has yet produced a fully satisfactory solution; so this is an area in which further work should be done.

First, since the chordal structure is so nice, one might write down the conditions associated with retaining partial positive definiteness as certain entries are added until a chordal pattern is achieved. Using the one-variable case and noting the *set* of maximal principal submatrices completed at each step, this may be done. The result is a list of "quadratic" (in each variable) inequalities that must be simultaneously satisfied. This gives a theorem that has two unfortunate drawbacks. First, minimal chordal supergraphs are not unique and, for a general graph, are not computationally easy to find. Second, in practice, feasibility of an arbitrary list of quadratic inequalities may be difficult to determine. More special structure of the problem needs to be exploited.

A second approach is to add to the list of "obvious" necessary conditions, hopefully until it becomes sufficient. The positive definiteness of specified principal submatrices is an especially obvious condition, but there are others. According to the Schur Product theorem **[HJ]**, the Hadamard, or entry-wise, product of any two positive definite matrices A and B, denoted $A \circ B$, must be positive definite. In particular, if H is a partial Hermitian matrix and P is any positive definite matrix with 0's in all positions in which H has unspecified entries, then $H \circ P$, an ordinary matrix, must be positive definite if H is to have a positive definite completion. This adds many necessary conditions, some of which are surely redundant. For example, any partial positive semidefinite matrix H with the minimal nonchordal pattern

$$\begin{bmatrix} * & * & ? & * \\ * & * & * & ? \\ ? & * & * & * \\ * & ? & * & * \end{bmatrix}$$

must produce a positive semidefinite matrix upon Hadamard multiplication by the positive semidefinite matrix

$$P = \begin{bmatrix} 1 & 1 & 0 & 1 \\ 1 & 2 & 1 & 0 \\ 0 & 1 & 1 & -1 \\ 1 & 0 & -1 & 2 \end{bmatrix}$$

in order that H have a positive semidefinite completion. In particular, the example

$$H = \begin{bmatrix} 1 & 1 & ? & -1 \\ 1 & 1 & 1 & ? \\ ? & 1 & 1 & 1 \\ -1 & ? & 1 & 1 \end{bmatrix}$$

has no positive semidefinite completion because $\det(H \circ P) < 0$. It is not yet clear whether this approach can be pursued to produce a finite set of necessary and sufficient conditions for any given pattern. It is enough to check the extremals of the cone of all positive semidefinite matrices with the corresponding zero-pattern, and these cones and their extremals have begun to be studied [**AHMR**, **HPR**]. But, generally there are infinitely many such extremals. This is a focus of current study.

We conclude this section with some remarks tangentially related to the general positive definite completion problem.

Theorem DG and theorem 1 guarantee a completion unique (among the positive definite completions) with regard to its inverse zero-pattern. The proof of theorem 1 makes this fact appear analytic, which, in part it is, and nothing is said about the possibility of (other than positive definite) invertible completions with a similar inverse zero pattern. However in the chordal case and under a regularity condition (that is automatic in the partial positive definite case), the universe of uniqueness may be extended completely, so that the inverse zero-pattern uniqueness is also significantly algebraic. If a square, combinatorially symmetric (the j, i entry is specified whenever the i, j is) partial matrix has a chordal graph for its specified entries and certain key specified principal minors are nonzero (those corresponding to maximal cliques and minimal separators to be defined in section III), then there is a unique invertible completion the graph of the nonzero entries of the inverse of which is contained in the graph of the specified entries of the original partial matrix. A formula for the entries of this completion may also be given **[L]**. Thus, in the chordal case of theorem 1, no qualification on the uniqueness (other than invertibility, of course) need be given. However, this broad uniqueness does not, in general, remain valid if *either* the regularity condition or the chordality requirement is dropped. Thus, the nonchordal case of theorem 1 is partly dependent on positivity and requires the qualification on uniqueness.

An interesting variation upon the combinatorial version of the positive definite completion problem remains open. Which sequences of bands $0, k_1, k_2, \dots, k_p$ (in which the main diagonal is numbered 0) for the specified entries of a Toeplitz partial positive definite matrix ensure the existence of a positive definite Toeplitz completion? The Caratheodory-Fejer case (which is easily proven using the methods of theorem DG or theorem 1) is that any sequence $0, 1, 2, \dots, p$ is a good one, and, more generally, any arithmetic sequence $0, t, 2t, \dots, pt$ also works. It is a working conjecture that this constitutes all such sequences; a number of unpublished (negative) partial results are consistent with such a conjecture.

The Toeplitz positive definite completion problem may be regarded as a special case of the general linearly constrained positive definite completion problem, in which a set of linear relations among the unspecified entries is required of a completion. Another special case of interest is one in which a completion A, of a given partial positive definite matrix, is desired that is not only positive definite but also satisfies $Ax = y$ for pre-specified vectors x and y. Initial study of the linearly constrained problem, especially with respect to determinant maximization, has begun in **[L]**.

There are, of course, other specialized, but natural, positive definite, or Hermitian, completion problems. We have mentioned the maximum minimum eigenvalue Hermitian completion problem, which has a simple solution in the chordal case (the minimum of all the smallest eigenvalues of specified principal submatrices) but is of interest in the nonchordal case. Minimum condition number and related completions are also of interest for numerical reasons. Fields of values of completions of general matrices and notions of numerical range for partial matrices have been studied in **[JL]**, and rank conditions on Hermitian completions have been studied in **[EL]** and by Jerry Dancis.

III. Maximum Determinant/Maximum Entropy Completions. If a partial positive definite matrix has positive definite completions, the one with maximum determinant has attracted the most attention. (We know it is unique and has a striking inverse zero pattern by Theorem 1.) This is due to an important role in applications and the attractive mathematical structure associated with it. The determinant maximizing completion is often called the *maximum entropy* completion because of its role in statistical entropy methods. Specific applications have come from seismic reconstruction and other missing data problems, image enhancement and the so-called model reduction problem among others. Discussion of some of these may be found among the references.

This author's interest in the determinant maximizing completion stems primarily from a long term interest in determinantal inequalities for positive definite matrices. Here we elaborate further on the determinant maximizer, formulae for its determinant and its role in determinantal inequalities. Many results still focus upon the chordal case in which the entries and determinant of the maximizing completion are rational in the specified entries. As before, we continue to assume that all diagonal entries of a partial positive definite matrix are specified.

We begin with a trivial observation and then show how it can be put to use. If the maximum determinant over all positive definite completions of a given partial positive definite matrix A may be expressed in terms of the specified entries, then a (sharp) upper bound for all determinants of positive definite matrices B that agree with A in its specified entries results. This idea may be used to show that either completion result (Theorem DG or Theorems 1 and 2) implies the classical determinantal inequality of Hadamard. Consider the simplest partial positive definite matrix

$$A = \begin{bmatrix} a_{11} & ? & \dots & ? \\ ? & a_{22} & \ddots & \vdots \\ \vdots & \ddots & \ddots & ? \\ ? & \dots & ? & a_{nn} \end{bmatrix}$$

in which only the diagonal entries $a_{11}, a_{22}, \dots, a_{nn} > 0$ are specified. According to either completion result, there is a unique determinant maximizing completion $\hat{A}$ of A, and its *inverse* must be diagonal (all off-diagonal entries 0, because the off-diagonal entries of A are unspecified). But, if $\hat{A}^{-1}$ is diagonal, then $\hat{A}$ must be also, and it follows that

$$\hat{A} = \begin{bmatrix} a_{11} & 0 & \dots & 0 \\ 0 & a_{22} & & \vdots \\ \vdots & & \ddots & 0 \\ 0 & \dots & 0 & a_{nn} \end{bmatrix}$$

is the determinant maximizing completion of A. Since $\det \hat{A} = a_{11}a_{22}\cdots a_{nn}$, we have

$$\det B \leq a_{11}a_{22}\cdots a_{nn}$$

if B is any positive definite matrix with diagonal entries $a_{11}, \ldots, a_{nn}$. This is just Hadamard's inequality, and equality occurs exactly when B is diagonal.

More complicated examples require more sophisticated methodology to express the maximum determinant in terms of the data. As chance would have it, such methodology was being developed in [**BJ1**, **BJ2**, **JB**] independent of the work on completions. This work focused upon simple formulae for the determinant of a nonsingular matrix A based upon some information about the 0-pattern of A^{-1}, and any 0-pattern could be at least partly exploited. To give an example we need some standard notation.

For index sets $\alpha \subseteq \{1, \ldots, m\}$, and $\beta \subseteq \{1, \ldots, n\}$, we denote the submatrix of an m-by-n (partial or ordinary) matrix lying in the rows indicated by α and the columns indicated by β as

$$A[\alpha, \beta].$$

Alternatively, the complementary submatrix in which the rows α and columns β are deleted is denoted by

$$A(\alpha, \beta).$$

When A is square ($m = n$), we abbreviate the principal submatrices $A[\alpha, \alpha]$ to

$$A[\alpha]$$

and $A(\alpha, \alpha)$ to

$$A(\alpha).$$

A simple, but good, example of the formulae for $\det A$ given the 0-pattern of A^{-1} is the following. Suppose that $A^{-1} = B = (b_{ij})$ is tridiagonal, i.e. $b_{ij} = 0$ whenever $|i - j| > 1$. Then, if $A = (a_{ij})$,

$$\det A = \frac{\det A[\{1,2\}] \det A[\{2,3\}] \cdots \det A[\{n-1,n\}]}{a_{22}a_{33}\cdots a_{n-1,n-1}}.$$

The numerator is the product of the 2-by-2 principal minors of A occurring consecutively down the main diagonal (there are $n - 1$ of these) and the denominator is the product of the principal minors in which the numerator terms overlap (the $n - 2$ diagonal entries 2 through $n - 1$, in this case), and all of this hinges upon the inverse B being tridiagonal. This is indicative of a fairly general, though more complicated theory developed in [**BJ1**, **BJ2**, **JB**]. It follows by the same logic used in our proof of Hadamard's inequality that for *any* n-by-n positive definite matrix $A = (a_{ij})$, we have the universal inequality

$$\det A \leq \frac{\det A[\{1,2\}] \det A[\{2,3\}] \cdots \det A[\{n-1,n\}]}{a_{22}a_{33}\cdots a_{n-1,n-1}}.$$

Moreover, equality occurs in this inequality exactly when A^{-1} is tridiagonal, and given the diagonal and sub- and super-diagonal, there is a unique matrix that achieves equality. The reasoning is this. Consider an n-by-n positive definite matrix A. Forget the entries outside the sub- and superdiagonals. The resulting partial positive definite matrix has positive definite completions. The (unique) determinant maximizing one has tridiagonal inverse (Theorem 2), but the determinant of this one (which bounds all others) is as indicated ([**BJ1**]). This idea was used to exhibit very broad generalizations of the so-called Hadamard-Fischer inequalities in [**JB**].

Instead of summarizing the ideas of [**BJ1**, **BJ2**], we describe three different ways of presenting the maximum determinant over all positive definite completions of a given partial positive definite matrix in the most general case in which we know such completions exist, namely when the graph of the specified entries is chordal. Though we do not pursue it here, bounds for the maximum determinant may be given in general, using chordal subgraphs and supergraphs. Simple, exact descriptions of the maximum determinant for a general partial positive definite matrix that has positive definite completions are yet to be found. It is clear that they are no longer rational in the specified entries, and it may be difficult to give such a nice description.

It should be kept in mind that any formula for the maximum determinant over positive definite completions gives another general determinantal inequality for positive definite matrices.

In order to proceed, we need a few more graph theoretical ideas. A *tree* is a connected (undirected) graph on n vertices that has the fewest possible edges, namely $n - 1$. A *spanning tree* T of a connected graph G is simply a tree on the same vertices as G whose set of edges is a subset of that of G. Given a family of subsets $C_1, C_2, \dots, C_m \subseteq \{1, 2, \dots, n\}$, the *intersection graph* of this family is just the graph whose vertices are $C_1, \dots, C_m$ with an (undirected) edge between C_i and C_j if and only if $C_i \cap C_j \neq \emptyset$. We say that a spanning tree of the intersection graph of $C_1, C_2, \dots, C_m$ satisfies the *intersection property* if whenever C_k lies on the (unique) path between C_i and C_j, then $C_i \cap C_j \subseteq C_k$. A natural intersection graph to associate with a given graph G is the one in which $C_1, \dots, C_m$ are the vertex sets of the maximal cliques of G. We call this graph $G_{\mathscr{C}}$. As with much of the development of completion problems for positive definite matrices, further understanding of the structure of chordal graphs is useful. An important idea is the following fact, which may be found in [**BJL**].

THEOREM 3. *Let G be a connected, undirected graph and let $G_{\mathscr{C}}$ be the intersection graph of the maximal cliques of G. Then, there is a spanning tree of $G_{\mathscr{C}}$ satisfying the intersection property if and only if G is chordal.*

As with the chordal orderings of the missing edges of a chordal graph, the spanning tree guaranteed by Theorem 3 need not be unique. However, the important expressions we derive from it are unique.

We may now mention the first of three (Theorems 4, 5, 6) presentations of the maximal determinant for completions of a partial positive definite matrix, the graph of whose specified entries is chordal. For further details and links with determinantal formulae and inequalities, see [**BJ1**, **BJ2**, **BJL**, **JB**].

THEOREM 4. *Let G be a connected, chordal graph and suppose A is a partial positive definite matrix, the graph of whose specified entries is G. The maximum determinant occurring among all positive definite completions of A is then given by*

$$\frac{\prod_{k=1}^{m} \det A[C_k]}{\prod \det A[C_i \cap C_j]}.$$

Here, $C_1, \ldots, C_m$ are the maximal cliques of G, and for any spanning tree T of the intersection graph $G_{\mathscr{C}}$ of the maximal cliques that satisfies the intersection property, the product in the denominator is taken over all pairs i, j such that $\{C_i, C_j\}$ is an edge of T. Furthermore, there is a unique positive definite completion of A that attains this maximum, and the undirected graph of its inverse is contained in G. This completion is the only one with the latter property.

The example given earlier in this section involving a tridiagonal inverse is the special case of Theorem 3 in which G is just a path from 1 to n.

Since virtually all computational tasks associated with chordal graphs may be carried out efficiently (including those necessary for Theorem 4), Theorem 4 does provide a computationally effective, as well as theoretically useful, description of the maximum determinant. The denominator terms turn out to be the minimal vertex separators, and, together with their multiplicities (as well as the numerator terms, which are maximal cliques) they are unique, even though the necessary spanning tree is not. However, an alternate presentation that requires no "external" structure (such as the special spanning tree) is a natural goal. It turns out that there is one that is attractive by its elementary nature. It has the flavor of inclusion/exclusion. If we begin, as in Theorem 4 with numerator terms corresponding to maximal cliques and then divide by terms corresponding to *all* pairwise intersections, we have put too much in the denominator. If we pay back by putting all 3-way intersections in the numerator, we have gone too far on top. Continuing with odd order intersections in the numerator and even order in the denominator turns out to terminate with a correct ratio. It can be shown that in the case of a chordal pattern, cancellation occurs to leave just the expression given in Theorem 4.

THEOREM 5. *Let G be a connected chordal graph and suppose A is a partial positive definite matrix, the graph of whose specified entries is G. Let $C_1, \ldots, C_m$ be the index sets of the maximal cliques of G. The maximum determinant occurring among all positive definite completion of A is then given* *by*

$$\frac{\prod_{k \text{ odd}} \prod_{|s|=k} \det A\left[\bigcap_{i\in s} C_i\right]}{\prod_{k \text{ even}} \prod_{|s|=k} \det A\left[\bigcap_{i\in s} C_i\right]}.$$

It is useful to illustrate Theorem 5 with an example that relates it to Theorem 4 by exhibitting the cancellation pattern that reduces the expression of the former to the latter.

Example: Consider the chordal graph

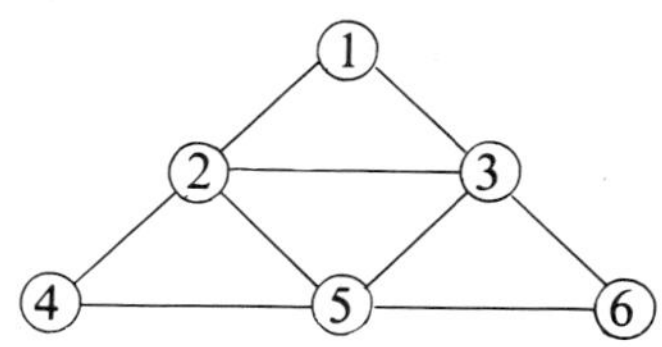

with maximal cliques $C_1 = \{1,2,3\}$, $C_2 = \{2,3,5\}$, $C_3 = \{2,4,5\}$ and $C_4 = \{3,5,6\}$. Then, in this case the expression of Theorem 4 is

$$\frac{\prod_{k \text{ odd}} \prod_{|s|=k} \det A\left[\bigcap_{i\in s} C_i\right]}{\prod_{k \text{ even}} \prod_{|s|=k} \det A\left[\bigcap_{i\in s} C_i\right]}$$

$$= \frac{\det A[C_1]\cdot \det A[C_2] \det A[C_3]\cdot \det A[C_4]}{\det A[C_1\cap C_2]\cdot \det A[C_1\cap C_3]\cdot \det A[C_1\cap C_4] \det A[C_2\cap C_3]\cdot \det A[C_2\cap C_4]\cdot \det A[C_3\cap C_4]}$$

$$\cdot\frac{\det A[C_1\cap C_2\cap C_3]\cdot \det A[C_1\cap C_2\cap C_4]\cdot \det A[C_1\cap C_3\cap C_4] \det A[C_2\cap C_3\cap C_4]}{\det A[C_1\cap C_2\cap C_3\cap C_4]}$$

$$= \frac{\det A[C_1]\cdot \det A[C_2]\cdot \det A[C_3]\cdot \det A[C_4]\cdot \det A[\{2\}]\cdot \det A[\{3\}] \det A[\phi]\cdot \det A[\{5\}]}{\det A[\{2,3\}]\cdot \det A[\{2\}]\cdot \det A[\{3\}]\cdot \det A[\{2,5\}]\cdot \det A[\{3,5\}] \det A[\{5\}]\cdot \det A[\phi]}$$

$$= \frac{\det A[C_1]\cdot \det A[C_2]\cdot \det A[C_3]\cdot \det A[C_4]}{\det A[C_1\cap C_2]\cdot \det A[C_2\cap C_3]\cdot \det A[C_2\cap C_4]}.$$

The latter is the expression of Theorem 5 corresponding to the unique (in this case) spanning tree of $G_{\mathscr{C}}$ satisfying the intersection property:

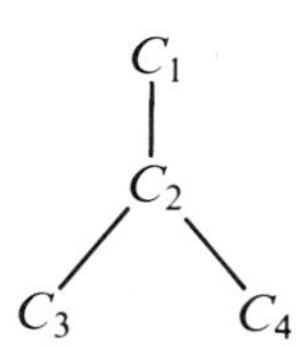

While there is great redundancy among numerator and denominator terms in Theorem 5, this expression has the virtue of being independent of the tree T; only a knowledge of the set of maximal cliques is needed.

A feature that the presentations of Theorems 4 and 5 lack is that they do not explicitly describe the entries of the determinant maximizing completion. A third, and perhaps even more elementary description of the maximum determinant, is a surprisingly simple algorithm that actually produces the maximizing matrix in the chordal case. The very nice surprise is that the sequence of one variable determinant maximizing problems associated with a chordal ordering of the missing edges actually produces the global maximum. This aspect of chordal structure likely has further implications in continuous optimization as well as those that are recognized in discrete optimization. That this local/global property holds for *some* chordal orderings was first noticed in **[JR3]** and that it actually holds for *any* chordal ordering was shown in **[BJL]**.

THEOREM 6. *Let G be a chordal graph and let A be a partial positive definite matrix the graph of whose specified entries is G. Let F be a chordal graph obtained by adding one edge u, v to G and let C be the unique maximal clique of F containing the new edge. If* $L = (l_{ij})$ *is the unique determinant maximizing positive definite completion of* $A[C]$ *and* $M = (m_{ij})$ *is the unique determinant maximizing completion of A, then*

$$l_{uv} = m_{uv}.$$

It follows that the global maximum M may be found by a sequence of problems of one variable type (described in detail in Section II). This has important implications. In the chordal case not only the maximum determinant but also the entries of the maximizing completion are rational in the specified entries. This means that if S is the real or rational field, instead of the complex field, the determinant maximizing completion will remain in the field. This does not remain valid beyond the chordal case.

Example **[JB]**. Consider the partial positive definite matrix

$$\begin{bmatrix} 2 & -1 & x & -1 \\ -1 & 2 & -1 & y \\ \overline{x} & -1 & 2 & -1 \\ -1 & \overline{y} & -1 & 2 \end{bmatrix}.$$

Here the graph of the specified entries is not chordal. Positive definite completions exist, and calculation based upon Theorem 2 yields that the maximum determinant (among positive definite completions) is $24\sqrt{3} - 36$, which is attained at the completion $x = y = \sqrt{3} - 1$, neither of which is rational in the data.

It is, however, a consequence of uniqueness (Theorem 2) that real data results in a real entried determinant maximizing matrix, regardless of pattern.

See **[JB]** for more details about the generality of determinantal inequalities for positive definite matrices resulting from the chordal case.

IV. Contraction Completions. Given an arbitrary rectangular complex matrix A, the *spectral norm* $|||A|||_2$ (or largest singular value) is just the square root of the largest eigenvalue of A^*A **[HJ]**. If $|||A|||_2 \leq 1$, then A is called a (spectral norm) *contraction*, or a strict contraction if $|||A|||_2 < 1$. For a rectangular partial matrix over the complex numbers, it is natural to ask if there is a completion that is a contraction, and this has been of interest from several points of view, including operator theory and numerical analysis **[DKW, JR2, P]**.

Since $|||\hat{A}|||_2 \leq |||A|||_2$ if $\hat{A}$ is a submatrix of A, it follows that each submatrix of a contraction is again a contraction and that the contraction completion problem also satisfies an inheritance property. If a partial matrix is to have a contraction completion, it is necessary that it be a *partial contraction*, i.e. every specified submatrix must be a contraction. Since A is a contraction if and only if the block matrix

$$\begin{bmatrix} I & A \\ A^* & I \end{bmatrix}$$

is positive semidefinite, it is clear that the contraction completion problem and the positive semidefinite completion problem are related.

What then may be said about the combinatorial version of the contraction completion problem: which patterns for the specified entires of a partial contraction ensure that there is a contraction completion? Certainly some patterns do not. For example, the 2–by–3 partial contraction

$$\begin{bmatrix} ? & \frac{1}{\sqrt{2}} & \frac{1}{\sqrt{2}} \\ \frac{1}{\sqrt{2}} & ? & \frac{1}{\sqrt{2}} \end{bmatrix}$$

has no contraction completion. Indeed, the ? in the $1, 1$ position lies in both partial submatrices

$$[? \;\; \tfrac{1}{\sqrt{2}} \;\; \tfrac{1}{\sqrt{2}}] \quad \text{and} \quad \begin{bmatrix} ? & \frac{1}{\sqrt{2}} \\ \frac{1}{\sqrt{2}} & \frac{1}{\sqrt{2}} \end{bmatrix}.$$

In order that the former be a contraction, the ? must be 0; in order that the latter be a contraction, the ? must be $-\frac{1}{\sqrt{2}}$. Since both conditions cannot be simultaneously met, there is no contraction completion. Some patterns, though, do admit a contraction completion if the "obvious" necessary condition is met. If

$$A(x) = \begin{bmatrix} a & x \\ b & c \end{bmatrix}$$

is a partial contraction with x the only unspecified entry, then

$$|||A(x)|||_2 = \max\left\{\left|\left|\left|\begin{bmatrix} a \\ b \end{bmatrix}\right|\right|\right|_2, |||[b \quad c]|||_2\right\}$$

for $x = -a(\frac{\overline{b}}{\overline{c}})$ if $\overline{c} \neq 0$ or $x = 0$ if $c = 0$. In fact the "single block case"

$$\begin{bmatrix} A & X \\ B & C \end{bmatrix},$$

in which all entries are specified, except those of the block X (all entries of which are unspecified), has received much attention **[DKW, P]**, and such a pattern always admits a contraction completion as long the obvious necessary condition is met. Formulae for the X in a minimum norm completion are available [**DKW**, **M**].

It is clear then that the combinatorial version of the contraction completion problem is the same as asking for which patterns of the specified entires is there a spectral norm nonincreasing completion of any partial matrix. We will describe a complete solution to this problem, but a more general problem remains very open: For a given pattern, what is the maximum norm over all minimum norm completions of partial contractions with this pattern for the specified entries?

Because the spectral norm is permutation equivalence invariant ($|||A|||_2 = |||PAQ|||_2$ for P and Q permutation matrices), it is natural to describe the pattern of specified entries of a rectangular partial matrix via an undirected *bipartite graph H*. If A is an m-by-n partial matrix, let $r_1, \dots, r_m$ and $c_1, \dots, c_n$ be the vertices of $H(A)$, and let $\{r_i, c_j\}$ be an edge of $H(A)$ if and only if the i, j entry of A is specified. Informally, we refer to the pattern P of the partial matrix A interchangeably with this graph. We call the pattern (or the partial matrix) *indecomposable* if its bipartite graph is connected. It suffices to consider the indecomposable case as the partial matrix

$$\begin{bmatrix} B_{11} & B_{12} \\ B_{21} & B_{22} \end{bmatrix}$$

has a contraction completion if and only if B_{11} and B_{22} do when B_{12} and B_{21} consist entirely of unspecified entries. This is so because B_{12} and B_{21} may be taken to be zero in finding a norm minimizing completion. In order to describe a complete solution [**JR2**] to the combinatorial version of the contraction completion problem, it is convenient to derive an ordinary undirected graph G from the bipartite graph H on vertices $r_1, \dots, r_m, c_1, \dots, c_n$. The graph G is simply the result of adding *all* edges of the form $\{r_i, r_j\}$ and $\{c_i, c_j\}$ to H. This graph is another equivalent description of the pattern P of specified entries in view of the block matrix

$$\begin{bmatrix} I & A \\ A^* & I \end{bmatrix}.$$

We then have

THEOREM 7. *For an m-by-n indecomposable pattern P (with associated bipartite graph H and undirected graph G) for the specified entries of an m-by-n partial matrix, the following statements are equivalent:*

(a) *every m-by-n partial contraction with the pattern P for its specified entries may be completed to an m-by-n contraction;*

(b) *the undirected graph G is chordal; and*

(c) *there do not exist two "rows" r_{i_1} and r_{i_2} and two "columns" c_{j_1} and c_{j_2} such that the subgraph of the bipartite graph H induced by the vertices $r_{i_1}, r_{i_2}, c_{j_1}, c_{j_2}$ contains exactly the edges $\{r_{i_1}, c_{j_1}\}$ and $\{r_{i_2}, c_{j_2}\}$.*

PROOF OUTLINE. Conditions (b) and (c) must be seen to be combinatorially equivalent. For the sufficiency of either, it is possible to either (i) complete one step at a time along a properly chosen sequence, exploiting the single block case at each step or (ii) use the positive semidefinite completion results of section II, exploiting the connection via

$$\begin{bmatrix} I & A \\ A^* & I \end{bmatrix}.$$

For necessity, it is possible to prescribe counterexamples (of the sort described earlier in this section) when the "forbidden subpatterns" occur (just as in the necessity portion of theorem 2). □

Note that condition (c) in theorem 7 simply means that no 2-by-2 partial matrix of the form $[\begin{smallmatrix} * & ? \\ ? & * \end{smallmatrix}]$ or $[\begin{smallmatrix} ? & * \\ * & ? \end{smallmatrix}]$ appears as a submatrix.

Matricially, in the general (possibly decomposable) case, the patterns for which all partial contractions may be completed to contractions may be described as follows. Up to permutation equivalence, the partial matrix A must be of the "block diagonal" form:

$$A = \begin{bmatrix} A_1 & ? & \cdots & ? \\ ? & A_2 & \ddots & \vdots \\ \vdots & \ddots & \ddots & ? \\ ? & \cdots & ? & A_r \end{bmatrix},$$

possibly bordered by rows and/or columns of question markes, in which

$$A_j = \begin{bmatrix} A_{j11} & ? & ? & \cdots & ? \\ A_{j21} & A_{j22} & \ddots & & \vdots \\ \vdots & & \ddots & & ? \\ A_{jp1} & A_{jp2} & \cdots & & A_{jpp} \end{bmatrix}, \qquad j = 1, \ldots, r,$$

and the (possibly rectangular) blocks A_{jst}, $1 \le t \le s \le p$, consist entirely of specified entries. The blocks A_j need not be square. As noted earlier, these are also all the patterns that guarantee a norm nonincreasing completion, and for most other patterns it is not known how much the minimum norm of a completion of a partial contraction may have to be.

Variants on the contraction completion problem and its combinatorial version have also been of interest. For example, partial results on the Toeplitz contraction completion problem may be found in **[JR4]**, although the general question of which sequences of specified bands for a Toeplitz partial contraction ensure existence of a Toeplitz contraction completion remains open.

The contraction completion question, when "contraction" is defined with respect to various other norms, has, thus far, received little attention. For norms such that the minimum norm completion is always given by using 0's, this question is uninteresting, but for several norms other than the spectral norm, it may be quite challenging and worthwhile.

V. Ranks of Completions. For a rectangular partial matrix over a field (usually the complex, real or rational numbers, but more general fields may be of interest), another natural question is what the possible ranks of completions are. For a given partial matrix A, it is clear that there is a maximum and minimum possible rank and that each will be attained. We denote these by $\mathrm{MR}(A)$ and $\mathrm{mr}(A)$, respectively. As changing a minimum rank completion into a maximum rank completion one unspecified entry at a time will change the rank of a completion by at most one at each step, it is also clear that each rank between $\mathrm{mr}(A)$ and $\mathrm{MR}(A)$ is attained by some completion of A. Thus, in order to know all possible ranks of completions, it suffices to know $\mathrm{mr}(A)$ and $\mathrm{MR}(A)$.

For any partial matrix over any field with any pattern for its specified entries, it is possible to give a formula for $\mathrm{MR}(A)$ in terms of ranks of specified submatrices. This is not surprising as most completions attain the maximum rank. Consider first a pattern in which the specified entries form a rectangular submatrix. Let A be an m–by–n partial matrix with rectangular pattern,

$$A = \begin{bmatrix} F & ? \\ ? & ? \end{bmatrix}$$

in which F is a fully specified p–by–q submatrix. If $p \leq m$ and $q \leq n$ are sufficiently large and $\operatorname{rank} F$ is sufficiently small, every completion of A will have to be rank deficient. Quantitatively, it is not difficult to show that $\mathrm{MR}(A) = \min\{m, n, (m-p) + (n-q) + \operatorname{rank}(F)\}$. For such a pattern J, define

$$\rho(A, J) = (m-p) + (n-q) + \operatorname{rank} F.$$

If J is a *trivial rectangular pattern* ($p = 0$ or $q = 0$), we understand $\operatorname{rank} F$ to be 0. Furthermore, by a *subpattern* of a given pattern for the specified entries of a partial matrix, we mean a pattern in which some specified entries have been replaced by ?'s. If A is now an arbitrary m–by–n partial matrix and J is any rectangular subpattern of the pattern of A, it is clear that

$$\mathrm{MR}(A) \leq \rho(A_J, J).$$

Here A_J means the partial matrix in which any specified entries other than those allowed by J have been replaced by ?'s. With this notation and proper interpretation the formula for maximum rank may now be given as:

THEOREM 8. *Let A be an m–by–n partial matrix, the pattern of whose specified entries is P. Then*

$$\mathrm{MR}(A) = \min \rho(A_J, J)$$

in which J runs over all the (maximal) rectangular subpatterns of P, including the trivial ones.

Theorem 8 may be found in **[CJRW]** and generalizes a result of **[HL]** (in which only matrices whose specified entries are 0 were considered), which, in turn, generalized a classical result of Frobenius/König/Egervary, which may be phrased in terms of matchings.

Generally, minimum rank is much more difficult to determine than maximum rank. The difficulty varies with the pattern, and, for some patterns, the minimum rank is not expressible in terms of ranks of specified submatrices. For example

$$\operatorname{mr}\begin{bmatrix} 1 & 1 & ? \\ 1 & ? & 1 \\ ? & 1 & 1 \end{bmatrix} = 1,$$

while

$$\operatorname{mr}\begin{bmatrix} 1 & 1 & ? \\ 2 & ? & 1 \\ ? & 1 & 1 \end{bmatrix} = 2,$$

although corresponding specified submatrices have the same rank.

Rank again satisfies an inheritance property: the rank of any submatrix is a lower bound for the rank of a given matrix. Thus, it is natural to ask which patterns for the specified entries permit a formula for $\operatorname{mr}(\cdot)$ just in terms of ranks of specified submatrices. As rank is again permutation equivalence invariant ($\operatorname{rank} PAQ = \operatorname{rank} A$ for permutation matrices P and Q), it is natural again to describe the pattern of specified entries of an m–by–n partial matrix via the bipartite graph H on vertices $r_1, \ldots, r_m, c_1, \ldots, c_n$, as in the last section. For which bipartite graphs do all partial matrices have the same mr when all corresponding specified submatrices have the same rank? One nice such class of patterns are the *block triangular* patterns. If a partial matrix is permutation equivalent to one of the form

$$A = \begin{bmatrix} A_{11} & ? & \cdots & ? \\ A_{21} & A_{22} & \cdots & ? \\ \vdots & & \ddots & ? \\ A_{p1} & A_{p2} & \cdots & A_{pp} \end{bmatrix}$$

then the partial matrix is called block triangular. If A is of this form, it was shown in **[W1]** that

THEOREM 9.

$$\operatorname{mr}(A) = \sum_{i=1}^{p} \operatorname{rank}\begin{bmatrix} A_{i1} & \cdots & A_{ii} \\ \vdots & & \vdots \\ A_{p1} & \cdots & A_{pi} \end{bmatrix} - \sum_{i=2}^{p} \operatorname{rank}\begin{bmatrix} A_{i1} & \cdots & A_{i,i-1} \\ \vdots & & \vdots \\ A_{p1} & \cdots & A_{p,i-1} \end{bmatrix}.$$

The triangular patterns are not the only ones for which $\operatorname{mr}(\cdot)$ is a function of the ranks of specified submatrices. A specific conjecture may be found in **[CJRW]**, and the notion of triangular minimum rank appears important. The triangular minimum rank, $\operatorname{tmr}(A)$, of a partial matrix A is the maximum of $\operatorname{mr}(A_T)$ as T runs over all triangular subpatterns of the pattern of A. A bipartite graph is called *bipartite chordal* if it has no minimal simple circuit of 6 or more edges. Note the distinction from chordal, which is defined for arbitrary undirected graphs. Several partial results (among them a proof of necessity) pointing to the following conjecture are given in **[CJRW]**.

CONJECTURE 10. A bipartite graph H is bipartite chordal if and only if every partial matrix A whose bipartite graph is H satisfies

$$\mathrm{mr}(A) = \mathrm{tmr}(A).$$

Other questions about minimum rank completions are organized in **[JW]**.

Acknowledgment. The author wishes to thank Michael Lundquist for helpful discussions during the preparation of this manuscript.

REFERENCES

[AHMR] J. Agler, W. Helton, S. McCullough and L. Rodman, Positive Semidefinite Matrices with a Given Sparsity Pattern, *Linear Algebra Appl.* **107** (1988), 101–149.

[B] J. P. Burg, Maximum Entropy Spectral Analysis, Ph.D. Thesis (Dept. of Geophysics), Stanford University, Stanford, CA, 1975.

[BEGL] A. Ben-Artzi, R. L. Ellis, I. Gohberg & D. C. Lay, The Maximum Distance Problem and Band Sequences, *Linear Algebra Appl.* **87** (1987), 93–112.

[BF] W. Barrett and P. Feinsilver, Inverses of Banded Matrices, *Linear Algebra Appl.* **41** (1981), 111–130.

[BJ1] W. Barrett and C. R. Johnson, Determinantal Formulae for Matrices with Sparse Inverses, *Linear Algebra Appl.* **56** (1984), 73–88.

[BJ2] W. Barrett and C. R. Johnson, Determinantal Formulae for Matrices with Sparse Inverses, II: Assymmetric Zero Patterns, *Linear Algebra Appl.* **81** (1986), 237–261.

[BJL] W. Barrett, C. R. Johnson and M. Lundquist, Determinantal Formulae for Matrix Completions Associated with Chordal Graphs, *Linear Algebra Appl.* **121** (1988), 265–289.

[BJOV] W. Barrett, C. R. Johnson, D. Olesky, and P. van den Driessche, Inherited Matrix Entries: Principal Submatrices of the Inverse, *SIAM J. Alg. Disc. Meth.* **8** (1987), 313–322.

[CJRW] N. Cohen, C. R. Johnson, L. Rodman and H. Woerdeman, Ranks of Completions of Partial Matrices, Operator Theory: Adv. and Appl. **40** (1989), 165–185.

[D] C. Davis, Completing a Matrix so as to Minimize its Rank, *Operator Theory: Advances and Appl.* **29** (1988), 87–95.

[DG1] H. Dym and I. Gohberg, Extensions of Band Matrices with Band Inverses, *Linear Algebra Appl.* **36** (1981), 1–24.

[DG2] H. Dym and I. Gohberg, A new class of contractive interpolents and maximum entropy principles, preprint.

[DKW] C. Davis, W. M. Kahan and H. F. Weinberger, Norm-preserving Dilations and their Applications to Optimal Error Bounds, *SIAM J. Numerical Anal.* **19** (1982), 445–469.

[EGL1] R. L. Ellis, I. Gohberg and D. C. Lay, Band Extensions, Maximum Entropy and the Permanence Principle, in *Maximum Entropy and Bayesian Methods in Applied Statistics*, ed. J. Justice, Cambridge University Press, Cambridge, 1986.

[EGL2] R. Ellis, I. Gohberg and D. Lay, Invertible Selfadjoint Extensions of Band Matrices and their Entropy, *SIAM J. Alg. Disc. Meth.* **8** (1987), 483–500.

[EGL3] R. Ellis, I. Gohberg and D. C. Lay, On Negative Eigenvalues of Self-adjoint Extensions of Band Matrices, *Linear and Multilinear Algebra* **24** (1988), 15–25.

[EL] R. Ellis and D. Lay, Rank Preserving Extensions of Band Matrices, *Linear and Multilinear Algebra*, to appear.

[G] M. Golumbic, *Algorithmic Graph Theory and Perfect Graphs*, Academic Press, New York, 1980.

[GJSW] R. Grone, C. R. Johnson, E. Sá and H. Wolkowicz, Positive Definite Completions of Partial Hermitian Matrices, *Linear Algebra Appl.* **58** (1984), 109–124.

[GKW] I. Gohberg, M. A. Kaashoek & H. Woerdeman, The Band Method for Positive and Contractive Extension Problems, preprint from the Summer Institute in Operator Theory, Operator Algebras and Applications, Durham, N.H., July 1988.

[GR] I. Gohberg, S. Rubinstein, Proper Contractions and their Unitary Minimal Completions, Operator Theory: Advances and Applications vol. 33.

[HL] D. Hartfiel and R. Loewy, A Determinant Version of the Frobenius-König Theorem, *Lin. and Multilin. Alg.* **16** (1984), 155–165.

[HPR] J. Helton, S. Pierce and L. Rodman, The Ranks of Extremal Positive Semidefinite Matrices with Given Sparsity Pattern, *SIAM J. Matrix Anal. Appl.* **10** (1989), 407–423.

[HJ] R. Horn and C. R. Johnson, *Matrix Analysis*, Cambridge University Press, New York, 1985.

[J] C. R. Johnson, Optimization, Matrix Inequalities and Matrix Completions, in *Operator Theory, Analytic Functions, Matrices and Electrical Engineering*, ed. J. W. Helton, CBMS Regional Conference Series in Mathematics 68, American Mathematical Society, Providence, R.I., 1987.

[J2] C. R. Johnson, Combinatorial Matrix Analysis: An Overview, *Linear Algebra Appl.* **107** (1988), 3–15.

[JB] C. R. Johnson and W. Barrett, Spanning Tree Extensions of the Hadamard-Fischer Inequalities, *Linear Algebra Appl.* **66** (1985), 177–193.

[JB2] C. R. Johnson and W. Barrett, Matrix Completions, Determinantal Maximization and Maximum Entropy, in *Proceedings of the Interface* 88 *Symposium*, American Statistical Association, Alexandria, VA, 1988.

[JB3] C. R. Johnson and W. Barrett, Determinantal Inequalities for Positive Definite Matrices, preprint.

[JL] C. R. Johnson and M. Lundquist, Numerical Ranges for Partial Matrices, submitted.

[JR1] C. R. Johnson and L. Rodman, Inertia Possibilities for Completions of Partial Hermitian Matrices, *Linear and Multilinear Algebra* **16** (1984), 179–195.

[JR2] C. R. Johnson and L. Rodman, Completion of Partial Matrices to Contractions, *J. Func. Anal.* **69** (1986), 260–267.

[JR3] C. R. Johnson and L. Rodman, Chordal Inheritance Principles and Positive Definite Completions of Partial Matrices over Function Rings, in *Contributions to Operator Theory and its Applications*, (Proc. of the Mesa Conference on Operator Theory and Functional Analysis), Birkhäuser-Verlag, Basel, 1988.

[JR4] C. R. Johnson and L. Rodman, Completion of Toeplitz Partial Contractions, *SIAM J. Matrix Anal. Appl.* **9** (1988), 159–167.

[JW] C. R. Johnson and G. Whitney, Minimum Rank Completions, submitted.

[KW] M. A. Kaashoek and H. J. Woerdeman, Unique Minimal Rank Extensions of Triangular Operators, *J. Math. Anal. Appl.* **131** (1988), 501–516.

[L] M. Lundquist, Ph.D. thesis, in preparation.

[M] J. Meinguet, On the *D–K–W* Solution of the Norm Preserving Dilation Problem, *Numer. Math.* **49** (1986), 331–341.

[N] H. Nelis, Sparse Approximations of Inverse Matrices, Delft University of Technology, (October) 1989.

[NDD] H. Nelis, E. Deprettere and P. Dewilde, Approximate Inversion of Positive Definite Matrices, Specified on a Multiple Band, preprint, 1988.

[P] S. Parrot, On a Quotient Norm and St.-Nagy-Foias Lifting Theorem, *J. Funct. Anal.* **30** (1978), 311–328.

[W] G. Whitney, Rank Maximin Problems, Senior Honors Thesis, Harvard University, 1989.

[W1] H. Woerdeman, The Lower Order of Lower Triangular Operators and Minimal Rank Extensions, *Integral Equations and Operator Theory* **10** (1987), 859–879.

[W2] H. Woerdeman, Strictly Contractive and Positive Definite Completions for Block Matrices, Rapport WS-337, Vrije Universiteit, Amsterdam, November, 1987.

[W3] H. Woerdeman, Minimal Rank Completions for Block Matrices, *Linear Algebra Appl.*, to appear.

[W4] H. Woerdeman, Matrix and Operator Extensions, Ph.D. Thesis, Vrije Universiteit Amsterdam, 1989.

Department of Mathematics
College of William and Mary
Williamsburg, VA 23185

Proceedings of Symposia in Applied Mathematics
Volume **40**, 1990

The Role Of Nonnegative Idempotent Matrices In Certain Problems In Probability

ARUNAVA MUKHERJEA

ABSTRACT. In this article we present results on structures of certain groups and semigroups of nonnegative matrices based on the structures of their idempotents, and discuss their implications in certain problems in probability.

1. Introduction

In this paper we present a number of interesting contexts in probability where we encounter nonnegative idempotent matrices and show how studying their structures can help us to better understand the problems involved.

In §2, we look into the structure of infinite dimensional nonnegative idempotent matrices and show how we can utilize this structure to characterize idempotent probability measures on discrete semigroups. Such measures play a major role in the study of random walks on semigroups. When we study random walks in nonnegative matrices, which form a semigroup under multiplication and a topological semigroup with a suitable topology, idempotent probability measures appear as weak limits of their distributions and the supports of these measures are what specialists in semigroups call completely simple semigroups. These semigroups are roughly special unions of groups. These motivate our study in sections 3 and 4 of groups and completely simple semigroups of nonnegative matrices. We show in these sections that these classes can be characterized in terms of the structures of the idempotent matrices in these classes.

In sections 5, 6 and 7 we apply the results obtained in earlier sections in problems involving random matrices, non-homogeneous Markov chains, and embeddable singular stochastic matrices.

1980 *Mathematics Subject Classification* (1985 *Revision*). Primary 60B10, 60J10.

Let us finally mention here that though idempotent matrices have been looked into by many mathematicians from various angles, we could not include all of the important references. But we hope to have included the ones relevant to our context, and these can lead to those we have omitted.

2. Nonnegative Idempotent Matrices

The structure of a stochastic idempotent matrix P with state space the set E of positive integers can be described as follows:

There is a unique partition of E given by

$$E = T \cup C_1 \cup C_2 \cup \cdots .$$

such that

$$\begin{aligned} P_{ij} &= 0, \text{ whenever } j \in T; \\ &= 0, \text{ whenever } i \text{ and } j \text{ belong to different } C-\text{classes}; \\ &= P_{kj}, \text{ whenever } i, j \text{ and } k \text{ are in the same } C-\text{class}; \\ &= a_s(i)P_{jj}, \text{ whenever } i \in T \text{ and } j \in C_s, \sum_s a_s(i) = 1. \end{aligned}$$

For nonnegative idempotent matrices we have the following analogous result given in [**22**].

PROPOSITION 2.1. *For an infinite dimensional nonnegative idempotent matrix P with state space E, there is a unique partition of E, called the basis of P, given by*

$$E = T \cup C_1 \cup C_2 \cup \cdots$$

such that the following conditions hold:

(i) *j is in T* iff *the jth row or the jth column or both consist of only zeros;*
(ii) *$P_{ij} = 0$ if i and j are in different C-classes;*
(iii) *in any $C_i \times C_i$ block of P, the rows are all proportional, each entry is positive, and the sum of the diagonal entries is 1;*
(iv) *if i is in T and the ith column (ith row, resp.) is a zero column (zero row, resp.), then for j and k both in C_s, a C-class, $P_{ij}/P_{jj} = P_{ik}/P_{jk}$ ($P_{ji}/P_{jj} = P_{ki}/P_{kj}$, resp.).* □

Note that in the finite dimensional case, the number of C-classes in the basis of P above is finite and equals the rank of P. The finite dimensional version of Prop. 2.1 was first obtained by Denny Brown. Later this result was rediscovered by other mathematicians.

One can push the result above a bit further to include nonnegative idempotent kernels $K(x, B)$, which are defined as follows:

Let X be a state space (possibly uncountable) and $\mathscr{B}$ a sigma algebra of subsets of X. Let $K\colon X \times \mathscr{B} \to [0, \infty]$ be such that

(i) for each $x, K(x, \ )$ is a nonnegative measure on $\mathscr{B}$;
(ii) for B in $\mathscr{B}, K(\ , B)$ is a $\mathscr{B}$-measurable nonnegative function;
(iii) for x in X and B in $\mathscr{B}, K(x, B) = \int K(y, B)K(x, dy)$.

Such a function K is a nonnegative idempotent kernel. Nonnegative idempotent kernels are natural generalizations of infinite dimensional nonnegative idempotent matrices. The simplest of such kernels are of the form $f(x)\cdot L(B)$, where f is a nonnegative Borel measurable function and L is a nonnegative Borel measure such that $\int f dL = 1$. The interesting nature of the structure of general nonnegative idempotent kernels is that they are often obtained by piecing together kernels of the above simplest type. Indeed it has been proven in **[23]** that when $\mathscr{B}$ is strictly separable (that is, generated by a denumerable subcollection of its elements) and X is the union of a sequence of A_n's such that for x in X and any n, $K(x, A_n)$ is finite, then there is a partition of X as follows:

$$X = S \cup N \cup (\cup S_t),$$

such that the following conditions hold:

(i) $S = \{x \colon K(x, X) = 0\}$;
(ii) N is a null set, that is, for x in X, $K(x, N) = 0$;
(iii) for each t, S_t is measurable and for x in S_t, $K(x, X - S_t) = 0$;
(iv) if x and y are in S_t, then there exists a positive number $u(x, y)$ such that for any B in $\mathscr{B}$, $K(x, B) = u(x, y) \cdot K(y, B)$.

One can use the above structure theory to determine the structure of idempotent probability measures on topological semigroups (such as multiplicative semigroups of nonnegative matrices) and we will see in section 5 that such measures come up naturally as weak limits in the context of weak convergence problems. It should be pointed out that the above structure in the stochastic case was first determined by David Blackwell (see **[2]**).

Before we close this section, let us show briefly how we can determine the structure of the support of an idempotent (with respect to convolution) probability measure on a countable discrete semigroup using Proposition 2.1 (in the stochastic case).

Suppose that Q is an idempotent probability measure on a countable discrete semigroup S. Since Q is idempotent, for any x in S we have:

$$Q(x) = \sum Q(xy^{-1})Q(y) = \sum Q(y^{-1}x)Q(y);$$

here, xy^{-1} means the set $\{z \colon zy = x\}$ and $y^{-1}x = \{z \colon yz = x\}$. This means that S_Q, the support of Q, is a subsemigroup and $S_Q S_Q = S_Q$. Thus, with no loss of generality, we can assume that

$$S = S_Q = \{x_1, x_2, x_3, \ldots\}.$$

Define the matrix P with state space S by $P_{ij} = Q(x_i^{-1}x_j)$. Then it can be verified easily that P is a stochastic idempotent matrix. By Proposition 2.1, there is a partition of S, $S = T \cup C_1 \cup C_2 \cup \ldots$, where T and the C-classes have the properties given in the proposition. Let x_j be in T. Then for any x_i in S, $Q(x_i^{-1}x_j) = P_{ij} = 0$ which means that x_j cannot be in $S \cdot S$. Thus, T must be empty. Since each C-block of P has identical strictly positive rows, it follows that if x_i and x_j are in some C_k, then $Q(x_i^{-1}z) = Q(x_j^{-1}z)$ for all z in S, and which is positive iff z is in C_k. This means that $x_i \cdot S = x_j \cdot S$. Also, $x_i C_k$ is a minimal right ideal of S. Since $x_i S = x_i C_k$, it follows that for any x in S, xS is a minimal right ideal of S. Similarly, considering the matrix P' given by $P'_{ij} = Q(x_j x_i^{-1})$, we can show that for each x in S, Sx is a minimal left ideal. By standard results in the theory of semigroups (see [**40**]), it follows that S is completely simple.

Let us remark that in the above case knowing that the support of an idempotent probability measure on a countable discrete semigroup is a completely simple subsemigroup, using the structure of such semigroups it is not difficult to show that such measures, restricted to their supports, are always product measures of the form $Q_1 \times Q_2 \times Q_3$, where Q_1 and Q_3 are probability measures on a left zero subsemigroup and a right zero subsemigroup respectively, and Q_2 is the uniform (Haar) probability measure on a finite subgroup. This result remains valid even in the general continuous case (with "compact" replacing "finite"). See [**40**]. The proof in this case is much more difficult.

Let us finally mention that understanding the structure of a general nonnegative idempotent kernel can help us understand why there cannot be INFINITE regular idempotent measures in a group. See [**23**].

3. Multiplicative Groups of Nonnegative Matrices

In this section we will characterize multiplicative groups of nonnegative matrices with state space E (the positive integers) satisfying the property:

$$\sup_{1} \sum_{j} A_{ij} < \infty.$$

(These results remain valid in the finite dimensional situation.)

Let G be such a group and e be its identity. By Proposition 2.1, there is a basis of e given by the following partition of E:

$$E = T \cup C_1 \cup C_2 \cup \cdots$$

such that for i and j in C_k, there exists $a_k(i,j) > 0$ such that

$$e_{is} = a_k(i,j) \cdot e_{js}, \quad s \text{ in } E. \tag{3.1}$$

Let x be in G, $x \neq e$. Then the following are true.

(i) The zero rows and zero columns of x and e are the same since $xe = ex = x$ and $xx^{-1} = x^{-1}x = e$.

(ii) Since $x = ex$, in the equation (3.1) e can be replaced by x. In other words, the rows of x (restricted to C_k) are proportional with the same strictly positive proportionality constant as holds for e.

(iii) For a given C-class C_k, there exists a unique k' and a unique k'' such that

$$x|_{C_s \times C_k},\ x|_{C_k \times C_t},\ x^{-1}|_{C_k \times C_s},\ x^{-1}|_{C_t \times C_k}$$

are all strictly positive blocks when $t = k'$ and $s = k''$, and are all blocks of zeros when $t \neq k'$ and $s \neq k''$.

Let us prove one of the assertions in (iii). The proofs of the others are similar. Let i be an arbitrary but fixed element of C_k. Then since $e = xx^{-1}$,

$$0 < e_{ii} = \sum_{u \notin T} x_{iu} x^{-1} ui$$

and therefore, there exists u in C_k, (for some k') such that $x_{iu} > 0$ and $x_{ui}^{-1} > 0$. Notice that $x_{ui}^{-1} x_{ij} \leq e_{uj}$ so that

(3.2) $$x_{ij} > 0 \text{ for } j \notin T => j \in C_{k'}.$$

Similarly,

(3.3) $$x_{uj}^{-1} > 0 \text{ for } j \notin T => j \in C_k.$$

Let $s \in C_{k'}$. Then x_{is} must be positive since $x_{is} = 0 => x_{vs} = 0$ for each v in C_k (because of (ii)), and then it follows from (3.3) that $e_{us} = \sum_{v \in C_k} x_{uv}^{-1} x_{vs} = 0$ and this contradicts Proposition 2.1.

Now it is clear from (iii) that for each x in G, there exists a permutation $p(x)$ on the set $E_o = \{j\colon$ there is a C_j in the basis of $e\}$, such that $p(x)[k] = k'$ iff the C_k by $C_{k'}$ block of x is a strictly positive block. The correspondence p from G into the set of permutations on E_o is a homomorphism in the sense that $p(xx') = p(x')op(x)$. The map p is not necessarily one-to-one, since for a positive, $p(ax) = p(x)$. However, $p(x) = p(x')$ implies the following:

For each C_k, there exists a positive number $a(k)$ such that

(3.4) $$x'|_{C_k \times C_{p(x)[k]}} = a(k) \cdot x|_{C_k \times C_{p(x)[k]}}.$$

To prove (3.4), let i be in C_k. Since for j in $C_{p(x)[k]} = C_{p(x')[k]'} x_{ij}$ and x'_{ij} are both positive, there exists $a(k, i, j) > 0$ such that

(3.5) $$x_{ij} = a(k, i, j) \cdot x'_{ij}.$$

By earlier observation (ii), it follows that

(3.6) $a(k, i, j)$ if independent of i in C_k.

Now notice that for j and s in $C_{p(x)[k]}$, there exists $b(j, s) > 0$ such that

(3.7) $$e_{vj} = b(j, s) \cdot e_{vs}, \qquad v \in C_{p(x)[k]}.$$

Using this and the fact that $x = xe, x' = x'e$, we have:

(3.8) $$x_{vj} = b(j, s) \cdot x_{vs}, \qquad x'_{vj} = b(j, s) \cdot x'_{vs'}$$

whenever j and s are in $C_{p(x)[k]}$ and v is in C_k. It follows from (3.5) and (3.8) that

$$b(j, s)x_{is} = x_{ij} = a(k, i, j)x'_{ij} = a(k, i, j)b(j, s)x'_{is}.$$

This means that $a(k, i, j)$ is the same for all i in C_k and all j in $C_{p(x)[k]}$. This establishes (3.4).

A simple example (in the finite dimensional case) illustrating (3.4) is the following: Consider the multiplicative group G given by

$$\left\{ \begin{vmatrix} 3^k & 3^{k+1} & 0 \\ 2(3^{k-1}) & 2(3^k) & 0 \\ 0 & 0 & 2^{k+1} \end{vmatrix} \quad : k \text{ is an integer} \right\}$$

Note that $k = 1$ gives the identity whose basis is $\{C_1, C_2\}$, where $C_1 = \{1, 2\}$ and $C_2 = \{3\}$ and $p(x)$ is the identity permutation on $\{1, 2\}$ for all x in G.

(iv) Suppose that G is compact (with respect to any topology where convergence implies pointwise convergence) and suppose that the basis of its identity has a finite number of C-classes. Then G is finite.

(Let us remark that this result in the finite dimensional case is due to Denny Brown [**4**]. Our proof here is different, elementary, and works for finite as well as infinite dimensional matrices.)

To prove (iv), let $x \in G$ such that $p(x) = p(e)$. Let $x \neq e$. Then for each C-class C_k, there exists $a(k)$ positive such that

$$x|_{C_k \times C_k} = a(k) \cdot e|_{C_k \times C_k}.$$

If $a(k) > 1$, then for i in C_k, $(x^m)_{ii} \geq [a(k)]^m \cdot e_{ii}$, which goes to infinity as m tends to infinity since e_{ii} is positive. This contradicts compactness of G. Thus, $a(k) \leq 1$. Similarly, since the $C_k \times C_k$ block of x^{-1} is $[1/a(k)]$ times that of e, it follows that $a(k)$ has to be exactly 1. This means that the map p in this case must be an isomorphism since $p(x) = p(x') => p(x^{-1}x') = p(x')op(x^{-1}) = p(x')o[p(x)]^{-1}$ = the identity permutation on E_o, so that $x^{-1}x' = e$ or $x = x'$.

Thus, p is an isomorphism from G into a group of permutations on E_o. Hence if E_o is finite, G is finite. Now we can state the following theorem.

THEOREM 3.1. *Let G be a group of finite or infinite dimensional nonnegative matrices. Let e be the identity of G with the basis*

$$\{T, C_1, C_2, C_3, \ldots\}$$

Then there is an (anti)-homomorphism $p\colon G \to G'$, where G' is the group of permutations on $E_o = \{1, 2, \ldots, m\}$, m being the number of C-classes in the basis of e, such that the following results hold:

(a) Each x in G has the same rank as that of e, and the same zero rows and zero columns as those of e. Also, it is completely determined by e and x_{ij} where i and j are not in T.

(b) For each x in G, the block $C_k \times C_s$ is a strictly positive block or a block of zeros according as $s = p(x)[k]$ or not.

(c) If G is compact (with a topology at least as strong as that of pointwise convergence) and if there are only a finite number of C-classes in the basis of e, then G must be finite (in fact, a singleton if G consists only of strictly positive matrices). □

It is clear that a multiplicative group of finite dimensional stochastic matrices must be finite since we can topologize it with usual topology to make it relatively compact. We also have that for a group of infinite or finite dimensional stochastic matrices, the map p in Theorem 3.1 is an isomorphism, since for i in C_k and j in $C_{p(x)[k]}$, we have

$$x_{ij} = (xe)_{ij} = \sum_{u \in C_{p(x)[k]}} x_{iu} e_{uj} = e_{jj} \cdot \sum x_{iu} = e_{jj}$$

so that x is completely determined by e and $p(x)$ (look at Theorem 3.1 (a)). Thus, G has to be finite if the number of C-classes in the basis of the identity of G is finite.

4. Completely Simple Semigroups of Nonnegative Matrices

A semigroup S is completely simple if it is simple (that is, has no proper two-sided ideals, i.e., subsets I of S such that $SI \subset I$ and $IS \subset I$) and if it has minimal left ideals (that is, minimal among subsets I such that $SI \subset I$) and minimal right ideals (that is, minimal among subsets I such that $IS \subset I$). Such semigroups are among the most widely studied semigroups (see [7]). They also come up naturally in the context of weak convergence problems as supports of idempotent probability measures on locally compact or complete metric topological semigroups.

Suppose that S is a completely simple semigroup. Let $x \in S$. Then Sx is a minimal left ideal, xS is a minimal right ideal and xSx is a group. Let e be the identity of xSx. Let X denote the set of idempotents in Se, Y the set of idempotents in eS and let $G = xSx$. Then X is a left zero semigroup, Y is a right zero semigroup and S can be identified with (or is isomorphis to) the semigroup $X \times G \times Y$, where the multiplication is given by $(x, g, y)(x', g', y') = (x, gyx'g', y')$. The isomorphism mapping is $(x, g, y) \to xgy$.

In this section, we discuss the structure of completely simple semigroups of nonnegative matrices. For simplicity we consider only matrices which have no zero rows or zero columns. (With a little more work, the theory can be presented without this assumption.) Also, all matrices are assumed to have the property: $\sup_i \sum_j A_{ij} < \infty$. Let S be a completely simple semigroup of such matrices.

Fix an idempotent e in S. Let the basis of e be $\{C_1, C_2, \dots\}$. Let $x \in S, x \neq e$. Consider the group eSe. Let p be the map of Theorem 3.1 corresponding to this group. Then the $C_k \times C_k$, block of the matrix exe is a strictly positive or a zero block according as $k' = p(exe)[k]$ or not. For $i \in C_k$ and $j \in C_{k'}$,

$$(exe)_{ij} = \sum_{u \in C_k, v \in C_{k'}} e_{iu} x_{uv} e_{vj},$$

which means that if $k' \neq p(exe)[k]$, then the $C_k \times C_{k'}$ block of x must be an all zero block and if $k' = p(exe)[k]$, then there exist u in C_k and v in $C_{k'}$ such that $x_{uv} > 0$. We claim that the $C_k \times C_{p(exe)[k]}$ block of x is a block with all its entries positive.

To prove this, we need to show first that if f is another idempotent element of S, then it must have the same basis as that of e. Consider C', a C-class in the basis of $f = f^2$, $f \neq e$. Let $i \in C'$. Then $i \in C_h$, where C_h is a C-class in the basis of e. As we have seen above, the $C_h \times C_t$ block of f is an all zero block if $t \neq p(efe)[h]$. For j in C', $f_{ij} > 0$. Therefore, $C' \subset C_{p(efe)[h]}$. Thus, each C-class of f is contained in a C-class of e. Reversing the roles of e and f, it is clear that e and f must have the same basis. Now since S is isomorphic to the product $X \times G \times Y$ as mentioned before with the mapping $(u, g, v) \to ugv$, it is clear that x can be written as $x = e'x'e''$, where e' and e'' are idempotent elements in X and Y respectively and x' is an element of the group eSe. Write: $p(x')[k] = r$. Then we know that the $C_k \times C_k$ block of e', the $C_k \times C_r$ block of x' and the $C_r \times C_r$ block of e'' are each strictly positive blocks. This means that the $C_k \times C_r$ block of x is a strictly positive block and $x' = exe$. Thus, we can now state the following theorem.

THEOREM 4.1. *Suppose that S is a completely simple semigroup (under usual matrix multiplication) of d by d (where d is either finite or d is the cardinality of the integers) nonnegative matrices with no zero rows or zero columns. Then there is a unique partition $\{C_1, C_2, \dots\}$ of the set $\{1, 2, \dots, d\}$ such that the following statements hold:*

(i) *For x in S, $\operatorname{rank}(x|_{C_k \times E}) = 1$ for each C_k.*

(ii) *There is a map p from S into the set of permutations on the set $E_o = \{1, 2, \dots, m\}$, where m is the number (possible, infinity) of C-classes above such that (a) it is a homomorphism in the sense $p(xy) = p(y) o p(x)$ and $p(e)$ is the identity permutation whenever e is an idempotent in S, and (b) for each x in S, $x_{ij} > 0$ or $= 0$ according as $j \in C_{p(x)[k]}$ or $j \notin C_{p(x)[k]}$, whenever $i \in C_k$.*

(iii) *With usual topology (from R^{d^2}) and d finite, S is relatively compact iff its group factor G (which is described earlier and can be taken as eSe, e being an idempotent in S) is relatively compact with relative topology from S.*

PROOF. The proof of (i) and (ii) has been given already. To prove (iii), let x be in S. (Here S is allowed to have nonnegative matrices with zero rows or zero columns. We only need for (iii) the condition that S has at least one matrix with no zero rows or columns.) Suppose x has no zero rows or columns. Then xSx is a group of nonnegative matrices none of which has a zero row or zero column. Let e be the identity of xSx. Consider X and Y as described earlier. Then X, the set of idempotents in Se, is a left zero semigroup so that for any z in $X, ez = e$. Let $1 \leq i, j \leq d$. Since e has no zero column, there exists k such that $e_{ki} > 0$. Since $ez = e$, $e_{ki} z_{ij} \leq e_{kj}$ or for all z in X, z_{ij} is bounded above. (So far we have not needed d to be finite.) Since X is a closed subset of S, X is compact when d if finite, and in this case, it follows similarly that Y, the set of idempotents in eS, is compact. Since the mapping $(x, g, y) \to xgy$ from $X \times G \times Y$ (with produce topology and relative topologies for X, $G = eSe$ and Y from S) to S is a continuous bijection with inverse map $s \to (s(ese)^{-1}, ese, (ese)^{-1}s)$, which is also continuous with G being a topological group, assertion (iii) follows easily.

COROLLARY 4.2. *A completely simple multiplicative semigroup of $d \times d$ (where d is finite or infinite) bistochastic matrices (that is, nonnegative matrices with each row sum as well as each column sum one) is a group, and when d is finite, it is a finite group.*

PROOF. Note that for an infinite dimensional idempotent bistochastic matrix each C-class in its basis must be finite since the rows of the matrix restricted to this class are identical with all positive entries. This means that corresponding to a partition $\{C_1, C_2, \dots\}$ of $\{1, 2, \dots, d\}$, where each C_i is finite, there is exactly one bistochastic idempotent matrix with this partition as its basis. It follows from Theorem 4.1 that a completely semigroup of bistochastic matrices has exactly one idempotent element and consequently, it is a group. When d is finite, with usual topology, the group is compact and by Theorem 3.1 (c), it is finite. □

It is well known that a compact (Hausdorff) topological semigroup has a smallest two-sided ideal, called its kernel, which is the nonempty intersection of all such ideals, closed and completely simple. (See **[40]**.) It is also known that if a multiplicative semigroup S of finite dimensional matrices (with real entries) has a smallest two-sided ideal (that is, a kernel) which is completely simple, then this kernel must be $m(S)$, the set of all matrices in S with minimal rank. (See **[6]**.) Thus, the kernel of all $d \times d$, d finite, stochastic matrices is the set of all $d \times d$ stochastic matrices with identical rows. Let us now show how some of these ideas can be extended to the infinite dimensional case.

Consider a multiplicative semigroup S of infinite dimensional real matrices with state space E = the set of positive integers such that $\sup_i \sum_{j \in E} |x_{ij}| < \infty$, for each x in S. Each x in S can be regarded as a linear operator on l_∞ defined by $(T_x(v))_i = \sum_{j \in E} x_{ij} v_j$. Let $R(x)$ denote the range of T_x. We can now present the following theorem.

THEOREM 4.3. *Suppose that S has a (nonempty) kernel K which is completely simple. Then, $K = \{x \in S\colon R(xy) = R(x)$ for every y in $S\}$. If S has also the property that it contains an element y such that dimension $R(y)$ is finite and minimal (among all the elements of S), then $K = \{z \in S\colon \dim R(z) = \dim R(y)\}$.*

PROOF. Let K be the completely simple kernel of S. Let us write: $I = \{x \in S\colon R(xz) = R(x)$ for each z in $S\}$. Let us first prove that I is nonempty and I contains K. Let $x \in K$ and $y \in S$. Then there exists an idempotent e in K such that $x = xe$. This means that $xy = (xe)y = x(ey) \in xK$. It follows from the structure theory of completely simple semigroups that there exists w in K such that $xyw = x$ so that $R(x) \subset R(xy)$. Since $R(xy) \subset R(x)$ trivially, $x \in I$. Thus, $K \subset I$. To prove the converse inclusion, let $w \in I$.

Let $w' \in K$. Then $ww' \in K$. Since K, being completely simple, is a union of groups, there exist an idempotent f in K and an element w'' in K such that $f = (ww')w''$. This implies that

$$R(f) = R(w(w'w'')) = R(w), \qquad \text{since } w \in I.$$

Thus, for any v in l_∞, there exists v' in l_∞ such that $fv' = wv$ or $fwv = ffv' = fv' = wv$. This means that $w = fw$ which belongs to K. □

We will now close this section with a few examples. (It may be relevant to point out that in Example 4.1 below, if S denotes the set of all 1-1, but not onto maps from the countable set V into itself, then S is a semigroup (with respect to composition of maps) with no completely simple kernel, since for this S, the set $\{x \in S\colon R(xy) = R(x)$ for every y in $S\}$ is empty. The reason is: for x in S and y in S, let w be in the complement of $R(y)$; then, if $v = x(w), v \notin R(xy)$, while $v \in R(x)$.)

EXAMPLE 4.1. Consider V^V, the set of all mappings from a countable discrete set V into itself. Then they form a semigroup under composition of maps. Note that each f in V^V corresponds to a unique infinite dimensional stochastic matrix A_f with state space V and all its entries 0 or 1 given by: $(A_f)_{ij} = 1$ iff $f(i) = j$ and $= 0$ otherwise. Note that for f and g in V^V, $A_{fg} = A_g A_f$. Suppose that S is a multiplicative semigroup of such infinite stochastic matrices such that S contains an element y such that the number of nonzero columns of y is finite and minimal (among all elements of S). Then S has a kernel which is completely simple and it is given by $\{z \in S\colon$ the number of nonzero columns of $z =$ the number of nonzero columns of $y\}$. (For its proof, see [**12**].)

EXAMPLE 4.2. Consider a multiplicative semigroup of 3×3 bistochastic matrices. Then since the usual topology makes it compact (assuming it is closed in R^{3^2}), it is clear that it has a kernel which is a finite group. The kernel K consists of those matrices in the semigroup with the minimal rank. It this rank is one, then K is the singleton each of whose entry is $1/3$. If this rank is two, then again K must be a singleton, this time it has to be one of the following three idempotents:

$$\begin{bmatrix} \frac12 & \frac12 & 0 \\ \frac12 & \frac12 & 0 \\ 0 & 0 & 1 \end{bmatrix}, \qquad \begin{bmatrix} \frac12 & 0 & \frac12 \\ 0 & 1 & 0 \\ \frac12 & 0 & \frac12 \end{bmatrix}, \qquad \begin{bmatrix} 1 & 0 & 0 \\ 0 & \frac12 & \frac12 \\ 0 & \frac12 & \frac12 \end{bmatrix}.$$

If the above rank is three, then K must be all of S.

EXAMPLE 4.3. Consider $S = \{e_0, e_1, e_2, e_3\}$, where

$$e_0 = \begin{bmatrix} u \\ u \\ u \end{bmatrix}, \; e_1 = \begin{bmatrix} v \\ v \\ v \end{bmatrix}, \; e_2 = \begin{bmatrix} \frac12 w \\ w \\ \frac14 w \end{bmatrix}, \; e_3 = \begin{bmatrix} \frac12 t \\ t \\ \frac14 t \end{bmatrix},$$

where $u = (1/2\ 1/6\ 1/3), v = (3/4\ 1/12\ 1/6), w = (1\ 1/3\ 2/3)$ and $t = (3/2\ 1/6\ 1/3)$. Then S is a completely simple semigroup where each of its element has rank one and is an idempotent.

5. Convergence in Distribution of Products of Random Matrices

As usual, all matrices in this section are assumed to satisfy the condition

(5.1) $$\|A\| = \sup \sum_j |A_{ij}| < \infty.$$

Let (A_i) be a sequence of independent and identically distributed (iid, for short) $d \times d, 2 \leq d \leq \infty$, random matrices with real or nonnegative entries. We consider, in this section, questions concerning the limiting distribution of the products $M_n = A_1 A_2 \cdots A_n$, as n tends to infinity. [Note that the backward product $A_n A_{n-1} \cdots A_1$ has the same distribution as M_n.] Such problems have been studied, among others, in [3,12,28,29,35,45,50]. Weak convergence in PR(S), the space of regular probability measures on the Borel sets of S, a locally compact Hausdorff (or complete metric) second countable topological space, has the usual meaning here, namely that $Q_n \to Q$ weakly in PR(S) iff $\int f(x)Q_n(dx) \to \int f(x)Q(dx)$ for each bounded continuous function f on S. If S is further a topological semigroup (that is, has an associative jointly continuous multiplication), then it is well known that PR(S) is a topological semigroup with respect to weak topology and convolution $(\star)$ operation as its multiplication defined by

$$Q_1 \star Q_2(B) = \int Q_1(Bx^{-1})Q_2(dx) = \int Q_2(x^{-1}B)Q_1(dx),$$

where

$$Bx^{-1} = \{y \in S \colon yx \in B\} \quad \text{and} \quad x^{-1}B = \{y \in S \colon xy \in B\}.$$

In this section, we are mainly concerned with the problem of weak convergence of the sequence of probability measures $P^n = P \star P \star \cdots \star P$($n$ times), which is the distribution of M_n above when A_i has distribution P and S is a complete metric topological semigroup of $d \times d$ matrices, the metric being induced by the norm in (5.1) and the semigroup operation being the usual matrix multiplication. (It can be verified easily that any semigroup of $d \times d$ matrices, when given the metric from the norm above, becomes a complete metric separable topological semigroup. For a special S and d infinite, a weaker topology is often used.)

There are many contexts where one comes across problems of this type. Let us consider a simple example.

EXAMPLE 5.1. Suppose that V is a countable set with discrete topology. Note that the set of mappings from V into itself, denoted by $S = V^V$, with product topology can be identified with infinite dimensional 0-1 stochastic matrices with pointwise convergence topology with the following correspondence: f corresponds to A_f, where $(A_f)_{ij} = 1$ if $f(i) = j$, $= 0$ otherwise.

Note that $A_f A_g = A_{gof}$. Let $X = \{0,1\}^V$. Then S is a complete separable metric topological semigroup with respect to usual composition of mappings or, if viewed as a semigroup of matrices, then usual matrix multiplication; moreover, S acts on X on the right by $(V, f) \to V o f$. For Q' in PR(X) and Q in PR(S), define:

$$Q' \star Q(B) = \int Q'\{v : v \circ f \in B\} Q(df).$$

Then it follows that

(1) $Q' \star (Q_1 \star Q_2) = (Q' \star Q_1) \star Q_2$;

(2) $Q_n \to Q$ weakly $\Rightarrow Q' \star Q_n \to Q' \star Q$ weakly.

Note that for v in X and f in S, $(v \circ f)(i) = (A_f \cdot v)(i)$. Let $T_1, T_2, \ldots$ be a sequence of iid with values in S and let w be a random variable with values in X and independent of the T_i's. Then the problem of studying the limiting (in the weak sense) distribution of the products $wT_1T_2 \ldots T_n$ is really a problem of weak convergence of the distributions of the random matrix products $A_{T_n} A_{T_{n-1}} \cdots A_{T_1} \cdot w$.

[One can give an interesting voter interpretation of this problem. Think of V as a collection of individuals, each of which has one of two possible positions denoted by 0 or 1 on a political issue. At each unit of time, an individual reassesses his position by choosing another individual at random according to certain probability distribution and then adopting his position. This model, in continuous time, has been studied in depth in [**37**]. The discrete time model has been studied in [**13**].] We will come back to this example later. □

The following result is fundamental in the context of weak convergence in topological semigroups. (Note that it is motivated by the corresponding result for discrete semigroups given in §2.)

THEOREM 5.1. *Let S be a locally compact Hausdorff or complete metric second countable topological semigroup and $Q \in$ PR(S) such that $Q = Q \star Q$. Then S_Q, the support of Q, defined by $S_Q = \{x \in S : Q(V_x) > 0$ for any open set V_x containing $x\}$, is a closed completely simple subsemigroup of S, with a compact group factor.* □

For a proof, see [**40**]. It follows from Theorem 5.1 and Theorem 4.1 that when S is the semigroup of finite dimensional nonnegative matrices with usual topology, then the support of an idempotent probability measure on S is a compact completely simple subsemigroup if it contains at least one matrix with no zero rows or columns. Also note that the weak limits of a convolution sequence Q^n, when it exists, is always idempotent.

To discuss weak convergence results in matrices, we need the following basic result in a slightly different form in [**41**, Theorem 4.3].

THEOREM 5.2. *Let S be a locally compact Hausdorff or complete metric second countable topological semigroup. Suppose that $Q \in$ PR(S) and S is the* *smallest closed subsemigroup containing* S_Q. *Suppose also that the convolution sequence* (Q^n) *is tight (that is, for each positive integer m, there is a compact set* K_m *such that for all n,* $Q^n(K_m) > 1 - (1/m)$). *Then the weak closure of* $\{Q^n : n \geq 1\}$ *is weakly compact in* PR(S). *Let G be the set of all weak cluster points of* (Q^n) *and let* cl(supp G) *be the closure of the union of supports of all measures in G. Then the following results hold:*

(i) *The sequence* $(1/n)\sum_{k=1}^{n} Q^k$ *converges weakly to some* Q_o *in* PR(S). Q_o *is idempotent,* $Q \star Q_o = Q_o \star Q = Q_o$ *and* S_{Q_o} *is the closed completely simple kernel of S. Moreover,* $S_{Q_o} = \text{cl}(\text{supp}\, G)$.

(ii) *G is a group with respect to convolution.*

(iii) *Let* Q' *be the identity of G. Then in the usual product representation of completely simple semigroups,* S_Q, *as a subsemigroup of* S_{Q_o} *can be expressed as* $S_{Q'} = X \times H_o \times Y$, $S_{Q_o} = X \times H \times Y$, *where* $YX \subset H_o$, H_o *a normal subgroup of the group H,* $H = eS_{Q_o}e$, $X =$ *the set of idempotents in* $S_{Q_o}e$, $Y =$ *the set of idempotents in* $eS_{Q'_o}$, *e being an idempotent element in* S_{Q_o}.

(iv) *If* Q'' *is an element of G different from* Q', *then* $S_{Q''}$ *is of the form* $X \times gH_o \times Y$ *for some g in* $H - H_o$.

(v) *The sequence* Q^n *converges weakly to* Q_o *iff there does not exist any normal subgroup* H_o *of H such that* $eS_Qe \subset gH_o$ *for some g in* $H - H_o$, *e being as in* (iii).

Let us indicate briefly an idea of the proof.

PROOF. By the tightness assumption, the weak closure of $\{Q^n : n \geq 1\}$ is weakly compact. (See [**49**].) Since $(1/n)\sum_{k=1}^{n} Q^k$ and $(1/n)\sum_{k=1}^{n} Q^{k+1}$ have the same weak cluster points, the cluster points must satisfy $Q \star T = T \star Q = Q$. It follows easily that any two weak cluster points must be identical. Thus, most of (i) follows from Theorem 5.1. Also, it follows from $Q^n \star Q_o = Q_o \star Q^n = Q_o$ that S_{Q_o} is an ideal of S and for any open set W containing S_{Q_o}, $\lim_n Q^n(W) = 1$. This then implies that $\text{cl}(\text{supp}\, G) \subset S_{Q_o}$. Since cl(supp G) is an ideal of S and S_{Q_o} is simple, it follows that $\text{cl}(\text{supp}\, G) = S_{Q_o}$.

To prove (ii), we observe that the weak closure of $\{Q^n : n \geq 1\}$ is a compact abelian topological semigroup (with respect to convolution as multiplication and weak topology) and its kernel is G. Thus, G, being an abelian completely simple semigroup, is a group. Using the group property, (iii) and (iv) can be established. To prove (v), notice that Q^n converges weakly iff G is a singleton, that is, iff $Q \star Q' = Q', Q'$ being the identity of G. Thus, if Q^n does not converge weakly, then $Q \star Q' \neq Q'$ so that there exists a normal subgroup H_o such that for the identity e in $H_o, S_Q \cdot e \subset \text{cl}(S_Q \cdot S_{Q'}) = S_{Q \cdot Q'}$ or $eS_Qe \subset e \cdot S_{Q \cdot Q'} \cdot e = gH_o$ for some g in $H - H_o$, using (iv). Conversely, if there is such a normal subgroup H_o of H such that for some idempotent e in S_{Q_o}, $eS_Qe \subset gH_o, g \in H - H_o$, then for $n \geq 1$, $e \cdot S_Q^n \cdot e \subset g^nH_o$ and $eSe = H = \text{cl}(\cup_{n=1}^{\infty} g^nH_o)$. This means that there are two disjoint cosets g_1H_o and g_2H_o such that using compactness of H, one can find open sets W_1 and W_2 with disjoint closures such that $g_1H_o \subset W_1$, $g_2H_o \subset W_2$ and each of

W_1 and W_2 contains infinitely many of the sets $eS_Q^n e, n \geq 1$. This means that $d_e \star Q^n \star d_e$, where d_e is the unit mass at e, cannot converge weakly so that Q^n cannot converge weakly. □

Let us now see what all these mean when S is a multiplicative semigroup of nonnegative matrices. In what follows, J will always represent matrices with at least one zero row or one zero column or both.

THEOREM 5.3. *Let S be the closed (multiplicative) semigroup generated by the support of a probability measure Q on $d \times d$ (d finite) nonnegative matrices with usual topology. Then the convolution sequence $Q_n = (1/n) \sum_{k=1}^{n} Q^k$ converges weakly to Q_o in* PR(S) *and $S_{Q_o} \cap J^c$ is nonempty iff S is compact and $m(S) \cap J^c$ is nonempty. (Recall that $m(S)$ is the set of all matrices in S with minimal rank.) If S contains the zero matrix, then Q^n converges weakly to the unit mass at the zero matrix iff (Q^n) is tight.*

PROOF. The last assertion follows easily from Theorem 5.2 Suppose that Q_n converges weakly to Q_o. Then $Q \star Q_o = Q_o \star Q = Q_o = Q_o \star Q_o$ and S_{Q_o} is the closed completely simple kernel of S. Then, by **[6]**, $S_{Q_o} = m(S)$. Let x be in $S_{Q_o} \cap J^c$. Since S_{Q_o} has a compact group factor (see Theorem 5.1), $xS_{Q_o}x$ is a compact group of nonnegative matrices. Since $x \in J^c \cap xS_{Q_o}x, xS_{Q_o}x \subset J^c$. Let e be the identity of $xS_{Q_o}x$, which is then the same as $eS_{Q_o}e$ and this set is finite by Theorem 3.1. Since $e \in J^c, e$ has a strictly positive diagonal. Since $eSe = eS_{Q_o}e$ is finite, there is a positive number m such that for $1 \leq i, j \leq d$, and for any s in S, we have: $e_{ii}s_{ij}e_{jj} \leq (ese)_{ij} \leq m$. This means that S is compact. The rest of the theorem follows from Theorem 5.2. □

A natural companion to the above theorem is the following theorem. (This theorem was given in **[45]**. The conditions (i) and (ii) along with parts of this theorem were first considered in **[35]**.)

THEOREM 5.4. *Let Q and S be as in the previous theorem. Suppose also that*

(i) *$Q(J) = 0$;*
(ii) *$Q^m(I) > 0$ for some positive integer m,*

where I is the set of all strictly positive matrices. Then the following statements are equivalent:

(a) *(Q^n) is tight and $0 \notin S$;*
(b) *Q^n converges weakly to Q_o in* PR(S) *and $Q_o(I \cap S) = 1$;*
(c) *S is compact and $0 \notin S$;*
(d) *$m(S)$ is bounded and $m(S) \neq \{0\}$.*

PROOF. The proof of the equivalence of (a), (b) and (c) follows from Theorems 5.2 and 5.3. To prove that $(d) \Rightarrow (c)$, notice that $S \cap I \neq \emptyset$, by (ii). If $m(S)$ is bounded, then it, being closed, is compact. This means that $m(S)$ must be the closed completely simple kernel of S. If $m(S) \neq \{0\}$, then $0 \notin S$, and since $(I \cap S)m(S)(I \cap S) \subset I \cap m(S)$, it follows that $I \cap m(S) \neq \emptyset$.

Let x be in $I \cap m(S)$. Then $xm(S)x$ is a group. Let e be the identity of the group. Then, $eSe = em(S)e = xm(S)x$ which is a compact group of strictly positive matrices, and therefore, by Theorem 3.1(c), a singleton. As in Theorem 5.3, it follows that S is compact. □

Let us now see what happens in the infinite dimensional situation. The results here are not as complete as we would like. Let us present first two results in this direction.

THEOREM 5.5. *Let Q be a probability measure on the Borel sets of S', the complete metric multiplicative semigroup of infinite dimensional real matrices where the topology is induced by the norm $\|A\| = \sup_i \sum_{j=1}^{\infty} |A_{ij}|$. Let $S = \mathrm{cl}(\cup_{n=1}^{\infty} S_Q^n)$, the closed semigroup generated by the support of Q. Suppose that (Q^n) is tight. Then, the sequence $(1/n)\sum_{k=1}^{n} Q^k$ converges weakly to Q_o in $\mathrm{PR}(S)$ such that $S_{Q_o} = \{A \in S\colon R(A) = R(AB)$ for all B in $S\}$. If S contains a matrix B such that $\dim R(B)$ is finite, then S_{Q_o} consists of all those matrices in S whose ranges have minimal finite dimension. If in this case, $S_{Q_o} \cap J^c \neq \varnothing$ and the matrices in S are nonnegative, then for each i, j, there exists a positive number m_{ij} such that for all A in S, $A_{ij} \leq m_{ij}$.* □

The proof of this theorem can be given easily using Theorem 4.3 and Theorem 5.2. The next theorem also follows easily from Theorems 5.1 and 5.2 and Corollary 4.2.

THEOREM 5.6. *Let the semigroup S' in the previous theorem consist of finite or infinite dimensional bistochastic matrices with the topology induced by the norm. Let Q and S be as in the previous theorem. Suppose that (Q^n) is tight. Then (i) S has a kernel K which is a compact group and the probability measure Q_o, the weak limit of $(1/n)\sum_{k=1}^{n} Q^k$, is the normed Haar measure on K. In case of finite dimensional matrices, K is a finite group and Q_o is the uniform distribution on K. (ii) the sequence Q^n converges weakly to Q_o when the support of Q contains the identity of K.* □

It is clear that the main question in the context of weak convergence of (Q^n) is first whether the sequence (Q^n) is tight. So far we are not aware of any result that settles this problem completely even for finite dimensional nonnegative matrices via an easily verifiable condition. The question of tightness has been studied specifically in **[3]**, **[35]**, **[45]**. For the special case of 2×2 nonnegative matrices, this question has been somewhat reasonably settled in **[35]**. On questions of tightness, let us first present this result in a slightly more general form (with fewer assumptions) and with a different proof (taken from **[45]**) that also allows the result to be generalized to matrices with real entries. Kesten and Spitzer's theorem **[35]** is essentially the following.

THEOREM 5.7. *Let Q be a probability measure on the (multiplicative) semigroup of 2×2 nonnegative matrices with usual topology. Let S be the closed subsemigroup generated by S_Q. Suppose that $Q(J) = 0$ and for some positive integer m, $Q^m(I) > 0$, where I is the set of all strictly positive matrices. Then,* *the sequence Q^n converges weakly to Q_o in* PR(S) *and* $Q_o(I \cap S) = 1$ *iff there exists a strictly positive vector* $v = (v_1, v_2), v_1 > 0, v_2 > 0$, *such that either* $vS_Q = v$ *or* $S_Q v^T = v^T$.

PROOF. If $vS_Q = v$ or $S_Q v^T = v^T$ for some strictly positive v, then $vS = v$ or $Sv^T = v^T$ so that S is compact and does not contain the zero matrix. Then the "if" part of the theorem follows from Theorem 5.4.

For the "only if" part, let Q^n converge weakly to Q_o. Then Q_o is idempotent and $Q \star Q_o = Q_o \star Q = Q_o$ so that S_{Q_o} is the completely simple kernel K of S, with a compact group factor (by Theorem 5.1). By Theorem 5.4, $0 \notin S$. By assumption, $I \cap S_{Q_o}$ is nonempty. Let $x \in I \cap S_{Q_o}$. Then, $xSx \subset I$. If e is the identity of the group $xS_{Q_o}x$, then $eS_{Q_o}e (= xS_{Q_o}x = xSx)$, being a compact group of strictly positive matrices, must be a singleton. Notice that every element of S_{Q_o} is then an idempotent since S_Q is a completely simple semigroup whose group factor is a singleton. Let us now write:

$$e = \begin{vmatrix} a_1 e_1 & a_1 e_2 \\ a_2 e_1 & a_2 e_2 \end{vmatrix}, \; a_1 > 0, \; a_2 > 0, \; e_1 > 0, \; e_2 > 0, \; a_1 e_1 + a_2 e_2 = 1;$$

let x be in Se and y be in eS such that

$$x = \begin{bmatrix} b_1 x_1 & b_1 x_2 \\ b_2 x_1 & b_2 x_2 \end{bmatrix}, \quad y = \begin{bmatrix} c_1 y_1 & c_1 y_2 \\ c_2 y_1 & c_2 y_2 \end{bmatrix}$$

where (since $yx = e$) $c_1 > 0, c_2 > 0, x_1 > 0, x_2 > 0$, and since $x = x^2$, $y = y^2$, $b_1 x_1 + b_2 x_2 = 1 = c_1 y_1 + c_2 y_2$.

Then the following results follow by looking into the matrix equation $yx = e$:

(i) If $b_1 = 0$ for some x in Se, then $yx = e \Rightarrow y = e$.
(ii) If $y_2 = 0$ for some y in eS, then $yx = e \Rightarrow x = e$
(iii) Suppose that $b_1 > 0$ for each x in Se and $y_2 > 0$ for each y in eS.

Then $yx = e$ and $x \neq e \Rightarrow y = e$. Thus, in every case, either $eS = e$ or $Se = e$. □

An analogous result with an exactly similar proof holds for matrices (2×2) with real entries. See **[45]**.

It is clear that an easy, though very restrictive, sufficient condition for the tightness of (Q^n) in the case of finite dimensional $(d \times d, d \geq 2)$ nonnegative matrices is the existence of a strictly positive vector v such that

$$vS_Q = v \text{ or } S_Q v^T = v^T, \qquad v = (v_1, v_2, \ldots, v_d), \tag{5.2}$$

since this implies that the closed semigroup generated by S_Q is compact (under the usual topology). However, this condition is not necessary for $d > 2$ and certainly not sufficient to imply the compactness of S for matrices not necessarily nonnegative. Take, for example, the semigroup of strictly positive matrices in Example 4.3. Define the probability measure Q on this semigroup by $Q(e_i) = 1/4, i = 0, 1, 2, 3$. Since every subgroup of S here is a singleton, by Theorem 5.2, Q^n converges weakly. However, it can be verified that there does not exist a nonzero vector x such that $xS = x$ or $Sx^T = x^T$. However, we show below that condition (5.2) is often sufficient for the weak convergence of Q^n even for $d > 2$.

First, we need to observe the following:

(1) If x is a finite or infinite dimensional idempotent real matrix with no zero row such that

$$xy = x \quad \text{and} \quad xz = x, \tag{5.3}$$

where y and z are rank one real matrices, then $yz = y, y^2 = y$ and $z^2 = z$. Similarly, if x is idempotent real with no zero column and $yx = x, zx = x$ where y and z have rank one, then y and z are idempotent and $yz = z$.

We omit the proof. See Lemma 5.1 in **[45]**. The proof there works here too.

(2) A compact (with respect to any topology where convergence implies pointwise convergence) group (multiplicative) of finite or infinite dimensional rank one matrices is either a singleton or has exactly two elements.

To see this, notice that if A is an element in this group different from the identity, then A having rank one must have its ith row A_i of the following form: $A_1 = (b_i x_1, b_i x_2, \ldots)$ where not all the x_i's are zeros. Notice that then we have

$$A^{n+1} = b^n A, \text{ where } b = \sum_{i=1}^{\infty} b_i x_i \neq 0.$$

This means that b is either 1 or -1, since the group is compact. Then since A is not the identity, $A^2 = -A$. Then A must be $-$(the identity).

Now we can state and prove the following theorem.

THEOREM 5.8. *Let S' be a semigroup (under multiplication) of finite or infinite dimensional real matrices with the metric topology induced by the usual norm. Let Q be a probability measure and S be the closed subsemigroup generated by S_Q, the support of Q. Suppose that the following two conditions hold:*

(i) *there exists a nonzero vector v in l_1 such that*

$$vS_Q = v \quad \text{or} \quad S_Q v^T = v^T;$$

(ii) *$Q^m(I') > 0$ for some positive integer m, where I' is the set of all rank one matrices in S.*

Then there is a Q_o in PR(S)*, $S_{Q_o} \subset I'$ such that Q^n converges to Q_o weakly as n tends to infinity.*

PROOF. Suppose that there exists $v \neq 0$ such that $v \in l_1$ and $vS_Q = v$. Then, $vS = v$. It is clear then that there is an idempotent real matrix x (whose first row is v) such that $xS = x$. Hence, whenever y and z are rank one matrices in S, we must have by our earlier observation $yz = y$. This means that for each y in I', $My = M$ for any subset M of I'. Note that

$vS = v$ and v nonzero implies that 0 is not in S and consequently, I' is an ideal of S. It is not difficult to see then that condition (ii) implies that $Q^n(I')$ converges to 1 as n tends to infinity. (See p. 38 in **[40]**.) Now notice that for any subset M of I' and any positive integer N, we have:

$$Q^{n+N}(M) \geq \int_{I'} Q^N(Mz^{-1})Q^n(dz) \geq Q^N(M)Q^n(I')$$

and this implies that the sequence (Q^n) is tight. By Theorem 5.2, the sequence $(1/n)\sum_{k=1}^n Q^k$ converges weakly to some Q_o, where S_{Q_o} is a completely simple semigroup (actually the kernel of S) with a compact group factor. Since I' is a closed subset of S and since $Q^n(I')$ converges to 1 as $n \to \infty$, it is clear that $S_{Q_o} \subset I'$ and therefore, the group factor of S_{Q_o} is either a singleton or of the form $\{e, -e\}$. Since $xS = x$, this group factor must be a singleton. It follows from Theorem 5.2 that Q^n converges weakly to Q_o. □

Let us mention that in the finite dimensional case, the condition (ii) in Theorem 5.8 can be replaced by the weaker condition (but not very easy to check): $\lim_{n\to\infty} Q^n(W) = 1$ for every open set W containing I'.

Let us now illustrate the results by some examples.

EXAMPLE 5.2. Consider the semigroup S of 2×2 nonnegative matrices with discrete topology given by

$$S = \left\{ \begin{vmatrix} 1 & m \\ 0 & n \end{vmatrix} : m, n \text{ are nonnegative integers} \right\}.$$

Notice that S has a completely simple kernel given by

$$K = \left\{ \begin{vmatrix} 1 & m \\ 0 & 0 \end{vmatrix} : m \text{ is a nonnegative integer} \right\}.$$

If $v = (1\ 0)$, then $Sv^T = v^T \cdot$ (K is a right zero semigroup.) Define a probability measure Q on S such that

$$S_Q \supset \left\{ \begin{bmatrix} 1 & 1 \\ 0 & 1 \end{bmatrix}, \begin{bmatrix} 1 & 1 \\ 0 & 0 \end{bmatrix}, \begin{bmatrix} 1 & 0 \\ 0 & 0 \end{bmatrix} \right\}$$

Note that the semigroup generated by S_Q contains K. By Theorem 5.8, the sequence Q^n converges weakly to some Q_o and $S_{Q_o} \subset K$. The point of this example is that neither S_{Q_o} nor S is compact (unlike what we saw in Theorem 5.3).

EXAMPLE 5.3. Let $d_o = 1$ and for each $i \geq 1, d_i > 0$ such that $\sum_{i=0}^{\infty} 2^{-i-1} d_i = c$ is finite. Consider the matrices A and B such that $A_{ij} = 2^{-j}c^{-1}d_{i-1}, i, j \geq 1$, and

$$B = \begin{vmatrix} z_1 & w_1 & & & \\ 2(1-z_1) & 1-2w_1 & & & 0's \\ & & 1 & & \\ & 0's & & 1 & \\ & & & & 1 \end{vmatrix}$$

where $0 < z_1 < 1$, $0 < w_1 < 1/2$. Define the probability measure Q such that $Q(A) = a$ and $Q(B) = 1 - a$, where $0 < a < 1$. Notice that for

$$v = (1/2, 1/2^2, 1/2^3, \dots), vS_Q = v.$$

By Theorem 5.8, Q^n converges weakly to a probability measure Q_o, whose support is contained in the set of all rank one matrices.

EXAMPLE 5.4. Suppose that the support S_Q of a probability measure is given by:

$$S_Q = \left\{ \begin{pmatrix} s & rs \\ s'/r & s' \end{pmatrix}, \begin{pmatrix} t & r(1-p) \\ t' & p \end{pmatrix} \right\},$$

where $s' = 1 - s, t' = (1 - t)/r$, and r, s, t, p are any real numbers with r nonzero. Notice that if $v = (1, r)$, then $vS_Q = v$. By Theorem 5.8, the sequence Q^n converges weakly to a probability measure whose support is contained in the set of rank one matrices.

Before we close this section, let us consider Example 5.1. Let Q be the distribution of each random transformation T_i there so that $Q \in \mathrm{PR}(S)$ and let Q' in $\mathrm{PR}(X)$ be the distribution of the random variable w there. Then Q induces a Markov chain on V, called the one point motion on V, whose transition probability is given by $p_{ij} = Q\{f\colon V \to V\colon f(i) = j\}$. Then using Theorem 5.2, we can prove the following theorem given in **[13]**.

THEOREM 5.9. *Suppose that the one point motion on V induced by Q is irreducible and positive recurrent (that means $\lim_{n\to\infty} \sup p_{ij}^n$ is positive for all i, j in V). Suppose also that S_Q, the support of Q, contains an idempotent e in V^V such that for all f in S, the closed subsemigroup of V^V generated by S_Q, $e(V) = e(f(V))$. Then for any Q' in $\mathrm{PR}(X)$, the sequence $Q' \star Q^n$ converges weakly in $\mathrm{PR}(X)$.* □

We omit the proof. However, we point out that irreducibility and positive recurrence of the one point motion will guarantee the tightness of the sequence (Q^n).

6. Nonhomogeneous Markov Chains

In this section we consider convergence of nonhomogeneous stochastic chains with finite or countably many states. Here again we will see that idempotent matrices and their structures play an important role.

Let (P_n) be a sequence of finite or infinite dimensional stochastic matrices such that for each positive integer k, the sequence

$$P_{k,n} = P_{k+1}P_{k+2}\cdots P_n \qquad (n > k)$$

converges to a stochastic matrix Q_k. Such a sequence (P_n) will be called a convergent forward stochastic chain (f. s. c.). If instead of the forward products $P_{k,n}$, we consider the backward products $P_{n,k} = P_nP_{n-1}\cdots P_{k+1}$ ($n > k$), then we will call (P_n) a convergent backward stochastic chain (b.s.c.).

Let us present our first theorem for general nonnegative matrices.

THEOREM 6.1. *Let (P_n) be a sequence of nonnegative matrices with state space E (the set of positive integers) such that*

(i) *the products $P_{k,n}$ are all well-defined and for all i, j, k and n, $(P_{k,n})_{ij} < M$ (some positive integer);*
(ii) *there exists a subsequence (n_t) of E such that for each k in E, P_{k,n_t} converges to a nonnegative matrix Q_k as t tends to infinity pointwise and for every $i \geq 1, \sum_{j=1}^{\infty}(Q_k)_{ij} < \infty$ and*

$$\lim_{t\to\infty} \sum_{j=1}^{\infty} |(P_{k,n_t})_{ij} - (Q_k)_{ij}| = 0.$$

Let Q' be a pointwise limit of the Q_{n_t}'s and let $T \subset E$ be the set of its zero columns, that is, $T = \{j: Q'_{ij} = 0$ for all $i\}$. Suppose that for some k, Q_k is not the zero matrix. Then $E - T$ is nonempty, and Q', restricted to $E - T$, is a nonnegative idempotent matrix. If Q'' is another pointwise limit of the Q_{n_t}'s, then Q'' has the same set of zero columns as Q', Q'' is also idempotent when restricted to $E - T$, and $Q''Q' = Q''$. □

This theorem, given in **[22]**, leads to the following interesting result.

THEOREM 6.2. *Consider a convergent f.s.c.(P_n). Let Q' and Q'' be any two pointwise limits of the Q_k's. Then Q' and Q'' have the same set T of zero columns; moreover, Q' and Q'', when restricted to T^c, the complement of T, are equal and idempotent.* □

We omit the proof, which uses Theorem 6.1 and Theorem 2.1. This theorem allows us to define the BASIS of a convergent f. s. c.

DEFINITION 6.1. Suppose that (P_n) is a convergent f. s. c. such that for each k in E, $\lim_{n\to\infty} P_{k,n} = Q_k$ (pointwise). Then there exists a unique partition $\{T, C_1, C_2, \dots\}$ of E such that for any (pointwise) limit Q' of the Q_k's, the following holds:

(i) $T = \{j: Q'_{ij} = 0$ for each $i\}$;
(ii) for i, j in different C-classes, $Q'_{ij} = 0$;
(iii) for i, j and k in the same C-class, $Q'_{ij} = Q'_{kj}$.

The partition $\{T, C_1, C_2, \dots\}$ is called the basis of the chain. □

The concept of the basis was first introduced by Maximov **[39]** for convergent bistochastic chains and then for convergent finite-dimensional and infinite dimensional stochastic chains by Mukherjea and Nakassis (see **[42]** and Springer-Verlag Lecture Notes in Math. Vol. 928, 1981, pages 315–336).

The following theorem is one reason why the basis plays an important role in describing the convergence of the chain.

THEOREM 6.3. *Let (P_n) be a finite-dimensional convergent f.s.c. with basis $\{C_1, C_2, \dots, C_p\}$. (Note that the "T" set is assumed here to be empty as, for example, in the case of a bistochastic chain.) Then the following hold:*

(i) *for i, j in different C-classes, $\sum_{n=1}^{\infty}(P_n)_{ij} < \infty$;*

(ii) *for each* C_i, $1 \leq i \leq p$, *the sequence* $P_n(C_i)$ *defined by*

$$(P_n(C_i))_{jk} = (P_n)_{jk} / \left[\sum_{s \in C_i} (P_n)_{js} \right], \qquad j \in C_i, k \in C_i,$$

is strongly ergodic, that is, for each k *in* E, *the sequence* $P_k(C_i)P_{k+1}(C_i) \cdots P_n(C_i)$, *as* n *tends to infinity, converges to a stochastic matrix with identical rows.*

Conversely, if there is a partition $\{C_1, C_2, \ldots, C_p\}$ *of the state space such that conditions* (*i*) *and* (*ii*) *hold for a stochastic chain* (P_n), *then the chain is convergent.*

Also, for a convergent f.s.c. the condition (*i*) *above continues to hold even when the state space is infinite and the "T" set is non-empty.* □

This theorem was first discovered by Maximov for finite-dimensional bi-stochastic chains in **[39]**. For the present extension, see **[42]**, **[46]**.

Let us now make a few remarks about b.s.c. Let (P_n) be a convergent b.s.c. such that $R_k = \lim_{n\to\infty} P_{n,k}$. Note that unlike in the case of forward products, for backwards products (in the infinite-dimensional case), a pointwise limit of the R_k's can be a zero matrix. For instance, let

$$\begin{aligned}(P_n)_{ij} &= 0 \quad \text{for } j = 1, 2, \ldots, n \text{ and } i \geq 1;\\ &= 1/2^k \quad \text{for } j = n + k, i \geq 1.\end{aligned}$$

Then for $n > k, P_{n,k} = P_{k+1}$ and $\lim P_{k+1} = 0$. Suppose that all the (pointwise) limit points of the R_k's are stochastic. Any two such limit points R' and R'' then satisfy $R'R'' = R''$, and therefore, each such limit point is idempotent. If, furthermore, none of these limit points has a zero column, then it can be shown that they all have the same basis. If, however, some of these limit points have zero columns, we can then define the basis of the convergent b.s.c. as follows.

DEFINITION 6.2. A partition $\{S_1, S_2, \ldots\}$ of E is called an S-basis of a convergent b.s.c. if for each k in E, i and j are both in S_k iff the ith and the jth rows of each limit point R' of the R_k's are identical.

It is clear that any C-class in the basis of a limit point R' is completely contained in a S-class and each such S-class can contain one and only one such C-class. However, we can say more. Suppose that the bases of two limit points R' and R'' are given by R': $\{T, C_1, C_2, \cdots\}$ and R'': $\{T', C_1', C_2', \ldots\}$. Then, if $S_i \subset T$ for some i, it must also be contained in T'.

DEFINITION 6.3. Let (P_n) be a convergent b.s.c. such that $R_k = \lim_{n\to\infty} P_{n,k}$ and the limit points (pointwise) of the R_k's are all stochastic. A state i in E is called strongly (resp. weakly) recurrent if $R'_{ii} > 0$ for each (resp. some) limit point R' of the R_k's.

For a convergent b.s.c. with a S-basis $\{S_1, S_2, \ldots\}$, $\sum (P_n)_{ij} < \infty$, whenever i is strongly recurrent, j is weakly recurrent, and i, j are in different S-classes. Also, Theorem 6.3 holds word for word for a convergent b.s.c. where $R =$

$\lim R_k$ exists. This result as well as Theorem 6.3 holds even when the "T" set is nonempty provided we also have the condition

$$\sum_{n=1}^{\infty}(P_n)_{ij} < \infty, \qquad \text{for } i \notin T \text{ and } j \in T.$$

See [**46**].

Let us consider two examples.

EXAMPLE 6.1. Consider

$$P_n = \begin{pmatrix} 0 & a_n & 1-a_n \\ 0 & 1 & 0 \\ a_n & 0 & 1-a_n \end{pmatrix}, \qquad 0 < a_n < 1.$$

Then (P_n) is a convergent b.s.c. If $\sum_{n=1}^{\infty} a_n a_{n+1} < \infty$, then the R_k's converge and the basis of the chain is: $\{T = 1, C_1 = 2, C_2 = 3\}$, while the S-basis of the chain is $\{S_1 = \{1,3\}, S_2 = 2\}$. However, if $\sum_{n=1}^{\infty} a_n a_{n+1} = \infty$, then each R_k has identical rows, each row being (0 1 0). □

EXAMPLE 6.2. Here is an example of a convergent f. s. c. Consider the following sequence of infinite dimensional stochastic matrices (P_n) given by

$$(P_n)_{11} = (P_n)_{22} = 1 - (1/n); \qquad (P_n)_{1,n+2} = (P_n)_{2,n+2} = 1/n;$$
$$(P_n)_{ii} = 1 \ \text{ for } 2 < i \le n; \qquad (P_n)_{i1} = (P_n)_{i2} = 1/2 \ \text{ for } i > n.$$

Then it can be shown that $P_{k,n} \to Q_k$ and $Q_k \to Q$ (pointwise), where

(i) $(Q_k)_{11} = (Q_k)_{12} = (Q_k)_{21} = (Q_k)_{22} = 1/2; (Q_k)_{ii} = 1$ for $2 < i \le k+1; (Q_k)_{i1} = (Q_k)_{i2} = 1/2$ for $i > k+1$.

(ii) $Q_{11} = Q_{12} = Q_{21} = Q_{22} = 1/2; Q_{ii} = 1$ for $i > 2$.

The basis of this convergent chain is:

$$\{C_1 = 1, 2, C_2 = 3, \ldots, C_n = n+1, \ldots\}$$

Note that the "T" set is empty, 1 is in $C_1, \sum_{n=1}^{\infty}(P_n)_{1s} < \infty$ for s not in C_1, but

$$\sum_{n=1}^{\infty}\sum_{s \notin C_1}(P_n)_{1s} = \infty \cdot \qquad (6.1)$$
□

Results like (6.1) do not hold for convergent f. s. c. satisfying the following uniformity condition: $\lim_{k\to\infty} \sup_{i \notin T} |(Q_k)_{ij} - Q'_{ij}| = 0$ for $j \notin T$, where Q' is a limit point of the Q_k's, $Q_k = \lim_{n\to\infty} P_{k,n}$ (pointwise). For a proof, see [**46**].

Let us now discuss briefly what all these mean in the probabilistic sense. It is known, see [**5**], that given any sequence (P_n) of $d \times d$ (d possibly infinite) stochastic matrices and given a probability vector $v = (v_1, \ldots, v_d)$, there exists a non-homogeneous Markov chain $(X_n), n \ge 0$, with state space $\{1, 2, \ldots, d\}$ such that

(i) $\Pr(X_o = i) = v_i, \qquad 1 \le i \le d;$

(ii) $\Pr(X_n = j | X_k = i) = (P_{k,n})_{ij}, \quad$ whenever $k < n$ and $\Pr(X_k = i) > 0$.

In the discussion to follow, we will need the following fact (due originally to Csiszá r, see **[11]**): For any given sequence (P_n) of finite-dimensional stochastic matrices and given any sequence (m_t) of positive integers, there exists a subsequence $(n_t) \subset (m_t)$ such that for each $k \geq 0$,

$$P_{k,n_t} \to Q_k, \; Q_{n_t} \to Q = Q^2 \text{ and } Q_k Q = Q_k.$$

Let $\{T, C_1, \dots, C_p\}$ be the basis of Q here (the tail idempotent of the chain (P_n) corresponding to the sequence (n_t) introduced above). Let (X_n) be the Markov chain corresponding to the chain (P_n). Then the following results (see **[43]**) hold:

(i) If $\Pr(X_k = i) > 0$ and $j \notin T$, then

$$\Pr(X_{n_t} = j \text{ i. o.} \,|\, X_k = i) = [(Q_k)_{ij}]/Q_{jj}.$$

(i.o. means "for infinitely many t")

(ii) For j in C_k, $1 \leq k \leq p$,

$$\Pr(X_{n_t} = j \text{ i. o.}) = \sum_{u \in C_k} [\lim_{t\to\infty} \Pr(X_{n_t} = u)],$$

which is positive iff for all u in $C_k, \lim_{t\to\infty} \Pr(X_{n_t} = u) > 0$.

The following theorem (given in this form in **[43]**, looked into earlier by many probabilists including J. F. C. Kingman and specially H. Cohn) shows how the basis of the tail idempotent Q is related to the tail sigma- field T_∞ of the Markov chain (X_n), defined by $T_\infty = \cap_{n=1}^{\infty} F_n$, where F_n is the sigma-algebra of events generated by $\{X_m \colon m \geq n\}$.

THEOREM 6.4. *Suppose (P_n) is a sequence of $d \times d$ (d finite) stochastic matrices, Q be a tail idempotent (corresponding to (n_t)) as introduced above with basis $\{T, C_1, \dots, C_p\}$. Let T_∞ be the tail sigma-field of the Markov chain (X_n) corresponding to (P_n). Suppose that $p_o = \operatorname{card}(I_o)$, where $I_o = \{m \colon \lim_{t\to\infty} \Pr(X_{n_t} = i) > 0$, whenever i is in $C_m\}$. Then for each m in I_o and j, k in C_m, the sets $\{X_{n_t} = j \text{ i. o.}\}$ and $\{X_{n_t} = k\text{i.o.}\}$ are equivalent, each is non-null and an atom in T_∞; moreover, these atoms generate T_∞.*

Also, for i in T, the event $\{X_{n_t} = i\text{io}\}$, whenever non-null, is also an atom in T_∞, and in this case, is equivalent to $\{X_{n_t} = j \text{ i. o.}\}$ for some j not in T. In fact, for any i and j such that $Q_{ij} > 0$, $\{X_{n_t} = i\text{i.o.}\}$ is contained in $\{X_{n_t} = j \text{ i. o.}\}$ except perhaps for a null subset. □

Let us consider some examples. As usual, a state i is called recurrent if $\Pr(X_n = i \text{ i. o.}) > 0$. (Here, i.o. means "for infinitely many n".) The state i, when not recurrent, is called transient.

EXAMPLE 6.3. Consider the chain (P_n) defined by

$$P_1 = \begin{pmatrix} 1 & 0 & 0 \\ 0 & 1 & 0 \\ 1 & 0 & 0 \end{pmatrix}, \qquad P_n = \begin{pmatrix} 0 & 1 & 0 \\ 0 & 1 & 0 \\ 0 & 0 & 1 \end{pmatrix} \quad \text{for each } n \geq 2.$$

Then notice that for all $n > k \geq 2, P_{k,n} = P_n$, but $P_{0,n}$ is a matrix where each row is (0 1 0). Thus, the chain is a convergent f. s. c. with basis $\{T = 1, C_1 = 2, C_2 = 3\}$. The state 3 (though a C-state in the basis) is a transient state here since for each $n \geq 1$, $\Pr(X_n = 3) = 0$. However, if we start the chain at a later time (by dropping P_1), then 3 will be a recurrent state (with a suitable initial distribution).

EXAMPLE 6.4. The point of this example is that a "T" state can be recurrent. Consider the f. s. c. given by

$$P_n = \begin{pmatrix} 0 & \frac{1}{2} & \frac{1}{2} \\ \frac{1}{n} & 0 & 1 - \frac{1}{n} \\ \frac{1}{n} & 1 - \frac{1}{n} & 0 \end{pmatrix}, \qquad n \geq 1.$$

It can be shown that for each $k \geq 1, P_{k,n}$ converges to a matrix where each row is $(0 \ \frac{1}{2} \ \frac{1}{2})$, so that the chain is strongly ergodic with basis $\{T = 1, C_1 = 2, 3\}$. Let us compute $\Pr(X_n = 1 \text{ i. o.} \mid X_k = 3)$, where (X_n) is the Markov chain corresponding to (P_n). For fixed k, write $(P_{k,n})_{31} = a_n$. Note that for $n > k, a_n > 0$ (though $\lim a_n = 0$). Now we have:

$$\begin{aligned}
&\Pr(X_n = 1 \text{ i. o.} \mid X_k = 3) \\
&= \lim_{n\to\infty} \Pr\left(\bigcup_{m=n}^{\infty} X_m = 1 | X_k = 3\right) \\
&= \lim_{n\to\infty} [\Pr(X_n = 1|X_k = 3) + \Pr(X_{n+1} = 1, X_n \neq 1|X_k = 3) + \cdots] \\
&= \lim_{n\to\infty} \left[a_n + (1 - a_n)(1/(n+1)) + (1 - a_n)\left(1 - \frac{1}{n+1}\right)\frac{1}{n+2} + \cdots\right] \\
&= \lim_{n\to\infty} \left[a_n + (1 - a_n)\left\{1 - \lim_{r\to\infty} \prod_{i=n+1}^{n+r} \left(1 - \frac{1}{i}\right)\right\}\right] = 1. \quad \square
\end{aligned}$$

In the above example, note that for j not in T and i in $T, \sum_{n=1}^{\infty}(P_n)_{ji} = \infty$. However, since $\lim_{k\to\infty} \Pr(X_k = i) = 0$ for each i in T (for any initial distribution) and since then for i in T,

$$\begin{aligned}
&\Pr(X_n = i \text{ i. o.}) \\
&= \lim_{m\to\infty} \sum_{j\notin T} \Pr(X_n = i \text{ i. o.}, X_m = j) \\
&\leq \lim_{m\to\infty} \sum_{j\notin T} \sum_{r=m}^{\infty} \Pr(X_{r+1} \in T, X_r \notin T),
\end{aligned}$$

which means that a "T" state is transient for a finite-dimensional convergent f. s. c.(P_n), whenever $\sum_{n=1}^{\infty}(P_n)_{ji} < \infty$ for $j \notin T$ and i in T. (See **[9]**.)

For a convergent f. s. c. one will expect probabilistically the total flow of probability from the C-states to the T-states to be small compared to that from the T-states to the C-states. Indeed the following result **[48]** holds.

THEOREM 6.5. *Let (P_n) be a finite dimensional convergent f.s.c. with basis $\{T, C_1, \ldots, C_p\}$. Let $L \subset T$. Write: $(P_n)_{AB} = \sum_{i\in A, j\in B}(P_n)_{ij}$. Then for each* *positive integer k, we have:*

$$\lim_{m\to\infty}\left[\sum_{n=k}^{m}(P_n)_{T C_L}\right] \Big/ \left[\sum_{n=k}^{m}(P_n)_{L L C}\right] = 0. \qquad \square$$

Before we close this section, let us mention that readers interested in ergodicity results for nonnegative matrices should look into **[31]**, **[52]**; in this context, the paper of Hajnal **[25]** is important. This work of Hajnal given for finite dimensional nonnegative matrices has been recently extended in **[21]** to the infinite dimensional case. Let us finally mention the following open problem: Given a convergent f. s. c.(P_n), is there a simple method (that does not involve computing the products $P_{k,n}$ for large n) to determine the basis of the chain? For some partial results in this direction, see **[48]**.

7. Embeddable Stochastic Matrices

In this final section we present another problem in probability where we will come across idempotent matrices and need Theorem 3.1. This is the problem of embedding a finite state discrete time Markov chain in a continuous time Markov chain. The problem can be stated as follows:

The transition function of a continuous time Markov chain with d states, d finite, is a map $P(t)$ from $(0,\infty)$ into S, the set of all $d \times d$ stochastic matrices such that

(7.1)
(i) $P(s)P(t) = P(s+t)$;
(ii) $P_{ij}(t)$ is a continuous function of t, $1 \le i, j \le d$.

It is known (see **[5]**) that $P_o = \lim_{t\to 0+} P(t)$ exists and $P_o = P_o^2$. The problem is then to find the class of $d \times d$ stochastic matrices Q for which there exists a transition function $P(t)$, satisfying (7.1), such that $P(1) = Q$.

Note that if $P(t)$ is as in (7.1) and $P(1) = Q$, then

$$P_o \text{ is singular iff } Q \text{ is singular}.$$

The following result was proven by Kingman **[36]**.

THEOREM 7.1. *A non-singular stochastic matrix Q is embeddable in a continuous time Markov chain* iff *it is infinitely divisible, that is, for each positive n, there exists a stochastic matrix Q_n such that $Q_n^n = Q$, Iff $Q = \exp A$, where $\exp A = I + \sum_{n=1}^{\infty}(1/n!)A^n$, I is the usual identity matrix and A is an intensity matrix, that is, A satisfies:*

(i) $A_{ij} \ge 0 \quad (i \ne j)$
(ii) $\sum_{j=1}^{d} A_{ij} = 0 \quad$ *(each i).* $\square$

An easily verifiable simple condition for embeddability can be given for 2×2 stochastic matrices, namely that a non-singular 2×2 stochastic matrix Q is embeddable iff $\det Q > 0$ iff trace $Q > 1$. This result is given in **[36]** and credited to D. G. Kendall. For results in this direction for the $d = 3$ case, see **[18]**, **[34]**. To the best of our knowledge, the corresponding question for a general finite d is still open.

In what follows, we present a theorem similar to Theorem 7.1 for a singular stochastic matrix. The ideas we use are not original. They are similar to those used by Martin-Löf [**38**] and earlier by others in the context of studying infinitely divisible probability measures on finite semigroups.

Suppose that Q is a $d \times d$ infinitely divisible stochastic matrix. Let Q_n be a stochastic matrix such that $Q_n^n = Q$. For each positive integer m, consider the sequence $Q_{(m!)n}^{m!}$. Let Q_n' be a limit point of this sequence. Then $Q_n'^n = Q$ and Q_n' is itself infinitely divisible, since for any positive integer r and a limit point R_n of a suitable subsequence of $Q_{(m!)n}^{m!/r}$, $R_n^r = Q'$. Now we define a sequence of infinitely divisible stochastic matrices (E_n) such that $Q = E_1$ and for $n \geq 1$, $E_{n+1}^{n+1} = E_n$. Define $Q(t)$ for each rational number t, $t = p/q$, p, q natural numbers, as the stochastic matrix $E_k^{(p/q)k!}, k \geq q$. Note that then for any two positive rational numbers t_1 and t_2, $Q(t_2)Q(t_1) = Q(t_1)Q(t_2) = Q(t_1+t_2)$, and also, $Q(1) = Q$.

We claim that

$$(7.2) \qquad \lim_{\substack{t \to 0+ \\ t \text{ rational}}} Q(t) \text{ exists and equals } Q_o, \text{ an idempotent stochastic matrix.}$$

To prove this, we will use Theorem 3.1. Let (t_i) and (s_i) be two sequences of positive rationals such that $t_i \to 0+, s_i \to 0+$ and

$$\lim Q(t_i) = Q' \quad \text{and} \quad \lim Q(s_i) = Q''.$$

Choosing a subsequence $(s_i') \subset (s_i)$ such that $t_i > s_i'$ (each i), we have:

$$\begin{aligned} &Q(t_1 - s_i')Q(s_i') = Q(s_i')Q(t_i - s_i') = Q(t_i) \text{ so that} \\ &\qquad Q''X = Q' = XQ'', X \text{ being a limit point of } (Q(t_i - s_i')). \end{aligned}$$

This means that the set S' of all such limit points is an abelian semigroup under matrix multiplication such that $S'x = S'$ for each x in S'. Thus, S' is a group of finite dimensional stochastic matrices. By the remarks following Theorem 3.1, S' is finite. Let s be the number of elements of S' and Q_o be the identity of this group. Then $x^s = Q_o$, whenever x is a limit point of the sequence $(Q(t_i/s))$, where (t_i) is a sequence of positive rationals converging to zero. Since $Q(t_i) = [Q(t_i/s)]^s$, it is clear that (7.2) holds.

Now let (p_n) and (q_n) be two sequences of positive rationals converging to a positive real number a such that $Q(p_n) \to Q_1$ and $Q(q_n) \to Q_2$ as n goes to ∞. As before, considering a suitable subsequence of $(q_n), (q_n')$, we have:

$$Q(p_n) = Q(q_n')Q(p_n - q_n') = Q(p_n - q_n')Q(q_n')$$

which means that $Q_1 = Q_2Q_o = Q_oQ_2$; similarly, $Q_2 = Q_1Q_o = Q_oQ_1$. Since Q_o is idempotent, it follows that $Q_2 = Q_oQ_1 = Q_oQ_2 = Q_1$. This means that we can define $Q(a) = \lim_{n\to\infty} Q(p_n)$, for any real $a > 0$ whenever (p_n) is a sequence of positive rationals converging to a. Thus, we can embed any given $d \times d$ infinitely divisible stochastic matrix Q in a continuous parameter

Markov chain with transition function $Q(t)$ such that

$$\text{(7.3)}\qquad \begin{aligned} &\text{(i) } Q(t+s) = Q(t)Q(s),\ t>0, s>0;\\ &\lim_{t\to 0+} Q(t) = Q_o = Q_o^2.\\ &\text{(ii) } Q(1) = Q \quad \text{and for } t>0,\quad Q(t)Q_o = Q_oQ(t) = Q(t).\end{aligned}$$

Suppose now that a transition function $Q(t), t > 0$, satisfying (7.3) is given. Then, as in Theorem 7.1, we show here that $Q(t)$ has also a similar exponential representation.

First, notice that for t sufficiently small $(0 < t \leq t_o)$, we can define:

$$\text{(7.4)}\qquad \operatorname{Log} Q(t) = -\sum_{n=1}^{\infty}(1/n)(Q_o - Q(t))^n.$$

For any $d \times d$ matrix C such that $Q_oC = CQ_o = C$, we define:

$$\text{(7.5)}\qquad \operatorname{Exp} C = Q_o + \sum_{n=1}^{\infty}(1/n!)C^n.$$

[Notice that if D denotes the matrix represented by the series in (7.4), then taking $A = I - (Q_o - Q(t))$, I the usual identity matrix, we have: $\log A = D$ (with usual definition of log) so that $A = \exp D = \operatorname{Exp} D + I - Q_o$ or $Q(t) = \operatorname{Exp} D$. Also, for C in (7.5), whenever $\operatorname{Log}(\operatorname{Exp} C)$ is defined,

$$\operatorname{Log}\ (\operatorname{Exp}\ C) = \operatorname{Log}\ (Q_o - I + \exp C) = \log\ (\exp C) = C.$$

It is also clear that $(\operatorname{Exp} C)(\operatorname{Exp} D) = \operatorname{Exp}(C + D)$ when $CD = DC$.]

Because of these remarks, it follows that for positive integers m and n, we can write:

$$\operatorname{Log} Q(t_o) = n \operatorname{Log} Q(t_o/n), \qquad \operatorname{Log} Q\left(\frac{m}{n}\cdot t_o\right) = \frac{m}{n}\cdot \operatorname{Log} Q(t_o).$$

Using continuity, for $t > 0$ (t small),

$$\operatorname{Log} Q(t) = (t/t_o)\cdot \operatorname{Log} Q(t_o).$$

Write: $G = (1/t_o)\cdot \operatorname{Log} Q(t_o)$. Then for small positive t,

$$\text{(7.6)}\qquad Q(t) = \operatorname{Exp}(tG).$$

Because of (7.3 (i)), (7.6) holds for all $t > 0$. Notice that G satisfies:

$$\text{(7.7)}\qquad GQ_o = Q_oG = G \quad \text{and} \quad G = \lim_{t\to 0+}[Q(t) - Q_o]/t,$$

and (7.7) implies

$$\text{(7.8)}\qquad \sum_{j=1}^{d} G_{ij} = 0 \quad \text{(each i)}; \qquad (Q_o)_{ij} = 0 \Rightarrow G_{ij} \geq 0.$$

Now suppose that we are given a $d \times d$ idempotent stochastic matrix Q_o and a $d \times d$ matrix G satisfying (7.8) and the first property in (7.7). (Such a matrix G will be called an intensity matrix corresponding to the idempotent

Q_o.) Then we claim that $\text{Exp}(tG)$, defined as in (7.5), defines a stochastic matrix for all $t > 0$. Once this is established, it will then be clear that $\text{Exp}(tG)$ defines a transition function of a continuous parameter Markov chain.

To prove our claim, choose a so large that

$$a \cdot (Q_o)_{ij} > |G_{ij}|, \quad \text{whenever } (Q_o)_{ij} > 0.$$

Define the stochastic matrix R by $aR = aQ_o + G$. Then $RQ_o = Q_oR = R$. Taking $G^o = Q_o$, we have:

$$\begin{aligned}\text{Exp}(tG) &= \sum_{n=0}^{\infty} t^n G^n/n! = \sum_{n=0}^{\infty} t^n a^n (R - Q_o)^n/n! \\ &= \sum_{n=0}^{\infty} (t^n a^n/n!) \cdot \sum_{k=0}^{n} \binom{n}{k} R^{n-k}(-1)^k \\ &= \sum_{k=0}^{\infty} (-a.t)^k/k! \cdot \sum_{n=k}^{\infty} [(at)^{n-k} R^{n-k}]/(n-k)! \\ &= \exp(a.t) \cdot \text{Exp}(a.tR)\end{aligned}$$

It is now clear that $\text{Exp}(tG)$ is a stochastic matrix for all $t > 0$. Thus we have proven the following theorem.

THEOREM 7.2. *For a $d \times d$, d finite, stochastic matrix Q (possibly singular) the following conditions are equivalent:*

(i) *Q is infinitely divisible;*
(ii) *Q is embeddable in a continuous time Markov chain with transition function $Q(t)$ satisfying (7.3);*
(iii) *there exists an idempotent matrix Q_o and an intensity matrix G corresponding to Q_o (that is, a matrix satisfying (7.8) and $GQ_o = Q_oG = G$) such that $Q = \text{Exp}\, G$ (that is, $Q = Q_o + \sum_{n=1}^{\infty} G^n/n!$).* □

EXAMPLE 7.1. Let us use our Theorem 7.2 to find all embeddable 3×3 stochastic matrices with rank 2. (Note that the rank three case is the non-singular case and the rank one stochastic matrices are idempotent which are, of course, embeddable.)

Consider

$$Q = \begin{pmatrix} a & b & 1-a-b \\ a & b & 1-a-b \\ c & d & 1-c-d \end{pmatrix}, \quad Q \text{ stochastic.}$$

(Here we have taken the first two rows identical. The other cases can be dealt with exactly as this one.)

CASE 1. a is positive.

In this case, Q is infinitely divisible iff

$$Q = \begin{pmatrix} a & 1-a & 0 \\ a & 1-a & 0 \\ c & \frac{c}{a} - c & 1 - \frac{c}{a} \end{pmatrix}, \qquad 0 \le c < a.$$

In this case, the matrix

$$Q_o = \begin{pmatrix} a & 1-a & 0 \\ a & 1-a & 0 \\ 0 & 0 & 1 \end{pmatrix}$$

and the intensity matrix G is

$$\begin{pmatrix} 0 & 0 & 0 \\ 0 & 0 & 0 \\ \frac{c}{q_o} & \frac{1-a}{a}\cdot\frac{c}{q_o} & \frac{-c}{aq_o} \end{pmatrix}$$

where $q = \ln(a/[a-c])$ and $qq_o = c/a$.

Here Q is embeddable in the Markov chain with transition function

$$\operatorname{Exp}(tG) = \begin{vmatrix} a & 1-a & 0 \\ a & 1-a & 0 \\ \frac{cf(t)}{q_o} & \frac{(1-a)cf(t)}{aq_o} & \frac{-cf(t)}{aq_o}+1 \end{vmatrix}$$

where $f(t) = [1-\exp(-tq)]/q$.

CASE 2. $a = 0$.

In this case, Q is infinitely divisible iff

$$Q = \begin{pmatrix} 0 & b & 1-b \\ 0 & b & 1-b \\ 0 & d & 1-d \end{pmatrix} \qquad \text{with } d < b.$$

In this case, Q_o is as above with $a = 0$ and G is given by

$$G = \begin{vmatrix} 0 & \frac{1-b}{q_o} & \frac{1-b}{q_o} \\ 0 & \frac{1-b}{q_o} & \frac{1-b}{q_o} \\ 0 & \frac{d}{q_o} & \frac{-d}{q_o} \end{vmatrix}$$

where $qq_o = 1-b+d = 1-\exp(-q)$, $q = \ln(1/[b-d])$. □

The following is a result of W. E. Clark, W. C. Holland and G. Szekely (personal communication from Clark). We present it below as an application of our Theorem 7.2.

COROLLARY 7.3. *Let S be a finite semigroup and x be an element of S such that for each positive integer n, there is x_n in S such that $x_n^n = x$. Then x must be an idempotent.*

PROOF. With no loss of generality, we can assume that S has a (two-sided) identity. Let us write $S = \{s_1, s_2, \ldots, s_n\}$. Suppose that the element s_1 is infinitely divisible. For each s_i, consider the n by n stochastic matrix A_i defined by $(A_i)_{jk} = 1$ iff $s_i s_j = s_k, = 0$, otherwise. Then $M = \{A_1, A_2, \ldots, A_n\}$ is a matrix semigroup (anti)isomorphic to S and A_1 is an infinitely divisible stochastic matrix. By Theorem 7.2, there is a continuous time Markov chain $A(t)$ such that

$$A(t) = \operatorname{Exp}(tG),\ A(1) = A_1,\ \lim_{m\to\infty} A(l/m) = A_o = A_o^2$$

where G is an intensity matrix corresponding to A_o. Since the semigroup M is finite, it is clear that there are arbitrarily small positive and distinct t and t' such that $A(t) = A(t')$ so that taking Log, $tG = t'G$ implying that G must be the zero matrix. This means that A_1 must be A_o. The rest is clear. □

Our final theorem in this section connects singular infinitely divisible stochastic matrices with the corresponding nonsingular ones. Though the proof could be shorter using certain standard results such as given in books like **[5]**, for completeness we give a self-contained proof.

THEOREM 7.4. *A $d \times d$ singular stochastic matrix Q with rank p is infinitely divisible iff there exists an idempotent stochastic matrix Q_o with rank $r (p \leq r < d)$ and a non-singular $r \times r$ infinitely divisible stochastic matrix R satisfying the following condition:*

If $\{T, C_1, C_2, \ldots, C_r\}$ is the basis of Q_o, then

$$\begin{aligned} Q_{ij} &= (Q_o)_{jj} \cdot R_{uv}, \quad \textit{if } i \in C_u \textit{ and } j \in C_v; \\ &= 0, \quad \textit{if } j \in T \textit{ (each } i\textit{)}; \\ &= (Q_o)_{jj} \cdot (MR)_{iv}, \quad \textit{if } i \in T \textit{ and } j \in C_v, \end{aligned}$$

where M is a $|T| \times r$ matrix such that $M_{iv} = \sum_{k \in C_v} (Q_o)_{ik}$.

PROOF. Suppose that Q is a $d \times d$ singular infinitely divisible stochastic matrix with rank p. Then by Theorem 7.2, there is a singular idempotent stochastic matrix Q_o with rank r such that

$$QQ_o = Q_oQ = Q \tag{7.9}$$

and

$$Q = \mathrm{Exp}(G), \tag{7.10}$$

where G is an intensity matrix corresponding to Q_o. Let $\{T, C_1, C_2, \ldots, C_r\}$ be the basis of Q_o. [The basis of Q_o is not uniquely determined by (7.9). However, if Q has no zero columns and if the basis is given, then Q_o is uniquely determined by (7.9), since then for j in C_v,

$$(Q_o)_{jj} = Q_{ij}/[\sum_{s \in C_v} Q_{is}] \text{ if } Q_{ij} > 0.$$

We also make the following observation: If the jth column of Q is not a zero column, then $j \notin T$ and the C-class in the basis of Q_o that contains j is contained in $\{k : Q_{ik} > 0$ and the kth row of Q is identical with the jth row of $Q\}$, where i is so chosen that $Q_{ij} > 0$.] By definition, G satisfies

$$GQ_o = Q_oG = G \tag{7.11}$$

$$\sum_{j=1}^{d} G_{ij} = 0 \quad \text{(each } i\text{)} \tag{7.12}$$

$$G_{ij} \geq 0, \quad \text{whenever } (Q_o)_{ij} = 0. \tag{7.13}$$

Now we define the $r \times r$ matrix H by $H_{uv} = \sum_{j \in C_v} G_{ij}$ (i in C_u). (Note that for each i in C_u, the ith row of G is the same.) Then H is an $r \times r$ intensity matrix in the sense that

(7.14)
$$\begin{aligned} H_{ij} &\leq 0, \qquad \text{if } i = j; \\ &\geq 0, \qquad \text{if } i \neq j; \\ \sum_{j=1}^{r} H_{ij} &= 0 \qquad \text{(each } i\text{)} . \end{aligned}$$

For i in C_u and j in C_v,

(7.15)
$$G = GQ_o => G_{ij} = (Q_o)_{jj} \cdot H_{uv},$$

and for $n \geq 1$, $G^n_{ij}(Q_o)_{jj} \cdot (H^n)_{uv}$. Define the $d \times r$ matrix M by

(7.16)
$$M_{iv} = \sum_{k \in C_v} (Q_o)_{ik}.$$

Then for $n \geq 1$, i in C_u and j in C_v, we have:

(7.17)
$$(M \cdot H^n)_{iv} = H^n_{uv}.$$

Now let i be any state and j be in C_u. We claim:

(7.18)
$$G^n_{ij} = (Q_o)_{jj} \cdot (MH^n)_{iu}.$$

To prove (7.18), observe first that

$$\begin{aligned} \sum_{s \in C_u} G_{is} &= \sum_{s \in C_u} \sum_{v=1}^{r} \sum_{k \in C_v} (Q_o)_{ik} G_{ks} \\ &= \sum_{s \in C_u} \sum_{v=1}^{r} G_{k_v,s} \sum_{k \in C_v} (Q_o)_{ik} \qquad (k_v \in C_v) \\ &= \sum_{v=1}^{r} M_{iv} \cdot \sum_{s \in C_u} G_{k_v,s} = (MH)_{iu}; \end{aligned}$$

(7.19)
$$\begin{aligned} G_{ij} &= \sum_{k \in C_u} G_{ik}(Q_o)_{kj} \qquad \text{(since } j \in C_u\text{)} \\ &= (Q_o)_{jj} \cdot \sum_{k \in C_u} G_{ik} = (Q_o)_{jj} \cdot (MH)_{iu} \end{aligned}$$

Now using (7.17) and (7.19), we have:

$$\begin{aligned} G^{n+1}_{ij} &= \sum_{v=1}^{r} \sum_{k \in C_v} G_{ik} G^n_{kj} \qquad \text{(since } k \in T => G_{ik} = 0\text{)} \\ &= \sum_{v=1}^{r} \sum_{k \in C_v} (Q_o)_{kk}(MH)_{iv} \cdot (Q_o)_{jj} \cdot H^n_{vu} \qquad \text{(since } j \in C_u\text{)} \\ &= (Q_o)_{jj} \cdot \sum_{v=1}^{r} (MH)_{iv} \cdot H^n_{vu} \cdot \sum_{k \in C_v} (Q_o)_{kk} \\ &= (Q_o)_{jj} \cdot (MH^{n+1})_{iu}. \end{aligned}$$

This establishes (7.18). Let $j \in C_u$. Then we have from (7.10) that

$$\begin{aligned} Q_{ij} &= (Q_o)_{ij} + \sum_{n=1}^{\infty} G_{ij}^n / n! \\ &= (Q_o)_{ij} + (Q_o)_{jj} \cdot \sum_{n=1}^{\infty} [MH^n]_{iu} / n! \\ &= (Q_o)_{ij} + (Q_o)_{jj} \cdot [M(\exp H - I)]_{iu}. \end{aligned}$$

Simplifying, we have for all i, j: $1 \leq i, j \leq d$,

$$\begin{aligned} Q_{ij} &= (Q_o)_{jj} \cdot (M(\exp H))_{iu}, && \text{if } j \in C_u; \\ &= 0, && \text{if } j \in T. \end{aligned}$$

The rest of the proof follows easily. □

Let us finally mention that Professor H. Heyer of the University of Tübingen has recently informed the author that in the context of this section a Tübingen dissertation by G. Maroudas is relevant.

References

1. A. Berman, and R. J. Plemmons, Nonnegative matrices in the mathematical sciences, Academic Press, 1979.

2. D. Blackwell, Idempotent Markoff chains, Ann. of Math. 43, No. 3 (1942), 560–567.

3. P. Bougerol, Tightness of products of random matrices and stability of linear stochastic systems, Ann. Probab. 15 (1987), 40–74.

4. D. R. Brown, On clans of nonnegative matrices, Proc. Amer. Math. Soc. (1964), 671–674.

5. K. L. Chung, Markov chains with stationary transition probabilities (2nd Edition), Springer-Verlag, 1967.

6. W. E. Clark, Remarks on the kernel of a matrix semigroup, Czechoslovak Math. J. 15 (90) (1965), 305–309.

7. A. H. Clifford and G. B. Preston, The Algebraic Theory of Semigroups I, Math. Surveys Monogr., 7, Amer. Math. Soc., 1961.

8. H. Cohn, On the tail events of a Markov chain, Wahrsch. Verw. Gebiete 29 (1974), 65–72.

9. ——, On a class of non-homogeneous Markov chains, Math. Proc. Camb. Philo. Soc. (1982), 528–534.

10. J. E. Cohen, Ergodic theorems in demography, Bull. Amer. Math. Soc. (N. S.) 1, No. 2 (1979), 275–295.

11. I. Csiszar, On infinite products of random elements and infinite convolutions of probability distributions on locally compact groups, Wahrsch. Verw. Gebiete 5 (1965), 279–295.

12. R. Darling, and A. Mukherjea, Stochastic flows on a countable set, J. Theoret. Prob. 1, No. 2 (1988), 121–147.

13. ——, *Stochastic automata* (Preprint).

14. R. Demarr, Nonnegative idempotent matrices, Proc. Amer. Math. Soc. 45, No. 2 (1974), 185–188.

15. P. Flor, Groups of nonnegative matrices, Comp. Math. 21 (1969), 376–382.

16. F. J. Fritz, B. Huppert und W. Willems, Stochastische Matrizen, Springer-Verlag, 1979.

17. H. Frydman, A structure of the bang bang representation for 3×3 embeddable matrices, Wahrsch. Verw. Gebiete 53 (1980), 305–316.

18. ——, The embedding problem for Markov chains with 3 states, Math. Proc. Camb. Philo. Soc. 87 (1980), 285–294.

19. ——, A structure of doubly stochastic Markov chains, Linear Algebra Appl. 67 (1985), 51–66.
20. H. Furstenberg, and H. Kesten, Products of random matrices, Ann. Statist. 31 (1960), 457–469.
21. S. Gibert and A. Mukherjea, Products of infinite dimensional nonnegative matrices: extension of Hajnal's results, Math. Z. 196 (1987), 485–490.
22. ——, Tail idempotents and their structures for infinite dimensional nonnegative matrices in Proc. of an Oberwolfach Conf. in 1983, Springer-Verlag Lecture Notes in Math. Vol. 1064, 147–160.
23. ——, Nonnegative idempotent kernels, J. Math. Anal. Appl. 135, No. 1 (1988), 326–341.
24. W. Hazod, Bemerkungen zum bang-bang problem fur stochastische matrizen, Z. Wahrsch. Verw. Gebiete 35 (1976), 39–43.
25. J. Hajnal, On products of nonnegative matrices, Math. Proc. Camb. Philo. Soc. 79 (1976), 521–530.
26. G. Hognas, Random semigroup acts on a finite set, J. Austral. Math. Soc. 23 (Series A) (1977), 481–497.
27. ——, A note on products of random matrices, Statist. Probab. Lett. 5 (1987), 367–370.
28. ——, On products of random projections, Acta Acad. Abo., Series B, 44, No. 5, 1984, 18 pages.
29. G. Hognas and A. Mukherjea, Recurrent random walks and invariant measures on semigroups of $n \times n$ matrices, Math. Z. 173 (1980), 69–94.
30. M. Iosifescu, Finite Markov processes and their applications, J. Wiley & Sons, 1980.
31. D. Isaacson, and R. Madsen, Markov chains: Theory and Applications, J. Wiley & Sons, 1976.
32. S. Johansen, A central limit theorem for finite semigroups and its application to the embedding problem for finite state Markov chains, Z. Wahrsch. Verw. Gebiete 26 (1973), 171–190.
33. ——, The bang-bang problem for stochastic matrices, Z. Wahrsch. Verw. Gebiete 26 (1973), 191–195.
34. S. Johansen, and F. L. Ramsey, A bang-bang representation for 3×3 embeddable stochastic matrices, Z. Wahrsch. Verw. Gebiete 47 (1979), 107–118.
35. H. Kesten, and F. Spitzer, Convergence in distribution of products of random matrices, Z. Wahrsch. Verw. Gebiete 67 (1984), 363–386.
36. J. F. C. Kingman, The imbedding problem for finite Markov chains, Z. Wahrsch. Verw. Gebiete 1 (1962), 14–24.
37. T. Liggett, Interacting particle systems, Springer-Verlag, New York, Heidelberg Berlin, 1985.
38. P. Martin-Lof, Probability theory on discrete semigroups, Z. Wahrsch. Verw. Gebiete 4 (1965), 78–102.
39. V. M. Maximov, Convergence of non-homogeneous bistochastic Markov chains, Theory of Probab. & Appl. 15, No. 4 (1970), 604–618.
40. A. Mukherjea, and N. A. Tserpes, Measures on topological semigroups: Convolution products and random walks, Springer-Verlag Lecture Notes Vol. 547 (1976), 195 pages.
41. A. Mukherjea, Limit theorems: Stochastic matrices, ergodic Markov chains and measures on semigroups, Probab. Analysis and Related Topics, Vol. 2, Academic Press (1979), 143–203.
42. ——, A new result on the convergence of non-homogeneous Markov chains, Trans. Amer. Math. Soc. 262, No. 2 (1980), 505–520.
43. ——, Non-homogeneous Markov chains: tail-idempotents, tail sigma-fields and basis, Math. Z. 183 (1983), 293–309.
44. ——, Completely simple semigroups of matrices, Semigroup Forum 33 (1986), 405–429.
45. ——, Convergence in distribution of products of random matrices, Trans. Amer. Math. Soc. 303, No. 1 (1987), 395–411.
46. A. Mukherjea, and A. Nakassis, Convergence of non-homogeneous stochastic chains with countable states, J. Multivariate Anal. 16, No. 1 (1985), 85–117.

47. ——, The combinatorial structure of non-homogeneous Markov chains with countable states, Internat. J. Math. Math. Sci. 6, No. 2 (1983), 371–386.

48. A. Mukherjea, A. Nakassis, and D. Issacson, Determination of the basis of a convergent non-homogeneous Markov chain, Statist. Decisions 2 (1984), 363–375.

49. K. R. Parthasarathy, Probability measures on metric spaces, Academic Press, New York, 1967.

50. M. Rosenblatt, Products of independent identically distributed stochastic matrices, J. Math. Anal. Appl. 11 (1965), 1–10.

51. ——, Markov processes: Structure and asymptotic behavior, Springer-Verlag, New York, 1971.

52. E. Seneta, Nonnegative matrices and Markov chains (second edition), Springer-Verlag, Heidelbert, Berlin, 1981.

Department of Mathematics
University of South Florida
Tampa, Fla 33620-5700

Proceedings of Symposia in Applied Mathematics
Volume **40**, 1990

Interface Between Statistics and Linear Algebra

INGRAM OLKIN

ABSTRACT. The goal of this paper is to illustrate the strong interplay between statistics and linear algebra arising from problems in statistics. The areas that will be covered relate to a variety of extremal problems and integrals arising from the multivariate normal distribution, independence of linear and quadratic forms in normal variates, probability inequalities and their relation to eigenvalue inclusion theorems, partial orders, and simulations.

This lecture is designed to illustrate the interplay between statistics and linear algebra. We will focus on several areas of statistics that have a strong linear algebra component. These areas do not exhaust the list of connections between statistics and linear algebra, and different authors would no doubt choose different areas of emphasis. The topics that I have chosen have a strong personal component—they are the ones that I have worked on. In addition to deciding on which topics might be of interest, there is a choice as to whether to delve deeply into one or two areas or whether to attempt to communicate the flavor of the connections with linear algebra. I have chosen the latter course primarily because it provides a broader scope and touches more aspects of linear algebra. Because the emphasis is on breadth, we will not offer proofs except where it provides some illumination to the connection with linear algebra. More specifically, I will start with a statistical model and show the role that linear algebra plays.

The general topic headings are

A: Extremal problems in multivariate distributions

B: Integrals and distributions

C: Independence properties and characterization of distributions

D: Probability inequalities

E: Orderings

F: Simulations

1980 *Mathematics Subject Classification* (1985 *Revision*). Primary 15-02, 62H99.

Key words and phrases. Extremals, multivariate integrals, independence properties, probability inequalities, eigenvalue inclusion theorems, simulations.

For each topic I will provide some preliminaries together with historical comments to set the stage for the discussion. A full list of references would constitute a long paper in itself. I have been selective in the choice of references, but a perusal of these references should uncover a list that will permit the reader to generate a more complete bibliography.

Commentary on Notation: It is customary in the linear algebra literature to use lower case Roman letters to denote elements or vectors and capital letters to denote matrices, whereas in the statistical literature we use capital Roman letters to denote random variables and lower case letters to denote realizations. When we combine the two fields we are led to an inconsistency. Some authors have introduced bold face type to permit additional distinctions. This curse of notation has to be weighed against clarity. I have chosen not to add notation, but instead to depend on the context or to use the descriptor "random variable," "random vector," "random matrix" when the context is unclear.

We need to distinguish between random variables and parameters, and do so by using Greek letters (lower case for vectors, capitals for matrices) to denote parameters.

The term "random sample" or simply "sample" is a synonym for saying that the observations are independently and identically distributed.

Definiteness is often used, and we write $A > 0$ ($A \geq 0$) to mean that A is positive definite (semi-definite), and $A > B$ to mean that $A - B > 0$.

If $Y = (y_{ij})$ is a $p \times n$ matrix we write dY to mean $\prod_{i,j} dy_{ij}$, whereas if $S = (s_{ij})$ is a $p \times p$ symmetrix matrix, we write dS to mean $\prod_{i \leq j} ds_{ij}$.

We write $p(\cdot)$ to denote a generic density function. The expectation is $\mathscr{E} h(x) = \int h(x) p(x)\, dx$.

If $h(x) = (h_{ij}(x))$ is a matrix then $\mathscr{E} h(x) = (\mathscr{E} h_{ij}(x))$ is the matrix of elementwise expectations.

A. The Multivariate Normal Distribution and its Relatives.

A.1. Preliminaries. The most fundamental statistical distribution is the Gaussian or normal distribution with density function

$$p(x) = \frac{1}{\sqrt{2\pi}\sigma} \exp -\frac{1}{2} \frac{(x-\mu)^2}{\sigma^2}, \tag{A.1}$$

$-\infty < x < \infty$, $-\infty < \mu < \infty$, $\sigma > 0$. We write $x \sim \mathscr{N}(\mu, \sigma^2)$ as shorthand to denote that the random variable x has a normal distribution with the density function given by (A.1). The parameters μ and σ^2 represent the mean and variance of x.

If $x_1, \ldots, x_p$ are independent, identically distributed random variables, with $x_i \sim \mathscr{N}(0, 1)$, $i = 1, \ldots, p$, then the joint distribution of $x = (x_1, \ldots, x_p)$ is

$$p(x_1, \ldots, x_p) = \frac{1}{(2\pi)^{p/2}} \exp\left(-\frac{1}{2} x x'\right). \tag{A.2}$$

If A is a non-singular matrix and $y = xA + \mu$, then the joint distribution of $y = (y_1, \ldots, y_p)$ is

$$p(y_1, \ldots, y_p) = \frac{1}{(2\pi)^{p/2}} |A|^{-1} \exp -\frac{1}{2}(y - \mu)(A'A)^{-1}(y - \mu)' = \frac{1}{(2\pi)^{p/2}} |\Sigma|^{-\frac{1}{2}} \exp -\frac{1}{2}(y - \mu)\Sigma^{-1}(y - \mu)', \tag{A.3}$$

where $\Sigma = A'A$. To denote that $y = (y_1, \ldots, y_p)$ has the distribution (A.3) we write $y \sim \mathscr{N}_p(\mu, \Sigma)$, where p denotes the dimension of y, $\mu = (\mu_1, \ldots, \mu_p)$ represents the mean vector, and $\Sigma = (\sigma_{ij})$ represents the covariance matrix. That is, $\mathscr{E} y = \mu$ and $\mathscr{E}(y-\mu)'(y-\mu) = \Sigma$. Thus a random vector y with an arbitrary mean and covariance matrix can be obtained from a standard normal distribution (with means 0 and variances 1) by an affine transformation.

To simulate observations from a distribution with a given covariance matrix Σ, let $\Sigma = A'A$ denote any factorization, generate $x = (x_1, \ldots, x_p)$ where each x_i has a $\mathscr{N}(0, 1)$ distribution, and let $y = xA$. Since any factorization will serve, it behooves us to use a simple one, namely the Cholesky decomposition.

Various mechanisms for efficiently generating standard normal random variables are described in Devroye (1986), Thisted(1988), and Bratley, Fox, Schrage (1987).

Let $(y_{1\alpha}, \ldots, y_{p\alpha})$, $\alpha = 1, \ldots, n$, denote a random sample of size n from the p-variate normal distribution $\mathscr{N}_p(0, \sigma)$, and let

$$Y = \begin{pmatrix} y_{11} & \cdots & y_{1n} \\ \vdots & \vdots & \vdots \\ y_{p1} & \cdots & y_{pn} \end{pmatrix}$$

denote the sample matrix. Each column of Y represents a single observation on the p-variables, whereas each row represents the n observations on a single variable.

A major distributional result was obtained by Wishart (1928). The elements of $V = YY'$ represent inner products between variables, that is, inner products of $(y_{i1}, \ldots, y_{in})$ and $(y_{j1}, \ldots, y_{jn})$; $i, j = 1 \ldots, p$. This joint distribution is called the Wishart distribution, and has the joint density function

$$p(V \mid n, p, \Sigma) = c(p, n)|\Sigma|^{-\frac{n}{2}}|V|^{\frac{n-p-1}{2}} e^{-\frac{1}{2} tr\, \Sigma^{-1} V}, \qquad V > 0, \tag{A.4}$$

where

$$c(p, n) = \left[2^{np/2}\pi^{p(p-1)/4} \prod_1^p \Gamma\left(\frac{n - i + 1}{2}\right)\right]^{-1}$$

is determined so that

$$\int p(V \mid n, p, \Sigma)\, dV = 1.$$

The expression $V \sim \mathscr{W}_p(\Sigma, n)$ denotes that the elements of the symmetric matrix V have the joint distribution (A.4).

The case $p = 1$ in (A.4) yields the well-known gamma distribution

$$p(v \mid n, 1, \sigma^2) = cv^{\frac{n}{2}-1} e^{-\frac{1}{2}\frac{v}{\sigma^2}}, \qquad v > 0,$$

so that the Wishart distribution represents a multivariate version of the gamma distribution. This suggests the possibility of generating other multivariate versions of distributions related to the gamma distributions. We shall see several such generalizations in Section B.

A number of alternative derivations of the Wishart distribution appear in the literature. These are based on a geometric argument, on induction, on the inversion of the characteristic function, or on a functional equation. For a detailed discussion see Anderson (1984, Chapter 7). We here sketch the derivation of Rasch (1948) which is based on a functional equation, and provides an interesting connection with linear algebra. Functional equations with matrix arguments arise in other contexts, as in the case of obtaining Jacobians of matrix transformations. See Olkin and Sampson (1972).

Suppose the joint density function $p(Y)$ of Y is a known function $f(\cdot)$ of YY':

$$p(Y) = f(YY').$$

Consider two alternative sequences of transformations:

(a) $V = YY'$, $\quad W = AVA' \quad = AYY'A'$;

(b) $X = AY$, $\quad W = XX' \quad = AYY'A'$,

where A is nonsingular. In each sequence we start with the joint distribution of the elements of Y and end with the joint distribution of the elements of W. Because the routes taken are different we are able to relate the two results.

With the sequence (a) we have with $V = YY'$:

$$p(V) = f(V)\, J(V), \tag{A.5}$$

where $J(V)$ is an unknown function obtained from the Jacobian and an integration. Now let $W = AVA'$ be a transformation from V to W, the Jacobian of which is $|A|^{-(p+1)}$. (See Deemer and Olkin (1951) for a general discussion of Jacobians of matrix transformations.) Then

$$p(W) = f(A^{-1}WA'^{-1})\, J(A^{-1}WA'^{-1})\, |A|^{-(p+1)}. \tag{A.6}$$

With the sequence (b), let $X = AY$ be a transformation from Y to X, the Jacobian of which is $|A|^{-n}$. Then

$$p(X) = f(A^{-1}XX'A'^{-1})\, |A|^{-n}.$$

Now let $W = XX'$, in which case the function $J(\cdot)$ again enters:

$$p(W) = f(A^{-1}WA'^{-1})\, |A|^{-n} J(W). \tag{A.7}$$

Equating the two expressions (A.6) and (A.7) for the density $p(W)$ of W, and eliminating the common term $f(A^{-1}WA'^{-1})$ yields the functional equation

$$|A|^{-n} J(W) = J\left(A^{-1}WA'^{-1}\right) |A|^{-p+1},$$

which holds for all positive definite matrices W and all nonsingular matrices A. With $W = I$ and $S = A^{-1}A'^{-1}$ the functional equation becomes

$$J(S) = J(I)\,|S|^{\frac{n-p-1}{2}}.$$

Since $J(I)$ is a constant, we have from (A.5) that

$$p(V) = f(V)\,J(V) = c(p,n)\,e^{-\frac{1}{2}trV}\,|V|^{\frac{n-p-1}{2}},$$

where the constant $c(p,n)$ is determined so that

$$c(p,n)\int_{V>0} |V|^{\frac{n-p-1}{2}} e^{-\frac{1}{2}trV}\,dV = 1.$$

The evaluation of this integral can be accomplished in a variety of ways. An inductive argument provides an elementary proof and also provides a mechanism in more complicated settings. Partition

$$V = \begin{pmatrix} V_{11} & v' \\ v & v_{pp}, \end{pmatrix}$$

where $V_{11} = (v_{ij})$, $i, j = 1, \dots, p-1$, $v = (v_{p1}, \dots v_{p,p-1})$. Then $|V| = |V_{11}|\left(v_{pp} - vV_{11}^{-1}v'\right)$, so that

$$\int_{V>0} |V|^{\frac{n-p-1}{2}} e^{-\frac{1}{2}\operatorname{tr} V}\,dV$$
$$= \int_{\substack{V_{11}>0 \\ v_{pp}>vV_{11}^{-1}v'}} |V_{11}|^{\frac{n-p-1}{2}} (v_{pp} - vV_{11}^{-1}v')^{\frac{n-p-1}{2}} e^{-\frac{1}{2}v_{pp}-\frac{1}{2}\operatorname{tr} V_{11}}\,dV.$$

Let $w = v_{pp} - vV_{11}^{-1}v'$ be a transformation from v_{pp} to w to yield

$$\int_{\substack{w>0 \\ V_{11}>0}} |V_{11}|^{\frac{n-p-1}{2}} w^{\frac{n-p-1}{2}} e^{-\frac{1}{2}w-\frac{1}{2}vV_{11}^{-1}v'-\frac{1}{2}\operatorname{tr} V_{11}}\,dV_{11}\,dw\,dv.$$

Integration with respect to v yields $|V_{11}|^{\frac{1}{2}}(2\pi)^{\frac{p-1}{2}}$, which permits a breakdown of the result into the product of two integrals, one of which yields the gamma function:

$$(2\pi)^{\frac{p-1}{2}}\int_{V_{11}>0} |V_{11}|^{\frac{n-p}{2}} e^{-\frac{1}{2}\operatorname{tr} V_{11}} dV_{11} \int_{w>0} w^{\frac{n-p-1}{2}} e^{-\frac{1}{2}w} dw$$
$$= (2\pi)^{\frac{p-1}{2}}\Gamma(\frac{n-p+1}{2})2^{\frac{n-p+1}{2}} \int_{V_{11}>0} |V_{11}|^{\frac{n-p}{2}} e^{-\frac{1}{2}\operatorname{tr} V_{11}} dV_{11}.$$

Thus

$$c(p,n) = \left[2^{\frac{n}{2}}\pi^{\frac{p-1}{2}}\Gamma\left(\frac{n-p+1}{2}\right)\right]^{-1} c(p-1,n),$$

which now can be solved recursively to yield (A.4).

Note that the Wishart distribution (A.4) with $\Sigma = I$ is orthogonally invariant. That is, the distribution of GVG' is the same as the distribution of V for all orthogonal matrices G. Consequently, the joint moment generating function

$$\mathscr{E}e^{\operatorname{tr}\Theta V} = \int_{V>0} e^{\operatorname{tr}\Theta V} p(V \mid n,p,I)\,dV$$

has the property that

(A.8) $$\mathscr{E} e^{\operatorname{tr}\Theta V} = \mathscr{E} e^{\operatorname{tr}\Theta G V G'}.$$

Differentiation in (A.8) with respect to θ_{ij} evaluated at $\Theta = 0$ yields

$$\mathscr{E} v_{ij} = \mathscr{E} g_i V g_j'$$

for all unit vectors g_i, g_j with $g_i g_j' = \delta_{ij}$, the Kronecker delta. The choice $g_i = (1, 0, \ldots, 0)$, $g_j = (0, -1, 0, \ldots, 0)$ yields $\mathscr{E} v_{12} = -\mathscr{E} v_{12}$, so that $\mathscr{E} v_{12} = 0$, and with permutations $\mathscr{E} v_{ij} = 0$ for all $i \neq j$. For $g_i = g_j = (0, 1, 0, \ldots, 0)$ we obtain $\mathscr{E} v_{ii} = \mathscr{E} v_{22}$, so that with permutations $\mathscr{E} v_{ii} = c$. Consequently,

$$\mathscr{E} v_{ij} = c\delta_{ij}.$$

In a similar manner by using permutations and reflections we obtain

$$\mathscr{E} v_{ij} v_{k\ell} = c_1 \delta_{ij} \delta_{k\ell} + c_2(\delta_{ik}\delta_{j\ell} + \delta_{i\ell}\delta_{jk}).$$

The constants c, c_1, c_2 depend on the particular distribution $p(V)$. The above argument is used by Olkin and Rubin (1962) in resolving a characterization of the Wishart distribution.

A.2. Extremals for Maximum Likelihood Estimation. One method for estimating the parameters of a distribution is the method of maximum likelihood estimation (MLE) obtained as the maximum of the joint density function of the sample. That is, if $f(x; \theta)$ is a family of density functions depending on an unknown parameter θ, and $x_1, \ldots, x_n$ is a sample from this population, then the maximum likelihood estimator $\hat{\theta}$ of θ is obtained from $\max_\theta \prod_1^n f(x_i, \theta)$. This criterion is the basis of a large class of extremal problems, some of which are illustrated in the following examples.

To obtain the maximum likelihood estimate of Σ based on an observation of the $p \times n$ matrix Y, and thereby of the $p \times p$ matrix $V = YY'$, we require

$$\max_{\Sigma > 0} c(p, n) |\Sigma|^{-\frac{n}{2}} |V|^{\frac{n-p-1}{2}} e^{-\frac{1}{2}\operatorname{tr}\Sigma^{-1}V}.$$

Taking the logarithm of the Wishart density and omitting constants leads to the problem of maximizing the function.

$$g(\Sigma) = -n \log|\Sigma| - \operatorname{tr}\Sigma^{-1}V,$$

over the region $\Sigma > 0$. Since $\log|\Sigma|$ is a concave function of Σ and $\operatorname{tr}\Sigma^{-1}V$ is a convex function of Σ, the function $g(\Sigma)$ is neither concave nor convex. However, if we let $\Lambda = \Sigma^{-1}$ and consider instead maximizing

$$g^*(\Lambda) = n \log|\Lambda| - \operatorname{tr}\Lambda V,$$

over $\Lambda > 0$, then we have gained the advantage that $g^*(\Lambda)$ is strictly concave in Λ. Further, $g^*(\Lambda)$ approaches $-\infty$ as Λ approaches the boundary, that is, as any characteristic root approaches 0 or ∞. Consequently, the maximum is achieved at an interior point, which can be obtained by differentiating $g^*(\Lambda)$. This yields the matrix equation

$$n\Lambda^{-1} - V = 0,$$

or equivalently,

$$n\Sigma - V = 0,$$

so that the maximum likelihood estimator $\hat{\Sigma} = V/n$.

The Wishart distribution is the basis for a variety of extremal problems. We give several illustrations; discussions of alternative derivations for the solution in each example are contained in Anderson and Olkin (1985).

A.2.a. *A Model with Missing Observations.* In this model we have n observations on p-variates and m observations on $k \leq p$ variates. In effect we have m observations for which there is no data on $p - k$ variables. The data matrices are

$$Y_{(1)} = \begin{bmatrix} y_{11} & \cdots & y_{1n} \\ \vdots & \vdots & \vdots \\ y_{k1} & \cdots & y_{kn} \\ \vdots & \vdots & \vdots \\ y_{p1} & \cdots & y_{pn} \end{bmatrix}, \quad Y_{(2)} = \begin{bmatrix} y_{1,n+1} & \cdots & y_{1,n+m} \\ \vdots & \vdots & \vdots \\ y_{k,n+1} & \cdots & y_{k,n+m} \end{bmatrix}.$$

The first set of observations generates a $p \times p$ matrix $V = Y_{(1)} Y_{(1)}' \sim \mathscr{W}_p(\Sigma, n)$, and the second set of observations generates a $k \times k$ matrix $\mathscr{W} = Y_{(2)} Y_{(2)}' \sim \mathscr{W}_k(\Sigma_{11}, m)$, where the $p \times p$ covariance matrix Σ is partitioned

$$\Sigma = \begin{pmatrix} \Sigma_{11} & \Sigma_{12} \\ \Sigma_{21} & \Sigma_{22}, \end{pmatrix} \qquad \Sigma_{11} : k \times k.$$

The joint distribution of the independent random matrices V and W is the product of the respective densities:

$$p(V, W \mid \Sigma) = \left(c(p, n) |\Sigma|^{-\frac{n}{2}} |V|^{\frac{n-p-1}{2}} e^{-\frac{1}{2} \operatorname{tr} \Sigma^{-1} V} \right)$$
$$\left(c(k, m) |\Sigma_{11}|^{-\frac{m}{2}} |W|^{\frac{m-k-1}{2}} e^{-\frac{1}{2} \operatorname{tr} \Sigma_{11}^{-1} W} \right).$$

The maximum likelihood estimator of Σ is obtained by maximizing $p(V, W \mid \Sigma)$ over $\Sigma > 0$. If we take the logarithm of $p(V, W \mid \Sigma)$ and ignore constants, then the maximization problem becomes

$$\max_{\Sigma > 0} \left[-n \log |\Sigma| - m \log |\Sigma_{11}| - \operatorname{tr} \Sigma^{-1} V - \operatorname{tr} \Sigma_{11}^{-1} W \right].$$

This model and variants is studied in Gleser and Olkin (1970).

A.2.b. *A Model with Means.* In addition to observing a Wishart matrix V we observe an independent normal matrix X. This yields the joint distribution

$$p(V, X | \Sigma, B) \propto |\Sigma|^{-\frac{n}{2}} \exp -\frac{1}{2} \operatorname{tr} \Sigma^{-1} [V + (X - BZ)(X - BZ)'],$$

where we have omitted terms not involved in the extremal problem. The $p \times p$ matrix V and the $p \times n$ matrix X are the matrices of observations. The $k \times n$ matrix Z is a design matrix of constants and is of rank k. The unknown matrices of parameters are Σ: $p \times p$ and $B = (\beta_{ij})$: $p \times k$. The required maximization is of $p(V, X \mid \Sigma, B)$ over the region $\Sigma > 0$ and $-\infty < \beta_{ij} < \infty$ to yield maximum likelihood estimators $\hat{\Sigma}$ and $\hat{B}$ of Σ and B, respectively.

A.2.c. *A Linear Model with Constraints.* This model is similar to the previous one in that there is a Wishart component V and a normal component X:

$$p(V, X \mid \Sigma, B) \propto |\Sigma|^{-\frac{n}{2}} \exp -\frac{1}{2} \operatorname{tr} \Sigma^{-1}[V + (X - Z_1 B Z_2)(X - Z_1 B Z_2)'].$$

The $p \times p$ matrix V and the $p \times n$ matrix X are the matrices of observations. The $p \times q$ matrix Z_1 and the $m \times n$ matrix Z_2 are design matrices of constants of rank q and m respectively. The unknown matrices of parameters are Σ: $p \times p$ and $B = (\beta_{ij})$: $q \times m$. The maximization is of $p(V, X \mid \Sigma, B)$ over $\Sigma > 0$ and $-\infty < \beta_{ij} < \infty$ satisfying the constraint equation

$$Z_3 B Z_4 = B_0,$$

where the $v \times q$ matrix Z_3 of rank v, the $m \times k$ matrix Z_4 of rank k, and the $v \times u$ matrix B_0 are known. The maximization yields the maximum likelihood estimators $\hat{\Sigma}$ and $\hat{B}$ of Σ and B, respectively.

A.2.d. *Patterned Covariance Matrices.* Physical models often suggest symmetries in the data. If $x_1, \ldots, x_p$ are interchangeable random variables, then the means and the variances will be interchangeable so that $\mu_1 = \cdots = \mu_p$ and $\sigma_{11} = \cdots = \sigma_{pp}$, and the covariances will be interchangeable so that $\sigma_{12} = \sigma_{13} = \cdots = \sigma_{p-1,p}$. Consequently, the covariance matrix is of the form

$$\Sigma = \sigma^2 \begin{bmatrix} 1 & \rho & \cdots & \rho \\ & 1 & \cdots & \rho \\ & & \vdots & \\ & & & 1 \end{bmatrix}.$$

This matrix is called the intraclass covariance matrix. Interchangeability often arises in practice from conditions of the model.

Measurements at the vertices of a regular polygon or on the petals of a flower generate a spatial pattern in which the relation between adjacent petals is the same, relation between petals spaced two apart is the same, etc. so that

$$\Sigma = \sigma^2 \begin{bmatrix} 1 & \rho_1 & \rho_2 & \cdots & \rho_2 & \rho_1 \\ & 1 & \rho_1 & \cdots & \rho_3 & \rho_2 \\ & & 1 & \cdots & \rho_4 & \rho_3 \\ & & & \vdots & & \\ & & & & 1 & \rho_1 \\ & & & & & 1 \end{bmatrix}.$$

This matrix is a circulant. As the name implies, this model arises when observations are taken on a circle. A discussion of applications and statistical results are given in Olkin and Press (1969), Olkin (1973).

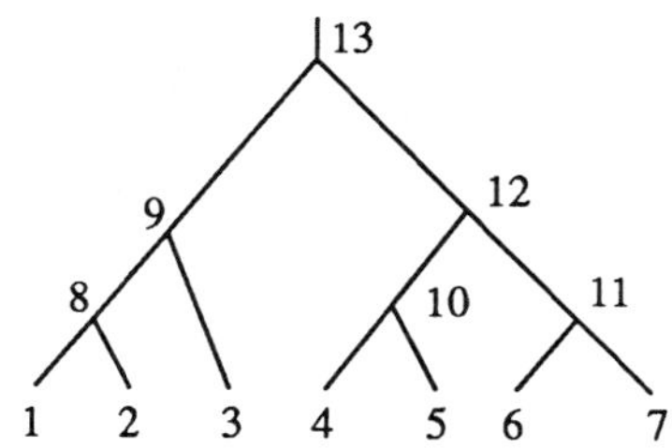

FIGURE 1. A tree representation of a patterned covariance matrix.

Observations at the terminals of a tree such as in Figure 1 generate a patterned matrix of the form

$$\Sigma = \left[\begin{array}{cc|c|cc|cc} \alpha & \beta & \delta & \tau & \tau & \tau & \tau \\ & \alpha & \delta & \tau & \tau & \tau & \tau \\ \hline & & \gamma & \tau & \tau & \tau & \tau \\ \hline & & & \varepsilon & \eta & \mu & \mu \\ & & & & \varepsilon & \mu & \mu \\ \hline & & & & & \psi & \nu \\ & & & & & & \psi \end{array}\right] \equiv \begin{pmatrix} \Sigma_{11} & \Sigma_{12} & \Sigma_{13} & \Sigma_{14} \\ \Sigma_{21} & \Sigma_{22} & \Sigma_{23} & \Sigma_{24} \\ \Sigma_{31} & \Sigma_{32} & \Sigma_{33} & \Sigma_{34} \\ \Sigma_{41} & \Sigma_{42} & \Sigma_{43} & \Sigma_{44} \end{pmatrix}.$$

Here the nodes show how to partition the matrix; within the diagonal matrices Σ_{11}, Σ_{22}, Σ_{33} the variances are interchangeable, and the covariances are interchangeable. Further, every element in one branch is interchangeable with every element in another branch. Such trees have been used as genetic and migration models by Piazza and Cavalli-Sforza (1983), and by Cavalli-Sforza and Piazza (1985).

Maximum likelihood estimates require the determination of

$$\max_{\Omega} |\Sigma|^{-\frac{n}{2}} e^{-\frac{1}{2}\operatorname{tr}\Sigma^{-1}V}$$

where $\Omega = \{\Sigma\colon \Sigma > 0,\ \Sigma \text{ patterned}\}$. In the three cases above a closed-form solution of this maximization can be obtained. However, in the case of a pattern in the form of a Toeplitz matrix

$$\Sigma = \sigma^2 \begin{pmatrix} 1 & \rho_1 & \rho_2 & \cdots & \rho_{p-1} \\ & 1 & \rho_1 & \cdots & \rho_{p-2} \\ & & & \vdots\vdots\vdots & \\ & & & & \rho_1 \\ & & & & 1 \end{pmatrix},$$

or a pattern

$$\Sigma = \left(\begin{array}{ccc|cc} \alpha & \beta & \beta & \gamma & \delta \\ & \alpha & \beta & \gamma & \delta \\ & & \alpha & \gamma & \delta \\ \hline & & & \alpha & \beta \\ & & & & \alpha \end{array}\right),$$

closed form solutions cannot be given. Of course, numerical solutions are available.

A natural candidate for an estimate of Σ in the last pattern is to average corresponding elements of V/n. Thus, for the above example

$$\begin{aligned}\hat{\alpha} &= (v_{11}+v_{22}+v_{33}+v_{44}+v_{55})/5n,\\ \hat{\beta} &= (v_{12}+v_{13}+v_{23}+v_{45})/4n,\\ \hat{\gamma} &= (v_{14}+v_{24}+v_{34})/3n,\\ \hat{\delta} &= (v_{15}+v_{25}+v_{35})/3n\end{aligned}$$

and

$$\hat{\Sigma} = \left(\begin{array}{ccc|cc} \hat{\alpha} & \hat{\beta} & \hat{\beta} & \hat{\gamma} & \hat{\delta} \\ & \hat{\alpha} & \hat{\beta} & \hat{\gamma} & \hat{\delta} \\ & & \hat{\alpha} & \hat{\gamma} & \hat{\delta} \\ \hline & & & \hat{\alpha} & \hat{\beta} \\ & & & & \hat{\alpha} \end{array}\right).$$

Although this is a reasonable estimation procedure, it has a potential deficiency in that the $\hat{\Sigma}$ formed in this way need not be positive definite. Thus we will want to modify the definition of $\hat{\Sigma}$, but still maintain the pattern, and move closer to the maximum likelihood estimator. At this time there does not appear to be a theoretical approach to solving this type of problem. However, we can employ numerical methods to obtain a solution.

When Σ is a linear function of the parameters in the sense that $\Sigma = \alpha_1 G_1 + \cdots + \alpha_m G_m$, where the matrices $G_1, \ldots, G_m$ are symmetric and known, Anderson (1984) provides an iterative procedure that yields the MLE. Applications that generate patterns are discussed in Olkin (1973), Gleser and Olkin (1973), Olkin (1974), Goodenough, Gandini, Olkin, Pizzamiglio, Thayer and Witkin (1977). Patterned matrices arising from group theoretic structure are studied by Diaconis (1989).

A.3. Other Extremals. Although maximum likelihood estimation provides a rich source of extremal problems, other models also lead to extremal problems. We discuss several such.

A.3.a. *A Distance Problem.* Let X and Y be p-dimensional random vectors having a covariance matrix

$$\Sigma = \begin{pmatrix} \Sigma_{11} & \Psi \\ \Psi' & \Sigma_{22} \end{pmatrix},$$

where the diagonal within-covariance matrices Σ_{11} and Σ_{22} are known, but the between-covariance, $\mathrm{Cov}(X, Y) = \Psi$, between X and Y is unknown. The problem is to determine Ψ subject to $\Sigma > 0$, for which X and Y are closest in the sense

$$\min_{\Psi} \mathscr{E}\, \mathrm{tr}(X - Y)'(X - Y) = \min_{\Psi}(\mathrm{tr}\,\Sigma_{11} + \mathrm{tr}\,\Sigma_{22} - 2\,\mathrm{tr}\,\Psi).$$

This and several related problems are resolved by Olkin and Pukelsheim (1982).

A.3.b. *Estimating the Cholesky Decomposition.* We noted at the outset of this section that V/n is the maximum likelihood estimate of Σ. If we factor $\Sigma = \Psi\Psi'$ and similarly $V/n = TT'$ by the Cholesky decomposition, will T be a good estimator of Ψ? It turns out that TD_a, where D_a is a diagonal matrix with elements $a_1, \ldots, a_p$, yields an "optimal" estimator of Ψ. The elements of the diagonal matrix depend on the particular loss function used. For example, to obtain an unbiased estimator the diagonal elements are

$$a_i = \Gamma((p-i+1)/2) \Big/ \sqrt{2}\Gamma((p-i+2)/2), \qquad i = 1, \ldots, p.$$

However, criteria other than unbiasedness lead to different solutions for the diagonal elements. See Olkin (1985), and Eaton and Olkin (1987).

This suggests several related questions. Suppose that we factor Σ and V/n by the characteristic root decomposition

$$\Sigma = \Gamma D_\lambda \Gamma', \qquad V/n = GD_\ell G',$$

where Γ and G are orthogonal, $D_\lambda = \operatorname{diag}(\lambda_1, \ldots, \lambda_p)$, $D_\ell = \operatorname{diag}(\ell_1, \ldots, \ell_p)$, and λ_i and ℓ_i, $i = 1, \ldots, p$ are the characteristic roots of Σ and V/n respectively. In what sense are G and ℓ good estimators of Γ and λ? This problem is one of current interest, and is, as yet, not completely resolved.

Another factorization that is of general interest is

$$\Sigma = D_\sigma P D_\sigma, \qquad V/n = D_v R D_v,$$

where $D_\sigma = \operatorname{diag}(\sigma_{11}^{1/2}, \ldots, \sigma_{pp}^{1/2})$, $D_v = \frac{1}{\sqrt{n}} \operatorname{diag}(v_{11}^{1/2}, \ldots, v_{pp}^{1/2})$, and the matrices $P = (\rho_{ij})$ and $R = (r_{ij})$ are correlation matrices, (that is, are positive definite with diagonal elements equal to one). Here we know that $v_{ii}^{1/2}/n$ and r_{ij} are maximum likelihood estimates of $\sigma_{ii}^{1/2}$ and ρ_{ij}, respectively. The problem that is difficult is to find the maximum likelihood estimates of P when the σ_{ii} are known. This problem requires the maximization

$$\max_P |P|^{-\frac{1}{2}} \exp -\frac{n}{2} \operatorname{tr} D_\sigma^{-1} P^{-1} B_\sigma^{-1} V,$$

over the region P a correlation matrix. This problem is studied extensively by Styan (1969).

B. Integrals and Distributions

We have noted that the Wishart distribution is a multivariate version of the gamma distribution. This suggests other related multivariate versions, and we describe several such.

B.1. Multivariate Beta Distributions. If U and V are independently distributed, each having a Wishart distribution with $\Sigma = I$, then the joint density is

$$p(U, V) = c(p, n)\, c(p, m)\, |U|^{\frac{n-p-1}{2}} e^{-\frac{1}{2}\operatorname{tr} U} |V|^{\frac{m-p-1}{2}} e^{-\frac{1}{2}\operatorname{tr} V}.$$

Following the univariate version, let

$$Z = U^{-1/2} V U^{-1/2},$$

and

$$W = (U+V)^{-1/2} V (U+V)^{-1/2}.$$

Here the key issue is how to generalize a univariate ratio. For example, we could define $Z_* = \frac{1}{2}\left(VU^{-1} + U^{-1}V\right)$ as a matrix version of a ratio, but this formulation does not lead to a natural generalization of the beta distribution. The resulting distributions of Z and W are

$$\text{(B.1)} \qquad p(Z) = \frac{1}{B_p(m,n)} |Z|^{\frac{m-p-1}{2}} / |I+Z|^{\frac{n+m-p-1}{2}}, \qquad Z > 0,$$

$$p(W) = \frac{1}{B_p(m,n)} |W|^{\frac{m-p-1}{2}} |I-W|^{\frac{n-p-1}{2}}, \qquad 0 < W < I,$$

respectively. The constants $B_p(2m, 2n)$ need to be evaluated so that the integrals are equal to 1. These constants turn out to be generalized beta functions defined by

$$\text{(B.2)} \qquad \begin{aligned} B_p(a_1, \dots, a_k) &= \prod_1^k \Gamma_p(a_i) / \Gamma_p(\Sigma a_i), \\ \Gamma_p(s) &= \pi^{p(p-1)/4} \prod_1^p \Gamma\left(\frac{s-i+1}{2}\right), \end{aligned}$$

Olkin (1959), Olkin and Rubin (1964). Note that

$$B_1(a_1, \dots, a_k) = \prod_1^k \Gamma(a_i) / \Gamma(\Sigma a_i)$$

is the ordinary beta function when $k = 2$.

B.2. Multivariate Liouville-Dirichlet Integrals. The univariate Liouville-Dirichlet integral is

$$\text{(B.3)} \qquad \int_\Omega \prod_1^k x_i^{a_i - 1} f(x_1 + \cdots + x_k) \, \Pi \, dx_i = B(a_1, \dots, a_k) \int_{\Omega_*} z^{a-1} f(z) \, dz,$$

where $a = \Sigma a_i$, $a_i > 0$, $i = 1, \dots, k$ and

$$\Omega = \{x_1, \dots, x_k \colon x_i \geq 0, \ 0 < s \leq \Sigma x_i \leq t\}, \qquad \Omega_* = \{z \colon 0 < s < z < t\}.$$

A multivariate version of (B.3) is

(B.4)

$$\int_\Omega \prod_1^k |V_i|^{a_i - \frac{p+1}{2}} f(V_1 + \cdots + V_k) \, \Pi dV_i = B_p(a_1, \dots, a_k) \int_{\Omega_*} |Z|^{a - \frac{p+1}{2}} f(Z) \, dZ,$$

where $a = \Sigma a_i$, $a_i > (p-1)/2$, and

$$\Omega = \{V_1, \dots, V_k \colon V_i \geq 0, \ 0 < A < \Sigma V_i < B\},$$

$$\Omega_* = \{z \colon z > 0, \ 0 < A < z < B\},$$

Olkin (1979).

The functions $f(Z) = |I - Z|^b$ and $|I + Z|^{-b}$ yield familiar results. Indeed, the joint distribution of $V_1, \ldots, V_k$ given by

$$p(V_1, \ldots, V_k) = \text{const} \prod_1^k |V_i|^{a_i - \frac{p+1}{2}} \Big/ |I + V_1 + \cdots + V_k|^b \tag{B.5}$$

and

$$p(V_2, \ldots, V_k) = \text{const} \prod_1^k |V_i|^{a_i - \frac{p+1}{2}} |I - V_1 - \cdots - V_k|^b \tag{B.6}$$

arise in a number of contexts. This distribution for $p = 1$ and $f(Z) = 1 - Z$ or $(1 + Z)^{-1}$ is used in evaluating the tails of a multinomial distribution.

B.3. Distribution of Characteristic Roots. If V has a Wishart distribution $\mathscr{W}_p(I, n)$, and if $V = GD_\ell G'$ is the characteristic root decomposition, where G is orthogonal and $D_\ell = \text{diag}\,(\ell_1, \ldots, \ell_p)$, the joint distribution of the characteristic roots $\ell_1, \ldots, \ell_p$ is

$$p(\ell_1, \ldots, \ell_p) = c \prod_1^p \ell_i^{\frac{n-p-1}{2}} \prod_{i<j} |\ell_i - \ell_j|\, e^{-\frac{1}{2}\Sigma \ell_i}. \tag{B.7}$$

This distribution was first obtained in 1939 by a number of authors. For a more complete history, see Anderson (1984). It has since served as the basis for many tantalizing results. The derivation of non-central distributions has led to the development of results about zonal polynomials. For a general discussion see Muirhead (1982, Chapter 7).

Recently A. Selberg reviewed his earlier results, Selberg (1944), on the integrals

$$\int_\Omega \prod_1^p t_i^{a-1} \prod_{i<j} |t_i - t_j|^{2c} (a - \Sigma t_i)^{b-1} \prod dt_i$$

over the region $\Omega = \{t_1, \ldots, t_p \colon t_i > 0, \Sigma t_i < 1\}$. Here $a, b, c > 0$. A transformation and limiting argument leads to the integral

$$\int \prod_1^p t_i^{a-1} \prod_{i<j} |t_i - t_j|^{2c} e^{-\frac{1}{2}\Sigma t_i} \Pi\, dt_i$$

over the region $\Omega = \{t_q, \ldots, t_p; t_i > 0\}$. This integral with $c = 1/2$ reduces to the case of the distribution of the characteristic roots.

Related integrals are studied by Askey and Richards (1989) and Gross and Richards (1989), Richards (1989).

C. Independence Properties and Characterizations

There is a long history of different ways that quadratic forms play a central role for the normal distribution. One of the earliest developments relates to the chi-square distribution:

$$f(z) = z^{\frac{m}{2}-1} e^{-\frac{1}{2}z} \Big/ 2^{\frac{n}{2}} \Gamma(n/2), \qquad 0 < x < \infty,$$

which is referred to as the chi-square distribution with m degrees of freedom, and is denoted $z \sim \chi_m^2$. The chi-square distribution arises as the sum of squares $z = x_1^2 + \cdots + x_m^2$, of independently distributed, $\mathcal{N}(0,1)$ variates.

Let $x_1, \ldots, x_n$ be a sample from a $\mathcal{N}(0,1)$ distribution, let $x = (x_1, \ldots, x_n)$ and let $z = xAx'$ be a quadratic form. When $A = I$, then z has a chi-square distribution with n degrees of freedom. This raises the question under what conditions will z have a chi-square distribution with m degrees of freedom. The characteristic function of a chi-square distribution with m degrees of freedom is $(1 - 2it)^{m/2}$, whereas the characteristic function of z is $|I - 2itA| = \prod_{k=1}^n (1 - 2it\lambda_k)$, where $\lambda_1, \ldots, \lambda_n$ are the characteristic roots of A. Consequently, it follows that $\lambda_1 = \cdots = \lambda_m = 1$, $\lambda_{m+1} = \cdots = \lambda_n = 0$, so that A is idempotent.

A remarkable result is due to Cochran (1934). Suppose that $x_1, \ldots, x_n$ are independent $\mathcal{N}(0,1)$ random variables, that

$$Q_i = xA_ix', \qquad i = 1, \ldots, k$$

are k quadratic forms, where rank $A_i = r_i$. Then the following are equivalent:

(i) $r_1 + \cdots + r_k = n$,

(ii) $A_1, \ldots, A_k$ are idempotent,

(iii) $A_iA_j = 0$ for all $i \neq j$.

There are many extensions and variations of this result. For a general discussion see Marsaglia and Styan (1974), and Anderson and Styan (1982).

C.1. Independence Properties. If $x_1, \ldots, x_n$ are independent $\mathcal{N}(0,1)$ random variables, and $x = (x_1, \ldots, x_n)$, then two linear forms $L_1 = xa'$ and $L_2 = xb'$, where $a = (a_1, \ldots, a_n)$, $b = (b_1, \ldots, b_n)$, are independent provided $ab' = 0$, that is, a and b are orthogonal.

Under what conditions are a linear form and a quadratic form independent? Let A be a symmetric matrix, and let

$$L = xa', \qquad Q = xAx'$$

denote a linear and quadratic form. Then L and Q are independent if and only if the joint characteristic function factors:

$$\mathscr{E} e^{isL+itQ} = \mathscr{E} e^{isL} \mathscr{E} e^{itQ}.$$

A direct calculation yields the equation

$$aa' = a(I - 2itA)^{-1}a' \tag{C.1}$$

for all t for which the inverse exists. But (C.1) holds if and only if

$$aA = 0. \tag{C.2}$$

This can be seen by expanding (C.1) in a power series.

When are two quadratic forms independent? Let A and B be two positive semi-definite symmetric matrices and

$$Q_1 = xAx', \qquad Q_2 = xBx'$$

be two quadratic forms. Then Q_1 and Q_2 are independent if and only if

$$|I - \alpha A|\,|I - \beta B| = |I - \alpha A - \beta B| \tag{C.3}$$

for all α, β, which holds if and only if

$$AB = 0. \tag{C.4}$$

That (C.4) implies (C.3) is immediate. There are now many proofs of the converse, most of which are based on an expansion of (C.3). For example, using the fact that

$$\log|I - Q| = \operatorname{tr} Q - \frac{1}{2}\operatorname{tr} Q^2 + \frac{1}{3}\operatorname{tr} Q^3 - \dots \tag{C.5}$$

we can expand both sides of (C.3) to yield

$$\begin{aligned}\sum_{j=1}^{\infty}\frac{(-\alpha)^j}{j}\operatorname{tr} A^j + \sum_{j=1}^{\infty}\frac{(-\beta)^j}{j}\operatorname{tr} B^j \\ = \sum_{j=1}^{\infty}\frac{(-1)^j}{j}\operatorname{tr}(\alpha A + \beta B)^j, \qquad \text{for all } \alpha, \beta.\end{aligned} \tag{C.6}$$

Equating powers on both sides of (C.6) yields

$$\begin{aligned}&\operatorname{tr} A^k B = 0, \qquad \operatorname{tr} AB^k = 0, \qquad k = 1, 2, \dots, \\ &2\operatorname{tr} ABBA + \operatorname{tr} ABAB = 0.\end{aligned} \tag{C.7}$$

The last equation implies that $AB = 0$. To see this note that $\operatorname{tr}(AB)(AB) \ge 0$ so that we must have $\operatorname{tr}(AB)(BA) = \Sigma\lambda_i^2(AB) \le 0$. Since $A \ge 0$, $B \ge 0$, $\lambda(AB) = \lambda\left(A^{\frac{1}{2}}BA^{\frac{1}{2}}\right) \ge 0$, which implies that $A^{\frac{1}{2}}B^{\frac{1}{2}} = 0$, and yields the result.

An alternative but similar argument is obtained by expanding

$$1 = |I + (I - \alpha A - \beta B)^{-1}\alpha\beta AB|$$

to yield

$$0 = \operatorname{tr}(I - \alpha A - \beta B)^{-1}AB + \frac{\alpha\beta}{2}\operatorname{tr}\left[(I - \alpha A - \beta B)^{-1}AB\right]^2 + \dots .$$

An examination of the coefficients of $\alpha\beta$ yields a similar analysis.

C.2. Characterizations. An early conjecture of Lévy was proved by Cramér in 1936 to the effect that if x_1 and x_2 are independent, and $x_1 + x_2$ is normal, that x_1 and x_2 must be normal. In effect, Cramér showed that the only solution of the functional equation

$$\varphi_1(s)\,\varphi_2(s) = \exp -\frac{1}{2}s^2,$$

for $\varphi_j(s) = \mathscr{E}\, e^{isx_j}$, $j = 1, 2$, characteristic functions, is that $\log\varphi_j(s)$ is quadratic.

This result set the stage for a large class of characterizations. One of the most famous is a result of Darmois and Skitovič. For references on characterizations and for a general discussion see Lukacs (1956), Kagan, Linnik, and Rao (1973), and Galambos and Kotz (1978). The Darmois-Skitovič result states that if $x_1, \ldots, x_n$ are independent random variables, $a_1, \ldots, a_n,\quad b_1, \ldots, b_n$ are non-zero with $\sum a_i b_i = 0$, then

$$L_1 = \sum a_i x_i \text{ independent of } L_2 = \sum b_i x_i \Longrightarrow x_i \text{ normal.}$$

The proof is based on showing that the solutions of the functional equation

$$\prod_1^n \varphi_j(sa_j + tb_j) = \prod_1^n \varphi_j(sa_j)\varphi_j(tb_j),$$

where $\varphi_j(s) = \mathscr{E} e^{itx_j}$ is the characteristic function of x_j, are those of a normal distribution.

A multivariate version can be obtained by letting x_i be vectors. However, this does not lead to a new result. To obtain a non-trivial extension let $A_1, \ldots, A_n,\quad B_1, \ldots, B_n$ be non-singular matrices. Then the independence of $L_1 = \sum_1^n x_i A_i$ and $L_2 = \sum_1^n x_i B_i$ is equivalent to the functional equation

$$\prod_1^n \varphi_j(sA_j + tB_j) = \prod_1^n \varphi_j(sA_m)\varphi_m(tB_j), \tag{C.8}$$

where $\varphi_j(s) = \mathscr{E} e^{isx_j'}$ is the characteristic function of the p-dimensional vector x_j. The solution of (C.8) is the normal distribution; see Ghurye and Olkin (1962).

D. Chebychev Inequalities. The prototype Chebychev inequality is that we want to find a bound for the probability that a random vector X lies in a region $\mathscr{X}$. What is known are the means of certain functions of X:

$$\mathscr{E} g_j(x) = \mu_j, \qquad j = 1, \ldots, m.$$

(We include the constant function by choosing $g_1(x) \equiv 1$.)

It is remarkable that an extraordinarily simple argument leads to sharp results. Let $g = (g_1(x), \ldots, g_m(x))$, and $a = (a_1, \ldots, a_m)$, Then

$$\mathscr{E} ag' = \int_{x \in \mathscr{X}} (ag')dF(x) + \int_{x \notin \mathscr{X}} (ag')dF(x). \tag{D.1}$$

Let

$$\mathscr{A} = \{a_1, \ldots, a_m \colon ag' \geq 0 \text{ for } x \notin \mathscr{X}, \quad ag' \geq 1 \text{ for } x \in \mathscr{X}\},$$

then from (D.1)

$$\mathscr{E} ag' \geq \int_{x \in \mathscr{X}} (ag')dF(x) \geq \int_{x \in \mathscr{X}} dF(x) = P\{x \in \mathscr{X}\}, \tag{D.2}$$

which holds for all $a \in \mathscr{A}$. The left-hand side of (D.2) is known: $\mathscr{E} ag' = a\mu'$, where $\mu = (\mu_1, \ldots, \mu_m)$. Consequently, the best inequality is obtained from

$$\inf_{a \in \mathscr{A}} a\mu' \geq P\{x \in \mathscr{X}\}. \tag{D.3}$$

A general discussion of Chebyshev inequalities is provided by Karlin and Studden (1966, Chapters 12, 13).

D.1. A Multivariate Inequality. We now show for a particular example that the determination of the infimum yields a new matrix factorization. This model is studied in detail in Olkin and Pratt (1958).

Let $x_1, \ldots, x_p$ be random variables with $\mathscr{E} x_i = 0$, $\mathscr{E} x_i x_j = \sigma_{ij}$. We seek a bound for

$$P\{|x_i| \geq k_i \sigma_i \text{ for some } i\}.$$

A simplification is achieved by letting $z_i = x_i / k_i \sigma_i$ in which case $\mathscr{E} z_i = 0$, $\mathscr{E} z_i z_j = \sigma_{ij}/(k_i k_j \sigma_i \sigma_j) \equiv \pi_{ij}$ and we seek a bound for

$$P\{|z_i| \geq 1 \text{ for some } i\}.$$

Following the format in (D.1) each $g(x)$ is quadratic so we consider

$$\mathscr{E}\, zAz' = \mathscr{E} \operatorname{tr} zAz' = \operatorname{tr} A(\mathscr{E} z'z) = \operatorname{tr} A\Pi,$$

where $\Pi = (\pi_{ij})$. Consequently, the best bound is obtained as

$$\text{(D.4)} \qquad \inf_{\mathscr{A}} \operatorname{tr} A\pi \geq P\{|Z_i| \geq 1 \text{ for some } i\},$$

where $\mathscr{A} = \{A\colon zAz' \geq 0 \text{ for } z \notin \mathscr{Z},\ zAz' \geq 1 \text{ for } z \in \mathscr{Z}\}$, $\mathscr{Z} = \{z_1, \ldots, z_p\colon |z_i| \geq 1 \text{ for some } i\}$.

Suppose $z_1 = 1$, $\ddot{z} = (z_2, \ldots, z_p)$, and partition $A = \begin{pmatrix} a_{11} & a \\ a' & A_{22} \end{pmatrix}$, $B = A^{-1} = \begin{pmatrix} b_{11} & b \\ b' & B_{22} \end{pmatrix}$. Then

$$(1, \ddot{z})A(1, \ddot{z})' = b_{11}^{-1} + (\ddot{z} - b/b_{11})A_{22}(\ddot{z} - b/b_{11})',$$

with equality at $\ddot{z} = b/b_{11}$. This leads to

i) $A \in \mathscr{A}$ if and only if $B > 0$ and $b_{jj}^{-1} \geq 1$, $j = 1, \ldots, p$.

ii) A is extreme in $\mathscr{A}$ if and only if $B > 0$ and $b_{jj} = 1$, $j = 1, \ldots, p$.

As a consequence, (D.4) becomes

$$\text{(D.5)} \qquad \inf_{a \in \mathscr{A}} \operatorname{tr} A\Pi = \inf_{b \in \mathscr{B}} \operatorname{tr} B^{-1}\Pi \geq P\{|z_i| \geq 1 \text{ for some } i\},$$

where $\mathscr{B} = \{B\colon B > 0,\ b_{ii} = 1, i = 1, \ldots, p\}$.

The problem in terms of $\mathscr{B}$ is more manageable than (D.4). The function $\operatorname{tr} B^{-1}\Pi$ is strictly convex and approaches infinity on the boundary; it has a unique minimum at an interior point $\bar{B} \in \mathscr{B}$ obtained by setting the first derivatives equal to zero. This yields that $\bar{B}^{-1}\Pi\bar{B}^{-1}$ is a diagonal matrix, say D_θ. Consequently, we have the following factorization: Given a positive definite matrix Π, there exists a correlation matrix $\bar{B}$ (that is, $\bar{B} > 0$ and $\bar{b}_{ii} = 1$, $i = 1, \ldots, p$) and a diagonal matrix D_θ with positive elements, such that

$$\Pi = \bar{B} D_\theta \bar{B}.$$

In terms of the original problem, the bound is $tr\bar{B}\Pi = trD_\theta\bar{B} = \Sigma\theta_i$. This means that for any problem of this form, the covariance matrix will generate the diagonal matrix D_θ, which in turn yields the bound.

A discussion of other multivariate versions is given in Marshall and Olkin (1960).

D.2. A Connection Between Chebychev Inequalities and Inclusion Theorems. That a connection exists between Chebychev inequalities and inclusion theorems can most easily be seen from the famous Bienaymé-Chebychev inequality:

If X is a random variable with mean $\mathscr{E}X = \mu$ and variance $\mathscr{E}(X-\mu)^2 = \sigma^2$, then

$$P\{\mu - k\sigma \leq X \leq \mu + k\sigma\} \geq 1 - 1/k^2.$$

In effect this says that there is some probability in the interval $[\mu - \sigma, \mu + \sigma]$.

For any symmetric matrix A with eigenvalue decomposition $A = GD_\lambda G'$, where G is orthogonal, $D_\lambda = \operatorname{diag}(\lambda_1, \ldots, \lambda_n)$, and $\lambda_1, \ldots, \lambda_n$ are the eigenvalues of A, and any unit vector u:

$$uAu' = uGD_\lambda G'u' \equiv vD_\lambda v' = \Sigma\lambda_i v_i^2 \equiv \Sigma\lambda_i p_i,$$

where $v = uG$ is a unit vector, $p_i = v_i^2 \geq 0$, $\Sigma p_i = 1$. Thus uAu' is the mean of a distribution on $\lambda_1, \ldots, \lambda_n$ with respective probabilities $p_1, \ldots, p_n$.

Similarly, $uA^m u' = \Sigma\lambda_i^m p_i$ is the m-th moment of this distribution. Thus, for any unit vector u let $uAu' = \mu$ and $u(A - \mu I)^2 u' = \sigma^2$, then there exists at least one eigenvalue in the interval $[\mu - \sigma, \mu + \sigma]$.

Each Chebychev inequality yields an inclusion theorem. For example, a result of Selberg (1940) is that there must be some probability—or there must exist at least one eigenvalue—in the interval

$$\left[\mu + a - \sqrt{\sigma^2 + a^2},\ \mu + a + \sqrt{\sigma^2 + a^2}\right].$$

We note that many of the proof of inclusion theorems follow along the same lines as the proofs of Chebychev inequalities. For more details on the connection between Chebychev inequalities and inclusion theorems, see Marshall and Olkin (1964).

E. Orderings

A variety of "orderings" have been introduced in the literature. The usefulness of an ordering, $x \succ y$, depends in part on the number of applications for which the ordering is natural, and also on the richness of the class of order-preserving functions, that is, functions f for which $x \succ y$ implies $f(x) \geq f(y)$.

Not all orderings prove to be very useful. For example define $(x_1, \ldots, x_n) \succ (y_1, \ldots, y_n)$ to mean $x_i \geq y_i$, $i = 1, \ldots, n$. The order-preserving functions $f(x_1, \ldots, x_n)$ are those that are monotone increasing in each argument. But in practice most inequalities are not obtained from such a simple monotonicity argument.

Probability has been a source for a variety of orderings because there are many ways to compare distribution functions. For example, a distribution function $F(x)$ is called "more peaked" than a distribution $G(x)$ if

$$\int_{-t}^{t} dF(x) \geq \int_{-t}^{t} dG(x) \qquad \text{for all } t > 0.$$

In effect, this states that more mass is centered about the origin in the F distribution than in the G distribution.

In another example we compare the whole distribution function. We say that $F(x, y)$ is "more concordant" than $G(x, y)$ if the marginal distributions are equal: $F(x, \infty) = G(x, \infty)$, $F(\infty, y) = G(\infty, y)$ and $F(x, y) \geq G(x, y)$ for all x, y.

E.1. The Loewner Ordering. One of the most familiar partial orders in linear algebra is the Loewner ordering $A > B (A \geq B)$ between two symmetric matrices A, B to mean that $A - B$ is positive definite (positive semi-definite). A key result arising from this ordering is that if $\alpha_1 \geq \cdots \geq \alpha_n$ and $\beta_1 \geq \cdots \geq \beta_n$ are the real characteristic roots of the symmetric matrices A and B, respectively, then $A \geq B$ implies that $\alpha_i \geq \beta_i$, $i = 1, \ldots, n$. Consequently, a variety of inequalities involving characteristic roots follow immediately from the Loewner ordering. In addition, the order-preserving functions for the Loewner order have been characterized. Although this class is not too rich, the ordering has proved to be a very useful tool.

One of the most useful aspects of the Loewner ordering is its connection with matrix-convexity. See Loewner (1950), and Bendat and Sherman (1955).

Definition: A function f defined on a convex set $\mathscr{A}$ of matrices and taking values on the set $\mathscr{H}_n$ of Hermitian matrices is said to be matrix-convex if

$$f(aA + (1 - \alpha)B) \leq \alpha f(A) + (1 - \alpha) f(B), \tag{E.1}$$

for all $\alpha \in [0, 1]$ and $A, B \in \mathscr{A}$.

The matrix inequality in $(E.1)$ is the Loewner ordering. An equivalent definition to (E.1)is that f is matrix-convex on $\mathscr{A}$ if, for all fixed $A, B \in \mathscr{A}$, $f(\alpha A + (1 - \alpha)B)$ is convex in $\alpha \in [0, 1]$.

The function $f(A) = A^{-1}$, $A > 0$, is matrix-convex; the function $f(A) = A^{1/2}$, $A > 0$, is matrix concave; the function $f(A) = (MA^{-1}M^*)^{-1}$, $A > 0$, $M \colon m \times n$ complex, rank $M = m$, is matrix concave.

E.2. Majorization. One of the most useful orderings that appears to have an increasingly wide class of applications is that of "majorization." We say that $x = (x_1, \ldots, x_n)$ majorizes $y = (y_1, \ldots, y_n)$, denoted $x \succ y$ if $x_1 \geq \cdots \geq x_n$, $y_1 \geq \cdots \geq y_n$,

$$\sum_1^k x_i \geq \sum_1^k y_i, \qquad k = 1, \ldots, n-1,$$

$$\sum_1^n x_i = \sum_1^n y_i.$$

An equivalent definition is that the vector y lies in the convex hull of the vectors $(x_{\Pi_1}, \ldots, x_{\Pi_n})$, over all permutations $\{\Pi_1, \ldots, \Pi_n\}$ of $\{1, \ldots, n\}$. Finally, $x \succ y$ if and only if $y = xP$, where $P = (p_{ij})$ is a doubly stochastic matrix. For a detailed discussion of majorization and further references see Marshall and Olkin (1979). See also Arnold (1987).

That majorization enters in a natural way in linear algebra can be seen from the following key result of Schur. Let A be a symmetric matrix with characteristic roots $(\lambda_1, \ldots, \lambda_n)$ and diagonal elements $(a_{11}, \ldots, a_{nn})$. Then $(\lambda_1, \ldots, \lambda_n) \succ (a_{11}, \ldots, a_{nn})$.

There are now a number of proofs of this result. For example, the result is a direct consequence of the extremal property

$$\max \operatorname{tr} XAX' = \sum_1^k \lambda_{[i]},$$

$\lambda_{[1]} \geq \cdots \geq \lambda_{[n]}$, where the maximum is taken over all $k \times n$ matrices X satisfying $XX' = I_k$. Consequently,

$$\max \operatorname{tr} XAX' \geq \operatorname{tr}(I_k, O)A(I_k, O)' = \sum_1^k a_{ii}.$$

That majorization is a "natural" ordering in this context is exhibited by the converse obtained by Horn in 1954: given $\alpha_1, \ldots, \alpha_n$, $\beta_1, \ldots, \beta_n$ with $\beta \succ \alpha$, then there exists a symmetric matrix A with $\beta = \lambda(A)$ and $\alpha = \operatorname{diag}(A)$.

There are many results that relate majorization with matrix theory, and we mention but a few. For other examples and for proofs and references see Marshall and Olkin (1979, Chapters 9, 10).

(E.2) If A and B are symmetric matrices, then

$$(\lambda_1(A+B), \ldots, \lambda_n(A+B)) \prec \left(\lambda_{[1]}(A) + \lambda_{[1]}(B), \ldots, \lambda_{[n]}(A) + \lambda_{[n]}(B)\right),$$

where $\lambda_{[1]} \geq \cdots \geq \lambda_{[n]}$ are the ordered roots.

(E.3) For any $n \times n$ complex matrix A with characteristic roots ordered $|\lambda_1(A)| \geq \cdots \geq |\lambda_n(A)| > 0$ and singular values $\sigma_1, \ldots, \sigma_n$,

$$(\log|\lambda_1(A)|, \ldots, \log|\lambda_n(A)|) \prec (\log \sigma_1(A), \ldots, \log \sigma_n(A)).$$

(E.4) For any $n \times n$ complex matrix A,

$$(\mathscr{R}\lambda_1(A), \ldots, \mathscr{R}\lambda_n(A)) \prec \left(\lambda_1\left(\frac{A+A^*}{2}\right), \ldots, \lambda_n\left(\frac{A+A^*}{2}\right)\right).$$

F. Simulations

We have already mentioned methods for simulating a multivariate normal distribution. Even in the normal distribution case many simulation problems remain unresolved. The case of multivariate non-normal distributions opens up a variety of new problems. In this section we review but a few simulation methods to indicate the type of problems that have been studied.

F.1. Random Orthogonal Matrices. The problem here is to generate random orthogonal matrices distributed according to Haar measure over the orthogonal group.

One method is to generate p^2 independent standard normal random variables x_{ij}, $i, j = 1, \ldots, p$, and to factor (using the Gram-Schmidt orthogonalization) the matrix $X = (x_{ij}) = QR$, where R is an upper triangular matrix. The matrix Q is the desired orthogonal matrix.

Other methods use the relation between an orthogonal matrix and skew-symmetrix matrix or Householder transformation. These methods are discussed in Anderson, Olkin and Underhill (1987). In this paper another method based on a factorization of an orthogonal matrix into a product of simple orthogonal matrices is proposed. More explicitly let

$$G_{ij}(\theta_{ij}) = \begin{array}{c} \\ i \\ \\ j \\ \\ \end{array} \!\!\begin{array}{c} \qquad\quad i \qquad\qquad\quad j \\ \begin{pmatrix} I & 0 & 0 & 0 & 0 \\ 0 & \cos\theta_{ij} & 0 & -\sin\theta_{ij} & 0 \\ 0 & 0 & I & 0 & 0 \\ 0 & \sin\theta_{ij} & 0 & \cos\theta_{ij} & 0 \\ 0 & 0 & 0 & 0 & I \end{pmatrix} \end{array},$$

$\pi/2 < \theta_{ij} \le \pi/2$. Then every orthogonal matrix G can be written as

$$G = (G_{12}G_{13}\cdots G_{1n})(G_{23}\cdots G_{2n})(G_{n-1,n})D_\epsilon,$$

where $D_\epsilon =$ diag $(\epsilon_1, \ldots, \epsilon_n)$, $\epsilon_i = \pm 1$, $i = 1, \ldots, n$. The θ_{ij} represent the angles of rotation and the ϵ_i represent reflections.

To obtain the desired random orthogonal matrix the ϵ_i are chosen as $+1$ or -1, each with probability $1/2$, and the $n(n-1)/2$ random angles are chosen according to a beta distribution. For more details see Anderson, Olkin, Underhill (1987); further discussion of this problem is given by Holmes (1988).

F.2. Characteristic Roots of a Wishart Matrix. Suppose that the $p \times p$ random matrix V has a Wishart distribution $\mathscr{W}_p(I, n)$, and let

$$V = GD_\ell G'$$

denote the characteristic root decomposition, where G is orthogonal, $D_\ell =$ diag $(\ell_1, \ldots, \ell_p)$, and $\ell_1 \ge \cdots \ge \ell_p$ are the characteristic roots of V. Recall that the joint distribution of $\ell_1, \ldots, \ell_p$ is given by (B.7).

The problem is to determine

$$E\ell_1 = c \int \max\{\ell_1, \ldots, \ell_p\} \prod_{i \ne j} |\ell_i - \ell_j| \prod \ell_i^{\frac{n-p-1}{2}} e^{-\frac{1}{2}\Sigma \ell_i} \prod d\ell_i.$$

One procedure to evaluate this integral is to use numerical integration. Another is to generate random variables according to the exponential distribution or the gamma distribution:

$$E\ell_1 = c^* \int \max\{\ell_1, \ldots, \ell_p\} \prod_{i \ne j} |\ell_i - \ell_j| \prod \ell_i^{\frac{n-p-1}{2}} g(\ell_1, \ldots, \ell_p) \prod d\ell_i,$$

or

$$E\ell_1 = c^* \int \max\{\ell_1, \dots, \ell_p\} \prod_{i \neq j} |\ell_i - \ell_j| \, h(\ell_1, \dots, \ell_p) \prod d\ell_i$$

where

$$g(\ell_1, \dots, \ell_p) = c_1 \prod_1^p \left(e^{-\frac{1}{2}\ell_i}\right),$$

$$h(\ell_1, \dots, \ell_p) = c_2 \prod_1^p \left(\ell_i^{\frac{n-p-1}{2}} e^{-\frac{1}{2}\ell_i}\right)$$

represent the densities corresponding to the exponential and gamma distribution, and where $c^* = c/c_1$ or c/c_2, respectively. The problem is that each of these methods, which is called "importance sampling," can lead to instability. The types of instability and the use of regression methods to remove the instability are discussed in Luzar and Olkin (1988). Numerical methods for evaluating these integrals are discussed in Edelman (1988, 1989).

References

Anderson, T.W. (1984). *An Introduction to Multivariate Statistical Analysis.* Second Edition. New York: Wiley.

Anderson, T.W. and Olkin, I. (1985). Maximum-likelihood estimation of the parameters of a multivariate normal distribution. *Linear Algebra and Its Applications*, 70, 147–171.

Anderson, T.W., Olkin, I., and Underhill, L. (1987). Generation of random orthogonal matrices. *SIAM Journal of Scientific and Statistical Computing*, 8, 625–629.

Anderson, T.W. and Styan, G.P.H. (1982). Cochran's theorem, rank additivity and tripotent matrices. Pp. 1–23 in *Statistics and Probability: Essays in Honor of C.R. Rao* (ed. by G. Kallianpur, P.R. Krishnaiah, J.K. Ghosh) North-Holland Publishing Co.

Arnold, B. (1987). *Majorization and the Lorenz order: A brief introduction.* Lecture Notes in Statistics 43. Berlin: Springer-Verlag.

Askey, R. and Richards, D. (1989). Selberg's second beta integral and an integral of Mehta. in *Probability, Statistics, and Mathematics: Papers in Honor of Samuel Karlin.* (ed. by T. W. Anderson, K. B. Athreys, and D. L. Iglehart) New York: Academic Press, pp. 27–39.

Bendat, J. and Sherman, S. (1955). Monotone and convex operator functions. *Transactions of the American Mathematical Society*, 79, 58–71.

Bratley, P., Fox, B.L. and Schrage, L.E. (1987). *A Guide to Simulation.* Second Edition. New York: Springer-Verlag.

Cavalli-Sforza, L.L. and Piazza, A. (1985). Analysis of evolution: evolutionary rates, independence and treeness. *Theoretical Population Biology*, 8, 127–165.

Cochran, W.G. (1934). The distribution of quadratic forms in a normal system, with applications to the analysis of covariance. *Proceedings of the Cambridge Philosophical Society*, 30, 178–191.

Deemer, W.L. and Olkin, I. (1951). The Jacobians of certain matrix transformations useful in multivariate analysis. *Biometrika*, 38, 345–367.

Devroye, L. (1986). *Non-uniform Random Variate Generation.* New York: Springer-Verlag.

Diaconis, P. (1989). Patterned Matrices. *Technical Report No. 320*, Department of Statistics, Stanford University.

Eaton, M.L. and Olkin, I. (1987). Best invariant estimators of a Cholesky decomposition. *Annals of Statistics*, 155, 1639–1650.

Edelman, A. (1988). Eigenvalues and condition numbers of random matrices. *SIAM Journal on Matrix Analysis and Applications*, 9, 543–560.

Edelman, A. (1989). The distribution and moments of the smallest eigenvalue of a random matrix of Wishart type. *Numerical Analysis Report 89-4*, Massachusetts Institute of Technology.

Galambos, J. and Kotz, S. (1978). *Characterizations of probability distributions.* Lecture Notes in Mathematics 675. Berlin: Springer-Verlag.

Ghurye, S.G. and Olkin, I. (1962). A characterization of the multivariate normal distribution. *Annals of Mathematical Statistics*, 33, 533–541.

Gleser, L.J. and Olkin, I. (1970). Linear models in multivariate analysis. Pp. 267–292 in *S.N. Roy Memorial Volume* (ed. by R.C. Bose, I.M. Chakravarti, P.C. Mahalanobis, C.R. Rao, and K.J.C. Smith) University of North Carolina Press, Chapel Hill.

Gleser, L.J. and Olkin, I. (1973). Multivariate statistical inference under marginal structure, I. *British Journal of Mathematical and Statistical Psychology*, 26, 98–123.

Goodenough, D.R., Gandini, E., Olkin, I., Pizzamiglio, L., Thayer, D., and Witkin, H.A. (1977). A study of X chromosome linkage with field dependence and spatial visualization. *Behavior Genetics*, 7, 373–387.

Gross, K.I. and Richards, D.St.P. (1989). Total positivism, spherical series, and hypergeometric functions of matrix arguments. *Journal of Approximation Theory*, to appear.

Holmes, R. (1988). On random correlation matrices. *Technical Report TR-803*. Lexington, MA: MIT Lincoln Laboraties.

Kagan, A.M., Linnik, Yu.V. and Rao, C.R. (1973). *Characterization Problems in Mathematical Statistics.* New York: Wiley.

Karlin, S.J. and Studden, W.J. (1966). *Tchebycheff Systems: With Applications in Analysis and Statistics.* New York: Wiley Interscience.

Loewner, C. (1950). Some classes of functions defined by difference or differential inequalities. *Bulletin of the American Mathematical Society*, 56, 308–319.

Lukacs, E. (1956). Characterization of populations by properties of suitable statistics. *Proceedings of the Third Berkeley Symposium in Probability and Statistics*, 2, 195–214. Berkeley: University of California Press.

Luzar, V. and Olkin, I. (1988). Comparison of Simulation Methods in the Estimation of the Ordered Characteristic Roots of a Random Covariance Matrix. *Technical Report No. 249*, Department of Statistics, Stanford University.

Marsaglia, G. and Styan, G.P.H. (1974). Equalities and inequalities for ranks of matrices. *Linear and Multilinear Algebra*, 2, 269–292.

Marshall, A.W. and Olkin, I. (1960). Multivariate Chebyshev inequalities. *Annals of Mathematical Statistics*, 31, 1001–1014.

Marshall, A.W. and Olkin, I. (1964). Inclusion theories for eigenvalues from probability inequalities. *Numerische Mathematik*, 6, 98–102.

Marshall, A.W. and Olkin, I. (1979). *Theory of Majorization and its Applications.* New York: Academic Press.

Muirhead, R.J. (1982). *Aspects of Multivariate Statistical Theory.* New York: Wiley.

Olkin, I. (1959). A class of integral identities with matrix argument. *Duke Mathematical Journal*, 26, 209–213.

Olkin, I. (1973). Testing and estimation for structures which are circularly symmetric in blocks. Pp. 183–195 in *Proceedings of the Symposium on Multivariate Statistical Inference*, (ed. by D. G. Kabe and R. P. Gupta) Dalhousie, Nova Scotia, North-Holland, Amsterdam.

Olkin, I. (1974). Inference for a normal population when the parameters exhibit some structure. Pp. 759–773 in *Reliability and Biometry: Statistical Analysis of Lifelength* (ed. by F. Proschan and R.J. Serfling), SIAM, Philadelphia, PA.

Olkin, I. (1979). Matrix extensions of Liouville-Dirichlet type integrals. *Linear Algebra and Its Applications*, 28, 155–160.

Olkin, I. (1985). Estimating the Cholesky decomposition. *Linear Algebra and Its Applications*, 67, 201–205.

Olkin, I. and Pratt, J. (1958). On a multivariate Tchebycheff inequality. *Annals of Mathematical Statistics*, 29, 226–234.

Olkin, I. and Press, S.J. (1969). Testing and estimation for a circular stationary model. *Annals of Mathematical Statistics*, 40, 1358–1373.

Olkin, I. and Pukelsheim, F. (1982). The distance between two random vectors with given dispersion matrices. *Linear Algebra and Its Applications*, 48, 257–263.

Olkin, I. and Rubin, H. (1962). A characterization of the Wishart distribution. *Annals of Mathematical Statistics*, 33, 1272–1280.

Olkin, I. and Rubin, H. (1964). Multivariate beta distributions and independence properties of the Wishart distribution. *Annals of Mathematical Statistics*, 35, 261–269.

Olkin, I. and Sampson, A.R. (1972). Jacobians of matrix transformations and induced functional equations. *Linear Algebra and Its Applications*, 5, 257–276.

Piazza, A. and Cavalli-Sforza, L.L. (1983). Treeness and the problem of variable evolutionary rates. Pp. 451–463 in *Numerical Taxonomy* (ed by J. Felsenstein). Berlin: Springer-Verlag.

Rasch, G. (1948). A functional equation for Wishart's distribution. *Annals of Mathematical Statistics*, 19, 262–266.

Richards, D. St. P. (1989). Analogs and extensions of Selberg's integral. *Proceedings of the Workshop on q-series and Combinatorics.* (ed. by D. Stanton) Institute for Mathematical Analysis, University of Minnesota, New York, Springer.

Selberg, H.L. (1940). Zwei Ungleichungen zur Ergänzung des Tchebycheffen lemmas. *Skandinavisk Aktuarietidskrift*, 23, 121–125.

Selberg, A. (1944). Bemerkninger om et multipelt integral. *Nordisk Matematish Tidskrift*, 26, 71–78.

Styan, G.P.H. (1969). *Multivariate normal inference with correlation structure.* Doctoral dissertation, Columbia University

Thisted, R. (1988). *Elements of Statistical Computing: Numerical Computation.* New York: Chapman & Hall.

Wishart, J. (1928). The generalised product moment distribution in samples from a normal multivariate population. *Biometrika*, 20A, 32–52.

Department of Statistics, Stanford University, California 94305

Subject Index